ANNUAL REVIEW OF
CELL BIOLOGY

EDITORIAL COMMITTEE (1987)

ANNUAL REVIEW OF CELL BIOLOGY

VOLUME 3, 1987

GEORGE E. PALADE, *Editor*
Yale University School of Medicine

BRUCE M. ALBERTS, *Associate Editor*
University of California, San Francisco

JAMES A. SPUDICH, *Associate Editor*
Stanford University School of Medicine

ANNUAL REVIEWS INC.   4139 EL CAMINO WAY   P.O. BOX 10139   PALO ALTO, CALIFORNIA 94303-0897

ANNUAL REVIEWS INC.
Palo Alto, California, USA

International Standard Serial Number : 0743-4634
International Standard Book Number : 0-8243-3103-6

Annual Review and publication titles are registered trademarks of Annual Reviews Inc.

Annual Reviews Inc. and the Editors of its publications assume no responsibility for the statements expressed by the contributors to this Review.

TYPESETTING BY AUP TYPESETTERS (GLASGOW) LTD., SCOTLAND
PRINTED AND BOUND IN THE UNITED STATES OF AMERICA

PREFACE

The first two volumes of the *Annual Review of Cell Biology* were received with considerable interest by the scientific community, as judged by the number of copies acquired by individuals and libraries. Moreover, they proved useful, as indicated by the frequency with which their chapters were cited in the recent literature. These developments encourage the Editorial Committee to continue its current policies, which are based on the premise that cell biology is part of a continuous body of knowledge properly defined as cellular and molecular biology. These policies also proceed from the assumption that principles of cellular organization and function apply to all living organisms and represent—in our times—a common denominator for all biological sciences, basic or applied.

Over the last decades, few fields of scientific research have advanced as rapidly as cellular and molecular biology, and few have enjoyed the advantages inherent in acquiring, in a generation, broad vistas over apparently inexhaustible territories. But few have had to face the problems generated by onrushing, often spectacular advances achieved with sustained vigor in many directions. Broad vistas and vast territories imply great diversity, which in itself invites divergence and encourages fragmentation. The generation of researchers responsible for the opening of those vistas now has the problem of keeping the central areas of the field in focus. In-depth research in such areas will undoubtedly continue to generate new concepts and new technologies, which in time will animate and illuminate other areas. Hence the Editorial Committee proposes to continue to concentrate on central topics in cellular and molecular biology. Yet it also intends to cover significant, specific developments in broad fields of primordial importance, such as plant cell biology, or in active fields currently undergoing impressive and exciting developments, such as immunology and developmental biology.

Keeping all these desiderata in proper balance is not an easy task. Over the last three years, the Editorial Committee has done its best to achieve this goal and will continue to do so. In the process, it welcomes comments and suggestions from other members of the scientific research community, especially members of the American Society for Cell Biology.

The Editorial Committee believes that a vigorous Annual Review, judiciously selective in its coverage and demanding in the quality of its content, is particularly timely in cellular and molecular biology. It can become a unifying factor clearly needed in a large field undergoing rapid expansion in diverse directions. It can keep many of us well informed or reasonably educated in fields adjacent to our individual areas of interest. We trust that our readers will continue to benefit from this exercise in communication.

GEORGE E. PALADE
EDITOR

ANNUAL REVIEWS INC. is a nonprofit scientific publisher established to promote the advancement of the sciences. Beginning in 1932 with the *Annual Review of Biochemistry*, the Company has pursued as its principal function the publication of high quality, reasonably priced *Annual Review* volumes. The volumes are organized by Editors and Editorial Committees who invite qualified authors to contribute critical articles reviewing significant developments within each major discipline. The Editor-in-Chief invites those interested in serving as future Editorial Committee members to communicate directly with him. Annual Reviews Inc. is administered by a Board of Directors, whose members serve without compensation.

Annual Review of Cell Biology
Volume 3, 1987

CONTENTS

UBIQUITIN-MEDIATED PATHWAYS FOR INTRACELLULAR PROTEOLYSIS,
Martin Rechsteiner   1

CELL TRANSFORMATION BY THE VIRAL *src* ONCOGENE, *Richard
Jove and Hidesaboro Hanafusa*   31

LAMININ AND OTHER BASEMENT MEMBRANE COMPONENTS, *George
R. Martin and Rupert Timpl*   57

REPLICATION OF PLASMIDS DERIVED FROM BOVINE PAPILLOMA
VIRUS TYPE 1 AND EPSTEIN-BARR VIRUS IN CELLS IN CULTURE,
Joan Mecsas and Bill Sugden   87

EARLY EVENTS IN MAMMALIAN FERTILIZATION, *Paul M.
Wassarman*   109

MOLECULAR ASPECTS OF B-LYMPHOCYTE ACTIVATION, *Anthony
L. DeFranco*   143

CELL SURFACE RECEPTORS FOR EXTRACELLULAR MATRIX MOLECULES,
Clayton A. Buck and Alan F. Horwitz   179

GENERATION OF PROTEIN ISOFORM DIVERSITY BY ALTERNATIVE
SPLICING: Mechanistic and Biological Implications, *Athena
Andreadis, Maria E. Gallego, and Bernardo Nadal-Ginard*   207

CONSTITUTIVE AND REGULATED SECRETION OF PROTEINS, *Teresa
Lynn Burgess and Regis B. Kelly*   243

OLIGOSACCHARIDE SIGNALLING IN PLANTS, *Clarence A. Ryan*   295

CELL ADHESION IN MORPHOGENESIS, *David R. McClay and
Charles A. Ettensohn*   319

INTRACELLULAR TRANSPORT USING MICROTUBULE-BASED MOTORS,
Ronald D. Vale   347

MYOSIN STRUCTURE AND FUNCTION IN CELL MOTILITY, *Hans
M. Warrick and James A. Spudich*   379

GROWTH AND DIFFERENTIATION IN THE HEMOPOIETIC SYSTEM,
T. M. Dexter and E. Spooncer   423

POLYPEPTIDE GROWTH FACTORS: Roles in Normal and Abnormal
Cell Growth, *Thomas F. Deuel*   443

viii CONTENTS (*continued*)

INDEXES

Subject Index 493
Cumulative Index of Contributing Authors, Volumes 1–3 499
Cumulative Index of Chapter Titles, Volumes 1–3 500

SOME RELATED ARTICLES IN OTHER *ANNUAL REVIEWS*

From the *Annual Review of Biochemistry*, Volume 56 (1987):

Dynamics of Membrane Lipid Metabolism and Turnover, E. A. Dawidowicz

Topography of Glycosylation in the Rough Endoplasmic Reticulum and Golgi Apparatus, C. B. Hirschberg and M. D. Snider

Inositol Trisphosphate and Diacylglycerol: Two Interacting Second Messengers, M. J. Berridge

Intracellular Proteases, J. S. Bond and P. E. Butler

The Structure and Function of the Hemagglutinin Membrane Glycoprotein of Influenza Virus, D. C. Wiley and J. J. Skehel

Alternative Splicing: A Ubiquitous Mechanism for the Generation of Multiple Protein Isoforms from Single Genes, R. E. Breitbart, A. Andreadis, and B. Nadal-Ginard

Inhibitors of the Biosynthesis and Processing of N-Linked Oligosaccharide Chains, A. D. Elbein

The Nucleus: Structure, Function, and Dynamics, J. W. Newport and D. J. Forbes

Protein Serine/Threonine Kinases, A. M. Edelman, D. K. Blumenthal, and E. G. Krebs

G Proteins: Transducers of Receptor-Generated Signals, A. G. Gilman

Interferons and their Actions, S. Pestka, J. A. Langer, K. C. Zoon, and C. E. Samuel

Biosynthetic Protein Transport and Sorting by the Endoplasmic Reticulum and Golgi, S. R. Pfeffer and J. E. Rothman

Receptors for Epidermal Growth Factor and Other Polypeptide Mitogens, G. Carpenter

From the *Annual Review of Biophysics and Biophysical Chemistry*, Volume 16 (1987):

Structure and Assembly of Coated Vesicles, B. M. F. Pearse and R. A. Crowther

Structure and Dynamics of Water Surrounding Biomolecules, W. Saenger

The Structural Basis of Antigen-Antibody Recognition, R. A. Mariuzza, S. E. V. Phillips, and R. J. Poljak

An Introduction to Molecular Architecture and Permeability of Ion Channels, G. Eisenman and J. A. Dani

Molecular Properties of Ion Permeation Through Sodium Channels, T. Begenisich

Calcium Channels: Mechanisms of Selectivity, Permeation, and Block, R. W. Tsien, P. Hess, E. W. McCleskey, and R. L. Rosenberg

Absorption, Scattering, and Imaging of Biomolecular Structures with Polarized Light, I. Tinoco, Jr., W. Mickols, M. F. Maestre, and C. Bustamante

Measurement of Metal Cation Compartmentalization in Tissue by High-Resolution Metal Cation NMR, C. S. Springer, Jr.

Real-Time Spectroscopic Analysis of Ligand-Receptor Dynamics, L. A. Sklar

Peptides With Affinity for Membranes, E. T. Kaiser and F. J. Kézdy

From the *Annual Review of Genetics*, Volume 21 (1987):

Oncogene Activation by Chromosome Translocation in Human Malignancy, F. G. Haluska, Y. Tsujimoto, and C. M. Croce

Regulation of DNA Replication During Drosophila *Development*, A. Spradling and T. Orr-Weaver

Arabidopsis thaliana, E. M. Meyerowitz

DNA Methylation in Escherichia coli, M. G. Marinus

Genetic Analysis of the Yeast Cytoskeleton, T. C. Huffaker, M. A. Hoyt, and D. Botstein

RNA 3′ End Formation in the Control of Gene Expression, D. Friedman, M. J. Imperiale, and S. Adhya

From the *Annual Review of Immunology*, Volume 5 (1987):

The Role of Somatic Mutation of Immunoglobulin Genes in Autoimmunity, A. Davidson, R. Shefner, A. Livneh, and B. Diamond

Disorders of Phagocyte Function, D. Rotrosen and J. I. Gallin

Molecular Mechanisms of Transmembrane Signaling in B Lymphocytes, J. C. Cambier and J. T. Ransom

Viruses Perturb Lymphocyte Functions: Selected Principles Characterizing Virus-Induced Immunosuppression, M. B. McChesney and M. B. A. Oldstone

B-Cell Stimulatory Factor-1/Interleukin 4, W. E. Paul and J. Ohara

The Structure, Function, and Serology of the T-Cell Antigen Receptor Complex, J. P. Allison and L. I. Lanier

Genes of the T-Cell Antigen Receptor in Normal and Malignant T Cells, B. Toyonaga and T. Mak

From the *Annual Review of Microbiology*, Volume 41 (1987):

Compartmentation of Carbohydrate Metabolism in Trypanosomes, F. R. Opperdoes

The Mitochondrial Genome of Kinetoplastid Protozoa: Genomic Organization, Transcription, Replication, and Evolution, L. Simpson

Export of Protein: A Biochemical View, L. L. Randall, S. J. S. Hardy, and J. R. Thom

High-Resolution NMR Studies of Saccharomyces cerevisiae, S. L. Campbell-Burk and R. G. Shulman

Genetic Research with Photosynthetic Bacteria, P. A. Scolnik and B. L. Marrs

From the *Annual Review of Neuroscience*, Volume 11 (1988):

Modulation of Ion Channels in Neurons and Other Cells, I. B. Levitan

Excitatory Amino Acid Neurotransmission: NMDA Receptors and Hebb-Type Synaptic Plasticity, C. W. Cotman, D. T. Monaghan, and A. H. Ganong

Microtubule-Associated Proteins: Their Potential Role in Determining Neuronal Morphology, A. Matus

Probing the Molecular Structure of the Voltage-Dependent Sodium Channel, R. L. Barchi

From the *Annual Review of Physiology*, Volume 49 (1987):

Proton Transport by Hepatocyte Organelles and Isolated Membrane Vesicles, B. F. Scharschmidt and R. W. Van Dyke

Lateral Diffusion of Proteins in Membranes, K. Jacobson, A. Ishihara, and R. Inman

Intracellular Lipid Transport in Eukaryotes, R. G. Sleight

Lipid Modulation of Transport Proteins in Vertebrate Cell Membranes, B. Deuticke and C. W. M. Haest

Gastroenteropancreatic Peptides and the Central Nervous System, D. P. Figlewicz, F. Lacour, A. Sipols, D. Porte, Jr., and S. C. Woods

Functions of Angiotensin in the Central Nervous System, M. I. Phillips

Mechanisms of Angiogenesis, P. A. D'Amore and R. W. Thompson

Kinetics of the Actomyosin ATPase in Muscle Fibers, Y. E. Goldman

(*continued*)

Mechanical and Structural Approaches to Correlation of Cross-Bridge Action in Muscle with Actomyosin ATPase in Solution, B. Brenner

Spectroscopic Probes of Muscle Cross-Bridge Rotation, D. D. Thomas

From the *Annual Review of Plant Physiology*, Volume 38 (1987):

Photochemical Reaction Centers: Structure, Organization, and Function, A. N. Glazer and A. Melis

Membrane-Proton Interactions in Chloroplast Bioenergetics: Localized Proton Domains, R. A. Dilley, S. M. Theg, and W. A. Beard

Phosphorylation of Proteins in Plants: Regulatory Effects and Potential Involvement in Stimulus/Response Coupling, R. Ranjeva and A. M. Boudet

The Plant Cytoskeleton: The Impact of Fluorescence Microscopy, C. W. Lloyd

Regulation of Gene Expression in Higher Plants, C. Kuhlemeier, P. J. Green, and N.-H. Chua

Evolution of Higher-Plant Chloroplast DNA-Encoded Genes: Implications for Structure-Function and Phylogenetic Studies, G. Zurawski and M. T. Clegg

Ann. Rev. Cell Biol. 1987. 3 : 1–30
Copyright © 1987 by Annual Reviews Inc. All rights reserved

UBIQUITIN-MEDIATED PATHWAYS FOR INTRACELLULAR PROTEOLYSIS

Martin Rechsteiner

Department of Biochemistry, School of Medicine, University of Utah, Salt Lake City, Utah 84132

CONTENTS

INTRODUCTION.. 1
GENERAL PROPERTIES OF INTRACELLULAR PROTEOLYSIS 2
   Rates 2
   Nucleotide Dependence... 3
   Sites of Proteolysis... 3
   Protein Structure and Intracellular Stability 6
   Complete Versus Partial Proteolysis 7
   Significance 8
UBIQUITIN-MEDIATED PROTEOLYTIC PATHWAYS.. 9
   History.. 9
   Properties of Ubiquitin and Its Genes 9
   Models for Ubiquitin-Dependent Proteolysis 12
   Enzymes of Ubiquitin Metabolism.. 16
   Stable Ubiquitin Conjugates... 21
   Ubiquitin Pool Dynamics 22
   Ubiquitin and Developing Tissues.. 24
SUMMARY.. 24

INTRODUCTION

Proteolysis occurs in nearly all cellular compartments, where it serves a number of purposes ranging from removal of protein targeting sequences to complete hydrolysis of polypeptide chains. The cytosolic process is particularly fascinating because individual proteins are destroyed at vastly different rates. We are just beginning to understand the mechanisms

1

0743–4634/87/1115–0001$02.00

involved in this process, particularly the importance of ubiquitin. This remarkable eucaryotic protein has scarcely changed during the 3 billion years of evolution that separate yeast and man. Besides its evolutionary conservation, ubiquitin is unusual in that it can be covalently attached to histones, to various cytoplasmic proteins, and to an external domain on the lymphocyte homing receptor. Although the physiological significance of histone and homing receptor ubiquitination remains unclear, there is abundant evidence that conjugation of ubiquitin to cytosolic proteins can mark those proteins for destruction. Ubiquitin also plays a key role in the heat-shock response.

This review concentrates on intracellular proteolysis and ubiquitin's function in that process. Space limitations require a selective review of the literature. Therefore, I have focused on experiments that use rabbit reticulocyte lysate or microinjection. For other perspectives see Hershko & Ciechanover (1986), Beynon & Bond (1986), Mayer & Doherty (1986), and *Ubiquitin*, a multiauthored book forthcoming from Plenum Press.

GENERAL PROPERTIES OF INTRACELLULAR PROTEOLYSIS

Rates

Metabolic labeling experiments have revealed two sets of proteins with stabilities that differ more than tenfold. When rat livers or cultured mammalian cells are exposed to radioactive amino acids for a few minutes, 20–40% of the newly synthesized proteins are degraded within the following hour (Poole & Wibo 1973). After cells are exposed to radioactive amino acids for 24 hr or longer, degradation rates are between 2 and 4% per hour, values similar to those measured by nonisotopic methods (Steinberg & Vaughan 1956). These two distinct classes of proteins are designated "short-lived" and "long-lived" proteins.

Turnover rates of individual proteins do not necessarily fall within the ranges just presented. In fact, half-lives of proteins vary from several minutes to weeks, irrespective of their cellular location. The term "half-life" implies that loss of an individual protein is a first-order process. As Schimke (1973) pointed out, this kinetic behavior has two important implications: (a) "aging" of a protein molecule does not lead to an increased probability of its being degraded; (b) all molecules of a specific protein are present in a common pool. Exponential loss has been clearly established for some enzymes. However, the degradation of other proteins, especially in muscle or nerve, does not conform to first-order kinetics. For example, the turnover of vimentin and actin displays biphasic kinetics in astrocytes; about 40% of each protein is degraded with a half-life of 12–18 hr, whereas more than half of the remaining molecules are still present 8 days later (Chiu & Goldman 1984). The stabilization of a fraction of each protein population may reflect its assembly into macromolecular complexes (Siekevitz 1972). Alternatively, nerve processes may be proteolytically privileged sites since proteins exhibit remarkable stability during slow axonal transport (Brady & Lasek 1981).

Basal rates of proteolysis can double when mammalian cells are deprived of polypeptide hormones or essential nutrients. It is well established that most of the enhanced proteolysis results from autophagy, a process whereby portions of the cytoplasm are encapsulated in membrane vesicles that subsequently fuse with lysosomes. Several studies have shown that, except for more rapid degradation of mitochondria (Chandler & Ballard 1983), cytoplasmic components are randomly included within autophagic vacuoles during enhanced degradation (Amenta & Brocher 1981). Little is known about the cellular and molecular mechanisms responsible for autophagy, and there is no evidence for or against ubiquitin's participation in the process.

Nucleotide Dependence

Simpson (1953) demonstrated that ATP is required for the degradation of intracellular proteins. This observation has since been confirmed for individual enzymes (Hershko & Tomkins 1971), for proteins microinjected into cultured mammalian cells (Katznelson & Kulka 1983), and for both short- and long-lived proteins in general (Gronostajski et al 1985). Recently, however, two papers reported intracellular proteolysis in the absence of ATP. The degradation of chick reticulocyte β-spectrin was unaffected by ATP depletion, whereas breakdown of α-spectrin ceased under similar conditions (Woods & Lazarides 1985). Likewise, abnormal globin chains were degraded in the absence of high-energy compounds, albeit at an incredibly low rate (Fagan et al 1986). Since removal of ATP produces an abnormal metabolic condition, depletion may activate proteases normally inhibited by ATP-dependent processes, e.g. phosphorylation. For this reason, results from ATP-depletion experiments may not apply to normal physiological conditions.

Sites of Proteolysis

Proteolytic activities have been identified in virtually every cellular compartment. Although the physiological significance of some, e.g. leader peptidase, transit peptide protease, and hormone processing enzymes, seems apparent, we do not know the contribution of most cellular proteases to overall protein metabolism, nor do we know which enzymes participate in the degradation of specific proteins. There is evidence that cytoskeletal proteins are the principal substrates for calcium-activated proteases (Vorgias & Traub 1986). Still, to my knowledge, we have yet to identify the protease(s) responsible for the degradation of any specific cytoplasmic protein. Clearly this is an important task for the future.

A general discussion of cellular proteases is outside the scope of this essay, and the recent review by Bond & Butler (1987) provides a guide to the literature. However, two issues relating to possible sites of proteolysis do warrant further consideration. It has been suggested that the degradation of transit peptides in mitochondria and leader peptides in the endoplasmic reticulum can account for the class of short-lived proteins (Kominami et al 1983; Hough & Rechsteiner 1984). Temperature studies support the notion that different pathways are responsible for the destruction of short-lived and long-lived proteins. Degradation of short-lived proteins exhibits a temperature dependence typical of many enzyme reactions; the Q_{10} is about two (Neff et al 1979). In contrast, the Q_{10} for turnover of long-lived proteins is often greater than four (Hough & Rechsteiner 1984). Nevertheless, several observations challenge the hypothesis that the existence of short-lived proteins reflects co- or post-translational proteolytic processing. Proteins containing transit or leader peptides do not account for 20–40% of newly synthesized polypeptide chains. Moreover, targeting sequences are removed rapidly after synthesis (Reid & Schatz 1982). Also, many proteins with short half-lives, e.g. *myc*, *fos*, *p53*, *e1a*, are found in the nucleus. Thus it appears that most short-lived proteins inhabit the nuclear and cytosolic compartments. Their instability surely has regulatory significance, and a common degradative pathway, characterized by a low activation energy, may be responsible for their rapid breakdown.

The second issue concerns the possible role of lysosomes in the selective degradation of cytosolic proteins. The diverse half-lives of cytosolic proteins would require differential stabilities within lysosomes or selective transfer into these organelles, and mechanisms to accomplish this have been proposed (Dean 1984). It has been suggested that lysosomes can directly take up proteins, a process known as microautophagy (Ahlberg & Glaumann 1985). However, recent studies indicate that lysosomes do not participate in the turnover of short-lived proteins, but rather they contribute to the degradation of long-lived proteins.

In 1973, Poole & Wibo demonstrated that cells contain two mechanistically distinct proteolytic pathways. They showed that while fresh serum inhibited the degradation of long-lived proteins in cultured rat fibroblasts, it had no effect on the degradation of short-lived proteins. A year later, Wibo & Poole demonstrated that the organic amine, chloroquine, accumulated in lysosomes and inhibited lysosomal proteolysis. A number of experiments to assess lysosomal involvement in proteolysis have been based on this observation. The intracellular degradation of a protein is measured in the presence or absence of chloroquine or ammonia, and the extent to which these agents inhibit its proteolysis is taken to be proportional to the fraction of the protein hydrolyzed within lysosomes. In view of the various cell lines examined, e.g. hepatoma, muscle, and HeLa (Knowles & Ballard 1976; Libby & Goldberg 1981; Rote & Rechsteiner 1983), surprisingly consistent results have been observed. The degradation of long-lived proteins is inhibited 20–40%, while the degradation of short-lived proteins, when measured, is barely affected.

The conclusion that lysosomes are minor contributors to basal proteolysis is valid only if chloroquine or ammonia completely inhibit lysosomal proteolysis. Although this assumption has been questioned (Mortimore 1982; Ahlberg et al 1985), the degradation of membrane proteins, proteins internalized by pinocytosis, and autophagy-induced proteolysis resulting from nutrient deprivation (Libby & Goldberg 1981; J. L. Goldstein et al 1975; Amenta & Brocher 1981, respectively) all cease in the presence of these agents. Clearly, they inhibit the intralysosomal degradation of certain proteins; microinjection experiments show that they efficiently inhibit proteolysis of cytosolic proteins as well.

Rogers & Rechsteiner (1985) measured the stabilities of 35 proteins injected into HeLa cells. Half-lives of the injected proteins varied from 16 to 220 hr, and proteolysis of a specific protein was inhibited by ammonia and chloroquine only if its half-life was greater than 45 hr. The extent of inhibition correlated well with increasing half-life, such that the degradation of injected proteins with half-lives greater than 200 hr was inhibited by 60%. The observed relationship, which is consistent with complete inhibition of lysosomal proteolysis by ammonia and chloroquine, indicated that HeLa cells randomly sequester 0.3% of their cytoplasm per hour into autophagic vacuoles. This value agrees with measurements by Hendil (1981), who showed that injected [^{3}H]dextran accumulated in BHK lysosomes at a rate between 0.2 and 0.4% per hour. Both results indicate that HeLa or hamster proteins impervious to cytosolic proteases would still exhibit half-lives less than 350 hr as a result of basal autophagy.

Based on evidence obtained with lysosomatotropic agents alone, one might still hesitate to eliminate lysosomes as agents of selective proteolysis. Fortunately, the results of inhibitor studies are supported by those of microinjection experiments of distinctly different design. Bigelow et al (1981) injected [^{14}C]sucrose-labeled proteins into mouse cells or allowed the proteins to be taken up by endocytosis and found that the degradation products, sucrose peptides, were cytosolic in the former case and intralysosomal in the latter. These authors concluded that a major proteolytic system exists in the cytosol. While confirming these results for [³H]raffinose bovine serum albumin (BSA) injected into human fibroblasts, McElligott et al (1985) found that raffinose peptides arising from injected RNase A were located within lysosomes. Conversely, Rote & Rechsteiner (1986) observed proteolysis of RNase A to be unaffected by lysosomotropic agents in HeLa cells. At present, there is no satisfactory explanation for the apparent discrepancy in the site of RNase degradation.

Temperature studies also indicate major involvement of cytosolic pathways in selective protein degradation. Hough & Rechsteiner (1984) measured degradation rates of several injected proteins at temperatures between 6 and 37°C and found no "breaks" in the resulting Arrhenius plots. However, in nutrient-poor medium, in which HeLa cell autophagy is increased, enhanced degradation was only observed above 20°C. These investigators suggested that the formation of autophagic vacuoles or their fusion to lysosomes was inhibited below 20°C, and they interpreted the smooth Arrhenius plots as evidence against protein degradation in lysosomes under normal physiological conditions.

Microinjection and inhibitor experiments thus combine to reinforce the idea that lysosomes play a nonselective role in the turnover of cytosolic proteins. These organelles appear to be reserved mainly for the destruction of exogenous proteins and membrane proteins. Further, lysosomes may not be exclusive sites for the turnover of membrane proteins. Half of the basal degradation and most of the enhanced degradation of hydroxymethyl glutaryl (HMG) CoA reductase, an integral endoplasmic reticulum protein and a key enzyme in cholesterol biosynthesis, is insensitive to lysosomotropic agents (Tanaka et al 1986).

Protein Structure and Intracellular Stability

Half-lives of cytoplasmic proteins can differ more than a thousandfold, and this raises intriguing questions about the causes of this variable stability. Early studies documented the rapid degradation of proteins with altered conformations due to mutation, incorporation of amino acid analogs, denaturation, or premature chain termination (Goldberg & St. John 1976). Although one might expect major disruptions of protein structure to result in rapid breakdown of the protein, normal cellular proteins also display a wide range of half-lives. This fact has generated speculation about the structural features of proteins that elicit proteolysis, and correlations have been proposed between half-life and subunit molecular weight, isoelectric point, hydrophobicity, thermal stability, and surface negative-charge density (Rechsteiner et al 1987). Unfortunately, few hypotheses have received consistent experimental support.

Some hypotheses have focused on primary sequence rather than structural features. For example, deamidation of asparagine and glutamine has been proposed to enhance the turnover of proteins rich in these residues (Robinson 1974). Oxidation may also cause rapid degradation (Stadtman 1986), and it is worth noting that the oxidizable amino acids (his, cys, met, tyr, and trp) are the least abundant residues in proteins. Several groups have attached particular importance to the N terminus (Hershko et al 1984; Bachmair et al 1986); others have speculated that local enrichment for certain amino acids, e.g. pro, ser, glu, and thr (Rogers et al 1986) or the presence of specific sequences, e.g. lys-phe-glu-arg-gln (Dice et al 1986) can result in rapid degradation. In a later section, we reconsider two of these newer sequence hypotheses since it appears that E3, a key factor in the ubiquitination pathway, specifically recognizes free α-amino groups and oxidized methionine residues.

Complete Versus Partial Proteolysis

The degradation of leader peptides and the processing of polypeptide hormones from their precursors attest to the prevalence of partial proteolytic events in the secretory pathway. Because many cellular proteins consist of multiple domains joined by hinges or transmembrane helices, cytosolic proteases might produce partial degradation products by simply cleaving flexible regions between domains. If partial degradation of cytosolic proteins normally occurs, it would have mechanistic as well as regulatory implications. A proteolytic process generating diffusible fragments could impair normal functions if the released peptides were adsorbed to or melted into important cellular structures. For this reason, one might expect cytoplasmic proteases to be processive rather than distributive enzymes.

Detection of intermediates is difficult after biosynthetic labeling because fragments of one protein are likely to be obscured by other labeled proteins. Using antibody techniques, Reznick et al (1985) and Toda & Ohashi (1986) have reported intracellular fragments derived from aldolase and phosphofructokinase, respectively. However, proteolytic fragments are usually not observed (Beynon et al 1985). When they are present, special precautions, such as direct extraction of cellular monolayers in trichloroacetic acid (TCA), markedly reduce their concentration (Chandler & Ballard 1986). Proteolytic intermediates are also not apparent in cells receiving injections of various radioiodinated proteins.

The apparent absence of intermediates suggests that once degradation is initiated, the protein is hydrolyzed completely. Microinjection experiments address several interesting aspects of the "*all-or-none*" question. McGarry et al (1983) injected IgG molecules into HeLa cells and found that heavy and light chains were degraded at the same rate. In contrast, Fab fragments were more stable than intact IgGs or Fc fragments. One can conclude that although the Fc fragment contains the primary recognition site(s), once proteolysis of an IgG is initiated, all domains are degraded. Two experiments show that even noncovalently linked domains are degraded in an all-or-none fashion. Yamaizumi et al (1982) found that whereas the half-life of diphtheria toxin fragment A injected alone was more than 48 hr, its stability decreased sevenfold to that of anti–fragment A IgG in L929 cells injected with both proteins. Likewise, Rote & Rechsteiner (1986) found that both components in trypsin/trypsin inhibitor complexes were degraded at the same rate in HeLa cells even though their half-lives differed when the proteins were injected separately. The simultaneous degradation of components in antibody-antigen or enzyme-inhibitor complexes indicates that all-or-none proteolysis is characteristic of cytosolic pathways. Most evidence argues against transfer of soluble proteins to lysosomes. Hence, the mechanism for sequestering multiple domains during the actual cleavage process remains an intriguing problem. Conceivably, cells construct molecular cages (or perhaps vaults; see Kedersha & Rome 1986) around proteins to be degraded.

Significance

Intracellular proteolysis plays a key role in the assembly of cellular structures. Leader and transit peptides are degraded after they have targeted proteins to their proper compartments. Construction of multicomponent complexes within the cytosol is also accompanied by proteolysis. Cells generally produce unequal amounts of subunits and then destroy the excess. The fates of unassembled ribosomal proteins, mitochondrial precursors, and unincorporated globin chains (Abovich et al 1985; Reid & Schatz 1982; Shaeffer 1983) clearly illustrate this strategy. The rapid destruction of unassembled proteins, like that of abnormal or damaged proteins noted in an earlier section, presumably prevents damage from "unstructured" polypeptides.

Selective proteolysis also serves as an important regulatory mechanism. The maximum rate at which protein levels can change is determined by the half-life of the protein in question (Schimke 1973), and key metabolic enzymes are often rapidly degraded. It is now evident that regulatory proteins without known enzymatic activity can also be short-lived. Many of these proteins contain regions rich in pro, glu, ser, and thr (PEST sequences), and they include steroid receptors, heat-shock proteins, phytochrome, cyclins, and various oncogene products, e.g. *myc*, *myb*, *fos*, *ela* (Rogers et al 1986). Some regulatory proteins may even self-destruct (HSP 70, Mitchell et al 1985; TF111A, Kmiec & Worcel 1986; Lex A,

Slilaty et al 1986). The mRNAs that encode short-lived proteins are also rapidly degraded, which emphasizes the regulatory significance of fast turnover (Dani et al 1984).

A number of *Drosophila* homeotic proteins contain PEST sequences, and there are strong suggestions that several are rapidly degraded (Carroll & Scott 1986; Edgar et al 1986). Indeed, the presumed product of *snake*, an important determinant of dorsal-ventral axis in *Drosophila* embryos, appears to be a serine protease that itself contains a PEST region (DeLotto & Spierer 1986). Thus, proteolytic cascades in the embryonic syncytium may be a central mechanism for body-pattern specification in insects (North 1986).

UBIQUITIN-MEDIATED PROTEOLYTIC PATHWAYS

History

Ubiquitin (Ub) was isolated by Gideon Goldstein and colleagues in 1975 as a protein capable of inducing B-lymphocyte differentiation. Because the antibodies prepared against Ub reacted with similar proteins in organisms as diverse as mammals, yeast, and celery, the protein was named ubiquitous immunopoietic peptide. Later that year, Schlesinger et al published the sequence for ubiquitin, as it was now called. Ub reappeared in the literature in 1977. Two groups reported that A24, a rat liver chromosomal protein identified by two-dimensional electrophoresis, contained histone H2A covalently linked to Ub (Goldknopf & Busch 1977; Hunt & Dayhoff 1977). The two proteins were joined by an isopeptide bond between the ε-amino group of lysine 119 on histone H2A and the carboxyl terminus of Ub. The sequence of Ub in protein A24 varied from that originally reported; it was 76 amino acids with a C terminus composed of arg-gly-gly rather than arg. We now know that active Ub contains 76 amino acids and that the two C-terminal glycines can be rapidly lost by proteolysis during purification. In 1980, Ub surfaced for a third time in yet another guise. Hershko and his colleagues showed that the ATP-dependent proteolytic factor I (APFI), a 9-kDa polypeptide necessary for ATP-dependent proteolysis in reticulocyte lysates, was covalently linked to various proteins, including exogenous proteolytic substrates (Ciechanover et al 1980; Hershko et al 1980). It was soon demonstrated that APFI was, indeed, ubiquitin (Wilkinson & Audhya 1981).

Properties of Ubiquitin and Its Genes

Although it lacks cysteine and tryptophan, there is nothing unusual about the sequence of Ub other than its extreme conservation. The identical sequences of animal Ubs differ at three positions from the yeast and plant proteins (see Figure 1A). The x-ray structure of Ub, which has been resolved to 2.8 Å, reveals a compact globular protein with the carboxyl terminal arg-gly-gly extended into the solvent (Vijay-Kumar et al 1985). The molecule contains four strands of β-sheet plus a single α-helix with three and one half turns; all sequence differences are located on a small portion of ubiquitin's surface (see Figure 2). NMR studies have shown that ubiquitin remains folded at pHs 1 to 13 and below 80°C (Cary et al 1980). A distinct hydrophobic core and extensive hydrogen bonding are present, which may account for the molecule's exceptional stability.

Wilkinson & Mayer (1986) report increased α-helicity in Ub in response to decreased solvent polarity, and they speculate that Ub may undergo conformational changes following conjugation to a target protein. This is an attractive idea for several reasons. In some cells more than half of the Ub molecules are unconjugated, so one might expect that conformational differences between free and conjugated Ub would reduce competition for components in the activating and proteolytic pathways. The existence of several physiologically important conformations could also explain the

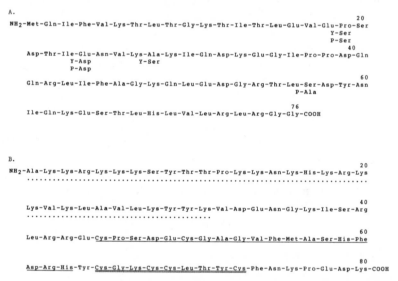

Figure 1 Amino acid sequences of ubiquitin and a carboxyl extension. (A) The sequence of animal ubiquitin is presented using the three-letter amino acid code. Substitutions at positions 19, 24, 28, and 57 are shown below the animal sequence; Y denotes yeast and P denotes plant. (B) The hypothetical amino acid sequence of a carboxyl terminal extension of ubiquitin obtained from a human cDNA (Lund et al 1985). The sequence contains an N-terminal basic region reminiscent of histones (*dotted underline*), a cys/hys region similar to the Zn finger motif (*solid underline*) and a cys cluster like those found in metallothionein (*double underline*).

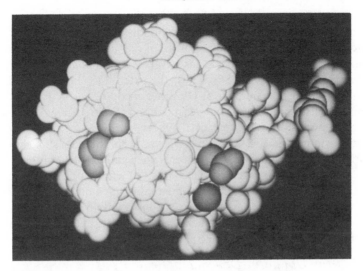

Figure 2 A space-filling model of ubiquitin showing residues altered in oat and yeast ubiquitin. The side-chain atoms of residues that differ from animal ubiquitin are colored grey. This view of ubiquitin is almost directly opposite the carboxyl-terminal extension. (Courtesy of Keith Wilkinson, Emory University.)

severe constraints on ubiquitin's sequence. Support for the existence of distinct conformations is provided by antibodies that preferentially bind free or conjugated forms of the protein (Haas & Bright 1985). Future studies using two-dimensional NMR should provide additional evidence on this important issue.

Ub is encoded by multiple genes in all organisms examined to date. Some ubiquitin genes are located in tandem arrays with repeat lengths varying from five in yeast to twelve in *Xenopus* (Finley & Varshavsky 1985). Since stop signals are not present between Ub coding regions, cleavage at the gly-met bonds separating each Ub molecule presumably generates monomers from a polyubiquitin translation product. Human genomes contain several tandem clusters, whereas yeast contains just one. Deletion of the polyubiquitin locus from yeast does not produce a lethal phenotype under all growth conditions. Rather, the cells die only at sporulation or upon heat stress (Finley et al 1987), which indicates that tandemly arrayed Ub genes enable cells to produce large quantities of the protein during periods of acute need.

Single Ub coding regions are also present. Lund et al (1985) sequenced a human cDNA that would code for ubiquitin and a carboxyl terminal extension of 80 amino acids. The extended sequence contains a 32–amino acid region that is 50% basic residues; other regions exhibit patterns of cysteine and histidine similar to the zinc fingers proposed to exist in nucleic acid binding proteins (Figure 1B). Antibodies produced to a synthetic peptide corresponding to residues 30–42 of the human extension identify a 16-kDa protein in organisms as diverse as *Acanthameba* and man (Redman & Rechsteiner 1986). Thus, the extensions, like Ub itself, are highly conserved. The number of Ub genes encoding carboxyl extensions in animal cells is not known, but yeast contains three such loci, each with a different extension (Ozkaynak et al 1987).

Recent studies indicate that Ub is cleaved from the extension shown in Figure 1 (K. L. Redman & M. C. Rechsteiner, unpublished observation). While the unusual arrangement of coding sequences for Ub and its carboxyl extensions seems designed to maintain stoichiometry between the two molecules, the function of the extensions is unknown. The zinc finger motif in the extension suggests nucleic acid binding, but the arrangement of cysteine residues is also similar to that found in some protease inhibitors.

Models for Ubiquitin-Dependent Proteolysis

Early biochemical studies on intracellular proteolysis were hampered by lysosome breakage. The released cathepsins, present in high concentrations, also removed the C-terminal glycines from Ub, thereby inactivating it (Haas et al 1985). Thus, the 1977 report by Etlinger & Goldberg that rabbit reticulocyte lysates degrade abnormal hemoglobins in the presence of ATP was an important advance. This system exhibited two key features of intracellular proteolysis, selectivity and energy dependence. Subsequent studies by Hershko and his colleagues implicated Ub in ATP-dependent proteolysis. They chromatographed reticulocyte lysate on DEAE-cellulose and showed that degradation required the flow-through proteins (Ub) and proteins eluting at 0.5 M KCl (fraction II). When iodinated Ub was added to fraction II, it became covalently bound to various proteins (Ciechanover et al 1980). This finding led Hershko et al (1980) to propose that attachment of Ub signals degradation of the conjugated protein.

This marking hypothesis, shown schematically in Figure 3, has received considerable support. Five studies document a good correlation between rapid degradation of a protein and its ubiquitination. First, when hemoglobin is injected into cultured mammalian cells and then denatured with phenylhydrazine, it is rapidly degraded. The concentration of globin-ubiquitin conjugates that form upon denaturation is proportional to the rate of globin degradation (Chin et al 1982). Second, proteins that incorporate amino acid analogs are generally degraded rapidly, and the concentration of Ub conjugates increases in Ehrlich ascites cells exposed to analogs (Hershko et al 1982). Third, *Dictyostelium* calmodulin is ubiqui-

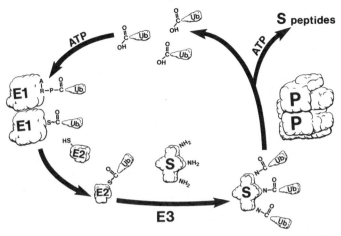

Figure 3 Schematic representation of ubiquitin activation and ATP-dependent proteolysis of conjugated substrates. According to our present understanding, the carboxyl terminus of ubiquitin (Ub) is activated by dimeric enzyme (E1) and transferred to one of several small carrier proteins (E2s) in the form of a reactive thiolester. Ubiquitin is then transferred to lysine amino groups on the proteolytic substrate (S) by a third enzyme (E3). The conjugated substrate is subsequently hydrolyzed by a large ATP-dependent protease (P), and ubiquitin is recycled.

tinated at lysine 119 and subsequently degraded after being added to reticulocyte lysate; bovine calmodulin, which contains a methylated lysine 119, is not conjugated to Ub and is more stable (Gregori et al 1985). Fourth, phytochrome, a cytoplasmic light receptor in plants, exists in two interconvertible forms that differ in half-life more than 100-fold. When dark-grown oat seedlings receive a light flash, rapid degradation of phytochrome follows. At the same time, a portion of the phytochrome molecules become multiply ubiquitinated (Shanklin et al 1987). Fifth, Bachmair et al (1986) constructed ubiquitin/β-galactosidase fusion proteins, and upon their expression in yeast, ubiquitin was rapidly removed from the N terminus of β-galactosidase. When site-directed mutagenesis was used to produce enzymes with different residues at the N-terminus, the resulting proteins varied considerably in stability. Those with met, ser, ala, thr, val, or gly were stable; those with N-terminal phe, leu, asp, lys, or arg were degraded with half-lives of less than 3 min. Western blots showed that the latter proteins were ubiquitinated, some with as many as eleven attached ubiquitins.

Supporting evidence of a different kind has been obtained in studies on a mutant mouse lymphoma line, ts-85. Originally isolated as temperature sensitive in G traverse, the ts lesion was subsequently shown to be a thermolabile E1, the ubiquitin activating enzyme (Finley et al 1984). Incubation of mutant cells or extracts at 39°C, the nonpermissive temperature, led to inactivation of ubiquitination reactions in vitro. The intracellular turnover of abnormal or short-lived proteins was also inhibited at nonpermissive temperatures (Ciechanover et al 1984). Although these studies are cited as evidence for the marking hypothesis, they do not address the model per se. Strictly speaking, they only show that ubiquitin conjugation is necessary for intracellular proteolysis. The ubiquitinated target could be a protease, kinase, inhibitor, or activator. In fact, ts-85 cells synthesize heat-shock proteins at 39°C, and several studies report that decreased proteolysis is a characteristic of the heat-shock response (see Carlson et al 1987). Hence, the inability of ts-85 cells to degrade short-lived proteins at 39°C could reflect a regulated metabolic response rather than the failure to conjugate ubiquitin to proteolytic substrates.

Recognition of ubiquitinated substrates by a specific protease(s) is implicit in the Ub marking hypothesis. Recently, this central feature of the model was confirmed. Haas & Rose 1981 showed that addition of hemin to reticulocyte lysates inhibits proteolysis but not ubiquitin conjugation. Hough & Rechsteiner (1986) exploited this observation to generate ubiquitin-lysozyme conjugates in hemin-inhibited lysates. Upon return to fresh lysate, the partially purified conjugates were degraded ten times faster than free lysozyme. More importantly, these unique substrates allowed Hough et al (1986) to identify a large ATP-dependent protease that degrades ubiquitin-lysozyme conjugates; unmodified lysozyme, however, was not hydrolyzed by the enzyme.

A review of the evidence reveals that the central features of the marking hypothesis have received support during the past six years. Abnormal, denatured, or short-lived proteins are ubiquitinated at levels consistent with their enhanced rates of degradation. In addition, a protease that preferentially degrades ubiquitinated proteins has been purified (Hough et al 1987). The hypothesis should not, however, be considered proved in every detail. Some ubiquitinated substrates are larger than expected from the model; other excellent proteolytic substrates do not appear to form conjugates. Moreover, there are alternate interpretations for certain observations.

Although Ub is attached to cysteines during activation, in conjugates it has only been found linked to lysine ε-amino groups. Since lysozyme contains six lysine residues, according to the simplest form of the marking hypothesis, ubiquitin-lysozyme conjugates should not have molecular weights greater than 65,000 (8,500 for each ubiquitin and 14,000 from lysozyme). Yet, experiments using SDS-PAGE, gel filtration, and sedimentation demonstrate that such conjugates have molecular weights greater than 200,000 (Hough & Rechsteiner 1986). The basis for their anomalous size is not known. Rechsteiner et al (1984a) suggested that Ub might conjugate to itself to produce polyubiquitin chains. This explanation is favored by Hershko & Heller (1985), who showed that large conjugates did not form from Ub whose lysines were converted to homoarginine by guanidination. However, Hough & Rechsteiner (1986) found that disassembly of high molecular weight conjugates in ATP-depleted lysates produced intermediates varying in molecular mass by 34, 22, and 12 kDa, values unexpected from a simple hyperubiquitination model. Moreover, the ratio of [^{125}I]ubiquitin to [^{131}I]lysozyme in double-labeled conjugates was not consistent with hyperubiquitination. It is possible that transglutaminase cross-links ubiquitinated substrates to other proteins. Alternatively, Ub may introduce reactive thiolesters into other proteins (Rechsteiner 1985), and these proteins could, like α-2-macroglobulin or complement components C3 and C4, form cross-links with ubiquitinatedsubstrates. Whatever the eventual explanation, the large size of ubiquitin lysozyme conjugates is not consistent with a simple model for Ub marking.

Several groups have demonstrated that substrates with blocked amino groups are degraded in an ATP-dependent process (Tanaka et al 1983; Katznelson & Kulka 1983). By itself, this finding does not affect the model; ubiquitin-mediated proteolysis need not be the only pathway for degrading cytosolic proteins. In fact, the turnover of ornithine decarboxylase or HMG CoA reductase is unimpaired in ts-85 cells at nonpermissive temperatures (Glass & Gerner 1986; Tanaka et al 1986). If Ub conjugation is abolished in ts-85 cells at 39°C, there must be alternate pathways.

One result difficult to reconcile with the substrate marking model was obtained by Chin et al 1986. They found that normal lysozyme and guanidinated lysozyme were degraded at similar rates in reticulocyte lysates. Whereas Ub conjugates of normal lysozyme were readily apparent, covalent adducts between guanidinated lysozyme and Ub could not be detected even when care was taken to preserve labile bonds, such as thiolesters. Nevertheless, degradation of guanidinated lysozyme was *ubiquitin dependent*. They proposed that Ub has essential functions for proteolysis in addition to direct marking of the substrate.

Speiser & Etlinger (1983) treated reticulocyte lysate with ammonium sulfate and obtained a fraction that rapidly degraded casein in the absence of ATP. Adding back an inhibitor fraction obtained by ammonium sulfate precipitation resulted in a marked decrease in casein degradation, both in the presence and absence of ATP. However, when Ub was added to the inhibited protease and ATP, casein was degraded in an ATP-dependent reaction. These investigators proposed that Ub represses an endogenous protease inhibitor. Recently, Murakami & Etlinger (1986) have reported purification of the inhibitor, which consists of 6 subunits of M_r 40,000 each. Interestingly, the hexamer inhibited both a high molecular weight reticulocyte protease and calcium-activated proteases, both noncompetitively.

The inhibitor hypothesis has been challenged by Eytan & Hershko (1984). They contend that inhibition is due to competition by endogenous substrates. Clearly, this is a controversial issue that will not be easily resolved without purified components in hand. Nevertheless, aspects of both views may be correct. One can imagine that ubiquitinated proteolytic substrates compete with ubiquitinated inhibitors for the active site of an ATP-dependent protease. Hough et al (1986) treated lysate using methods similar to those described by Speiser & Etlinger (1983) and observed the appearance of an ATP-independent protease with properties much like those of the ATP/ubiquitin-dependent enzyme. In view of this result and the substantial evidence for endogenous inhibitors of the calcium-activated proteases (Murachi 1985), it seems premature to dismiss inhibitor hypotheses.

Enzymes of Ubiquitin Metabolism

UBIQUITINATION   Using affinity chromatography on ubiquitin-Sepharose and gel filtration, Hershko et al (1983) reported the isolation of three enzymes required for conjugating Ub to proteins: E1, the Ub activating enzyme; E2s, Ub carrier proteins; and E3, a factor that recognizes proteolytic substrates (see Figure 3). All three components plus Ub, a protease fraction, and ATP are required for the degradation of proteolytic substrates. However, as discussed below, the ubiquitination of certain proteins does not require E3.

Ub is activated in a two-step reaction by E1, an enzyme composed of one or two 100-kDa chains. The carboxyl terminus of ubiquitin is first adenylated and then transferred to a cysteine on E1, with energy being conserved as a highly reactive thiolester. The essential reactions are:

$$\text{Ub}-\text{COOH} + \text{ATP} \; \underset{\text{E1}}{\rightleftharpoons} \; \text{Ub}-\overset{\text{O}}{\underset{||}{\text{C}}}-\text{OAMP} + \text{PP}_i$$

$$\text{Ub}-\overset{\text{O}}{\underset{||}{\text{C}}}-\text{OAMP} + \text{E1}-\text{SH} \; \rightleftharpoons \; \text{E1}-\text{s}-\overset{\text{O}}{\underset{||}{\text{C}}}-\text{Ub} + \text{AMP}$$

The reversibility of the reactions allows E1 to be substantially purified on ubiquitin-Sepharose (Ciechanover et al 1982). However, complete purification of E1 has not been reported, and antibodies are apparently not available. Consequently, we know little about the physical properties of the enzyme or its location in nucleated cells.

The Ub carrier molecules, E2s, transfer activated Ub from E1 to various protein substrates. The essential reactions are:

$$E1_{-S-\overset{O}{\overset{\|}{C}}-}Ub \; + \; E2_{-SH} \; \rightleftharpoons \; E1_{-SH} \; + \; E2_{-S-\overset{O}{\overset{\|}{C}}-}Ub$$

$$E2_{-S-\overset{O}{\overset{\|}{C}}-}Ub \; + \; H_{-NH_2} \; \longrightarrow \; H_{-\underset{H}{N}-\overset{O}{\overset{\|}{C}}-}Ub \; + \; E2_{-SH}$$

$$E2_{-S-\overset{O}{\overset{\|}{C}}-}Ub \; + \; S_{-NH_2} \; \xrightarrow{\text{E3}} \; S_{-\underset{H}{N}-\overset{O}{\overset{\|}{C}}-}Ub \; + \; E2_{-SH}$$

Here H denotes histone H2A, and S denotes a proteolytic substrate.

Hershko et al (1983) originally identified four separate E2 molecules; in more recent studies Pickart & Rose (1985a) have isolated five. The Ub carriers, termed $E2_1$–$E2_5$ have subunit molecular masses of 32, 24, 20, 17, and 14 kDa, respectively. In solution, E2s 1, 3, 4, and 5 behave as homodimers, whereas $E2_2$ is apparently monomeric. Surprisingly, just two of the five carrier proteins ligated Ub to protein substrates. In the presence of E3, the smallest carrier, $E2_5$, was the only one that transferred Ub to the proteolytic substrates, RNase or BSA. Both $E2_3$ and $E2_5$ were able to transfer Ub to histones or cytochrome C in the absence of E3. Thus, Ub carrier proteins are functionally heterogeneous: $E2_5$ participates in the ubiquitination of BSA or RNase; $E2_3$ is capable of directly ubiquitinating histones; and E2s 1, 2, and 4 are assayable only by transfer of Ub to small primary amines in vitro. The latter three carriers surely transfer to larger molecules in vivo, perhaps to specific cellular proteins or to proteolytic substrates other than BSA or RNase.

Of all the components in the conjugation pathway, E3 has proved the most elusive and the most interesting. This factor is required for ubiquitination of proteins as diverse as lysozyme ($M_r = 14,000$; pI $\approx$ 11) and bovine serum albumin ($M_r = 68,000$; pI $= 5.4$). The obvious question is: What is recognized by E3? The best substrates for ubiquitin-mediated proteolysis, lysozyme, RNase, and BSA, have unblocked amino termini and disulfide bonds in common. Earlier studies by Hershko et al (1984) using blocking reagents with differing selectivity for α- or ε-amino groups demonstrated the importance of an unblocked α-amino terminus; they suggested that ubiquitination of the amino terminus was critical. However, none of the identified E2s transfers Ub to amines on secondary carbons, and α-ubiquitinated substrates have not been reported.

While recent studies reaffirm the importance of the amino terminus, they now implicate E3 binding as the critical event. Using chemical cross-linking and pulse-chase experiments, Hershko et al (1986) have identified a 180-kDa protein as E3, the substrate binding factor. Proteins with a free α-NH$_2$ terminus bind better to E3 than do corresponding proteins with a blocked terminus. In addition, oxidation of methionines increased the susceptibility of certain proteins to ubiquitination in parallel with their increased binding to E3. Thus, in selecting proteolytic substrates, E3 appears to recognize oxidized methionines and unblocked α-NH$_2$ termini. However, the ligand need not be capable of accepting Ub. For example, reductively methylated lysozyme was a strong inhibitor of lysozyme binding even though the methylated protein is not conjugated to Ub. Similar results were obtained by Breslow et al (1986), who suggested that E3 and/or E2s recognize the unfolded polypeptide backbone.

These biochemical experiments complement the genetic studies on ubiquitin/β-galactosidase fusion proteins. Bachmair et al (1986) found that the half-life of β-galactosidase depends on the N-terminal residue exposed after Ub cleavage. Since the stabilizing residues are those normally acetylated during chain synthesis, one can imagine that unacetylated nascent polypeptides are recognized by E3 and targeted for destruction. There are two reasons to suspect that "N-end" is a cotranslational rule. The observed half-lives for the most rapidly degraded proteins in the series (2–3 min) are on the same time scale as chain synthesis itself. Furthermore, micro-injection studies in two laboratories have shown that RNase A with its well-exposed, amino-terminal lysine has a half-life longer than 60 hr (Rote & Rechsteiner 1986; McElligott et al 1985). Thus E3 may specifically recognize the N terminus of unfolded polypeptides.

Whereas ubiquitination of histone H2A is restricted to lysine 119, proteolytic substrates are often multiply ubiquitinated. This is clearly illustrated by rapidly degraded forms of β-galactosidase or phytochrome (Bachmair et al 1986; Shanklin et al 1987). On SDS-PAGE both proteins generate distinct ladders with steps of 5 kDa. The larger species, comprising only 1–3% of the protein, stain positively with anti-ubiquitin antibodies. The presence of unmodified and hyperubiquitinated molecules implies that conjugation is processive. Possibly, E3 is limiting. Since high molecular weight conjugates of lysozyme are preferentially degraded, attachment of multiple Ubs may be the critical event that targets a protein for destruction.

Ciechanover et al (1985) showed that tRNA was required for the degradation of [^{125}I]-bovine serum albumin in reticulocyte lysate. Recent studies indicate that Ub conjugation is the step specifically inhibited by RNase and restored by tRNA. Ferber & Ciechanover (1986) found that conjugation of

[^{125}I]ubiquitin to proteins with acidic N termini, including bovine serum albumin, α-lactalbumin, and soybean trypsin inhibitor, was RNase sensitive. However, conjugation of labeled ubiquitin to lysozyme, α-casein, or β-lactalbumin was actually stimulated by RNase. They propose that there are at least two ubiquitin conjugation systems sharing common components since inhibition of the tRNA-dependent pathway stimulates the independent pathway. The role of tRNA has not yet been elucidated.

In summary, ubiquitination is an unusual and important protein modification that targets proteins for destruction and also serves regulatory functions (see below). Whereas the activation reaction is understood reasonably well, the respective roles of E2s and E3 in substrate selection are not. tRNAs are required for ubiquitination of some proteolytic substrates, and E3 may be essential for all. However, some E2s transfer Ub to small, basic proteins in the absence of E3, so the Ub carriers could play a direct role in marking proteins for destruction.

PROTEOLYTIC PATHWAY   Hough et al (1986, 1987) have identified and purified a large (26S) protease that degrades ubiquitin-lysozyme conjugates (see Figure 3). ATP stimulates the breakdown of conjugates fourfold and hydrolysis of a fluorogenic peptide twofold; the protease does not degrade unconjugated lysozyme. Based on sedimentation, gel filtration, and non-denaturing PAGE, the ATP/ubiquitin dependent protease has a sedimentation coefficient of 26S and a molecular weight of about 1,000,000. As might be expected, it is a multisubunit complex containing major subunits between 34,000 and 110,000; several polypeptide chains with M_r between 21,000 and 32,000 are also present.

The 26S protease copurifies through four chromatographic steps with a 20S protease that does not require nucleotides for degradation of protein or peptide substrates. The smaller enzyme is composed of 8–10 subunits with M_r between 21,000 and 32,000 and has a molecular weight of 700,000. It is similar, if not identical, to the "multicatalytic proteinase complex" first described by Wilk & Orlowski (1983).

Both enzymes hydrolyze peptides with tyr, phe, or arg at the P1 position, and both are inhibited by hemin, thiol reagents, chymostatin, and leupeptin. They differ, however, by other criteria. The 26S ATP-dependent protease is inactivated by low levels of sodium dodecyl sulfate or oleic acid, whereas the 20S enzyme is stimulated. Moreover, the 20S enzyme shows little preference for ubiquitinated substrates.

Since these two high molecular weight protease complexes have only recently been identified, we do not know much about them. Their large size and complex subunit structure presumably reflect the presence of regulatory subunits that recognize appropriately marked proteins. They may even be involved in RNA metabolism. The 20S protease bears striking resemblance to "prosomes," 19S ribonucleoprotein particles (RNPs) purified from a number of sources (Martins de Sa et al 1986). Prosomes have recently been implicated in tRNA processing (Castano et al 1986). Future studies on the 26S and 20S protease complexes should prove illuminating.

REGENERATION PATHWAY   Proteolysis is not the only fate of ubiquitinated proteins. Eucaryotic cells contain enzymes that can hydrolyze the isopeptide bond linking Ub to conjugated proteins. Unfortunately, with one exception, isopeptidases have not been extensively studied. Matsui et al (1982) identified an activity in hamster or mouse culture cell extracts that converts ubiquitinated histone H2A (μH2A) to H2A and Ub. The enzyme contains an SH group essential for activity, is inhibited by divalent cations, and copurifies with a polypeptide of M_r 38,000 by SDS-PAGE. More recent studies by Kanda et al (1986) on a similar activity from calf thymus show the enzyme to be specific for glycyl-lysine isopeptide bonds; the isopeptidase would not cleave transglutaminase-catalyzed isopeptides between glutamine and putrescine nor would it cleave glycyl-lysine dipeptides. Hough et al (1986) observed widespread distribution of conjugate disassembly activities upon DEAE-fractionation of reticulocyte lysate, so there are probably a number of isopeptidases.

Figure 3 shows peptides arising from the ubiquitinated substrate. In principle, the immediate products of the ATP-dependent protease could be ubiquitinated peptides or free peptides. A hypothetical protease-isopeptidase complex might start at one end of a multiply ubiquitinated protein and release free Ub as it processively cleaved the polypeptide chain. A distributive protease would hydrolyze peptide bonds between ubiquitinated lysines and thus release Ub peptides. The latter reaction can explain the existence of Ub carboxyl-terminal hydrolase, an enzyme first described by Rose & Warms (1983) and studied in more detail by Pickart & Rose (1985b). The enzyme catalyzes the following reaction:

$$\text{Ub--}\overset{\overset{\text{O}}{\|}}{\text{C}}\text{--R} \longrightarrow \text{Ub--COOH + R--H}$$

where R = lysine ε-NH$_2$, spermidine, glutathione-SH. It has been purified to homogeneity and is a monomer of $M_r = 30,000$. The hydrolase is selective for small ubiquitin derivatives. While it may regenerate free ubiquitin from Ub-peptides, it does not appear to have substantial isopeptidase activity on larger protein conjugates.

The enzyme that processes the polyubiquitin precursor has not yet been described. Although it seems unlikely that Ub-carboxyl-hydrolase has such an activity, the 26S ATP-dependent protease complex might.

Stable Ubiquitin Conjugates

Several pieces of evidence indicate that ubiquitination serves functions besides marking proteins for destruction. SDS-PAGE reveals that injected [^{125}I]Ub labels histones and a characteristic set of polypeptides in addition to a general background "smear" of presumed proteolytic intermediates (Atidia & Kulka 1982; Carlson & Rechsteiner 1987). These discrete conjugates, also evident on Western blots stained with anti-Ub antibodies, probably represent cellular proteins whose activities are reversibly regulated by ubiquitination.

The histone conjugates, μH2A and μH2B, provide clear examples that ubiquitination need not signal proteolysis. Modified forms can comprise a significant portion of the total histone pool even though histone turnover is barely detectable. For example, in L1210 cells, 11% of H2A and 1.5% of H2B are ubiquinated (West & Bonner 1980); in *Physarum* about 6% of each histone is conjugated (Mueller et al 1985). The role of this histone modification is not well understood. Levinger & Varshavsky (1982) found μH2A to be enriched in transcribed chromatin in *Drosophila* and, based on the changes in HSP 70 observed upon heat shock, suggested that limited destruction of nucleosomal proteins might be important for transcription. But, μH2A is not present in the actively transcribed κ-chain genes (Huang et al 1986). Moreover, as Ub is now known to be a heat-shock protein, changes in μH2A concentration at the HSP 70 locus may be specifically related to the heat-shock response. Mezquita et al (1982) report that Ub stimulates histone deacetylase in vitro. Thus, μH2A and μH2B may affect transcription indirectly.

Whatever their role in interphase chromatin, ubiquitinated histones do not appear compatible with chromatin condensation at mitosis. μH2A and μH2B are absent from metaphase chromosomes in all organisms so far examined. This rapid and transient removal of Ub is particularly striking in *Physarum*, in which as many as 10^9 nuclei divide synchronously every 9 hr; ubiquitinated histones are absent for a 7–10 min period at metaphase (Mueller et al 1985).

The lymphocyte homing receptor provides another example of a stable ubiquitin conjugate, and a surprising one at that. In attempts to clone the epitope recognized by MEL 14, a monoclonal antibody that prevents lymphocyte homing, Siegelman et al (1986) and St. John et al (1986) found positive λgt11 clones to contain ubiquitin genes. Radiochemical sequencing of the antibody-precipitated receptor provided additional evidence that Ub was conjugated to it, and because MEL 14 binds the cell surface, it was concluded that Ub is attached to an extracellular domain on the receptor.

The genesis of surface Ub conjugates presents a problem. Few enzymes are located both in the secretory pathway and the cytosol. So unless El and E2 are exceptions, it is unclear how Ub could be activated within the endoplasmic reticulum (ER) or Golgi apparatus. One might propose cotranslational insertion of the ubiquitinated receptor, but given the inhibition of dihydrofolate reductase (DHFR) uptake into mitochondria after its presumed folding by methotrexate (Eilers & Schatz 1986), transfer of a conjugated protein through the ER membrane seems improbable.

In view of these difficulties, it is worth considering other possible pathways to the cell surface. Indeed, alternate routes have been advanced to explain cell surface SV40 large T antigen. Sharma et al (1985) proposed that vesicles of the nuclear envelope fuse with the plasma membrane during mitosis. The presentation of viral nucleoprotein antigens to cytotoxic T lymphocytes also seems to require a novel mechanism for externalization (Townsend 1987). Finally, some puzzling results from microinjection studies can be explained by a second pathway. Rechsteiner et al (1984b) injected a large number of different proteins into HeLa cells and found that the proteins were both degraded within cells and released from them, apparently intact. Since cells can encapsulate cytosol within membrane vesicles, Rechsteiner et al speculated that such vesicles might fuse with endosomes, and proteins within the vesicles could shuttle to the cell surface. Cells can release surface blebs, so as an alternative possibility those features of proteins that result in rapid degradation might also promote their interaction with the cytoskeleton and subsequent release to the medium. Release and turnover were positively correlated, i.e. rapidly degraded proteins were also rapidly released. This observation suggests that the two processes may have a common step (Ub conjugation?).

Ubiquitin Pool Dynamics

NORMAL PHYSIOLOGICAL CONDITIONS   Ub rivals actin, tubulin, and histones as one of the most abundant intracellular proteins. Using immunological methods, Haas & Bright (1985) determined that IMR 90 fibroblasts and green monkey kidney cells contain, respectively, 8×10^7 and 1.8×10^8 Ub molecules per cell. They also reported that anywhere from 50 to 80% of the Ub was conjugated. Both estimates have been confirmed by microinjecting [^{125}I]Ub into cultured cells. Carlson & Rechsteiner (1987) found that under normal physiological conditions HeLa cells contain 2×10^8 Ub molecules of which 10% are conjugated to histones, 40% are present in conjugates with M_rs greater than 30,000, 15% are present in thioester linkage to activating enzymes, and the remainder are free.

Microinjection experiments also show that Ub is in rapid equilibrium with intracellular protein acceptors. Atidia & Kulka (1982) used red blood cell (RBC)-mediated injection to introduce radiolabeled Ub into hepatoma cells and found the high molecular weight conjugates to be maximally labeled within minutes. Similar results were obtained by Carlson & Rechsteiner (1987), who noted, however, that histone ubiquitination was slower, requiring 60–120 min to reach equilibrium. Ubiquitin is degraded with a half-life of 10–30 hr (Wu et al 1981; Carlson & Rechsteiner 1987; Haas & Bright 1987). Hence, long-term studies on labeled conjugate pools are difficult. Nevertheless, Carlson & Rechsteiner (1987) were able to compare the patterns of HeLa conjugates over a 30-hr period postinjection and found them unchanged. This indicates that stable Ub conjugates are rare or absent since the concentration of such conjugates should increase with time. The absence of stable conjugates and the rapid redistribution of Ub to denatured globin observed in the same study imply that most Ub molecules are in dynamic equilibrium with HeLa proteins.

HEAT SHOCK   After heat treatment or exposure to amino acid analogs, cells respond by increasing the synthesis of a small set of proteins, called heat-shock proteins (HSPs). This response is found in all organisms from bacteria to man and confers the ability to withstand a second exposure to heat (see Lindquist 1986 for review). In 1980, Hightower suggested that HSPs are synthesized in response to the generation of abnormal proteins and that HSPs may be involved in the degradation of these newly formed substrates. This hypothesis has received increasing support over the past seven years. HSPs are synthesized in *Drosophila* cells producing mutant actins (Hiromi & Hotta 1985); HSP synthesis can be induced in oocytes after injection of denatured proteins (Ananthan et al 1986); and two known HSPs, *Escherichia coli* protease La and ubiquitin, are components of proteolytic pathways (Phillips et al 1984; Bond & Schlesinger 1985). Finley et al (1984) and Munro & Pelham (1985) have proposed that ubiquitination of transcription factors may control the heat-shock response. The hypothetical factor would be ubiquitinated under normal conditions, thereby making it a substrate for Ub-dependent proteolysis or otherwise inactivating it. After heat-shock, this transcription factor would be stabilized by competition with other substrates or activated by conversion to its nonubiquitinated form.

Two recent experiments provide support for such speculations. Parag et al (1987) exposed rat hepatoma cells to a 43°C heat shock and observed a transient acceleration of proteolysis concomitant with a fall in Ub and μH2A. Carlson et al (1987) introduced labeled Ub into HeLa cells and then incubated them at 45°C for 5 min (reversible heat shock). This produced dramatic changes in the levels of Ub conjugates. Under normal culture conditions, $\sim 10\%$ of the injected ubiquitin is linked to histones, 40% is found in conjugates with molecular weights greater than 25,000, and the rest is unconjugated. After heat-shock, free Ub and μH2A decreased rapidly, while high molecular weight conjugates predominated. Concomitant with the loss of free ubiquitin, the degradation of both endogenous proteins and injected hemoglobin, BSA, and ubiquitin was reduced in heat-shocked HeLa cells. The shift of Ub to high molecular weight conjugates of heat-denatured cytoplasmic proteins could spare a transcription factor from inactivation. Similarly, depletion of μH2A might promote expression of heat-shock genes since ubiquitinated histones alone may be sufficient to prevent transcription. Production of HSPs would then be sensitive to levels of abnormal proteins, as suggested by Hightower, and could be reversed by increasing ubiquitin levels or by degrading the abnormal proteins, thereby restoring the pool of free ubiquitin.

Ubiquitin and Developing Tissues

Lysosomal proteolysis plays a central role in amphibian tail resorption, and in an analogous fashion, Ub-mediated proteolysis might have specific developmental roles. Rapoport (1986) has stressed the importance of the Ub-dependent pathway in RBC maturation. He proposes that RBC mitochondria are damaged by lipoxygenase and degraded by the nonlysosomal Ub pathway. RBCs are unusual in another respect. They produce large amounts of Hb, a protein that requires unblocked amino termini for physiological reasons (Bohr effect). These two features of erythroid maturation may result in enhanced expression of the Ub proteolytic pathway. Whole-scale destruction of cellular organelles also occurs during lens development, and ubiquitin-mediated proteolysis is prevalent in this tissue as well (Jahngen et al 1986).

SUMMARY

Ubiquitination is one of several ways in which cells modify their proteins. As for phosphorylation or acetylation, there are distinct enzymes for adding and removing Ub from the surfaces of protein substrates. The dynamic equilibration of Ub with cellular proteins is also typical of most posttranslational modifications. Ubiquitination differs, however, in that the added group is large compared to acetate or phosphate. Its size must provide great potential for recognition by other cellular proteins. Ub may be the cell's reversible cross-linking reagent, covalently bound to protein substrates at one end and noncovalently associated with various Ub binding proteins at the other.

It is likely that one ubiquitin binding protein is a component of the 26S ATP-dependent protease. The presence of Ub on histones and on the lymphocyte homing receptor suggests that ubiquitination does not serve exclusively to mark proteins for degradation. There are probably various ubiquitin binding proteins since Ub appears to be a multifunctional protein that affects chromatin structure, intracellular proteolysis, cellular interactions, and the stress response. This abundant protein may serve as an intracellular barometer whose distribution among several pools regulates a variety of processes.

ACKNOWLEDGMENTS

I would like to thank Kristen Ballantyne, Ron Hough, Nancy Johnson, Greg Pratt, Kevin Rote, and Rod Wells for helpful suggestions on the manuscript. Special thanks go to Florence Rechsteiner and Mary Beckerle for stylistic changes that improved the final essay considerably. A special thanks is also due Kristen Ballantyne for her cheerful attitude during numerous revisions. Research from my laboratory was supported by grants from the National Institutes of Health.

Literature Cited

Abovich, N., Gritz, L., Tung, L., Rosbash, M. 1985. Effect of RP51 gene dosage alterations on ribosome synthesis in *Saccharomyces cerevisiae. Mol. Cell. Biol.* 5: 3429–35

Ahlberg, J., Berkenstam, A., Henell, F., Glaumann, H. 1985. Degradation of short and long lived proteins in isolated rat liver lysosomes: Effects of pH, temperature, and proteolytic inhibitors. *J. Biol. Chem.* 260: 5847–54

Ahlberg, J., Glaumann, H. 1985. Uptake—microautophagy—and degradation of exogenous proteins by isolated rat liver lysosomes: Effects of pH, ATP, and inhibitors of proteolysis. *Exp. Mol. Pathol.* 42: 78–88

Amenta, J. S., Brocher, S. C. 1981. Mechanisms of protein turnover in cultured cells. *Life Sci.* 28: 1195–1208

Ananthan, J., Goldberg, A. L., Voellmy, R. 1986. Abnormal proteins serve as eukaryotic stress signals and trigger the activation of heat shock genes. *Science* 232: 522–24

Atidia, J., Kulka, R. G. 1982. Formation of conjugates by ^{125}I-labelled ubiquitin microinjected into cultured hepatoma cells. *FEBS Lett.* 142: 72–76

Bachmair, A., Finley, D., Varshavsky, A. 1986. In vivo half-life of a protein is a function of its amino-terminal residue. *Science* 234: 179–86

Beynon, R. J., Bond, J. S. 1986. Catabolism of intracellular protein: Molecular aspects. *Am. J. Physiol.* 251: C142–52

Beynon, R. J., Cookson, E. J., Butler, P. E. 1985. Intracellular proteolysis: The elusive intermediates. *Biochem. Soc. Trans.* 13: 1005–7

Bigelow, S., Hough, R., Rechsteiner, M. 1981. The selective degradation of injected proteins occurs principally in the cytosol rather than in lysosomes. *Cell* 25: 83–93

Bond, J. S., Butler, P. E. 1987. Intracellular proteases. *Ann. Rev. Biochem.* 56: 333–64

Bond, U., Schlesinger, M. J. 1985. Ubiquitin is a heat shock protein in chicken embryo fibroblasts. *Mol. Cell. Biol.* 5: 949–56

Brady, S. T., Lasek, R. J. 1981. Nerve-specific enolase and creatine phosphokinase in axonal transport: Soluble proteins and the axoplasmic matrix. *Cell* 23: 515–23

Breslow, E., Daniel, R., Ohba, R., Tate, S. 1986. Inhibition of ubiquitin-dependent proteolysis by nonubiquitinatable proteins. *J. Biol. Chem.* 261: 6530–35

Carlson, N., Rechsteiner, M. 1987. Microinjection of ubiquitin: Intracellular distribution and metabolism in HeLa cells maintained under normal physiological conditions. *J. Cell Biol.* 104: 537–46

Carlson, N., Rogers, S., Rechsteiner, M. 1987. Microinjection of ubiquitin: Changes in protein degradation in HeLa cells subjected to heat-shock. *J. Cell Biol.* 104: 547–55

Carroll, S. B., Scott, M. P. 1986. Zygotically active genes that affect the spatial expression of the *fushi tarazu* segmentation gene during early *Drosophila* embryogenesis. *Cell* 45: 113–26

Cary, P. D., King, D. S., Crane-Robinson, C., Bradbury, E. M., Rabbini, L., et al. 1980. Structural studies on two high-mobility-group proteins from calf thymus, HMG-14 and HMG-20 (ubiquitin), and their interaction with DNA. *Eur. J. Biochem.* 112: 577–80

Castano, J. G., Ornberg, R., Koster, J. G., Tobian, J. A., Zasloff, M. 1986. Eukaryotic pre-tRNA 5′ processing nuclease: Copurification with a complex cylindrical particle. *Cell* 46: 377–87

Chandler, C. S., Ballard, F. J. 1983. Inhibition of pyruvate carboxylase degradation and total protein breakdown by lysosomotropic agents in 3T3-L1 cells. *Biochem. J.* 210: 845–53

Chandler, C. S., Ballard, F. J. 1986. Multiple biotin-containing proteins in 3T3-L1 cells. *Biochem. J.* 237: 123–30

Chin, D. T., Carlson, N., Kuehl, L., Rechsteiner, M. 1986. The degradation of guanidinated lysozyme in reticulocyte lysate. *J. Biol. Chem.* 261: 3883–90

Chin, D. T., Kuehl, L., Rechsteiner, M. 1982. Conjugation of ubiquitin to denatured hemoglobin is proportional to the rate of hemoglobin degradation in HeLa cells. *Proc. Natl. Acad. Sci. USA* 79: 5857–61

Chiu, F. C., Goldman, J. E. 1984. Synthesis and turnover of cytoskeletal proteins in cultured astrocytes. *J. Neurochem.* 42: 166–74

Ciechanover, A., Elias, S., Heller, H., Hershko, A. 1982. "Covalent affinity" purification of ubiquitin-activating enzyme. *J. Biol. Chem.* 257: 2537–42

Ciechanover, A., Finley, D., Varshavsky, A. 1984. Ubiquitin dependence of selective protein degradation demonstrated in the mammalian cell cycle mutant ts85. *Cell* 37: 57–66

Ciechanover, A., Heller, H., Elias, S., Haas, A. L., Hershko, A. 1980. ATP-dependent conjugation of reticulocyte proteins with the polypeptide required for protein degradation. *Proc. Natl. Acad. Sci. USA* 77: 1365–68

Ciechanover, A., Wolin, S. L., Steitz, J. A., Lodish, H. F. 1985. Transfer RNA is an essential component of the ubiquitin- and ATP-dependent proteolytic system. *Proc. Natl. Acad. Sci. USA* 82: 1341–45

Dani, C., Blanchard, J. M., Piechaczyk, M., El Sabouty, S., Marty, L., et al. 1984. Extreme instability of myc mRNA in normal and transformed human cells. *Proc. Natl. Acad. Sci. USA* 81: 7046–50

Dean, R. T. 1984. Modes of access of macromolecules to the lysosomal interior. *Biochem. Soc. Trans.* 12: 911–13

DeLotto, R., Spierer, P. 1986. A gene required for the specification of dorsal-ventral pattern in *Drosophila* appears to encode a serine protease. *Nature* 323: 688–92

Dice, J. F., Chiang, H.-L., Spencer, E. P., Backer, J. M. 1986. Regulation of catabolism of microinjected ribonuclease A. *J. Biol. Chem.* 261: 6853–59

Edgar, B. A., Weir, M. P., Schubiger, G., Kornberg, T. 1986. Repression and turnover pattern of *fushi tarazu* RNA in the early *Drosophila* embryo. *Cell* 47: 747–54

Eilers, M., Schatz, G. 1986. Binding of a specific ligand inhibits import of a purified precursor protein into mitochondria. *Nature* 322: 228–32

Etlinger, J. D., Goldberg, A. L. 1977. A soluble ATP-dependent proteolytic system responsible for the degradation of abnormal proteins in reticulocytes. *Proc. Natl. Acad. Sci. USA* 74: 54–58

Eytan, E., Hershko, A. 1984. Relevance of protease "inhibitor" to the ATP-ubiquitin proteolytic system. *Biochem. Biophys. Res. Commun.* 122: 116–23

Fagan, J. M., Waxman, L., Goldberg, A, L. 1986. Red blood cells contain a pathway for the degradation of oxidant-damaged hemoglobin that does not require ATP or ubiquitin. *J. Biol. Chem.* 261: 5705–13

Ferber, S., Ciechanover, A. 1986. Transfer RNA is required for conjugation of ubiquitin to selective substrates of the ubiquitin- and ATP-dependent proteolytic system. *J. Biol. Chem.* 261: 3128–34

Finley, D., Ciechanover, A., Varshavsky, A. 1984. Thermolability of ubiquitin-activating enzyme from the mammalian cell cycle mutant ts85. *Cell* 37: 43–55

Finley, D., Ozkaynak, E., Varshavsky, A. 1987. The yeast polyubiquitin gene is essential for resistance to high temperatures, starvation and other stresses. *Cell* 48: 1035–46

Finley, D., Varshavsky, A. 1985. The ubiquitin system: Functions and mechanisms. *TIBS* 10: 343–47

Glass, J. R., Gerner, E. W. 1986. Spermidine mediates degradation of ornithine decarboxylase by a nonlysosomal, ubiquitin-independent mechanism. *J. Cell Biol.* 103: 327a

Goldberg, A. L., St. John, A. C. 1976. Intracellular protein degradation in mammalian and bacterial cells: Part 2. *Ann. Rev. Biochem.* 45: 747–803

Goldknopf, I. L., Busch, H. 1977. Isopeptide linkage between nonhistone and histone 2A polypeptides of chromosomal conjugate-protein A24. *Proc. Natl. Acad. Sci. USA* 74: 864–68

Goldstein, G., Scheid, M., Hammerling, U., Boyse, E. A., Schlesinger, D. H., et al. 1975. Isolation of a polypeptide that has lymphocyte-differentiating properties and is probably represented universally in living cells. *Proc. Natl. Acad. Sci. USA* 72: 11–15

Goldstein, J. L., Brunschede, G. Y., Brown, M. S. 1975. Inhibition of the proteolytic degradation of low density lipoprotein in human fibroblasts by chloroquine, concanavalin A, and Triton WR 1339. *J. Biol. Chem.* 250: 7854–62

Gregori, L., Marriott, D., West, C. M., Chau, V. 1985. Specific recognition of calmodulin from *Dictyostelium discoideum* by the ATP, ubiquitin-dependent degradative pathway. *J. Biol. Chem.* 260: 5232–35

Gronostajski, R. M., Pardee, A. B., Goldberg, A. L. 1985. The ATP dependence of the degradation of short- and long-lived proteins in growing fibroblasts. *J. Biol. Chem.* 260: 3344–49

Haas, A. L., Bright, P. M. 1985. The immunochemical detection and quantitation of intracellular ubiquitin-protein conjugates. *J. Biol. Chem.* 260: 12464–73

Haas, A. L., Bright, P. M. 1987. The dynamics of ubiquitin pools within cultured human lung fibroblasts. *J. Biol. Chem.* 262: 345–51

Haas, A. L., Murphy, K. E., Bright, P. M. 1985. The inactivation of ubiquitin accounts for the inability to demonstrate ATP, ubiquitin-dependent proteolysis in liver extracts. *J. Biol. Chem.* 260: 4694–4703

Haas, A. L., Rose, I. A. 1981. Hemin inhibits ATP-dependent ubiquitin-dependent proteolysis: Role of hemin in regulating ubiquitin conjugate degradation. *Proc. Natl. Acad. Sci. USA* 78: 6845–48

Hendil, K. B. 1981. Autophagy of metabolically inert substances injected into fibroblasts in culture. *Exp. Cell Res.* 135: 157–66

Hershko, A., Ciechanover, A. 1986. The ubiquitin pathway for the degradation of intracellular proteins. *Progr. Nucl. Acid Res. Mol. Biol.* 33: 19–56

Hershko, A., Ciechanover, A., Heller, H., Haas, A. L., Rose, I. A. 1980. Proposed role of ATP in protein breakdown: Conjugation of proteins with multiple chains of the polypeptide of ATP-dependent proteolysis. *Proc. Natl. Acad. Sci. USA* 77: 1783–86

Hershko, A., Eytan, E., Ciechanover, A., Haas, A. L. 1982. Immunochemical analysis of the turnover of ubiquitin-protein conjugates in intact cells. *J. Biol. Chem.* 257: 13964–70

Hershko, A., Heller, H. 1985. Occurrence of a polyubiquitin structure in ubiquitin-protein conjugates. *Biochem. Biophys. Res. Commun.* 128: 1079–86

Hershko, A., Heller, H., Elias, S., Ciechanover, A. 1983. Components of ubiquitin-protein ligase system: Resolution, affinity purification, and role in protein breakdown. *J. Biol. Chem.* 258: 8206–14

Hershko, A., Heller, H., Eytan, E., Kaklij, G., Rose, I. A. 1984. Role of the α-amino group of protein in ubiquitin-mediated protein breakdown. *Proc. Natl. Acad. Sci. USA* 81: 7021–25

Hershko, A., Heller, H., Eytan, E., Reiss, Y. 1986. The protein substrate binding site of the ubiquitin-protein ligase system. *J. Biol. Chem.* 261: 11992–99

Hershko, A., Tomkins, G. M. 1971. Studies on the degradation of tyrosine aminotransferase in hepatoma cells in culture: Influence of the composition of the medium and adenosine triphosphate dependence. *J. Biol. Chem.* 246: 710–14

Hightower, L. E. 1980. Cultured animal cells exposed to amino acid analogs or puromycin rapidly synthesize several polypeptides. *J. Cell Physiol.* 102: 407–27

Hiromi, Y., Hotta, Y. 1985. Actin gene mutations in *Drosophila*; heat shock activation in the indirect flight muscles. *EMBO J.* 4: 1681–87

Hough, R., Pratt, G., Rechsteiner, M. 1986. Ubiquitin-lysozyme conjugates: Identification and characterization of an ATP-dependent protease from rabbit reticulocyte lysates. *J. Biol. Chem.* 261: 2400–8

Hough, R., Pratt, G., Rechsteiner, M. 1987. Purification of two high molecular weight proteases from rabbit reticulocyte lysate. *J. Biol. Chem.* In press

Hough, R., Rechsteiner, M. 1984. Effects of temperature on the degradation of proteins in rabbit reticulocyte lysates and after injection into HeLa cells. *Proc. Natl. Acad. Sci. USA* 81: 90–94

Hough, R., Rechsteiner, M. 1986. Ubiquitin-lysozyme conjugates: Purification and susceptibility to proteolysis. *J. Biol. Chem.* 261: 2391–99

Huang, S.-Y., et al. 1986. The active immunoglobulin κ chain gene is packaged by non-ubiquitin-conjugated nucleosomes. *Proc. Natl. Acad. Sci. USA* 83: 3738–42

Hunt, L. T., Dayhoff, M. O. 1977. Amino-terminal sequence identity of ubiquitin and the nonhistone component of nuclear protein A24. *Biochem. Biophys. Res. Commun.* 74: 650

Jahngen, J. H., et al. 1986. The eye lens has an active ubiquitin-protein conjugation system. *J. Biol. Chem.* 261: 13760–67

Kanda, F., Sykes, D. E., Yasuda, H., Sandberg, A. A., Matsui, S. 1986. Substrate recognition of isopeptidase: Specific cleavage of the ε-(α-glycyl)lysine linkage in ubiquitin-protein conjugates. *Biochim. Biophys. Acta* 870: 64–75

Katznelson, R., Kulka, R. G. 1983. Degradation of microinjected methylated and unmethylated proteins in hepatoma tissue culture cells. *J. Biol. Chem.* 258: 9597–99

Kedersha, N. L., Rome, L. H. 1986. Isolation and characterization of a novel ribonucleoprotein particle: Large structures contain a single species of small RNA. *J. Cell Biol.* 103: 699–709

Kmiec, E. B., Worcel, A. 1986. The positive transcription factor of the 5S RNA gene proteolyses during direct exchange between 5S DNA sites. *J. Cell Biol.* 103: 673–81

Knowles, S. E., Ballard, F. J. 1976. Selective control of the degradation of normal and aberrant proteins in Reuber H35 hepatoma cells. *Biochem. J.* 156: 609–17

Kominami, E., Hashida, S., Khairallah, E. A., Katunuma, N. 1983. Sequestration of cytoplasmic enzymes in an autophagic vacuole-lysosomal system induced by injection of leupeptin. *J. Biol. Chem.* 258: 6093–6100

Levinger, L., Varshavsky, A. 1982. Selective arrangement of ubiquitinated and D1 protein-containing nucleosomes within the *Drosophila* genome. *Cell* 28: 375–85

Libby, P., Goldberg, A. L. 1981. Comparison of the control and pathways for degradation of the acetylcholine receptor and average protein in cultured muscle cells. *J. Cell. Physiol.* 107: 185–94

Lindquist, S. 1986. The heat-shock response. *Ann. Rev. Biochem.* 55: 1151–91

Lund, P. K., Moats-Staats, B. M., Simmons, J. G., Hoyt, E., D'Ercole, A. J., et al. 1985. Nucleotide sequence analysis of a cDNA encoding human ubiquitin reveals that ubiquitin is synthesized as a precursor. *J. Biol. Chem.* 260: 7609–13

Martins de Sa, C. M., Grossi de Sa, M.-F., Akyakat, O., Broders, F., Scherrer, K., et al. 1986. Prosomes: Ubiquity and interspecies structural variation. *J. Mol. Biol.* 187: 479–93

Matsui, S., Sandberg, A. A., Negoro, S., Seon, B. K., Goldstein, G. 1982. Isopeptidase: A novel eukaryotic enzyme that cleaves isopeptide bonds. *Proc. Natl. Acad. Sci. USA* 79: 1535–39

Mayer, R. J., Doherty, F. 1986. Intracellular protein catabolism: State of the art. *FEBS Lett.* 198: 181–93

McElligott, M. A., Miao, P., Dice, J. F. 1985. Lysosomal degradation of ribonuclease A and ribonuclease S-protein microinjected into the cytosol of human fibroblasts. *J.*

Biol. Chem. 260: 11986–93

McGarry, T., Hough, R., Rogers, S., Rechsteiner, M. 1983. Intracellular distribution and degradation of immunoglobulin G and immunoglobulin G fragments injected into HeLa cells. *J. Cell Biol.* 96: 338–46

Mezquita, J., Chiva, M., Vidal, S., Mezquita, C. 1982. Effect of high mobility group nonhistone proteins HMG-20 (ubiquitin) and HMG-17 on histone deacetylase activity assayed in vitro. *Nucl. Acids Res.* 10: 1781–97

Mitchell, H. K., Petersen, N. S., Buzin, C. H. 1985. Self-degradation of heat shock proteins. *Proc. Natl. Acad. Sci. USA* 82: 4969–73

Mortimore, G. E. 1982. Mechanisms of cellular protein catabolism. *Nutr. Rev.* 40: 1–12

Mueller, R. D., Yasuda, H., Hatch, C. L., Bonner, W. M., Bradbury, E. M. 1985. Identification of ubiquitinated histones 2A and 2B in *Physarum polycephalum*: Disappearance of these proteins at metaphase and reappearance at anaphase. *J. Biol. Chem.* 260: 5147–53

Munro, S., Pelham, H. 1985. What turns on heat shock genes? *Nature* 317: 477–78

Murachi, T. 1985. The proteolytic system involving calpains. *Biochem. Soc. Trans.* 13: 1015–18

Murakami, K., Etlinger, J. D. 1986. Endogenous inhibitor of nonlysosomal high molecular weight protease and calcium-dependent protease. *Proc. Natl. Acad. Sci. USA* 83: 7588–92

Neff, N. T., DeMartino, G. N., Goldberg, A. L. 1979. The effect of protease inhibitors and decreased temperature on the degradation of different classes of proteins in cultured hepatocytes. *J. Cell. Physiol.* 101: 439–58

North, G. 1986. Pattern formation: Descartes and the fruitfly. *Nature* 322: 404–5

Ozkaynak, E., Finley, D., Solomon, M. J., Varshavsky, A. 1987. The yeast ubiquitin genes: A family of natural gene fusions. *EMBO J.* 6: 1429–39

Parag, H. A., Raboy, B., Kulka, R. G. 1987. Effect of heat shock on protein degradation in mammalian cells: Involvement of the ubiquitin system. *EMBO J.* 6: 55–61

Phillips, T. T., VanBogelen, R., Neidhardt, F. 1984. *Lon* gene product of *Escherichia coli* is a heat-shock protein. *J. Bacteriol.* 159: 283–87

Pickart, C. M., Rose, I. A. 1985a. Functional heterogeneity of ubiquitin carrier proteins. *J. Biol. Chem.* 260: 1573–81

Pickart, C. M., Rose, I. A. 1985b. Ubiquitin carboxyl-terminal hydrolase acts on ubiquitin carboxyl-terminal amides. *J. Biol. Chem.* 261: 7903–10

Poole, B., Wibo, M. 1973. Protein degradation in cultured cells: The effect of fresh medium, fluoride, and iodoacetate on the digestion of cellular protein of rat fibroblasts. *J. Biol. Chem.* 248: 6221–26

Rapoport, S. M. 1986. Functions of the ubiquitin system. *TIBS* 11: 67

Rechsteiner, M. 1985. Ubiquitin-dependent proteolysis in eucaryotic cells. *Curr. Top. Plant Biochem. Physiol.* 4: 15–24

Rechsteiner, M., ed. 1988. *Ubiquitin.* New York: Plenum. In press

Rechsteiner, M., Rogers, S., Rote, K. V. 1987. Protein structure and intracellular stability. *TIBS.* In press

Rechsteiner, M., Carlson, N., Chin, D., Hough, R., Rogers, S., et al. 1984a. On the role of covalent protein modification and protein aggregation in intracellular proteolysis. In *Protein Transport and Secretion,* ed. D. L. Oxender, pp. 391–402. New York: Liss

Rechsteiner, M., Chin, D., Hough, R., McGarry, T., Rogers, S., et al. 1984b. What determines the degradation rate of an injected protein? *CIBA Found. Symp.* 103: 181–201

Redman, K. L., Rechsteiner, M. C. 1986. Extended ubiquitin may function as more than a precursor. *J. Cell Biol.* 103: 326a

Reid, G. A., Schatz, G. 1982. Import of proteins into mitochondria: Yeast cells grown in the presence of carbonyl cyanide *m*-chlorophenylhydrazone accumulate massive amounts of some mitochondrial precursor polypeptides. *J. Biol. Chem.* 257: 13056–61

Reznick, A. Z., Rosenfelder, L., Shpund, S., Gershon, D. 1985. Identification of intracellular degradation intermediates of aldolase B by antiserum to the denatured enzyme. *Proc. Natl. Acad. Sci. USA* 82: 6114–18

Robinson, A. B. 1974. Evolution and distribution of glutaminyl and asparaginyl residues in proteins. *Proc. Natl. Acad. Sci. USA* 71: 885–88

Rogers, S., Wells, R., Rechsteiner, M. 1986. Amino acid sequences common to rapidly degraded proteins: The PEST hypothesis. *Science* 234: 364–68

Rogers, S. W., Rechsteiner, M. C. 1985. Degradation rates and intracellular distributions of structurally characterized proteins injected into HeLa cells. In *Intracellular Protein Catabolism,* ed. E. Khairallah, J. S. Bond, J. W. C. Bird, 405–16 pp. New York: Liss

Rose, I. A., Warms, J. V. B. 1983. An enzyme with ubiquitin carboxy-terminal esterase

activity from reticulocytes. *Biochemistry* 22: 4234–37

Rote, K. V., Rechsteiner, M. 1983. Degradation of microinjected proteins: Effects of lysosomotropic agents and inhibitors of autophagy. *J. Cell. Physiol.* 116: 103–10

Rote, K. V., Rechsteiner, M. 1986. Degradation of proteins microinjected into HeLa cells: The role of substrate flexibility. *J. Biol. Chem.* 261: 15430–36

Schimke, R. T. 1973. Control of enzyme levels in mammalian tissues. *Adv. Enzymol.* 37: 135–87

Schlesinger, D. H., Goldstein, G., Niall, H. D. 1975. The complete amino acid sequence of ubiquitin, an adenylate cyclase stimulating polypeptide probably universal in living cells. *Biochemistry* 14: 2214–18

Shaeffer, J. R. 1983. Turnover of excess hemoglobin α chains in β-thalassemic cells is ATP-dependent. *J. Biol. Chem.* 258: 13172–77

Shanklin, J., Jabben, M., Vierstra, R. D. 1987. Red light–induced formation of ubiquitin-phytochrome conjugates: Identification of possible intermediates of phytochrome degradation. *Proc. Natl. Acad. Sci. USA* 84: 359–63

Sharma, S., Rodgers, L., Brandsma, J., Gething, M.-J., Sambrook, J. 1985. SV40 T antigen and the exocytotic pathway. *EMBO J.* 4: 1479–89

Siegelman, M., Bond, M. W., Gallatin, W. M., St. John, T., Smith, H. T., et al. 1986. Cell surface molecule associated with lymphocyte homing is a ubiquitinated branched-chain glycoprotein. *Science* 231: 823–29

Siekevitz, P. 1972. The turnover of proteins and the usage of information. *J. Theor. Biol.* 37: 321–34

Simpson, M. V. 1953. The release of labeled amino acids from the proteins of rat liver slices. *J. Biol. Chem.* 201: 143–54

Slilaty, S. N., Rupley, J. A., Little, J. W. 1986. Intramolecular cleavage of LexA and Phage λ repressors: Dependence of kinetics on repressor concentration, pH, temperature, and solvent. *Biochemistry* 25: 6866–75

Speiser, S., Etlinger, J. D. 1983. ATP stimulates proteolysis in reticulocyte extracts by repressing an endogenous protease inhibitor. *Proc. Natl. Acad. Sci. USA* 80: 3577–80

Stadtman, E. R. 1986. Oxidation of proteins by mixed-function oxidation systems: Implication in protein turnover, ageing and neutrophil function. *TIBS* 11: 11–12

Steinberg, D., Vaughan, M. 1956. Observations on intracellular protein catab-

30   RECHSTEINER

olism studied in vitro. *Arch. Biochem. Biophys.* 65: 93–105

St. John, T., Gallatin, W. M., Siegelman, M., Smith, H. T., Fried, V. A., et al. 1986. Expression cloning of lymphocyte homing receptor cDNA: Ubiquitin is the reactive species. *Science* 231: 845–50

Tanaka, K., Waxman, L., Goldberg, A. L. 1983. ATP serves two distinct roles in protein degradation in reticulocytes, one requiring and one independent of ubiquitin. *J. Cell Biol.* 96: 1580–85

Tanaka, R. D., Li, A. C., Fogelman, A. M., Edwards, P. A. 1986. Inhibition of lysosomal protein degradation inhibits the basal degradation of 3-hydroxy-3-methylglutaryl coenzyme A reductase. *J. Lipid Res.* 27: 261–73

Toda, T., Ohashi, M. 1986. Purification and identification of intermediate catabolic products in the in vivo degradation of pig liver phosphofructokinase. *J. Biol. Chem.* 261: 12455–61

Townsend, A. R. M. 1987. Recognition of influenza virus proteins by cytotoxic T lymphocytes. In *Immunologic Research*, ed. J. Coligan. Basel: Karger. In press

Vijay-Kumar, S., Bugg, C. E., Wilkinson, K. D., Cook, W. J. 1985. Three-dimensional structure of ubiquitin at 2.8 Å resolution (crystal structure/protein conformation). *Proc. Natl. Acad. Sci. USA* 82: 3582–85

Vorgias, C. E., Traub, P. 1986. Efficient degradation in vitro of all intermediate filament subunit proteins by the Ca^{2+}-activated neutral thiol proteinase from

Ehrlich ascite tumor cells and porcine kidney. *Biosci. Rep.* 6: 57–63

West, M. H. P., Bonner, W. M. 1980. Histone 2B can be modified by the attachment of ubiquitin. *Nucl. Acids Res.* 8: 4671–80

Wibo, M., Poole, B. 1974. Protein degradation in cultured cells. *J. Cell Biol.* 63: 430–40

Wilk, S., Orlowski, M. 1983. Evidence that pituitary cation-sensitive neutral endopeptidase is a multicatalytic protease complex. *J. Neurochem.* 40: 842–49

Wilkinson, K. D., Audhya, T. 1981. Stimulation of ATP-dependent proteolysis requires ubiquitin with the C-terminal sequence arg-gly-gly. *J. Biol. Chem.* 256: 9235–41

Wilkinson, K. D., Mayer, A. N. 1986. Alcohol-induced conformational changes of ubiquitin. *Arch. Biochem. Biophys.* 250: 390–99

Woods, C. M., Lazarides, E. 1985. Degradation of unassembled α- and β-spectrin by distinct intracellular pathways: Regulation of spectrin topogenesis by β-spectrin degradation. *Cell* 40: 959–69

Wu, R. S., Kohn, K. W., Bonner, W. M. 1981. Metabolism of ubiquitinated histones. *J. Biol. Chem.* 256: 5916–20

Yamaizumi, M., Uchida, T., Takamatsu, K., Okada, Y. 1982. Intracellular stability of diphtheria toxin fragment A in the presence and absence of anti-fragment A antibody. *Proc. Natl. Acad. Sci. USA* 79: 461–65

Ann. Rev. Cell Biol. 1987. 3 : 31–56
Copyright © 1987 by Annual Reviews Inc. All rights reserved

CELL TRANSFORMATION BY THE VIRAL *src* ONCOGENE

Richard Jove and Hidesaburo Hanafusa

The Rockefeller University, 1230 York Avenue, New York, New York 10021

CONTENTS

PERSPECTIVES AND SUMMARY 31
OVERVIEW OF CELL TRANSFORMATION 32
ORIGIN OF VIRAL *src* 33
PROTEIN–TYROSINE KINASE ACTIVITY 35
   Identification of p60$^{v\text{-}src}$ Kinase 35
   Catalytic Domain 36
   Mutations in the Catalytic Domain 37
SUBCELLULAR LOCALIZATION 39
   Interaction with Cell Membranes 39
   Membrane-Binding Domain 40
   Biogenesis of p60$^{v\text{-}src}$ 41
   Unusual src *Proteins* 42
MODULATION OF *src* FUNCTION... 42
   Mutations in the Amino-Terminal Half 42
   Proposed Modulatory Domain 43
PHOSPHORYLATION OF *src* PROTEIN 44
CELLULAR PROTEIN SUBSTRATES 45
ACTIVATION OF CELLULAR *src* 46
   *Mutations in c-*src 46
   Biochemical Properties of p60$^{c\text{-}src}$ 47
PROSPECTS 49

PERSPECTIVES AND SUMMARY

Even though the processes involved in the development of cancers are complex, many aspects of oncogenesis can be studied under simplified conditions in cell culture. Transformation of cells by Rous sarcoma virus (RSV) is one system that has received considerable attention because of

31

0743–4634/87/1115–0031$02.00

the rapid and dramatic cellular response. Remarkable progress has been made during the past decade in understanding the origin, structure, and function of the RSV oncogene responsible for cell transformation. It is now well established that the viral *src* gene (v-*src*) was originally derived by transduction of homologous cellular sequences (c-*src*). This system gained further interest when v-*src* was discovered to encode a protein tyrosine kinase that is solely responsible for initiation and maintenance of the many diverse phenotypic changes accompanying transformation. One major goal of current research is to understand the molecular mechanisms governing cell transformation. The particular emphasis of this review is on genetic and molecular analysis of the structure and function of both the v-*src* gene product, $p60^{v-src}$, and its cellular homolog, $p60^{c-src}$. We discuss contributions of domains and posttranslational modifications of the *src*-encoded proteins to transforming activity. We also point out areas that are not well understood and that warrant further investigation.

OVERVIEW OF CELL TRANSFORMATION

Cell transformation is the acquisition, by cells cultured in vitro, of abnormal phenotypic properties that reflect characteristics of tumor cells in vivo. Expression of the viral *src* oncogene induces three major phenotypic changes generally observed in cells transformed by diverse agents (reviewed in Hanafusa 1977; Teich et al 1982). (*a*) Transformed cells have an altered morphology, typically round, and are more refractile than cultured normal flat cells. Disorganized clusters of such cells on a background of normal cells form foci that are readily recognized visually. The disruption and reorganization of cytoskeletal structures observed in transformed cells apparently contribute to this morphological alteration mentioned above. (*b*) Transformed cells proliferate under conditions in which normal cells do not, such as in media containing low serum concentrations. This reduced requirement for exogenous mitogenic stimuli may result from abnormal expression of growth factors or growth factor receptors. Alternatively, oncogenes encoding protein kinases, including the *src* gene, may directly intervene in signal transduction pathways involved in normal growth control. (*c*) Transformed cells acquire anchorage independence, the ability to grow without attachment to a substrate. This property is manifest as the growth of colonies from single cells suspended in semisolid media. Although the basis for anchorage independence is unclear, transformation results in an alteration of the extracellular matrix, which may subvert the requirement for substratum attachment.

Cell transformation by RSV is accompanied by changes in gene expression and in the synthesis of many cellular macromolecules, including hyaluronic acid, glycolipid, fibronectin, collagen, and plasminogen activator. RSV-transformed cells also exhibit changes in cellular physiology, such as increased glucose transport and glycolysis. Individual changes may either result from, or contribute to, the three major traits of the transformation phenotype. The ultimate criterion of cell transformation is tumorigenicity. In the case of transformation by RSV, a good correlation exists between the transformation parameters described above and tumorigenicity.

ORIGIN OF VIRAL *src*

Three avian sarcoma viruses have been independently isolated as *src*-containing retroviruses. One of these was the first virus shown to induce solid tumors. It was isolated from "chicken tumor number one" by Rous (1911) and later named RSV. Two other avian sarcoma viruses, S1 and S2, are recent isolates from a chicken flock deliberately infected with avian leukosis virus (ALV) (Hihara et al 1984). Unlike the later two isolates, RSV has had long passage histories in different laboratories, giving rise to various mutations that have accumulated in separate stocks, now called strains. Among the best-studied are the Schmidt-Ruppin (SR-RSV), Prague (PR-RSV), and Bryan high-titer (BH-RSV) RSV strains. SR-RSV and PR-RSV are nondefective in replication, whereas BH-RSV is defective (Hanafusa et al 1963; Wang 1978). S1 and S2 are also defective in replication (Yamagishi-Hagino et al 1984). The genome structures of several representative avian sarcoma viruses and a prototypic virus that lacks oncogene sequences (an ALV) are illustrated in Figure 1. Several lines of evidence strongly suggest that the original RSV, like BH-RSV, was replication defective (Lerner & Hanafusa 1984; Dutta et al 1985). The nondefective forms may have subsequently arisen via recombination between defective RSV and an ALV that was present and acted as a helper virus. Duplication of the flanking sequence *F*, located 5′ and 3′ to *src* in nondefective RSV genomes, may be the result of the latter recombination process.

The c-*src* sequence has been inserted into various sites within the ALV genome (Figures 1 and 2). In all cases the viral 5′ recombination junction preserves the c-*src* splice acceptor site located immediately upstream of the *src* translation initiation codon (Figure 2) (Schwartz et al 1983; Takeya & Hanafusa 1983). In RSV this splice acceptor site is used for the formation of subgenomic v-*src* mRNA by the splicing of genomic viral RNA (Varmus & Swanstrom 1982). Translation of v-*src* sequences in all viruses begins at the c-*src*–derived ATG codon indicated in Figure 2, yielding the same amino-terminal sequence in all *src* proteins.

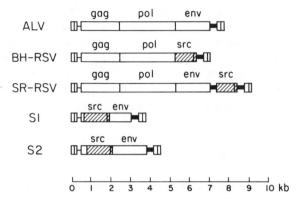

Figure 1 Structures of avian retrovirus genomes depicted as viral DNAs. Genomes are those of avian leukosis virus (ALV), the Bryan high-titer (BH-RSV) and Schmidt-Ruppin (SR-RSV) strains of Rous sarcoma virus (RSV), and two independently isolated avian sarcoma viruses (S1 and S2). One copy of the long terminal repeat is shown at either end of each genome, and the thick solid line denotes flanking *F* sequences. Viral replicative genes, *gag*, *pol*, and *env*, are represented by open boxes. All of the v-*src* genes, indicated by the hatched boxes, contain carboxy-terminal coding sequences that are not present in chicken c-*src* (kb = kilobase pairs).

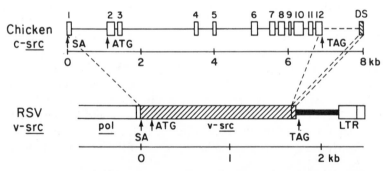

Figure 2 Structure of chicken c-*src* genomic DNA compared to that of the RSV v-*src* oncogene. Exons 1–12 of c-*src* are denoted by the open boxes, and v-*src* coding sequences derived from cellular DNA are represented by the hatched box. The splice acceptor site of v-*src*, which is used to generate a subgenomic *src* mRNA, is derived from exon 1 of c-*src*. Part of the carboxy-terminal coding sequences in exon 12 of c-*src* (encoding 19 amino acids) has been deleted in v-*src*. The extreme carboxy-terminal coding sequences of v-*src* (encoding 12 novel amino acids) are derived from the DS region located approximately 900 base pairs downstream from the c-*src* termination codon and from viral flanking *F* sequences (*thick solid line*). (Abbreviations are: splice acceptor site, SA; translation initiation codon, ATG; translation termination codon, TAG; downstream sequence, DS; viral polymerase gene, *pol*; long terminal repeat, LTR; kilobase pairs, kb.)

At the viral 3′ recombination junction, c-*src* sequences are joined to the middle of *env* in S1 and to the end of *pol* in S2 (Figure 1) (Ikawa et al 1986). In both cases, however, recombination occurred at a site upstream from the c-*src* termination codon. As a result, 8 carboxy-terminal amino acids of c-*src* protein are replaced by 43 unrelated residues coded for by an incorrect reading frame of *env* in S1. In S2, 14 carboxy-terminal amino acids of c-*src* protein are replaced by 38 residues derived from *pol* (Ikawa et al 1986). In contrast, the carboxy-terminal coding sequence of RSV v-*src*, which is present in all RSV strains, appears to have been generated by two separate recombination events. One event replaced 57 base pairs at the 3′ end of c-*src* by a 31–base pair sequence located approximately 900 base pairs downstream from the c-*src* termination codon (Figure 2) (Takeya & Hanafusa 1983). Another event joined this 31–base pair sequence with the *F* region of ALV, thereby providing a new TAG termination codon for the v-*src* gene (Lerner & Hanafusa 1984; Bizub et al 1984). The net result of these two recombination events is the substitution of 19 carboxy-terminal amino acids in c-*src* protein with 12 novel amino acids in RSV v-*src* protein.

In summary, all three independently isolated avian sarcoma viruses encode v-*src* proteins of approximately 60 kDa that are similar to the c-*src* gene product but contain different carboxy termini. This is also the case for isolates of recovered avian sarcoma virus (rASV), which were formed in chickens under experimental conditions by recombination between c-*src* and RSV deletion mutants lacking the majority of v-*src* (Hanafusa et al 1977; Wang et al 1984; Wang 1987).

PROTEIN–TYROSINE KINASE ACTIVITY

Identification of $p60^{v\text{-src}}$ Kinase

Development of antisera raised in rabbits bearing tumors induced by RSV allowed identification of the *src* gene product in immunoprecipitates from RSV-transformed cells and from in vitro translation reactions programmed with virion RNA (Brugge & Erikson 1977; Purchio et al 1978). Immunoprecipitates of $p60^{v\text{-src}}$ with sera from tumor-bearing rabbits were subsequently shown to possess an associated protein kinase activity that phosphorylates the immunoglobulin heavy chain in the presence of ATP and Mg^{2+} (Collett & Erikson 1978; Levinson et al 1978). Many lines of evidence, including the observation that recombinant viral *src* protein expressed in *Escherichia coli* functions as a protein kinase (Gilmer & Erikson 1981), demonstrated that this kinase activity is intrinsic to $p60^{v\text{-src}}$. An unusual and important feature of the $p60^{v\text{-src}}$ protein kinase is its strict specificity for tyrosine residues. The phosphotyrosine generated in this rare protein modification represents less than 0.1% of total phosphoamino acids in cells (Hunter & Sefton 1980; Collett et al 1980; Levinson et al 1980). Cell transformation by RSV results in a tenfold elevation in total cellular protein phosphotyrosine (Hunter & Sefton 1980; Sefton et al 1980). Phosphorylation of cellular substrates on tyrosine residues by the p60^{v-src} kinase is presumed to be responsible for transformation of RSV-infected cells (Sefton et al 1980; Iba et al 1985a).

Studies of *src* proteins have been facilitated by the availability of monoclonal and polyclonal antibodies directed against p60^{v-src} produced in bacteria (Gilmer & Erikson 1983; Lipsich et al 1983; S. J. Parsons et al 1984). The *src* protein in immune complexes formed with many of these antibodies and highly purified p60^{v-src} undergo autophosphorylation and phosphorylate exogenous protein substrates at tyrosine residues (Purchio 1982; Lipsich et al 1983; Cooper et al 1984). Even though in vitro studies have contributed greatly to our understanding of *src* function, the *src* kinase phosphorylates many proteins in vitro that are not considered to be bona fide substrates in vivo (Cooper & Hunter 1983). Evidence also has been presented that purified p60^{v-src} can phosphorylate diacylglycerol and phosphatidylinositol in vitro (Sugimoto et al 1984). This finding raised the possibility that lipid kinase activity of p60^{v-src} could directly contribute to the previously reported increase in phospholipid turnover in RSV-transformed cells (Diringer & Friis 1977). Subsequent studies, however, demonstrated that the p60^{v-src} kinase activity does not significantly contribute to overall phosphatidylinositol kinase activity in transformed cells (MacDonald et al 1985; Sugano & Hanafusa 1985; Sugimoto & Erikson 1985). How phospholipid levels are regulated by p60^{v-src} expression in transformed cells and whether elevated phospholipid turnover is causally related to transformation are interesting questions that deserve further investigation.

Catalytic Domain

The catalytic activity of p60^{v-src} resides within a carboxy-terminal 30-kDa proteolytic fragment (Levinson et al 1981; Brugge & Darrow 1984). This region of the *src* protein shares a high degree of amino acid homology with other tyrosine and serine/threonine kinases (Hunter & Cooper 1985; Van Beveren & Verma 1986). By the criteria of biochemical function and structural homology, this region is designated the tyrosine kinase catalytic domain (Figure 3). Discernible homology with other protein kinases extends from amino acid 250 to 516 of p60^{v-src}, and there are many short blocks of highly conserved sequence scattered throughout this region (Van Beveren & Verma 1986). Interestingly, the sequence leu-ala-ala-arg-asn or the related sequence leu-arg-ala-ala-asn (amino acids 387–391 in p60^{v-src}) is

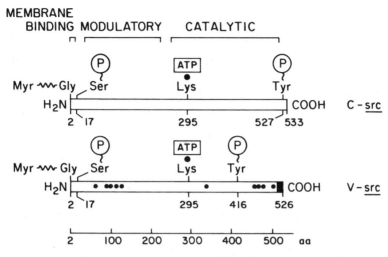

Figure 3   Structural features of chicken p60$^{c\text{-}src}$ and SR-RSV p60$^{v\text{-}src}$. Suggested locations of the membrane binding, modulatory, and kinase catalytic domains are shown at top. In the mature proteins, initiator methionine is absent and amino-terminal glycine (Gly) is fatty acylated with myristic acid (Myr). The major sites of serine (Ser) and tyrosine (Tyr) phosphorylation and the lysine residue (Lys) that forms part of the ATP-binding site are indicated. Small solid dots denote single amino acid substitutions in p60$^{v\text{-}src}$ compared to p60$^{c\text{-}src}$. The solid box represents the substitution of 12 carboxy-terminal amino acids in p60$^{v\text{-}src}$ for 19 different residues in p60$^{c\text{-}src}$. Numbers indicate amino acid (aa) positions.

highly conserved among all known tyrosine kinases, whereas an unrelated sequence is invariably present in kinases that phosphorylate serine or threonine residues (Foster et al 1986). Perhaps this sequence is important for determining amino acid substrate specificity, a possibility that could be addressed by site-directed mutagenesis. Examination of the corresponding region in novel protein kinase sequences may prove useful for preliminary assignment to either the tyrosine or serine/threonine kinase category.

Mutations in the Catalytic Domain

NONCONDITIONAL MUTANTS   Studies using the ATP analog *p*-fluoro-sulfonyl 5′-benzoyl adenosine, which reacts with the ATP binding site and inactivates the kinase, indicate that lysine 295 forms part of the ATP binding site in p60$^{v\text{-}src}$ (Kamps et al 1984). This lysine residue is absolutely conserved within the catalytic domain of all known protein kinases (Hunter & Cooper 1985; Van Beveren & Verma 1986). Direct evidence that lysine 295 is essential for the kinase activity was obtained by oligonucleotide-directed mutagenesis. Substitution of lysine 295 with methionine, glutamic acid, arginine, or histidine abolished both kinase and transforming activities (Snyder et al 1985; Kamps & Sefton 1986). Based on the inability of these amino acids to substitute for lysine 295, it has been proposed that the primary amino group of lysine carries out the specialized function of proton exchange in the phosphotransferase reaction (Kamps & Sefton 1986).

Bisulfite mutagenesis in vitro was used to obtain a series of mutations within a highly conserved region spanning amino acids 430–433 (the CHpm series of mutants) (Bryant & Parsons 1984). Single amino acid substitutions in this region severely diminished kinase activity (which suggests that these amino acids are important for catalytic activity) and concomitantly abolished cell transformation. In another series of mutants, amino acids 517–526 could be deleted or replaced by unrelated sequences of varying lengths (the pRS mutant series) without abolishing kinase activity or cell transformation (Yaciuk & Shalloway 1986). These results suggest that the extreme carboxy-terminal region, which is not present in the cellular *src* protein, does not provide a specific function in p60$^{v\text{-}src}$. This conclusion is further supported by the finding that two independently isolated avian sarcoma viruses, S1 and S2, encode *src* proteins containing entirely different carboxy termini (Ikawa et al 1986). However, deletion or substitution of leucine 516, which is highly conserved among protein kinases (Van Beveren & Verma 1986), severely destabilized the *src* protein and caused loss of transforming activity (Yaciuk & Shalloway 1986). In retrospect, it is probable that *src* protein instability at least partly contributed to the loss of function of carboxy-terminal mutants analyzed in an earlier study, all of which lacked leucine 516 (Wilkerson et al 1985). From these results, the carboxy-terminal boundary of the kinase catalytic domain appears to be amino acid 517 (Yaciuk & Shalloway 1986).

TEMPERATURE-SENSITIVE MUTANTS   Temperature-sensitive (ts) transformation mutants selected in vivo have proven valuable in many studies of *src* gene function (reviewed in Hanafusa 1977; Linial & Blair 1982; Bishop & Varmus 1982; Wyke & Stoker 1987). Molecular cloning and DNA sequence analysis revealed that the variant tsNY68, which is derived from SR-RSV, encodes a *src* protein containing two mutations in the kinase domain (Nishizawa et al 1985). One mutation is the substitution of valine 461 with methionine, and the other is deletion of amino acids 352–354; both mutations are required for expression of the ts property. Surprisingly, analysis of two other independent ts variants derived from SR-RSV, tsNY72-4 and PA104, revealed that these *src* genes encode the identical valine 461 to methionine substitution present in tsNY68 (Mayer et al 1986). In addition, each of these two variants contains a second, variant-specific mutation: leucine 325 is replaced with proline in PA104, and proline 503 with serine in tsNY72-4. In each case both the common and the variant-specific mutation are required for temperature sensitivity, which is consistent with the tsNY68 results. A temperature-sensitive BH-RSV mutant, PA101, also was found to encode two amino acid substitutions within the carboxy-terminal portion of *src* protein at positions 328 and 524 (Mayer et al 1986), although the contribution of the individual mutations to temperature sensitivity has not been evaluated. The lesions within the kinase domains of all four of these ts variants affect kinase activity in vitro as well as in vivo (Jove et al 1986a; Garber et al 1987).

In contrast to the mutants described above, DNA sequence analysis of two molecularly cloned ts mutants derived from PR-RSV revealed that single amino acids substitutions in the kinase domain are responsible for temperature sensitivity: arginine 480 to histidine in tsLA24, and glycine 478 to aspartic acid in tsLA31 (Fincham & Wyke 1986). Interestingly, revertants of both variants that transform at high temperatures contain second-site mutations at either amino acid 377 or 492, which suggests an interaction between these compensating mutations and the original ts lesions.

Three general features have emerged concerning the lesions involved in temperature sensitivity of the biologically selected kinase domain mutants. (*a*) Lesions throughout the kinase domain, spanning amino acids 325–503, can influence expression of the ts phenotype. (*b*) Although many lesions coincide with highly conserved sequences among protein kinases and some are predicted to have significant effects on protein secondary structure, this is not the case for all lesions in the kinase domain that have a profound effect on *src* protein function. (*c*) Analysis of these mutants has revealed many long-range interactions between distant residues, separated by up to 135 amino acids, that might not otherwise have been discovered. It is anticipated that an explanation of the effect of these ts lesions on protein conformation will become apparent when the three-dimensional structure of p60$^{v\text{-}src}$ is solved.

Taken together, the studies described above of nonconditional and conditional mutants harboring lesions in the *src* kinase domain provide strong support for the hypothesis that tyrosine kinase activity is critical for all manifestations of cell transformation induced by p60$^{v\text{-}src}$.

SUBCELLULAR LOCALIZATION

Interaction with Cell Membranes

Greater than 80% of p60$^{v\text{-}src}$ is associated with plasma membranes, as determined by biochemical subcellular fractionation experiments (Krueger et al 1980; Courtneidge et al 1980; Krzyzek et al 1980). The *src* protein is localized primarily at the cytoplasmic surface of the plasma membrane (reviewed in Krueger et al 1983), where it is enriched in regions of cell-to-cell contact and adhesion plaques (Willingham et al 1979; Rohrschneider 1980; Nigg et al 1982). Subpopulations of p60$^{v\text{-}src}$ also appear to be associated with intracellular membranes, including perinuclear membranes (Krueger et al 1983; Resh & Erikson 1985).

Membrane-Binding Domain

Early biochemical analyses suggested that p60$^{v\text{-}src}$ is an integral membrane protein anchored by its amino terminus (Krueger et al 1983), even though DNA sequence analysis of cloned viral *src* DNA demonstrated the lack of any long hydrophobic stretches at the amino terminus (Takeya et al 1982; Takeya & Hanafusa 1982). Further analyses revealed that p60$^{v\text{-}src}$ has lipid attached to its amino terminus (Sefton et al 1982; Garber et al 1983), which was subsequently shown to be the rare 14-carbon saturated fatty acid myristic acid (Schultz et al 1985; Buss & Sefton 1985). In mature p60$^{v\text{-}src}$, the initiator methionine is cleaved off and myristic acid is covalently linked via an amide bond to the α-amino group of the amino-terminal glycine. Evidence presented below suggests that the amino-terminal fatty acylation of p60$^{v\text{-}src}$ is essential for interaction with lipid bilayers.

Amino-terminal mutants of p60$^{v\text{-}src}$ were constructed by in vitro mutagenesis of cloned viral *src* DNA in order to delineate sequences required for myristylation and membrane association (Cross et al 1984; Pellman et al 1985a). These studies demonstrated that amino acid substitutions or deletions affecting residues 2 through 14 abolished myristylation, membrane association, and gross cell transformation. Replacement of the amino-terminal glycine with either alanine or glutamic acid by oligonucleotide-directed mutagenesis also yielded nonmyristylated, soluble *src*-proteins that are transformation-defective (Kamps et al 1985; Buss et al 1986). In addition, fusion of DNA encoding the first 14 amino acids of p60$^{v\text{-}src}$ to sequences encoding heterologous soluble, previously non-myristylated proteins converted the latter to myristylated, membrane-associated proteins (Pellman et al 1985b). Because all of the above results are consistent with the suggestion that amino acids 2 through 14 are necessary and sufficient for myristylation and membrane localization of p60$^{v\text{-}src}$, this amino-terminal region is designated the membrane-binding domain (Figure 3).

Despite the inability of nonmyristylated *src* proteins to induce morphological alteration and anchorage-independent growth in soft-agar suspension, these proteins retain high levels of kinase activity in vitro and in vivo (Cross et al 1984; Kamps et al 1986). This suggests that cellular substrates involved in cell transformation are localized at or near the plasma membrane and are inaccessible to phosphorylation by non-myristylated *src* proteins. Interestingly, however, the nonmyristylated proteins do promote growth of cells to high densities and stimulate the rate of cell proliferation compared to normal monolayer cultures (Kamps et al 1986; Calothy et al 1987). These observations indicate that cellular targets involved in regulation of certain growth properties remain accessible to nonmyristylated *src* proteins, which raises the possibility that targets for different transformation traits may reside in distinct subcellular compartments.

Biogenesis of p60$^{v\text{-src}}$

p60$^{v\text{-src}}$ is synthesized on free polyribosomes in the cytosol (Lee et al 1979; Purchio et al 1980), where it immediately becomes associated with two cellular proteins, p50 and p90, to form a soluble complex that has a half-life of less than 15 min (Brugge et al 1981; Courtneidge & Bishop 1982; Brugge et al 1983). Within 15 min, however, p60$^{v\text{-src}}$ is localized at the plasma membrane (Levinson et al 1981) and dissociated from the complex. These observations led to the proposal that complex formation with p50 and p90 might be involved in processing or transporting newly synthesized p60$^{v\text{-src}}$ to the plasma membrane, although true understanding of the function of this complex remains elusive (reviewed in Brugge 1986).

Kinetic and cell fractionation experiments indicate that myristylation is an extremely early step in the biogenesis of p60$^{v\text{-src}}$, preceding its association with the plasma membrane (Buss et al 1984; Garber et al 1985). Several interesting aspects of protein myristylation have emerged from the studies of *src* mutants described above (Cross et al 1984; Pellman et al 1985a; Kamps et al 1985; Garber et al 1985; Buss et al 1986). (*a*) Prior removal of the initiator methionine is required for myristylation, although cleavage of this methionine can occur without myristylation, which demonstrates that these events are not obligatorily linked. (*b*) Aminopeptidase cleaves the initiator methionine when either glycine or alanine is the second residue, but not if glutamic acid, aspartic acid, or asparagine is present at this position. These results suggest that the aminopeptidase has a strong preference for small, uncharged amino acids in the second position. (*c*) Amino-terminal alanine is not myristylated, which suggests that the myristyltransferase has a strict specificity for amino-terminal glycine. (*d*) In certain mutants, amino-terminal glycine is not myristylated, which indicates that adjacent amino acids are important for myristylation. It is presently not clear whether the myristyltransferase recognizes primary or secondary structural features of the amino terminus. (*e*) Some myristylated *src* proteins do not associate with membranes, apparently because of internal mutations that stabilize complex formation with the soluble proteins p50 and p90; thus myristylation alone is not sufficient to ensure membrane association.

Unusual src Proteins

There is an exceptional class of *src* proteins encoded by recovered avian sarcoma viruses rASV157 and rASV1702 that are not myristylated and yet are active in cell transformation (Karess & Hanafusa 1981; Garber et al 1983). These mutant proteins are localized primarily in adhesion plaques and regions of cell-to-cell contact and behave as peripheral membrane proteins during subcellular fractionation (Krueger et al 1982, 1984). Molecular cloning and DNA sequence analysis revealed that the membrane binding domains of these rASV *src* proteins have been replaced by different amino termini (Garber & Hanafusa 1987). In the case of rASV157, 30 amino acids of *env* signal sequence are fused to serine 6 of the *src* sequence, whereas in rASV1702 there are 45 amino acids of *env* signal sequence fused to alanine 76 of the *src* sequence. These findings suggest that the novel amino-terminal sequences are responsible for the unusual biochemical and biological properties of these proteins, including induction in chickens of tumors that eventually regress.

MODULATION OF *src* FUNCTION

Mutations in the Amino-Terminal Half

An interesting and diverse group of *src* variants contain mutations within the amino-terminal portion of $p60^{v-src}$ located between the membrane binding domain and the kinase catalytic domain. Many lesions within this region result in an elongated, fusiform cell morphology, in contrast to the typical round morphology of cells transformed by wild-type $p60^{v-src}$. Four independent, biologically selected mutants that induce a fusiform morphology (ST529, WO101, WO201, and WO401) were shown to encode *src* proteins containing gross structural alterations within the amino-terminal 18 kDa (Fujita et al 1981), which suggests that a similar alteration in these mutant proteins has a common effect on cell morphology. Another fusiform mutant, d15, has been analyzed by molecular cloning and DNA sequencing (Kitamura & Yoshida 1983). In this spontaneous mutant, amino acids 135–236 have been altered by deletion and frameshift mutations, resulting in a substitution of 33 amino acids and deletion of 69 amino acids. The d15 mutant *src* protein induces anchorage-independent growth of cells with approximately the same efficiency as does the wild type, although the in vitro and in vivo kinase levels are slightly reduced compared to wild type (Iwashita et al 1983; Kitamura & Yoshida 1983).

The ts mutant PA101, which was found to contain lesions in the kinase domain responsible for temperature sensitivity (described above), also contains amino-terminal substitutions (located at amino acid positions 53, 58, 85, and 105) that induce a fusiform morphology independent of temperature when combined with a wild-type kinase domain (Jove et al 1986b; Mayer et al 1986). These amino-terminal mutations do not significantly affect the ability of the PA101 *src* protein to promote anchorage independence or cell proliferation, but they do cause a partial reduction in the levels of in vitro and in vivo kinase activities (Jove et al 1986a).

An extensive series of amino-terminal deletion mutants was constructed by in vitro mutagenesis of the cloned wild-type viral *src* gene (Cross et al 1984, 1985). Biological analysis of these mutants revealed that deletion of amino acids 15–81 (NY309) or 15–149 (NY311) induced a fusiform morphology, whereas deletion of amino acids 15–169 (NY310) or 149–169 (NY320) induced only subtle alterations in cell morphology compared to normal cells. Moreover, the ability of the mutant *src* proteins to promote anchorage independence and cell proliferation was generally more severely impaired as the deletion was extended (Cross et al 1985; Calothy et al 1987). Deletion of amino acids 15–49 (NY308), on the other hand, yielded a protein indistinguishable from wild-type p60^{v-src} with respect to all properties examined.

Temperature-sensitive transformation mutants containing aminoterminal alterations have also been generated by in vitro mutagenesis of cloned viral *src* DNA (Bryant & Parsons 1982; J. T. Parsons et al 1984). One mutant has a deletion of amino acids 173–227 (CHd1119), and the other lacks amino acids 169–225 (CHd1120). Both variants induce a wildtype transformation phenotype at permissive temperature and a normal cell phenotype at restrictive temperature, accompanied by only a partial reduction in kinase levels as assayed in vitro and in vivo. Another ts mutant, tsLA32, retains high levels of kinase activity at the nonpermissive temperature for cell transformation and has no other obvious functional defects (Stoker et al 1984, 1986).

Proposed Modulatory Domain

All of the above studies support the notion that there is a functionally important domain in p60^{v-src} that is separate from the membrane binding and kinase domains. Despite the diversity among this group of mutants, several distinctive patterns have become evident from these analyses. (*a*) Lesions throughout the region between amino acids 53–227 apparently have an influence on cell morphology; in many cases, the effect on cell morphology is not accompanied by a significant effect on cell growth properties. (*b*) There appears to be a general correlation between extent of *src* deletions and impairment of cell growth properties. However, whether this observation reflects loss of specific functions or gross conformational alteration of the *src* proteins as a result of the large deletions is unclear. (*c*) The mutants retain kinase activity, but in most cases reduced levels of in vitro and in vivo kinase activity have been demonstrated, which suggests an interaction between an amino-terminal region and the kinase domain.

Because the one important characteristic all mutations in this aminoterminal portion have in common is that they appear to modulate p60$^{v\text{-}src}$ function, this region is tentatively designated the modulatory domain (Figure 3). Comparison of protein kinase sequences reveals a region located between amino acids 137 and 241 of p60$^{v\text{-}src}$ that is conserved among those tyrosine protein kinases that do not span the plasma membrane (Sadowski et al 1986). In vitro mutagenesis experiments suggest that the corresponding region in Fujinami sarcoma virus transforming protein P130$^{gag\text{-}fps}$ is not required for catalytic activity, but it does modulate the kinase activity and is important in cell transformation (Sadowski et al 1986). The apparent conservation of both structural and functional homology in this amino-terminal region of tyrosine protein kinases suggests a common interaction with the kinase domain or cellular components. Taken together, the available data suggest that the modulatory domain could play an important role in regulating levels of kinase activity or substrate specificity. It is important to emphasize that representation of *src* protein as a linear arrangement of discrete domains (Figure 3) may be a topological oversimplification, reflecting the lack of information on complex interactions in the native conformation. More refined and extensive analyses of *src* mutants will be required to define the exact location and functions of this putative domain.

PHOSPHORYLATION OF *src* PROTEIN

The major sites of phosphorylation of p60$^{v\text{-}src}$ in vivo are serine 17 and tyrosine 416 (Figure 3) (Smart et al 1981; Patschinsky et al 1982, 1986; Takeya et al 1982; Cross & Hanafusa 1983). Serine 17 appears to be phosphorylated by cAMP-dependent protein kinase (Collett et al 1979a), whereas tyrosine 416 phosphorylation appears to be the result of autophosphorylation by p60$^{v\text{-}src}$ itself (Purchio 1982). Deletion of either serine 17 or tyrosine 416 in p60$^{v\text{-}src}$ by in vitro mutagenesis demonstrated that the major sites of phosphorylation are dispensable for kinase activity and cell transformation, although deletion of tyrosine 416 resulted in reduced tumorigenicity in chickens (Cross & Hanafusa 1983; Cross et al 1984). Replacement of tyrosine 416 with phenylalanine by oligonucleotide-directed mutagenesis also showed that this residue is not required for cell transformation or kinase activity, even though it is required for the full oncogenic potential of p60^{v-src} in vivo (Snyder et al 1983; Snyder & Bishop 1984).

Mutagenesis of the major sites of phosphorylation is not equivalent to alteration of the phosphorylation state at these sites in wild-type p60^{v-src}; therefore, the question remains whether phosphorylation of the wild-type protein can regulate its function. In this regard, it is interesting that treatment of RSV-transformed cells with the phosphatase inhibitor orthovanadate stimulates p60^{v-src} kinase activity in vitro, coincident with the appearance of novel amino-terminal tyrosine phosphorylation (Brown & Gordon 1984; Collett et al 1984). Phosphorylation of novel amino-terminal tyrosine residues is also observed when p60^{v-src} is incubated with high concentrations of ATP in vitro (Collett et al 1983; Purchio et al 1983). In addition, protein kinase C, the major cellular receptor for tumor-promoting phorbol esters, phosphorylates p60^{v-src} at serine 12 and serine 48 (Purchio et al 1985; Gould et al 1985). Since many transient amino-terminal phosphorylations appear to occur in the modulatory domain, these modifications may regulate p60^{v-src} function. However, demonstration of the physiological significance of these and other phosphorylations in p60^{v-src} awaits further investigation.

CELLULAR PROTEIN SUBSTRATES

Many presumptive cellular substrates of p60^{v-src} that are phosphorylated on tyrosine have been identified in transformed cells. Among these are three cytoskeleton-associated proteins: vinculin, p36, and p81; three soluble glycolytic enzymes: enolase, lactate dehydrogenase, and phosphoglycerate mutase; and two additional cytosolic proteins: p42 and p50 (reviewed in Cooper & Hunter 1983; Krueger et al 1983; Sefton 1986; Wyke & Stoker 1987). Recent studies have identified several more candidate substrates that are also associated with cytoskeleton. One of these is the fibronectin receptor complex, which is enriched in adhesion plaques and appears to provide a link between extracellular matrix proteins and the cytoskeleton (Hirst et al 1986). Two other interesting candidates are talin, a component of the cytoskeleton localized in adhesion plaques (Pasquale et al 1986; DeClue & Martin 1987), and the Ca^{2+} binding protein calmodulin (Fukami et al 1986). The activity of p60^{v-src} on the submembranous cytoskeleton is compatible with its plasma membrane localization; moreover, the majority of p60^{v-src} is apparently associated with cytoskeletal components resistant to nonionic detergent extraction (Burr et al 1980). A good correlation has been found between association with the detergent-resistant structure and cell transformation by mutant *src* proteins (Hamaguchi & Hanafusa 1987). Thus phosphorylation of cytoskeleton-associated proteins may be responsible for the observed alteration of cellular architecture and other transformation phenotypes.

Despite the progress in identifying cellular proteins phosphorylated on tyrosine, there have been two major impediments in the continuing search for primary cellular targets of $p60^{v-src}$ involved in cell transformation. First, it has been difficult to demonstrate that cellular proteins phosphorylated on tyrosine in transformed cells are direct substrates of $p60^{v-src}$. Second, it has not yet been possible to firmly establish a causal relationship between tyrosine phosphorylation of any particular cellular protein and cell transformation. This raises the possibility that many of the known or suspected cellular substrates could be simply adventitious substrates whose phosphorylation is not physiologically relevant. The existence of *src* mutants that are partially transformation-defective, i.e. that induce certain transformation parameters but not others, suggests that there may be multiple primary targets of $p60^{v-src}$ involved in cell transformation (Becker et al 1977; Weber & Friis 1979; Calothy et al 1980; Tanaka et al 1980; Anderson et al 1981; Jove et al 1986b). Availability of many such mutants, which differentially phosphorylate multiple candidate substrates (Cooper et al 1983; Kamps et al 1986), may prove indispensable for identifying critical cellular targets using combined biochemical and genetic approaches.

ACTIVATION OF CELLULAR *src*

Mutations in *c*-src

In normal chicken cells, the level of c-*src* expression is at least 50-fold lower than that of v-*src* in RSV-transformed cells (Collett & Erikson 1978; Karess et al 1979). Transduction of c-*src* sequences into retroviruses places these sequences under the control of viral regulatory elements, thereby increasing the level of expression. Nevertheless, mere overexpression of chicken c-*src* using SV40 or RSV promoters is not sufficient to induce cell transformation, even when $p60^{c-src}$ is expressed at levels comparable to that of $p60^{v-src}$ in RSV-transformed cells (Parker et al 1984; Iba et al 1984, 1985b; Shalloway et al 1984; Wilhelmsen et al 1984). These results demonstrate that structural changes are important in activating the transforming potential of c-*src*. In the case of the SR-RSV $p60^{v-src}$, there are ten internal amino acid substitutions relative to chicken $p60^{c-src}$, in addition to the extreme carboxy-terminal alteration (Figure 3) (Takeya & Hanafusa 1982, 1983; Czernilofsky et al 1983; Levy et al 1986).

To determine which sequence differences between chicken c-*src* and SR-

RSV v-*src* contribute to transforming activity, chimeric coding DNA sequences were constructed by exchanging various portions of the two genes. Reciprocal recombinants in which v-*src* and c-*src* sequences coding for the 431 amino-terminal residues were exchanged induced transformation (Iba et al 1984; Shalloway et al 1984). A recombinant encoding the carboxy-terminal 19 amino acids of p60^{c-src} in place of the 12 unique carboxy-terminal amino acids of p60^{v-src} was also transforming (Iba et al 1984; Wilkerson et al 1985). Further analysis demonstrated that a single substitution of threonine 338 with isoleucine activates the transforming potential of p60^{c-src} (Kato et al 1986). This is the only amino acid substitution outside of the carboxy terminus that is common among the *src* proteins of all RSV strains (but not S1 or S2) (Mayer et al 1986). Substitution of amino acids located at positions 63, 95, and 96 in p60^{c-src} with the corresponding residues in p60^{v-src}, which are located in the proposed modulatory domain (Figure 3), also yields a transforming protein (Kato et al 1986). In addition, two spontaneous transforming variants derived from an RSV expression vector that originally contained normal chicken c-*src* were found to encode one amino acid substitution each, at positions 378 and 441 (Iba et al 1984; Levy et al 1986).

Taken together, the above results demonstrate that the transforming potential of p60^{c-src} can be activated by mutations throughout the protein, including single amino acid substitutions. However, the only structural modification of p60^{c-src} that is common to all of the RSV, S1, and S2 *src* proteins is substitution of extreme carboxy-terminal sequences. Replacement of only the 14 carboxy-terminal amino acids in chicken p60^{c-src} with the 38 carboxy-terminal residues of reverse transcriptase, which occurred naturally by transduction in avian sarcoma virus S2, yields a transforming protein (Ikawa et al 1986). Similarly, studies carried out with the human c-*src* gene showed that replacement of sequences encoding the carboxy-terminal 19 amino acids with those encoding the 12 carboxy-terminal amino acids of RSV p60^{v-src} activates human p60^{c-src} (Tanaka & Fujita 1986). These observations indicate that substitution of the p60^{c-src} carboxy terminus by various unrelated sequences is sufficient to activate the protein. In addition to structural changes, other studies have demonstrated that a threshold level of p60^{v-src} expression is required for cell transformation (Jakobovits et al 1984). Therefore, activation of the transforming potential of c-*src* requires regulatory as well as structural alterations (Hanafusa 1986).

Biochemical Properties of p60^{c-src}

Cellular and viral *src* proteins are similar in many respects: both are myristylated, membrane-associated phosphoproteins that possess intrinsic tyrosine protein kinase activity (Oppermann et al 1979; Collett et al 1979b; Karess & Hanafusa 1981; Iba et al 1985a; Buss & Sefton 1985). Nevertheless, as discussed above, the transforming viral protein and its normal cellular progenitor are not identical in either structure or function. Differences in certain notable biochemical properties arising from structural alterations may determine the distinct biological functions of p60$^{v\text{-}src}$ and p60$^{c\text{-}src}$. Of particular interest is the finding that the specific activity of the p60$^{c\text{-}src}$ kinase is at least tenfold lower than that of p60$^{v\text{-}src}$ as assayed in vitro or in vivo (Iba et al 1985a; Coussens et al 1985). Moreover, kinase activity is elevated in all of the p60$^{c\text{-}src}$ mutants described above, which is consistent with the notion that unrestrained activation of the src kinase induces cell transformation. The reduced level of p60$^{c\text{-}src}$ kinase activity is concomitant with phosphorylation of its tyrosine 527 residue, in contrast to the phosphorylation of tyrosine 416 observed in p60$^{v\text{-}src}$ (Figure 3) (Cooper et al 1986; Laudano & Buchanan 1986).

Several lines of evidence suggest that phosphorylation of tyrosine 527 regulates the kinase activity of p60$^{c\text{-}src}$. (a) Transforming mutants derived from p60$^{c\text{-}src}$ are phosphorylated on tyrosine 416 instead of on tyrosine 527 and have elevated kinase activity (Iba et al 1985a). (b) p60$^{c\text{-}src}$ associated in a complex with polyoma virus middle T antigen exhibits elevated kinase activity and is phosphorylated on tyrosine 416 but not tyrosine 527 (Bolen et al 1984; Courtneidge 1985; Cartwright et al 1986). (c) Dephosphorylation of tyrosine 527 in p60$^{c\text{-}src}$ by phosphatase elevates its in vitro kinase activity (Courtneidge 1985; Cooper & King 1986). (d) All independent isolates of avian sarcoma virus containing src sequences encode transforming proteins that lack the extreme carboxy-terminal region of p60$^{c\text{-}src}$, which includes tyrosine 527 (Takeya et al 1982; Ikawa et al 1986). (e) Site-directed mutagenesis of tyrosine 527 so that it can no longer be phosphorylated activates both the kinase and transforming activity of the mutant c-src protein (Kmiecik & Shalloway 1987).

The above results support the hypothesis that phosphorylation of tyrosine 527 suppresses the kinase activity and transforming potential of p60$^{c\text{-}src}$. Since tyrosine 416 is not essential for either kinase or transforming activity (Snyder et al 1983; Cross & Hanafusa 1983), perhaps the shift to phosphorylation of tyrosine 416 in activated src proteins is the consequence, rather than the cause, of elevated kinase activity. Recent studies in animal cells revealed that mutation of the ATP binding site in p60$^{c\text{-}src}$ expressed from a retrovirus vector abolishes its kinase activity but not its tyrosine 527 phosphorylation (R. Jove, S. Kornbluth, and H. Hanafusa, unpublished results). These results raise the possibility that another protein kinase phosphorylates tyrosine 527. In view of the implications concerning regulation of p60$^{c\text{-}src}$ function, it is of interest to identify the cellular tyrosine protein kinase and phosphatase that modify tyrosine 527.

The normal functions of p60$^{c\text{-}src}$ remain enigmatic. The fact that p60$^{c\text{-}src}$ is expressed at high levels in postmitotic, terminally differentiated cells suggests that the normal cellular protein may not be involved in regulating cell proliferation (reviewed in Rohrschneider 1986; Golden & Brugge 1987). Whether p60$^{c\text{-}src}$ and p60$^{v\text{-}src}$ are in any way related in their biological functions will most likely become apparent when the cellular targets of these kinases are known.

PROSPECTS

What are the molecular events that connect the kinase activity of p60$^{v\text{-}src}$ with the myriad and complex changes collectively termed cell transformation? The answer will undoubtedly be found at many different levels. One critical step is identification of primary cellular substrates involved in cell transformation. Phosphorylation of some substrates may directly trigger gross phenotypic alterations. Other phosphorylated substrates may represent an initial step in a cascade or a series of reactions that are involved in signal transduction and amplification. Many cellular phenotypes also are likely to result directly or indirectly from regulation of gene expression. Extracellular structural components and diffusible factors need to be considered as well. At each of these levels, other oncogene-related products may interact to promote cell transformation. More subtle, but perhaps no less crucial, are mechanisms that may suppress the transformed phenotype. Related and equally important problems for future investigation pertain to the normal function of p60$^{c\text{-}src}$ and what role, if any, cellular *src* plays in the etiology of human tumors.

ACKNOWLEDGMENTS

We thank our colleagues Marius Sudol and Bruce Mayer for their comments on this review. Work from this laboratory was supported by grants from the National Cancer Institute (CA14935 and CA18213) and from the American Cancer Society (MV128). Richard Jove was supported by the Damon Runyon–Walter Winchell Cancer Fund (DRG-786).

Literature Cited

Anderson, D. D., Beckmann, R. P., Harms, E. H., Nakamura, K., Weber, M. J. 1981. Biological properties of "partial" transformation mutants of Rous sarcoma virus and characterization of their pp60^{src} kinase. *J. Virol.* 37: 445–58

Becker, D., Kurth, R., Critchley, D., Friis, R., Bauer, H. 1977. Distinguishable trans-

50   JOVE & HANAFUSA

formation-defective phenotypes among temperature-sensitive mutants of Rous sarcoma virus. *J. Virol.* 21: 1042–55

Bishop, J. M., Varmus, H. 1982. Functions and origins of retroviral transforming genes. In *RNA Tumor Viruses*, ed. R. Weiss, N. Teich, H. Varmus, J. Coffin, pp. 999–1108. Cold Spring Harbor, NY: Cold Spring Harbor Lab. 2nd ed.

Bizub, D., Katz, R. A., Skalka, A. M. 1984. Nucleotide sequence of noncoding regions in Rous-associated virus-2: Comparisons delineate conserved regions important in replication and oncogenesis. *J. Virol.* 49: 557–65

Bolen, J. B., Thiele, C. J., Israel, M. A., Yonemoto, W., Lipsich, L. A., Brugge, J. S. 1984. Enhancement of cellular *src* gene product associated tyrosyl kinase activity following polyoma virus infection and transformation. *Cell* 38: 767–77

Brown, D. J., Gordon, J. A. 1984. The stimulation of pp60^v-src kinase activity by vanadate in intact cells accompanies a new phosphorylation state of the enzyme. *J. Biol. Chem.* 259: 9580–86

Brugge, J. S. 1986. Interaction of the Rous sarcoma virus protein pp60^src with the cellular proteins pp50 and pp90. *Curr. Top. Microbiol. Immunol.* 123: 1–22

Brugge, J. S., Darrow, D. 1984. Analysis of the catalytic domain of phosphotransferase activity of two avian sarcoma virus transforming proteins. *J. Biol. Chem.* 259: 4550–57

Brugge, J. S., Erikson, E., Erikson, R. L. 1981. The specific interaction of the Rous sarcoma virus transforming protein, pp60^src, with two cellular proteins. *Cell* 25: 363–72

Brugge, J. S., Erikson, R. L. 1977. Identification of a transformation-specific antigen induced by an avian sarcoma virus. *Nature* 269: 346–48

Brugge, J. S., Yonemoto, W., Darrow, D. 1983. Interaction between the Rous sarcoma virus transforming protein and two cellular phosphoproteins: Analysis of the turnover and distribution of this complex. *Mol. Cell. Biol.* 3: 9–19

Bryant, D., Parsons, J. T. 1982. Site-directed mutagenesis of the *src* gene of Rous sarcoma virus: Construction and characterization of a deletion mutant temperature sensitive for transformation. *J. Virol.* 44: 683–91

Bryant, D., Parsons, J. T. 1984. Amino acid alterations within a highly conserved region of the Rous sarcoma virus *src* gene product pp60^src inactivate tyrosine protein kinase activity. *Mol. Cell. Biol.* 4: 862–66

Burr, J. G., Dreyfuss, G., Penman, S., Buchanan, J. M. 1980. Association of the *src* gene product of Rous sarcoma virus with cytoskeletal structures of chicken embryo fibroblasts. *Proc. Natl. Acad. Sci. USA* 77: 3484–88

Buss, J. E., Kamps, M. P., Gould, K., Sefton, B. M. 1986. The absence of myristic acid decreases membrane binding of p60^src but does not affect tyrosine protein kinase activity. *J. Virol.* 58: 468–74

Buss, J. E., Kamps, M. P., Sefton, B. M. 1984. Myristic acid is attached to the transforming protein of Rous sarcoma virus during or immediately after synthesis and is present in both soluble and membrane-bound forms of the protein. *Mol. Cell. Biol.* 4: 2697–2704

Buss, J. E., Sefton, B. M. 1985. Myristic acid, a rare fatty acid, is the lipid attached to the transforming protein of Rous sarcoma virus and its cellular homolog. *J. Virol.* 53: 7–12

Calothy, G., Laugier, D., Cross, F. R., Jove, R., Hanafusa, T., Hanafusa, H. 1987. The membrane-binding domain and myristylation of p60^v-src are not essential for stimulation of cell proliferation. *J. Virol.* 61: 1678–81

Calothy, G., Poirier, F., Dambrine, G., Mignatti, P., Combes, P., Pessac, B. 1980. Expression of viral oncogenes in differentiating chick embryo neuroretinal cells infected with avian tumor viruses. *Cold Spring Harbor Symp. Quant. Biol.* 44: 983–90

Cartwright, C. A., Kaplan, P. L., Cooper, J. A., Hunter, T., Eckhart, W. 1986. Altered sites of tyrosine phosphorylation in pp60^c-src associated with polyoma virus middle tumor antigen. *Mol. Cell. Biol.* 6: 1562–70

Collett, M. S., Belzer, S. K., Purchio, A. F. 1984. Structurally and functionally modified forms of pp60^v-src in Rous sarcoma virus-transformed cell lysates. *Mol. Cell. Biol.* 4: 1213–20

Collett, M. S., Erikson, E., Erikson, R. L. 1979a. Structural analysis of the avian sarcoma virus transforming protein: Sites of phosphorylation. *J. Virol.* 29: 770–81

Collett, M. S., Erikson, E., Purchio, A. F., Brugge, J. S., Erikson, R. L. 1979b. A normal cell protein similar in structure and function to the avian sarcoma virus transforming gene product. *Proc. Natl. Acad. Sci. USA* 76: 3159–63

Collett, M. S., Erikson, R. L. 1978. Protein kinase activity associated with the avian sarcoma virus *src* gene product. *Proc. Natl. Acad. Sci. USA* 75: 2021–24

Collett, M. S., Purchio, A. F., Erikson, R. L. 1980. Avian sarcoma virus transforming protein, pp60^src, shows protein kinase activity specific for tyrosine. *Nature* 285: 167–69

Collett, M. S., Wells, S. K., Purchio, A. F. 1983. Physical modification of purified Rous sarcoma virus pp60^{v-src} protein after incubation with ATP/Mg^{2+}. *Virology* 128: 285–97

Cooper, J. A., Esch, F. S., Taylor, S. S., Hunter, T. 1984. Phosphorylation sites in enolase and lactate dehydrogenase utilized by tyrosine protein kinases in vivo and in vitro. *J. Biol. Chem.* 259: 7835–41

Cooper, J. A., Gould, K. L., Cartwright, C. A., Hunter, J. 1986. Tyr 527 is phosphorylated in pp60^{c-src}: Implications for regulation. *Science* 231: 1431–34

Cooper, J. A., Hunter, T. 1983. Regulation of cell growth and transformation by the tyrosine-specific protein kinases: The search for important cellular substrate proteins. *Curr. Top. Microbiol. Immunol.* 107: 125–62

Cooper, J. A., King, C. S. 1986. Dephosphorylation or antibody binding to the carboxy terminus stimulates pp60^{c-src}. *Mol. Cell. Biol.* 6: 4467–77

Cooper, J. A., Nakamura, K. D., Hunter, T., Weber, M. J. 1983. Phosphotyrosine-containing proteins and expression of transformation parameters in cells infected with partial transformation mutants of Rous sarcoma virus. *J. Virol.* 46: 15–28

Courtneidge, S. A. 1985. Activation of the pp60^{c-src} kinase by middle T antigen binding or by dephosphorylation. *EMBO J.* 4: 1471–77

Courtneidge, S. A., Bishop, J. M. 1982. Transit of pp60^{v-src} to the plasma membrane. *Proc. Natl. Acad. Sci. USA* 79: 7117–21

Courtneidge, S. A., Levinson, A. D., Bishop, J. M. 1980. The protein encoded by the transforming gene of avian sarcoma virus (pp60^{src}) and a homologous protein in normal cells (pp60$^{proto-src}$) are associated with the plasma membrane. *Proc. Natl. Acad. Sci. USA* 77: 3783–87

Coussens, P. M., Cooper, J. A., Hunter, T., Shalloway, D. 1985. Restriction of the in vitro and in vivo tyrosine protein kinase activities of pp60^{c-src} relative to pp60^{v-src}. *Mol. Cell. Biol.* 5: 2753–63

Cross, F. R., Garber, E. A., Hanafusa, H. 1985. N-terminal deletions in Rous sarcoma virus p60^{src}: Effects on tyrosine kinase and biological activities and on recombination in tissue culture with cellular *src* gene. *Mol. Cell. Biol.* 5: 2789–95

Cross, F. R., Garber, E. A., Pellman, D., Hanafusa, H. 1984. A short sequence in the p60^{src} N terminus is required for p60^{src} myristylation and membrane association and for cell transformation. *Mol. Cell. Biol.* 4: 1834–42

Cross, F. R., Hanafusa, H. 1983. Local mutagenesis of Rous sarcoma virus: The major sites of tyrosine and serine phosphorylation of p60^{src} are dispensable for transformation. *Cell* 34: 597–607

Czernilofsky, A. P., Levinson, A. D., Varmus, H. E., Bishop, J. M., Tischer, E., Goodman, H. 1983. Corrections to the nucleotide sequence of the *src* gene of Rous sarcoma virus. *Nature* 301: 736–38

DeClue, J. E., Martin, G. S. 1987. Phosphorylation of talin at tyrosine in Rous sarcoma virus–transformed cells. *Mol. Cell. Biol.* 7: 371–78

Diringer, H., Friis, R. R. 1977. Change in phosphatidylinositol metabolism correlated to growth state of normal and Rous sarcoma virus-transformed Japanese quail cells. *Cancer Res.* 37: 2979–84

Dutta, A., Wang, L.-H., Hanafusa, T., Hanafusa, H. 1985. Partial nucleotide sequence of Rous sarcoma virus-29 provides evidence that the original Rous sarcoma virus was replication defective. *J. Virol.* 55: 728–35

Fincham, V. J., Wyke, J. A. 1986. Localization of temperature-sensitive transformation mutations and back mutations in the Rous sarcoma virus *src* gene. *J. Virol.* 58: 694–99

Foster, D. A., Levy, J. B., Daley, G. Q., Simon, M. C., Hanafusa, H. 1986. Isolation of chicken cellular DNA sequences with homology to the region of viral oncogenes that encodes the tyrosine kinase domain. *Mol. Cell. Biol.* 6: 325–31

Fujita, D. J., Bechberger, J., Nedic, I. 1981. Four Rous sarcoma virus mutants which affect transformed cell morphology exhibit altered *src* gene products. *Virology* 114: 256–60

Fukami, Y., Nakamura, T., Nakayama, A., Kanehisa, T. 1986. Phosphorylation of tyrosine residues of calmodulin in Rous sarcoma virus-transformed cells. *Proc. Natl. Acad. Sci. USA* 83: 4190–93

Garber, E. A., Cross, F. R., Hanafusa, H. 1985. Processing of p60^{v-src} to its myristylated membrane-bound form. *Mol. Cell. Biol.* 5: 2781–88

Garber, E. A., Hanafusa, H. 1987. NH$_2$-terminal sequences of two *src* proteins that cause aberrant transformation. *Proc. Natl. Acad. Sci. USA* 84: 80–84

Garber, E. A., Krueger, J. G., Hanafusa, H., Goldberg, A. R. 1983. Only membrane-associated RSV *src* proteins have amino-terminally bound lipid. *Nature* 302: 161–63

Garber, E. A., Mayer, B. J., Jove, R., Hanafusa, H. 1987. Analysis of p60^{v-src} mutants carrying lesions involved in temperature sensitivity. *J. Virol.* 61: 354–60

Gilmer, T. M., Erikson, R. L. 1981. Rous sarcoma virus transforming protein, p60*src*, expressed in *E. coli* functions as a protein kinase. *Nature* 294: 771–73

Gilmer, T. M., Erikson, R. L. 1983. Development of anti-pp60*src* serum with antigen produced in *Escherichia coli. J. Virol.* 45: 462–65

Golden, A., Brugge, J. S. 1987. The cellular and viral *src* genes. In *Oncogene Handbook*, ed. A. M. Skalka, T. Curran. Amsterdam: Elsevier. In press

Gould, K. L., Woodgett, J. R., Cooper, J. A., Buss, J. E., Shalloway, D., Hunter, T. 1985. Protein kinase C phosphorylates pp60*src* at a novel site. *Cell* 42: 849–57

Hamaguchi, M., Hanafusa, H. 1987. Association of p60*src* with Triton-resistant cellular structure correlates with morphological transformation. *Proc. Natl. Acad. Sci. USA* 84: 2312–16

Hanafusa, H. 1977. Cell transformation by RNA tumor viruses. In *Comprehensive Virology*, ed. H. Fraenkel-Conrat, R. R. Wagner, 10: 401–83. New York: Plenum

Hanafusa, H. 1986. Activation of the c-*src* gene. In *Oncogenes and Growth Control*, ed. P. Kahn, T. Graf, pp. 100–5. Berlin: Springer-Verlag

Hanafusa, H., Halpern, C. C., Buchhagen, D. L., Kawai, S. 1977. Recovery of avian sarcoma virus from tumors induced by transformation-defective mutants. *J. Exp. Med.* 146: 1735–47

Hanafusa, H., Hanafusa, T., Rubin, H. 1963. The defectiveness of Rous sarcoma virus. *Proc. Natl. Acad. Sci. USA* 49: 572–80

Hihara, H., Shimizu, T., Yamamoto, H., Yoshino, T. 1984. Two strains of avian sarcoma virus induced by lymphatic leukemia virus subgroup A in two lines of chickens. *J. Natl. Cancer Inst.* 72: 631–35

Hirst, R., Horwitz, A., Buck, C., Rohrschneider, L. 1986. Phosphorylation of the fibronectin receptor complex in cells transformed by oncogenes that encode tyrosine kinases. *Proc. Natl. Acad. Sci. USA* 83: 6470–74

Hunter, T., Cooper, J. A. 1985. Protein-tyrosine kinases. *Ann. Rev. Biochem.* 54: 897–930

Hunter, T., Sefton, B. M. 1980. Transforming gene product of Rous sarcoma virus phosphorylates tyrosine. *Proc. Natl. Acad. Sci. USA* 77: 1311–15

Iba, H., Cross, F. R., Garber, E. A., Hanafusa, H. 1985a. Low level of cellular protein phosphorylation by nontransforming overproduced p60*c-src*. *Mol. Cell. Biol.* 5: 1058–66

Iba, H., Jove, R., Hanafusa, H. 1985b. Lack of induction of neuroretinal cell proliferation by Rous sarcoma virus variants that carry the c-*src* gene. *Mol. Cell. Biol.* 5: 2856–59

Iba, H., Takeya, T., Cross, F. R., Hanafusa, T., Hanafusa, H. 1984. Rous sarcoma virus variants that carry the cellular *src* gene instead of the viral *src* gene cannot transform chicken embryo fibroblasts. *Proc. Natl. Acad. Sci. USA* 81: 4424–28

Ikawa, S., Yamagishi-Hagino, K., Kawai, S., Yamamoto, T., Toyoshima, K. 1986. Activation of the cellular *src* gene by transducing retrovirus. *Mol. Cell. Biol.* 6: 2420–28

Iwashita, S., Kitamura, N., Yoshida, M. 1983. Molecular events leading to fusiform morphological transformation by partial *src* deletion mutant of Rous sarcoma virus. *Virology* 125: 419–31

Jakobovits, E. B., Majors, J. E., Varmus, H. E. 1984. Hormonal regulation of the Rous sarcoma virus *src* gene via a heterologous promoter defines a threshold dose for cellular transformation. *Cell* 38: 757–65

Jove, R., Garber, E. A., Iba, H., Hanafusa, H. 1986a. Biochemical properties of p60*v-src* mutants that induce different cell transformation parameters. *J. Virol.* 60: 849–57

Jove, R., Mayer, B. J., Iba, H., Laugier, D., Poirier, F., et al. 1986b. Genetic analysis of p60*v-src* domains involved in the induction of different cell transformation parameters. *J. Virol.* 60: 840–48

Kamps, M. P., Buss, J. E., Sefton, B. M. 1985. Mutation of NH₂-terminal glycine of p60*src* prevents both myristylation and morphological transformation. *Proc. Natl. Acad. Sci. USA* 82: 4625–28

Kamps, M. P., Buss, J. E., Sefton, B. M. 1986. Rous sarcoma virus transforming protein lacking myristic acid phosphorylates known polypeptide substrates without inducing transformation. *Cell* 45: 105–12

Kamps, M. P., Sefton, B. M. 1986. Neither arginine nor histidine can carry out the function of lysine-295 in the ATP-binding site of p60*src*. *Mol. Cell. Biol.* 6: 751–57

Kamps, M. P., Taylor, S. S., Sefton, B. M. 1984. Direct evidence that oncogenic tyrosine kinases and cyclic AMP-dependent protein kinase have homologous ATP-binding sites. *Nature* 310: 589–92

Karess, R. E., Hanafusa, H. 1981. Viral and cellular *src* genes contribute to the structure of recovered avian sarcoma virus transforming protein. *Cell* 24: 155–64

Karess, R. E., Hayward, W. S., Hanafusa, H. 1979. Cellular information in the genome of recovered avian sarcoma virus directs the synthesis of transforming protein. *Proc. Natl. Acad. Sci. USA* 76: 3154–58

Kato, J.-Y., Takeya, T., Grandori, C., Iba, H., Levy, J. B., Hanafusa, H. 1986. Amino acid substitutions sufficient to convert the nontransforming $p60^{c-src}$ protein to a transforming protein. *Mol. Cell. Biol.* 6: 4155–60

Kitamura, N., Yoshida, M. 1983. Small deletion in *src* of Rous sarcoma virus modifying transformation phenotypes: Identification of 207-nucleotide deletion and its smaller product with protein kinase activity. *J. Virol.* 46: 985–92

Kmiecik, T. E., Shalloway, D. 1987. Activation and suppression of $pp60^{c-src}$ transforming ability by mutation of its primary sites of tyrosine phosphorylation. *Cell.* In press

Krueger, J. G., Garber, E. A., Chin, S. S.-M., Hanafusa, H., Goldberg, A. R. 1984. Size-variant $p60^{src}$ proteins of recovered avian sarcoma virus interact with adhesion plaques as peripheral membrane proteins: Effects on cell transformation. *Mol. Cell. Biol.* 4: 454–67

Krueger, J. G., Garber, E. A., Goldberg, A. R. 1983. Subcellular localization of $pp60^{src}$ in RSV-transformed cells. *Curr. Top. Microbiol. Immunol.* 107: 51–124

Krueger, J. G., Garber, E. A., Goldberg, A. R., Hanafusa, H. 1982. Changes in aminoterminal sequences of $pp60^{src}$ lead to decreased membrane association and decreased in vivo tumorigenicity. *Cell* 28: 889–96

Krueger, J. G., Wang, E., Goldberg, A. R. 1980. Evidence that the *src* gene product of Rous sarcoma virus is membrane associated. *Virology* 101: 25–40

Krzyzek, R. A., Mitchell, R. L., Lau, A. F., Faras, A. J. 1980. Association of $pp60^{src}$ and *src* protein kinase activity with the plasma membrane of nonpermissive and permissive avian sarcoma virus-infected cells. *J. Virol.* 36: 805–15

Laudano, A. P., Buchanan, J. M. 1986. Phosphorylation of tyrosine in the carboxyl-terminal tryptic peptide of $pp60^{c-src}$. *Proc. Natl. Acad. Sci. USA* 83: 892–96

Lee, J. S., Varmus, H. E., Bishop, J. M. 1979. Virus-specific messenger RNAs in permissive cells infected by avian sarcoma virus. *J. Biol. Chem.* 254: 8015–22

Lerner, T. L., Hanafusa, H. 1984. DNA sequence of the Bryan high-titer strain of Rous sarcoma virus: Extent of *env* deletion and possible genealogical relationship with other viral strains. *J. Virol.* 49: 549–56

Levinson, A. D., Courtneidge, S. A., Bishop, J. M. 1981. Structural and functional domains of the Rous sarcoma virus transforming protein ($pp60^{src}$). *Proc. Natl. Acad. Sci. USA* 78: 1624–28

Levinson, A. D., Oppermann, H., Levintow, L., Varmus, H. E., Bishop, J. M. 1978. Evidence that the transforming gene of avian sarcoma virus encodes a protein kinase associated with a phosphoprotein. *Cell* 15: 561–72

Levinson, A. D., Oppermann, H., Varmus, H. E., Bishop, J. M. 1980. The purified protein product of the transforming gene of avian sarcoma virus phosphorylates tyrosine. *J. Biol. Chem.* 255: 11973–80

Levy, J. B., Iba, H., Hanafusa, H. 1986. Activation of the transforming potential of $p60^{c-src}$ by a single amino acid change. *Proc. Natl. Acad. Sci. USA* 83: 4228–32

Linial, M., Blair, D. 1982. Genetics of retroviruses. In *RNA Tumor Viruses*, ed. R. Weiss, N. Teich, H. Varmus, J. Coffin, pp. 649–783. Cold Spring Harbor, NY: Cold Spring Harbor Lab. 2nd ed.

Lipsich, L. A., Lewis, A. J., Brugge, J. S. 1983. Isolation of monoclonal antibodies that recognize the transforming proteins of avian sarcoma viruses. *J. Virol.* 48: 352–60

MacDonald, M. L., Kuenzel, E. A., Glomset, J. A., Krebs, E. G. 1985. Evidence from two transformed cell lines that the phosphorylation of peptide tyrosine and phosphatidylinositol are catalyzed by different proteins. *Proc. Natl. Acad. Sci. USA* 82: 3993–97

Mayer, B. J., Jove, R., Krane, J. F., Poirier, F., Calothy, G., Hanafusa, H. 1986. Genetic lesions involved in temperature sensitivity of the *src* gene products of four Rous sarcoma virus mutants. *J. Virol.* 60: 858–67

Nigg, E. A., Sefton, B. M., Hunter, T., Walter, G., Singer, S. J. 1982. Immunofluorescent localization of the transforming protein of Rous sarcoma virus with antibodies against a synthetic *src* peptide. *Proc. Natl. Acad. Sci. USA* 79: 5322–26

Nishizawa, M., Mayer, B. J., Takeya, T., Yamamoto, T., Toyoshima, K., et al. 1985. Two independent mutations are required for temperature-sensitive cell transformation by a Rous sarcoma virus temperature-sensitive mutant. *J. Virol.* 56: 743–49

Oppermann, H., Levinson, A. D., Varmus, H. E., Levintow, L., Bishop, J. M. 1979. Uninfected vertebrate cells contain a protein that is closely related to the product of the avian sarcoma. virus transforming gene (*src*). *Proc. Natl. Acad. Sci. USA* 76: 1804–8

54   JOVE & HANAFUSA

Parker, R. C., Varmus, H. E., Bishop, J. M. 1984. Expression of v-src and chicken c-src in rat cells demonstrates qualitative differences between pp60^{v-src} and pp60^{c-src}. *Cell* 37: 131–39

Parsons, J. T., Bryant, D., Wilkerson, V., Gilmartin, G., Parsons, S. J. 1984. Site-directed mutagenesis of Rous sarcoma virus pp60^{src}: Identification of functional domains required for transformation. In *Cancer Cells*, ed. G. F. Vande Woude, A. J. Levine, W. C. Topp, J. D. Watson, 2: 37–42. Cold Spring Harbor, NY: Cold Spring Harbor Lab.

Parsons, S. J., McCarley, D. J., Ely, C. M., Benjamin, D. C., Parsons, J. T. 1984. Monoclonal antibodies to Rous sarcoma virus pp60^{src} react with enzymatically active cellular pp60^{src} of avian and mammalian origin. *J. Virol.* 51: 272–82

Pasquale, E. B., Maher, P. A., Singer, S. J. 1986. Talin is phosphorylated on tyrosine in chicken embryo fibroblasts transformed by Rous sarcoma virus. *Proc. Natl. Acad. Sci. USA* 83: 5507–11

Patschinsky, T., Hunter, T., Esch, F. S., Cooper, J. A., Sefton, B. M. 1982. Analysis of the sequence of amino acids surrounding sites of tyrosine phosphorylation. *Proc. Natl. Acad. Sci. USA* 79: 973–77

Patschinsky, T., Hunter, T., Sefton, B. M. 1986. Phosphorylation of the transforming protein of Rous sarcoma virus: Direct demonstration of phosphorylation of serine 17 and identification of an additional site of tyrosine phosphorylation in p60^{v-src} of Prague Rous sarcoma virus. *J. Virol.* 59: 73–81

Pellman, D., Garber, E. A., Cross, F. R., Hanafusa, H. 1985a. Fine structural mapping of critical NH$_2$-terminal region of p60^{src}. *Proc. Natl. Acad. Sci. USA* 82: 1623–27

Pellman, D., Garber, E. A., Cross, F. R., Hanafusa, H. 1985b. An N-terminal peptide from p60^{src} can direct myristylation and plasma membrane localization when fused to heterologous proteins. *Nature* 314: 374–77

Purchio, A. F. 1982. Evidence that pp60^{src}, the product of the Rous sarcoma virus src gene, undergoes autophosphorylation. *J. Virol.* 41: 1–7

Purchio, A. F., Erikson, E., Brugge, J. S., Erikson, R. L. 1978. Identification of a polypeptide encoded by the avian sarcoma virus src gene. *Proc. Natl. Acad. Sci. USA* 75: 1567–71

Purchio, A. F., Jovanovich, S., Erikson, R. L. 1980. Sites of synthesis of viral proteins in avian sarcoma virus-infected chicken cells. *J. Virol.* 35: 629–36

Purchio, A. F., Shoyab, M., Gentry, L. E. 1985. Site-specific increased phosphorylation of pp60^{v-src} after treatment of RSV-transformed cells with a tumor promoter. *Science* 229: 1393–95

Purchio, A. F., Wells, S. K., Collett, M. S. 1983. Increase in the phosphotransferase specific activity of purified Rous sarcoma virus pp60^{v-src} protein after incubation with ATP plus Mg^{2+}. *Mol. Cell. Biol.* 3: 1589–97

Resh, M. D., Erikson, R. L. 1985. Highly specific antibody to Rous sarcoma virus src gene product recognizes a novel population of pp60^{v-src} and pp60^{c-src} molecules. *J. Cell Biol.* 100: 409–17

Rohrschneider, L. R. 1980. Adhesion plaques of Rous sarcoma virus-transformed cells contain the src gene product. *Proc. Natl. Acad. Sci. USA* 77: 3514–18

Rohrschneider, L. R. 1986. Tissue-specific expression and possible functions of pp60^{c-src}. In *Oncogenes and Growth Control*, ed. P. Kahn, T. Graf, pp. 27–31. Berlin: Springer-Verlag

Rous, P. 1911. A sarcoma of the fowl transmissible by an agent separable from the tumor cells. *J. Exp. Med.* 13: 397–411

Sadowski, I., Stone, J. C., Pawson, T. 1986. A noncatalytic domain conserved among cytoplasmic protein-tyrosine kinases modifies the kinase function and transforming activity of Fujinami sarcoma virus P130$^{gag-fps}$. *Mol. Cell. Biol.* 6: 4396–4408

Schultz, A. M., Henderson, L. E., Oroszlan, S., Garber, E. A., Hanafusa, H. 1985. Amino terminal myristylation of the protein kinase p60^{src}, a retroviral transforming protein. *Science* 227: 427–29

Schwartz, D. E., Tizard, R., Gilbert, W. 1983. Nucleotide sequence of Rous sarcoma virus. *Cell* 32: 853–69

Sefton, B. M. 1986. The viral tyrosine protein kinases. *Curr. Top. Microbiol. Immunol.* 123: 39–72

Sefton, B. M., Hunter, T., Beemon, K., Eckhart, W. 1980. Evidence that the phosphorylation of tyrosine is essential for cellular transformation by Rous sarcoma virus. *Cell* 20: 807–16

Sefton, B. M., Trowbridge, I. S., Cooper, J. A., Scolnick, E. M. 1982. The transforming proteins of Rous sarcoma virus, Harvey sarcoma virus and Abelson virus contain tightly bound lipid. *Cell* 31: 465–74

Shalloway, D., Coussens, P. M., Yaciuk, P. 1984. Overexpression of the c-src protein does not induce transformation of NIH 3T3 cells. *Proc. Natl. Acad. Sci. USA* 81: 7071–75

Smart, J. E., Oppermann, H., Czernilofsky, A. P., Purchio, A. F., Erikson, R. L., Bishop, J. M. 1981. Characterization of sites for tyrosine phosphorylation in the transforming protein of Rous sarcoma virus (pp60^{v-src}) and its normal cellular homologue (pp60^{c-src}). *Proc. Natl. Acad. Sci. USA* 78: 6013–17

Snyder, M. A., Bishop, J. M. 1984. A mutation at the major phosphotyrosine in pp60$^{(v-src)}$ alters oncogenic potential. *Virology* 136: 375–86

Snyder, M. A., Bishop, J. M., Colby, W. W., Levinson, A. D. 1983. Phosphorylation of tyrosine-416 is not required for the transforming properties and kinase activity of pp60$^{(v-src)}$ alters oncogenic potential.

Snyder, M. A., Bishop, J. M., McGrath, J. P., Levinson, A. D. 1985. A mutation at the ATP-binding site of pp60^{v-src} abolishes kinase activity, transformation, and tumorigenicity. *Mol. Cell. Biol.* 5: 1772–79

Stoker, A. W., Enrietto, P. J., Wyke, J. A. 1984. Functional domains of the pp60^{v-src} protein as revealed by analysis of temperature-sensitive Rous sarcoma virus mutants. *Mol. Cell. Biol.* 4: 1508–14

Stoker, A. W., Kellie, S., Wyke, J. A. 1986. Intracellular localization and processing of pp60^{v-src} proteins expressed by two distinct temperature-sensitive mutants of Rous sarcoma virus. *J. Virol.* 58: 876–83

Sugano, S., Hanafusa, H. 1985. Phosphatidylinositol kinase activity in virus-transformed and nontransformed cells. *Mol. Cell. Biol.* 5: 2399–2404

Sugimoto, Y., Erikson, R. L. 1985. Phosphatidylinositol kinase activities in normal and Rous sarcoma virus-transformed cells. *Mol. Cell. Biol.* 5: 3194–98

Sugimoto, Y., Whitman, M., Cantley, L. C., Erikson, R. L. 1984. Evidence that the Rous sarcoma virus transforming gene product phosphorylates phosphatidylinositol and diacylglycerol. *Proc. Natl. Acad. Sci. USA* 81: 2117–21

Takeya, T., Feldman, R. A., Hanafusa, H. 1982. DNA sequence of the viral and cellular *src* gene of chickens. I. Complete nucleotide sequence of an *Eco*RI fragment of recovered avian sarcoma virus which codes for gp37 and pp60^{src}. *J. Virol.* 44: 1–11

Takeya, T., Hanafusa, H. 1982. DNA sequence of the viral and cellular *src* gene of chickens. II. Comparison of the *src* genes of two strains of avian sarcoma virus and of the cellular homolog. *J. Virol.* 44: 12–18

Takeya, T., Hanafusa, H. 1983. Structure and sequence of the cellular gene homologous to the RSV *src* gene and the mechanism for generating the transforming virus. *Cell* 32: 881–90

Tanaka, A., Fujita, D. J. 1986. Expression of a molecularly cloned human c-*src* oncogene by using a replication-competent retroviral vector. *Mol. Cell. Biol.* 6: 3900–9

Tanaka, A., Parker, C., Kaji, A. 1980. Stimulation of growth rate of chondrocytes by Rous sarcoma virus is not coordinated with other expressions of the *src* gene phenotype. *J. Virol.* 35: 531–41

Teich, N., Wyke, J., Mak, T., Bernstein, A., Hardy, W. 1982. Pathogenesis of retrovirus-induced disease. In *RNA Tumor Viruses*, ed. R. Weiss, N. Teich, H. Varmus, J. Coffin, pp. 785–998. Cold Spring Harbor, NY: Cold Spring Harbor Lab. 2nd ed.

Van Beveren, C., Verma, I. M. 1986. Homology among oncogenes. *Curr. Top. Microbiol. Immunol.* 123: 73–98

Varmus, H., Swanstrom, R. 1982. Replication of retroviruses. In *RNA Tumor Viruses*, ed. R. Weiss, N. Teich, H. Varmus, J. Coffin, pp. 369–512. Cold Spring Harbor, NY: Cold Spring Harbor Lab. 2nd ed.

Wang, L.-H. 1978. The gene order of avian RNA tumor viruses derived from biochemical analysis of deletion mutants and viral recombinants. *Ann. Rev. Microbiol.* 32: 561–92

Wang, L.-H. 1987. The mechanism of transduction of proto-oncogene c-*src* by avian retroviruses. *Mutation Res.* In press

Wang, L.-H., Beckson, M., Anderson, S. M., Hanafusa, H. 1984. Identification of the viral sequence required for the generation of recovered avian sarcoma viruses and characterization of a series of replication-defective recovered avian sarcoma viruses. *J. Virol.* 49: 881–91

Weber, M. J., Friis, R. R. 1979. Dissociation of transformation parameters using temperature-conditional mutants of Rous sarcoma virus. *Cell* 16: 25–32

Wilhelmsen, K. C., Tarpley, W. G., Temin, H. M. 1984. Identification of some of the parameters governing transformation by oncogenes in retroviruses. In *Cancer Cells*, ed. G. F. Vande Woude, A. J. Levine, W. C. Topp, J. D. Watson, 2: 303–8. Cold Spring Harbor, NY: Cold Spring Harbor Lab.

Wilkerson, V. W., Bryant, D. L., Parsons, J. T. 1985. Rous sarcoma virus variants that encode *src* proteins with an altered carboxy terminus are defective for cellular transformation. *J. Virol.* 55: 314–21

Willingham, M. C., Jay, G., Pastan, I. 1979. Localization of the ASV *src* gene product to the plasma membrane of transformed cells by microscopic immunocyto-chemistry. *Cell* 18: 125–34

Wyke, J. A., Stoker, A. W. 1987. Genetic analysis of the form and function of the viral *src* oncogene product. *Biochim. Biophys. Acta.* In press

Yaciuk, P., Shalloway, D. 1986. Features of the pp60^{v-src} carboxyl terminus that are required for transformation. *Mol. Cell. Biol.* 6: 2807–19

Yamagishi-Hagino, K., Ikawa, S., Kawai, S., Hihara, H., Yamamoto, T., Toyoshima, K. 1984. Characterization of two strains of avian sarcoma virus isolated from avian lymphatic leukosis virus induced-sarcomas. *Virology* 137: 266–75

Ann. Rev. Cell Biol. 1987. 3 : 57–85

LAMININ AND OTHER BASEMENT MEMBRANE COMPONENTS

George R. Martin

Laboratory of Developmental Biology and Anomalies, National Institute of Dental Research, National Institutes of Health, Bethesda, Maryland 20892

Rupert Timpl

Max-Planck-Institut für Biochemie, D-8033 Martinsried/Munich, Federal Republic of Germany

CONTENTS

INTRODUCTION... 57
STRUCTURE 58
INTERACTIONS WITH OTHER BASEMENT MEMBRANE PROTEINS.. 65
BIOSYNTHESIS AND REGULATION 69
CELL BINDING AND OTHER BIOLOGICAL PROPERTIES OF LAMININ...................................... 71
MOLECULAR MODELS OF BASEMENT MEMBRANE STRUCTURE 76

INTRODUCTION

Laminin, the most abundant glycoprotein in basement membranes, is both a structural and a biologically active component. It is only found in significant quantities in basement membranes, the thin extracellular matrices that surround epithelial tissues, nerves, fat cells, and smooth, striated, and cardiac muscle. All basement membranes contain a common set of proteins that includes laminin, collagen IV, various heparan sulfate proteoglycans, and entactin/nidogen. Basement membrane is the first extra-

57

0743–4634/87/1115–0057$02.00

cellular matrix to appear during embryogenesis, and laminin is the first matrix protein to be detected. Current concepts suggest that basement membranes create barriers that allow embryonic cells to segregate and differentiate into specific tissues. In the adult, basement membranes serve as molecular filters in capillaries and glomeruli preventing the passage of proteins, and they provide the scaffolding that maintains normal tissue form during regeneration and growth.

The laminin molecule ($M_r = 850,000$), as isolated from tumor tissue or from placenta, is visualized as a cross-shaped structure with three short and one long arm. The laminin from these sources contains three distinct chains (A, B1, and B2). These have been cloned, and the nucleotide sequences for the B1 and B2 chains indicate that these chains are homologous structures containing several distinct domains arranged sequentially along their length.

Laminin binds to collagen IV, heparan sulfate proteoglycan, to entactin/nidogen, and to itself to create an integrated structure within the basement membrane. Additionally, due to its size and shape, it is able to span the basement membrane and bind to various substances on the surface of cells. These substances include a specific high-affinity membrane receptor ($M_r = 67,000$) and various other ligands (heparan sulfate, gangliosides, and sulfatides). Additional receptors on the cells may also be present and involved in these responses but are not yet well defined. Binding to laminin elicits cell-specific responses, causing secretory cells to become polarized, neural cells to extend axonlike processes, various cells to migrate, and a variety of cells to differentiate. These findings indicate that laminin is an important structural and regulatory molecule.

There are already a considerable number of reports on the occurrence, structure, and cellular interactions of laminin. Limitations of space prohibit a comprehensive discussion of every publication involving laminin. Rather we cover areas related to the structure and function of laminin in basement membranes and to its interaction with cells that allow their attachment, migration, and the formation of neurites and that induce invasive behavior in malignant tumor cells. Additional material can be found in other articles in this Annual Reviews series and in a recent comprehensive treatment of basement membranes (Timpl & Dziadek 1986).

STRUCTURE

Laminin is a ubiquitous basement membrane component, as demonstrated in numerous immunohistological and cell-culture studies (reviewed in Timpl et al 1983a; Timpl & Dziadek 1986). It was initially isolated in intact form from neutral extracts of a mouse tumor that produces large quantities of basement membrane, the Engelbreth-Holm-Swarm (EHS) tumor (Timpl et al 1979), and its constituent chains from cultured terato-carcinoma cells (Chung et al 1979). Laminin from these sources consists of three large polypeptide chains designated B1, B2, and A (Cooper et al 1981), which by many inter- and intrachain disulfide bonds form a unique cross-shaped component (Figure 1a). Electron microscopical analyses of laminin molecules (Engel et al 1981) demonstrated several rodlike segments and globular structures within this cruciform shape (Figure 2), indicating that laminin is a multidomain protein. Each molecule has three very similar short arms ~ 37 nm long, and each arm contains two globular and two rodlike segments. In addition, laminin contains a long arm consisting of a 77-nm-long rod that terminates in a large globular domain. This globule appears to have a complex structure with three smaller globules apparent in negatively stained preparations (Paulsson et al 1985a). The rodlike

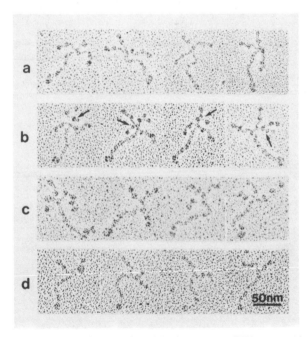

Figure 1   Rotary-shadowing images of laminin from mouse EHS tumor (*a*), sea urchin embryo (*c*), and rat Schwannoma cell culture (*d*), and of particles of the laminin-nidogen complex from EHS tumor (*b*). The probable position of the noncovalently attached, dumbbell-shaped nidogen is indicated. The individual particles are typical examples of mouse laminin (Engel et al 1981), laminin-nidogen complex (Paulsson et al 1987), sea urchin laminin (McCarthy et al 1987), and Schwannoma laminin (Davis et al 1985a; D. Edgar et al, unpublished). The bar indicates a length of 50 nm.

elements have a high mass-per-length ratio, which indicates that they contain compactly folded peptide segments (Engel et al 1981).

Protease-Derived Fragments

Digestion of laminin by various proteases and characterization of the large fragments by electron microscopical and immunochemical methods were useful approaches for dissecting the multidomain structure of laminin (Rao et al 1982a,b; Ott et al 1982; Timpl et al 1983b; Palm et al 1985; Paulsson et al 1985a). These studies demonstrated a higher protease resistance of the short arms compared to the long arm and provided fragments useful for mapping domains within the laminin molecule with diverse biological activities (see below). The locations of some important fragments are illustrated in Figure 2. A remarkably high stability was observed for the disulfide-rich fragment 1 generated by pepsin, which consists of three rodlike segments (Rohde et al 1980; Engel et al 1981) and originates from the center of the cross. More limited proteolysis yielded α3, a fragment with three apparently intact short arms (α3, fragment 1–4) (Ott et al 1982; Rao et al 1982a). Other interesting structures were the major heparin binding domain (Ott et al 1982), and fragment 8, derived from the end of

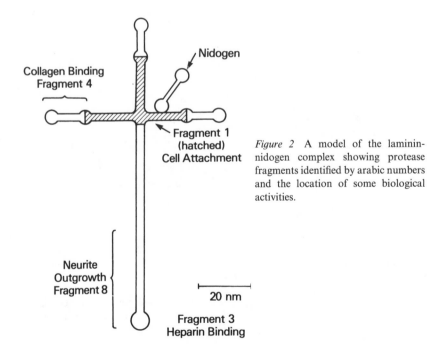

Figure 2 A model of the laminin-nidogen complex showing protease fragments identified by arabic numbers and the location of some biological activities.

the long arm, which consists of the globular domain plus a 32-nm-long rod (Paulsson et al 1985a). This fragment contains peptides originating from all three chains of laminin (Paulsson et al 1987).

Studies with laminin fragments also demonstrated complex conformational features (Ott et al 1982; Paulsson et al 1985a). Most of the short-arm fragments show no well-defined circular dichroism (CD) spectra. In contrast, fragment 8 from the long arm and some related fragments are 50–100% α-helical and have a high thermal stability ($T_m = 56$–$58°C$). The α-helix is found in the rodlike element of the long arm and exists in a coiled-coil configuration formed by portions of the B1 and B2 chains. The presence of a coiled-coil helix was originally predicted from cDNA sequences (Barlow et al 1984), which demonstrated that the C termini of the B1 and B2 chains are composed of a characteristic heptad repeat of hydrophobic and polar amino acids typical for proteins with a coiled-coil configuration (i.e. tropomyosin, myosin, cytokeratins). The conformation of the globular domains of the long arm may include a substantial portion of β structure (Ott et al 1982).

Carbohydrate

Laminin contains about 13% carbohydrate, attached mainly in N-linked form. These appear to be clustered in certain regions of the molecule, as judged by the carbohydrate content of various proteolytic fragments. The amino acid sequence deduced for the B1 chain from recombinant DNA clones shows 13 Asn-X-Thr/Ser sequences, which are potential carbohydrate acceptor sites (Sasaki et al 1987), in agreement with predictions from biosynthetic studies (Howe 1984). About 15 different oligosaccharides with rather complex structures have been isolated from digests of laminin (Arumugham et al 1986). Unique features of these structures are the frequent extension of the oligosaccharide structures by lactosamine oligomers and the addition of α-glycosidicly linked galactose at the nonreducing end. The latter fact was suggested earlier based on binding studies with an α-galactose-specific lectin (Shibata et al 1982; Rao et al 1983a). On tissue sections the same lectin shows strong binding to murine but not human basement membranes (Peters & Goldstein 1979). The role of oligosaccharides on laminin is unknown, but they do not appear to confer stability against proteases (Howe 1984).

Studies with cDNA Clones

Most of the progress made in understanding the covalent structure of laminin has come from recent analysis of cDNA clones that established the amino acid sequence of the mouse B1 chain (Sasaki et al 1987). The chain consists of 1765 amino acid residues. Computer analysis of the sequence revealed a multidomain structure (Figure 3) for the B1 chain, as expected from other morphological and structural data. The N-terminal domains, VI and IV, each of which is about 250 residues and has a low cysteine content, are believed to form globular structures. Domains V and III are formed of homologous, cysteine-rich segments of ~50 amino acids, which are repeated five times in domain V and eight times in domain III. These repeats can be aligned, utilizing gaps, to show eight cysteine residues at regular intervals. Such a cysteine-rich repeat structure is also found in precursor EGF, TGF$_\alpha$, coagulation factors, and thrombospondin, although their cysteine motif is different from that of the B1 chain (Lawler & Hynes 1986). The sequence of the C-terminal domains II and I (about 600 residues) predicts an α-helical structure. The heptad repeat is more perfect for domain I than domain II, but it is likely that both can participate

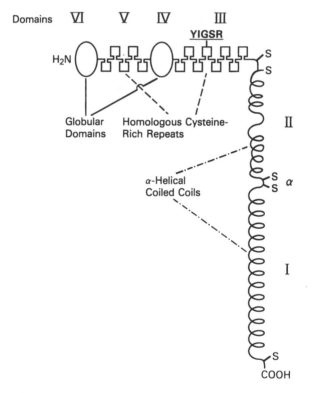

Figure 3 Domain model of the B1 chain of laminin based on computer analysis of its amino acid sequence as deduced from its nucleotide sequence (Sasaki et al 1987). The location and nature of its domains are indicated. Domain III contains a sequence (YIGSR) that supports cell attachment, receptor binding, and chemotaxis (Graf et al 1987).

in a coiled coil. These α-helical domains are interrupted by a 30-residue segment ("α") containing six cysteines, whose function is not known.

The B2 chain is shorter by about 100 amino acid residues than the B1 chain but shows basically the same arrangement of domains (M. Sasaki & Y. Yamada, personal communication). Differences between the two chains include a lower number of EGF-like repeats and absence of the interrupting segment "α" and its complement of cysteines. The characterization and analysis of the A chain is still incomplete (M. Sasaki & Y. Yamada; R. Deutzmann, personal communications) but suggests that a large N-terminal segment of the A chain shares homology to comparable domains in the B1 and B2 chains. Sequences at its C-terminal end are different and presumably correspond to the large globular domain at the end of the long arm of laminin.

A Model for Laminin

Electron microscopic evidence, protein studies, and the data obtained by cloning the B1 and B2 chains, suggest a model for the arrangement of the chains within the molecule as well as the assignment of specific domains (Figure 4). In this model, the N terminus of each chain is placed at an outer globule with domains VI through III forming the globules and adjacent structures observed in the short arms. Subsequently, the three chains intersect at the center of the cross, are linked by disulfide bonds and then run together, in parallel, to form the rodlike segment of the long arm (Hogan et al 1985). A significant portion of the rod contains a two-

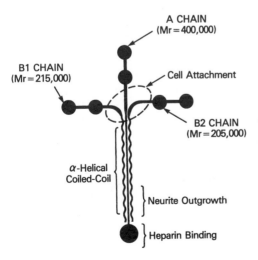

Figure 4 Arrangement of the A, B1, and B2 chains in the laminin molecule.

or three-chain coiled coil, which creates a rigid and extended structure. The B1 and B2 chains terminate at the end of the rod and are linked together in this region by a single disulfide bond (Paulsson et al 1985a, 1987). The carboxyl terminus of the A chain forms the large globule at the end of the long arm.

Peptide sequencing (R. Deutzmann, personal communication) shows that the B1 chain starts with a blocked N terminus and that this region corresponds to protease fragment 4 (Figure 2). As expected, protease fragment 1, from the center of the cross, contains sequences from all three constituent chains of laminin.

Immunochemistry

Specific polyclonal antibodies have been raised against intact laminin and its constituent chains (Chung et al 1979; Timpl et al 1979; Rohde et al 1979; Foidart et al 1980). Such antisera have been extensively used for detecting laminin in biological specimens (see a summary of these studies in Timpl & Dziadek 1986) and for developing sensitive radioimmunoassays applicable to measurements on human serum (Brocks et al 1986). It was also possible to generate antibodies specific to various fragments of laminin that block certain biological properties of laminin (see below) (Ott et al 1982; Timpl et al 1983b; Paulsson et al 1985a; Charonis et al 1986; Graf et al 1987; Aumailley et al 1987). Monoclonal antibodies specific for A or B chains and for the long arm of laminin have also been described (Chung et al 1983; Wewer et al 1983; Palm et al 1985). Some of the antibodies to specific regions of the molecule give a differential staining of basement membranes in tissues (Jaffe et al 1984; Wan et al 1984); in contrast poly-clonal antibodies to laminin itself stain all basement membranes. This finding either indicates molecular heterogeneity of laminin in different sites or unequal masking of epitopes. Antibodies that react with mouse laminin have been also detected in patients with Chagas disease (Szarfman et al 1982) and with American cutaneous leishmaniasis (Avila et al 1984). These antibodies have been shown to react with α-galactose-containing carbo-hydrate epitopes of laminin (Towbin et al 1987), which indicates that these epitopes are shared by the rodent protein and *Trypanosomatidae* parasites.

Isoforms

Laminin with similar chain composition and shape has been obtained from parietal yolk-sac carcinomas (Engvall et al 1983; Martinez-Hernandez et al 1982) and teratocarcinoma cells (Sakashita & Ruoslahti 1980; Ozawa et al 1983). Laminin from normal human tissues appears to be less soluble and must be extracted with denaturing solvents (Ohno et al 1983, 1985; Dixit 1985). The human material also shows a cruciform structure but contains, in addition to A and B chains, a component with intermediate electrophoretic mobility, the M chain. Whether the M component represents a new constituent chain or a degradation artefact is unknown. Laminin has also been purified from invertebrate tissues, such as *Drosophila* (Fessler et al 1984) and sea urchin embryos (McCarthy et al 1987). These materials also exhibit a typical cross-shape (Figure 1c), but the A and B chains and the rodlike domain of the long arm appear to be 30 nm longer than those of mouse laminin.

The occurrence of additional forms of laminin was also suggested by the finding that the oocyte and the two-cell-stage embryo produce only the B chains of laminin (Cooper & MacQueen 1983), as do Schwann cells (Cornbrooks et al 1983; Palm & Furcht 1983). Particularly interesting were the observations that Schwann cells and myoblasts secrete substances with neurogenic activity that are precipitated by antibodies to tumor laminin but consist only of B chains and lack fragment 3 epitopes (Davis et al 1985a; Lander et al 1985; Dziadek et al 1986; D. Edgar & M. Paulsson, unpublished). Partially purified material from such cultures contains Y-shaped particles by electron microscopy rather than the cruciform structure of tumor laminin (Figure 1d). The possibility that some tissues produce laminins with a different chain structure is also supported by the finding that kidney and liver contain disproportionately low levels of mRNA for the A relative to the B chains of laminin (Kleinman et al 1987). Thus a variety of observations suggests the existence of isoforms of laminin.

INTERACTIONS WITH OTHER BASEMENT MEMBRANE PROTEINS

Laminin binds to several structural components of the matrix. Indeed, extracts of EHS tumor matrix, which contain laminin and a variety of other basement membrane components, reassociate under physiological conditions into a gel-like matrix containing the specific components of the matrix, including laminin, collagen IV, heparan sulfate proteoglycan, and nidogen. Reconstitution of the gel is dependent on the presence of laminin, collagen IV, and heparan sulfate proteoglycan (Kleinman et al 1983, 1986). Several of the components that bind to laminin are well characterized, and their properties are summarized in Table 1.

Self-Assembly

Laminin has the ability to self-assemble into polymers with a defined structure. Aggregation of laminin has been detected by ultracentrifugation

Table 1 Properties of basement membrane components with potential for interaction with laminin[a]

Component	Molecular mass (kDa)	Chain composition	Special features	References
Laminin	850	B1 (220 kDa), B2 (210 kDa), and A (350–400 kDa) chains; disulfide bonded	Major cell binding protein	Terranova et al 1980; Engel et al 1981; Timpl et al 1983a
Collagen IV	550	Two α1(IV) and one α2(IV) chain (each about 180 kDa); disulfide bonded	Major structural component; network formation	Timpl et al 1981; Yurchenco & Furthmayr 1984
Heparan sulfate proteoglycans	LD, 600–700	Single compact polypeptide core (400–450 kDa); 3 HS chains	Multiglobular; HS chains at one pole	Hassell et al 1985; Paulsson et al 1985b
	HD, 130	Single small peptide core (5–30 kDa); 3–4 HS chains	Star-shaped particles	Kanwar et al 1981; Fujiwara et al 1984
Nidogen/entactin	150	Single polypeptide	Dumbbell shape; tyrosine sulfate	Carlin et al 1981; Paulsson et al 1986a
BM-40 (SPARC, osteonectin)	35	Single polypeptide	Calcium binding	Dziadek et al 1986b; Mason et al 1986

[a] Abbreviations: LD, low-density form; HD, high-density form; HS, heparan sulfate.

(Engel et al 1981) and in heat gelation experiments in which the formation of large polymers occurs (Yurchenco et al 1985). In the latter system polymerization depends on temperature and a critical minimum protein concentration and is enhanced by calcium. Aggregation is reversible and can also be arrested at the level of laminin dimers and trimers by chelating agents. Analysis of laminin oligomers demonstrated interaction between outer globular domains of the short arms and between globular domains at the end of the long arm. The importance of the latter interaction was shown in inhibition experiments with antibodies against laminin fragment 3 (Charonis et al 1986).

Binding to Nidogen

Among the various complexes laminin can form with proteins, those formed with nidogen/entactin are particularly stable with $K_d < 10$ nM (Dziadek et al 1985; Paulsson et al 1987). A fragment of nidogen was originally isolated from extracts of the EHS tumor (Timpl et al 1983c) and later shown to be a dumbbell-shaped protein ($M_r = 150,000$) (Paulsson et al 1986a). Entactin was discovered as a sulfated polypeptide of similar molecular mass produced by cultured teratocarcinoma cells (Carlin et al 1981). These proteins coelute in various chromatographic systems, which indicates that they are similar or even identical proteins (Paulsson et al 1985b). Structural studies of nidogen demonstrated a single polypeptide chain that is highly sensitive to proteolytic digestion, which creates multiple fragments differing in their N termini (Paulsson et al 1986a). Immuno-logical analysis demonstrated ubiquitous occurrence of nidogen/entac-tin in basement membranes (reviewed in Timpl & Dziadek 1986) and that laminin and nidogen occur in equimolar proportions in a variety of tissue extracts (Dziadek & Timpl 1985).

The existence of stable complexes between laminin and nidogen/entactin was initially indicated by the coprecipitation of a noncovalently associated 150-kDa polypeptide with laminin by specific antibodies from cell-culture media (Hogan et al 1980; Carlin et al 1983; Palm & Furcht 1983; Dziadek & Timpl 1985). Such complexes can also be extracted from the EHS tumor using chelating agents (Paulsson et al 1987). A model of the complex was proposed based on electron microscopy (Figure 1b) and binding studies indicating that a single nidogen molecule binds through one of its globular domains to the center of the laminin cross (Figure 2). The biological function of nidogen and of this complex is unknown, although the complex may modulate the activity of a cell binding site present in laminin fragment 1 (Aumailley et al 1987). The data also indicate that bivalent cations, such as calcium, contribute to the stabilization of basement membranes.

Binding to Collagen IV

Collagen IV is considered the major structural component in the basement membrane, which anchors the cell-laminin complex (Terranova et al 1980; Rao et al 1982a). This collagen is unique to basement membranes and forms, by specific interactions and cross-linking of its terminal domains, large open or polygonal networks (Timpl et al 1981; Yurchenco & Furthmayr 1984). It is thought that this network forms a scaffolding to which other proteins bind at specific sites as demonstrated for laminin and heparan sulfate proteoglycan (Woodley et al 1983; Charonis et al 1985; Laurie et al 1986). The binding of laminin to collagen IV requires an intact triple-helical conformation, and one or two major binding sites have been localized along its major triple-helical domain. The collagen binding sites on laminin have been mapped to globular domains at the end of the short arms in cell attachment studies (Rao et al 1982a) and to the globular domain at the end of the long arm by electronmicroscopy and by inhibition with antibodies of known specificity (Laurie et al 1986; Charonis et al 1985, 1986).

Binding to Heparan Sulfate Proteoglycan

Basement membranes contain polyanionic sites composed of heparan sulfate and possibly chondroitin sulfate chains, which serve as selective charge barriers and make the matrix impermeable to proteins (Farquhar 1981). Heparan sulfate binds directly to laminin (Woodley et al 1983; Fujiwara et al 1984), probably via the major heparin binding domain (fragment 3; see Figure 2) (Sakashita et al 1980; Ott et al 1982). Two heparan sulfate proteoglycans have been isolated from extracts of the EHS tumor; they differ in size and in the ratio of protein to glycosaminoglycan and thus in buoyant density (Hassell et al 1980, 1985; Fujiwara et al 1984). The small, high-density proteoglycan has a starlike shape and a low protein content (Fujiwara et al 1984). The large, low-density proteoglycan has a multidomain protein core 80 nm long with heparan sulfate side chains attached only at one end (Paulsson et al 1986b; Ledbetter et al 1987). Heparan sulfate proteoglycans from glomerular basement membranes are intermediate in size (Kanwar et al 1981, 1984) and may represent a distinct but related species. Other heparan sulfate (Parthasarathy & Spiro 1984) and chondroitin sulfate proteoglycans (Couchman et al 1984; Paulsson et al 1985b) may exist in basement membranes but have not yet been fully characterized.

The high-density heparan sulfate proteoglycan shows weak binding ($K_d > 1$ μM) to laminin via its heparan sulfate chains; this binding is eliminated by exposure to moderate salt concentrations (Fujiwara et al

1984). The low-density form also binds to laminin, but the strength and domain specificity of this interaction are not known (Woodley et al 1983). In the basement membrane itself, it is likely that strong binding of the large protein core to other matrix components exists, since concentrated guanidine solutions are required for its extraction (Hassell et al 1985; Paulsson et al 1986b).

Other Components

A small basement membrane protein (BM-40) has been isolated (Dziadek et al 1986b). Based on sequence data, this protein is likely to be identical to the cell-culture protein SPARC (Mason et al 1986) and bone osteonectin (M. Bolander, personal communication). The protein shows distinct calcium binding, which changes its α-helical conformation (J. Engel, personal communication). Since BM-40 is extracted together with the laminin-nidogen complex by EDTA containing neutral buffer (R. Timpl, unpublished), it may also participate in a calcium-mediated, supramolecular complex with laminin in the basement membranes.

Laminin was also reported to interact with proteins of extracellular fluids, such as the complement components C1q (Bohnsack et al 1985) and C3 (Leivo & Engvall 1986), and with plasminogen and plasminogen activator (Salonen et al 1984). The biological significance of these interactions is not clear, but they may be involved in homeostatic control or pathological alterations.

BIOSYNTHESIS AND REGULATION

Laminin is synthesized by many different cells (see Timpl & Dziadek 1986 for a partial listing), and studies on cultured cells have helped to define the structure, biosynthesis, and functions of the protein. Particularly useful models have been differentiated teratocarcinoma cells that secrete laminin and other basement membrane proteins in large amounts (Chung et al 1979; Hogan 1980; Leivo 1983; Pierce et al 1982). Production of basement membrane constituents occurs constitutively or, in the case of F9 and certain other teratocarcinoma cells, can be induced by treating the cells with retinoic acid plus cyclic AMP (cAMP) analogues (Strickland et al 1980). Following a one- and two-day lag, there is a greater than 20-fold increase in laminin synthesis in F9 cells treated with retinoic acid and dibutyryl cAMP, and the levels of the chains increase coordinately (Strickland et al 1980; Prehm et al 1982; Cooper et al 1983; Grover & Adamson 1986).

Studies on cultured cells showed that the laminin they produce contains three distinct chains, A, B1, and B2 (Cooper et al 1981); these were particularly well resolved by electrophoresis when glycosylation reactions were inhibited by tunicamycin (Howe & Dietzschold 1983; Kurkinen et al 1983). Variable glycosylation increases the mass of the B chains and causes them to migrate as a single diffuse band. The fact that the A, B1, and B2 chains are distinct proteins was demonstrated (*a*) by pulse-chase studies, which did not indicate interconversion; (*b*) by specific immunoassays, which showed A and B chains to be distinct; and (*c*) by obtaining different peptide maps of the three chains (Chung et al 1979; Cooper et al 1981; Kurkinen et al 1983; Howe & Dietzschold 1983). In vitro translation of mRNA from cells synthesizing laminin produced B1 and B2 chains (Kurkinen et al 1983; Wang & Gudas 1983) as well as A chain polypeptides (Hogan et al 1984; Durkin et al 1986).

The size of the nonglycosylated chains of laminin is estimated to be: A chain, $M_r = 300,000$; B1 chain, $M_r = 205,000$, and B2 chain, $M_r = 185,000$. The values for the B chains are in close agreement with the size of the B1 ($M_r = 194,668$) and B2 ($M_r = 174,527$) chains predicted from sequencing recombinant clones (Sasaki et al 1987; M. Sasaki & Y. Yamada, unpublished observations). The A chain has been cloned (Durkin et al 1986; Kleinman et al 1987) but has not yet been sequenced in its entirety. Northern hybridization indicates that the mRNAs for the A, B1, and B2 chains are 10, 6, and 8 kilobases (kb), respectively (see references above). Genetic analysis showed that the B1 and B2 chains of laminin are tightly linked in the central region of chromosome 1 of the mouse (Elliott et al 1985). Genomic clones for the B1 chain have been isolated and examined by R-looping. These studies indicate that the mouse B1 gene is greater than 60 kb and contains more than 35 exons, each of 100–400 base pairs (bp) (Yamada et al 1985, 1987). Current evidence suggests that the increase in laminin synthesis in F9 cells following treatment with retinoic acid and dibutyryl cAMP is preceded by proportional increases in mRNA for the laminin chains. This finding suggests that in this system synthesis is regulated at the level of transcription (Carlin et al 1983; Kurkinen et al 1983; Wang & Gudas 1983; Durkin et al 1986; Kleinman et al 1987).

Current concepts suggest that the B1 and B2 chains form a disulfide-linked dimer that is an intermediate in the assembly of laminin (Cooper et al 1981; Morita et al 1985; Peters et al 1985). The synthesis of A chains appears to be rate limiting, and the addition of an A chain to the B1-B2 dimer is rapid and precedes secretion, as do various glycosylation reactions. These glycosylation reactions are not required, however, either for assembly or secretion of the molecule (Howe & Dietzschold 1983). Secretion of labeled laminin molecules can be detected within an hour.

CELL BINDING AND OTHER BIOLOGICAL PROPERTIES OF LAMININ

Laminin has diverse effects on cultured cells; it can alter their growth, survival, morphology, differentiation, and motility (Kleinman et al 1985). Such effects may be related to and proceed from the binding of the cells to a laminin substrate. The type of morphological response depends on the cell examined. Glandular cells, such as Sertoli cells (Hadley et al 1985), assume a columnar form, while neural cells extend axonlike processes (Baron van Evercooren et al 1982). Considerable progress has been made in defining the molecular basis of the interaction of cells with laminin, including defining a major cellular receptor for laminin (reviewed in von der Mark et al 1984; von der Mark & Kühl 1985; Liotta et al 1985a) and identifying a major cell-attachment sequence on laminin (Graf et al 1987). However, there are likely to be other receptors and domains active in cell binding (Timpl et al 1983b; Aumailley et al 1987) since laminin is a very large molecule with many functional domains. Further, laminin resides in basement membranes in combination with nidogen, collagen IV, and heparan sulfate proteoglycan, each of which may also have an affinity for cells or modulate the cell binding activity of laminin. Thus it is likely that multiple interactions occur between the cell and the basement membrane (Kleinman et al 1985; Mai & Chung 1984).

Laminin Receptors and Attachment Sites in Laminin

Studies with tumor cells first demonstrated that cells contain high-affinity binding sites for laminin (Terranova et al 1983). Subsequently, a specific laminin receptor ($M_r = 67,000$) was isolated by affinity chromatography on immobilized laminin from muscle cells and from various tumor cell lines (Rao et al 1983b; Lesot et al 1983; Malinoff & Wicha 1983; Barsky et al 1984a). The isolated receptor retains a high affinity for laminin ($K_d \approx 2$ nM), does not bind fibronectin, requires detergent for solubilization, can be inserted into liposomes, and is associated with the membrane. Monoclonal antibodies have been raised to the receptor and have been used to study its distribution in normal and tumor tissue (Liotta et al 1985b; Hand et al 1985). These results indicate that the receptor is normally displayed along the basal surface of epithelial cells, while a less organized distribution is observed in malignant tumor cells. One of these monoclonal antibodies blocks the binding of laminin to cells (Liotta et al 1985b).

The laminin receptor has been cloned and sequenced in part (Wewer et al 1986). These studies suggest that it has a typical transmembrane segment and predict extracellular and intracellular domains. It is likely that the laminin receptor binds, either directly or indirectly, to cytoskeletal components, as judged by its effects on cell shape and by its ability to stabilize cell membranes (Sugrue & Hay 1981; Couchman et al 1983). The binding of actin to the laminin receptor has been demonstrated in isolated in vitro systems, which suggests a possible direct linkage of the laminin receptor to actin cables (Brown et al 1983). However, since many proteins bind to actin in such isolated systems, more detailed information is needed on the intracellular interaction of this receptor.

A major cell binding site was initially identified on laminin fragment 1 (Rao et al 1982a; Timpl et al 1983b) and was shown to interact with the 67-kDa high-affinity receptor (Terranova et al 1983; Graf et al 1987). A second high-affinity ($K_d = 1–4$ nM) cell binding site was more recently localized to fragment 8 (Aumailley et al 1987). Competition studies indicated different binding structures on fragments 1 and 8, which implies the existence of a second ubiquitous laminin receptor used by cells for attachment. Whether this putative receptor corresponds to that promoting neurite outgrowth (Edgar et al 1984; see below) is unknown. Some other small laminin fragments have been found to promote adhesion of hepatocytes (Timpl et al 1983b) but are presumably of low affinity. Proteins with strong laminin binding activity have also been identified on the surface of several bacteria (Switalsky et al 1984; Lopes et al 1985) and could play a role in the invasion of host tissues.

Possible Relation Between Fibronectin and Laminin Binding

Fibronectin is another extracellular, cell binding protein and many, but not all, cells show a comparable attachment to both laminin and fibronectin (reviewed in Timpl & Dziadek 1986). These observations raise the possibility that both interactions may involve similar or even identical mechanisms. Two independently raised monoclonal antibodies (SCAT and JG22) inhibit the attachment of cells to fibronectin and to laminin (Knudsen et al 1985) but do not appear to recognize the previously described laminin receptor. These antibodies recognize a complex of proteins (CSAT) whose most prominent component ($M_r = 145,000$) has been identified as the fibronectin receptor. The CSAT antigen has been reported to bind to laminin and fibronectin (Horwitz et al 1985). Further work is needed to demonstrate that the CSAT antigen is an actual laminin receptor and that the monoclonal antibodies are acting directly on the receptor to inhibit laminin binding by cells.

Other Cell Surface Ligands Binding to Laminin

Heparan sulfate (see section on interaction with other basement membrane proteins), sulfatides (Roberts et al 1985), and gangliosides (Kennedy et al

1983) bind to laminin. The physiological significance of these interactions is not established, but they could help to stabilize the binding of laminin to the cell surface via multiple interactions. They could also be involved in erythrocyte agglutination by laminin (Kennedy et al 1983).

A Defined Attachment Sequence in the B1 Chain of Laminin

As noted above, the B1 chain of laminin has been cloned and sequenced at the nucleotide level (Sasaki et al 1987). This information has been used to deduce the amino acid sequence of this protein and to predict its structure. Synthetic peptides were prepared to sequences from the various domains of the B1 chain, and specific antibodies were also prepared to conjugates of the peptides. The antibody to a peptide from domain III, a cysteine-rich region of homologous repeats, inhibited cell attachment to laminin, although the peptide itself was inactive (Graf et al 1987). Additional peptides were synthesized to sequences in domain III, and a nonapeptide (CDPGYIGSR) was found to be directly active in cell attachment and in chemotaxis. In addition, this peptide eluted the laminin receptor from columns of laminin-Sepharose to which detergent extracts of cellular membranes had been applied. This peptide exhibits only 0.5–1% of the activity of laminin in cell attachment on a molar basis. However, the peptide was able to inhibit attachment of the cells to laminin by up to 80%, which indicates that it is one of the major sites of cell attachment in laminin. Preliminary structure-function studies suggest that a pentapeptide (YIGSR) contains sufficient information to allow cell attachment, migration, and receptor binding and identify this sequence as a major receptor binding and attachment site in laminin (J. Graf, F. Robey, Y. Iwamoto, unpublished).

It is noteworthy that a tetrapeptide (RGDS) (Pierschbacher & Ruoslahti 1984; Suzuki et al 1985; Oldberg et al 1986) has been identified as a cell attachment site in fibronectin and in certain other proteins. The YIGSR sequence does not occur in fibronectin, and the RGDS sequence has not so far been found in the chains of laminin. Possibly these distinctive sequences determine the cellular receptors to which these proteins bind and thereby establish the cells with which they can interact.

Effects on Cell Migration

Laminin has the ability to stimulate cell movement (McCarthy et al 1983; McCarthy & Furcht 1984). Investigations in this area have been prompted by the knowledge that tumor cells migrate as they invade normal tissues and that certain matrix molecules, including collagen I and fibronectin, are chemotactic. Although tumor cells have received the most attention

(McCarthy et al 1985; Situ et al 1984; Varani et al 1985), epithelial cells, neural cells, leucocytes (Donaldson & Mahan 1984; Goodman & Newgreen 1985; Liesi 1985b; Selak et al 1985; Terranova & Lyall 1986; Terranova et al 1986a), and probably other cells also respond to laminin. In these studies, laminin was tested in soluble form, but in vivo laminin is largely insoluble and concentrated in basement membranes. For this reason, the response of tumor cells on filters coated with laminin was also studied (McCarthy & Furcht 1984). Indeed, surface-bound laminin also stimulates the cells to migrate, especially when it is applied to the lower surface of the filter. Such results indicate that laminin is chemotactic when it is in solution and haptotactic when on surfaces, and in either form it is able to guide cell movement. Proteolytic fragments were also chemotactic (Furcht et al 1984). Studies with the synthetic peptide CDPGYIGSR, which corresponds to the attachment site in the B1 chain of laminin, demonstrated that this peptide is a chemoattractant (Graf et al 1987). The peptide is not as active as laminin on a molar basis nor does it induce as large a response. However, these results indicate that laminin's chemotactic activity is mediated at least in part via the 67-kDa laminin receptor.

Cells of the Nervous System, Laminin, and the Formation of Neural Processes

Neural cells can attach to a variety of substrates, including polyornithine, polylysine, collagen, fibronectin, and laminin (reviewed in Thoenen & Edgar 1985). Laminin stimulates the attachment of neural cells and induces these cells to produce long thin processes that resemble axons (Baron van Evercooren et al 1982; Rogers et al 1983; Manthorpe et al 1983; Edgar et al 1984; Davis et al 1985b,c; Liesi et al 1984a; Adler et al 1985). The number of cells with processes and the number of processes per cell are proportional to the amount of laminin to which the cells are exposed. The first such observations were made with organ cultures of spinal ganglia and showed that laminin is more potent than either fibronectin or collagen in inducing neurite formation and in promoting the association of Schwann cells with these neural processes (Baron van Evercooren et al 1982). Such interactions are not entirely unexpected since peripheral neural processes, but not those in central nerves, are encased in basement membranes, and this matrix supports their growth and regeneration (Anglister & McMahan 1984). However, central nervous system (CNS) neurons also respond to laminin by producing neurites (Rogers et al 1983; Manthorpe et al 1983; Liesi et al 1984a; Smalheiser et al 1984). In addition, laminin is produced in the central nervous system after injury, where it appears to be involved in regeneration (Liesi et al 1984b; Liesi 1985a,b). Laminin is present in the basement membranes of the neural tube and around neurite processes and is clearly involved in the development of peripheral, and possibly central, neurites. This trophic activity might have a practical application since tubes containing laminin have been found to stimulate optic and sciatic nerve regeneration in rodents (Madison et al 1985).

Other studies suggest that laminin may be an important component in media that promote neurite process formation in a variety of cultured cells (Lander et al 1985; Davis et al 1985a,c; Tomaselli et al 1986; Dziadek et al 1986a). It appears that the laminin in such preparations is not free but is associated with proteoglycans and other proteins such as nidogen/entactin. Such large soluble complexes have been termed adherons and have been found to promote neural retinal cell adhesion (Shubert et al 1983).

Attempts have been made to localize the domain in laminin responsible for neurite outgrowth (Edgar et al 1984; Engvall et al 1986). Studies with proteolytic fragments of laminin, with antibodies to defined fragments, and with monoclonal antibodies whose site of reaction in laminin is known, place the neurite-promoting site next to the heparin binding globule at the end of the long arm (fragment 8). Peptides from the cell attachment site, including the sequence CDPGYIGSR, support the attachment of neural cells but not the formation of neurites (Graf et al 1987). These findings indicate that the formation of neurites is a process distinct from cell attachment and that it involves different sites on laminin. Such observations suggest that there also may be at least two distinct laminin receptors on nerve cells that remain to be isolated and identified. It has also been demonstrated that antibody to the CSAT antigen is able to block neurite outgrowth, which suggests a role for this membrane complex in the process (Bozyczko & Horwitz 1986).

It should be noted that basement membrane and laminin act on other cellular elements of neural tissue. Schwann cells on laminin or on a laminin-rich gel elongate and produce myelin-specific proteins (McGarvey et al 1984; Carey et al 1986; Kleinman et al 1985). In addition, Schwann cells in the presence of laminin spread on and ensheathe the nerve fibers. The domains in laminin that mediate these activities have been located to fragments 1 and 8 using Schwannoma cells (Aumailley et al 1987), but the cellular ligands with which these domains interact have not been identified. The same domain on laminin, fragment 8, seems to be responsible for neurite outgrowth and the stimulation of tyrosine hydroxylase in nervelike chromaffin cells from the adrenal medulla (Acheson et al 1986).

Metastasis of Tumor Cells and Basement Membranes

Metastasis, the process by which secondary lesions arise from the original tumor, is a major factor in the morbidity of cancer. It is thought that the ability to metastasize is a complex process requiring activities that only a small proportion of tumor cells acquire (Fidler 1978). These activities include the ability to disseminate from the primary tumor, to survive in the circulation, and to penetrate into and proliferate at a distant site. In this process, the metastatic cell encounters and must cross various basement membranes (Liotta 1984). Basement membranes represent significant barriers whose passage by the cells occurs in stages involving (*a*) the binding of the cells to its surfaces, (*b*) the secretion by the cells of enzymes that degrade the basement membrane, and (*c*) the movement of the cells into adjacent tissue in response to various chemotactic or haptotactic factors.

Metastatic cells bind preferentially to laminin substrates (Vlodavsky & Gospodarowicz 1981; Terranova et al 1982), have more available membrane receptors for laminin (Terranova et al 1983), and are attracted to laminin (McCarthy & Furcht 1984). A high affinity for laminin would be expected to assist cell binding to basement membranes, for example, in the capillary walls of their target tissue. Indeed, it has been found in animal studies that antibodies to laminin (Terranova et al 1982) as well as fragments of laminin (fragment 1) (Barsky et al 1984b) and the YIGSR synthetic peptide (Graf et al 1987), which bind to the laminin receptor, reduce the formation of metastases when injected together with the cells, presumably by interfering with their arrest on the vessel wall. In addition, exposure to laminin induces metastatic tumor cells in culture to be more invasive (Terranova et al 1984) and to produce increased amounts of collagenase IV (Turpeeniemi-Hujanen et al 1986), an enzyme required to degrade the collagen IV network in the basement membrane (Liotta et al 1979). Also, as noted above, laminin has the ability to activate the migration of tumor cells by haptotaxis and by chemotaxis. These are probably not random but are concerted responses initiated by the interaction of the tumor cell's laminin receptors with laminin. The metastatic activity of tumor cells can be increased by prior exposure in culture to laminin; however, prior fibronectin exposure has the opposite effect (Terranova et al 1984). Apparently, the malignant cells can express either a malignant or nonmalignant phenotype, which is controlled in part by cell surface receptors. Since good systems are available for measuring the invasive activity of tumor cells in vitro (Terranova et al 1986b), novel therapeutic approaches may be developed to prevent tumor cell metastasis by interference with these molecular events.

MOLECULAR MODELS OF BASEMENT MEMBRANE STRUCTURE

Advances in isolating and characterizing laminin and the other proteins discussed here have helped to establish more detailed models for the supramolecular structure of basement membranes (summarized by Furth-mayr et al 1985). It is generally agreed that basement membranes are composed of a dense network of fine cords (Inoue et al 1983). The basic scaffolding forming the cords is a highly cross-linked network of collagen IV, which endows the matrix with mechanical stability. This network can be either open and planar or condensed into polygonal structures formed by the lateral alignment of collagen triple helices (Yurchenco et al 1986).

Three general models of assembly have been suggested for basement membranes. The layered model, based in part on earlier proposals (Spiro 1970; Kefalides 1970), suggests that layers of collagen and noncollagenous proteins alternate in the matrix and are separately deposited but inter-connected at discrete sites by covalent and noncovalent bonds to produce a three-dimensional structure (Schwartz & Veis 1980). This model emphasizes the homopolymeric association of components such as col-lagen IV, laminin, and heparan sulfate proteoglycan. It is supported by the observation that the anionic groups in the basement membrane are arrayed on the surface of basement membrane (Kanwar et al 1980) and that differences in reactivity are observed within the regions of the basement membrane using antibodies to laminin, collagen IV, and heparan sulfate proteoglycan (reviewed in Martinez-Hernandez & Amenta 1983).

The matrisome model (Martin et al 1984) proposes that the major components of basement membranes form stable complexes of defined stoichiometry, which are deposited as such when the matrix is formed but with subsequent rearrangement of collagen IV molecules into the network structure. This model is supported by the isolation of soluble complexes such as that between laminin and nidogen/entactin and between laminin, nidogen/entactin, and heparan sulfate proteoglycan, by the strong and specific affinity these components have for one another, and by the obser-vation that these components are codistributed in the cords within the basement membrane as visualized with immunoelectron microscopy (Inoue et al 1983).

A third possibility, the polymorphic polymerization model (Furthmayr et al 1985), is based on the potential of the various components to form either hetero- or homopolymers. This model suggests that there may be differences in the structure of the basement membranes at different ana-tomical locations generated by variations in the synthesis and secretion of individual components. That such a selective secretion and deposition of components can occur has been observed in animals with an induced autoimmune disease in which spikes of laminin without collagen IV are deposited along the basement membrane (Matsuo et al 1986).

Our present knowledge does not allow us to distinguish between these possibilities. The actual basement membrane architecture probably integrates features of each of these models. A full understanding will also certainly depend on the discovery and characterization of quite a few more constituents, which very likely are required to form an intact basement membrane.

ACKNOWLEDGMENT

Thanks are due Mrs. H. Wiedemann for preparing the electron micrographs and Dr. K. Beck for supplying pictures of sea urchin laminin.

Literature Cited

Acheson, A., Edgar, D., Timpl, R., Thoenen, H. 1986. Laminin increases levels and activity of tyrosine hydroxylase in calf adrenal medullary cells. *J. Cell Biol.* 102: 151–59

Adler, R., Jerdan, J., Hewitt, A. T. 1985. Responses of cultured neural retinal cells to substratum-bound laminin and other extracellular matrix molecules. *Dev. Biol.* 112: 100–14

Anglister, T., McMahan, U. J. 1984. Extracellular matrix components involved in neuromuscular transmission and regeneration. *Ciba Symp.* 108: 163–78

Arumugham, R. G., Hsieh, T. C. Y., Tanzer, M. L., Laine, R. A. 1986. Structure of asparagine-linked sugar chains of laminin. *Biochim. Biophys. Acta* 883: 112–26

Aumailley, M., Nurcombe, V., Edgar, D., Paulsson, M., Timpl, R. 1987. Laminin is a bivalent cell-binding protein which implicates the existence of two distinct laminin receptors. *J. Biol. Chem.* In press

Avila, J. L., Rojas, M., Rieber, M. 1984. Antibodies to laminin in American cutaneous Leishmaniasis. *Infect. Immunol.* 43: 402–6

Barlow, D. P., Green, N. M., Kurkinen, M., Hogan, B. L. M. 1984. Sequencing of laminin B chain cDNAs reveals C-terminal regions of coiled-coil alpha-helix. *EMBO J.* 3: 2355–62

Baron van Evercooren, A., Kleinman, H. K., Ohno, S., Marangos, P., Schwartz, J. P., Dubois-Dalcq, M. E. 1982. Nerve growth factor, laminin and fibronectin promote neurite growth in human fetal sensory ganglia cultures. *J. Neurosci. Res.* 8: 179–94

Barsky, S. H., Rao, C. N., Hyams, D., Liotta, L. A. 1984a. Characterization of a laminin receptor from human breast carcinoma tissue. *Breast Cancer Res. Treat.* 4: 181–88

Barsky, S. H., Rao, C. N., Williams, J. E., Liotta, L. A. 1984b. Laminin molecular domains which alter metastasis in a murine model. *J. Clin. Invest.* 74: 843–48

Bohnsack, J. F., Tenner, A. J., Laurie, G. W., Kleinman, H. K., Martin, G. R., Brown, E. J. 1985. C1q binds to laminin: A mechanism for the deposition and retention of immune complexes in basement membrane. *Proc. Natl. Acad. Sci. USA* 82: 3824–28

Bozyczko, D., Horwitz, A. 1986. The participation of a putative cell surface receptor for laminin and fibronectin in peripheral neurite extension. *J. Neurosci.* 6: 1241–51

Brocks, D. G., Strecker, H., Neubauer, H. P., Timpl, R. 1986. Radioimmunoassay of laminin in serum and its application to cancer patients. *Clin. Chem.* 32: 787–91

Brown, S. S., Malinoff, H. L., Wicha, M. S. 1983. Connectin: Cell surface protein that binds both laminin and actin. *Proc. Natl. Acad. Sci. USA* 80: 5927–30

Carey, D. J., Todd, M. S., Rafferty, C. M. 1986. Schwann cell myelination: Induction by exogenous basement membrane-like extracellular matrix. *J. Cell Biol.* 102: 2254–63

Carlin, B., Jaffe, R., Bender, B., Chung, A. E. 1981. Entactin, a novel basal lamina associated sulfated glycoprotein. *J. Biol. Chem.* 256: 5209–14

Carlin, B., Durkin, M. E., Bender, B., Jaffe, R., Chung, A. E. 1983. Synthesis of laminin and entactin by F9 cells induced with retinoic acid and dibutyryl cyclic AMP. *J. Biol. Chem.* 258: 7729–37

Charonis, A. S., Tsilibary, E. C., Yurchenco, P. D., Furthmayr, H. 1985. Binding of laminin to type IV collagen: A morphological study. *J. Cell Biol.* 100: 1848–53

Charonis, A. S., Tsilibary, E. C., Saku, T., Furthmayr, H. 1986. Inhibition of laminin self-assembly and interaction with type IV collagen by antibodies to the terminal domain of the long arm. *J. Cell Biol.* 103: 1689–97

Chung, A. E., Jaffe, R., Freeman, I. L., Vergnes, J. P., Braginski, J. E., Carlin, B. 1979. Properties of a basement membrane related glycoprotein synthesized in culture by a mouse embryonal carcinoma-derived cell line. *Cell* 16: 277–87

Chung, A. E., Jaffe, R., Bender, B., Lewis, M., Durkin, M. 1983. Monoclonal antibodies against the GP-2 subunit of laminin. *Lab. Invest.* 49: 576–81

Cooper, A. R., Kurkinen, M., Taylor, A., Hogan, B. L. M. 1981. Studies on the biosynthesis of laminin by murine parietal endoderm cells. *Eur. J. Biochem.* 119: 189–97

Cooper, A. R., MacQueen, H. A. 1983. Subunits of laminin are differentially synthesized in mouse eggs and early embryos. *Dev. Biol.* 96: 467–71

Cooper, A. R., Taylor, A., Hogan, B. L. M. 1983. Changes in the rate of laminin and entactin synthesis in F9 embryonal carcinoma cells treated with retinoic acid and cyclic AMP. *Dev. Biol.* 99: 510–16

Cornbrooks, C. J., Carey, D. J., McDonald, J. A., Timpl, R., Bunge, R. P. 1983. In vivo and in vitro observations on laminin productions by Schwann cells. *Proc. Natl. Acad. Sci. USA* 80: 3850–54

Couchman, J. R., Höök, M., Rees, D. A., Timpl, R. 1983. Adhesion, growth and matrix production by fibroblasts on laminin substrates. *J. Cell Biol.* 96: 177–83

Couchman, J. R., Caterson, B., Christner, J. E., Baker, J. R. 1984. Mapping by monoclonal antibody of glycosaminoglycans in connective tissues. *Nature* 307: 650–52

Davis, G. E., Manthorpe, M., Engvall, E., Varon, S. 1985a. Isolation and characterization of rat Schwannoma neurite promoting factor: Evidence that the factor contains laminin. *J. Neurosci.* 5: 2662–71

Davis, G. E., Manthorpe, M., Varon, S. 1985b. Parameters of neuritic growth from ciliary ganglion neurons in vitro: Influence of laminin, Schwannoma polyornithine-binding neurite promoting factor and ciliary neuronotrophic factor. *Dev. Brain Res.* 17: 75–84

Davis, G. E., Varon, S., Engvall, E., Manthorpe, M. 1985c. Substratum binding neurite-promoting factors: Relationships to laminin. *Trends Neurosci. Res.* 8: 528–32

Dixit, S. N. 1985. Isolation, purification and characterization of intact and pepsin-derived fragments of laminin from human placenta. *Connect. Tissue Res.* 14: 31–40

Donaldson, D. J., Mahan, J. T. 1984. Epi-dermal cell migration on laminin-coated substrates. Comparison with other extracellular matrix and non-matrix proteins. *Cell Tissue Res.* 235: 221–24

Durkin, M. E., Phillips, S. L., Chung, A. E. 1986. Control of laminin synthesis during differentiation of F9 embryonal carcinoma cells. *Differentiation* 32: 260–66

Dziadek, M., Paulsson, M., Timpl, R. 1985. Identification and interaction repertoire of large forms of the basement membrane protein nidogen. *EMBO J.* 4: 2513–18

Dziadek, M., Timpl, R. 1985. Expression of nidogen and laminin in basement membranes during mouse embryogenesis and in teratocarcinoma cells. *Dev. Biol.* 111: 372–82

Dziadek, M., Edgar, D., Paulsson, M., Timpl, R., Fleischmajer, R. 1986a. Basement membrane proteins produced by Schwann cells and in neurofibromatosis. *Ann. NY Acad. Sci.* 486: 248–59

Dziadek, M., Paulsson, M., Aumailley, M., Timpl, R. 1986b. Purification and tissue distribution of a small protein (BM-40) extracted from a basement membrane tumor. *Eur. J. Biochem.* 161: 455–64

Edgar, D., Timpl, R., Thoenen, H. 1984. The heparin-binding domain of laminin is responsible for its effects on neurite outgrowth and neuronal survival. *EMBO J.* 3: 1463–68

Elliott, R. W., Barlow, D., Hogan, B. L. M. 1985. Linkage of genes for laminin B1 and B2 subunits on chromosome 1 in mouse. *In Vitro* 21: 474–84

Engel, J., Odermatt, E., Engel, A., Madri, J. A., Furthmayr, H., et al. 1981. Shapes, domain organizations and flexibility of laminin and fibronectin, two multifunctional proteins of the extracellular matrix. *J. Mol. Biol.* 150: 97–120

Engvall, E., Krusius, T., Wewer, U., Ruoslahti, E. 1983. Laminin from rat yolk sac tumor: Isolation, partial characterization, and comparison with mouse laminin. *Arch. Biochem. Biophys.* 222: 649–56

Engvall, E., Davis, G. E., Dickerson, K., Ruoslahti, E., Varon, S., Manthorpe, M. 1986. Mapping of domains in human laminin using monoclonal antibodies: Localization of the neurite promoting site. *J. Cell Biol.* 103: 2457–65

Farquhar, M. G. 1981. The glomerular basement membrane: A selective macromolecular filter. In *Cell Biology of Extracellular Matrix*, ed. E. D. Hay, pp. 335–78. New York: Plenum

Fessler, J. H., Lunstrum, G., Duncan, K. G., Campbell, A. G., Sterne, R., et al. 1984. Evolutionary constancy of basement membrane components. In *The Role of Extracellular Matrix in Development*, ed.

R. L. Trelstad, pp. 207–19. New York: Liss

Fidler, I. J. 1978. Tumor heterogenicity and the biology of cancer invasion and metastasis. *Cancer Res.* 38: 2651–60

Foidart, J. M., Bere, E. W., Yaar, M., Rennard, S. I., Gullino, M., et al. 1980. Distribution and immunoelectron microscopic localization of laminin, a non-collagenous basement membrane glycoprotein. *Lab. Invest.* 42: 336–42

Fujiwara, S., Wiedemann, H., Timpl, R., Lustig, A., Engel, J. 1984. Structure and interactions of heparan sulfate proteoglycans from a mouse tumor basement membrane. *Eur. J. Biochem.* 143: 145–57

Furcht, L. T., McCarthy, J. B., Palm, S. L., Basara, M. L., Enenstein, J. 1984. Peptide fragments of laminin and fibronectin promote migration (haptotaxis and chemotaxis) of metastatic cells. *Ciba Symp.* 108: 130–45

Furthmayr, H., Yurchenco, P. D., Charonis, A. S., Tsilibary, E. C. 1985. Molecular interactions of type IV collagen and laminin: Models of basement membrane assembly. In *Basement Membranes*, ed. S. Shibata, pp. 169–80. Amsterdam: Elsevier

Goodman, S. L., Newgreen, D. 1985. Do cells show an inverse locomotory response to fibronectin and laminin substrates? *EMBO J.* 4: 2769–71

Graf, J., Iwamoto, Y., Sasaki, M., Martin, G. R., Kleinman, H. K., et al. 1987. Identification of an amino acid sequence in laminin mediating cell attachment, chemotaxis and receptor binding. *Cell* 48: 989–96

Grover, A., Adamson, E. 1986. Roles of extracellular matrix components in differentiating teratocarcinoma cells. *J. Biol. Chem.* 260: 12242–58

Hadley, M. A., Byers, S. W., Suarez-Quian, C. A., Kleinman, H. K., Dym, M. 1985. Extracellular matrix regulates Sertoli cell differentiation, testicular cord formation and germ cell development in vitro. *J. Cell Biol.* 101: 1511–22

Hand, P., Thor, A., Schlom, J., Rao, C. N., Liotta, L. A. 1985. Expression of laminin in normal and carcinomatous human tissue as defined by a monoclonal antibody. *Cancer Res.* 45: 2713–19

Hassell, J. R., Gehron Robey, P., Barrach, H. J., Wilczek, J., Rennard, S. I., Martin, G. R. 1980. Isolation of a heparan sulfate containing proteoglycan from basement membrane. *Proc. Natl. Acad. Sci. USA* 77: 4494–98

Hassell, J. R., Leyshon, W. C., Ledbetter, S. R., Tyree, B., Suzuki, S., et al. 1985. Isolation of two forms of basement membrane proteoglycan. *J. Biol. Chem.* 260: 8098–8105

Hogan, B. L. M. 1980. High molecular weight extracellular proteins synthesized by endoderm cells derived from mouse teratocarcinoma cells and normal extraembryonic membranes. *Dev. Biol.* 76: 275–85

Hogan, B. L. M., Cooper, A. R., Kurkinen, M. 1980. Incorporation into Reichert's membrane of laminin-like extracellular proteins synthesized by parietal endoderm cells of the mouse embryo. *Dev. Biol.* 80: 289–300

Hogan, B. L. M., Barlow, D. P., Kurkinen, M. 1984. Reichert's membrane as a model for studying the biosynthesis and assembly of basement membrane components. *Ciba Symp.* 108: 60–74

Hogan, B. L. M., Barlow, D. P., Green, N. M., Elliott, R. W., McVey, J., et al. 1985. Laminin: Towards the structure of the protein and its genes. In *Basement Membranes*, ed. S. Shibata, pp. 147–54. Amsterdam: Elsevier

Horwitz, A., Duggan, K., Greggs, R., Decker, C., Buck, C. 1985. The cell substrate attachment (CSAT) antigen has properties of a receptor for laminin and fibronectin. *J. Cell Biol.* 101: 2134–44

Howe, C. C. 1984. Functional role of laminin carbohydrate. *Mol. Cell Biol.* 4: 1–7

Howe, C. C., Dietzschold, B. 1983. Structural analysis of three subunits of laminin from teratocarcinoma-derived parietal endoderm cells. *Dev. Biol.* 98: 385–91

Inoue, S., Leblond, C. P., Laurie, G. W. 1983. Ultrastructure of Reichert's membrane, a multilayered basement membrane of the parietal wall of the rat yolk sac. *J. Cell Biol.* 97: 1524–37

Jaffe, R., Bender, B., Santamaria, M., Chung, A. E. 1984. Segmental staining of the murine nephron by monoclonal antibodies directed against the GP-2 subunit of laminin. *Lab. Invest.* 51: 88–96

Kanwar, Y. S., Linker, A., Farquhar, M. G. 1980. Increased permeability of the glomerular basement membrane to ferritin after removal of glycosaminoglycans (heparan sulfate) by enzyme digestion. *J. Cell Biol.* 86: 688–93

Kanwar, Y. S., Hascall, V. C., Farquhar, M. G. 1981. Partial characterization of newly synthesized proteoglycans isolated from the glomerular basement membrane. *J. Cell Biol.* 90: 527–32

Kanwar, Y. S., Veis, A., Kimura, J. H., Fikubowski, M. L. 1984. Characterization of heparan sulfate proteoglycans of glomerular basement membranes. *Proc. Natl. Acad. Sci. USA* 81: 762–66

Kefalides, N. A. 1970. Comparative biochemistry of mammalian basement membranes. In *Chemistry and Molecular*

Biology of the Intracellular Matrix, ed. E. A. Balazs, Vol. 1, pp. 535–73. London: Academic

Kennedy, D. W., Rohrbach, D. H., Martin, G. R., Mosmoi, T., Yamada, K. M. 1983. The adhesive glycoprotein laminin is an agglutinin. *J. Cell Physiol.* 114: 257–62

Kleinman, H. K., McGarvey, M. L., Hassell, J. R., Martin, G. R. 1983. Formation of a supramolecular complex is involved in the reconstitution of basement membrane components. *Biochemistry* 22: 4969–74

Kleinman, H. K., Cannon, F. B., Laurie, G. W., Hassell, J. R., Aumailley, M., et al. 1985. Biological activities of laminin. *J. Cell Biochem.* 27: 317–25

Kleinman, H. K., McGarvey, M. L., Hassell, J. R., Star, V. L., Cannon, F. B., et al. 1986. Basement membrane complexes with biological activity. *Biochemistry* 25: 312–18

Kleinman, H. K., Ebihara, I., Killen, P. D., Sasaki, M., Cannon, F. B., et al. 1987. Genes for basement membrane proteins are coordinately expressed in differentiating F9 cells but not in normal adult murine tissues. *Dev. Biol.* In press

Knudsen, K. A., Horwitz, A. F., Buck, C. A. 1985. A monoclonal antibody identifies a glycoprotein complex involved in cell substratum adhesion. *Exp. Cell Res.* 157: 218–26

Kurkinen, M., Barlow, D. P., Jenkins, J. R., Hogan, B. L. M. 1983. In vitro synthesis of entactin and laminin polypeptides. *J. Biol. Chem.* 258: 6543–48

Lander, A. D., Fujii, D. K., Reichardt, L. F. 1985. Purification of a factor that promotes neurite outgrowth: Isolation of laminin and associated molecules. *J. Cell Biol.* 101: 898–913

Laurie, G. W., Bing, J. T., Kleinman, H. K., Hassell, J. R., Aumailley, M., et al. 1986. Localization of binding sites for laminin, heparan sulfate proteoglycan and fibronectin on basement membrane (type IV) collagen. *J. Mol. Biol.* 189: 205–16

Lawler, J., Hynes, R. O. 1986. The structure of human thrombospondin, an adhesive glycoprotein with multiple calcium-binding sites and homologies with several different proteins. *J. Cell Biol.* 103: 1635–48

Ledbetter, S. R., Fisher, L. W., Hassell, J. R. 1987. Domain structure of the basement membrane heparan sulfate proteoglycan. *Biochemistry* 26: 989–95

Leivo, I. 1983. Structure and composition of early basement membranes: Studies with early embryos and teratocarcinoma cells. *Med. Biol.* 61: 1–30

Leivo, I., Engvall, E. 1986. C3d fragment of complement interacts with laminin and binds to basement membranes of glomerulus and trophoblast. *J. Cell Biol.* 103: 1091–1100

Lesot, H., Kuhl, U., von der Mark, K. 1983. Isolation of a laminin binding protein from muscle cell membranes. *EMBO J.* 2: 861–65

Liesi, P. 1985a. Laminin-immunoreactive glia distinguish regenerative adult CNS systems from non-regenerative ones. *EMBO J.* 4: 2505–11

Liesi, P. 1985b. Do neurons in the vertebrate CNS migrate on laminin? *EMBO J.* 4: 1163–70

Liesi, P., Koakkola, S., Dahl, D., Vaheri, A. 1984a. Laminin is induced in astrocytes of adult brain by injury. *EMBO J.* 3: 683–86

Liesi, P., Dahl, D., Vaheri, A. 1984b. Neurons cultured from developing rat brain attach and spread preferentially on laminin. *J. Neurosci. Res.* 11: 241–51

Liotta, L. A. 1984. Tumor invasion and metastases: Role of the basement membrane. *Am. J. Pathol.* 117: 339–48

Liotta, L. A., Abe, S., Gehron-Robey, P., Martin, G. R. 1979. Preferential digestion of a basement membrane collagen by an enzyme derived from a metastatic murine tumor. *Proc. Natl. Acad. Sci. USA* 76: 2268–72

Liotta, L. A., Weiner, U. M., Rao, C. N., Bryant, G. 1985a. Laminin receptors. In *The Cell In Contact*, ed. G. M. Edelman, J. P. Thiery, pp. 333–44. New York: Wiley

Liotta, L. A., Hand, P. H., Rao, C. N., Bryant, G., Barsky, S. H., Schlom, J. 1985b. Monoclonal antibodies to the human laminin receptor recognize structurally distinct sites. *Exp. Cell Res.* 156: 117–26

Lopes, J. D., dos Reis, M., Brentani, R. R. 1985. Presence of laminin receptors in *Staphylococcus aureus*. *Science* 229: 275–77

Madison, R., da Silva, C. F., Dikkes, P., Chiu, T. H., Sidman, R. L. 1985. Increased rate of peripheral nerve regeneration using bioresorbable nerve guides and a laminin-containing gel. *Exp. Neurol.* 88: 767–72

Mai, S., Chung, A. E. 1984. Cell attachment and spreading on extracellular matrix-coated beads. *Exp. Cell Res.* 152: 500–9

Malinoff, H. L., Wicha, M. S. 1983. Isolation of a cell surface receptor protein for laminin from murine fibrosarcoma cells. *J. Cell Biol.* 96: 1475–79

Manthorpe, M., Engvall, E., Ruoslahti, E., Longo, F. M., Davis, G. E., Varon, S. 1983. Laminin promotes neuritic regeneration from cultured peripheral and central neurons. *J. Cell Biol.* 97: 1882–90

Martin, G. R., Kleinman, H. K., Terranova, V. P., Ledbetter, S., Hassell, J. R. 1984.

The regulation of basement membrane formation and cell-matrix interactions by defined supramolecular complexes. *Ciba Symp.* 108: 197–212

Martinez-Hernandez, A., Miller, E. J., Damjanov, I., Gay, S. 1982. Laminin secreting yolk sac carcinoma of the rat. Biochemical and electron immuno-histochemical studies. *Lab. Invest.* 47: 247–57

Martinez-Hernandez, A., Amenta, P. S. 1983. The basement membrane in pathology. *Lab. Invest.* 48: 656–77

Mason, I. J., Taylor, A., Williams, J. G., Sage, H., Hogan, B. L. M. 1986. Evidence from molecular cloning that SPARC, a major product of mouse embryo parietal endoderm is related to an endothelial cell "culture shock" glycoprotein of M_r 43,000. *EMBO J.* 5: 1465–72

Matsuo, S., Brentzens, J. R., Andres, G., Martin, G. R., Martinez-Hernandez, A. 1986. Distribution of basement membrane antigens in glomeruli of mice with autoimmune glomerulonephritis. *Am. J. Pathol.* 122: 36–49

McCarthy, J. B., Palm, S. L., Furcht, L. T. 1983. Migration by haptotaxis of a Schwann cell tumor line to the basement membrane glycoprotein laminin. *J. Cell Biol.* 97: 772–77

McCarthy, J. B., Furcht, L. T. 1984. Laminin and fibronectin promote the haptotactic migration of B16 mouse melanoma cells in vitro. *J. Cell Biol.* 98: 1474–80

McCarthy, J. B., Basara, M. L., Palm, S. L., Sas, D. F., Furcht, L. T. 1985. The role of cell adhesion proteins—laminin and fibronectin in the movement of malignant and metastatic cells. *Cancer Metasis Rev.* 4: 125–52

McCarthy, R. A., Beck, K., Burger, M. M. 1987. Laminin is structurally conserved in the sea urchin basal lamina. *EMBO J.* In press

McGarvey, M. L., Baron van Evercooren, A., Kleinman, H. K., Dubois-Dalcq, M. E. 1984. Synthesis and effects of basement membrane components in cultured rat Schwann cells. *Dev. Biol.* 105: 18–28

Morita, A., Sugimoto, E., Kitagawa, Y. 1985. Post-translational assembly and glycosylation of laminin subunits in parietal endoderm-like F9 cells. *Biochem. J.* 229: 259–64

Ohno, M., Martinez-Hernandez, A., Ohno, N., Kefalides, N. A. 1983. Isolation of laminin from human placental basement membranes: Amnion, chorion and chorionic micro vessels. *Biochem. Biophys. Res. Commun.* 112: 1091–98

Ohno, M., Martinez-Hernandez, A., Ohno, N., Kefalides, N. A. 1985. Comparative study of laminin found in normal placental membrane with laminin of neoplastic origin. In *Basement Membranes*, ed. S. Shibata, pp. 3–11. Amsterdam: Elsevier

Oldberg, A., Franzen, A., Heinegard, D. 1986. Cloning and sequence analysis of rat bone sialoglycoprotein (osteopontin) cDNA reveals an Arg-Gly-Asp cell-binding sequence. *Proc. Natl. Acad. Sci. USA* 83: 8819–23

Ott, U., Odermatt, E., Engel, J., Furthmayr, H., Timpl, R. 1982. Protease resistance and conformation of laminin. *Eur. J. Biochem.* 123: 63–72

Ozawa, M., Sato, M., Muramatsu, T. 1983. Basement membrane glycoprotein laminin is an agglutinin. *J. Biochem.* 94: 479–85

Palm, S. L., Furcht, L. T. 1983. Production of laminin and fibronectin by Schwannoma cells: Cell-protein interactions in vitro and protein localization in peripheral nerve in vivo. *J. Cell Biol.* 96: 1218–26

Palm, S. L., McCarthy, J. B., Furcht, L. T. 1985. Alternative model for the internal structure of laminin. *Biochemistry* 24: 7753–60

Parthasarathy, N., Spiro, R. G. 1984. Isolation and characterization of heparan sulfate proteoglycan of the bovine glomerular basement membrane. *J. Biol. Chem.* 259: 12749–55

Paulsson, M., Deutzmann, R., Timpl, R., Dalzoppo, D., Odermatt, E., Engel, J. 1985a. Evidence for coiled-coil α-helical regions in the long arm of laminin. *EMBO J.* 4: 309–16

Paulsson, M., Dziadek, M., Suchanek, C., Huttner, W. B., Timpl, R. 1985b. Nature of sulfated macromolecules in mouse Reicherts membrane. Evidence for tyrosine-O-sulfate in basement membrane proteins. *Biochem. J.* 231: 571–79

Paulsson, M., Deutzmann, R., Dziadek, M., Nowack, H., Timpl, R., et al. 1986a. Purification and properties of intact and degraded nidogen obtained from a tumor basement membrane. *Eur. J. Biochem.* 156: 467–78

Paulsson, M., Fujiwara, S., Dziadek, M., Timpl, R., Pejlar, G., et al. 1986b. Structure and function of basement membrane proteoglycans. *Ciba Symp.* 124: 189–203

Paulsson, M., Aumailley, M., Deutzmann, R., Timpl, R., Beck, K., Engel, J. 1987. Laminin-nidogen complex: Extraction with chelating agents and structural characterization. *Eur. J. Biochem.* In press

Peters, B. P., Goldstein, I. 1979. The use of fluorescein-conjugated *Bandiraea simplificolia* B4-isolectin as a histochemical reagent for the detection of α-D-galactopyranosyl groups. Their occurrence in basement membranes. *Exp. Cell Res.* 120: 321–34

Peters, B. P., Hartle, R. J., Krzesick, R. F., Kroll, T. G., Perini, F., et al. 1985. The biosynthesis, processing and secretion of laminin by human choriocarcinoma cells. *J. Biol. Chem.* 260: 14732–42

Pierce, G. B., Jones, A., Orfanakis, N. G., Nakane, P. K., Lustig, L. 1982. Biosynthesis of basement membrane by parietal yolk sac cells. *Differentiation* 3: 60–72

Pierschbacher, M. D., Ruoslahti, E. 1984. Cell attachment activity of fibronectin can be duplicated by small synthetic fragments of the molecule. *Nature* 309: 30–33

Prehm, P., Dessau, W., Timpl, R. 1982. Rates of synthesis of basement membrane proteins by differentiating teratocarcinoma stem cells and their modulation by hormones. *Connect. Tissue Res.* 10: 275–85

Rao, C. N., Margulies, I. M. K., Tralka, T. S., Terranova, V. P., Madri, J. A., Liotta, L. A. 1982a. Isolation of a subunit of laminin and its role in molecular structure and tumor cell attachment. *J. Biol. Chem.* 257: 9740–44

Rao, C. N., Margulies, M. K., Goldfarb, R. H., Madri, J. A., Woodley, D. T., Liotta, L. A. 1982b. Differential proteolytic susceptibility of laminin alpha and beta subunits. *Arch. Biochem. Biophys.* 219: 65–70

Rao, C. N., Goldstein, I. J., Liotta, L. A. 1983a. Lectin-binding domains on laminin. *Arch. Biochem. Biophys.* 227: 118–24

Rao, C. N., Barsky, S. H., Terranova, V. P., Liotta, L. A. 1983b. Isolation of a tumor cell laminin receptor. *Biochem. Biophys. Res. Commun.* 111: 804–8

Roberts, D. D., Rao, C. N., Magnani, J. L., Spitalnik, S. L., Liotta, L. A., Ginsburg, V. 1985. Laminin binds specifically to sulfated glycolipids. *Proc. Natl. Acad. Sci. USA* 82: 1306–10

Rogers, S. L., Letourneau, P. C., Palm, S. L., McCarthy, J., Furcht, L. T. 1983. Neurite extension by peripheral and central nervous system neurons in response to substratum-bound fibronectin and laminin. *Dev. Biol.* 98: 212–20

Rohde, H., Wick, G., Timpl, R. 1979. Immunochemical characterization of the basement membrane glycoprotein laminin. *Eur. J. Biochem.* 102: 195–201

Rohde, H., Bachinger, H. P., Timpl, R. 1980. Characterization of pepsin fragments of laminin in a tumor basement membrane. Evidence for the existence of related proteins. *Hoppe-Seylers Z. Physiol. Chem.* 361: 1651–60

Sakashita, S., Ruoslahti, E. 1980. Laminin-like glycoproteins in extracellular matrix of endodermal cells. *Arch. Biochem. Biophys.* 205: 283–90

Sakashita, S., Engvall, E., Ruoslahti, E. 1980. Basement membrane glycoprotein laminin binds to heparin. *FEBS Lett.* 116: 243–46

Salonen, E. M., Zitting, A., Vaheri, A. 1984. Laminin interacts with plasminogen and its tissue type activator. *FEBS Lett.* 172: 29–32

Sasaki, M., Kato, S., Kohno, K., Martin, G. R., Yamada, Y. 1987. Sequence of cDNA encoding the laminin B1 chain reveals a multidomain protein containing cysteine-rich repeats. *Proc. Natl. Acad. Sci. USA* 84: 935–39

Schwartz, D., Veis, A. 1980. Characterization of bovine anterior-lens capsule basement membrane collagen. *Eur. J. Biochem.* 103: 29–37

Selak, I., Foidart, J. M., Moonen, G. 1985. Laminin promotes cerebellar granule cells migration in vitro and is synthesized by cultured astrocytes. *Dev. Neurosci.* 7: 285–88

Shibata, S., Peters, B. P., Roberts, D. D., Goldstein, I. J., Liotta, L. A. 1982. Isolation of laminin by affinity chromatography on immobilized *Griffonia simplicifolia* I lectin. *FEBS Lett.* 142: 194–98

Shubert, D., LaCorbiere, M., Klier, F. G., Birdwell, C. 1983. A role for adherons in neural retina cell adhesion. *J. Cell Biol.* 96: 990–98

Situ, R., Lee, E. C., McCoy, J. P. Jr., Varani, J. 1984. Stimulation of murine tumor cell motility by laminin. *J. Cell Sci.* 70: 167–76

Smalheiser, N. R., Crain, S. M., Reid, L. M. 1984. Laminin as a substrate for retinal axons in vitro. *Dev. Brain Res.* 12: 136–40

Spiro, R. G. 1970. Biochemistry of basement membranes. In *Chemistry and Molecular Biology of the Intercellular Matrix*, ed. E. A. Balazs, Vol. 1, pp. 511–34. London: Academic

Strickland, S., Smith, K. K., Marotti, K. R. 1980. Hormonal induction of differentiation in teratocarcinoma stem cells: Generation of parietal endoderm by retinoic acid and dibutyryl cAMP. *Cell* 21: 347–55

Sugrue, S. P., Hay, E. D. 1981. Response of basal epithelial cell surface and cytoskeletal to solubilized extracellular matrix molecules. *J. Cell Biol.* 91: 45–54

Suzuki, S., Oldberg, A., Hayman, E. G., Pierschbacher, M. D., Ruoslahti, E. 1985. Complete amino acid sequence of human vitronectin deduced from cDNA. Similarity of cell attachment sites in vitronectin and fibronectin. *EMBO J.* 4: 2519–24

Switalsky, L., Speciale, P., Höök, M., Wadström, T., Timpl, R. 1984. Binding of *Streptococcus pyogenes* to laminin. *J. Biol. Chem.* 259: 3734–38

Szarfman, A., Terranova, V. P., Rennard, S. I., Foidart, J. M., de Fatima Lima, M., et al. 1982. Antibodies to laminin in Chagas disease. *J. Exp. Med.* 155: 1161–71

Terranova, V. P., Rohrbach, D. H., Martin, G. R. 1980. Role of laminin in the attachment of PAM 212 (epithelial) cells to basement membrane collagen. *Cell* 22: 719–26

Terranova, V. P., Liotta, L. A., Russo, R. G., Martin, G. R. 1982. Role of laminin in the attachment and metastasis of murine tumor cells. *Cancer Res.* 42: 2265–69

Terranova, V. P., Rao, C. N., Kalebic, T., Margulies, M. K., Liotta, L. A. 1983. Laminin receptor on human breast carcinoma cells. *Proc. Natl. Acad. Sci. USA* 80: 444–48

Terranova, V. P., Williams, J. E., Liotta, L. A., Martin, G. R. 1984. Modulation of the metastatic activity of melanoma cells by laminin and fibronectin. *Science* 226: 982–85

Terranova, V. P., DiFlorio, R., Hujanen, E. S., Lyall, R. M., Liotta, L. A., et al. 1986a. Laminin promotes rabbit neutrophil motility and attachment. *J. Clin. Invest.* 77: 1180–86

Terranova, V. P., Hujanen, E. S., Martin, G. R. 1986b. Basement membranes and the invasive activity of metastatic tumor cells. *J. Natl. Cancer Inst.* 77: 311–16

Terranova, V. P., Lyall, R. M. 1986. Chemotaxis of human gingival epithelial cells to laminin: A mechanism for epithelial cell apical migration. *J. Periodontol.* 57: 311–17

Thoenen, H., Edgar, D. 1985. Neurotrophic factors. *Science* 229: 238–42

Timpl, R., Rohde, H., Gehron Robey, P., Rennard, S. I., Foidart, J. M., Martin, G. R. 1979. Laminin—a glycoprotein from basement membranes. *J. Biol. Chem.* 254: 9933–37

Timpl, R., Wiedemann, H., van Delden, V., Furthmayr, H., Kuhn, K. 1981. A network model for the organization of type IV collagen molecules in basement membranes. *Eur. J. Biochem.* 120: 203–11

Timpl, R., Engel, J., Martin, G. R. 1983a. Laminin a multifunctional protein of basement membranes. *Trends Biochem. Sci.* 8: 207–9

Timpl, R., Johansson, S., van Delden, V., Oberbäumer, I., Höök, M. 1983b. Characterization of protease resistant fragments of laminin mediating attachment and spreading of rat hepatocytes. *J. Biol.*

Chem. 258: 8922–27

Timpl, R., Dziadek, M., Fujiwara, S., Nowack, H., Wick, G. 1983c. Nidogen: A new self-aggregating basement membrane protein. *Eur. J. Biochem.* 137: 455–65

Timpl, R., Dziadek, M. 1986. Structure, development and molecular pathology of basement membranes. *Int. Rev. Exp. Pathol.* 29: 1–112

Tomaselli, K. J., Reichardt, L. F., Bixby, J. L. 1986. Distinct molecular interactions mediate neuronal process outgrowth on non-neural cell surfaces and extracellular matrices. *J. Cell Biol.* 103: 2659–72

Towbin, H., Rosenfelder, G., Wieslander, J., Avila, J. L., Rojas, M., et al. 1987. Circulating antibodies to mouse laminin in Chagas disease, American cutaneous Leishmaniasis and normal individuals recognize terminal galactosyl [α-3]galactose epitopes. *J. Exp. Med.* In press

Turpeeniemi-Hujanen, T., Thorgeisson, U. P., Rao, C. N., Liotta, L. A. 1986. Laminin increases the release of type IV collagenase. *J. Biol. Chem.* 261: 1883–89

Varani, J., Fligiel, S. E., Perone, P. 1985. Directional motility in strongly malignant murine tumor cells. *Int. J. Cancer* 35: 559–64

Vlodavsky, I., Gospodarowicz, D. 1981. Respective roles of laminin and fibronectin in adhesion of human carcinoma and sarcoma cells. *Nature* 289: 304–6

von der Mark, K., Mollenhauer, J., Kühl, U., Bee, J., Lesot, H. 1984. In *The Role of Extracellular Matrix in Development*, ed. R. Trelstad, pp. 67–87. New York: Liss

von der Mark, K., Kühl, U. 1985. Laminin and its receptor. *Biochem. Biophys. Acta* 823: 147–60

Wan, Y. J., Wu, T. C., Chung, A. E., Damjanov, I. 1984. Monoclonal antibodies to laminin reveal the heterogeneity of basement membranes in developing and adult mouse tissues. *J. Cell Biol.* 98: 971–79

Wang, S., Gudas, L. J. 1983. Isolation of cDNA clones specific for collagen IV and laminin from mouse teratocarcinoma cells. *Proc. Natl. Acad. Sci. USA* 80: 5880–84

Wewer, U., Albrechtsen, R., Manthorpe, M., Varon, S., Engvall, E., Ruoslahti, E. 1983. Human laminin isolated in a nearly intact, biologically active form from placenta by limited proteolysis. *J. Biol. Chem.* 258: 12654–60

Wewer, U. M., Liotta, L. A., Jaye, M., Ricca, G. A., Drohan, W. N., et al. 1986. Altered levels of laminin receptor mRNA in various human carcinoma cells that have different abilities to bind laminin. *Proc. Natl. Acad. Sci. USA* 83: 7137–41

Woodley, D. T., Rao, C. N., Hassell, J. R., Liotta, L. A., Martin, G. R., Kleinman, H. K. 1983. Interactions of basement membrane components. *Biochim. Biophys. Acta* 761: 278–83

Yamada, Y., Sasaki, M., Kohno, K., Kleinman, H. K., Kato, S., Martin, G. R. 1985. A novel structure for the protein and gene of the mouse laminin B1 chain. In *Basement Membranes*, ed. S. Shibata, pp. 139–46. Amsterdam: Elsevier

Yamada, Y., Albini, A., Ebihara, I., Graf, J., Kato, S., et al. 1987. Structure expression and function of mouse laminin. In *Mesenchymal-Epithelial Interactions in Neural Development*, ed. J. R. Wolff, J. Sievers, M. Berry, pp. 31–43. Heidelberg/New York: Springer-Verlag. In press

Yurchenco, P. D., Furthmayr, H. 1984. Self-assembly of basement membrane collagen. *Biochemistry* 23: 1839–50

Yurchenco, P. D., Tsilibary, E. C., Charonis, A. S., Furthmayr, H. 1985. Laminin polymerization in vitro: Evidence for a two step assembly with domain specificity. *J. Biol. Chem.* 260: 7636–44

Yurchenco, P. D., Tsilibary, E. C., Charonis, A. S., Furthmayr, H. 1986. Models for the self-assembly of basement membrane. *J. Histochem. Cytochem.* 34: 93–102

Ann. Rev. Cell Biol. 1987. 3:87–108

REPLICATION OF PLASMIDS DERIVED FROM BOVINE PAPILLOMA VIRUS TYPE 1 AND EPSTEIN-BARR VIRUS IN CELLS IN CULTURE

Joan Mecsas and Bill Sugden

McArdle Laboratory for Cancer Research, University of Wisconsin, Madison, Wisconsin 53706

CONTENTS

COMPARISON OF BPV-1 AND EBV DNA REPLICATION 87
   Introduction to BPV-1 and EBV 88
   Replication of BPV-1 and EBV as Plasmids in Cells in Culture 90
   Trans-Acting Factors Required for Replication of BPV-1 and EBV Plasmids 93
   Segregation of BPV-1 and EBV Plasmids in Proliferating Cell Populations 96
   Control of the Copy Number of BPV-1 and EBV Plasmids in Cells in Culture 97
EXPERIMENTAL APPROACHES TO FUTURE STUDIES ON THE REPLICATION OF BPV-1 AND EBV
   DNA 98
   Studies of the Replication of mtDNA and SV40 DNA as Guides for Analysis of the
      Plasmid Replication of BPV-1 and EBV 98
   Sites of Initiation of DNA Synthesis within BPV-1 and EBV Origins of Replication. 99
   The Roles of Viral Trans-Acting Factors in the Replication of BPV-1 and EBV 102
   Cellular Factors Required for the Replication of BPV-1 and EBV DNAs 103
   The Replication of BPV-1 and EBV Plasmids in Cell-Free Extracts 104
   Summary 104

COMPARISON OF BPV-1 AND EBV DNA REPLICATION

This article focuses on recently acquired knowledge on the replication of bovine papilloma virus type 1 (BPV-1) and Epstein-Barr virus (EBV) DNAs as plasmids in cells in culture. Interest in these plasmids is increasing for three reasons. (*a*) Papilloma viruses and EBV are associated with

0743–4634/87/1115–0087$02.00

human cancers (reviewed in Fields 1985). (*b*) Vectors derived from BPV-1 and EBV are used to express and maintain DNAs as plasmids in mammalian cells (reviewed by Baichwal & Sugden 1986). (*c*) These DNAs are presumed to share some steps in their pathways of replication with chromosomal DNA. After extensively reviewing the plasmid replication of BPV-1 and EBV, we describe selected findings from the studies of mitochondrial DNA (mtDNA) and simian virus 40 (SV40) DNA that may provide insights and directions for the further study of BPV-1 and EBV. The replication of mtDNA and SV40 DNA has been studied for a longer period of time and in greater depth than has the replication of BPV-1 DNA and EBV DNA. The study of the replication of all of these DNAs promises to shed light on the replication of mammalian chromosomes.

Introduction to BPV-1 and EBV

BPV-1 currently is the most studied papilloma virus because its introduction into some established rodent cells induces a detectable phenotype, morphological transformation (Dvoretzky et al 1980). The recognition that some human papilloma viruses (HPV) are likely to be causally associated with cervical carcinoma (zur Hausen 1986) and that their DNAs can also induce morphological transformation of established rodent cells (Tsunokawa et al 1986; Yasumoto et al 1986) ensures that HPVs will also be studied intensively. This review focuses on DNA replication in the papilloma virus BPV-1. BPV-1 and HPV have similar genomic organizations, and we expect that many of the results described below may also apply to the replication of HPVs.

BPV-1 virions, which can be isolated only from bovine warts, contain double-stranded, covalently closed, superhelical DNA of approximately 8000 bp (base pairs). The viral DNA has been sequenced (Chen et al 1982). The largest open reading frames (ORFs) are all on one strand of the viral DNA (Chen et al 1982), and all transcription probably occurs on that strand (Figure 1) (Amtmann & Sauer 1982; Heilman et al 1982). Most *cis*-acting elements that have been identified within BPV-1 are clustered within 1000 bp of DNA, which is referred to as the upstream regulatory region (URR). Transcription initiation from within the URR or adjacent to it leads to production of the RNAs expressed during latent infection by BPV-1 (Figure 1). Multiple splicing of these pre-mRNAs leads to the synthesis of complex molecules, the structures of which are not all known (Stenlund et al 1985; Yang et al 1985). Latent infection occurs in mouse cells in culture and is characterized by the replication of the viral DNA as a plasmid and by the expression of genes encoded by the early ORFs but not those encoded by the late ORFs. This latent infection in culture presumably reflects that phase of the life cycle of BPV-1 in animals in

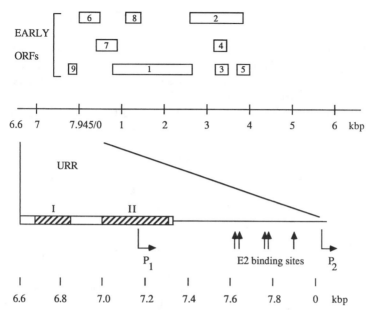

Figure 1   Genomic organization of BPV-1. The circular DNA of BPV-1 has been arbitrarily linearized at 6.6 kbp and is represented by the horizontal line in the middle of the figure. The numbering of the nucleotides on that DNA and its early open reading frames (E1–E9 ORFs) are derived from Chen et al (1982), Danos et al (1983), and Stenlund et al (1987). The three rows of early ORFs correspond to each of the reading frames in which the early ORFs reside. The upstream regulatory region (URR) is shown below the complete, linearized genome in an expanded version. The enlarged box between 6.6 and 7.4 kbp denotes the approximate position of the origin of replication of BPV-1 (Waldeck et al 1984; Lusky & Botchan 1984, 1986a). The two cross-hatched domains within this box indicate sequences identified by mutational analysis as being required in *cis* for plasmid replication (Lusky & Botchan 1986a). For plasmid replication, domain I can be replaced by a transcription-enhancing element (Lusky & Botchan 1986a). Two promoters within URR have been identified. The transcription start site of one promoter, P_1, has been localized to position 7186 by analysis of RNAs isolated from cells in culture by primer extension and by analysis of RNAs synthesized in cell-free extracts (Stenlund et al 1987). These transcripts may encode products of the E1 ORF (Stenlund 1987). The transcriptional start site of the second promoter within URR, P_2, has been mapped to position 89 by analysis of cDNA clones (Stenlund et al 1985; Yang et al 1985). These transcripts may encode products of the E6, E7, E2, E3, E4, and E5 ORFs. The vertical arrows between 7.6 and 0 kbp indicate some of the sites to which the E2 gene product binds within URR in vitro (Androphy et al 1987). A region of DNA containing these five sites acts as a transcription-enhancing element whose function is dependent on the product of the E2 ORF (Spalholz et al 1985).

which the virus latently infects basal cells of stratified epithelia. In animals the lytic phase of BPV-1 is thought to be confined to the differentiated progeny of the basal cells, in which virion proteins are synthesized from late ORFs and viral DNA is amplified, and from which virions are released.

EBV is the only herpesvirus that latently infects B lymphocytes and some epithelial cells in humans and in culture. The virus is associated with several human cancers, including Burkitt's lymphoma and nasopharyngeal carcinoma (Miller 1985). In culture it latently infects and transforms (that is, immortalizes) human B lymphocytes (Pope et al 1968; Pattengale et al 1973). The majority of adult human B lymphocytes can be transformed by EBV in culture (Sugden & Mark 1977), and up to 20% of peripheral B-lymphoid cells are infected in patients with EBV-associated infectious mononucleosis (Robinson et al 1981). Some primary epithelial cells can be infected in culture, but the infected cells are not transformed and have not been maintained in culture (Sixbey et al 1983). A variety of epithelial cells are likely to be infected by EBV in humans (Wolf et al 1984; Greenspan et al 1985; Sixbey et al 1986), but only those that are derived from naso-pharyngeal carcinomas have been isolated, and those only by propagation in athymic mice (Klein et al 1974).

Unfortunately, as with BPV-1, no cell lines have been identified that support a productive infection to yield large amounts of progeny EBV. Latently infected B lymphoblasts in culture do release infectious viral progeny with varying frequencies. The rate of this shift in the life cycle of EBV varied from approximately $1/1000$ cells per 24 hr to less than $1/10^6$ cells per 24 hr in 30 independently cloned EBV-transformed lymphoblasts (Sugden 1984). It is not known whether infected lymphoblasts or infected epithelial cells or both are the source of the cell-free EBV detected in vivo (Yao et al 1985; Sixbey et al 1986).

In transformed lymphoblasts EBV DNA is usually maintained as mul-tiple, complete, circularized copies of virion DNA (Adams et al 1979; Kintner & Sugden 1981). The viral DNA contains 172,000 bp and, for-tunately, has been sequenced (Baer et al 1984). Transcription from viral plasmids is complex. Approximately 100,000 bp of EBV DNA are tran-scribed and spliced to yield at least five mRNAs that are translated in the transformed cells (Fennewald et al 1984; Bodescot et al 1984, 1986; Bodescot & Perricaudet 1986; Sample et al 1986; Speck et al 1986). One region of EBV DNA contains an origin of replication and a transcrip-tion enhancer. The transcription enhancer may affect the synthesis of some or all of these latent viral transcripts (Yates et al 1984; Reisman & Sugden 1986 (Figure 2).

Replication of BPV-1 and EBV as Plasmids in Cells in Culture

Both BPV-1 and EBV DNAs are present as plasmids in the nuclei of infected cells (Waldeck et al 1984; Shaw et al 1979). After latent infection has been established by BPV-1 and EBV, the copy number of the plasmids

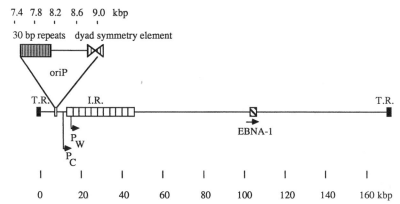

Figure 2   Genomic organization of EBV. The DNA of EBV is depicted in its linear form as it is found in the virion. In transformed lymphoblasts it is found as a circular plasmid molecule (Lindahl et al 1976) whose terminal repeats (T.R.) are joined (Kintner & Sugden 1979). The DNA molecule is 172 kbp in length and contains 10 or 11 identical copies of an internal repeat (I.R.) (Baer et al 1984). The *cis*-acting element of EBV required for plasmid replication, termed *oriP*, maps between positions 7333 and 9109 (Yates et al 1984) and is depicted in an expanded form at the top of the figure. *oriP* contains two components, a family of 30-bp repeats with 20 members and a dyad symmetry element, both of which are required for plasmid replication (Reisman et al 1985). The dyad symmetry element is composed of four repeats that are partial copies of the repeats comprising the family of 30-bp repeats (Hudson et al 1985). The only viral protein required in *trans* for the replication of *oriP* is EBNA-1 (Yates et al 1985; Lupton & Levine 1985), whose coding sequences map between positions 107 and 110 kbp (Summers et al 1982; Baer et al 1984), indicated by the cross-hatched box. Two possible promoters for the expression of the EBNA-1 protein in latently infected cells have been located by studying cDNA clones. One promoter lies within the Bam C fragment of EBV DNA (Bodescot et al 1986). Its "TATAA" box is at 11305 bp and is designated P_C. The second candidate promoter is in the Bam W fragment of EBV (each copy of I.R. is cleaved once by the *Bam*HI endonuclease to yield 10–11 copies of the Bam W fragment) (Sample et al 1986). Its "TATAA" box lies at 14352 bp and is designated P_W. Both P_C and P_W yield rightward transcripts that are multiply spliced. Only cDNAs of transcripts initiating near P_C have so far been found to contain coding sequences for EBNA-1; however, these cDNAs are isolated from a population of cells in which some cells are supporting the expression of late viral RNAs (Bodescot et al 1986).

is maintained at approximately 100 for BPV-1 and between 1 and 500 for EBV when measured in different clones (Lusky & Botchan 1984; Sugden et al 1979). BPV-1 and EBV plasmids replicate during the S phase of the cell cycle (Botchan et al 1986a; Hampar et al 1974; B. Sugden & N. Warren, unpublished observations). These observations distinguish BPV-1 and EBV from the better-characterized mammalian plasmids of mitochondria and SV40. mtDNAs replicate throughout the cell cycle and within their own organelles in the cytoplasm of the cell (Bogenhagen & Clayton 1977). SV40 DNA replicates in the nucleus and is amplified exponentially in a single S phase such that it usually kills its host cell (reviewed in Tooze 1980).

The origins of replication of BPV-1 and EBV differ structurally, but both are associated with transcriptional enhancers. The origin of replication of BPV-1 has been defined by studying replicative intermediates with a combination of restriction endonucleases and electron microscopy (Waldeck et al 1984). This work localized the origin to the region 6940 ± 400 bp within the URR (Waldeck et al 1984). The origin has also been localized between positions 6670 and 7330 by the identification of a *cis*-acting element of BPV-1 that permits plasmid replication in cells that provide BPV-1 gene products in *trans* (Lusky & Botchan 1984, 1986a; Waldeck et al 1984). Sites at which DNA synthesis initiates within origins of replication are often associated with dyad symmetry elements (see below). The 660 bp comprising the plasmid origin of replication of BPV-1 has no long dyad symmetry elements. (A search of this sequence with the "stemloop" program of the University of Wisconsin Genetics Computer Group failed to identify dyad symmetry elements that had more hydrogen bonds than those identified in a randomized version of the same sequence.)

The origin of replication of EBV has also been localized by the identification of the only *cis*-acting sequence within the 172,000 bp of EBV that permits plasmid replication in cells that provide EBV gene products in *trans* (Yates et al 1984). This element is termed *oriP* for origin of plasmid replication and contains two striking structural features within its 1800 bp (Figure 2) (Reisman et al 1985). EBV *oriP* contains a 20-member family of 30-bp repeats and, 1000 bp away, a dyad symmetry element that can be depicted as a stem of 31 bp with three mismatches and a loop of three bases. Although binding of proteins to this dyad symmetry element may favor its forming a stem and a loop, there is no evidence that such a structure forms during replication. Both the family of 30-bp repeats and the dyad symmetry element are required for the replication of plasmids containing *oriP*, although their relative orientations and the number of base pairs separating them can be changed while still preserving their function (Reisman et al 1985).

The origins of replication of BPV-1 and EBV are associated with transcription-enhancing elements, as is that of SV40 (Spalholz et al 1985; Lusky & Botchan 1986a; Reisman et al 1985; Reisman & Sugden 1986). The region between positions 6670 and 6850 (domain I in Figure 1) in BPV-1 may function as a transcription enhancer since the requirement for this region in *cis* for replication can be met by substituting a known transcription-enhancing element for it (Lusky & Botchan 1986a). However, this region has not been tested directly for transcription-enhancing activity. A bona fide transcription enhancer does lie between positions 7620 and 7906

(Androphy et al 1987), yet it does not sustain replication of mutants with deletions in domain I of BPV-1. These mutants do replicate when linked to an enhancer found between positions 4173 and 4451 (Lusky & Botchan 1986a). The family of repeated sequences within *oriP* of EBV is also a transcription enhancer and is absolutely required for the replication of *oriP* (Reisman et al 1985; Reisman & Sugden 1986). A transcription enhancer is required for the efficient replication of SV40 DNA (Hertz & Mertz 1986).

The role that the transcription enhancers perform for these origins of replication is unknown. Initiation of DNA synthesis from the origins of replication of BPV-1 and EBV may be primed with RNA, as is the case for mt and SV40 DNAs (see below), and the synthesis of these RNA primers could be affected by the transcription enhancers associated with the BPV-1 and EBV origins.

Although the origins of replication of both BPV-1 and EBV have been defined functionally, the exact sites within the origins at which DNA synthesis is initiated are not known. The work of Waldeck et al (1984) indicates that BPV-1 DNA synthesis initiates within an 800-bp region. The site of initiation is elusive in that there are no dyad symmetry elements within this sequence that might distinguish an origin of DNA synthesis. The recently described promoter P_1 within URR (Stenlund et al 1987) may serve to prime both pre-mRNA synthesis and DNA synthesis, as is the case for O_H in mtDNA (see below). Another possibility is that a site distinct from this promoter between positions 7068 and 7330 (domain II, Figure 1) is used to initiate DNA synthesis. As stated above, this region is necessary in *cis* for the replication of BPV-1 plasmids (Lusky & Botchan 1986a). Although it is not known whether DNA synthesis initiates within *oriP* of EBV, it seems likely that it does and that it does so at the dyad symmetry element of *oriP*. This supposition is based simply on the observation that the origin of DNA synthesis of SV40 and one of the origins of mtDNA are closely associated with dyad symmetry elements (see below).

Trans-*Acting Factors Required for Replication of BPV-1 and EBV Plasmids*

The description of the *trans*-acting factors encoded by BPV-1 and EBV required for their plasmid replication initially looks like a study in contrast. BPV-1 encodes multiple factors that appear to act in a complex manner, whereas EBV, which has a genome size twenty times that of BPV-1, encodes only one.

At least three genes encoded by BPV-1 have been implicated in its plasmid replication. Mutations at the 3′ end of the E1 ORF eliminate plasmid replication of BPV-1 (Sarver et al 1984; Lusky & Botchan 1985,

1986b; Groff & Lancaster 1986; Rabson et al 1986). Two nonsense mutations mapping only within the E2 ORF prevent plasmid replication of these mutants (DiMaio 1986; Rabson et al 1986). A mutant containing a deletion within the E2 ORF prevented plasmid replication when tested in three studies (DiMaio et al 1985; Groff & Lancaster 1986; Rabson et al 1986), but the same mutant behaved variably in another study (Lusky & Botchan 1985). Another deletion mutant that has no E4 or E5 ORF and lacks most of the E2 ORF replicates as a plasmid (Lusky & Botchan 1984). A frame-shift mutation within the E5 ORF, however, prevents plasmid replication (Groff & Lancaster 1986). How can these observations, some of which are contradictory, be resolved?

Firstly, the 3' end of the E1 ORF is required for plasmid replication. This BPV-1 gene product may be a site-specific DNA binding protein that binds within the origin of replication of BPV-1 and aids in the initiation of replication, as does SV40 T antigen (see below). This possibility is supported by the amino acid homology found between the 3' end of the E1 ORF and SV40 T antigen (Clertant & Seiff 1984). Secondly, the E2 ORF encodes a protein that binds to several sites within the URR and *trans*-activates a transcription enhancer within the URR (Figure 1) (Spalholz et al 1985; Androphy et al 1987). If E2 function is eliminated, expression of other BPV-1 genes (e.g. that encoded by the 3' E1 ORF) is expected to be affected. Thus different experimental protocols could lead to the detection either of cells that allow sufficient expression of E1 to maintain the mutants as plasmids or to cells that can maintain the mutant DNA only in an integrated form. Thirdly, since a mutant with a deletion encompassing most of E2 and all of the E4 and E5 ORFs can replicate, the E5 gene product must not be required for the plasmid replication of BPV-1. Perhaps the protein product derived from the frame-shift mutation within E5 is toxic to cells. According to the above view, the only BPV-1 product known now to be directly required in *trans* for plasmid replication is that derived from the 3' end of the E1 ORF. This idea could be tested by constructing cell lines that express single gene products of BPV-1 and then assaying plasmids that contain only the *cis*-acting sequences required for replication in these cell lines for extrachromosomal replication. Such an approach has been used successfully to identify the virally encoded protein required for replication of *oriP* in EBV (Yates et al 1985).

The only *trans*-acting factor encoded by EBV required for the replication of *oriP* is the EBNA-1 protein (Yates et al 1985; Lupton & Levine 1985). Both *cis*-acting components of *oriP*, the family of 30-bp repeats and the dyad symmetry element, appear to bind EBNA-1. The carboxy-terminal third of the EBNA-1 protein has been synthesized in *Escherichia coli* and shown to bind to each 30-bp member of the family of repeats and to the four sites within and beside the dyad symmetry element that share homology with the family of repeats (Rawlins et al 1985). This in vitro binding was monitored with a nuclease-protection assay (Rawlins et al 1985), and it presumably reflects the capacity of intact EBNA-1 to bind to these same sites in cells in culture. In the presence of EBNA-1, the family of 30-bp repeats of *oriP* can act as a transcription enhancer for heterologous promoters (Reisman et al 1985; Reisman & Sugden 1986) and for at least one EBV promoter (B. Sugden & N. Warren, unpublished observations). Thus the binding of EBNA-1 to the family of 30-bp repeats is probably required for EBNA-1 to *trans*-activate this enhancer.

Binding of EBNA-1 to the four sites within and beside *oriP*'s dyad symmetry element (only three of these sites are required for plasmid replication) (Yates et al 1984) is reminiscent of the binding of SV40 T antigen to three sites beside and within the dyad symmetry element of the SV40 *ori* (see below). Binding of EBNA-1 to the dyad symmetry element may be required for the initiation of *oriP* replication, just as binding of SV40 T antigen is required for the initiation of SV40 DNA replication (see below); but there are as yet no experiments that address this possibility.

It is important to underscore the contrast between the roles of the *trans*-activated enhancers of BPV-1 and EBV in plasmid replication. The E2 ORF of BPV-1 encodes a *trans*-activator that activates an enhancer adjacent to the origin of replication (Spalholz et al 1985; Androphy et al 1987). This enhancer can be deleted from a plasmid containing the BPV-1 URR (mutant *D41-H2-1* of Lusky & Botchan 1986a), and the plasmid replicates as a plasmid when introduced into cells producing BPV-1 gene products (Lusky & Botchan 1986a). This *trans*-activated enhancer is, therefore, not necessary for the plasmid replication of BPV-1. EBNA-1 *trans*-activates the family of 30-bp repeats (Reisman & Sugden 1986). A mutant DNA must have no less than half of this family to be able to replicate as a plasmid (Reisman et al 1985). Substitution of the EBV *trans*-activated enhancer by a heterologous enhancer prevents detectable plasmid replication (Reisman & Sugden 1986). The *trans*-activated enhancer within *oriP* is, therefore, required for the plasmid replication of EBV.

Two features of the replication of BPV-1 and EBV plasmids are now being addressed for which there are no precedents in the study of mtDNA and SV40 DNA replication. These features are the modes by which these viral plasmids segregate into daughter cells in dividing cell populations and the methods by which these viruses affect the copy number of their plasmid molecules in latently infected cells. It is not known to what extent these features of BPV-1 and EBV plasmid replication are novel or to what extent they are similar to the mechanisms for the segregation and copy-number control of chromosomal origins of replication.

Segregation of BPV-1 and EBV Plasmids in Proliferating Cell Populations

BPV-1 and EBV genomes are maintained as plasmids in cells proliferating in culture. Their maintenance could result either directly from viral elements that mediate segregation or indirectly from selective growth advantages they provide their host cells. There is circumstantial evidence that both BPV-1 and EBV segregate nonrandomly into daughter cells in a proliferating cell population. This circumstantial evidence is consistent with BPV-1 and EBV DNAs containing elements that behave as centromeres do in chromosomes.

Cells that contain a mutation in the E7 ORF of BPV-1 maintain the mutant viral DNA as plasmids at one or two copies per cell (Lusky & Botchan 1985). Random loss of plasmids from cells having only two copies of the plasmid would lead to more than 25% of the cells losing their plasmids after two cell generations. However, these cells containing mutated BPV-1 can be cloned without apparent selection and amplified, and they still maintain the mutant viral DNA as plasmids (Lusky & Botchan 1985). That these mutants of BPV-1 are faithfully maintained as plasmids indicates that they are not lost randomly.

Two established cell lines contain one to three EBV plasmids, and the EBV plasmids are maintained stably. This observation is consistent with EBV also having a centromerelike function. One is a Burkitt's lymphoma cell line, which originally lacked EBV DNA but was subsequently infected in culture. It contains one or two copies of the viral DNA as a plasmid per cell and maintains those plasmid copies after cloning (Reisman & Sugden 1984). The maintenance of the viral DNA in this cell line indicates that the EBV plasmid either contains a centromerelike function or provides the cell with a selective growth advantage. Since the parental cell prior to infection was an immortalized tumor cell, it is not clear what this putative selective growth advantage might be. The second cell line contains two or three copies of EBV DNA as plasmids and one complete copy integrated into the host DNA (Matsuo et al 1984). It seems likely that the integrated copy of EBV DNA could provide viral functions required for cell growth. If so, the stable maintenance of the EBV genome as plasmids is also consistent with EBV having a centromerelike function. All of these arguments are indirect ones; if in any of the cited examples the viral plasmids confer selective growth advantages to their host cells, the arguments are unfounded.

If there is a classical centromeric element in *oriP*, it is not one with the classical properties of yeast centromeres (Mann & Davis 1983). Plasmids that contain two copies of *oriP* behave like plasmids with one copy of *oriP* and can be easily maintained intact in cells under selection (B. Sugden & N. Warren, in preparation). In contrast, dicentromeric molecules are unstable at mitosis. In addition, plasmids derived from EBV that contain *oriP* and are ~ 10,000 bp in size are lost at an appreciable rate from cells in the absence of selection (Yates et al 1984, 1985; Reisman et al 1985). In cell populations that contain an average of ten copies of these plasmids per cell, 3–4% of the cells lose the plasmids per cell generation (Yates et al 1984, 1985; Reisman et al 1985). These small *oriP* plasmids derived from EBV may be lost in a dividing cell population because they are small (Hieter et al 1985), because they lack a centromerelike element located elsewhere on the EBV genome, or because they do not provide the selective growth advantage to their host cells provided by the complete viral genome.

Control of the Copy Number of BPV-1 and EBV Plasmids in Cells in Culture

There is evidence that both BPV-1 and EBV exert some kind of control over the average number of molecules of plasmid DNA per cell. If BPV-1 and EBV lack centromerelike functions, then maintenance of a high copy number would aid in ensuring that viral plasmids would consistently segregate to both daughter cells in a proliferating cell population. Cells transformed by BPV-1 maintain an average of 100–200 viral plasmids per cell, as measured in different populations of cells at various times in different laboratories. Similarly, during the last ten years several laboratories have determined that the Burkitt lymphoma cell line Raji maintains an average of 25–100 EBV plasmids per cell. These measurements indicate that after BPV-1 and EBV establish latent infections of cells, the average copy number of viral plasmids per infected cell does not fluctuate widely. This homeostasis is affected directly by BPV-1; how EBV affects the number of its plasmids in a cell is unknown.

Two complementation groups in BPV-1 appear to influence the copy number of viral plasmids in their host cells. Mutations in the 5′ end of the E1 ORF are lethal (Berg et al 1986; Lusky & Botchan 1986b). Upon their introduction into cells, these mutant plasmids replicate extensively. This amplification may exceed that of wild-type BPV-1 and thereby be toxic to the host cell (Lusky & Botchan 1986b). If this proves true, the 5′ end of the E1 ORF may affect the copy number of BPV-1 plasmids by limiting their extent of replication prior to establishment of the latent state. Mutants with altered E6 and E7 ORFs initially amplify their DNAs to approximately ~ 100–200 copies per cell upon introduction into cells (Berg et al 1986). However, after latent infection is established these mutants maintain only one or two copies of plasmid DNA per cell (Lusky &

Botchan 1985). This observation indicates that their defect lies in their inability to maintain the initial high copy number of plasmids. Thus the wild-type BPV-1 gene product derived from the E6 and E7 ORFs may be required to maintain a high copy number of BPV-1 plasmids. However, newly introduced wild-type BPV-1 does not amplify in cells latently infected with BPV-1 DNA containing the above mutations in the E6 and E7 ORFs (Berg et al 1986). These mutants are not known to be null mutants, and it is therefore difficult to use their phenotypes to predict the wild-type function of the E6 and E7 ORF product.

In summary, products of BPV-1 genes affect plasmid copy number such that: (*a*) within two to four days after infection or transfection of cells, BPV-1 amplifies its DNA; (*b*) this amplification is limited to yield $\sim$100–200 copies of plasmid DNA per cell; and (*c*) the 100–200 copies of plasmid DNA per cell are maintained in the latently infected, proliferating cell population.

The EBV genome is also amplified upon infection of B lymphocytes in culture. Infection of human B lymphocytes with one particle of EBV per cell yields transformed lymphoblasts that accumulate $\sim$5–500 EBV plasmids per cell in different clones (Sugden et al 1979). This accumulation, however, occurs during 25–50 cell generations (Sugden et al 1979) and therefore differs fundamentally from the rapid amplification observed for BPV-1 plasmids. The long time course required for the accumulation of EBV plasmids indicates that the amplication may result from occasional replication of EBV DNA twice in one cellular S phase. If this explanation is correct, then the level of expression of cellular and/or viral factors that affect the initiation of replication in a cell may determine the ultimate average copy number of EBV plasmids in a cell.

EXPERIMENTAL APPROACHES TO FUTURE STUDIES ON THE REPLICATION OF BPV-1 AND EBV DNA

Studies of the Replication of mtDNA and SV40 DNA as Guides for Analysis of the Plasmid Replication of BPV-1 and EBV

The replication of mtDNA and SV40 DNA has been studied extensively. The success of these studies is due in part to the high copy number of these plasmids in cells. There are approximately 10^4 copies of mtDNA per cell, and they replicate throughout the cell cycle (Bogenhagen & Clayton 1977). There are more than 10^5 copies of the SV40 genome per cell late in the cycle of infection; these genomes replicate exponentially in one protracted S phase (reviewed in Tooze 1980). A second reason for the success of these studies is that cell-free extracts have been developed that permit initiation of replication at one of the two origins of mtDNA (Wong & Clayton 1985) and at the origin of SV40 DNA (Ariga & Sugano 1983; Li & Kelly 1984). The results from studies both in cells and in cell-free extracts of mtDNA and SV40 DNA replication allow testable predictions to be made about several unknown aspects of the plasmid replication of BPV-1 and EBV.

Sites of Initiation of DNA Synthesis Within BPV-1 and EBV Origins of Replication

As stated previously, the sites at which DNA synthesis is initiated within the origins of replication of BPV-1 and EBV DNAs are unknown; the origin of replication of BPV-1 lacks a long dyad symmetry element that would be a candidate for a site of initiation of DNA synthesis. Replication initiating at the O_H origin of mtDNA provides a precedent for DNA synthesis initiating near sites other than dyad symmetry elements. DNA synthesis in mtDNA is unidirectional and initiates asynchronously at two distant sites on opposite strands of the DNA (Figure 3) (reviewed in Clayton 1982). Although the O_H initiation site of the heavy strand of mtDNA is not associated with a dyad symmetry element, it occurs at three adjacent sites in the DNA whose sequences are conserved among mammalian mtDNAs (CSB, Figure 3) (Chang & Clayton 1985). DNA synthesis from O_H of mtDNA is apparently primed by the synthesis of pre-mRNA; this possible mode of priming can be readily tested for BPV-1 DNA synthesis (see below). It may be that DNA synthesis is initiated near a short dyad symmetry element in the origin of replication of BPV-1. A minor site of initiation of DNA synthesis of the light strand of mtDNA, O_L, is near a dyad symmetry element that can be depicted as a stem and loop in which the stem contains only four bp (Figure 3) (Tapper & Clayton 1981). There are many such short dyad symmetry elements within the origin of replication of BPV-1. If these sites are used for the initiation of BPV-1 DNA synthesis, the 3' end of the E1 ORF of BPV-1 may encode a protein that binds to those sites (see below).

EBV *oriP* contains a long dyad symmetry element to which EBNA-1, the only *trans*-acting viral protein required for replication, binds. These findings so closely parallel known facets of the replication of SV40 DNA that they lead easily to a prediction. The origin of replication of SV40 DNA contains a dyad symmetry element to which SV40 T antigen, which is required for the initiation of replication, binds (Figure 4) (Tegtmeyer

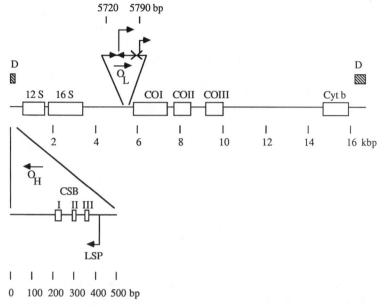

Figure 3 Genomic organization of human mitochondrial DNA. The circular DNA of human mitochondria is depicted as a linear form opened at a site recognized by the Mbol endonuclease (Anderson et al 1981). The DNA contains 16.5 kbp. Locations of the genes for 12S and 16S rRNAs, for cytochrome c oxidase subunits (CO I, II, and III), and for cytochrome b are depicted as boxes to orient the map (Anderson et al 1981). The locations of 22 tRNA genes and 8 ORFs are not shown (Anderson et al 1981). The location of the 7S DNA, which forms a displacement loop (D loop) often present in isolated mitochondrial DNA, is shown as the cross-hatched box labeled D (Brown et al 1978; reviewed in Clayton 1982). The replication of mitochondrial DNA initiates at two distant sites, O_H for origin of the heavy strand and O_L for origin of the light strand, and proceeds unidirectionally. The direction of DNA synthesis is indicated by arrows above O_H and O_L (reviewed in Clayton 1982). DNA synthesis first initiates in O_H, shown in an expanded form at the bottom of the figure (reviewed in Clayton 1982). It is apparently primed by a pre-mRNA molecule that itself initiates at the promoter marked LSP (light-strand promoter) (Chang & Clayton 1985). The switch from polyribonucleotide synthesis to DNA synthesis occurs at three conserved sequence blocks (CSB I, II, and III) that are common to mammalian mitochondrial DNAs (Chang & Clayton 1985). When one strand of DNA in the region of O_L (shown in an expanded form at the top of the figure) is rendered single-stranded by the passage of DNA synthesis initiated at O_H, second-strand synthesis initiates at O_L. There are two dyad symmetry elements in O_L at which synthesis initiates in vivo (Tapper & Clayton 1981): one at 5761 bp is associated with a stem and loop sequence in which the stem contains 11 bp, and the other at 5800 bp is associated with a stem and loop sequence in which the putative stem is only 4 bp (Tapper & Clayton 1981). The former of these two start sites functions in cell-free extracts and is primed by a ribonucleotide 17–30 nucleotides long (Wong & Clayton 1985; Hixson et al 1986).

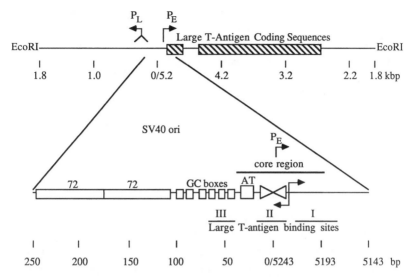

Figure 4 Genomic organization of SV40. The circular DNA of SV40 containing 5243 bp is shown as a linear molecule cleaved at its unique *Eco*R1 recognition site. The two arrows above the line show the location and direction of the late promoter region (P_L), which contains the multiple start sites for transcription of the late strand, and of the early promoter (P_E), from which transcription of the early strand initiates (reviewed in Tooze 1980). The coding sequences for SV40 T antigen are shown in the cross-hatched boxes following the early promoter (P_E). The *cis*-acting control region of SV40 that contains the SV40 origin of DNA synthesis is shown in an expanded form at the bottom of the figure. The three sites to which T antigen binds specifically are indicated by lines below the expanded figure (Tjian 1978). Binding to site II is required for initiation of DNA replication (Margolskee & Nathans 1984). Both the 72-bp repeats (shown as two boxes marked with 72) and the GC boxes, also known as 21-bp repeats (adjacent to the 72-bp repeats), positively affect transcription from P_E and DNA replication (reviewed by Fried & Prives 1986). The core region of the SV40 origin of DNA synthesis contains a stretch of A + T nucleotides (boxed) and a dyad symmetry element containing 13 nucleotides in each half of the dyad (depicted as tip-to-tip arrowheads). Both elements are required in *cis* for DNA synthesis (DiMaio & Nathans 1980). DNA synthesis is bidirectional and is primed with short oligoribonucleotides within the SV40 origin (Hay & DePamphilis 1982). The conversion of polyribonucleotide synthesis to DNA synthesis for the continuous or leading strands occurs within the core region at position 5211 (Hay & DePamphilis 1982) and is marked by the leftward- and rightward-pointing arrows.

1972; Tjian 1978; reviewed by Fried & Prives 1986). Discontinuous DNA synthesis initiates from multiple sites in the vicinity of the dyad symmetry element of SV40 DNA, but at one site continuous, bidirectional DNA synthesis begins (Figure 4) (Hay & DePamphilis 1982; Dean et al 1987). These observations with SV40 DNA suggest that the binding of EBNA-1 to the dyad symmetry element of *oriP* facilitates the initiation of bidirectional DNA synthesis near that element. It should be possible to test this prediction crudely by using the approach of Waldeck et al (1984) to localize the origin of DNA synthesis of EBV plasmids. The family of 30-bp repeats is 1000 bp away from the dyad symmetry element (Figure 2) and can be moved an additional 1500 bp away and still function (Reisman et al 1985), so it should be possible to distinguish whether the origin of DNA synthesis is associated with the family of 30-bp repeats or with the dyad symmetry element of *oriP* (or neither).

The Roles of Viral Trans-*Acting Factors in the Replication of BPV-1 and EBV*

Recent observations on SV40 T antigen lead to conjecture about the roles of the product of the 3′ end of the E1 ORF of BPV-1 and of EBNA-1 in DNA replication. SV40 T antigen is a DNA helicase, i.e. it can unwind duplex DNA containing a single-stranded region using ATP as an energy source (Stahl et al 1986). This activity, which presumably aids in the initial site-specific unwinding of circular SV40 DNA, initiates DNA synthesis in conjunction with the DNA polymerase α (polα)-primase complex (Dean et al 1987). The helicase activity may also unwind the parental DNA strands as polα elongates daughter strands (Dean et al 1987). This presumption is supported by a detectable association between polα and SV40 T antigen (Smale & Tjian 1986).

These findings with SV40 T antigen raise the question of whether the product of either the 3′ end of the E1 ORF of BPV-1 and/or EBNA-1 of EBV is a site-specific DNA helicase. To answer this question each of these proteins must be isolated. Unfortunately, neither protein is expressed efficiently in latently infected cells. Perhaps these proteins can be purified from cells in which they are overexpressed. [EBNA-1 has been overexpressed using a SV40 vector in which the SV40 late genes are replaced by EBNA-1 (Hammarskjöld et al 1986).] It is also possible that sufficient quantities of these proteins can be synthesized in vitro to test for their capacities to bind to DNA, to bind to DNA site-specifically, and to act as a helicase [25 ng of T antigen is sufficient to detect helicase activity (Stahl et al 1986).] EBNA-1 synthesized in vitro should bind to specific sites within *oriP*; whether or not it acts as a helicase remains to be determined. Negative results in these tests would be uninformative because proteins synthesized in vitro could lack crucial posttranslational modifications. If all the results are positive, then these two viral proteins are likely to be site-specific DNA helicases required for the initiation of DNA synthesis of BPV-1 and EBV.

Cellular Factors Required for the Replication of BPV-1 and EBV DNAs

With the single exception of SV40 T antigen, all of the factors required for the priming, elongation, and resolution of the replication of mt and SV40 plasmids are encoded by the cellular DNA. The priming of DNA synthesis within O_H of mtDNA appears to be mediated by the synthesis of pre-mRNA from a promoter termed LSP (light-strand promoter) (Figure 3) (Chang & Clayton 1985). A promoter within URR of BPV-1 (P_1 in Figure 1) may be a site of initiation for pre-mRNA (Stenlund et al 1987) that in turn could serve as a primer for DNA synthesis. This conjecture is easily tested because it predicts that this priming of DNA synthesis would be sensitive to α-amanitin, a toxin that inhibits cellular RNA polymerase II. RNA polymerase II is probably the enzyme involved in synthesis of the pre-mRNA at promoter P_1 (Stenlund et al 1987). In this scheme, the second strand of BPV-1 DNA would be primed by a primase similar to that used in the replication of SV40 DNA.

If the conjecture that pre-mRNA primes DNA synthesis within URR of BPV-1 is wrong, then SV40's mode of initiating DNA synthesis may be a model for that of BPV-1. In addition to the helicase activity provided by T antigen, SV40 requires polα and its associated primase to prime and elongate its DNA (Murakami et al 1986). Polα is inhibited by aphidicolin. Studies with this inhibitor first indicated that polα is required for the replication of SV40 plasmids (Otto & Fanning 1978; Edenberg et al 1978). A similar approach could be followed to test if aphidicolin inhibits the replication of BPV-1 and EBV plasmids in nuclei or in synchronized whole cells. Inhibition of replication of these two plasmids by this drug would be consistent with polα elongating their DNAs. Failure of aphidicolin to inhibit the replication of BPV-1 and EBV plasmids would indicate indirectly that neither polα nor, perhaps, its associated primase were required for BPV-1 and EBV replication. Experiments using inhibitors are indirect and can serve only as guides for direct experiments. For example, it could be that an additional DNA polymerase resistant to aphidicolin may also be required for SV40 and, perhaps, BPV-1 and EBV DNA replication (Prelich et al 1987).

A topoisomerase activity is required for the replication of SV40 DNA in vitro (Dean et al 1987). Studies using inhibitors of topoisomerases in cells indicate that both a topoisomerase type I and type II activity are required for SV40 plasmid replication (Snapka 1986). Camptothecin apparently inhibits a type I topoisomerase required during the elongation of SV40 DNA, while ellipticine and adriamycin appear to inhibit a type II topoisomerase required to decatenate replicated, intertwined molecules

(Snapka 1986). Similar inhibitor studies in whole cells could indicate if the same topoisomerase activities are likely to be required for the replication of BPV-1 and EBV plasmids.

The Replication of BPV-1 and EBV Plasmids in Cell-Free Extracts

Much insight into the details of the replication of mtDNA and SV40 DNA has been gained by analyzing them in cell-free extracts that support their replication (Wong & Clayton 1985; Ariga & Sugano 1963; Li & Kelly 1984; Dean et al 1987). Similarly rewarding insights are expected to result from the development of cell-free extracts that permit analysis of the replication of BPV-1 and EBV plasmids. The major hurdle confronting those who try to develop such extracts is the lack of a ready supply of the product of the 3' end of the E1 ORF and of EBNA-1. The cell-free extracts used for the replication of SV40 DNA contain SV40 T antigen in approximately 30-fold molar excess compared with DNA, i.e. at a concentration of approximately 20 μg/ml. If the product of the 3' end of the E1 ORF and EBNA-1 is required at similar concentrations, these proteins will have to be produced in large quantities and purified in order to develop cell-free extracts that support the replication of the corresponding DNAs. The effort to obtain these proteins is warranted. Analysis of plasmid DNA replication in cell-free extracts should aid in the identification of the functions that the product of the 3' end of the E1 ORF and EBNA-1 contribute to DNA replication. In addition, the extracts should permit the identification of the mode and sites of RNA priming and the start of DNA synthesis in the replication of BPV-1 and EBV plasmids. The extracts should also allow the positive identification of the cellular enzymatic activities required for these synthetic pathways.

Summary

The major components encoded by BPV-1 and EBV that act in plasmid replication of these viral DNAs in latently infected cells are now known. The minimal DNA sequences required in cis have been delineated, and the genes whose products are required in trans have been identified. The only required trans-acting gene of EBV, EBNA-1, has been shown to bind specifically to the cis-acting element, oriP. The current advanced understanding of plasmid replication of mtDNA and SV40 DNA is likely to aid in the selection of experiments that will assign functions to these components. In particular, it is exciting to contemplate the possible roles that transcription enhancers and the proteins that bind to these enhancers may perform in plasmid replication. Moreover, one can ask whether or not a virally encoded, site-specific DNA helicase participates in the replication of BPV-1 and EBV plasmids, as it does in SV40 DNA. Continued studies of the regulation of the segregation and the copy-number control of BPV-1 and EBV plasmids are likely to reveal other types of mechanisms, for which our knowledge of mtDNA and SV40 DNA replication does not provide precedents. Findings from these studies may aid not only in our understanding of the regulation of BPV-1 and EBV plasmid replication, but also in the regulation of chromosomal origins of replication.

ACKNOWLEDGMENTS

We thank H. Ariga, M. R. Botchan, and W. Waldeck, who provided us with manuscripts of their published and unpublished work. We thank Vijay Baichwal, Joyce Knutson, Stan Metzenberg, Jeff Ross, and Lars Sternas for their constructive criticisms of this review. We thank Stan Metzenberg for his scanning URR of BPV-1 for potential stem and loop sequences. Finally, we thank Ilse Riegel for her editorial guidance, Noreen Warren for her help in preparing the figures, and Mary Jo Markham for her careful and quick typing of our numerous drafts. We were supported by NIH grants CA-17175 and CA-22443. JM was supported by a National Research Service Award GM07215-12.

Literature Cited

Adams, A., Bjursell, G., Gussander, E., Koliais, S., Falk, L., Lindahl, T. 1979. Size of the intracellular circular Epstein-Barr virus DNA molecules in infectious mononucleosis-derived human lymphoid cell lines. *J. Virol.* 29: 815–17

Amtmann, E., Sauer, G. 1982. Bovine papillomavirus transcription: Polyadenylated RNA species and assessment of the direction of transcription. *J. Virol.* 43: 59–66

Anderson, S., Bankier, A. T., Barrell, B. G., de Bruijn, M. H. L., Coulson, A. R., et al. 1981. Sequence and organization of the human mitochondrial genome. *Nature* 290: 457–65

Androphy, E. J., Lowy, D. R., Schiller, J. T. 1987. Bovine papillomavirus E2 *trans*-activating gene product binds to specific sites in papilloma virus DNA. *Nature* 325: 70–73

Ariga, H., Sugano, S. 1983. Initiation of simian virus 40 DNA replication in vitro. *J. Virol.* 48: 481–91

Baer, R., Bankier, A. T., Biggin, M. D., Deininger, P. L., Farrell, P. J., et al. 1984. DNA sequence and expression of the B95-8 Epstein-Barr virus genome. *Nature* 310: 207–11

Baichwal, V. R., Sugden, B. 1986. Vectors for gene transfer derived from animal DNA viruses: Transient and stable expression of transferred genes. In *Gene Transfer*, ed. R. Kucherlapati, pp. 117–48. Plenum: New York. 447 pp.

Berg, L., Lusky, M., Stenlund, A., Botchan, M. R. 1986. Repression of bovine papilloma virus replication is mediated by a virally encoded *trans*-acting factor. *Cell* 46: 753–62

Bodescot, M., Brison, O., Perricaudet, M. 1986. An Epstein-Barr virus transcription unit is at least 84 kilobases long. *Nucleic Acids Res.* 14: 2611–20

Bodescot, M., Chambraud, B., Farrell, P., Perricaudet, M. 1984. Spliced RNA from the 1R1-U2 region of Epstein-Barr virus: Presence of an open reading frame for a repetitive polypeptide. *EMBO J.* 3: 1913–17

Bodescot, M., Perricaudet, M. 1986. Epstein-Barr virus mRNAs produced by alternative splicing. *Nucleic Acids Res.* 14: 7103–14

Bogenhagen, D., Clayton, D. A. 1977. Mouse L cell mitochondrial DNA molecules are selected randomly for replication throughout the cell cycle. *Cell* 11: 719–27

Botchan, M. R., Berg, L. J., Reynolds, J., Lusky, M. 1986a. The bovine papillomavirus replicon. *Ciba Found. Symp.* 120: 53–67

Botchan, M., Grodzicker, T., Sharp, P. A., eds. 1986b. *Cancer Cells—4. DNA Tumor Viruses: Control of Gene Expression and Replication.* Cold Spring Harbor, NY: Cold Spring Harbor Lab. 620 pp.

Brown, W. M., Shine, J., Goodman, H. M. 1978. Human mitochondrial DNA: Analysis of 7S DNA from the origin of replication. *Proc. Natl. Acad. Sci. USA* 75: 735–39

Chang, D. D., Clayton, D. A. 1985. Priming of human mitochondrial DNA replication occurs at the light strand promoter. *Proc. Natl. Acad. Sci. USA* 82: 351–55

Chen, E. Y., Howley, P. M., Levinson, A. D., Seeburg, P. H. 1982. The primary structure and genetic organization of the bovine papillomavirus type 1 genome. *Nature* 299: 529–34

Clayton, P. A. 1982. Replication of animal mitochondrial DNA. *Cell* 28: 693–705

Clertant, P., Seiff, I. 1984. A common function for polyomavirus large-T and papillomavirus E1 proteins. *Nature* 311: 276–79

Danos, O., Engel, L. W., Chen, E. Y., Yaniv, M., Howley, P. M. 1983. Comparative analysis of the human type 1a and bovine type 1 papillomavirus genomes. *J. Virol.* 46: 557–66

Dean, F. B., Bullock, P., Murakami, Y., Wobbe, C. R., Weissbach, L., Hurwitz, J. 1987. Simian virus 40 (SV40) DNA replication: SV40 large T antigen unwinds DNA containing the SV40 origin of replication. *Proc. Natl. Acad. Sci. USA* 84: 16–20

DiMaio, D. 1986. Nonsense mutation in open reading frame E2 of bovine papillomavirus DNA. *J. Virol.* 57: 475–80

DiMaio, D., Metherall, J., Neary, K., Guralski, D. 1985. Genetic analysis of cell transformation by bovine papillomavirus. In *Papillomaviruses: Molecular and Clinical Aspects,* ed. P. M. Howley, T. R. Broker, pp. 437–56. NY: Liss. 575 pp.

DiMaio, D., Nathans, D. 1980. Cold sensitive regulatory mutants of SV40. *J. Mol. Biol.* 129: 129–42

Dvoretzky, I., Shober, R., Chattopadhy, S. K., Lowy, D. 1980. A quantitative in vitro focus assay for bovine papillomavirus. *Virology* 103: 369–75

Edenberg, H. J., Anderson, S., DePamphilis, M. L. 1978. Involvement of DNA polymerase α in simian virus 40 DNA replication throughout the cell

cation. *J. Biol. Chem.* 253: 3273–80

Fennewald, S., van Santen, V., Kieff, E. 1984. Nucleotide sequence of an mRNA transcribed in latent growth-transforming virus infection indicates that it may encode a membrane protein. *J. Virol.* 51: 411–19

Fields, B. N., ed. 1985. *Virology.* NY: Raven. 1614 pp.

Fried, M., Prives, C. 1986. The biology of simian virus 40 and polyomavirus. See Botchan et al 1986b, pp. 1–16

Greenspan, J. S., Greenspan, D., Lennette, E. T., Abrams, D. I., Conant, M. A., et al. 1985. Replication of Epstein-Barr virus within the epithelial cells of oral "hairy" leukoplakia, an AIDS-associated lesion. *N. Engl. J. Med.* 313: 1564–71

Groff, D. E., Lancaster, W. D. 1986. Genetic analysis of the 3′ early region transformation and replication functions of bovine papillomavirus type 1. *Virology* 150: 221–30

Hammarskjöld, M. L., Wang, S. C., Klein, G. 1986. High-level expression of the Epstein-Barr virus EBNA-1 protein in CV1 cells and human lymphoid cells using a SV40 late replacement vector. *Gene* 43: 41–50

Hampar, B., Tanaka, A., Nonoyama, M., Derge, J. G. 1974. Replication of the resident repressed Epstein-Barr virus genome during the early S phase (S-1 period) of nonproducer Raji cells. *Proc. Natl. Acad. Sci. USA* 71: 631–33

Hay, R. T., DePamphilis, M. L. 1982. Initiation of SV40 DNA replication in vivo: Localization and structure of 5′ ends of DNA synthesized in the ori region. *Cell* 28: 767–79

Heilman, C. A., Engel, L., Lowy, D. R., Howley, P. M. 1982. Virus-specific transcription in bovine papillomavirus-transformed cells. *Virology* 119: 22–34

Hertz, G. Z., Mertz, J. E. 1986. Bidirectional promoter elements of simian virus 40 are required for efficient replication of the viral DNA. *Mol. Cell. Biol.* 6: 3513–22

Hieter, P., Mann, C., Snyder, M., Davis, R. W. 1985. Mitotic stability and yeast chromosomes: A colony color assay that measures nondisjunction and chromosome loss. *Cell* 40: 381–92

Hixson, J. E., Wong, T. W., Clayton, D. A. 1986. Both the conserved stem-loop and divergent 5′-flanking sequences are required for initiation at the human mitochondrial origin of light-strand DNA replication. *J. Biol. Chem.* 261: 2384–90

Hudson, G. S., Bankier, A. T., Satchwell, S. C., Barrell, B. G. 1985. The short unique region of the B95-8 Epstein-Barr virus genome. *Virology* 147: 81–98

Kintner, C. R., Sugden, B. 1979. The structure of the termini of the DNA of Epstein-Barr virus. *Cell* 17: 661–71

Kintner, C., Sugden, B. 1981. Conservation and progressive methylation of Epstein-Barr viral DNA sequences in transformed cells. *J. Virol.* 38: 305–16

Klein, G., Giovanella, B. C., Lindahl, T., Fialkow, P. J., Singh, S., Stehlin, J. S. 1974. Direct evidence for the presence of Epstein-Barr virus DNA and nuclear antigen in malignant epithelial cells from patients with poorly differentiated carcinoma of the nasopharynx. *Proc. Natl. Acad. Sci. USA* 71: 4737–41

Li, J. J., Kelly, T. J. 1984. Simian virus 40 DNA replication in vitro. *Proc. Natl. Acad. Sci. USA* 81: 6973–77

Lindahl, T., Adams, A., Bjursell, G., Bornkamm, G. W., Kaschka-Dierich, C., Jehn, U. 1976. Covalently closed circular duplex DNA of Epstein-Barr virus in a human lymphoid cell line. *J. Mol. Biol.* 102: 511–30

Lupton, S., Levine, A. J. 1985. Mapping genetic elements of Epstein-Barr virus that facilitate extrachromosomal persistence of Epstein-Barr virus-derived plasmids in human cells. *Mol. Cell. Biol.* 5: 2533–42

Lusky, M., Botchan, M. R. 1984. Characterization of the bovine papilloma virus plasmid maintenance sequences. *Cell* 36: 391–401

Lusky, M., Botchan, M. R. 1985. Genetic analysis of bovine papillomavirus: Type 1 *trans*-acting replication factors. *J. Virol.* 53: 955–65

Lusky, M., Botchan, M. R. 1986a. Transient replication of bovine papilloma virus type 1 plasmids: *cis* and *trans* requirements. *Proc. Natl. Acad. Sci. USA* 83: 3609–13

Lusky, M., Botchan, M. R. 1986b. A bovine papillomavirus type 1-encoded modulator function is dispensable for transient viral replication but required for the establishment of the stable plasmid state. *J. Virol.* 60: 729–42

Mann, C., Davis, R. W. 1983. Instability of dicentric plasmids in yeast. *Proc. Natl. Acad. Sci. USA* 80: 228–32

Margolskee, R. F., Nathans, D. 1984. Simian virus 40 mutant T antigens with relaxed specificity for the nucleotide sequence at the viral DNA origin of replication. *J. Virol.* 49: 386–93

Matsuo, T., Heller, M., Petti, L., O'Shiro, E., Kieff, E. 1984. The entire Epstein-Barr virus genome integrates into human lymphocyte DNA. *Science* 226: 1322–25

Miller, G. 1985. Epstein-Barr virus. See Fields 1985, pp. 563–89

Murakami, Y., Wobbe, C. R., Weissbach, L., Dean, F. B., Hurwitz, J. 1986. Role of DNA polymerase α and DNA primase in

simian virus 40 DNA replication in vitro. *Proc. Natl. Acad. Sci. USA* 83: 2869–73

Otto, B., Fanning, E. 1978. DNA polymerase α is associated with replicating SV40 nucleoprotein complexes. *Nucleic Acids Res.* 5: 1715–28

Pattengale, P. K., Smith, R. W., Gerber, P. 1973. Selective transformation of B lymphocytes by E.B. virus. *Lancet* ii: 93–94

Pope, J. H., Horne, M. K., Scott, W. 1968. Transformation of foetal human leukocytes in vitro by filtrates of a human leukaemic cell line containing herpes-like virus. *Int. J. Cancer* 3: 857–66

Prelich, G., Tan, C.-K., Kostura, M., Mathews, M. B., So, A. G., et al. 1987. Functional identity of proliferating cell nuclear antigen and a DNA polymerase-δ auxiliary protein. *Nature* 326: 517–20

Rabson, M. S., Yang, Y. C., Howley, P. M. 1986. A genetic analysis of bovine papillomavirus type-1 transformation and plasmid-maintenance functions. See Botchan et al 1986b, pp. 235–43

Rawlins, D. R., Milman, G., Hayward, S. D., Hayward, G. S. 1985. Sequence-specific DNA binding of the Epstein-Barr virus nuclear antigen (EBNA-1) to clustered sites in the plasmid maintenance region. *Cell* 42: 859–68

Reisman, D., Sugden, B. 1984. An EBNA-negative, EBV-genome positive human lymphoblast cell line in which superinfecting EBV DNA is not maintained. *Virology* 137: 113–26

Reisman, D., Sugden, B. 1986. *Trans*-activation of an Epstein-Barr viral (EBV) transcriptional enhancer by the EBV nuclear antigen-1. *Mol. Cell. Biol.* 6: 3838–46

Reisman, D., Yates, J., Sugden, B. 1985. A putative origin of replication of plasmids derived from Epstein-Barr virus is composed of two *cis*-acting components. *Mol. Cell. Biol.* 5: 1822–32

Robinson, J., Smith, D., Niederman, J. 1981. Plasmacytic differentiation of circulating Epstein-Barr virus–infected B lymphocytes during acute infectious mononucleosis. *J. Exp. Med.* 153: 235–44

Sample, J., Hummel, M., Braun, D., Birkenbach, M., Kieff, E. 1986. Nucleotide sequences of mRNAs encoding Epstein-Barr virus nuclear proteins: A probable transcriptional initiation site. *Proc. Natl. Acad. Sci. USA* 83: 5096–5100

Sarver, N., Rabson, M. S., Yang, Y.-C., Byrne, J. C., Howley, P. M. 1984. Localization and analysis of bovine papillomavirus type 1 transforming functions. *J. Virol.* 52: 377–88

Shaw, J. E., Levinger, L. F., Carter, C. W. Jr. 1979. Nucleosomal structure of Epstein-

Barr virus DNA in transformed cell lines. *J. Virol.* 29: 657–65

Sixbey, J. W., Lemon, S. M., Pagano, J. S. 1986. A second site for Epstein-Barr virus shedding: The uterine cervix. *Lancet* ii: 1122–24

Sixbey, J. W., Vesterinen, E. H., Nedrud, J. G., Raab-Traub, N., Walton, L. A., Pagano, J. S. 1983. Replication of Epstein-Barr virus in human epithelial cells infected in vitro. *Nature* 306: 480–83

Smale, S. T., Tjian, R. 1986. T-antigen-DNA polymerase α complex implicated in simian virus 40 DNA replication. *Mol. Cell. Biol.* 6: 4077–87

Snapka, R. M. 1986. Topoisomerase inhibitors can selectively interfere with different stages of simian virus 40 DNA replication. *Mol. Cell. Biol.* 6: 4221–27

Spalholz, B. A., Yang, Y., Howley, P. M. 1985. Transactivation of a bovine papilloma virus transcriptional regulatory element by the E2 gene product. *Cell* 42: 183–91

Speck, S. H., Pfitzer, A., Strominger, J. L. 1986. An Epstein-Barr virus transcript from a latently infected, growth-transformed B-cell line encodes a highly repetitive polypeptide. *Proc. Natl. Acad. Sci. USA* 83: 9298–9302

Stahl, H., Droge, P., Knippers, R. 1986. DNA helicase activity of SV40 large tumor antigen. *EMBO J.* 5: 1939–44

Stenlund, A., Bream, G. L., Botchan, M. R. 1987. *Science.* In press

Stenlund, A., Zabielski, J., Ahola, H., Moreno-Lopez, J., Pettersson, U. 1985. Messenger RNAs from the transforming region of bovine papilloma virus type 1. *J. Mol. Biol.* 182: 541–54

Sugden, B. 1984. Expression of virus-associated functions in cells transformed in vitro by Epstein-Barr virus: Epstein-Barr virus cell surface antigen and virus-release from transformed cells. In *Immune Deficiency and Cancer, Epstein-Barr Virus and Lymphoproliferative Malignancies*, ed. D. T. Purtillo, p. 165. New York: Plenum

Sugden, B., Mark, W. 1977. Clonal transformation of adult human leukocytes by Epstein-Barr virus. *J. Virol.* 23: 503–8

Sugden, B., Phelps, M., Domoradzki, J. 1979. Epstein-Barr virus DNA is amplified in transformed lymphocytes. *J. Virol.* 31: 590–95

Summers, W. P., Grogan, E. A., Shedd, D., Robert, M., Liu, C.-R., Miller, G. 1982. Stable expression in mouse cells of nuclear neoantigen following transfer of a 3.4 megadalton cloned fragment of Epstein-Barr virus DNA. *Proc. Natl. Acad. Sci. USA* 79: 5688–92

Tapper, D. P., Clayton, D. A. 1981. Mechanism of replication of human mitochondrial DNA. *J. Biol. Chem.* 256: 5109–15

Tegtmeyer, P. 1972. Simian virus 40 deoxyribonucleic acid synthesis: The viral replicon. *J. Virol.* 10: 591–98

Tjian, R. 1978. The binding site on SV40 DNA for a T antigen-related protein. *Cell* 13: 165–79

Tooze, J., ed. 1980. *Molecular Biology of Tumor Viruses. II. DNA Tumor Viruses.* Cold Spring Harbor, NY: Cold Spring Harbor Lab. 958 pp. 2nd ed.

Tsunokawa, Y., Takebe, N., Kasamatsu, T., Terada, M., Sugimura, T. 1986. Transforming activity of human papillomaviruses type 16 DNA sequences in a cervical cancer. *Proc. Natl. Acad. Sci. USA* 83: 2200–3

Waldeck, W., Rösl, F., Zentgraf, H. 1984. Origin of replication in episomal bovine papilloma virus type 1 DNA isolated from transformed cells. *EMBO J.* 3: 2173–78

Wolf, H., Haus, M., Wilmes, E. 1984. Persistence of Epstein-Barr virus in the parotid gland. *J. Virol.* 51: 795–98

Wong, T. W., Clayton, D. A. 1985. In vitro replication of human mitochondrial DNA: Accurate initiation of the origin of light-strand synthesis. *Cell* 42: 951–58

Yang, Y. C., Okayama, H., Howley, P. M. 1985. Bovine papillomavirus contains multiple transforming genes. *Proc. Natl. Acad. Sci. USA* 82: 1030–34

Yao, Q. Y., Rickinson, A. B., Epstein, M. A. 1985. A re-examination of the Epstein-Barr virus carrier state in healthy seropositive individuals. *Int. J. Cancer* 35: 35–42

Yasumoto, S., Burkhardt, A. L., Doniger, J., DiPaolo, J. A. 1986. Human papillomavirus type 16 DNA-induced malignant transformation of NIH 3T3 cells. *J. Virol.* 57: 572–77

Yates, J., Warren, N., Reisman, D., Sugden, B. 1984. A *cis*-acting element from the Epstein-Barr viral genome that permits stable replication of recombinant plasmids in latently infected cells. *Proc. Natl. Acad. Sci. USA* 81: 3806–10

Yates, J. L., Warren, N., Sugden, B. 1985. Stable replication of plasmids derived from Epstein-Barr virus in a variety of mammalian cells. *Nature* 313: 812–15

zur Hausen, H. 1986. Papillomaviruses in human urogenital cancer: Established results and prospects for the future. In *Viral Etiology of Cervical Cancer*, ed. R. Peto, H. zur Hausen, pp. 327–32. Banbury Report 21. Cold Spring Harbor, NY: Cold Spring Harbor Lab. 362 pp.

Ann. Rev. Cell Biol. 1987. 3 : 109–42

EARLY EVENTS IN MAMMALIAN FERTILIZATION

Paul M. Wassarman

Department of Cell Biology, Roche Institute of Molecular Biology,
Roche Research Center, Nutley, New Jersey 07110

CONTENTS

PREFACE 109
INTRODUCTION 110
CAPACITATION OF SPERM 111
THE ACROSOME REACTION 111
SITE OF THE ACROSOME REACTION 113
BINDING OF SPERM TO THE ZONA PELLUCIDA 118
NATURE OF THE SPERM RECEPTOR 124
NATURE OF THE EGG BINDING PROTEIN 132
SUMMARY AND FINAL COMMENTS 134

PREFACE

During the first half of this century, studies of fertilization mechanisms in mammals generally lagged behind those in nonmammals, such as marine animals. To a large extent, this can be ascribed to the difficulties inherent in studying organisms that carry out fertilization within (internally) rather than outside (externally) the adult female. As pointed out by Austin & Walton (1960), "The advancement of knowledge on mammalian fertilization would undoubtedly be aided if the process could be studied in vitro with the ease that it can with the eggs of some marine invertebrates." Consequently, in the late 1950s, introduction of in vitro culture systems that supported mammalian fertilization and preimplantation development, and refinement of these systems in succeeding years, overcame many of the

109

0743–4634/87/1115–0109$02.00

limitations associated with research on fertilization in mammals (Bedford 1971; Greep et al 1976; Gwatkin 1977; Blandau 1980; Bavister 1981; Brackett 1981; Oliphant & Eng 1981). As a result, during the past 20 years or so, significant progress has been made in our understanding of the cellular and molecular details of the mammalian fertilization process. With respect to human reproduction, some of the practical consequences of this progress are realized daily in in vitro fertilization (IVF) clinics around the world.

The subject of mammalian fertilization is so broad as to preclude comprehensive coverage of all its aspects here. Consequently, this review is limited largely to relatively recent developments in certain aspects of the subject. It particularly emphasizes events following the encounter of sperm and egg in the oviduct and preceding fusion of sperm and egg to form a zygote. This portion of the mammalian fertilization process has received considerable attention of late, resulting in new insights into the molecular basis of sperm-egg interactions and their consequences.

Many excellent reviews and monographs are available that address mammalian fertilization. To obtain a broader, more historical perspective of the field, readers are encouraged to refer to some of the more comprehensive treatments of this subject (Chang & Pincus 1951; Austin & Bishop 1957; Austin & Walton 1960; Austin 1965; Metz & Monroy 1967, 1969, 1985; Gwatkin 1977; Yanagimachi 1977, 1978, 1981; Bedford 1982; Hartmann 1983b; Hedrick 1986).

INTRODUCTION

Of the 50 (mouse) to 300 (human) million sperm ejaculated during coitus, only 100 (mouse) to 200 (human) actually reach the site of fertilization, the ampulla of the oviduct. There, in order to form a zygote, a sperm must (a) penetrate layers of cumulus cells surrounding each ovulated egg, (b) bind to and (c) penetrate the egg extracellular coat (zona pellucida; ZP), and (d) upon reaching the area between the ZP and plasma membrane (perivitelline space), fuse with the plasma membrane (i.e. fertilize the egg). The egg responds to fertilization, in part, by modifying its investments such that additional sperm are prevented from fusing with the zygote, which would result in polyspermic fertilization, a lethal condition. This review deals primarily with steps (a) and (b) of the mammalian fertilization pathway, and most experiments described here were carried out in vitro with gametes from mice and hamsters. Among the topics considered are the site of the acrosome reaction (AR), and the nature of the AR-inducer, the sperm receptor, and the egg binding protein. The traditional views of some of these topics are being reconsidered in light of recent findings.

CAPACITATION OF SPERM

It has been known for some time that to be able to fertilize ovulated eggs ejaculated mammalian sperm must reside in the female reproductive tract for several hours; the precise time required varies from one mammal to another. The term capacitation was coined to describe this phenomenon (Austin 1951, 1952; Chang 1951). Similarly, sperm removed from the cauda epididymes for in vitro fertilization must be incubated under conditions that promote capacitation (Bedford 1971; Inoue & Wolf 1975b; Wolf & Inoue 1976; Wolf et al 1977; Rogers 1978; Oliphant & Eng 1981). Perhaps the most up-to-date definition of capacitation (Moore & Bedford 1983) is "the process in the female (or in vitro) that prepares the spermatozoon to undergo the acrosome reaction and also quite probably to develop a whiplash or hyperactivated motility that may enhance ability to penetrate the zona pellucida" (Yanagimachi 1970). Because of the importance of capacitation during mammalian fertilization, this subject has been studied and reviewed extensively (Bedford, 1970, 1972b; Barros 1974; Austin 1975; Bedford & Cooper 1978; Rogers 1978; Yanagimachi 1981; Clegg 1983; Farooqui 1983; Moore & Bedford 1983; Langlais & Roberts 1985) and will not be covered here. Clegg (1983) presents a thoughtful, relatively up-to-date treatment of mechanisms involved in capacitation. Currently, considerable attention is being paid to dissociation and topographical rearrangements of sperm surface components that accompany capacitation (Aonuma et al 1973; Friend & Rudolf 1974; Gordon et al 1974; Koehler 1976, 1978; Friend et al 1977; Oliphant 1976; O'Rand 1977, 1979; Koehler et al 1980; Davis et al 1980; Langlais et al 1981; Heffner & Storey 1982; Shur & Hall 1982a; Ward & Storey 1984; Myles & Primakoff 1984). Apparently, such alterations, together with increased sperm metabolism, are part of the conditioning that enables sperm to bind to eggs properly and undergo the AR. Although numerous experiments involving the use of uncapacitated sperm have been reported, they are not considered here.

THE ACROSOME REACTION

The acrosome is a membrane-bound organelle that appears during spermiogenesis as a product of the Golgi complex (Leblond & Clermont 1952; Fawcett 1975; Fawcett & Bedford 1979; Bellvé 1979; Setchell 1982; Bellvé & O'Brien 1983). It may be considered biochemically analogous to a lysosome. The size and morphology of the acrosome vary considerably from one mammal to another. However, it always occupies the anterior region of the sperm head, just above the nucleus and beneath the plasma membrane. Acrosomal membranes underlying the plasma membrane and overlying the nuclear membrane are referred to as the outer and inner acrosomal membrane, respectively.

The AR is an exocytotic event involving fusion of sperm plasma and outer acrosomal membranes at many sites, formation of hybrid membrane vesicles that are eventually sloughed from the sperm head, and exposure of acrosomal contents (including a variety of hydrolases, such as hyaluronidase, proteinases, glycosidases, lipases, and phosphatases) and inner acrosomal membrane (Austin & Bishop 1958; Pikó & Tyler 1964; Barros et al 1967; Bedford 1968; Allison & Hartree 1970; Yanagimachi & Noda 1970a,b; McRorie & Williams 1974; Fawcett 1975; Morton 1977; Stambaugh 1978; Green 1978; Bedford & Cooper 1978; Meizel 1978, 1984; Russell et al 1979; Langlais & Roberts 1985). Among mammals, only sperm that have undergone the AR can penetrate the ZP and, upon reaching the perivitelline space, fuse with egg plasma membrane via plasma membrane overlying the postacrosomal region of the sperm head (Yanagimachi & Noda 1970a,b; Yanagimachi 1977, 1981). The AR has an absolute requirement for extracellular Ca^{2+}; involves a Ca^{2+}-dependent phospholipase, guanine nucleotide binding regulatory proteins, and altered cyclic nucleotide metabolism; and is characterized by Na^+ and Ca^{2+} influx and H^+ efflux through plasma membrane surrounding the sperm head. The latter involves an ATP-dependent H^+ pump and leads to an increase in intracellular pH (Yanagimachi & Usui 1974; Bedford & Cooper 1978; Garbers & Kopf 1980; Shapiro et al 1981; Meizel 1978, 1984; Yanagimachi 1981; Langlais & Roberts 1985; Trimmer & Vacquier 1986; Kopf et al 1986; Endo et al 1987). Many of these characteristics suggest that induction of the AR may be analogous to receptor-mediated signal transduction in somatic cells (Garbers & Kopf 1980; Gilman 1984; Birnbaumer et al 1984; Berridge 1985; Stryer & Bourne 1986).

Among the many substances reported to induce mammalian sperm to undergo the AR in vitro are serum albumin, β-lactoglobulin, lysolecithin, ionophores, catecholamines, glycosaminoglycans, cyclic nucleotides, steroids, as well as components originating from blood, oviductal fluid, follicular fluid, cumulus cells, oocytes, and the ZP (Gwatkin 1977; Green 1978; Meizel 1978, 1984; Yanagimachi 1981). This diversity has led to considerable confusion and controversy about both the nature of the AR inducer and the site of the AR in vivo. Since only sperm that have undergone the AR are capable of penetrating the ZP and fertilizing eggs, these are important issues.

Finally, it should be noted that a variety of techniques have been used to detect the AR in vitro. Although light microscopy is sufficient to assess the status of the acrosome of some mammalian sperm (e.g. guinea pig sperm; Huang et al 1981), in many cases it is necessary to employ transmission or scanning electron microscopy, or fluorescence microscopy (e.g. Saling & Storey 1979; Florman et al 1984) to make an unambiguous assessment.

SITE OF THE ACROSOME REACTION

Ovulated eggs of most mammals are surrounded by several layers of follicle cells that constitute the cumulus oophorus. For many years it has been generally held that only sperm that have undergone the AR can penetrate cumulus matrix (i.e. cumulus cells and extracellular matrix; Talbot & DiCarlantonio 1984a,b) surrounding ovulated eggs. This view stems primarily from observations of sperm that had undergone the AR (i.e. sperm with either an exposed inner acrosomal membrane or hybrid vesicles associated with the head) within cumulus matrix and associated with the ZP of fertilized and unfertilized eggs in vivo. In addition, this view stems from the belief that hyaluronidase, which is present in acrosomes, enables sperm to traverse cumulus matrix by degrading the hyaluronic acid–containing extracellular matrix (Austin & Bishop 1958; Bedford 1972a,b; Talbot & Franklin 1974a,b; McRorie & Williams 1974; Perreault et al 1979; Huang et al 1981; Cummins & Yanagimachi 1982; Yanagimachi & Phillips 1984; Talbot 1985). Although it has remained an open question whether sperm that have undergone the AR bind to the ZP by inner acrosomal membrane or by an acrosomal "ghost" (i.e. hybrid vesicles composed of plasma and outer acrosomal membranes) that remains associated with the sperm head (Moore & Bedford 1983; Yanagimachi & Phillips 1984; Cummins & Yanagimachi 1986), in either case, fusion of plasma and outer acrosomal membranes has been thought to occur in the cumulus matrix.

However, using static morphological methods, it is difficult to establish unambiguously whether sperm associated with the ZP in vivo underwent the AR before or after binding to the extracellular coat. Similarly, it is difficult to establish unambiguously whether sperm that have undergone the AR and are found in cumulus matrix in vivo are actually capable of fertilizing an egg. These determinations would require continuous monitoring of individual sperm in vivo from the time they first associated with cumulus of ovulated eggs until they fused with egg plasma membrane. Thus the site of the AR in vivo remains an unresolved and contested issue.

Recently, an alternative view of the fertilization pathway, in which sperm with an intact acrosome penetrate cumulus matrix, bind to the ZP, and only then undergo the AR, has received considerable attention (Gulyas & Schmell 1981; Wassarman & Bleil 1982; Florman et al 1982; Hartmann 1983a; Wassarman 1983, 1987a,b; Yanagimachi & Phillips 1984; Talbot 1984, 1985; Wassarman et al 1986a,b). The evidence for this sequence of events comes exclusively from experiments carried out in vitro and, consequently, its relevance to the in vivo situation is as yet unclear. Furthermore, since the experiments were performed with only mouse and hamster gametes, extrapolation to all other mammals may not be warranted. However, despite these reservations, since the more traditional view of the mammalian fertilization pathway (i.e. sperm undergo the AR *before* binding to the ZP) has been reviewed extensively (referenced above), recent evidence supporting the alternative scheme (i.e. sperm undergo the AR *after* binding to the ZP) is emphasized here (Figure 1).

Acrosome-Intact Sperm Can Penetrate Cumulus Matrix and Bind to the Zona Pellucida

Until relatively recently, it was considered likely that only sperm that had undergone the AR and released hyaluronidase would be able to traverse cumulus matrix surrounding ovulated eggs. However, although in the minority, some investigators have reported that acrosome-intact sperm penetrate cumulus-enclosed eggs and bind to the ZP during the course of in vitro fertilization. In all cases sperm were considered to have an intact acrosome when no signs of membrane vesiculation were detected shortly after binding of sperm to the ZP. For example, based on their electron microscopic analyses, Anderson et al (1975) concluded that both acrosome-intact mouse sperm and sperm that had undergone the AR penetrated cumulus matrix but that only the former became associated with the ZP and ultimately fertilized eggs. Storey et al (1984) reached the same conclusion using a chlortetracycline (CTC) fluorescence procedure (Saling & Storey 1979; Ward & Storey 1984) to score acrosome-intact mouse sperm and sperm that had undergone the AR during in vitro fertilization. Furthermore, Cherr et al (1986), using high-speed videomicrography to study hamster gamete interactions in vitro, found that few sperm ($\simeq 10\%$) underwent the AR within cumulus matrix and that those that did appeared to be immobilized due to adherence of their heads to cumulus cells (also, Suarez et al 1984). In these experiments, virtually all sperm that reached the ZP and bound to it had an intact acrosome. These recent findings are consistent with an earlier report that only acrosome-intact hamster sperm were able to penetrate cumulus-free eggs in vitro, whereas sperm induced to undergo the AR by cumulus were unable to bind to the ZP of hamster eggs (cited in Gwatkin 1977). Finally, in this context, Talbot et al (1985)

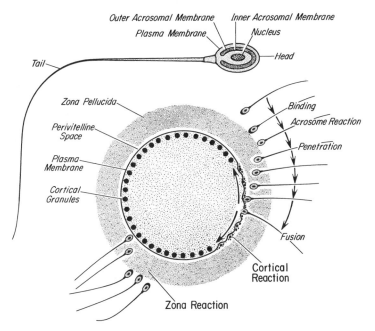

Figure 1   Diagrammatic representation of mouse gametes and the fertilization pathway in mice. The sequence of events includes attachment of sperm to eggs, followed by binding of sperm to eggs, completion of the AR, penetration of the ZP, sperm-egg fusion, cortical reaction, and zona reaction. (From Wassarman 1987a.)

found that sea urchin and frog sperm, motile cells lacking hyaluronidase, were able to penetrate hamster egg cumulus matrix in vitro and reach, but not penetrate, the ZP. Unlike the ZP, cumulus matrix did not appear to be a significant barrier to movement of either sea urchin or frog sperm. Naturally, these experiments must be interpreted cautiously, since the physical state of cumulus matrix surrounding ovulated eggs in vivo may differ significantly from that of cumulus matrix subjected to in vitro culture conditions.

These observations and others (e.g. reports that hyaluronidase is associated with plasma membrane of acrosome-intact sperm; Talbot & Franklin 1974a,b; Zao et al 1985; Talbot 1985) at least raise doubts about the necessity for sperm of some mammals to undergo the AR in order to release hyaluronidase and penetrate egg cumulus matrix. Interestingly, ovulated eggs of certain mammals (e.g. marsupials) are not surrounded by a cumulus matrix. Nevertheless, their sperm contain acrosomal hyaluronidase (Morton 1977; Rodger & Young 1981; Talbot & DiCarlantonio 1984a).

Acrosome-Intact Sperm Bind to the Zona Pellucida and Undergo the Acrosome Reaction

In addition to experiments noted above, performed with cumulus-enclosed eggs, mammalian gamete interactions have been studied extensively in vitro using cumulus-free ("denuded") eggs. Under these conditions, the ZP represents the sole barrier between advancing sperm and the egg plasma membrane. The same molecular interactions are likely responsible for binding of sperm to the ZP of either cumulus-enclosed or denuded eggs. Gamete components responsible for species-specific binding of sperm to the ZP (i.e. sperm receptors and egg binding proteins; discussed below) are expected to continue to function even after removal of cumulus cells from eggs. Certainly, removal of cumulus cells does not alter the ability of eggs to be fertilized and to develop in vitro.

Results of several investigations suggest that, as with cumulus-enclosed eggs (Anderson et al 1975; Storey et al 1984; Cherr et al 1986), only acrosome-intact sperm bind to the ZP of denuded eggs in vitro. For example, in their ultrastructural studies of mouse gamete interactions in vitro, Saling et al (1979) and Bleil & Wassarman (1983) found that approximately 10 min after insemination 90% or more of the sperm bound to eggs were acrosome-intact. Few, if any, of the sperm that had undergone the AR and were present in the free-swimming sperm population were able to bind to the eggs. (Note: It is clear that mouse sperm induced to undergo the AR with ionophore A23187 do not bind to eggs; however, this result must be interpreted cautiously since the ionophore may have deleterious effects on sperm motility and other properties of sperm.) Furthermore, as a result of their scanning electron microscopic studies, Phillips & Shalgi (1980) concluded that only acrosome-intact sperm were able to bind to ZP that were mechanically isolated from unfertilized hamster eggs. Sperm that had undergone the AR spontaneously (about 40% of the population) failed to bind to isolated ZP. These observations, which confirm results reported earlier by Wolf & Inoue (1976), are supported by results of a complementary study that employed CTC as a fluorescent probe to differentiate between acrosome-intact sperm and sperm that had undergone the AR (Saling & Storey 1979).

Once bound to the surface of the ZP by plasma membrane overlying their head, mouse sperm subsequently undergo the AR (Saling et al 1979; Bleil & Wassarman 1983). [Cherr et al (1986) made similar observations with cumulus-enclosed hamster eggs.] Florman & Storey (1982) carefully documented the time course for induction of the AR using the CTC assay and transmission electron microscopy. They found that following a lag time of 20–25 min postinsemination the proportion of ZP-bound sperm that underwent the AR increased steadily, reaching a plateau value of 80% of bound sperm after approximately 240 min ($t_{1/2}$ of 47 ± 12 min after the lag period); the $t_{1/2}$ for fertilization in these experiments was about 180 min. Interestingly, similar results were obtained with sperm bound to intact ZP that had been isolated from ovulated eggs ($t_{1/2}$ of 36 ± 6 min for the AR after a lag period). Thus it appears that for some mammals acrosome-intact sperm can penetrate cumulus matrix, bind to the ZP, and then undergo the AR in vitro. This scheme implies that a ZP component induces sperm to undergo the AR.

In addition to the use of light and electron microscopy, other means have been employed to distinguish between sperm binding and induction of the AR during in vitro fertilization. For example, Florman & Storey (1982) found that 3-quinuclidinyl benzylate (QNB; a high-affinity antagonist for the muscarinic class of cholinergic receptors), an inhibitor of in vitro fertilization (Florman & Storey 1981), did not prevent binding of mouse sperm to eggs but inhibited sperm bound to the ZP from undergoing the AR. Consequently, sperm were unable to penetrate the ZP and fertilize eggs. QNB had no effect on unbound sperm that underwent the AR either spontaneously in culture medium or in the presence of ionophore A23187. Thus, QNB specifically inhibits interactions between sperm plasma membrane and the ZP that are required for induction of the AR. Similarly, Saling (1986) described a monoclonal antibody (M42; directed against a mouse sperm plasma membrane protein) that inhibits in vitro fertilization (Saling & Lakoski 1985) by preventing ZP-bound sperm from undergoing the AR. Neither QNB (Florman & Storey 1982) nor M42 was able to inhibit the ionophore A23187–induced AR undergone by free-swimming sperm. These and other observations (discussed below) provide additional evidence that induction of the AR follows binding of sperm to the ZP in vitro.

Solubilized Zonae Pellucidae Induce Sperm to Undergo the Acrosome Reaction

Consistent with observations that mouse and hamster sperm undergo the AR at the ZP surface in vitro, are reports that preparations of acid-solubilized egg ZP (i.e. ZP dissolved at pH ≈ 3 and then neutralized) induce free-swimming sperm to undergo the AR in vitro. Florman & Storey (1982) found that exposure of mouse sperm to such preparations resulted in a rapid enhancement in the rate of the AR. In these experiments, the lag phase observed with sperm bound to intact ZP was not seen; however, QNB did inhibit induction of the AR by solubilized ZP. Bleil & Wassarman (1983) also found that acid-solubilized egg ZP induced mouse sperm to undergo the AR in vitro and identified the ZP glycoprotein responsible for inducing the reaction (Wassarman & Bleil 1982; Florman et al 1984; discussed below). Their experiments also demonstrated that acid-solubilized embryo ZP failed to induce sperm to undergo the AR (discussed below). Finally, Cherr et al (1986) reported that preparations of solubilized hamster ZP induced hamster sperm to undergo the AR in vitro but were significantly less effective at inducing mouse sperm to undergo the AR. The latter observation suggests a certain degree of species specificity for the interaction of sperm and the AR inducer. These, as well as other observations, strongly suggest that interactions between acrosome-intact sperm and a component of the ZP result in induction of the AR.

Summary

Results of experiments carried out with mouse and hamster gametes in vitro strongly suggest that only acrosome-intact sperm bind to the ZP of unfertilized eggs. Bound sperm are then induced to undergo the AR by a component of the ZP. Results obtained with QNB suggest alternative mechanisms for the ZP-induced and spontaneous AR in vitro, and there is some evidence indicating a degree of species specificity for the ZP-induced AR.

BINDING OF SPERM TO THE ZONA PELLUCIDA

Homologous sperm-egg interactions have been examined in a systematic manner in vitro using capacitated mouse and hamster gametes. Unfortunately, the methods used to assess such interactions are relatively imprecise, and consequently, interpretation of the results remains somewhat subjective. Most often, the strength of interactions between sperm and eggs has been evaluated by repeatedly pipetting the eggs, with their associated sperm, using broad-bore micropipettes (e.g. Hartmann et al 1972; Schmell & Gulyas 1980; Bleil & Wassarman 1980a). A gradient centrifugation assay has also been devised and used in a few studies (Saling et al 1978; Heffner et al 1980; Florman & Storey 1982). Sperm that remain associated with the ZP after pipetting or centrifugation are considered to be bound to the eggs. If fertilized eggs or embryos are included in these assays they serve as a control (Bleil & Wassarman 1980a).

Sperm Attach and Then Bind to the Zona Pellucida

When ovulated eggs and two-cell embryos are added to sperm in vitro, within seconds the ZP of both eggs and embryos are covered with motile sperm (Figure 2). These sperm are loosely associated with the ZP and can be removed by gentle pipetting with a broad-bore pipette. This state of adhesion is referred to as attachment (Hartmann et al 1972; Hartmann 1983a). Shortly thereafter, contact between sperm and egg ZP becomes more tenacious, such that gentle pipetting no longer dissociates the gametes. This state of adhesion is called binding (Hartmann et al 1972; Hartmann 1983a). The orientation of both gametes during the attachment phase probably influence whether or not sperm-egg interaction progresses to the binding stage. Although the initial, reversible attachment of sperm to embryo ZP is virtually indistinguishable from that observed with eggs, in the former case attachment does not proceed to the binding state. This pronounced difference in behavior provides an operational definition of bound sperm as those adhering to *egg* ZP under conditions that result in complete removal of sperm from *embryo* ZP. Bleil & Wassarman (1983) found that sperm attached to the ZP of unfertilized mouse eggs during a brief period ("pulse") in vitro, could be "chased" into bound sperm during a subsequent incubation period (~ 10 min) carried out in the absence of free-swimming sperm. This finding is consistent with the observation that attached sperm removed from one group of unfertilized eggs could subsequently bind to ZP of another group of eggs (Hartmann et al 1972; Hartmann & Hutchison 1974).

Many factors other than capacitation of sperm may influence binding of sperm to eggs in vitro. These include soluble peptide factors (Hartmann 1983b), temperature (Hartmann et al 1972; Heffner & Storey 1982), and Ca^{2+} (Saling et al 1978; Heffner et al 1980; Florman et al 1982). In the latter case, EGTA reversed binding of sperm to mouse eggs when added at early times postinsemination (< 15 min), but was ineffective at later times (> 20 min) (Florman et al 1982).

Binding of Sperm to the Zona Pellucida Is Relatively Species Specific

Most investigators agree that mammalian fertilization exhibits a certain degree of species specificity but not absolute specificity. [As pointed out by Schmell & Gulyas (1980), most in vitro experiments dealing with species specificity have been performed with gametes from different genera, families, and orders within the mammalian kingdom, not different species.] Clearly, a few mammalian hybrids, such as those resulting from crosses between horses and either donkeys (mule or hinny) or zebras (horsebras or zebrorses), exist and certain heterologous gamete combinations (e.g. gibbon and human; Bedford 1977; Schmell & Gulyas 1980) result in fertilization in vitro. Generally, however, in vitro fertilization experiments strongly suggest the existence of species-specific interactions between mam-

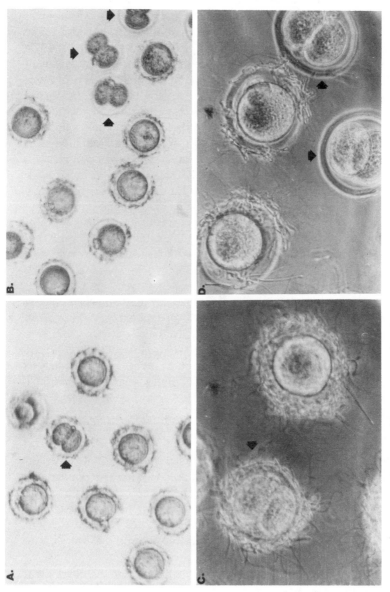

Figure 2 Light photomicrographs illustrating early interactions between mouse sperm and eggs in vitro. Attachment (A, C) and binding (B, D) of sperm to the ZP of unfertilized eggs and two-cell embryos are compared. Bound sperm were distinguished from attached sperm by repeated pipetting of eggs and embryos with broad-bore micropipettes, as described in the text. Note that sperm attach but do not bind to the ZP of two-cell embryos. Arrows indicate two-cell embryos. (From Wassarman et al 1986b.)

malian sperm and eggs (Adams 1974; Yanagimachi 1977, 1978, 1984; Barros & Leal 1980; Bedford 1981; Gulyas & Schmell 1981). When species specificity is observed it can usually be attributed to the ZP. This conclusion is drawn primarily from the finding that, whereas hybrid fertilization of ZP-intact eggs is rare in vitro, fertilization of ZP-free eggs by heterologous sperm is quite common. For example, ZP-free golden hamster eggs permit penetration by sperm from a wide variety of mammals, including humans; thus serving as a means to assess fertilizing capacity of human sperm in vitro (Yanagimachi 1984).

Unfortunately, it is not possible to draw a definitive conclusion about species specificity of sperm binding, since fertilization has usually been taken as the end point of in vitro experiments involving gametes from heterologous species. The inability of sperm to fertilize a heterologous egg could result from failure of any one of a number of steps in the fertilization pathway to take place, e.g. binding of sperm to eggs, induction of the AR, penetration of the ZP, or fusion of sperm with egg plasma membrane. However, on balance, results of in vitro fertilization experiments suggest a certain degree of species specificity for binding of sperm to eggs, and this is usually taken as evidence for the presence of sperm receptors in the ZP.

Sperm Bind to a Receptor Present in the Zona Pellucida

In addition to the relative species specificity of sperm-egg interactions, other observations are often taken as evidence for the presence of sperm receptors in the ZP of ovulated eggs. In particular, sperm do not bind to the ZP of homologous fertilized eggs or embryos in vitro (Barros & Yanagimachi 1972; Inoue & Wolf 1975b; Sato 1979; Bleil & Wassarman 1980a). This result implies that a ZP component of ovulated eggs that is recognized by, and permits binding of, sperm is modified shortly after fertilization, i.e. is inactivated. Modification of this sperm receptor is presumably attributable to the action of a cortical granule enzyme(s) released into the perivitelline space during the cortical reaction (Barros & Yanagimachi 1971; Gwatkin et al 1973; Nicosia et al 1977; Gwatkin 1977; Wolf & Hamada 1977; Schuel 1978, 1985; Gulyas 1980; Wassarman 1987a). The cortical reaction is triggered by fertilization of eggs and involves fusion of cortical granules, membrane-bound, lysosomelike organelles (200–600 nm in diameter) located in the egg cortex, with plasma membrane. Since the ZP is highly porous, making it permeable to large macromolecules and even small viruses (Gwatkin 1977), cortical granule contents (including various hydrolases) can enter the ZP and modify sperm receptors.

Additional evidence for the presence of sperm receptors in the ZP comes from in vitro experiments demonstrating that sperm-egg interactions are inhibited by solubilized ZP preparations. Exposure of sperm suspensions to preparations of either heat- or acid-solubilized ZP from unfertilized eggs has a marked inhibitory effect on the binding of sperm to eggs in vitro ("competition assay"; Gwatkin & Williams 1977; Bleil & Wassarman 1980a; Wassarman et al 1985b). For example, Bleil & Wassarman (1980a) found that treatment of mouse sperm with acid-solubilized egg ZP reduced the binding of sperm to eggs to as little as 10% of control values (Figure 3). Solubilized ZP from unfertilized eggs were shown to affect sperm rather than eggs, since mouse eggs that failed to bind sperm exposed to solubilized ZP could be removed from these cultures and were shown to bind untreated sperm to the same extent as control eggs. Sperm treated with acid-solubilized ZP from two-cell mouse embryos (Bleil & Wassarman 1980a) or with heat-solubilized ZP from either artificially activated hamster eggs or two-cell embryos (Gwatkin & Williams 1977) were able to bind to homologous unfertilized eggs to the same extent as untreated sperm (Figure 3). These results suggest that solubilized ZP from unfertilized eggs, but not from activated eggs or embryos, contains a component(s) that competes for binding sites on sperm and thus prevents their binding to eggs. This is the behavior expected for a sperm receptor.

Finally, a variety of circumstantial evidence also supports the idea that sperm receptors are present in egg ZP. This evidence is derived primarily from experiments in which treatment of eggs with proteinases, lectins, or antibodies directed against eggs or ZP resulted in inhibition of fertilization (reviewed in Gwatkin 1977; Dunbar 1983; Wassarman et al 1985b). Although it is tempting to suggest that these experiments demonstrate the presence of sperm receptors in the ZP, alternative explanations are as likely. For example, it is unlikely that either proteinases or lectins, under conditions employed thus far, modified sperm receptors exclusively. Rather, the inhibitory effect of these agents is probably attributable to global modification of the ZP surface, not necessarily including alteration of sperm receptors. Furthermore, while polyclonal antisera (Sacco 1978; Aitken et al 1981; Tsunoda et al 1981; Sacco et al 1981) and monoclonal antibodies (East et al 1984, 1985) directed against ZP antigens can clearly prevent sperm-egg interactions both in vivo and in vitro, it has not been established that their inhibitory effects are due exclusively to binding of antibody to sperm receptors.

Summary

Results of in vitro experiments suggest that sperm first loosely (attachment) and then tightly (binding) associate with the ZP of unfertilized eggs. Binding of sperm is relatively species-specific and is mediated by receptors

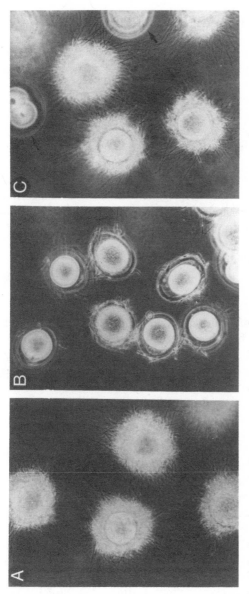

Figure 3 Light photomicrographs illustrating the effect of solubilized ZP on the binding of mouse sperm to eggs in vitro. Sperm exposed to buffer (A), solubilized ZP from unfertilized eggs (B), or solubilized ZP from two-cell embryos (C) were incubated with unfertilized eggs and two-cell embryos in vitro. Attached sperm were removed from eggs and embryos by repeated pipetting with broad-bore micropipettes. Only solubilized ZP from unfertilized eggs inhibited the binding of sperm to eggs (compare B with A and C). Arrows in C indicate two-cell embryos. (From Bleil & Wassarman 1980a.)

present in egg ZP. Sperm receptors are inactivated following the fusion of sperm and egg.

NATURE OF THE SPERM RECEPTOR

Based upon the fertilization pathway just described, a mammalian sperm receptor purified from the ZP would be expected to fulfill the following requirements in vitro: (*a*) Receptor purified from egg ZP should bind to sperm and prevent their binding to ovulated eggs (i.e. prevent fertilization). (*b*) Receptor purified from embryo ZP should not bind to sperm and not prevent their binding to ovulated eggs (i.e. not prevent fertilization). (*c*) Receptor purified from egg ZP should bind to the head of acrosome-intact sperm but not to the head of sperm that have undergone the AR. (*d*) Receptor purified from egg ZP should not bind to a wide variety of cell types. (*e*) Receptor purified from egg ZP should exhibit a certain degree of species specificity in binding to sperm. To date, only the mouse sperm receptor has been purified and characterized in this manner, although it should be noted that a porcine ZP glycoprotein has been reported to have sperm receptor activity (Sacco et al 1981). Results of experiments with the mouse sperm receptor are summarized below.

Zona Pellucida Glycoprotein ZP3 Is a Sperm Receptor

Three different glycoproteins, called ZP1 (200 kDa), ZP2 (120 kDa), and ZP3 (83 kDa), have been identified in the mouse egg ZP (Bleil & Wassarman 1980b; Wassarman et al 1985a; Wassarman 1987b). Bleil & Wassarman (1980a) tested each of the glycoproteins, purified to homogeneity, for sperm receptor activity by using an in vitro competition assay. In each case, the extent of binding of treated and untreated sperm to unfertilized eggs was compared (i.e. number of sperm bound per egg). Whereas neither ZP1 nor ZP2 had a significant effect on sperm binding (or fertilization), ZP3 was nearly as effective as solubilized egg ZP in reducing the number of sperm bound to eggs ($ID_{50} \approx 1$ ZP equivalent/μl) and preventing fertilization (Figures 4 and 5). Under identical conditions, serum albumin, submaxillary mucin, and transferrin, even at concentrations as high as 100 μg/ml (i.e. ten times the highest concentration of ZP3 tested), had no effect on binding of sperm to eggs. Comparisons of results obtained with purified ZP3 and solubilized ZP preparations indicated that ZP3 alone accounted for all sperm receptor activity present in egg ZP. Furthermore, as anticipated, ZP3 purified from embryo ZP (called ZP3$_f$) had no effect on sperm binding (or fertilization), even at concentrations at which egg ZP3 inhibited binding by 80% or more. [At concentrations 3–5 times higher receptor and AR-inducing activities were detected in preparations of embryo ZP3$_f$,

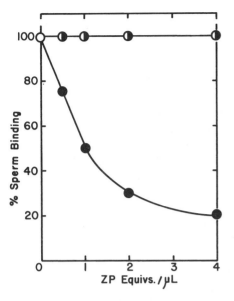

Figure 4 Effect of individual ZP glycoproteins on the binding of mouse sperm to eggs in vitro. Binding of sperm to eggs was compared after exposure of sperm to ZP1 (*half-filled circles*), ZP2 (*half-filled circles*), or ZP3 (*filled circles*) purified from egg ZP, or to culture medium alone (open circle). The number of sperm bound to eggs was determined as a function of ZP glycoprotein concentration (expressed as ZP equivalents per microliter). (From Wassarman et al 1985a.)

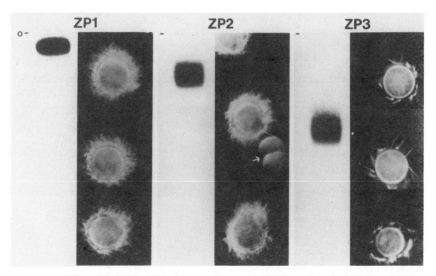

Figure 5 Effect of individual ZP glycoproteins on the binding of mouse sperm to eggs in vitro. Shown are autoradiograms of radiolabeled ZP1, ZP2, and ZP3, purified from egg ZP, following SDS-PAGE on nonreducing gels and photomicrographs of sperm bound to eggs after exposure of sperm to purified ZP1, ZP2, or ZP3. As seen in Figure 4, only ZP3 inhibited the binding of sperm to eggs. (From Wassarman et al 1985b.)

presumably due to the presence of residual ZP3 (Wassarman et al 1985a).] These results strongly suggest that ZP3 possesses the receptor activity responsible for binding of sperm to the ZP of unfertilized mouse eggs.

In subsequent studies, Bleil & Wassarman (1986) carried out whole-mount autoradiographic experiments to investigate the specificity of binding of ZP3 to sperm. Autoradiograms of whole-mount preparations of sperm incubated with radiolabeled ZP3 (^{125}I-ZP3) revealed that ZP3 was bound to heads of acrosome-intact sperm, but not to heads of sperm that had undergone the AR (Figures 6 and 7). Although both red blood cells and residual bodies were present in these sperm preparations, no silver grains due to radiolabeled ZP3 were found associated with these contaminants. Under the same conditions, radiolabeled fetuin was bound at only background levels to heads of both acrosome-intact sperm and sperm that had undergone the AR, and radiolabeled ZP2 bound preferentially to heads of sperm that had undergone the AR. Therefore, while ZP3 and fetuin share some common molecular features, only ZP3 was bound to sperm, and binding was specific with respect to cell type (i.e. only sperm), cellular localization (i.e. only the sperm head), and cellular state (i.e. only acrosome-intact sperm). Furthermore, although ZP2 and ZP3 also share common molecular features, they recognize different antigenic determinants on the sperm head. In the former case, determinants associated with inner acrosomal membrane and in the latter, determinants associated with plasma membrane. Quantitation of the autoradiographic data indicated that saturation of ZP3 binding sites occurred in the range 10–50 × 10^3 ZP3 molecules bound per sperm head. (Results of calculations suggest that the mouse sperm head can accommodate this number of receptor molecules, given the approximate dimensions of the ZP2-ZP3 complex that constitutes the structural repeat of mouse ZP filaments as discussed below.)

ZP3 Is Synthesized by Growing Oocytes and Assembled into the Zona Pellucida

ZP1, ZP2, and ZP3 are synthesized and secreted during the 2–3 week period of oocyte growth in mice (Bleil & Wassarman 1980c; Wassarman 1983; Shimizu et al 1983; Salzmann et al 1983; Wassarman et al 1985a). The

---

Figure 6 Autoradiographic visualization of radiolabeled sperm receptor bound to acrosome-intact sperm. Shown are photomicrographs of whole-mount autoradiograms of sperm incubated in the presence of ^{125}I-ZP3 (A, B) or ^{125}I-fetuin (C) and visualized by Nomarski differential interference contrast (DIC) optics using either a Zeiss 63X Planapo (*inset*, A) or 40X Plan-Neofluar (A, B, C) objective. Abbreviations: ar, acrosome-reacted sperm; rbc, red blood cell; rb, residual body. Bar = 10μm. (From Bleil & Wassarman 1986.)

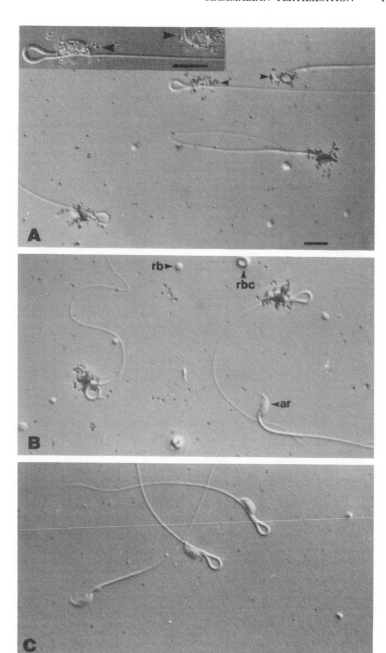

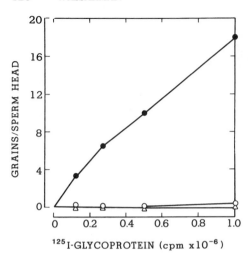

Figure 7 Binding of radiolabeled sperm receptor to acrosome-intact and -reacted sperm. Sperm were incubated in the presence of ^{125}I-ZP3 (*filled and open circles*) or ^{125}I-fetuin (*triangles*), subjected to whole-mount autoradiography and visualized by Nomarski DIC optics. Shown are the average number of silver grains per sperm head, plotted as a function of either ^{125}I-ZP3 or ^{125}I-fetuin concentration, for both acrosome-intact (*filled circles, triangles*) and -reacted (*open circles, triangles*) sperm. (From Bleil & Wassarman 1986.)

mature form of ZP3 (83 kDa) consists of a 44-kDa polypeptide chain, three or four complex-type N-linked oligosaccharides, and an undetermined number of O-linked oligosaccharides (Salzmann et al 1983; Florman & Wassarman 1985). The ZP3 gene is a low copy number (perhaps single-copy) gene: it is transcribed as a 1.7-kilobase (kb) mRNA that is found exclusively in ovarian tissue (Ringuette et al 1986) and is a major mRNA species in growing mouse oocytes (R. Kinloch, R. Roller, and P. Wassarman, unpublished results). The heterogeneity of mature ZP3, with respect to both M_r and isoelectric point, is attributable primarily to its oligosaccharide components and not to its polypeptide chain (Salzmann et al 1983; G. Salzmann & P. Wassarman, unpublished results). ZP glycoprotein synthesis ceases during or shortly after ovulation (i.e. when a fully grown oocyte becomes an unfertilized egg) and has not been detected in fertilized eggs or early embryos. There are more than a billion copies of ZP3 in the ZP of unfertilized eggs.

ZP3 is assembled, together with ZP2, into the filaments that make up the ZP (Greve & Wassarman 1985; Wassarman et al 1987; J. Greve & P. Wassarman, unpublished results). Isolated individual filaments are about 7 nm in width, can be 2–3 μm in length, and possess a structural repeat of about 15 nm. Apparently, this repeat reflects the presence of a ZP2-ZP3 complex ($\sim$ 180 kDa) every 15 nm or so along the filaments. ZP2 and ZP3 are present in approximately equimolar amounts in the ZP. ZP1 is present at irregular intervals along the filaments and serves as a cross-linker of individual filaments, thus giving rise to a three-dimensional matrix. Therefore, acrosome-intact sperm must interact with ZP3 that is part of a heterologous dimer (ZP2-ZP3) that, in turn, constitutes the structural repeat of ZP filaments.

Sperm Recognize O-Linked Oligosaccharides on ZP3

In order to evaluate the molecular basis of ZP3 sperm receptor activity, Florman et al (1984) and Florman & Wassarman (1985) subjected the purified glycoprotein to both chemical and enzymatic dissection. Firstly, to assess the role of the polypeptide chain, ZP3 was digested extensively with insolubilized pronase to give rise to glycopeptides of 1.5–6 kDa, which were then tested for sperm receptor activity in the in vitro competition assay. The glycopeptides were found to be virtually as effective as intact glycoprotein in inhibiting binding of sperm to eggs (i.e. receptor activity was present). Secondly, selective removal of N-linked oligosaccharides from ZP3 by digestion with endo-β-acetyl-D-glucosaminidase F (Endo F; Elder & Alexander 1982) had no effect on its sperm receptor activity. The O-glycosylated polypeptide chain was found to be virtually as effective as intact glycoprotein in inhibiting binding of sperm to eggs (i.e. receptor activity was present). Thirdly, extensive deglycosylation of ZP3 with tri-fluoromethanesulfonic acid (TFMS; Edge et al 1981), a reagent that removes both N- and O-linked oligosaccharides from glycoproteins, con-verted ZP3 to a species with an apparent M_r approximating that of the polypeptide chain (44 kDa); this species had no effect on the binding of sperm to eggs (i.e. receptor activity was absent). Fourthly, mild alkaline hydrolysis of ZP3 (β-elimination; Sharon 1975) selectively removed O-linked oligosaccharides without breakage of peptide bonds, resulting in an N-glycosylated polypeptide chain; this species had no effect on binding of sperm to eggs (i.e. receptor activity was absent). Fifthly, mild alkaline hydrolysis of ZP3 in the presence of NaBH$_4$ (Sharon 1975) resulted in release of intact O-linked oligosaccharides having N-acetyl-D-galac-tosaminitol at their reducing termini; these purified oligosaccharides inhibited binding of sperm to eggs (i.e. receptor activity was present). However, O-linked oligosaccharides released by alkaline-borohydride hydrolysis from either ZP2 or human chorionic gonadotropin had no effect on binding of sperm to eggs. Receptor activity was only found associated with ZP3 oligosaccharides having an apparent M_r of about 3900 on Bio-Gel P-6 columns (Figure 8), and this size class of oligosaccharides was significantly enriched following affinity purification of alkali-released oli-gosaccharides with sperm (Figure 9).

Collectively, results of these experiments strongly suggest that the sperm receptor activity of ZP3 is attributable not to the polypeptide chain, but to a specific class of O-linked oligosaccharides on the glycoprotein. This conclusion is consistent with the observation that ZP3 continues to exhibit

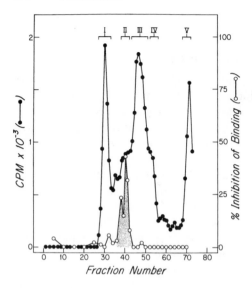

Figure 8 Gel-filtration and competition binding analyses of O-linked oligosaccharides released from ZP3. Shown are the Bio-Gel P-6 elution profile for radio-labeled (^{3}H-NaBH$_4$) oligosaccharides released from ZP3 by mild alkaline reduction (*filled circles*) and the sperm receptor activity profile for the released oligosaccharides (*open circles*). The region of the elution profile displaying sperm receptor activity (region II) is stippled. (From Florman & Wassarman 1985.)

sperm receptor activity in vitro even after exposure to denaturants, detergents, or high temperatures (Bleil & Wassarman 1980a; Wassarman et al 1985b) and with reports that various lectins, glycoconjugates, and monosaccharides prevent binding of sperm to eggs (Oikawa et al 1973; Huang et al 1982; Shur & Hall 1982b; Shalgi et al 1986).

ZP3 Induces Sperm to Undergo the Acrosome Reaction

ZP3 has been identified as the AR-inducer present in solubilized preparations of egg ZP (Bleil & Wassarman 1983; Florman et al 1984). Purified ZP3, at nanomolar concentrations, was found to be as effective as ionophore A23187 at inducing sperm to undergo the AR in vitro. [Mouse sperm induced to undergo the AR by ZP3 in vitro do not bind to unfertilized eggs, which provides further evidence that only acrosome-intact sperm bind to eggs (J. Bleil & P. Wassarman, unpublished results).] This suggests that binding of acrosome-intact mouse sperm to ZP3, through plasma membrane overlying the anterior region of the sperm head, is sufficient to alter the affected plasma membrane so that it becomes capable of fusing with outer acrosomal membrane. In view of the role of Ca^{2+} and the effect of ionophore A23187 (discussed above), it seems likely that binding of ZP3 to the sperm head affects the ion permeability of the plasma membrane.

Whereas the ability of ZP3 to serve as a sperm receptor depends solely on its O-linked oligosaccharides (discussed above), the ability of ZP3 to serve as an AR-inducer depends on its polypeptide chain as well (Florman et al

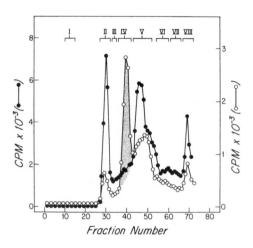

Figure 9   Gel-filtration, competition binding analyses, and determination of sperm association of O-linked oligosaccharides released from ZP3. These experiments were carried out as described for Figure 8. Shown are the Bio-Gel P-6 elution profile for radiolabeled oligosaccharides not incubated with sperm (*filled circles*) and those associated with sperm after a 1-hr incubation (*open circles*). The region of the elution profile that was significantly enriched following incubation of oligosaccharides with sperm (region IV; equivalent to region II in Figure 8) is stippled. Only region IV exhibited sperm receptor activity, reducing sperm binding by more than 50% as compared with controls. (From Florman & Wassarman 1985.)

1984; Wassarman et al 1985a, 1987). Although small (1.5–6 kDa) glycopeptides and O-linked oligosaccharides derived from ZP3 are able to bind to the head of acrosome-intact sperm and prevent them from binding to unfertilized eggs, neither the glycopeptides nor O-linked oligosaccharides induce sperm to undergo the AR in vitro. However, large proteolytic fragments of ZP3 (~ 40 kDa) are able to induce the AR in vitro (Wassarman et al 1985a). In view of this behavior, Wassarman (1987a) suggested that sperm bind to multiple O-linked oligosaccharides on each ZP3 molecule (i.e. via a multivalent interaction) and that these oligosaccharides must be joined by polypeptide chain for induction of the AR. The distance between these oligosaccharides on an individual ZP3 molecule may be an important factor during induction of the AR. Finally, results of experiments carried out thus far do not exclude the possibility that a portion of the ZP3 polypeptide chain itself functions as a "fusion peptide."

Summary

ZP glycoprotein ZP3 serves as both sperm receptor and AR-inducer in mice. Acrosome-intact sperm recognize and bind, via plasma membrane overlying their head, to specific O-linked oligosaccharides on ZP3. Apparently, these oligosaccharides alone are responsible for recognition and binding; however, the ZP3 polypeptide chain plays a role, either directly or indirectly, in inducing bound sperm to undergo the AR. Thus, ZP3 seems to have evolved so as to discharge the functions of polysaccharide located in the jelly coat (AR-inducer) and glycoprotein located in the vitelline envelope (species-specific sperm receptor) of sea urchin eggs (Wassarman 1987a).

NATURE OF THE EGG BINDING PROTEIN

The egg binding protein present on sperm is the counterpart of the sperm receptor present on eggs. In sea urchins, the egg binding protein is bindin, a 30.5-kDa hydrophobic protein localized within the acrosome. Bindin can be considered a lectin since it recognizes and binds to specific sequences of sugar residues (Vacquier & Moy 1977; Lopo et al 1982; Glabe et al 1982; Gao et al 1986). Among mammals, several different types of proteins are considered candidates for the role of egg binding protein (Peterson et al 1980, 1985; O'Rand et al 1985; Ahuja 1982; Saling & Lakoski 1985; Sullivan & Bleau 1985; Brown & Jones 1987; additional references below).

Glycosyltransferases

Glycosyltransferases have been implicated in mediating cell-cell interactions in a variety of biological systems (Pierce et al 1980; Schachter & Roseman 1980; Shur 1982). In this context, galactosyltransferase, an enzyme that normally catalyzes transfer of galactose from uridine 5'-diphosphate-galactose (UDP-galactose) to terminal N-acetylglucosamine residues to form N-acetyllactosamine, is a serious candidate for the role of egg binding protein in mice (Shur & Hall 1982a,b; Lopez et al 1985). Presumably, the enzyme recognizes and binds to specific N-acetylglucosamine residues on ZP3, a situation consistent with the proposed role of ZP3 O-linked oligosaccharides in sperm-egg interaction. Among the observations that suggest galactosyltransferase is an egg binding protein are the following. (a) Two different inhibitors of galactosyltransferase, α-lactalbumin and UDP-dialdehyde, inhibited both the activity of the sperm surface enzyme and binding of sperm to eggs in vitro. (Similarly, covalent modification of sperm surface galactosyltransferase by UDP-dialdehyde, under reducing conditions, inhibited binding of sperm to eggs.) (b) Both solubilized and intact ZP served as substrates for sperm surface galactosyltransferase. (c) A soluble form of galactosyltransferase, purified to homogeneity from bovine milk, catalyzed galactosylation of egg ZP and, coincidentally, inhibited binding of sperm to the eggs. (d) UDP-galactose, but not UDP-glucose, was able to dissociate bound sperm from egg zonae pellucidae in a time- and temperature-dependent manner. (e) Monospecific antibodies (or Fab fragments) raised against bovine milk galactosyltransferase inhibited the sperm surface enzyme, as well as binding of sperm to eggs, and were used to demonstrate that galactosyltransferase is localized to a plasma membrane domain on the dorsal surface of the anterior head overlying the acrosome. Collectively, these and other observations strongly suggest that a sperm surface galactosyltransferase plays a role during sperm-egg interaction in mice. Thus far, apparently neither the ZP glycoprotein(s) nor the oligosaccharide(s) galactosylated by the sperm surface galactosyltransferase has been identified, isolated, and characterized.

Proteinases

Based on the effects of proteinase inhibitors on sperm-egg interactions in vitro, Saling (1981) concluded that a trypsinlike activity associated with sperm is involved in binding of mouse sperm to the ZP of unfertilized eggs (i.e. serves as an egg binding protein). Exposure of sperm to inhibitors of trypsinlike proteinases prevented binding of sperm to the ZP in a concentration-dependent manner, whereas sperm binding to ZP-free eggs and fusion with egg plasma membrane was unaffected. The inhibitors prevented neither sperm penetration through the ZP nor membrane fusion (i.e. fertilization) when added to samples after sperm had tightly associated with egg ZP. Analogous observations were made by Hartmann & Hutchison (1974) with hamster gametes. The observations in mice were extended by Benau & Storey (1987), who found that the serine proteinase active-site titrant, 4-methylumbelliferyl-p-guanidinobenzoate (MUGB), was hydrolyzed by mouse sperm and inhibited binding of sperm to egg ZP. Furthermore, the presence of either trypsin inhibitors (e.g. soybean trypsin inhibitor) or solubilized ZP inhibited the hydrolysis of MUGB by sperm in a concentration-dependent manner. Surprisingly, lysyl and arginyl trypsin substrates (which are analogous to MUGB) were not hydrolyzed by sperm and did not inhibit either sperm binding to the ZP or hydrolysis of MUGB by sperm. The investigators concluded that the stereospecificity of the "trypsinlike" sperm site involved in binding is unique in that it reacts with trypsin inhibitors, but not substrates. These and other studies (e.g. Aarons et al 1984) suggest a role for sperm proteinases (perhaps acrosin or proacrosin; McRorie & Williams 1974) in mammalian sperm-egg interactions.

Summary

A variety of proteins are candidates for the role of egg binding protein. In at least two documented cases, involving glycosyltransferases and proteinases, recognition and binding would be accomplished through formation of an enzyme-substrate complex in which a ZP glycoprotein serves as substrate. Whether or not a lectinlike protein analogous to sea urchin bindin is involved in binding of sperm to the ZP remains to be determined. Whatever the nature of the egg binding protein, its structure must change in concert with that of the sperm receptor during evolution to provide species-specific binding of sperm to eggs.

SUMMARY AND FINAL COMMENTS

Here I have focused on events associated with the initial encounter of eggs and sperm during mammalian fertilization, with particular emphasis on experiments performed with mouse and hamster gametes in vitro. The extent to which the conclusions drawn may apply to other mammals and to in vivo fertilization remains to be determined. Certainly, it would not be surprising to find that the cellular and molecular details of the fertilization process differ to a degree among a wide range of mammals, just as they do among nonmammalian species. However, it is just as likely that, in time, certain principles will emerge that apply to virtually all mammals.

Results of experiments described here imply a pathway for fertilization of mouse eggs that consists of several steps that occur in a compulsory order. These steps include: (a) attachment and then (b) binding of acrosome-intact sperm, via plasma membrane overlying the acrosome, to the ZP of ovulated eggs; (c) induction of the AR at the surface of the ZP; (d) penetration of the ZP by sperm that have undergone the AR; and (e) fusion of sperm, via plasma membrane at the posterior region of the head, with egg plasma membrane. Involvement of the ZP in this process is attributable to two of its three glycoproteins, ZP2 and ZP3, that together constitute the repeating structural unit of ZP filaments. Acrosome-intact sperm recognize and bind to ZP3 (sperm receptor) via an egg binding protein (perhaps a glycosyltransferase, proteinase, or lectin) associated with plasma membrane overlying the sperm head. Binding of sperm to tens-of-thousands of ZP3 molecules causes them to undergo the AR and results in exposure of inner acrosomal membrane with its associated acrosomal contents. Bound sperm that have undergone the AR probably remain associated with the ZP by binding to ZP2, which may act as a secondary receptor. The strength of the interaction between ZP2 and sperm that have undergone the AR appears to be sufficient to maintain binding of sperm already associated with the ZP but insufficient to lead to binding of free-swimming sperm to the ZP. Since bound sperm that have undergone the AR must penetrate the ZP, relatively weak interactions could be advantageous for the sperm's progress through the extracellular coat. The possibility that binding of sperm that have undergone the AR to ZP2 is mediated by acrosin, the sperm proteinase credited with hydrolyzing a path through the ZP, should be considered.

The mouse sperm receptor, ZP3, illustrates a role for carbohydrates in supporting specific interactions between eukaryotic cells, in this case, between sperm and egg. A strong case can be made for involvement of carbohydrates in induction of the AR and species-specific binding of sperm to eggs in marine invertebrates as well. Apparently, only a small percentage of ZP3 O-linked oligosaccharides are actually recognized by acrosome-intact sperm, but each sperm binds to multiple O-linked oligosaccharides on each ZP3 molecule. It is possible that induction of the AR is dependent on multivalent interactions between the egg binding protein and ZP3. Whether ZP3 or O-linked oligosaccharides derived from ZP3 exhibit any species specificity has not yet been determined. The great diversity of known oligosaccharide structures is certainly compatible with generation of species specificity for mammalian sperm receptors. The variety of compositions, sequences, branching patterns, conformations, and other features of oligosaccharide structure provide for a staggering number of combinatorial possibilities. It will be of considerable interest to compare the structures of these ZP3 oligosaccharides with functionally analogous oligosaccharides isolated from eggs of other mammals.

The prospects for mammalian fertilization research are bright. Despite the practical limitations imposed on amounts of material available for experiments, significant progress is being made in elucidating various biochemical and molecular features of fertilization in mammals. In some respects, due in part to important advances in biotechnology, research on mammalian fertilization has achieved parity with research on nonmammals. In the near future, recombinant DNA technology will likely be used to produce large quantities of mammalian egg and sperm components, such as acrosomal enzymes, sperm receptors, and egg binding proteins, by transfection of cultured cells. With these materials in hand, together with knowledge of the primary structures of the proteins and the ability to apply site-directed mutagenesis, we can look forward to major advances in our understanding of the molecular basis of mammalian fertilization.

ACKNOWLEDGMENTS

Through the years I have enjoyed and benefited from discussions about fertilization with past and present members of my laboratory. Consequently, some of the ideas presented here may seem very familiar to them and may even have originated with them. Research performed in my laboratory was supported in part by the Rockefeller Foundation, the

National Institute of Child Health and Human Development, and the National Science Foundation.

Literature Cited

Aarons, D., Speake, J. L., Poirier, G. R. 1984. Evidence for a proteinase inhibitor binding component associated with murine spermatozoa. *Biol. Reprod.* 31: 811–17

Adams, C. E. 1974. Species specificity in fertilization. In *Physiology and Genetics of Reproduction*, ed. E. N. Coutinho, F. Fuchs, pp. 69–79. New York: Plenum

Ahuja, K. K. 1982. Fertilization studies in the hamster: The role of cell surface carbohydrates. *Exp. Cell Res.* 140: 353–62

Aitken, R. J., Rudak, E. A., Richardson, D. W., Dor, J., Djahanbakkeh, O., Templeton, A. A. 1981. The influence of anti-zona and antisperm antibodies on sperm-egg interactions. *J. Reprod. Fertil.* 62: 597–606

Allison, A. C., Hartree, E. F. 1970. Lysosomal enzymes in the acrosome and their possible role in fertilization. *J. Reprod. Fertil.* 21. 501–15

Anderson, E., Hoppe, P. C., Whitten, W. K., Lee, G. S. 1975. In vitro fertilization and early embryogenesis: A cytological analysis. *J. Ultrstruct. Res.* 50: 231–52

Aonuma, S., Mayumi, T., Suzuki, K., Noguchi, T., Iwai, M., Okabe, M. 1973. Studies on sperm capacitation. I. Relationships between a guinea pig sperm-coating antigen and a sperm capacitation phenomenon. *J. Reprod. Fertil.* 35: 425–32

Austin, C. R. 1951. Observations on the penetration of the sperm into the mammalian egg. *Aust. J. Sci. Res.* B4: 581–96

Austin, C. R. 1952. The "capacitation" of the mammalian sperm. *Nature* 170: 326

Austin, C. R. 1965. *Fertilization.* Englewood Cliffs, NJ: Prentice-Hall. 145 pp.

Austin, C. R. 1975. Membrane fusion events in fertilization. *J. Reprod. Fertil.* 44. 155–66

Austin, C. R., Bishop, M. W. H. 1957. Fertilization in mammals. *Biol. Rev.* 32: 296–349

Austin, C. R., Bishop, M. W. H. 1958. Role of the rodent acrosome and perforatorium in fertilization. *Proc. R. Soc. London Ser. B* 149: 241–48

Austin, C. R., Walton, A. 1960. Fertilisation. In *Marshall's Physiology of Reproduction*, ed. A. S. Parkes, Vol. 1, Pt. 2, pp. 310–416. London: Longmans, Green and Co.

Barros, C. 1974. Capacitation of mammalian spermatozoa. In *Physiology and Genetics of Reproduction*, ed. E. M. Coutinho, F. Fuchs, Pt. B, pp. 3–24. New York: Plenum

Barros, C., Bedford, J. M., Franklin, L. E., Austin, C. R. 1967. Membrane vesiculation as a feature of the mammalian acrosome reaction. *J. Cell Biol.* 34: C1–C5

Barros, C., Leal, J. 1980. In vitro fertilization and its use to study gamete interactions. In *In Vitro Fertilization and Embryo Transfer*, ed. E.S.E. Hafez, K. Semm, pp. 37–49. New York: Liss

Barros, C., Yanagimachi, R. 1971. Induction of the zona reaction in golden hamster eggs by cortical granule material. *Nature* 233: 268–69

Barros, C., Yanagimachi, R. 1972. Polyspermy-preventing mechanisms in the golden hamster egg. *J. Exp. Zool.* 180: 251–66

Bavister, B. D., 1981. Analysis of culture media for in vitro fertilization and criteria for success. In *Fertilization and Embryonic Development In Vitro*, ed. L. Mastroianni, J. D. Biggers, pp. 42–62. New York: Plenum

Bedford, J. M. 1968. Ultrastructural changes in the sperm head during fertilization in the rabbit. *Am. J. Anat.* 123: 329–58

Bedford, J. M. 1970. Sperm capacitation and fertilization in mammals. *Biol. Reprod.* 2: 128–58

Bedford, J. M. 1971. Techniques and criteria used in the study of fertilization. In *Methods in Mammalian Embryology*, ed. J. C. Daniel, Jr., pp. 37–63. San Francisco: Freeman

Bedford, J. M. 1972a. An electron microscopic study of sperm penetration into the rabbit egg after natural mating. *Am. J. Anat.* 133: 213–54

Bedford, J. M. 1972b. Sperm transport, capacitation and fertilization. In *Reproductive Biology*, ed. H. Balin, S. Glasser, pp. 338–92. Amsterdam: Excerpta Medica

Bedford, J. M. 1977. Sperm-egg interactions: The specificity of human spermatozoa. *Anat. Rec.* 188: 477–83

Bedford, J. M. 1981. Why mammalian gametes don't mix. *Nature* 291: 286–88

Bedford, J. M. 1982. Fertilization. In *Reproduction in Mammals: 1. Germ Cells and Fertilization*, ed. C. R. Austin, R. V. Short, pp. 128–63. Cambridge: Cambridge Univ. Press

Bedford, J. M., Cooper, G. W. 1978. Membrane fusion events in the fertilization of vertebrate eggs. In *Membrane Fusion*, ed. G. Poste, G. L. Nicholson, pp. 65–125. Amsterdam: Elsevier

Bellvé, A. R. 1979. The molecular biology of mammalian spermatogenesis. *Oxford Rev. Reprod. Biol.* 1: 159–261

Bellvé, A. R., O'Brien, D. A. 1983. The mammalian spermatozoon: Structure and temporal assembly. In *Mechanism and Control of Animal Fertilization*, ed. J. F. Hartmann, pp. 56–137. New York: Academic

Benau, D. A., Storey, B. T. 1987. Characterization of the mouse sperm plasma membrane zona-binding site sensitive to trypsin inhibitors. *Biol. Reprod.* 36: 282–92

Berridge, M. J. 1985. The molecular basis of communication within the cell. *Sci. Am.* 253 (Oct.): 142–152

Birnbaumer, L., Codina, J., Mattera, R., Cerione, R. A., Hildebrandt, J. D., et al. 1984. Regulation of hormone receptors and adenylyl cyclases by guanine nucleotide binding N proteins. *Recent Prog. Horm. Res.* 41: 41–94

Blandau, R. J. 1980. In vitro fertilization and embryo transfer. *Fertil. Steril.* 33: 3–11

Bleil, J. D., Wassarman, P. M. 1980a. Mammalian sperm-egg interaction: Identification of a glycoprotein in mouse egg zonae pellucidae possessing receptor activity for sperm. *Cell* 20: 873–82

Bleil, J. D., Wassarman, P. M. 1980b. Structure and function of the zona pellucida: Identification and characterization of the proteins of the mouse oocyte's zona pellucida. *Dev. Biol.* 76: 185–203

Bleil, J. D., Wassarman, P. M. 1980c. Synthesis of zona pellucida proteins by denuded and follicle-enclosed mouse oocytes during culture in vitro. *Proc. Natl. Acad. Sci. USA* 77: 1029–33

Bleil, J. D., Wassarman, P. M. 1983. Sperm-egg interactions in the mouse: Sequence of events and induction of the acrosome reaction by a zona pellucida glycoprotein. *Dev. Biol.* 95: 317–24

Bleil, J. D., Wassarman, P. M. 1986. Autoradiographic visualization of the mouse egg's sperm receptor bound to sperm. *J. Cell Biol.* 102: 1363–71

Brackett, B. G. 1981. In vitro culture of zygote and embryo. In *Fertilization and Embryonic Development In Vitro*, ed. L. Mastroianni, J. D. Biggers, pp. 63–81. New York: Plenum

Brown, C. R., Jones, R. 1987. Binding of zona pellucida proteins to a boar sperm polypeptide of M_r 53,000 and identification of zona moieties involved. *Development* 99: 333–39

Chang, M. C. 1951. Fertilizing capacity of spermatozoa deposited into the fallopian tube. *Nature* 168: 697–98

Chang, M. C., Pincus, G. 1951. Physiology of fertilization in mammals. *Physiol. Rev.* 31: 1–33

Cherr, G. N., Lambert, H., Meizel, S., Katz, D. F. 1986. In vitro studies of the golden hamster sperm acrosome reaction: Completion on the zona pellucida and induction by homologous soluble zonae pellucidae. *Dev. Biol.* 114: 119–31

Clegg, E. D. 1983. Mechanisms of mammalian sperm capacitation. In *Mechanism and Control of Animal Fertilization*, ed. J. F. Hartmann, pp. 178–212. New York: Academic

Cummins, J. M., Yanagimachi, R. 1982. Sperm-egg ratios and the site of the acrosome reaction during in vivo fertilization in the hamster. *Gamete Res.* 5: 239–56

Cummins, J. M., Yanagimachi, R. 1986. Development of ability to penetrate the cumulus oophorus by hamster spermatozoa capacitated in vitro, in relation to the timing of the acrosome reaction. *Gamets Res.* 15: 187–212

Davis, B. K., Byrne, R., Bedigian, K. 1980. Studies on the mechanism of capacitation: Albumin-mediated changes in plasma membrane lipids during in vitro incubation of rat sperm cells. *Proc. Natl. Acad. Sci. USA* 77: 1546–50

Dunbar, B. S. 1983. Morphological, biochemical, and immunochemical characterization of the mammalian zona pellucida. In *Mechanism and Control of Animal Fertilization*, ed. J. F. Hartmann, pp. 140–75. New York: Academic.

East, I. J., Gulyas, B. J., Dean, J. 1985. Monoclonal antibodies to the murine zona pellucida protein with sperm receptor activity: Effects on fertilization and early development. *Dev. Biol.* 109: 268–73

East, I. J., Mattison, D. R., Dean, J. 1984. Monoclonal antibodies to the major protein of the murine zona pellucida: Effects on fertilization and early development. *Dev. Biol.* 104: 49–56

Edge, A. S. B., Falftynek, C. R., Hof, L., Reichert, L. E., Weber, P. 1981. Deglycosylation of glycoproteins by trifluormethane-sulfonic acid. *Anal. Biochem.* 118: 131–37

Elder, J. H., Alexander, S. 1982. Endo-beta-N-acetyl-glucosaminidase F: Endoglycosidase from *Flavobacterium meningosepticum* that cleaves both high-mannose and complex glycoproteins. *Proc. Natl. Acad. Sci. USA* 79: 4540–44

Endo, Y., Lee, M. A., Kopf, G. S. 1987. Evidence for the role of a guanine nucleotide-binding regulatory protein in the

138   WASSARMAN

zona pellucida-induced mouse sperm acrosome reaction. *Dev. Biol.* 119. 210–16

Farooqui, A. A. 1983. Biochemistry of sperm capacitation. *Int. J. Biochem.* 15: 463–68

Fawcett, D. W. 1975. The mammalian spermatozoon. *Dev. Biol.* 44: 394–436

Fawcett, D. W., Bedford, J. M., eds. 1979. *The Spermatozoon.* Baltimore: Urban & Schwarzenberg

Florman, H. M., Bechtol, K. B., Wassarman, P. M. 1984. Enzymatic dissection of the functions of the mouse egg's receptor for sperm. *Dev. Biol.* 106: 243–55

Florman, H. M., Saling, P. M., Storey, B. T. 1982. Fertilization of mouse eggs in vitro. Time resolution of the reactions preceding penetration of the zona pellucida. *J. Androl.* 3: 373–81

Florman, H. M., Storey, B. T. 1981. Inhibition of in vitro fertilization of mouse eggs: 3-Quinuclidynyl benzilate specifically blocks penetration of zonae pellucidae by mouse spermatozoa. *J. Exp. Zool.* 216: 159–67

Florman, H. M., Storey, B. T. 1982. Mouse gamete interactions: The zona pellucida is the site of the acrosome reaction leading to fertilization in vitro. *Dev. Biol.* 91: 121–30

Florman, H. M., Wassarman, P. M. 1985. O-linked oligosaccharides of mouse egg ZP3 account for its sperm receptor activity. *Cell* 41: 313–24

Friend, D. S., Orci, L., Perrelet, A., Yanagimachi, R. 1977. Membrane particle changes attending the acrosome reaction in guinea pig spermatozoa. *J. Cell Biol.* 74: 561–77

Friend, D. S., Rudolf, I. 1974. Acrosomal disruption in sperm. *J. Cell Biol.* 63: 466–79

Gao, B., Klein, L. E., Britten, R. J., Davidson, E. H. 1986. Sequence of mRNA coding for bindin, a species-specific sea urchin sperm protein required for fertilization. *Proc. Natl. Acad. Sci. USA* 83: 8634–38

Garbers, D. L., Kopf, G. S. 1980. The regulation of spermatozoa by calcium and cyclic nucleotides. *Adv. Cyclic Nucleotide Res.* 13: 251–306

Gilman, A. G. 1984. G proteins and dual control of adenylate cyclase. *Cell* 36: 577–79

Glabe, C. G., Grabel, L. B., Vacquier, V. D., Rosen, S. D: 1982. Carbohydrate specificity of sea urchin sperm bindin: A cell surface lectin mediating sperm-egg adhesion. *J. Cell Biol.* 94: 123–28

Gordon, M., Dandekar, P. V., Bartoszewicz, W. 1974. Ultrastructural localization of surface receptors for concanavalin A on rabbit spermatozoa. *J. Reprod. Fertil.* 36: 211–14

Green, D. P. L. 1978. The mechanism of the acrosome reaction. In *Development in Mammals*, ed. M. H. Johnson, Vol. 3, pp. 65–81. Amsterdam: Elsevier

Greep, R. O., Koblinsky, M. A., Jaffe, F. S. 1976. *Reproduction and Human Welfare: A Challenge to Research.* Cambridge: MIT. 622 pp.

Greve, J. M., Wassarman, P. M. 1985. Mouse egg extracellular coat is a matrix of interconnecting filaments possessing a structural repeat. *J. Mol. Biol.* 181: 253–64

Gulyas, B. J. 1980. Cortical granules of mammalian eggs. *Int. Rev. Cytol.* 63: 357–92

Gulyas, B. J., Schmell, E. D. 1981. Sperm-egg recognition and binding in mammals. In *Bioregulators of Reproduction*, ed. G. Jagiello, H. Vogel, pp. 499–519. New York: Academic

Gwatkin, R. B. L. 1977. *Fertilization Mechanisms in Man and Mammals.* New York: Plenum. 161 pp.

Gwatkin, R. B. L., Williams, D. T. 1977. Receptor activity of the hamster and mouse solubilized zona pellucida before and after the zona reaction. *J. Reprod. Fertil.* 49: 55–59

Gwatkin, R. B. L., Williams, D. T., Hartmann, J. F., Kniazuk, M. 1973. The zona reaction of hamster and mouse eggs. Production in vitro by a trypsin-like protease from cortical granules. *J. Reprod. Fertil.* 32: 259–65

Hartmann, J. F. 1983a. Mammalian fertilization: Gamete surface interactions in vitro. In *Mechanism and Control of Animal Fertilization*, ed. J. F. Hartmann, pp. 325–64. New York: Academic

Hartmann, J. F., ed. 1983b. *Mechanisms and Control in Animal Fertilization.* New York: Academic. 561 pp.

Hartmann, J. F., Gwatkin, R. B., L., Hutchison, C. F. 1972. Early contact interactions between mammalian gametes in vitro: Evidence that the vitellus influences adherence between sperm and the zona pellucida. *Proc. Natl. Acad. Sci. USA* 69: 2767–69

Hartmann, J. F., Hutchison, C. F. 1974. Nature of the prepenetration contact interactions between hamster gametes in vitro. *J. Reprod. Fertil.* 36: 49–57

Hedrick, J. L., ed. 1986. *The Molecular and Cellular Biology of Fertilization.* New York: Plenum.

Heffner, L. J., Saling, P. M., Storey, B. T. 1980. Separation of calcium effects on motility and zona binding ability in mouse spermatozoa. *J. Exp. Zool.* 212: 53–59

Heffner, L. J., Storey, B. T. 1982. Cold

lability of mouse sperm binding to zona pellucida. *J. Exp. Zool.* 219: 155–61

Huang, T. T. F. Jr., Fleming, A. D., Yanagimachi, R. 1981. Only acrosome reacted spermatozoa can bind to and penetrate at the zona pellucida: A study using the guinea pig. *J. Exp. Zool.* 217: 287–90

Huang, T. T. F., Jr., Ohzu, E., Yanagimachi, R. 1982. Evidence suggesting that L-fucose is part of a recognition signal for sperm-zona pellucida attachment in mammals. *Gamete Res.* 5: 355–61

Inoue, M., Wolf, D. P. 1975a. Fertilization-associated changes in the murine zona pellucida: A time sequence study. *Biol. Reprod.* 13: 546–51

Inoue, M., Wolf, D. P. 1975b. Sperm binding characteristics of the murine zona pellucida. *Biol. Reprod.* 13: 340–46

Koehler, J. K. 1976. Changes in antigenic site distribution on rabbit spermatozoa after incubation in capacitating media. *Biol. Reprod.* 15: 444–56

Koehler, J. K. 1978. The mammalian sperm surface: Studies with specific labeling techniques. *Int. Rev. Cytol.* 54: 73–108

Koehler, J. K., Budelman, E. D., Hakamori, S. 1980. A collagen-binding protein on the surface of ejaculated rabbit spermatozoa. *J. Cell Biol.* 86: 529–36

Kopf, G. S., Woolkalis, M. J., Gerton, G. L. 1986. Evidence for a guanine nucleotide-binding regulatory protein in invertebrate and mammalian sperm: Identification by islet-activating protein-catalyzed ADP-ribosylation and immunochemical methods. *J. Biol. Chem.* 261: 7327–31

Langlais, J., Roberts, K. D. 1985. A molecular membrane model of sperm capacitation and the acrosome reaction of mammalian spermatozoa. *Gamete Res.* 12: 183–224

Langlais, J., Zoolinger, M., Plante, L., Chapdelaine, A., Bleau, G., Roberts, K. D. 1981. Localization of cholesteryl sulfate in human spermatozoa in support of a hypothesis for the mechanism of capacitation. *Proc. Natl. Acad. Sci. USA* 78: 7266–70

Leblond, C. P., Clermont, Y. 1952. Spermiogenesis of rat, mouse, hamster, and guinea pig as revealed by the "periodic acid-fuchsin sulfurous acid" technique. *Am. J. Anat.* 90: 167–216

Lopez, L. C.,., Bayna, E. M., Litoff, D., Shaper, N. L., Shaper, J. H., Shur, B. D. 1985. Receptor function of mouse sperm surface galactosyltransferase during fertilization. *J. Cell Biol.* 101: 1501–10

Lopo, A. C., Glabe, C. G., Lennarz, W. J., Vacquier, V. D. 1982. Sperm-egg binding events during sea urchin fertilization. *Ann. NY Acad. Sci.* 383: 405–25

McRorie, R. A., Williams, W. L. 1974. Biochemistry of mammalian fertilization. *Ann. Rev. Biochem.* 43: 777–804

Meizel, S. 1978. The mammalian sperm acrosome reaction: A biochemical approach. In *Development in Mammals*, ed. M. H. Johnson, Vol. 3, pp. 1–62. Amsterdam: Elsevier

Meizel, S. 1984. The importance of hydrolytic enzymes to an exocytotic event, the mammalian sperm acrosome reaction. *Biol. Rev.* 59: 125–57

Metz, C. B., Monroy, A., eds. 1967. *Fertilization: Comparative Morphology, Biochemistry and Immunology*, Vol. 1. New York: Academic

Metz, C. B., Monroy, A., eds. 1969. *Fertilization: Comparative Morphology, Biochemistry, and Immunology*, Vol. 2. New York: Academic

Metz, C. B., Monroy, A., eds. 1985. *Biology of Fertilization*, Vol. 3. New York: Academic

Moore, H. D. M., Bedford, J. M. 1983. The interaction of mammalian gametes in the female. In *Mechanism and Control of Animal Fertilization*, ed. J. F. Hartmann, pp. 453–97. New York: Academic

Morton, D. B. 1977. Immunoenzymic studies on acrosin and hyaluronidase in ram spermatozoa. In *Immunobiology of Gametes*, ed. M. Edidin, M. H. Johnson, pp. 115–55. Cambridge: Cambridge Univ. Press

Myles, D. G., Primakoff, P. 1984. Localized surface antigens of guinea pig sperm migrate to new regions prior to fertilization. *J. Cell Biol.* 99: 1634–41

Nicosia, S. V., Wolf, D. P., Inoue, M. 1977. Cortical granule distribution and cell surface characteristics in mouse ova. *Dev. Biol.* 57. 56–74

Oikawa, T., Yanagimachi, R., Nicolson, G. L. 1973. Wheat germ agglutinin blocks mammalian fertilization. *Nature* 241: 256–59

Oliphant, G. 1976. Removal of sperm-bound seminal plasma components as a prerequisite to induction of the rabbit acrosome reaction. *Fertil. Steril.* 27: 28–38

Oliphant, G., Eng, L. A. 1981. Collection of gametes in laboratory animals and preparation of sperm for in vitro fertilization. In *Fertilization and Embryonic Development In Vitro*, ed. L. Mastroianni, J. D. Biggers, pp. 11–27. New York: Plenum

O'Rand, M. G. 1977. Restriction of a sperm surface antigen's mobility during capacitation. *Dev. Biol.* 55: 260–70

O'Rand, M. G. 1979. Changes in sperm surface properties correlated with capacitation. In *The Spermatozoon*, ed. D. W.

140 WASSARMAN

Fawcett, J. M. Bedford, pp. 195–204. Baltimore: Urban & Schwarzenberg

O'Rand, M. G., Matthews, J. E., Welch, J. E., Fisher, S. J. 1985. Identification of zona binding proteins of rabbit, pig, human, and mouse spermatozoa on nitrocellulose blots. *J. Exp. Zool.* 25: 423–28

Perreault, S., Zanefeld, L., Rogers, B. J. 1979. Inhibition of fertilization by sodium aurothiomalate, a hyaluronidase inhibitor. *J. Reprod. Fertil.* 60: 461–67

Peterson, R. N., Henry, L., Hunt, W., Sazena, N., Russell, L. D. 1985. Further characterization of boar sperm plasma membrane proteins with affinity for the porcine zona pellucida. *Gamete Res.* 12: 91–100

Peterson, R. N., Russell, L., Bundman, D., Freund, M. 1980. Sperm-egg interaction: Direct evidence for boar plasma membrane receptors for porcine zona pellucida. *Science* 207: 73–74

Phillips, D. M., Shalgi, R. M. 1980. Surface properties of the zona pellucida. *J. Exp. Zool.* 213: 1–8

Pierce, M., Turley, E. A., Roth, S. 1980. Cell surface glycosyltransferase activities. *Int. Rev. Cytol.* 65: 1–47

Pikó, L., Tyler, A. 1964. Ultrastructure of the acrosome and the early events of sperm penetration in the rat. *Am. Zool.* 4: 287–96

Ringuette, M. J., Sobieski, D. A., Chamow, S. M., Dean, J. 1986. Oocyte-specific gene expression: Molecular characterization of a cDNA for ZP-3, the sperm receptor of the mouse zona pellucida. *Proc. Natl. Acad. Sci. USA* 83: 4341–45

Rodger, J. C., Young, R. J. 1981. Glycosidase and cumulus dispersal activities of acrosomal extracts from opossum (marsupial) and rabbit (eutherian) spermatozoa. *Gamete Res.* 4: 507–14

Rogers, B. J. 1978. Mammalian sperm, capacitation and fertilization in vitro: A critique of methodology. *Gamete Res.* 1: 165–223

Russell, L., Peterson, R. N., Freund, M. 1979. Direct evidence for formation of hybrid vesicles by fusion of plasma and outer acrosomal membranes during the acrosome reaction in boar spermatozoa. *J. Exp. Zool.* 208: 41–56

Sacco, A. G. 1978. Immunological specificity of anti-zona binding to zona pellucida. *J. Exp. Zool.* 204: 181–86

Sacco, A. G., Yurewicz, E. C., Subramanian, M. G., DeMayo, F. J. 1981. Zona pellucida composition: Species cross-reactivity and contraceptive potential of an antiserum to a purified pig zona antigen (PPZA). *Biol. Reprod.* 25: 997–1008

Saling, P. M. 1981. Involvement of trypsin-like activity in binding of mouse spermatozoa to zonae pellucidae. *Proc. Natl. Acad. Sci. USA* 78: 6231–35

Saling, P. M. 1986. Mouse sperm antigens that participate in fertilization. IV. A monoclonal antibody prevents zona penetration by inhibition of the acrosome reaction. *Dev. Biol.* 117: 511–19

Saling, P. M., Lakoski, K. A. 1985. Mouse sperm antigens that participate in fertilization. II. Inhibition of sperm penetration through the zona pellucida using monoclonal antibodies. *Biol. Reprod.* 33: 527–36

Saling, P. M., Sowinski, J., Storey, B. T. 1979. An ultrastructural study of epididymal mouse spermatozoa binding to zonae pellucidae in vitro. *J. Exp. Zool.* 209: 229–38

Saling, P. M., Storey, B. T. 1979. Mouse gamete interactions during fertilization in vitro. Chlortetracycline as a fluorescent probe for the mouse acrosome reaction. *J. Cell Biol.* 83: 544–55

Saling, P. M., Storey, B. T., Wolf, D. P. 1978. Calcium dependent binding of mouse epididymal spermatozoa to the zona pellucida. *Dev. Biol.* 65: 515–25

Salzmann, G. S., Greve, J. M., Roller, R. J., Wassarman, P. M: 1983. Biosynthesis of the sperm receptor during oogenesis in the mouse. *EMBO J.* 2: 1451–56

Sato, K. 1979. Polyspermy-preventing mechanisms in mouse eggs fertilized in vitro. *J. Exp. Zool.* 210: 353–59

Schachter, H., Roseman, S. 1980. In *The Biochemistry of Glycoproteins and Proteoglycans*, ed. W. J. Lennarz, pp. 85–160. New York: Plenum

Schmell, E., Gulyas, B. J. 1980. Mammalian sperm-egg recognition and binding in vitro. I. Specificity of sperm interactions with live and fixed eggs in homologous and heterologous inseminations of hamster, mouse, and guinea pig oocytes. *Biol. Reprod.* 23: 1075–85

Schuel, H. 1978. Secretory functions of egg cortical granules in fertilization and development: A critical review. *Gamete Res.* 1: 299–382

Schuel, H. 1985. Functions of egg cortical granules. In *Biology of Fertilization*, Vol. 3, ed. C. B. Metz, A. Monroy, pp. 223–51. New York: Academic

Setchell, B. P. 1982. Spermatogenesis and spermatozoa. In *Reproduction in Mammals: 1. Germ Cells and Fertilization*, ed. C. R. Austin, R. V. Short, pp. 63–101. Cambridge: Cambridge Univ. Press

Shalgi, R., Matityahu, A., Nebel, L. 1986. The role of carbohydrates in sperm-egg interactions in rats. *Biol. Reprod.* 34: 446–52

Shapiro, B. M., Shackmann, R. W., Gabel, C. A. 1981. Molecular approaches to the study of fertilization. *Ann. Rev. Biochem.* 50: 815–43

Sharon, N. 1975. *Complex Carbohydrates. Their Chemistry, Biosynthesis, and Functions.* Reading, Mass: Addison-Wesley

Shimizu, S., Tsuji, M., Dean, J. 1983. In vitro biosynthesis of three sulfated glycopoteins of murine zonae pellucidae by oocytes grown in culture. *J. Biol. Chem.* 258: 5858–63

Shur, B. D. 1982. Cell surface glycosyltransferase activities during fertilization and early embryogenesis. In *The Glycoconjugates*, ed. M. Horowitz, Vol. 3, pp. 145–85. New York: Academic

Shur, B. D., Hall, N. G. 1982a. Sperm surface galactosyltransferase activities during in vitro capacitation. *J. Cell Biol.* 95. 567–73

Shur, B. D., Hall, N. G. 1982b. A role for mouse sperm surface galactosyltransferase in sperm binding to the egg zona pellucida. *J. Cell Biol.* 95: 574–79

Stambaugh, R. 1978. Enzymatic and morphological events in mammalian fertilization. *Gamete Res.* 1: 65–85

Storey, B. T., Lee, M. A., Muller, C., Ward, C. R., Wirtshafter, D. F. 1984. Binding of mouse spermatozoa to the zonae pellucidae of mouse eggs in cumulus: Evidence that the acrosomes remain substantially intact. *Biol. Reprod.* 31: 1119–28

Stryer, L., Bourne, H. R. 1986. G poteins: A family of signal transducers. *Ann. Rev. Cell Biol.* 2: 391–419

Suarez, S. D., Katz, D. F., Meizel, S. 1984. Changes in motility that accompany the acrosome reaction in hyperactivated hamster spermatozoa. *Gamete Res.* 10: 253–66

Sullivan, R., Bleau, G. 1985. Interaction of isolated components from mammalian sperm and egg. *Gamete Res.* 12: 101–16

Talbot, P. 1984. Events leading to fertilization. In *Fertility and Sterility*, ed. R. F. Harrison, J. Bonnar, W. Thompson, pp. 121–31. Boston: MIT

Talbot, P. 1985. Sperm penetration through oocyte investments in mammals. *Am. J. Anat.* 174: 331–46

Talbot, P., DiCarlantonio, G. 1984a. The architecture of the hamster oocyte-cumulus complex. *Gamete Res.* 9: 261–72

Talbot, P., DiCarlantonio, G. 1984b. Ultrastructure of opposum oocyte investing coats and their sensitivity to trypsin and hyaluronidase. *Dev. Biol.* 103: 159–67

Talbot, P., DiCarlantonio, G., Zao, P., Penkala, J., Haimo, L. T. 1985. Motile cells lacking hyaluronidase can penetrate the hamster oocyte cumulus complex. *Dev. Biol.* 108: 387–98

Talbot, P., Franklin, L. E. 1974a. The release of hyaluronidase from guinea pig sperm during the course of the normal acrosome reaction in vitro. *J. Reprod. Fertil.* 39: 429–32

Talbot, P., Franklin, L. E. 1974b. Hamster sperm hyaluronidase. II. Its release from sperm in vitro in relation to the degenerative and normal acrosome reaction. *J. Exp. Zool.* 189: 321–32

Trimmer, J. S., Vacquier, V. D. 1986. Activation of sea urchin gametes. *Ann. Rev. Cell Biol.* 2: 1–26

Tsunoda, Y., Soma, T., Sugie, T. 1981. Inhibition of fertilization in cattle by passive immunization with anti-zona pellucida serum. *Gamete Res.* 4: 133–38

Vacquier, V. D., Moy, G. W. 1977. Isolation of bindin: The protein responsible for adhesion of sperm to sea urchin eggs. *Proc. Natl. Acad. Sci. USA* 74: 2456–60

Ward, C. R., Storey, B. T. 1984. Determination of the time course of capacitation in mouse spermatozoa using a chlortetracycline fluorescence assay. *Dev. Biol.* 104: 287–96

Wassarman, P. M. 1983. Fertilization. In *Cellular Interactions and Development: Molecular Mechanisms*, ed. K. M. Yamada, pp. 1–28. New York: Wiley-Interscience

Wassarman, P. M. 1987a. The biology and chemistry of fertilization. *Science* 235: 553–60

Wassarman, P. M. 1987b. The zona pellucida: A coat of many colors. *BioEssays* 6: 161–66

Wassarman, P. M., Bleil, J. D. 1982. The role of zona pellucida glycoproteins as regulators of sperm-egg interaction in the mouse. In *Cellular Recognition*, ed. W. A. Frazer, L. Glaser, D. Gottlieb, pp. 845–63. New York: Liss

Wassarman, P. M., Bleil, J. D., Fimiani, C. M., Florman, H. M., Greve, J. M., et al. 1987. Receptor mediated binding and membrane fusion during sperm-egg interaction in mice. In *Molecular Approaches to Developmental Biology*, ed. R. A. Firtel, E. H. Davidson, pp. 51–65. New York: Liss

Wassarman, P. M., Bleil, J. D., Florman, H. M., Greve, J. M., Roller, R. J., et al. 1985a. The mouse egg's receptor for sperm: What is it and how does it work? *Cold Spring Harbor Symp. Quant. Biol.* 50: 11–19

Wassarman, P. M., Bleil, J. D., Florman, H. M., Greve, J. M., Roller, R. J., Salzmann, G. S. 1986a. The mouse egg's extracellular coat: Synthesis, structure, and function. In

Gametogenesis and the Early Embryo, ed. J. Gall, pp. 371–88. New York: Liss

Wassarman, P. M., Bleil, J. D., Florman, H. M., Greve, J. M., Roller, R. J., Salzmann, G. S. 1986b. Nature of the mouse egg's receptor for sperm. In *The Molecular and Cellular Biology of Fertilization*, ed. J. L. Hedrick, pp. 55–77. New York: Plenum

Wassarman, P. M., Florman, H. M., Greve, J. M. 1985b. Receptor mediated sperm-egg interactions in mammals. In *Biology of Fertilization*, ed. C. B. Metz, A. Monroy, 2: 341–60. New York: Academic

Wolf, D. P., Hamada, M. 1977. Induction of zonal and oolemmal blocks to sperm penetration and with cortical granule exudate. *Biol. Reprod.* 21: 205–11

Wolf, D. P., Hamada, M., Inoue, M. 1977. Kinetics of sperm penetration into and the zona reaction of mouse eggs inseminated in vitro. *J. Exp. Zool.* 201: 29–36

Wolf, D. P., Inoue, M. 1976. Sperm concentration dependence in the fertilization and zonae sperm binding properties of mouse eggs inseminated in vitro. *J. Exp. Zool.* 196: 27–37

Yanagimachi, R. 1970. The movement of golden hamster spermatozoa before and after capacitation. *J. Reprod. Fertil.* 23: 193–96

Yanagimachi, R. 1977. Specificity of sperm-egg interaction. In *Immunobiology of Gametes*, ed. M. Edidin, M. Johnson, pp. 225–95. Cambridge: Cambridge Univ. Press

Yanagimachi, R. 1978. Sperm-egg associ-ation in mammals. *Curr. Top. Dev. Biol.* 12: 83–105

Yanagimachi, R. 1981. Mechanisms of fertilization in mammals. In *Fertilization and Embryonic Development In Vitro*, ed. L. Mastroianni, J. D. Biggers, pp. 81–182. New York: Plenum

Yanagimachi, R. 1984. Zona-free hamster eggs: Their use in assessing fertilizing capacity and examining chromosmes of human spermatozoa. *Gamete Res.* 10: 187–232

Yanagimachi, R., Noda, Y. D. 1970a. Ultra-structural changes in the hamster sperm head during fertilization. *J. Ultrastruct. Res.* 31: 465–85

Yanagimachi, R., Noda, Y. D. 1970b. Physiological changes in the post-nuclear cap region of mammalian spermatozoa: A necessary preliminary to the membrane fusion between sperm and egg cells. *J. Ultrastruct. Res.* 31: 486–93

Yanagimachi, R., Phillips, D. M. 1984. The status of acrosomal caps of hamster spermatozoa immediately before fertilization in vivo. *Gamete Res.* 9: 1–19

Yanagimachi, R., Usui, N. 1974. Calcium dependence of the acrosome reaction and activation of guinea pig spermatozoa. *Exp. Cell Res.* 89: 161–74

Zao, P., Meizel, S., Talbot, P. 1985. The release of hyaluronidase and *N*-acetylhexosaminidase from hamster sperm during in vitro incubation. *J. Exp. Zool.* 234: 63–74

Ann. Rev. Cell Biol. 1987. 3 : 143–78

MOLECULAR ASPECTS OF B-LYMPHOCYTE ACTIVATION

Anthony L. DeFranco

Department of Microbiology and Immunology and Department of Biochemistry, University of California, San Francisco, California 94143-0552

CONTENTS

INTRODUCTION.. 143

ROLE OF ANTIGEN RECEPTORS IN B-CELL ACTIVATION.................................... 145
    Transmembrane Signaling by B-Cell Antigen Receptors 146
    Interaction Between Fc Receptors and Antigen Receptors........................... 149
    Antigen Presentation by B Cells to Helper T Cells...................................... 151
    Antigen Uptake via Membrane Immunoglobulin 153
    Regulation of Antigen Presentation Efficiency 155

MECHANISMS OF ACTIVATION OF B CELLS BY HELPER T CELLS......................... 156
    Two Helper T Cell–Dependent Pathways of B-Cell Activation...................... 156
    Antigen Receptor–Independent B-Cell Activation....................................... 160

PRODUCTION OF ANTIBODIES IN RESPONSE TO BACTERIAL ANTIGENS.............. 161
    Polyclonal B-Cell Activators... 162
    Molecular Diversity of Polyclonal B-Cell Activators.................................... 163
    Additional Factors Promoting B-Cell Activation in Response to Bacteria...... 164
    Signal Transduction by the Putative LPS Receptor 165

ROLE OF c-*myc* PROTO-ONCOGENE IN B-CELL ACTIVATION 168

SUMMARY... 169

INTRODUCTION

Prior to the onset of an immune response, B lymphocytes are quiescent nondividing cells that circulate in the blood and reside in the secondary lymphoid organs. Upon the introduction of a foreign molecule that can elicit antibody production (an antigen), B cells capable of producing antibodies against this antigen become activated to proliferate, expanding the number of such cells, and to differentiate into "plasma" cells that secrete large amounts of antibody. These activation steps are under complicated

143

0743–4634/87/1115–0143$02.00

regulatory control from antigen, helper T lymphocytes, macrophages, complement components, and antibodies.

One key to understanding B-cell activation is the realization that the process can proceed by several distinct routes that involve different regulatory signals. One of the first suggestions of this diversity of activation mechanisms was the observation that different antigens require the presence of different elements to elicit antibody responses. For example, soluble protein antigens and red blood cell antigens do not elicit antibody production in animals depleted of T lymphocytes (e.g. athymic "nude" mice) and hence are referred to as T-dependent antigens. In contrast, antigens derived from certain bacterial cell wall components can induce strong antibody responses without the help of T cells. Such antigens contain moieties called polyclonal B-cell activators, which are capable of activating B cells without regard to antigen specificity. Antigens without this property but composed of regular repeating structures, such as polysaccharides, also are often able to elicit antibody responses without the aid of T cells.

The T-independent antigens can be subdivided according to their ability to induce antibodies in mice expressing a gene called *xid* (for X-linked immunodeficiency), which causes a defect in B cells (Scher 1982). In general, these mice respond to antigens derived from bacterial cell wall components that have polyclonal B-cell activator properties (type 1) but not to antigens with repeating structure (type 2) (Amsbaugh et al 1972; Mosier et al 1976). This latter type of antigen requires the presence of a small number of helper T cells to elicit responses in vitro, which suggests they may not be completely "T-independent" (Mond et al 1980). Similarly, T-dependent antigens can be divided into two types. For some antigens (e.g. red blood cell antigens), the helper T cells can be replaced in vitro by supernatants obtained from activated helper T cells (Dutton et al 1971). However, most soluble protein antigens require T cells to activate B cells.

The distinct requirements for B-cell activation by different antigens are not the result of different inherent properties of B cells with different specificities. Different requirements are observed when each type of antigen is derivatized with a small "hapten" such as trinitrophenyl (TNP) and the response analyzed by measuring anti-hapten antibodies (Table 1). Rather, some scaffolds or "carriers" to which haptens may be attached appear to be better or worse than others at promoting B-cell activation by the alternative pathways. The diversity in B-cell activation schemes presumably results from evolutionary pressure to produce antibodies to a wide variety of molecules, many of which are inherently better suited to one activation scheme than to another.

Table 1   Different types of antigens[a]

Type of antigen	Example	Property of carrier	Comments
T-independent type 1	TNP-LPS	polyclonal B-cell activator	response occurs in nude mice and in *xid* mice; probably requires growth factor signal from macrophage, C3, or B cell (autocrine)
T-independent type 2	TNP-Ficoll	repeating structure	response occurs in nude mice not in *xid* mice; in vitro response requires small number of T cells
T-dependent	TNP-SRBC	red blood cell based	T-cell requirement can be replaced by supernatants of activated T cells
T-dependent	TNP-ovalbumin	soluble protein	most stringent requirement for helper T cells, generally not replaceable by cell-free supernatants

[a] In all cases the responding B cells have the same specificity (anti-TNP).
Abbreviations: Trinitrophenyl hapten, TNP; bacterial lipopolysaccharide, LPS; sheep red blood cells, SRBC.

ROLE OF ANTIGEN RECEPTORS IN B-CELL ACTIVATION

In his theory of clonal selection, Burnet (1959) originally proposed that B lymphocytes would have on their surface a form of antibody that functions as an antigen receptor and that an antigen would act through its corresponding antigen receptor to select the appropriate cells for antibody production. It is now clear that antibody molecules can be made with one of two different carboxyl termini, one leading to secretion and the other leading to localization in the plasma membrane (Kehry et al 1980; Rogers et al 1980). Antibody of the latter type is referred to as membrane immunoglobulin (mIg); each of the antibody classes can be produced in a membrane form (Wall & Kuehl 1983). B cells that have not yet gone through an immune response express two forms of membrane immunoglobulins, mIgM and mIgD. The clonal selection theory requires that membrane immunoglobulins play critical roles at the onset of the antibody response. Indeed, these antigen receptors contribute to B-cell activation in two ways: by transducing a signal that informs the cell of the presence of the antigen and by concentrating the antigen on its specific B cell. Antigen binding may lead to the delivery of other activation signals coming, for example, from helper T cells or from the polyclonal activator moiety of type 1 T-independent antigens.

Transmembrane Signaling by B-Cell Antigen Receptors

Antibodies directed against membrane immunoglobulins are potent polyclonal activators of B cells; presumably they induce transmembrane signal transduction by mIg on all B cells simultaneously (Moller 1980). All resting, mature, splenic B cells enter the G_1 phase in response to anti-IgM or anti-IgD (DeFranco et al 1982a,b). Other early cellular responses observed in response to anti-IgM include increased expression of class II major histocompatibility complex (MHC) proteins (Mond et al 1981), a slow depolarization that peaks ~ 1 hr after anti-IgM binding (Monroe & Cambier 1983), and cell enlargement (DeFranco et al 1982a). A sizeable fraction of resting B cells proliferate when incubated at moderate cell density with a high concentration of anti-IgM (DeFranco et al 1982a). This proliferative response is stimulated considerably by the presence of several different B-cell growth factors (BCGFs) derived from helper T cells (Kishimoto & Ishizaka 1975; Howard et al 1982; Okada et al 1983; Muraguchi & Fauci 1982; Maizel et al 1982), the best characterized of which is now called interleukin 4 (IL-4, previously called B-cell stimulatory factor 1). Most of these B-cell growth factors do not stimulate B-cell proliferation in the absence of anti-IgM antibodies, which demonstrates a requirement for activation signals emanating from membrane IgM. Thus the B-cell antigen receptor appears capable of generating transmembrane signals that can promote a number of early activation events.

Recent work from several laboratories has shown that cross-linkage of mIgM with anti-IgM antibodies causes breakdown of phosphatidylinositol 4,5-bisphosphate (PIP_2) into the second messengers diacylglycerol, which activates protein kinase C (Nishizuka 1984), and inositol 1,4,5-trisphosphate [IP_3], which can cause an elevation of free calcium in the cytoplasm (Streb et al 1983; Berridge & Irvine 1984). The first evidence that mIgM could regulate phosphoinositide hydrolysis came from Maino et al (1975), who observed that anti-IgM caused enhanced labeling of phosphatidylinositol, an event now thought to reflect resynthesis of the phospholipid broken down by the initial signaling reaction. More recently, Coggeshall & Cambier (1984) observed that phosphatidic acid, a major metabolite of diacylglycerol, also increases rapidly in response to anti-IgM, which is consistent with the idea that diacylglycerol is being generated. Also consistent with this hypothesis are observations indicating that anti-IgM activates protein kinase C. Anti-IgM stimulation induces translocation of protein kinase C from the cytoplasmic fraction to the membrane fraction

(Chen et al 1986; Nel et al 1986), an event believed to be triggered by elevation of plasma membrane diacylglycerol. Also, phorbol esters, which are direct activators of protein kinase C, and anti-IgM can induce the phosphorylation of many of the same proteins in B cells (Hornbeck & Paul 1986). Recently, Bijsterbosch et al (1985) directly showed that PIP_2, rather than some other phospholipid, was broken down in response to anti-IgM by observing the rapid generation of IP_3 and diacylglycerol. Other classes of mIg (e.g. mIgD and mIgG) can also activate phosphoinositide breakdown (Mizuguchi et al 1986a). Thus, B-cell antigen receptors are members of the class of eukaryotic cell surface receptors that transduce signals via phosphoinositide breakdown.

Phosphoinositide breakdown is generally associated with elevation of the concentration of calcium in the cytoplasm (Michell 1975), and this is the case in B cells. Using $^{45}Ca^{2+}$, Braun et al (1979) observed mobilization of intracellular calcium in B cells upon treatment with anti-IgM, which led them to propose that cytoplasmic calcium concentration became elevated in the process. This was shown to be correct by Pozzan et al (1982), who used the fluorescent dye Quin2 to directly measure cytoplasmic calcium levels in B cells. The suggestion from the work of Braun et al (1979) that calcium is released from internal stores has received considerable support from the observation that cytoplasmic calcium concentration still increases if anti-IgM is added in the absence of extracellular calcium (Pozzan et al 1982; LaBaer et al 1986). In a number of different cell types IP_3, generated by breakdown of PIP_2, has been shown to cause release of calcium from internal stores, which are probably associated with the endoplasmic reticulum (Berridge & Irvine 1984). This has been directly demonstrated in B cells as well (Ransom et al 1986; LaBaer et al 1986). These results, combined with the fact that IP_3 levels rise in B cells stimulated with anti-IgM (Bijsterbosch et al 1985; Ransom et al 1986; Fahey & DeFranco 1987), support the idea that elevation of $[Ca^{2+}]$ is a consequence of phosphoinositide breakdown. The other well-characterized mechanism by which extracellular ligands cause elevation of cytoplasmic calcium concentration involves the depolarization-activated calcium channel of neurons and cardiac muscle. This has been ruled out as the mechanism responsible for elevating calcium concentration in response to anti-IgM in the B-cell lymphoma WEHI-231 (LaBaer et al 1986).

Elevation of cytoplasmic $[Ca^{2+}]$ may not be the sole mode of regulating cellular activity by the inositol phosphate side of the phosphoinositide signaling pathway. Recently it has been shown that $I(1,4,5)P_3$ can be phosphorylated to yield inositol 1,3,4,5-tetrakisphosphate (IP_4) (Irvine et al 1986; Stewart et al 1986). Subsequent removal of a phosphate by the 5-phosphatase, probably the most active enzyme acting on the inositol polyphosphates, is believed to generate the $I(1,3,4)P_3$ isomer. Concentrations of all three of these inositol polyphosphates, $I(1,4,5)P_3$, IP_4, and $I(1,3,4)P_3$, have been shown to rise rapidly in a B-cell line treated with anti-IgM (Fahey & DeFranco 1987). There is now evidence that IP_4 plays a discrete second-messenger role in sea urchin oocytes (Irvine & Moor 1986), so the inositol polyphosphates may turn out to have important biological effects in B cells in addition to causing cytoplasmic $[Ca^{2+}]$ elevation.

Thus three potential classes of intracellular regulators are generated by phosphoinositide breakdown in B cells (and many other cell types): diacylglycerol, inositol polyphosphates, and calcium. Two of these signals can be invoked artificially: $[Ca^{2+}]$ can be elevated with calcium ionophores, and protein kinase C, the only known target of diacylglycerol, can be potently activated by the tumor-promoting phorbol esters. If the effects of anti-IgM on B-cell activation are due to phosphoinositide breakdown, then the addition of phorbol diesters and calcium ionophores might also promote early activation events. Indeed, Monroe & Kass (1985) recently showed that these agents can drive resting B cells into the G_1 phase of the cell cycle. Similarly, Ransom & Cambier (1986) showed that these agents can promote class II MHC gene expression and cell depolarization. Thus it appears that many of the early events triggered by anti-IgM can also be triggered by elevation of cytoplasmic $[Ca^{2+}]$ and activation of protein kinase C. The implication is that these signaling events are at least partly responsible for the effects of anti-IgM on B cells.

Transmembrane signaling by membrane immunoglobulin has been studied to date mainly by stimulating the receptor with anti-immunoglobulin antibodies rather than with authentic antigens because cells specific for a given antigen are a small proportion of all B cells and hence signaling in response to antigen would be undetectable. However, since it is possible to purify antigen-specific B cells, it should be possible to measure signaling reactions in these cells by at least some of the available methods. Elevation of cytoplasmic $[Ca^{2+}]$ (Wilson et al 1987) and increased turnover of phosphatidylinositol (Grupp et al 1987) have been observed upon addition of hapten-carrier conjugates to hapten-specific B cells. Careful measurements of the magnitude and duration of signaling events in response to each of the major types of antigens (T-independent types 1 and 2; soluble protein T-dependent; and red blood cell T-dependent) should be made to see if there is a correlation between an antigen's ability to cause antigen receptor signaling and its ability to elicit antibody secretion under various circumstances. For example, T-independent type 2 antigens may have a relatively small requirement for helper T cells because they are more potent or more persistent in inducing antigen receptor signaling. Similarly, the antigens requiring the most help from T cells, the soluble protein-based antigens, may be intrinsically less efficient or less persistent at triggering antigen receptor signaling than are other types of antigens. In support of this idea, Stein et al (1986) recently demonstrated that purified TNP-specific B cells proliferate in response to TNP-Ficoll (a T-independent type 2 antigen) plus IL-4 but not in response to the soluble protein antigen TNP-ovalbumin plus IL-4. IL-4 was originally identified as a B-cell growth factor by virtue of its ability to act synergistically with anti-IgM to induce proliferation of resting B cells (Howard et al 1982). These results argue that TNP-ovalbumin is not as effective as TNP-Ficoll or anti-IgM at inducing signal transduction by membrane IgM.

The mechanism of action of IL-4 is poorly understood at this time. IL-4 does not induce phosphoinositide breakdown on its own, and it does not enhance the breakdown caused by anti-IgM, which suggests that it must generate a different signal (Mizuguchi et al 1986b; Justement et al 1986). Indeed, IL-4 has been found to induce phosphorylation of several B-cell proteins (Justement et al 1986).

It would be interesting to examine the transmembrane signaling efficacy of the T-independent type 2 antigens of carefully defined size studied by Dintzis and collaborators (1976, 1982) and their ability to act synergistically with IL-4 to induce B-cell proliferation. These antigens only induce antibody responses if they contain a critical number (~ 12–16) of appropriately spaced antigenic determinants. This finding led to the "immunon" hypothesis, which holds that a critical number of membrane IgM or IgD molecules must be clustered together by contact with one antigen molecule to activate a B cell. The simplest extension of this model would be that clustering of this many membrane Ig molecules leads to more signaling per membrane Ig molecule than does clustering of fewer mIg molecules and that a signal threshold must be reached to elicit a response in the T-independent type 2 mode.

Interaction Between Fc Receptors and Antigen Receptors

In addition to binding free or cell-bound antigen, B-cell membrane immunoglobulin can bind to immune complexes containing the appropriate antigen or to anti-idiotypic antibodies (e.g. antibodies that recognize the variable parts of a particular antibody molecule). The latter two types of ligand differ from the free or cell-bound antigens in that they bind not only to the antigen receptor but also to the Fc_γ receptor. Fc receptors are cell surface molecules that bind to the constant domains of the various heavy chains of antibodies, and the Fc_γ receptor is specific for the constant domains of IgG molecules.

Recent studies comparing the effects of anti-IgM antibodies, which bind to the B-cell Fc_γ receptor, with those of $F(ab')_2$ derivatives of these antibodies, which do not, provide a model for probing the consequences of Fc_γ receptor binding on transmembrane signaling. Anti-IgMs that bind to the Fc_γ receptor are poor activators of B cells compared to anti-IgMs that do not (Sidman & Unanue 1976; Phillips & Parker 1983; Klaus et al 1984). This effect appears to require the ligand to bring about the physical interaction of mIgM and the Fc_γ receptor, since the biological response to anti-IgM antibodies that do not bind to the Fc_γ receptor is not affected by addition of heat-aggregated IgG, which binds to Fc_γ receptors, but would not bring them into complexes with mIgM (Phillips & Parker 1984). Bijsterbosch & Klaus (1985) examined the effect of Fc_γ receptor interaction on the ability of anti-IgM to induce the phosphoinositide signaling reactions. They found that IP_3 production was very similar in response to either $F(ab')_2$ anti-IgM or intact anti-IgM for the first 30 sec. After 30 sec, however, IP_3 and total inositol phosphate production was dramatically less with intact anti-IgM, such that the rate of signaling in response to intact anti-IgM was only about 20% of that in response to $F(ab')_2$ anti-IgM. Thus combined interaction of the ligand with mIgM and Fc_γ was much less effective than interaction with only mIgM. Interestingly, there was very little difference in the concomitant increase of intracellular calcium concentration. This may mean that the amount of IP_3 generated is five times that needed to fully release intracellular stores of calcium.

It is not clear whether the decreased activation of B cells seen when the anti-IgM also binds to the Fc_γ receptor is due to the attenuation of the phosphoinositide signaling or whether it is due to additional effects of Fc_γ receptor signaling. Interestingly, Bijsterbosch & Klaus (1985) mention that they were unable to correct the poor proliferative response of B cells to intact anti-IgM by addition of phorbol esters. As the calcium signal was unaffected by the Fc_γ receptor interaction, and as phorbol esters would restore activation of protein kinase C, this result implies either that the Fc_γ receptor has additional important effects or that IP_3 and/or IP_4 are needed for B-cell proliferation in response to $F(ab')_2$ anti-IgM. One useful experiment would be to compare the effects on B-cell activation of calcium ionophores and phorbol esters (Monroe & Kass 1985; Ransom & Cambier 1986) with those of intact anti-IgM and phorbol esters. The results should indicate whether Fc_γ receptor signaling inhibits the effects of calcium elevation and protein kinase C activation. It should be noted that calcium ionophores A23187 or ionomycin and phorbol esters cause a number of early B-cell activation events but do not cause vigorous proliferation (Monroe & Kass 1985).

Although there are still issues that need resolution, such as how Fc_γ receptor interaction influences B-cell activation in response to helper T cells, it appears that antibody-antigen complexes may be poorer activators of B cells than are free antigens. This property may help terminate an immune response once sufficient antibody is made to complex most of the antigen.

Antigen Presentation by B Cells to Helper T Cells

It was recognized over a decade ago that to achieve a full-magnitude secondary response to a T-dependent antigen both the B-cell and the T-cell population must have been immunized against portions of the same antigen and that these antigen portions must be linked to each other during the second challenge. Introduction of the T-cell antigen and the B-cell antigen on separate molecules is insufficient to generate a full response (Mitchison 1971). These observations strongly argue that in a T-dependent antibody response the antigen-specific B cell binds antigen via its antigen receptor and then "presents" this antigen, together with the appropriate histocompatibility molecule, to the antigen-specific helper T cell. This presentation leads to activation of the helper T cell and consequent delivery of activation signals to the B cell. If the antigen molecule is split into two parts, such that the B cell recognizes one part and the T cell the other, then the antigen-specific B cell does not bind the T-cell antigen, does not present it to the helper T cell, and hence the response does not occur. Recently, this elegant explanation has received impressive experimental support. It is now clear that normal B cells are quite capable of presenting antigen to helper T lymphocytes (Chesnut et al 1982; Ashwell et al 1984). Furthermore, although B cells, like macrophages, will present soluble antigens that are simply incubated with them, presentation is much more efficient (10^3–10^4 times less antigen required) if the T-cell antigen can bind to the membrane immunoglobulin of the B cells, thereby focusing and concentrating the antigen (Kakiuchi et al 1983; Tony & Parker 1985; Rock et al 1984; Lanzavecchia 1985). This gain in efficiency is observed regardless of whether the antigen binds directly to the antigen-combining site on the membrane immunoglobulin (Rock et al 1984; Malynn et al 1985; Lanzavecchia 1985) or is bound experimentally by use of an anti-immunoglobulin antibody (Kakiuchi et al 1983; Tony & Parker 1985). Antigen presentation to T cells by this route appears to obey the rules of antigen processing previously described for macrophages (Ziegler & Unanue 1982; Casten et al 1985; Lanzavecchia 1985). This finding probably reflects the fact that most if not all T cells recognize fragments of proteins rather than intact proteins in their native conformation (Goodman & Sercarz 1983).

It has recently been possible to demonstrate direct binding between the oligopeptides that are recognized by T cells and the proteins encoded in the major histocompatibility complex (MHC) that are required for T-cell recognition (Babbitt et al 1985). The complexes formed by these interactions are very potent activators of T cells; their potency is presumed to reflect their biological importance (Buus et al 1986). Work with a peptide from lysozyme has demonstrated that the corresponding complex has an extremely slow rate of dissociation ($k_d = 3 \times 10^{-6}$ s^{-1}), only a modest affinity constant ($K_d = 2 \times 10^{-6}$ M) (Buus et al 1986), and thus a very slow association rate. The association and dissociation rates for this peptide are enhanced approximately tenfold by the low pH characteristic of a number of intracellular compartments. Thus perhaps antigen fragments bind to class II MHC molecules in a low-pH intracellular compartment following antigen fragmentation after endocytosis of the membrane immunoglobulin–antigen complex. The low pH of such compartments would speed up the association rate and provide a location where the concentration of the antigen fragment might be high enough for binding to class II MHC molecules to occur. A low pH could also facilitate dissociation of antigen from membrane Ig. Once the complex of class II MHC and antigen fragment reached the cell surface, the neutral pH would slow the dissociation rate, promoting the stability of the stimulatory complex. Indeed, agents that raise the pH of internal compartments interfere with antigen processing and/or presentation by macrophages (Ziegler & Unanue 1982) and B cells (Lanzavecchia 1985; Casten et al 1985). Furthermore, immunofluorescence suggests that capped and endocytosed membrane immunoglobulin reaches a compartment rich in class II MHC molecules (Pletscher & Pernis 1983). This conclusion is supported by recent biochemical evidence that during biosynthesis class II MHC molecules pass through a compartment accessible to recycling transferrin receptors (Cresswell 1985).

Although this model, illustrated in Figure 1, is attractive, many details are still lacking. It is unclear where within the cell fragmentation of the antigen occurs; whether membrane immunoglobulin releases the antigen before its fragmentation; and if so, how that is accomplished; and whether the pH in the relevant internal compartments is indeed acidic.

An equally intriguing issue is how class II MHC molecules recognize the peptide fragments they bind to. A number of peptides from unrelated proteins can be demonstrated to bind to a given class II MHC molecule in apparently identical ways, without having any obvious consensus amino acid sequence (Guillet et al 1986). These protein-protein interactions may be analogous to those involved in the recognition of signal sequences by signal recognition particle. In that case also, strict specificity is not apparent in the primary structure of the signal sequences.

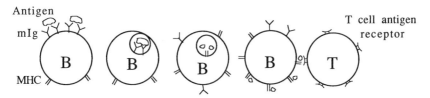

1.Binding 2.Internalization 3.Processing 4.Surface expression
of fragment + MHC

Figure 1 Mechanisms of helper T-cell/B-cell cooperation: antigen presentation by antigen-specific B cells. Antigen binds to membrane forms of immunoglobulin (mIg). Following antigen binding, antigen-mIg complexes are internalized. This is followed by degradation of the antigen into small fragments (e.g. proteolysis of a protein antigen). Small fragments capable of stimulating helper T cells then bind to class II major histocompatibility complex (MHC) proteins. The site of this interaction is unclear, but it may occur in a low pH endosome (see text). On the surface of the B cell, this complex has a very slow rate of dissociation. Helper T cells specific for the antigen fragment in question now interact with the B cell, are activated in response to recognition of antigen-MHC, and release B-cell growth and differentiation factors, which promote B-cell activation. This release of soluble mediators may be directed toward the antigen-presenting B cell. The antigen shown is bivalent, but monovalent antigens seem to proceed through this pathway as efficiently as do bivalent ones.

Antigen Uptake via Membrane Immunoglobulin

Perhaps better understood is the process by which membrane Ig is internalized after binding its cognate antigen. When membrane IgM is incubated with anti-immunoglobulin (used as a surrogate for a multivalent antigen), there is rapid formation of a "patch," an antibody-antigen complex between neighboring IgM molecules embedded in the plasma membrane and anti-immunoglobulins bound to them. Patching is a passive, energy-independent process. It is followed by an energy- and microfilament-dependent movement of these patches to one pole of the cell ("capping") (Taylor et al 1971; Schreiner & Unanue 1976). Capping of mIg has also been observed to occur in TNP-specific B cells incubated with TNP-KLH, a T-dependent antigen (Snow et al 1983). Currently, capping of membrane Ig is thought to occur by a mechanism similar to that involved in capping of a number of other cell surface proteins in lymphocytes (for a recent review, see Bourguignon & Bourguignon 1984).

The trigger for capping is not well understood; however, anti-IgM induces a rapid association of membrane IgM with the cytoskeleton, as demonstrated by the criterion of its association with the Triton X-100 insoluble cell fraction (Braun et al 1982). It is unclear whether the same events that lead to activation of the phosphoinositide signaling system also lead to IgM attachment to the cytoskeleton. Alternatively, patching of membrane IgMs may trap within the aggregate a still unidentified protein that mediates their attachment to the cytoskeleton (Bourguignon & Bourguignon 1984). A number of components of the cytoskeleton have been found to co-localize with mIg during the capping process, including actin (Gabbiani et al 1977), myosin (Schreiner et al 1977), α-actinin (Hoessli et al 1980), tubulin (Gabbiani et al 1977), and a spectrinlike protein that may be identical to fodrin (Levin & Willard 1983; Nelson et al 1983). The spectrinlike protein, possibly together with an ankyrinlike protein (Bourguignon et al 1986), may serve as the link between integral membrane proteins and the actin of microfilaments.

Another possible link between intracellular actin and mIgM is a 56-kilodalton (kDa) protein that binds to cross-linked mIg and to actin (Rosenspire & Choi 1982; Petrini et al 1983). This protein has been identified as group-specific component (Gc), a vitamin D_3 binding protein from serum (Petrini et al 1983). In B cells, the Gc protein is very tightly attached to membranes, from which it is released by detergents but not by 3-M KCl or by ammonium thiocyanate, which suggests that it is an integral membrane protein (Petrini et al 1983, 1984). However, purified serum Gc protein has been shown to bind to membrane Ig and to actin. Yet it is unclear whether Gc protein actually binds intracellular actin, as would be required for Gc to link mIg to microfilaments. At least some of the actin bound to Ig-Gc complexes comes from outside the cell, since it can be iodinated by surface labeling methods (Petrini et al 1983). As actin has been detected on the surface of B lymphocytes (Owen et al 1978), perhaps Gc protein binds actin on the outside of the membrane rather than on the inside, in which case it would not be the link to the cytoskeleton. The finding that Gc protein facilitates the depolymerization of F-actin (Korn 1982) is further evidence against a role for Gc in attachment of membrane Ig to microfilaments.

The calcium-regulatory protein calmodulin also co-localizes with membrane Ig caps (Salisbury et al 1981; Nelson et al 1982), which suggests that capping may be controlled in some way by elevation of cytoplasmic $[Ca^{2+}]$, an event triggered by anti-Ig, as described above. Indeed, several calmodulin inhibitors were found to block capping with a dose response in line with their inhibition of calmodulin activity in vitro (Nelson et al 1982). However, by depleting intracellular and extracellular Ca^{2+}, Pozzan et al (1982) were able to prevent the rise in intracellular $[Ca^{2+}]$ normally caused by anti-IgM without affecting capping detectably. This finding argues strongly against an obligatory role for calcium in capping. Interestingly, a moderate dose of trifluoperazine (a calmodulin inhibitor) inhibits endocytosis of clustered anti-IgM/IgM complexes but does not inhibit capping in the B-lymphoblastoid cell line WiL2 (Salisbury et al 1980). This result, coupled with careful electron microscopic analysis of capping, recruitment into coated pits, and endocytosis, led to the conclusion that endocytosis of cross-linked membrane Ig via clathrin-coated pits occurs independently of capping. Isolation of mIgM–containing coated vesicles revealed the presence of calmodulin as well as clathrin. Thus the calmodulin observed under caps may be involved in triggering or promoting endocytosis rather than in capping per se. This issue should be examined by the calcium-depletion protocols previously used by Pozzan et al (1982). If calcium is required for endocytosis, it will be interesting to see whether it comes from internal stores or from outside the cell.

One straightforward model of antigen uptake is as follows: A multivalent antigen binds to and cross-links several mIgM or mIgD molecules. This aggregation of mIgs is recognized by other components, leading to phosphoinositide breakdown, attachment of the antigen-receptor complex to the cytoskeleton, and their subsequent internalization by clathrin-coated pits. The relationships of these events to one another are presently unknown. Endocytosis then leads to antigen processing and association of antigen fragments with class II MHC molecules, which are involved in activating helper T cells.

If this were the main mechanism of antigen uptake via mIg, then monovalent antigens should be presented by B cells to T cells much less efficiently than bivalent antigens. However, in the one case examined, a comparison of dimeric rabbit anti-IgM F(ab')$_2$ to monomeric Fab for their ability to activate T-cell clones specific for rabbit IgG, B cells presented monomeric antigen to T cells as efficiently as they did bivalent antigen, which implies similar rates of uptake (Tony et al 1985). Thus either the anti-IgM Fab induces a conformational change that leads to internalization (without capping) or there is a brisk constitutive internalization of membrane IgM. In either case, it appears that B cells efficiently take up and present monomeric antigens that bind to mIgM. Although monovalent antigens can be presented satisfactorily to helper T cells, they appear to be unable to cause the B-cell transmembrane signaling reactions described above (Pozzan et al 1982; Bijsterbosch et al 1985). Furthermore, concentrations of antigen well below the range needed to activate phosphoinositide breakdown are sufficient to lead to maximal activation of antigen-specific helper T-cell clones (Tony et al 1985).

Regulation of Antigen Presentation Efficiency

Finally, it should be noted that antigen presentation to T cells by B cells is not a constitutive function but rather is up-regulated by early activation events. Anti-IgM and the T cell–derived B-cell growth factor IL-4 are each capable of inducing considerable increases in the expression of class II MHC gene products (Mond et al 1981; Roehm et al 1984; Noelle et al

1984), which should make B cells present antigen more efficiently (Matis et al 1983). Furthermore, there appear to be qualitative changes in the efficiency of antigen presentation by activated B cells that are not simply due to changes in the amount of class II MHC proteins (Krieger et al 1985, 1986; Frohman & Cowing 1985). The increased ability of antigen-activated B cells to elicit T-cell help would be expected to enhance further their activation and antibody production. In addition, by inducing a rise in the levels of class II MHC proteins, IL-4 may enhance the antigen-presenting ability of B cells that have taken up antigen but which have not generated efficient transmembrane signals (due to binding of monomeric antigen, for example). Presumably this regulation by IL-4 represents a positive feed-back loop whereby activation of T cells increases the ability of nearby B cells to present antigen to other T cells, hence increasing the overall response in that location.

MECHANISMS OF ACTIVATION OF B CELLS BY HELPER T CELLS

A major mechanism by which helper T cells participate in the activation of B cells is by secreting polypeptide growth and differentiation factors. The nature of these factors and the details of their effects on B cells are currently areas of intensive investigation. Table 2 summarizes some of the properties of these factors. This particular area has been reviewed recently in considerable detail (Kishimoto 1985; Melchers & Andersson 1986; Hamaoka & Ono 1986). Therefore the discussion here focuses on how these factors combine to generate an antibody response from resting B cells. Readers interested in the effects of the individual factors on B cells are urged to consult the reviews listed above.

Two Helper T Cell–Dependent Pathways of B-Cell Activation

There are two well understood systems in which combinations of T cell–derived B-cell growth and differentiation factors (BCGFs and BCDFs, respectively) can promote immunoglobulin production. Mishell & Dutton (1967) first developed an in vitro culture system for the production of an antigen-specific antibody response using sheep red blood cells (SRBCs) as the antigen. This antibody response was found to depend on T cells, as does in vivo immunization with sheep red blood cells. Furthermore, the T cells could be replaced by cell-free supernatants of activated T cells, which suggested that the action of T cells in this situation was mediated by secreted factors (Dutton et al 1971; Schimpl & Wecker 1972). This initial work was followed by chemical characterization of the soluble mediators involved and improvement in the methods for purifying B cells away from T cells (a critical advance). These developments culminated in the recent demonstration that highly purified B cells produce specific antibody when incubated with SRBCs and two purified T cell–derived factors, interleukin 2 (IL-2) and interferon-γ (IFN-γ) (Leibson et al 1984).

A second in vitro system capable of generating an antibody response is based on the use of anti-immunoglobulin antibodies in place of antigen. Anti-IgM–stimulated B cells proliferate but do not differentiate to the stage of antibody production (Sell & Gell 1965). Addition of supernatants from activated T cells enhances the B-cell proliferation caused by anti-IgM alone and induces high-level antibody production in rabbits (Kishimoto & Ishizaka 1975) and mice (Parker et al 1979). Three agents in T-cell supernatants appear to be responsible for promoting this response of murine B cells: IL-4, BCGF II, and a BCDF (Nakanishi et al 1983). IL-4 alone considerably enhances B-cell proliferation induced by anti-IgM (Howard et al 1982). A very similar activation scheme has been demonstrated with human B cells, in which anti-IgM, a low–molecular weight BCGF, and a high–molecular weight BCGF combine with a BCDF to stimulate antibody production (Kishimoto 1985). The high–molecular weight BCGF could be the human equivalent of murine BCGF II: their molecular properties and biological effects are very similar (Dutton et al 1984; Takatsu et al 1985; O'Garra et al 1986; Ambrus et al 1985). However, the low–molecular weight BCGF is not the human version of IL-4. Human IL-4 has been produced recently following the isolation of its gene by use of nucleic acid homology with the mouse gene (Yokata et al 1986) and biosynthesis from the cloned gene. It is clearly distinct from low–molecular weight BCGF, which was recently cloned by Sharma et al (1987). The simplest model that can reconcile the results in human and in mouse is that the low–molecular weight BCGF can replace IL-4 in this activation scheme and vice versa. More complicated explanations are also possible, so this must be considered an open issue. In any case, the anti-IgM activation scheme is clearly distinct from the one described above for the SRBC antigen.

Interestingly, many of the cloned T-cell lines that have the ability to help B-cell antibody responses can be subdivided clearly into two classes based on the spectrum of interleukins produced by these T cells upon antigenic stimulation (Mosmann et al 1986). Upon activation, one helper T-cell subset secretes IL-2, IFN-γ, lymphotoxin, and two growth factors for hematopoietic stem cells, interleukin 3 (IL-3) and granulocyte-mono-cyte colony stimulating factor (GM-CSF). The second helper T-cell subset also secretes IL-3 and GM-CSF, but instead of secreting IL-2 and IFN-γ it secretes IL-4 and BCGF II. If this second class of helper T cells also

Table 2 Major growth and differentiation factors for B cells identified to date[a,b]

Name	Cellular source	Species	Major activities on B cells	MW	State of characterization	References[e]
Interleukin 4 (IL-4) (BCGF I, BSF-1)	T cells	mouse	BCGF (+ anti-IgM); enhanced expression of MHC class II proteins; switch to IgG, IgE	15,000[c]	cloned; human equivalent identified	1, 2
B-cell growth factor II (BCGF II) (IL-5)	T cells	mouse	BCGF (for preactivated cells); switch to IgA	55,000[d] 18,000[c]	cloned; may correspond to human high-MW BCGF	3, 4
Low-molecular weight BCGF	T cells	human	BCGF (+ anti-IgM)	12,000	cloned; equivalent in mouse not yet identified	5, 6
B-cell stimulatory factor 2 (BSF-2)	T cells	human	BCDF	20,000	cloned; mouse BCDF and/or TRF-2 may be equivalent	7, 8
Interleukin 2 (IL-2)	T cells	mouse, human	BCGF; BCDF	15,500[c]	cloned; low-affinity receptor cloned	9
Interferon-γ (IFN-γ)	T cells	mouse, human	BCGF; BCDF; antagonist of IL-4 activity	17,000 and 18,000[c] (mouse); 21,000 and 25,000[c] (human)	cloned	10

Factor	Source	Species	Activity	Molecular weight	Comments	References[e]
B-cell activation factor (BCAF)	T cells	mouse, human	BCGF (acts alone on resting B cells); BCDF	—	partially purified; human and mouse factors could be different	11, 12
B-cell maturation factor (BMF)	T cells	mouse	BCDF acting directly on resting B cells; does not require proliferation to induce antibody secretion	50,000–55,000;[d] 16,000[c]	partially purified	13
Interleukin 1 (IL-1)	macrophages, B cells	mouse, human	BCGF; BCDF	15,000–20,000[c]	two different genes cloned (IL-1 α and β)	14, 15
Plasmacytoma growth factor	macrophages (fibroblasts)	mouse	BCGF for plasmacytoma cells; may be required for proliferation in response to LPS	25,000(?)	minimal characterization; may be the same as BCDF from helper T cells	16–18
B-BCGF	B cells	human	BCGF (+ anti-IgM)	32,000[d]	partially purified	19, 20
Neuroleukin	T cells	mouse, human	BCDF (?)	56,000[c,d]	cloned; antibodies against neuroleukin partially block helper T cell–dependent antibody response using pokeweed mitogen	21

[a] As of February, 1987.

[b] Abbreviations: interleukin (IL); B-cell growth factor (BCGF); B-cell differentiation factor (BCDF); B-cell stimulatory factor (BSF); interferon (IFN); B-cell activation factor (BCAF); B-cell maturation factor (BMF); B cell–derived BCGF (B-BCGF); T cell–replacing factor (TRF); bacterial lipopolysaccharide (LPS).

[c] Determined by SDS-polyacrylamide gel electrophoresis.

[d] Determined by nondenaturing gel filtration.

[e] (1) Howard et al 1982; (2) Noma et al 1986; (3) Swain et al 1983; (4) Kinashi et al 1986; (5) Maizel et al 1982; (6) Sharma et al 1987; (7) Hirano et al 1985; (8) Hirano et al 1986; (9) Smith 1984; (10) Gray & Goeddel 1983; (11) Leclerq et al 1984; (12) Bowen et al 1986; (13) Sidman et al 1984; (14) Booth et al 1983; (15) Durum et al 1985; (16) Corbel & Melchers 1984; (17) Nordin & Potter 1986; (18) Billiau 1986; (19) Jurgensen et al 1986; (20) Muraguchi et al 1986; (21) Gurney et al 1986.

secretes some form of BCDF, then the two main activation pathways described above are mediated by factors secreted by different subsets of T cells. It is not yet clear how these two types of helper T cells influence each other. Previous work suggested the existence of two cooperating types of helper T cells (Janeway et al 1980). However, IFN-γ blocks the major responses of B cells to IL-4, including the stimulation of proliferation in the presence of anti-IgM and the increased expression of class II MHC proteins (Rabin et al 1986; Mond et al 1986; Coffman & Carty 1986). Thus it is important to determine the extent to which these two types of helper T cells cooperate or interfere with each other in regulating B-cell antibody responses. Presumably, the two different types of helper T cells play different roles in promoting various types of immune responses.

Antigen Receptor–Independent B-Cell Activation

The two activation schemes described above are unlikely to explain fully the ability of helper T cells to activate B cells. Both of those systems are dependent on interaction of the relevant antigen or antigen surrogate (anti-IgM) with mIgM and/or mIgD. Furthermore, both anti-IgM and SRBCs are likely to induce stronger transmembrane signals from mIgM than are the soluble protein antigens, which have the strictest requirement for helper T cells. The nature of the antibody response to soluble protein antigens may be revealed by experiments in which B cells are activated in the absence of signal transduction from mIg. A substantial antibody response can be generated in the absence of antigen-mIg interaction by focusing T-cell help on B cells (Cammisuli et al 1978; Augustin & Coutinho 1980; Jones & Janeway 1981; Julius et al 1982; DeFranco et al 1984). For example, if cloned helper T-cell lines specific for a given antigen are incubated with sufficient amounts of that antigen and B cells expressing the appropriate class II MHC molecules, then often the B cells become polyclonally activated to proliferate and to produce antibody (Jones & Janeway 1981; DeFranco et al 1984). Similarly, if the cloned helper T cells react with a particular allogeneic class II MHC molecule, then they will activate B cells expressing this stimulatory MHC molecule (Augustin & Coutinho 1980). Typically, addition of anti-IgM to these systems enhances the B-cell response, especially with limiting numbers of T cells (Julius et al 1982), but often the anti-IgM is not absolutely required (DeFranco et al 1984). These results argue that helper T cells are able to deliver a signal to the B cell that replaces the signal from mIg. Some sort of mechanism of this type is needed to explain B-cell activation by antigens that cause little or no mIgM- or mIgD-induced phosphoinositide transmembrane signaling. Many soluble protein antigens may fall in this category.

    Two models for the nature of this signal have been proposed. The first is that T cells provide a signal via a direct cell-cell interaction, for example, by interaction of cell surface proteins on B cells and T cells. The second model is that this signal is transmitted through a soluble mediator, presumably secreted in limiting amounts to decrease the chance of activation of bystander B cells, i.e. B cells other than those directly activating the helper T cell by presenting antigen. In some but not all systems the bystander B cells also become activated, though to a lesser extent, in the presence of the helper T cells and the antigen-presenting B cells (Kaye et al 1983; DeFranco et al 1984; Tony & Parker 1985). This observation is most easily explained by the soluble mediator model. In further support of this idea, two groups have reported finding an activity in supernatants of activated T-cell lines that can cause proliferation of resting B cells, and they have partially purified this molecule (Leclerc et al 1984; Bowen et al 1986). The extent to which other B-cell growth and differentiation factors are required for this response is not known. This activity, termed B-cell activation factor, could promote antibody responses of B cells that have not contacted their antigen. This could decrease the specificity of the response. A mechanism that could limit this indiscriminant activation of bystander cells is suggested by the recent demonstration that helper T cells can orient their microtubule organizing center toward an antigen-presenting cell (Kupfer et al 1986). Therefore, the helper T cell may secrete soluble mediators directly toward the antigen-presenting B cell. Directed secretion would ensure that primarily those cells that had taken up antigen molecules through membrane Ig would be activated, with more limited activation of bystander B cells. If, however, the bystander B cells are binding an antigen that can induce strong antigen receptor signaling, then nonlimiting factors such as IL-4 and BCGF II may be able to promote B-cell activation without direct interaction of the helper T cell and the B cell.

PRODUCTION OF ANTIBODIES IN RESPONSE TO BACTERIAL ANTIGENS

The major mechanism for making antibodies against bacteria appears not to require helper T cells. For example, people with immunodeficiencies that compromise helper T cells, such as congenital thymic aplasia (Di George's syndrome) or acquired immunodeficiency syndrome (AIDS), generally have greater problems with opportunistic diseases than with pathogenic bacteria. Conversely, immunodeficiencies that affect B cells, macrophages or the complement pathway result in persistent and recurrent bacterial infections. One key to understanding antibody responses to infectious bacteria is the realization that a wide variety of bacterial cell surface components are potent activators of B cells and macrophages ("polyclonal

B-cell activators"). This activator property of many bacteria greatly increases the antibody response, and as such probably represents a system evolved by multicellular organisms for protection against bacteria.

Polyclonal B-Cell Activators

The T-independent type 1 antigens appear to generate antibody responses in the absence of growth and differentiation factors from helper T cells. The best-studied antigens of this type are based on bacterial lipopolysaccharide (LPS). Whereas the "carriers" described above, such as soluble proteins, carbohydrates, and red blood cells, may be considered scaffolds for antigenic determinants or for haptens such as TNP, LPS is clearly more than a carrier; it is a polyclonal B-cell activator. Exposure of B cells to LPS causes about one-third of them to proliferate and differentiate into antibody-secreting cells—a far greater fraction than that making antibodies directed against LPS itself (Andersson et al 1977). How then are immune responses to polyclonal activators kept from being nonspecific? There are probably two answers to this question. First, binding of haptenated LPS to hapten-specific membrane Ig apparently concentrates the polyclonal activator part of the antigen, causing it to be effective at much lower concentrations. For example, TNP-LPS can elicit an in vitro anti-TNP antibody response at concentrations 2–3 orders of magnitude below the concentration of LPS required for a vigorous polyclonal antibody response. A second mechanism increasing the specificity of the antibody response is a synergy between two activation signals, one conveyed through membrane Ig in response to binding of a specific antigen and the other through the putative LPS receptor, as demonstrated in experiments with anti-IgM and LPS (DeFranco et al 1982b; Zubler 1984).

LPS is also a powerful activator of macrophages, which in response to LPS release interleukin 1 (Lachman et al 1976), tumor necrosis factor (Beutler et al 1985), prostaglandins (Schade & Rietschel 1982), and a recently described B cell growth factor(s) that has activity on LPS-stimulated B cells (Corbel & Melchers 1983) and on plasmacytoma cells (Nordin & Potter 1986).

Macrophage-derived regulatory mediators may be required for the LPS-induced mitogenic and/or antibody-secreting responses of B cells. For example, only a small fraction of B cells plated at one cell per well proliferate in the presence of LPS alone (Wetzel & Kettman 1981). Similarly, Corbel & Melchers (1983) found that rigorous depletion of macrophages strongly decreased the proliferation of B cells stimulated with LPS. These cells did proliferate appropriately upon addition of either macrophages or a supernatant from LPS-stimulated P388D$_1$ macrophage cells. The exact nature of the B-cell growth factor in these supernatants is unclear. However, the same supernatants also are active in promoting the growth of plasmacytoma cells, which are tumor cells that correspond to the antibody-secreting stage of B-cell maturation. Nordin & Potter (1986) have partially characterized a factor from LPS-stimulated macrophages with the latter activity and termed it "plasmacytoma growth factor" (or PCGF). Recently it has been reported that a factor obtained from fibroblasts also has this activity and has the same amino acid sequence as B-cell stimulatory factor 2 (BSF-2) (Billiau 1986). The latter factor is a product of helper T cells, but so far there are no reports on its ability to stimulate the growth of plasmacytoma cells. It remains to be seen whether the macrophage-derived factor is actually BSF-2. Obviously, the details of the LPS activation scheme are less well understood than those of the helper T cell–dependent activation schemes described above.

One difficulty with analysis of this system is that macrophages are technically difficult to remove from B-cell populations. Since LPS induces macrophages to release B-cell growth factors, even a few contaminating macrophages may be sufficient to generate the necessary factors for B-cell proliferation and/or differentiation. Although macrophages probably play an important regulatory role in B-cell proliferation and antibody secretion in response to LPS, clearly LPS also has direct effects on B cells. In the experiments using a single B cell per well, as mentioned above (Wetzel & Kettman 1981), a majority of the B cells proliferated in the presence of LPS and dextran sulfate, a weak B-cell mitogen whose mechanism of action is unclear. However, since the single B cells did not proliferate in response to dextran sulfate unless LPS was added, the cells must be responsive to LPS. Additionally, a large number of B-cell lines, derived from lymphomas and leukemias, are responsive to LPS in a variety of ways (Paige et al 1981; Boyd et al 1981; Lanier 1982; Stavnezer et al 1984; Jakway et al 1986), which suggests an inherent ability of B cells to respond to LPS.

Molecular Diversity of Polyclonal B-Cell Activators

There is a considerable number of bacterial cell surface components other than LPS that have also been demonstrated to activate B cells and/or macrophages. Polyclonal B-cell activators are structurally quite diverse; they include LPS, which has attached fatty acids that are required for biological activity (Morrison & Ryan 1979); peptidoglycan (Damais et al 1975); capsular and released polysaccharides (Usinger et al 1985; Daley et al 1985); outer membrane proteins (Bessler & Henning 1979); and secreted bacterial proteins, such as toxic shock syndrome toxin (Parsonnet et al 1985). Even some viral membrane glycoproteins appear capable of triggering polyclonal B-cell proliferation (Goodman-Snitkoff et al 1981;

Butchko et al 1978). The human immunodeficiency virus (HIV) virion has also been reported to have this capacity (Pahwa et al 1985; Yarchoan et al 1986; Schnittman et al 1986), which may partially explain the B-cell hyperactivity seen in many AIDS patients. One attractive hypothesis is that B cells and macrophages express a series of different cell surface receptors, each specific for one or a few of these polyclonal B-cell activators, and that all of these receptors are capable of activating the same transmembrane signaling reactions and thus lead to identical cellular responses. Such a system might have evolved in multicellular organisms to enhance their responses to infectious bacteria. Interestingly, the standard bioassay for LPS involves a response of horseshoe crab amebocytes, which are presumably macrophagelike cells. Thus the ability to make protective responses to key bacterial surface components may have developed early in the evolution of multicellular organisms.

Additional Factors Promoting B-Cell Activation in Response to Bacteria

B-CELL GROWTH FACTOR PROPERTIES OF C3 FRAGMENTS   The B-cell growth factors apparently required for B-cell clonal expansion during the antibody response to bacteria may not come exclusively from macrophages. There is now good evidence that several forms of the third component of complement (C3) may play roles in B-cell proliferation. C3b or C3d (fragments of C3) when attached to beads and added to LPS-stimulated B cells promoted cell growth, apparently acting in the same manner as macrophage-generated BCGF (Erdei et al 1985; Melchers et al 1985). C3b is generated in two major ways: via the classical complement pathway, which is activated by antibodies, and via the alternative pathway, which is independent of antibody and may represent a first line of defense against many microorganisms. Once generated, C3b attaches covalently to nearby surfaces. The alternative pathway amplifies C3b expression because this bound C3b can participate in the cleavage of C3 to generate more C3b. Thus, particles with C3b and/or one of its inactivated forms, C3d, can bind to complement receptors on B cells and promote antibody responses to antigenic determinants on the particle surface.

B cells have two types of complement receptors, called CR1 and CR2 (Fearon & Wong 1983). CR2 is expressed primarily on a subset of B cells and is the cellular receptor for Epstein-Barr virus (EBV) (Fingeroth et al 1984; Nemerow et al 1985). EBV is a herpes virus that infects B cells, induces their long-term proliferation and antibody production, and participates in the origin of B-cell lymphomas of the Burkitt's type. The activation of B cells in the latter case seems to be directed by EBV gene products. The CR2 receptor has been implicated in B-cell activation in that anti-CR2 antibodies inhibit certain types of B-cell responses (Fearon & Wong 1983). Like most BCGFs, bead-attached C3b and C3d do not cause proliferation of resting B cells (Erdei et al 1985). Interestingly, treatment of resting B cells with anti-IgM induces phosphorylation of CR2 (Changelian & Fearon 1986), which could influence the ability of CR2 to signal the binding of C3 fragments in a productive way. In the experiments of Melchers et al (1985), LPS-activated B cells (depleted of macrophages) continued to proliferate either in the presence of LPS and C3b on beads or in the presence of anti-IgM, bead-attached C3b, and helper T cell–derived BCGFs. Thus the presence of C3-derived fragments on a particle, possibly resulting from the action of the alternate complement pathway, may provide a strong growth stimulus supporting a vigorous B-cell response prior to formation of specific antibody.

B CELL–DERIVED BCGF   A number of malignant B-cell lines and EBV-transformed B cells have been found to secrete B-cell growth factors (Gordon et al 1984a,b; Ambrus et al 1985; Blazer et al 1983; Brooks et al 1984; Nakajima et al 1985). In EBV-transformed B cells, the growth factor appears to be an autocrine requirement for the continued proliferation of the cells (Gordon et al 1984a). In some cases, the growth factor may be BCGF II (Nakajima et al 1985; Ambrus et al 1985); now that the gene for this latter factor has been cloned, it should be possible to demonstrate this directly.

These observations suggest that normal B cells may produce BCGFs under appropriate circumstances. Indeed, human B cells stimulated with fixed *Staphylococcus aureus* cells secrete a BCGF, called B-BCGF for B cell–derived BCGF. *Staphylococcus aureus* cells are mitogenic for human B cells, probably due to the interaction of protein A with membrane Ig and perhaps also due to polyclonal-activator properties of the cells. B-BCGF induces proliferation of B cells preactivated with fixed *Staphylococcus aureus* and acts synergistically with anti-IgM to induce proliferation of resting B cells (Jurgensen et al 1986; Muraguchi et al 1986). Biochemical and functional evidence suggests that B-BCGF is distinct from IL-4, IL-2, BCGF II, and BSF-2. Treatment of macrophages with fixed *Staphylococcus aureus* did not cause them to generate this factor. This result indicates that B-BCGF is derived from B cells rather than macrophages. The fact that its secretion is induced by a mitogenic bacterium suggests that B-BCGF may play a role in immune responses to bacteria.

Signal Transduction by the Putative LPS Receptor

The receptors for the polyclonal B-cell activators have not yet been identified conclusively. These compounds probably bind directly to cell surface receptors on macrophages and B cells, leading to generation of a trans-membrane signal and hence changes in cellular physiology. Alternatively, these agents may react with some component of serum, e.g. specific anti-bodies or complement components, and cells may recognize these moieties. However, this appears unlikely since LPS and peptidoglycan can cause excellent B-cell and macrophage responses in serum-free medium (Mosier 1981; Gold et al 1985). The effects of LPS and other polyclonal activators are more likely the result of their direct interaction with B cells and macrophages.

The response of B cells and macrophages to LPS appears to involve a G protein of the G_i type. The responses of the B-cell line WEHI-231 and of the macrophage line $P388D_1$ to LPS were blocked by pretreatment of the cells with pertussis toxin, a protein that ADP-ribosylates and thereby inactivates several members of the G-protein family (Jakway & DeFranco 1986). Evidence that the component blocked by pertussis toxin is a G protein comes from the biochemical demonstration that LPS can activate a G_i-type G protein in the membranes of these two cell lines, resulting in inhibition of adenylate cyclase (J. P. Jakway, M. R. Gold, S. J. Estey, and A. L. DeFranco, manuscript in preparation). As all of the known cases of activation of G proteins involve membrane protein receptors, such molecules are probably also involved in the LPS activation of G proteins in B cells and macrophages.

Unlike anti-IgM, LPS does not activate the phosphoinositide pathway in mature B cells (Grupp & Harmony 1985; Bijsterbosch et al 1985). This is consistent with the observations that LPS and anti-IgM have different effects on B cells but can act synergistically to induce B-cell proliferation (Zubler 1984). Although anti-IgM stimulation of phosphoinositide break-down has been reported to be insensitive to pertussis toxin (Jakway & DeFranco 1986; M. R. Gold, J. P. Jakway, and A. L. DeFranco, manu-script in preparation), stimulation of this pathway in macrophages by the chemotactic agent f-Met-Leu-Phe is blocked by pertussis toxin (Backlund et al 1985; Brandt et al 1985). As the responses of macrophages to LPS and to f-Met-Leu-Phe are clearly distinct (Hamilton & Adams 1984), macrophages most likely contain at least two pertussis toxin–sensitive G proteins: one that is stimulated by the f-Met-Leu-Phe receptor to activate breakdown of phosphoinositides and a second that is stimulated by the presumed LPS receptor to inhibit adenylate cyclase and possibly to activate other processes.

In the pre–B cell line 70Z/3, LPS induces transcription of the κ light-chain gene and consequently membrane IgM expression (Paige et al 1981). In this system, LPS causes a rise in cellular inositol phosphates and ele-vation of cytoplasmic calcium concentration (Rosoff & Cantley 1985), but these effects are modest compared to those of anti-IgM in mature B cells. It will be interesting to see whether these effects are mediated through a pertussis toxin–sensitive G protein.

The fact that LPS has covalently attached fatty acids, which are essential to all of its stimulatory effects on B cells and macrophages (Morrison & Ryan 1979), has led to the hypothesis that LPS may not bind to a cell surface receptor but rather may activate cells by inserting into membranes and directly activating processes there. This idea runs counter to the implication of the work showing that LPS must act through a G protein in a B-cell line and in a macrophage line, as G proteins are activated by receptor proteins in all known cases (Stryer & Bourne 1986). Nonetheless, it is possible that LPS also acts in this way and activates a lipid-regulated enzyme. The best-understood example of such an enzyme is the diacyl-glycerol-regulated protein kinase C (Nishizuka 1984). Interestingly, Wightman & Raetz (1984) have shown that LPS can activate protein kinase C. However, this enzyme requires phosphatidylserine and calcium in addition to diacylglycerol for the stimulation of its activity, and in the work mentioned LPS was used to replace phosphatidylserine rather than diacylglycerol. As phosphatidylserine would be expected to be continually present in B-cell and macrophage plasma membrane, with diacylglycerol being the regulated activator, the physiological relevance of these results is unclear. Certainly, it is important to determine whether LPS can replace diacylglycerol in the activation of the enzyme in vitro. Even if it can, one would still have to postulate that LPS reaches the inner leaflet of the plasma membrane to activate protein kinase C. Given the large amount of carbohydrate on LPS, it is not obvious how this could happen.

One way to examine this issue is to see if LPS induces a translocation of protein kinase C from the cytoplasm to the membrane in B cells, as this translocation has been associated with activation of the enzyme (Kraft & Anderson 1982). Indeed, LPS has been reported to cause a translocation of some of the protein kinase C from the cytosol to membranes in mature B cells (Chen et al 1986); but of course, this could be due to a receptor-mediated signaling event rather than a direct activation of protein kinase C by LPS.

Phorbol esters, which are potent activators of protein kinase C, cannot mimic many of the effects of LPS on B cells, which strongly argues that activation of protein kinase C is not solely responsible for the ability of LPS to activate B cells. Further, LPS cannot mimic the effects of phorbol esters on B cells. Activation of protein kinase C by phorbol esters causes a feedback inhibition that blocks membrane IgM–induced phosphoinositide breakdown in mature B cells and in several B-cell lines (Mizuguchi et al 1986c; Gold & DeFranco 1987), a form of regulation seen in many other cell types as well. LPS, however, does not block anti-IgM–induced phos-phoinositide breakdown (J. P. Jakway, K. A. Fahey, and A. L. DeFranco, manuscript in preparation). Thus LPS is not simply an activator of protein kinase C as are phorbol esters.

ROLE OF c-*myc* PROTO-ONCOGENE IN B-CELL ACTIVATION

An important role for the normal c-*myc* proto-oncogene in B-cell pro-liferation is suggested by the fact that genetic alterations in the c-*myc* gene are prominently associated with leukemias and lymphomas of the B-cell lineage in chickens, mice, and humans (Kelly & Siebenlist 1986). Furthermore, the expression of the normal c-*myc* gene is rapidly elevated in response to several growth-promoting stimuli in lymphocytes and fibroblasts. In B cells, LPS induces rapid elevation of c-*myc* mRNA levels (Kelly et al 1983), as do various anti-IgM antibodies (Smeland et al 1985).

The relationship between the ability of signals to stimulate B-cell growth and their ability to induce c-*myc* expression may not be a simple one. Smeland et al (1985) found that a monoclonal anti-IgM, which could not induce B-cell proliferation in the presence of a crude BCGF preparation, was able to induce c-*myc* mRNA production as well as did a polyclonal anti-IgM, which could stimulate proliferation under those circumstances. The monoclonal antibody did induce some early B-cell activation events, such as increased cell size. This antibody may be poorer at inducing phos-phoinositide breakdown than the polyclonal anti-IgM. Thus, in addition to elevating c-*myc* expression, the polyclonal anti-IgM must affect other processes to stimulate proliferation in the presence of BCGF. An additional argument that anti-IgM must stimulate B-cell activation by means other than elevation of c-*myc* is that phorbol esters induce increased production of c-*myc* mRNA (Smeland et al 1985), as does anti-IgM, but do not mimic many of anti-IgM's biological effects on B cells. These biological effects are mimicked reasonably well in B cells simultaneously exposed to phorbol esters and calcium ionophores, as discussed above. Thus elevation of $[Ca^{2+}]$ seems to be important for B-cell activation via anti-IgM but not for c-*myc* expression. These experiments argue that elevation of c-*myc* is not the only relevant effect of anti-IgM, although it may be required.

Recently, Snow et al (1986) have shown that addition of a haptenated soluble protein antigen (TNP-keyhole limpet hemocyanin) to purified hap-ten-specific B cells causes an elevation of c-*myc* mRNA similar to that induced by anti-IgM. Previous evidence indicates that such hapten-specific B cells do not proliferate in the presence of the corresponding haptenated soluble protein antigen and IL-4 (Snow et al 1983; Stein et al 1986). This is another indication that a signal from membrane Ig may induce elevation of c-*myc* mRNA without being of sufficient strength to act synergistically with a BCGF to induce B-cell proliferation.

Another way to address this issue is to introduce into B cells a constitutively expressed c-*myc* gene and examine its effects on B-cell activation. Adams and colleagues introduced a construct comprising c-*myc* and an Ig heavy-chain enhancer into a murine germ line. The resulting transgenic mice developed tumors of B lineage cells, which appeared to be monoclonal in origin (Adams et al 1985). The phenotype of B cells in the prelymphomatous state in these transgenic animals has now been analyzed (Langdon et al 1986). There is a considerable amplification of large, dividing pre–B cells in these animals. B cells are present in roughly normal numbers but are clearly larger in size (a parameter that correlates with activation) and proliferate more actively than B cells from normal mice. The authors did not determine whether any B cells were in the resting G_0 state, normally occupied by about two-thirds of splenic B cells (DeFranco et al 1982a), but the data are consistent with the idea that this state may be excluded when c-*myc* is expressed. Clearly a more detailed characterization of B cells from these mice is necessary to fully understand the effects of constitutive c-*myc* expression on B-cell physiology. For example, do B cells from these mice proliferate in response to the known growth factors, and to anti-IgM, LPS, etc? Are they capable of making normal or enhanced antibody responses to various types of antigens? Irrespective of remaining uncertainties, these experiments reinforce the idea that the c-*myc* gene plays an important role in the nuclear events underlying B-cell activation.

SUMMARY

The activation of B lymphocytes from the resting stage to the proliferating stage and then to the fully differentiated antibody-secreting stage is a highly regulated and complicated process. B-cell activation can clearly proceed by a number of different routes, each promoted by different regulatory cells (macrophages, two types of helper T cells) and each dependent upon the properties of the relevant antigen molecules. In most cases the nature of the antigen may simply control the magnitude and/or duration of antigen receptor signaling. An antigen that by itself generates inefficient signaling may require additional signals from helper T cells to induce B-cell activation. In contrast, antigens derived from bacterial cell surface components, such as LPS, can directly activate B cells and macrophages. This vigorous, polyclonal responsiveness to certain bacterial components probably represents a specialized system to enhance antibody responses to bacteria, whereas helper T cell–dependent responses may be primarily useful in making antibodies against soluble proteins and viruses. Thus multiple pathways of B-cell activation probably are needed to generate adequate antibody responses against the variety of pathogenic microorganisms encountered by vertebrates.

ACKNOWLEDGMENTS

I thank the members of my laboratory for helpful comments on the manuscript and colleagues who have shared their insights with me. The work in my laboratory was supported by NIH grant AI-20038 and by a grant from the School of Medicine, University of California, San Francisco (Hampton Fund).

Literature Cited

Adams, D. O., Hamilton, T. A. 1984. The cell biology of macrophage activation. *Ann. Rev. Immunol.* 2: 283–318

Adams, J. M., Harris, A. W., Pinkert, C. A., Corcoran, L. M., Alexander, W. S., et al. 1985. The *c-myc* oncogene driven by immunoglobulin enhancers induces lymphoid malignancy in transgenic mice. *Nature* 318: 533–38

Ambrus, J. L., Jr., Jurgensen, C. H., Brown, E. J., Fauci, A. S. 1985. Purification to homogeneity of a high molecular weight human B cell growth factor; demonstration of specific binding to activated B cells; and development of a monoclonal antibody to the factor. *J. Exp. Med.* 162: 1319–35

Amsbaugh, D. F., Hansen, C. T., Prescott, B., Stashak, P. W., Barthold, D. R., Baker, P. J. 1972. Genetic control of the antibody response to type III pneumococcal polysaccharide in mice. Evidence that an X-linked gene plays a decisive role in determining responsiveness. *J. Exp. Med.* 136: 931–49

Andersson, J., Coutinho, A., Lernhardt, W., Melchers, F. 1977. Clonal growth and maturation to immunoglobulin secretion in vitro of every growth-inducible B lymphocyte. *Cell* 10: 27–34

Ashwell, J. D., DeFranco, A. L., Paul, W. E., Schwartz, R. H. 1984. Antigen presentation by resting B cells: Radiosensitivity of the antigen-presenting function and two distinct pathways of T cell activation. *J. Exp. Med.* 159: 881–905

Augustin, A. A., Coutinho, A. 1980. Specific T helper cells that activate B cells polyclonally. *J. Exp. Med.* 151: 587–601

Babbitt, B. P., Allen, P. M., Matsueda, G., Haber, E., Unanue, E. R. 1985. Binding of immunogenic peptides to Ia histocompatibility molecules. *Nature* 317: 359

Backlund, P. S., Jr., Meade, B. D., Manclark, C. R., Cantoni, G. L., Aksamit, R. R. 1985. Pertussis toxin inhibition of chemotaxis and the ADP-ribosylation of a membrane protein in a human-mouse hybrid cell line. *Proc. Natl. Acad. Sci. USA* 82: 2637–41

Berridge, M. J., Irvine, R. F. 1984. Inositol trisphosphate, a novel second messenger in cellular signal transduction. *Nature* 312: 315–21

Bessler, W. G., Henning, U. 1979. Protein I and Protein II* from the outer membrane of *Escherichia coli* are mouse B-lymphocyte mitogens. *Z. Immunitaetsforsch.* 155: 387–98

Beutler, B., Mahoney, J., LeTrang, N., Pekala, P., Cerami, A. 1985. Purification of cachectin, a lipoprotein lipase-suppressing hormone secreted by endotoxin-induced RAW 264.7 cells. *J. Exp. Med.* 161: 984–95

Bijsterbosch, M. K., Klaus, G. G. B. 1985. Crosslinking of surface immunoglobulin and Fc receptors on B lymphocytes inhibits stimulation of inositol phospholipid breakdown via the antigen receptor. *J. Exp. Med.* 162: 1825–36

Bijsterbosch, M. K., Meade, C. J., Turner,

G. A., Klaus, G. G. B. 1985. B lymphocyte receptors and polyphosphoinositide degradation. *Cell* 41: 999–1006

Billiau, A. 1986. BSF-2 is not just a differentiation factor. *Nature* 324: 415

Blazar, B. A., Sutton, L. M., Strome, M. 1983. Self-stimulating growth factor production by B cell lines derived from Burkitt's lymphomas and other lines transformed in vitro by Epstein-Barr virus. *Cancer Res.* 43: 4562

Booth, R. J., Prestidge, R. L., Watson, J. D. 1983. Constitutive production by the WEHI-3 cell line of B cell growth and differentiation factor that co-purifies with interleukin 1. *J. Immunol.* 131: 1289–93

Bourguignon, L. Y. W., Bourguignon, G. J. 1984. Capping and the cytoskeleton. *Int. Rev. Cytol.* 87: 195–224

Bourguignon, L. Y. W., Walker, G., Suchard, S. J., Balazovich, K. 1986. A lymphoma plasma membrane–associated protein with ankyrin-like properties. *J. Cell Biol.* 102: 2115–24

Bowen, D. L., Ambrus, J. L., Jr., Fauci, A. S. 1986. Identification and characterization of a B cell activation factor (BCAF) produced by a human T cell line. *J. Immunol.* 136: 2158–63

Boyd, A. W., Goding, J. W., Schrader, J. W. 1981. The regulation of growth and differentiation of a murine B cell lymphoma. I. Lipopolysaccharide-induced differentiation. *J. Immunol.* 126: 2461–65

Brandt, S. J., Dougherty, R. W., Lapentina, E. G., Niedel, J. E. 1985. Pertussis toxin inhibits chemotactic peptide-stimulated generation of inositol phosphates and lysosomal enzyme secretion in human leukemic (HL-60) cells. *Proc. Natl. Acad. Sci. USA* 82: 3277–80

Braun, J., Hochman, P. S., Unanue, E. R. 1982. Ligand-induced association of surface immunoglobulin with the detergent-insoluble cytoskeletal matrix of the B lymphocyte. *J. Immunol.* 128: 1198–1204

Braun, J., Sha'afi, R. I., Unanue, E. R. 1979. Crosslinking by ligands to surface immunoglobulin triggers mobilization of intracellular $^{45}Ca^{2+}$ in B lymphocytes. *J. Cell Biol.* 82: 755–66

Brooks, K. H., Uhr, J. W., Vitetta, E. S. 1984. A B cell growth factor–like activity is secreted by cloned, neoplastic B cells. *J. Immunol.* 133: 3133

Burnet, F. M. 1959. *The Clonal Selection Theory of Acquired Immunity*, pp. 1–208. London: Cambridge Univ. Press

Butchko, G. M., Armstrong, R. B., Martin, W. J., Ennis, F. A. 1978. Influenza A viruses of the H2N2 subtype are lymphocyte mitogens. *Nature* 271: 66–67

Butler, J. L., Muraguchi, A., Lane, H. C., Fauci, A. S. 1983. Development of a human T-T cell hybridoma secreting B cell growth factor. *J. Exp. Med.* 157: 60–68

Buus, S., Sette, A., Colon, S. M., Jenis, D. M., Grey, H. M. 1986. Isolation and characterization of antigen-Ia complexes involved in T cell recognition. *Cell* 47: 1071–77

Cammisuli, S., Henry, C., Wofsy, L. 1978. Role of membrane receptors in the induction of in vitro secondary anti-hapten responses. I. Differentiation of B memory cells to plasma cells is independent of antigen-immunoglobulin receptor interaction. *Eur. J. Immunol.* 8: 656–62

Casten, L. A., Lakey, E. K., Jelachich, M. L., Margoliash, E., Pierce, S. K. 1985. Anti-immunoglobulin augments the B-cell antigen-presentation function independently of internalization of receptor-antigen complex. *Proc. Natl. Acad. Sci. USA* 82: 5890–94

Changelian, P. S., Fearon, D. T. 1986. Tissue-specific phosphorylation of complement receptors CR1 and CR2. *J. Exp. Med.* 163: 101–15

Chen, Z. Z., Coggeshall, K. M., Cambier, J. C. 1986. Translocation of protein kinase C during membrane immunoglobulin-mediated transmembrane signaling in B lymphocytes. *J. Immunol.* 136: 2300–4

Chesnut, R. W., Colon, S. M., Grey, H. M. 1982. Antigen presentation by normal B cells, B cell tumors and macrophages: Functional and biochemical comparison. *J. Immunol.* 128: 1764–68

Coffman, R. L., Carty, J. 1986. A T cell activity that enhances polyclonal IgE production and its inhibition by interferon γ. *J. Immunol.* 136: 949–54

Coggeshall, K. M., Cambier, J. C. 1984. B cell activation. VIII. Membrane immunoglobulins transduce signals via activation of phosphatidylinositol hydrolysis. *J. Immunol.* 133: 3382–86

Corbel, C., Melchers, F. 1983. Requirement for macrophages or for macrophage- or T-cell-derived factors in the mitogenic stimulation of murine B lymphocytes by lipopolysaccharides. *Eur. J. Immunol.* 13: 528–33

Corbel, C., Melchers, F. 1984. The synergism of accessory cells and of soluble factors derived from them in the activation of B cells to proliferation. *Immunol. Rev.* 78: 51–74

Cresswell, P. 1985. Intracellular class II HLA antigens are accessible to transferrin-neuraminidase conjugates internalized by receptor-mediated endocytosis. *Proc. Natl. Acad. Sci. USA* 82: 8188–92

Daley, L., Pier, G. B., Liporace, J. D.,

172   DeFRANCO

Eardley, D. D. 1985. Polyclonal B cell stimulation and interleukin 1 induction by the mucoid exopolysaccharide of *Pseudomonas aeruginosa* associated with cystic fibrosis. *J. Immunol.* 134: 3089–93

Damais, C., Bona, C., Chedid, L., Fleck, J., Nauciel, C., Martin, J. P. 1975. Mitogenic effect of bacterial peptidoglycans possessing adjuvant activity. *J. Immunol.* 115: 268–71

DeFranco, A. L., Ashwell, J. D., Schwartz, R. H., Paul, W. E. 1984. Polyclonal stimulation of resting B lymphocytes by antigen-specific T lymphocytes. *J. Exp. Med.* 159: 861–80

DeFranco, A. L., Kung, J. T., Paul, W. E. 1982b. Regulation of growth and proliferation in B cell subpopulations. *Immunol. Rev.* 64: 161–82

DeFranco, A. L., Raveche, E. S., Asofsky, R., Paul, W. E. 1982a. Frequency of B lymphocytes responsive to anti-immunoglobulin. *J. Exp. Med.* 155: 1523–36

Dintzis, H. M., Dintzis, R. Z., Vogelstein, B. 1976. Molecular determinants of immunogenicity: The immunon model of immune response. *Proc. Natl. Acad. Sci. USA* 73: 3671–75

Dintzis, R. Z., Vogelstein, B., Dintzis, H. M. 1982. Specific cellular stimulation in the primary immune response: Experimental test of a quantized model. *Proc. Natl. Acad. Sci. USA* 79: 884–88

Durum, S. K., Schmidt, J. A., Oppenheim, J. J. 1985. Interleukin 1: an immunological perspective. *Ann. Rev. Immunol.* 3: 263–87

Dutton, R. W., Falkoff, R., Hirst, J. A., Hoffmann, M., Kappler, J. W., et al. 1971. Is there evidence for a non-antigen specific diffusable chemical mediator from the thymus derived cell in the initiation of the immune response? *Prog. Immunol.* 1: 355–68

Dutton, R. W., Wetzel, G. D., Swain, S. L. 1984. Partial purification and characterization of a BCGFII from EL4 culture supernatants. *J. Immunol.* 132: 2451–56

Erdei, A., Melchers, F., Schulz, T., Dierich, M. 1985. The action of human C3 in soluble or cross-linked form with resting and activated murine B lymphocytes. *Eur. J. Immunol.* 15: 184–88

Fahey, K. A., DeFranco, A. L. 1987. Crosslinking membrane IgM induces production of inositol trisphosphate and inositol tetrakisphosphate in WEHI-231 B lymphoma cells. *J. Immunol.* 138: 3935–42

Fearon, D. T., Wong, W. W. 1983. Complement ligand-receptor interactions that mediate biological responses. *Ann. Rev.*

Immunol. 1: 243–71

Fingeroth, J. D., Weis, J. J., Tedder, T. F., Strominger, J. L., Biro, P. A., Fearon, D. T. 1984. Epstein-Barr virus receptor of human B-lymphocytes is the C3d receptor CR2. *Proc. Natl. Acad. Sci. USA* 81: 4510–14

Frohman, M., Cowing, C. 1985. Presentation of antigen by B cells: Functional dependence on radiation dose, interleukins, cellular activation, and differential glycosylation. *J. Immunol.* 134: 2269–75

Gabbiani, G., Chaponnier, C., Zumbe, A., Vassalli, P. 1977. Actin and tubulin co-cap with surface immunoglobulins in mouse B lymphocytes. *Nature* 269: 697–98

Gold, M. R., DeFranco, A. L. 1987. Phorbol esters and dioctanoylglycerol block anti-IgM-stimulated phosphoinositide hydrolysis in the murine B cell lymphoma WEHI-231. *J. Immunol.* 138: 868–76

Gold, M. R., Miller, C. L., Mishell, R. I. 1985. Soluble non-cross-linked peptidoglycan polymers stimulate monocyte-macrophage inflammatory functions. *Infect. Immun.* 49: 731–41

Goodman, J. W., Sercarz, E. E. 1983. The complexity of structures involved in T-cell activation. *Ann. Rev. Immunol.* 1: 465–98

Goodman-Snitkoff, G., Mannino, R. J., McSharry, J. J. 1981. The glycoprotein isolated from vesicular stomatitis virus is mitogenic for mouse B lymphocytes. *J. Exp. Med.* 153: 1489–1502

Gordon, J., Ley, S. C., Melamed, M. D., Aman, P., Hughes-Jones, N. C. 1984a. Soluble factor requirements for the autostimulatory growth of B lymphoblasts immortalized by Epstein-Barr virus. *J. Exp. Med.* 159: 1554–59

Gordon, J., Ley, S. C., Melamed, M. D., English, L. S., Hughes-Jones, N. C. 1984b. Immortalized B lymphocytes produce B-cell growth factor. *Nature* 310: 145–47

Gray, P. W., Goeddel, D. V. 1983. Cloning and expression of murine immune interferon cDNA. *Proc. Natl. Acad. Sci. USA* 80: 5842–46

Grupp, S. A., Harmony, J. A. K. 1985. Increased phosphatidylinositol metabolism is an important but not an obligatory early event in B lymphocyte activation. *J. Immunol.* 134: 4087–94

Grupp, S. A., Snow, E. C., Harmony, J. A. K. 1987. The phosphatidylinositol response is an early event in the physiologically relevant activation of antigen-specific B lymphocytes. *Cell. Immunol.* In press

Guillet, J.-G., Lai, M.-Z., Briner, T. J., Smith, J. A., Gefter, M. L. 1986. Interaction of peptide antigens and class II major histocompatibility complex antigens. *Nature* 324: 260–62

Gurney, M. E., Apatoff, B. R., Spear, G. T., Baumel, M. J., Antel, J. P., et al. 1986. Neuroleukin: A lymphokine product of lectin-stimulated T cells. *Nature* 234: 574–81

Hamaoka, T., Ono, S. 1986. Regulation of B cell differentiation: Interactions of factors and corresponding receptors. *Ann. Rev. Immunol.* 4: 167–204

Hirano, T., Taga, T., Nakano, N., Yasukawa, K., Kashiwamura, S., et al. 1985. Purification to homogeneity and characterization of human B-cell differentiation factor (BCDF or BSFp-2). *Proc. Natl. Acad. Sci. USA* 82: 5490–94

Hirano, T., Yasukawa, K., Harada, H., Taga, T., Watanabe, Y., et al. 1986. Complementary DNA for a novel human interleukin (BSF-2) that induces B lymphocytes to produce immunoglobulin. *Nature* 324: 73–76

Hoessli, D., Rungger-Brandle, E., Jockusch, B. M., Gabbiani, G. 1980. Lymphocyte α-actinin. Relationship to cell membrane and co-capping with surface receptors. *J. Cell Biol.* 84: 305–14

Hornbeck, P., Paul, W. E. 1986. Anti-immunoglobulin and phorbol ester induce phosphorylation of proteins associated with the plasma membrane and cytoskeleton in murine B lymphocytes. *J. Biol. Chem.* 261: 14817–24

Howard, M., Farrar, J., Hilfiker, M., Johnson, B., Takatsu, K., et al. 1982. Identification of a T cell–derived B cell growth factor distinct from interleukin 2. *J. Exp. Med.* 155: 914–23

Irvine, R. F., Letcher, A. J., Heslop, J. P., Berridge, M. J. 1986. The inositol tris/tetrakisphosphate pathway—demonstration of Ins(1,4,5)P_3 3-kinase activity in animal tissues. *Nature* 320: 631–34

Irvine, R. F., Moor, R. M. 1986. Microinjection of inositol 1,3,4,5-tetrakisphosphate activates sea urchin eggs by a mechanism dependent on external Ca^{2+}. *Biochem. J.* 240: 917

Jakway, J. P., DeFranco, A. L. 1986. Pertussis toxin inhibition of B cell and macrophage responses to bacterial lipopolysaccharide. *Science* 234: 743–46

Jakway, J. P., Usinger, W. R., Gold, M. R., Mishell, R. I., DeFranco, A. L. 1986. Growth regulation of the B lymphoma cell line WEHI-231 by anti-immunoglobulin, lipopolysaccharide, and other bacterial products. *J. Immunol.* 137: 2225–31

Janeway, C. A., Jr., Bert, D. L., Shen, F.-W. 1980. Cell cooperation during in vivo anti-hapten antibody responses. V. Two synergistic Ly-1$^+$ 23$^-$ helper T cells with distinctive specificities. *Eur. J. Immunol.* 10: 231–36

Jones, B., Janeway, C. A., Jr. 1981. Cooperative interaction of B lymphocytes with antigen-specific helper T lymphocytes is MHC restricted. *Nature* 292: 547–49

Julius, M. H., von Boehmer, H., Sidman, C. L. 1982. Dissociation of two signals required for activation of resting B cells. *Proc. Natl. Acad. Sci. USA* 79: 1989–93

Jurgensen, C. H., Ambrus, J. L., Jr., Fauci, A. S. 1986. Production of B cell growth factor by normal human B cells. *J. Immunol.* 136: 4542–47

Justement, L., Chen, Z., Harris, L., Ransom, J., Sandoval, V., et al. 1986. BSF1 induces membrane protein phosphorylation but not phosphoinositide metabolism, Ca^{2+} mobilization, protein kinase C translocation, or membrane depolarization in resting murine B lymphocytes. *J. Immunol.* 137: 3664–70

Kakiuchi, T., Chesnut, R. W., Grey, H. M. 1983. B cells as antigen-presenting cells: The requirement for B cell activation. *J. Immunol.* 131: 109–14

Kaye, J., Porcelli, S., Tite, J., Jones, B., Janeway, C. A., Jr. 1983. Both a monoclonal antibody and antisera specific for determinants unique to individual cloned helper T cell lines can substitute for antigen and antigen-presenting cells in the activation of T cells. *J. Exp. Med.* 158: 836–56

Kehry, M., Ewald, S., Douglas, R., Sibley, C., Raschke, W., et al. 1980. The immunoglobulin μ chains of membrane-bound and secreted IgM molecules differ in their C-terminal segments. *Cell* 21: 393–406

Kelly, K., Cochran, B. H., Stiles, C. D., Leder, P. 1983. Cell-specific regulation of the c-*myc* gene by lymphocyte mitogens and platelet-derived growth factor. *Cell* 35: 603–10

Kelly, K., Siebenlist, U. 1986. The regulation and expression of c-*myc* in normal and malignant cells. *Ann. Rev. Immunol.* 4: 317–38

Kinashi, T., Harada, N., Severinson, E., Tanabe, T., Sideras, P., et al. 1986. Cloning of complementary DNA encoding T-cell replacing factor and identity with B-cell growth factor II. *Nature* 324: 70–73

Kishimoto, T. 1985. Factors affecting B-cell growth and differentiation. *Ann. Rev. Immunol.* 3: 133–57

Kishimoto, T., Ishizaka, K. 1975. Regulation of antibody response in vitro. IX. Induction of secondary anti-hapten IgG antibody response by anti-immuno-

174 DeFRANCO

globulin and enhancing soluble factor. *J. Immunol.* 114: 585

Klaus, G. G. B., Hawrylowicz, C. M., Holman, M., Keeler, K. D. 1984. Activation and proliferation signals in mouse B cells. III. Intact (IgG) anti-immunoglobulin antibodies activate B cells but inhibit induction of DNA synthesis. *Immunology* 53: 693–701

Korn, E. D. 1982. Actin polymerization and its regulation by proteins from non-muscle cells. *Physiol. Rev.* 62: 672–737

Kraft, A. S., Anderson, W. B., Cooper, H. L., Sando, J. J. 1982. Decrease in cytosolic calcium/phospholipid-dependent protein kinase activity following phorbol ester treatment of EL4 thymoma cells. *J. Biol. Chem.* 257: 13193–96

Krieger, J. I., Chesnut, R. W., Grey, H. M. 1986. Capacity of B cells to function as stimulators of a primary mixed leukocyte reaction. *J. Immunol.* 137: 3117–23

Krieger, J. K., Grammer, S. F., Grey, H. M., Chesnut, R. W. 1985. Antigen presentation by splenic B cells: Resting B cells are ineffective, whereas activated B cells are effective accessory cells for T cell responses. *J. Immunol.* 135: 2937

Kupfer, A., Swain, S. L., Janeway, C. A., Jr., Singer, S. J. 1986. The specific direct interaction of helper T cells and antigen-presenting B cells. *Proc. Natl. Acad. Sci. USA* 83: 6080–83

LaBaer, J., Tsien, R. Y., Fahey, K. A., DeFranco, A. L. 1986. Stimulation of the antigen receptor of WEHI-231 B lymphoma cells results in a voltage-independent increase in cytoplasmic calcium. *J. Immunol.* 137: 1836–44

Langdon, W. Y., Harris, A. W., Cory, S., Adams, J. M. 1986. The c-*myc* oncogene perturbs B lymphocyte maturation in E_μ-*myc* transgenic mice. *Cell* 47: 11–18

Lanier, L. L. 1982. Activation of murine B cell lymphomas. I. Influence of lipopolysaccharide. *J. Immunol.* 129: 1130–37

Lanzavecchia, A. 1985. Antigen-specific interaction between T and B cells. *Nature* 314: 537–39

Leclercq, L., Bismuth, G., Theze, J. 1984. Antigen-specific helper T-cell clone supernatant is sufficient to induce both polyclonal proliferation and differentiation of small resting B lymphocytes. *Proc. Natl. Acad. Sci. USA* 81: 6491–95

Leibson, H. J., Gefter, M., Zlotnik, A., Marrack, P., Kappler, J. W. 1984. Role of interferon-γ in antibody-producing responses. *Nature* 309: 799–801

Levine, J., Willard, M. 1983. Redistribution of fodrin (a component of the cortical cytoplasm) accompanying capping of cell surface molecules. *Proc. Natl. Acad. Sci. USA* 80: 191–95

Maino, V. C., Hayman, M. J., Crumpton, M. J. 1975. Relationship between enhanced turnover of phosphatidylinositol and lymphocyte activation by mitogens. *Biochem. J.* 146: 247–52

Maizel, A., Sahasrabuddhe, C., Mehta, S., Morgan, J., Lachman, L., Ford, R. 1982. Biochemical separation of a human B cell mitogenic factor. *Proc. Natl. Acad. Sci. USA* 79: 5998–6002

Malynn, B. A., Romeo, D. T., Wortis, H. H. 1985. Antigen-specific B cells efficiently present low doses of antigen for induction of T cell proliferation. *J. Immunol.* 135: 980–88

Matis, L. A., Glimcher, L. H., Paul, W. E., Schwartz, R. H. 1983. Magnitude of response of histocompatibility-restricted T-cell clones is a function of the product of the concentrations of antigen and Ia molecules. *Proc. Natl. Acad. Sci. USA* 80: 6019

Melchers, F., Andersson, J. 1986. Factors controlling the B-cell cycle. *Ann. Rev. Immunol.* 4: 13–36

Melchers, F., Erdei, A., Schulz, T., Dierich, M. P. 1985. Growth control of activated, synchronized murine B cells by the C3d fragment of human complement. *Nature* 317: 264–67

Michell, R. H. 1975. Inositol phospholipids and cell surface receptor function. *Biochim. Biophys. Acta* 415: 81–140

Mishell, R. I., Dutton, R. W. 1967. Immunization of dissociated spleen cell cultures from normal mice. *J. Exp. Med.* 126: 423–42

Mitchison, N. A. 1971. The carrier effect in the secondary response to hapten-protein conjugates. II. Cellular cooperation. *Eur. J. Immunol.* 1: 18–27

Mizuguchi, J., Tsang, W., Morrison, S. L., Beavan, M. A., Paul, W. E. 1986a. Membrane IgM, IgD and IgG act as signal transmission molecules in a series of B lymphomas. *J. Immunol.* 137: 2162–67

Mizuguchi, J., Beavan, M. A., Ohara, J., Paul, W. E. 1986b. BSF-1 action on resting B cells does not require elevation of inositol phospholipid metabolism or increased $[Ca^{2+}]_i$. *J. Immunol.* 317: 2215–19

Mizuguchi, J., Beavan, M. A., Li, J. H., Paul, W. E. 1986c. Phorbol myristate acetate inhibits anti-IgM-mediated signaling in resting B cells. *Proc. Natl. Acad. Sci. USA* 83: 4474–78

Moller, G. 1980. Effects of anti-immunoglobulin sera on B lymphocyte function. *Immunol. Rev.* 52: 1

Mond, J. J., Carman, J., Sarma, C., Ohara, J., Finkelman, F. D. 1986. Interferon-γ suppresses B cell stimulation factor (BSF-1) induction of class II MHC determinants on B cells. *J. Immunol.* 137: 3534–37

Mond, J. J., Mongini, P. K. A., Sieckmann, D., Paul, W. E. 1980. Role of T lymphocytes in the response to TNP-AECM-Ficoll. *J. Immunol.* 125: 1066–70

Mond, J. J., Seghal, E., Kung, J., Finkelman, F. D. 1981. Increased expression of I-region associated antigen (Ia) on B cells after crosslinking of surface immunoglobulin. *J. Immunol.* 127: 881–88

Monroe, J. G., Cambier, J. C. 1983. B cell activation. I. Antiimmunoglobulin-induced receptor cross-linking results in a decrease in the plasma membrane potential of murine B lymphocytes. *J. Exp. Med.* 157: 2073–86

Monroe, J. G., Kass, M. J. 1985. Molecular events in B cell activation. I. Signals required to stimulate G_0 to G_1 transition of resting B lymphocytes. *J. Immunol.* 135: 1674–82

Morrison, D. C., Ryan, J. L. 1979. Bacterial endotoxins and host immune responses. *Adv. Immunol.* 28: 293–450

Mosier, D. E. 1981. Primary in vitro antibody responses by purified murine B lymphocytes in serum-free defined medium. *J. Immunol.* 127: 1490–93

Mosier, D. E., Scher, I., Paul, W. E. 1976. In vitro response of CBA/N mice: Spleen cells of mice with an X-linked defect that precludes immune response to several thymus-independent antigens can respond to TNP-lipopolysaccharide. *J. Immunol.* 117: 1336

Mosmann, T. R., Cherwinski, H., Bond, M. W., Giedlin, M. A., Coffman, R. L. 1986. Two types of murine helper T cell clones. I. Definition according to profiles of lymphokine activities and secreted proteins. *J. Immunol.* 136: 2348–57

Muraguchi, A., Fauci, A. S. 1982. Proliferative responses of normal human B lymphocytes: Development of an assay system for human B cell growth factor (BCGF). *J. Immunol.* 129: 1104–8

Muraguchi, A., Nishimoto, H., Kawamura, N., Hori, A., Kishimoto, T. 1986. B cell-derived BCGF functions as autocrine growth factor(s) in normal and transformed B lymphocytes. *J. Immunol.* 137: 179–86

Nakajima, K., Hirano, Y., Takatsuki, F., Sakaguchi, N., Yoshida, N., Kishimoto, T. 1985. Physicochemical and functional properties of murine B cell–derived B cell growth factor II (WEHI-231-BCGF-II). *J. Immunol.* 135: 1207–12

Nakanishi, K., Howard, M., Muraguchi, A., Farrer, J., Takatsu, K., et al. 1983. Soluble factors involved in B cell differentiation:

Identification of two distinct T cell–replacing factors (TRF). *J. Immunol.* 130: 2219–24

Nel, A. E., Wooten, M. W., Landreth, G. E., Goldschmidt-Clermont, P. J., Stevenson, H. C., et al. 1986. Translocation of phospholipid/Ca^{2+}-dependent protein kinase in B-lymphocytes activated by phorbol ester or cross-linkage of membrane immunoglobulin. *Biochem. J.* 233: 145–49

Nelson, G. A., Andrews, M. L., Karnovsky, M. J. 1982. Participation of calmodulin in immunoglobulin capping. *J. Cell Biol.* 95: 771–80

Nelson, W. J., Colaco, C. A. L. S., Lazarides, E. 1983. Involvement of spectrin in cell-surface receptor capping in lymphocytes. *Proc. Natl. Acad. Sci. USA* 80: 1626–30

Nemerow, G. R., Wolfert, R., McNaughton, M. E., Cooper, N. R. 1985. Identification and characterization of the Epstein-Barr virus receptor on human B lymphocytes and its relationship to the C3d complement receptor (CR2). *J. Virol.* 55: 347–51

Nishizuka, Y. 1984. The role of protein kinase C in cell surface signal transduction and tumor promotion. *Nature* 308: 693–98

Noelle, R., Krammer, P. H., Ohara, J., Uhr, J. W., Vitetta, E. S. 1984. Increased expression of Ia antigens on resting B cells: An additional role for B cell growth factor. *Proc. Natl. Acad. Sci. USA* 81: 6149–53

Noma, Y., Sideras, P., Naito, T., Bergstedt-Lindquist, S., Azuma, C., et al. 1986. Cloning of cDNA encoding the murine IgG1 induction factor by a novel strategy using SP6 promoter. *Nature* 319: 640–46

Nordin, R. P., Potter, M. 1986. A macrophage-derived factor required by plasmacytomas for survival and proliferation in vitro. *Science* 233: 566–69

O'Garra, A., Warren, D. J., Holman, M., Popham, A. M., Sanderson, C. J., Klaus, G. G. B. 1986. Interleukin 4 (B-cell growth factor II/eosinophil differentiation factor) is a mitogen and differentiation factor for preactivated murine B lymphocytes. *Proc. Natl. Acad. Sci. USA* 83: 5228–32

Okada, M., Sakaguchi, N., Yoshimura, N., Hara, H., Shimizu, K., et al. 1983. B cell growth factors and B cell differentiation factor from human T hybridomas: Two distinct kinds of B cell growth factors and their synergism in B cell proliferation. *J. Exp. Med.* 157: 583

Owen, M. J., Auger, J., Barber, B. H., Edwards, A. J., Walsh, F. S., Crumpton, M. J. 1978. Actin may be present on the lymphocyte surface. *Proc. Natl. Acad. Sci. USA* 75: 4484–88

176 DeFRANCO

Pahwa, S., Pahwa, R., Saxinger, C., Gallo, R. C., Good, R. A. 1985. Influence of the human T-lymphotropic virus/lymphadenopathy-associated virus on functions of human lymphocytes: Evidence for immunosuppressive effects and polyclonal B-cell activation by banded viral preparations. *Proc. Natl. Acad. Sci. USA* 82: 8198–8202

Paige, C. J., Kincade, P. W., Ralph, P. 1981. Independent control of immunoglobulin heavy and light chain expression in a murine pre-B cell line. *Nature* 292: 631–33

Parker, D. C., Fothergill, J. J., Wadsworth, D. C. 1979. B lymphocyte activation by insoluble anti-immunoglobulin: Induction of immunoglobulin secretion by a T cell-dependent soluble factor. *J. Immunol.* 123: 931–41

Parsonnet, J., Hickman, R. K., Eardley, D. D., Pier, G. B. 1985. Induction of human interleukin 1 by toxic-shock-syndrome toxin-1. *J. Infect. Dis.* 151: 514–22

Petrini, M., Emerson, D. L., Galbraith, R. M. 1983. Linkage between surface immunoglobulin and cytoskeleton of B lymphocytes may involve Gc protein. *Nature* 306: 73–74

Petrini, M., Galbraith, R. M., Werner, P. A. M., Emerson, D. L., Arnaud, P. 1984. Gc (vitamin D binding protein) binds to cytoplasm of all human lymphocytes and is expressed on B-cell membranes. *Clin. Immunol. Immunopathol.* 31: 282–95

Phillips, N. E., Parker, D. C. 1983. Fc-dependent inhibition of mouse B cell activation by whole anti-μ antibodies. *J. Immunol.* 130: 602–6

Phillips, N. E., Parker, D. C. 1984. Cross-linking of B lymphocyte Fc gamma receptors and membrane immunoglobulin inhibits antiimmunoglobulin-induced blastogenesis. *J. Immunol.* 132: 627–32

Pletscher, M., Pernis, B. 1983. Internalized membrane immunoglobulin meets intracytoplasmic DR antigen in human lymphoblastoid cells. *Eur. J. Immunol.* 13: 581–84

Pozzan, T., Arslan, P., Tsien, R. Y., Rink, T. J. 1982. Antiimmunoglobulin, cytoplasmic free calcium, and capping in B lymphocytes. *J. Cell Biol.* 94: 335–40

Rabin, E. M., Mond, J. J., Ohara, J., Paul, W. E. 1986. Interferon-γ inhibits the action of B cell stimulatory factor (BSF)-1 on resting B cells. *J. Immunol.* 137: 1573–76

Ransom, J. T., Cambier, J. C. 1986. B cell activation. VII. Independent and synergistic effects of mobilized calcium and diacylglycerol on membrane potential and I-A expression. *J. Immunol.* 136: 66–72

Ransom, J. T., Harris, L. K., Cambier, J. C. 1986. Anti-Ig induces release of inositol 1,4,5-trisphosphate, which mediates mobilization of intracellular Ca^{++} stores in B lymphocytes. *J. Immunol.* 137: 708–14

Rock, K. L., Benacerraf, B., Abbas, A. K. 1984. Antigen presentation by hapten-specific B lymphocytes. I. Role of surface immunoglobulin receptors. *J. Exp. Med.* 160: 1102–13

Roehm, N. W., Leibson, H. J., Zlotnik, A., Kappler, J., Marrack, P., Cambier, J. C. 1984. Interleukin-induced increase in Ia expression by normal mouse B cells. *J. Exp. Med.* 160: 679–94

Rogers, J., Early, P., Carter, C., Calame, K., Bond, M., et al. 1980. Two mRNAs with different 3' ends encode membrane-bound and secreted forms of immunoglobulin μ chain. *Cell* 20: 303–12

Rosenspire, A. J., Choi, Y. S. 1982. Relation between actin-associated proteins and membrane immunoglobulin in B cells. *Mol. Immunol.* 19: 1515–26

Rosoff, P. M., Cantley, L. C. 1985. Lipopolysaccharide and phorbol esters induce differentiation but have opposite effects on phosphatidylinositol turnover and Ca^{2+} mobilization in 70Z/3 pre-B lymphocytes. *J. Biol. Chem.* 260: 9209–15

Salisbury, J. L., Condeelis, J. S., Maihle, N. J., Satir, P. 1981. Calmodulin localization during capping and receptor-mediated endocytosis. *Nature* 294: 163–66

Salisbury, J. L., Condeelis, J. S., Satir, P. 1980. Role of coated vesicles, microfilaments, and calmodulin in receptor-mediated endocytosis by cultured B lymphoblastoid cells. *J. Cell Biol.* 87: 132–41

Schade, U., Rietschel, E. T. 1982. The role of prostaglandins in endotoxic activities. *Klin. Wochenschr.* 60: 743–45

Scher, I. 1982. The CBA/N mouse strain: An experimental model illustrating the influence of the X-chromosome on immunity. *Adv. Immunol.* 33: 1–71

Schimpl, A., Wecker, E. 1982. Replacement of T cell function by a T cell product. *Nature* 237: 15

Schnittman, S. M., Lane, H. C., Higgins, S. E., Folks, T., Fauci, A. S. 1986. Direct polyclonal activation of human B lymphocytes by the acquired immune deficiency syndrome virus. *Science* 233: 1084–86

Schreiner, G. F., Fujiwara, K., Pollard, T. D., Unanue, E. R. 1977. Redistribution of myosin accompanying capping of surface Ig. *J. Exp. Med.* 145: 1393–98

Schreiner, G. F., Unanue, E. R. 1976. Membrane and cytoplasmic changes in B lymphocytes induced by ligand surface interactions. *Adv. Immunol.* 24: 37–165

Sell, S., Gell, P. G. H. 1965. Studies on rabbit lymphocytes in vitro. I. Stimulation of blast transformation with an antiallotypic serum. *J. Exp. Med.* 122: 423

Sharma, S., Mehta, S., Morgan, J., Maizel, A. 1987. Molecular cloning and expression of a human B-cell growth factor gene in *Escherichia coli. Science* 235: 1489–92

Sidman, C. L., Paige, C. J., Schreier, M. H. 1984. B cell maturation factor (BMF): A lymphokine or family of lymphokines promoting the maturation of B lymphocytes. *J. Immunol.* 132: 209–22

Sidman, C. L., Unanue, E. R. 1976. Control of B-lymphocyte function. I. Inactivation of mitogenesis by interactions with surface immunoglobulin and Fc-receptor molecules. *J. Exp. Med.* 144: 882–96

Smeland, E., Godal, T., Ruud, E., Bleiske, K., Funderud, S., et al. 1985. The specific induction of *myc* protooncogene expression in normal human B cells is not a sufficient event for acquisition of competence to proliferate. *Proc. Natl. Acad. Sci. USA* 82: 6255–59

Smith, K. A. 1984. Interleukin 2. *Ann. Rev. Immunol.* 2: 319–33

Snow, E. C., Fetherston, J. D., Zimmer, S. 1986. Induction of the c-*myc* protooncogene after antigen binding to hapten-specific B cells. *J. Exp. Med.* 164: 944–49

Snow, E. C., Noelle, R. J., Uhr, J. W., Vitetta, E. S. 1983. Activation of antigen-enriched B cells. II. Role of linked recognition in B cell proliferation to thymus-dependent antigens. *J. Immunol.* 130: 614–18

Stavnezer, J., Abbott, J., Sirlin, S. 1984. Immunoglobulin heavy chain switching in cultured I.29 murine B lymphoma cells: Commitment to an IgA or IgE switch. *Curr. Top. Microbiol. Immunol.* 113: 109–16

Stein, P., Dubois, P., Greenblatt, D., Howard, M. 1986. Induction of antigen-specific proliferation in affinity-purified small B lymphocytes: Requirement for BSF-1 by type 2 but not type 1 thymus-independent antigens. *J. Immunol.* 136: 2080–89

Stewart, S. J., Prpic, V., Powers, F. S., Bocckino, S. B., Issaks, R. E., Exton, J. H. 1986. Perturbation of the human T-cell antigen receptor–T3 complex leads to the production of inositol tetrakisphosphate: Evidence for conversion from inositol trisphosphate. *Proc. Natl. Acad. Sci. USA* 83: 6098–6102

Streb, H., Irvine, R. F., Berridge, M. J., Schulz, I. 1983. Release of Ca^{2+} from a

nonmitochondrial intracellular store in pancreatic acinar cells by inositol-1,4,5-trisphosphate. *Nature* 306: 67–69

Stryer, L., Bourne, H. R. 1986. G-proteins: A family of signal transducers. *Ann. Rev. Cell Biol.* 2: 391–420

Swain, S. L., Howard, M., Kappler, J., Marrack, P., Watson, J., et al. 1983. Evidence for two distinct classes of murine B cell growth factors with activities in different functional assays. *J. Exp. Med.* 158: 822–35

Takatsu, K., Harada, N., Hara, Y., Takahama, Y., Yamada, G., et al. 1985. Purification and physicochemical characterization of murine T cell replacing factor (TRF). *J. Immunol.* 134: 382–89

Taylor, R. B., Duffus, P. H., Raff, M. C., dePetris, S. 1971. Redistribution and pinocytosis of lymphocyte surface immunoglobulin molecule induced by anti-immunoglobulin antibody. *Nature New Biol.* 233: 225–29

Tony, H.-P., Parker, D. C. 1985. Major histocompatability complex–restricted polyclonal B cell responses resulting from helper T cell recognition of anti-immunoglobulin presented by small B lymphocytes. *J. Exp. Med.* 161: 223–41

Tony, H.-P., Phillips, N. E., Parker, D. C. 1985. Role of membrane immunoglobulin (Ig) crosslinking in membrane Ig-mediated, major histocompatability-restricted T cell–B cell cooperation. *J. Exp. Med.* 162: 1695–1708

Usinger, W. R., Clark, G. C., Gottschalk, E., Holt, S., Mishell, R. I. 1985. Characteristics of bacterium GB-2, a presumptive Cytophaga species with novel immunoregulatory properties. *Curr. Microbiol.* 12: 203

Wall, R., Kuehl, M. 1983. Biosynthesis and regulation of immunoglobulins. *Ann. Rev. Immunol.* 1: 393–422

Wetzel, G. D., Kettman, J. R. 1981. Activation of murine B lymphocytes. III. Stimulation of B lymphocyte clonal growth with lipopolysaccharide and dextran sulfate. *J. Immunol.* 126: 723–28

Wightman, P. D., Raetz, C. R. H. 1984. The activation of protein kinase C by biologically active lipid moieties of lipopolysaccharide. *J. Biol. Chem.* 259: 10048–52

Wilson, H. A., Greenblatt, D., Poenie, M., Finkelman, F. D., Tsien, R. Y. 1987. *J. Exp. Med.* In press

Yarchoan, R., Redfield, R. R., Broder, S. 1986. Mechanisms of B cell activation in patients with acquired immunodeficiency syndrome and related disorders. *J. Clin. Invest.* 78: 439–47

Yokata, T., Otsuka, T., Mosmann, T., Banchereau, J., DeFrance, T., et al. 1986. Isolation and characterization of a human interleukin cDNA clone, homologous to mouse B-cell stimulatory factor 1, that expresses B-cell and T-cell stimulating activities. *Proc. Natl. Acad. Sci. USA* 83: 5894–98

Ziegler, H. K., Unanue, E. R. 1982. Decrease in macrophage antigen catabolism caused by ammonia and chloroquine is associated with inhibition of antigen presentation to T cells. *Proc. Natl. Acad. Sci. USA* 79: 175–78

Zubler, R. H. 1984. Polyclonal B cell responses in the presence of defined filler cells: Complementary effects of lipopolysaccharide and anti-immunoglobulin antibodies. *Eur. J. Immunol.* 14: 357–63

Ann. Rev. Cell Biol. 1987. 3 : 179–205
Copyright © 1987 by Annual Reviews Inc. All rights reserved

CELL SURFACE RECEPTORS FOR EXTRACELLULAR MATRIX MOLECULES

Clayton A. Buck

The Wistar Institute, 36th Street at Spruce, Philadelphia,
Pennsylvania 19104

Alan F. Horwitz[1]

Department of Biochemistry and Biophysics, The University of
Pennsylvania, Philadelphia, Pennsylvania 19104

CONTENTS

INTRODUCTION.. 179
INTEGRIN, THE AVIAN FIBRONECTIN RECEPTOR... 180
EXTRACELLULAR MATRIX RECEPTORS FROM MAMMALIAN CELLS...................................... 188
EXTRACELLULAR MATRIX–LIKE RECEPTORS ON PLATELETS... 190
EXTRACELLULAR MATRIX–LIKE RECEPTORS ON LYMPHOID CELLS 190
SUPERGENE FAMILY OF RECEPTORS 192
BIOSYNTHESIS AND PROCESSING OF INTEGRINS.. 193
EXTRACELLULAR MATRIX RECEPTORS NOT BELONGING TO THE INTEGRIN FAMILY 194
SUMMARY AND CONCLUSIONS... 195

INTRODUCTION

The ability of cells to adhere to the extracellular matrix is central to a
number of phenomena, including maintenance of tissue integrity, wound
healing, morphogenic movements, cellular migrations, and metastasis.
Sites of cell matrix adhesions have been identified ultrastructurally as
regions of cell surface thickening adjacent to extracellular matrix components (Abercrombie et al 1971). Actin filaments often appear to interact

[1] Present address: Department of Cell Biology, University of Illinois, Morrill Hall, Urbana,
Illinois 61801.

179

0743–4634/87/1115–0179$02.00

and terminate at these sites. The highly specialized nature of junctions of the adherens type and focal contacts, their analogous structures on fibroblasts in vitro, has made them easy to identify and subjects of intense investigation (Geiger et al 1985; Burridge 1987). A large number of molecules have been identified as components of these highly organized structures. They include actin, α-actinin, fibronectin, talin, vinculin, and other cytoskeleton-associated molecules thought to link actin filaments to the cell surface (Wehland et al 1979; Burridge & Feramisco 1980; Geiger et al 1980; Burridge & Connell 1983). Extracellular molecules in these junctions are less well characterized but include fibronectin and heparin sulfate proteoglycans (Singer & Paradiso 1982; Woods et al 1984). The codistribution of intracellular actin with extracellular fibronectin fibrils, along with the organizing effect of exogenous fibronectin on the cytoskeleton, has led to the hypothesis that transmembrane matrix receptors exist that link the cytoskeleton with the extracellular matrix (Heggeness et al 1978; Hynes & Destree 1978; Singer 1979; Birchmeier et al 1980).

Two approaches have been particularly productive in the search for such postulated cell surface receptors for extracellular matrix molecules. One is based on the early studies of slime mold aggregation (Gerisch 1977) in which antibodies capable of perturbing cellular adhesive events were used to identify by immunoprecipitation, biochemical purification, or antibody-affinity chromatography the molecules with which such antibodies interacted. This approach was extended to include cell matrix interactions (Wylie et al 1979) and became most useful with the advent of monoclonal antibodies (Kohler & Milstein 1975). The second approach evolved later and consists of affinity chromatography with ligands derived from the extracellular matrix (Malinoff & Wicha 1983; Rao et al 1983; Lesot et al 1983; Pyrela et al 1985). Convincing demonstrations that the molecules identified in these ways were cell-matrix receptors included the use of antibodies to inhibit cell adhesion and to localize the molecules in regions of cell-matrix interactions, and demonstrations that in vitro receptor-ligand binding was inhibited by adhesion-perturbing antibodies or small peptides derived from the cell-binding region of extracellular ligands. The purpose of this review is to summarize the work that has led to the isolation and characterization of molecules most likely to function as extracellular matrix receptors.

INTEGRIN, THE AVIAN FIBRONECTIN RECEPTOR

One of the first set of molecules to be identified and characterized extensively as possible receptors for extracellular matrix molecules was the complex from avian cells (Neff et al 1982; Greve & Gottlieb 1982). Avian integrin, known also as the CSAT antigen or the 140-kDa complex, was isolated using two adhesion-perturbing monoclonal antibodies, CSAT and JG22, produced in independent laboratories at about the same time (Greve & Gottlieb 1982; Horwitz et al 1982; Neff et al 1982). This multifunctional extracellular matrix receptor was found on muscle, fibroblasts, neurons, and many other cell types (Neff et al 1982; Decker et al 1984; Chen et al 1985a; Bozyczko & Horwitz 1986; Duband et al 1986; Krotoski et al 1986; Tomaselli et al 1986; Hall et al 1987). Three major lines of evidence suggest that integrin functions as a receptor for extracellular matrix molecules. (a) Monoclonal antibodies directed against it perturb cell-matrix adhesion specifically and reversibly; (b) the receptor is localized at adhesion sites; and (c) the purified receptor binds directly to the extracellular matrix molecules to which the monoclonal antibodies that inhibit cell adhesion bind.

The adhesion-inhibiting properties of the monoclonal antibodies CSAT and JG22 have been extensively studied in vitro. Although there is little effect of the antibodies on adhesion to poly-L-lysine, they do inhibit cellular adhesion to vitronectin (cell-spreading factor), fibronectin, laminin, and types I and IV collagen (Decker et al 1984; Chen et al 1985b; Horwitz et al 1985). Adhesion of migratory cells to these substrata is particularly sensitive to these antibodies. The addition of CSAT/JG22 antibodies to short-term cultures of skeletal muscle, somitic fibroblasts, neural crest cells, and neurons causes cell rounding and detachment from the substratum (Decker et al 1984; Bronner-Fraser 1985; Duband et al 1986; Bozyczko & Horwitz 1986). However, the adhesive properties of all cell types are not equally susceptible to the adhesion-perturbing effects of these antibodies (Neff et al 1982; Decker et al 1984; Horwitz et al 1985). For example, early cultures of myoblasts are more easily dissociated from their substrata by these antibodies than are older cultures; cardiac fibroblasts plated on fibronectin are highly resistant to dissociation by these antibodies, whereas the same cells plated on laminin are particularly susceptible to their action. In contrast, tendon fibroblasts are quite susceptible to the action of these antibodies under both conditions. These observations suggest that the role of integrin in cell-matrix adhesion may vary with the cell type and the matrix to which the cell is adhering. They also suggest that in addition to integrin other cell-matrix receptors are important in maintaining the interaction of cells with the various substrata to which they come in contact.

Monoclonal and polyclonal antibodies have been used to show the distribution of integrin on fibroblasts and other cell types (Chen et al 1985a,b; Damsky et al 1985). On well-spread, stationary fibroblasts integrin is found on regions of the membrane lying parallel to intracellular actin filament bundles and at the termini of these bundles near the cell periphery. Integrin is also present on the edge of the leading lamellae. The distribution of integrin along portions of stress fibers and at their termini in focal contacts corresponds to the distribution of areas enriched in cytoskeleton-associated molecules, such as vitronectin, talin, and α-actinin, thought to serve as linkers between the actin-containing stress fibers and the cell surface. Talin colocalizes with integrin at the leading lamellae. Similarly, the distribution of integrin along the surface of cells in contact with the substratum parallels that of extracellular fibronectin fibrils. These regions of the surface represent putative attachment sites where the substratum and the cell surface approach each other closely, as visualized by interference reflection microscopy and electron microscopy. The precise relative distribution of all these molecules within the adhesive sites is unclear and may depend upon the conditions under which cells are cultured. On some cells integrin and fibronectin are found at the periphery of focal contacts, which are shaped like the eye of a needle, and vinculin is at the center of the "eye." On other cells cultured under different conditions, or on other regions of the same cell, integrin, cytoskeleton-associated molecules, and fibronectin are colocalized to the same site on the cell surface. The reasons for this variability are not clear, but they may involve the properties of the different adhesive complexes. Another monoclonal antibody, 30B6, which may be directed against an integrinlike molecule, colocalizes with vinculin in focal contacts of fibroblasts and at other sites of actin-membrane linkage, e.g. the cleavage furrow of mitotic cells and the dense plaques of smooth and cardiac muscle. The antigen isolated by the 30B6 monoclonal antibody, called actosialin, on SDS-PAGE behaves as a dimer composed of a 170- and a 130-kDa subunit (165 and 107 kDa unreduced). It appears to be very similar to integrin, but a relationship remains to be established (Rogalski & Singer 1985; Rogalski 1987).

The distribution of integrin on nonfibroblastic cells has also been studied. On neurons integrin is localized along microspikes on the growth cone, along the axon, and on the cell body (Bozyczko & Horwitz 1986). The monoclonal antibody CSAT inhibits axonal extension and attachment of somas of newly plated neurons. Less well spread motile cells, such as skeletal myoblasts, neural crest cells, and short-term cultures of somitic fibroblasts, have been studied (Damsky et al 1985; Duband et al 1986). In general, these cells have neither highly organized bundles of actin filaments nor vinculin-rich adhesion plaques. Integrin is diffusely distributed on the surface of these cells. However, with continued time in culture, both muscle cells and fibroblasts develop more organized cell-matrix adhesion sites and coincidentally become more sensitive to the adhesion-perturbing properties of the CSAT monoclonal antibody. The less organized distribution may produce more labile contacts required for cell movement.

Integrin has been isolated by immunoaffinity chromatography of cell extracts obtained by lysis of cells with nonionic detergent. The receptor migrates on SDS-PAGE under reducing conditions as a broad band in the range of 140 kDa (Greve & Gottlieb 1982; Neff et al 1982; Chapman 1984; Hasegawa et al 1985; Knudsen et al 1985; Akiyama et al 1986). However, when electrophoresed under nonreducing conditions, integrin resolves into polypeptides migrating at 110, 120, and 160 kDa (Knudsen et al 1985; Akiyama et al 1986). The migration of band 3, the 110-kDa glycopeptide, is somewhat anomolous in that it runs slower on SDS-PAGE in its reduced form, which suggests it has a high degree of intramolecular disulfide bonding. Band 1 migrates more rapidly on reduced SDS-PAGE, presumably due in part to the release of a 25-kDa disulfide-linked component. Analysis of integrin as isolated by immunoaffinity reveals the following: the receptor consists of at least three distinct polypeptides having unique peptide maps. Antibodies prepared against the individual members of the complex do not show a high degree of cross-reactivity. All three polypeptides are glycosylated and appear to be integral membrane proteins. They cannot be extracted without detergent, they associate with liposomes, and sequence information available for two of the glycopeptides shows a single membrane-spanning region on each. Hydrodynamic data confirm that the individual peptides are associated and that the receptor exists as either a heterodimer or heterotrimer with an aggregate molecular mass of approximately 210–250 kDa (Buck et al 1985; Horwitz et al 1985).

Proof that the molecular complex composing integrin is indeed a receptor for extracellular matrix molecules has been complicated by the fact that the receptor-ligand interaction is of only moderate affinity (Akiyama et al 1985). In light of this, two methods have been devised to demonstrate the interaction of integrin with matrix molecules. One is a modification of conventional gel filtration in which the column is first equilibrated with the ligand to be assayed for binding and then a mixture of integrin and the ligand are applied to the column (Horwitz et al 1985). The receptor then elutes with the Stokes' radius of the receptor-ligand complex rather than that of the receptor alone. This technique avoids problems arising from receptor-ligand dissociation when, as is the case with fibronectin or laminin, the receptor-ligand complex tends to dissociate rapidly. The affinities of integrin for fibronectin and laminin are in the micromolar range and that for fibronectin agrees well with the affinity estimated by binding of fibronectin fragments to cells. In this case, the specificity of the ligand-receptor interaction was demonstrated by including the cell-binding fibronectin peptide arg-gly-asp-ser (RGDS) in the column at concentrations corresponding to those required to inhibit cell adhesion to these matrix molecules (Pierschbacher & Ruoslahti 1984b; Yamada & Kennedy 1985). A control peptide (RG*E*S) in which glutamate was substituted for aspartate failed to inhibit receptor-ligand interactions during equilibrium gel filtration. Another method to show integrin-fibronectin binding consists of comparing the elution profile of purified integrin from an affinity column made with the 75-kDa cell-binding fragment of fibronectin in the presence of either the cell-binding tetrapeptide RGDS or the control peptide RG*E*S (Akiyama et al 1986). Elution from the column in the presence of the control peptide was delayed.

The equilibrium gel filtration assay has been used to demonstrate that vitronectin, laminin, and fibronectin all bind directly to integrin. Vitronectin does not show the rapid dissociation rates seen with the other ligands. The RGDS peptide, but not control peptides, inhibits the adhesion of all ligands tested thus far, as do the adhesion-inhibiting monoclonal antibodies CSAT and JG22. Thus the binding properties of the ligand under conditions of equilibrium gel filtration resemble those of the cell-associated ligand. Interestingly, vitronectin binding competes with that of fibronectin, and fibronectin binding competes with that of laminin. Thus, each ligand binds to the same site or neighboring sites. Avian integrin appears to be a multifunctional, promiscuous receptor for at least three, and likely more, extracellular matrix molecules (Buck & Horwitz 1987).

A number of observations have suggested a transmembrane linkage between cytoskeletal elements and extracellular fibronectin (see Burridge 1987 for a recent review). The localization of integrin in regions of transmembrane linkage along with its integral association with the membrane points to its role as such a linkage molecule (Chen et al 1985a,b; Damsky et al 1985). Most data suggest that the interaction of cytoskeletal elements with the cell membrane occurs indirectly via cytoskeleton-associated molecules such as talin, α-actinin, or vinculin (Burridge 1987). The interaction of these molecules with avian integrin was tested by equilibrium gel filtration (Horwitz et al 1986). Of these, only talin showed a detectable interaction with integrin. This binding, also of a moderate affinity, was not inhibited by extracellular matrix ligands, such as vitronectin and fibronectin, nor was it inhibited by the RGD cell-binding peptide. Thus talin-integrin binding occurs at a site distinct from that required for integrin-fibronectin binding or binding to other matrix molecules. Also, the binding of integrin to talin occurs at a site distinct from that required for talin-vinculin binding (Buck & Horwitz 1987).

Peptides corresponding to amino acid sequences near the carboxyl terminus of band 3 of integrin have been synthesized. One of these peptides is a decamer that corresponds to a band 3 tryptic peptide containing the consensus tyrosine kinase substrate (Tamkun et al 1986); the other peptide corresponds to the last 20 amino acids of the cytoplasmic domain of band 3. Both peptides inhibit the binding of talin to integrin. Neither peptide has an effect on the binding of extracellular matrix ligands to integrin. In addition, recent observations indicate that the decamer serves as a kinase substrate and corresponds to the integrin peptide phosphorylated in vivo following Rous sarcoma virus transformation (see below). These findings taken together imply that this site on the receptor is involved in the binding of talin to integrin.

The structural features of integrin, particularly as they relate to its function, have been investigated (Buck et al 1986). The oligomeric nature of the receptor and the difficulty in purifying the individual subunits under nondenaturing conditions have made such studies difficult. However, a monoclonal antibody directed against the lower molecular weight, cysteine-rich subunit that has the property of dissociating the receptor has been quite useful for such studies. Using this antibody, integrin can be separated into two fractions, one containing the lower molecular weight band 3 subunit, and the second containing the two other subunits (SDS-PAGE bands 1 and 2). The structural and functional properties of these two fractions differ from those of the native receptor. When band 3 alone or bands 1 plus 2 are subjected to equilibrium gel filtration in the presence of fibronectin, laminin, or talin, no binding activity is seen. However, if these fractions are recombined, a typical integrin complex of three subunits is reconstituted, as seen by its behavior on gel filtration; this reconstituted complex is able to bind fibronectin, laminin, and talin. In addition, the monoclonal antibodies CSAT and JG22 bind only to band 3. These observations not only confirm the oligomeric structure of integrin, but suggest that it functions as an oligomer as well.

Further structural information has come from the recent sequencing of cDNA for band 3 glycoprotein from integrin (Tamkun et al 1986). It has several noteworthy features, including (a) a short (47–amino acid) cytoplasmic domain containing a consensus sequence for phosphorylation by tyrosine kinases; (b) four repeats of a cysteine-rich motif; (c) a single membrane spanning domain; and (d) several presumptive glycosylation sites. The entire sequence encodes an 803–amino acid peptide. A synthetic peptide corresponding to the consensus tyrosine phosphorylation sequence is phosphorylated in vitro by pp60 src, is identical to the tryptic peptide isolated from integrin phosphorylated in vivo, and inhibits talin binding.

Alterations in the function and distribution of integrin following viral transformation of chick cells have been examined. A major phenotypic change seen upon Rous sarcoma virus (RSV) transformation is an alteration in the adhesion and morphology of the cell (see Burridge 1987 for review). The localization of the RSV transforming gene product, pp60 src kinase, in sites of cell-matrix adhesion suggests that components of the adhesion plaque are likely substrates for this kinase (Rohrschneider 1980). Alterations in these substrates may contribute to the altered morphology and adhesiveness seen following viral transformation. As stated above, integrin is found localized in most cells at points of cell-matrix adhesion and beneath actin-containing microfilament bundles. Upon RSV transformation cellular fibronectin is lost, as are clearly organized actin bundles. The cytoskeleton-associated molecules α-actinin and vinculin are more diffusely distributed in transformed cells (David-Pfeuty & Singer 1980), as is integrin. These changes in the normal distribution of fibronectin receptors and cytoskeletal elements are seen in transformed mammalian cells and in chick cells transformed with temperature-sensitive viruses grown under permissive but not under nonpermissive conditions (Chen et al 1986b; Giancotti et al 1986a; Hirst et al 1986). Interestingly, the distributions of these elements in normal monocytes are similar to those in RSV-transformed cells (Marchisio et al 1987). The amount of integrin in both transformed and control cells is similar. There may be some differences in the migration of the individual bands on SDS-PAGE following transformation, but the origin of these differences remains to be studied. The glycoproteins composing bands 2 and 3 of integrin are phosphorylated on tyrosine, and to some extent serine, following RSV transformation (Hirst et al 1986). The ability of integrin from transformed cells to bind talin and extracellular matrix molecules is greatly reduced, which suggests that the changes in the distribution of cytoskeletal elements and the adhesive properties of transformed cells may be due in part to the phosphorylation of integrin. While previous explanations of changes in cellular behavior following transformation have focused upon the loss of matrix molecules from the cell surface and increased levels of cell-associated proteases, it seems clear that change in the functional properties of cell-matrix receptors must also be considered (Chen et al 1986b).

Integrin is clearly important as a receptor for extracellular matrix molecules in vitro. It also appears to function as such in vivo. The role of integrin in vivo has been investigated using CSAT and JG22 to reveal the distribution of integrin within various tissues and by studying the effect of these antibodies on embryonic development (Chen et al 1985a). The localization of the integrin complex in tissue has been determined using immunofluorescence and immunoelectron microscopy. In general, integrin is present in sites adjacent to junctions of the adherens type. In smooth muscle, it is concentrated adjacent to dense plaques and at membrane sites where fibrils of extracellular materials appear to contact the membrane.

On the columnar epithelium of the intestine, integrin is found primarily on the basolateral membrane in regions of close cell apposition.

Studies of neural crest and muscle development in vivo have demonstrated that integrin is important to adhesive interactions during development. Monoclonal antibodies and hybridomas producing the CSAT monoclonal antibody were injected into the mesenchyme of developing chick embryos, and the distribution of cranial neural crest cells was monitored by staining them with the neural crest–specific monoclonal antibody HNK-1 (Bronner-Fraser 1985). The distribution of neural crest cells within the embryo was notably altered. Some cells accumulated abnormally in the lumen of the neural tube, and others migrated along aberrant pathways. After longer periods of incubation (36–48 hr), the presence of the antibody resulted in formation of an open or otherwise malformed neural tube. In general, the effects of antibody on trunk neural crest cells were less apparent than those on cranial neural crest cells. Similar results are seen if the synthetic fibronectin cell-binding tetrapeptide RGDS is injected into early embryos (Boucaut et al 1984).

The long-term effects of the CSAT monoclonal antibody on embryonic muscle development have been studied (Jaffredo et al 1986). The injection of a CSAT-producing hybridoma into the body cavity of a 3- to 4-day-old embryo resulted, 7–9 days later, in anomolous development of the abdominal musculature. Abdominal muscles were either absent or disorganized, resulting in transparency of the ventral abdomen. Presumably, these anomalies arose from an inhibition of muscle cell migration.

The histotypic distribution of integrin in the embryo has been surveyed (Duband et al 1986; Krotoski et al 1986). Integrin can first be detected in the epiblast at the periphery of most cells. In older embryos it remains broadly distributed and is found on cells derived from all three primary germ layers. In general, integrin appears enriched near high concentrations of fibronectin and/or laminin. The intensity of integrin staining varies among tissues. Endocardiac, aortic, hematopoietic, and possibly vertebral cartilage are the only cell types displaying little or no receptor staining. In contrast, some tissues, like endothelium and limb cephalitic mesenchyme, show relatively high concentrations of integrin. During formation of epithelium, the receptor becomes regionalized along the basolateral surface and is generally enriched on the basal surface adjacent to the basement membrane. During the development of red cells, the receptor is entirely extinguished (Yamada et al 1986). Its concentration is also diminished during the development of lung epithelium, mesenchyme, and endothelium, and on certain skeletal muscles (Chen et al 1986a). In neural retinal cells changes in the molecular properties of the antigen that parallel alterations in adhesive phenomena have been reported (Hall et al 1987).

Receptors remaining on skeletal muscles appear regionalized at the myo-tendenous junctions and at neuromuscular junctions. Thus in vivo as well as in vitro, integrin is localized in avian cells at sites of important adhesive events, and reagents that alter the interaction of integrin with the extra-cellular matrix result in aberrant development.

EXTRACELLULAR MATRIX RECEPTORS FROM MAMMALIAN CELLS

Whereas adhesion-perturbing monoclonal antibodies were the primary tools in the detection and isolation of avian integrin, a different approach was instrumental in isolating the fibronectin binding receptors from mammalian cells. This approach was based upon the discovery that the cell-binding activity of fibronectin can be mimicked by the tetrapeptide RGDS (Pierschbacher & Ruoslahti 1984a,b). This peptide is found in the cell-binding fragment of fibronectin, promotes adhesion of cells to substrata upon which it is cross-linked, and inhibits the attachment of cells to fibronectin in a competition assay. Using this peptide to selectively elute material from a fibronectin affinity column, Pytela et al (1985a) identified a 140-kDa glycoprotein complex which, when incorporated into lipid vesicles, caused the vesicles to adhere to fibronectin but not laminin. The adhesion of these vesicles was inhibited specifically by peptides incor-porating the amino acid sequence RGDS. The evidence that this glyco-protein complex is a surface membrane constituent responsible for cell attachment to fibronectin is as follows: It is extractable only with deter-gents; it is labeled with ^{125}I following cell surface iodination; it binds to affinity columns containing the cell-binding fragment of fibronectin; it is specifically eluted with the cell-binding peptide, but not by chemically similar peptides in which the aspartate residue is replaced by glutamate; it can be incorporated into lipid vesicles which then bind to fibronectin-coated surfaces but not to surfaces coated with vitronectin or laminin.

As isolated by affinity chromatography, the fibronectin receptor from mammalian cells consists of two glycoproteins that when subjected to SDS-PAGE under reducing conditions migrate as a broad band of about 140 kDa but when electrophoresed under nonreducing conditions can be resolved into two components of approximately 160 and 120 kDa (Pytela et al 1985a). Structural studies have shown that the change in apparent M_r of the 160-kDa member of the complex upon reduction is the result of the loss of a disulfide-linked 25-kDa peptide. Thus, as with the cytoadherins the higher molecular weight subunit of the mammalian fibronectin receptor consists of two peptides, a heavy chain and a light chain, joined by a disulfide bridge. cDNA sequence analysis reveals that the light chain contains the cytoplasmic and membrane-spanning domain, which suggests that the light chain serves as the membrane anchor for the 160-kDa subunit (Argraves et al 1986; Plow et al 1986).

The receptor for the extracellular matrix molecule, vitronectin, was isolated similarly by affinity chromatography and eluted specifically with a hexapeptide containing the RGD sequence (Pytela et al 1985b). Like the mammalian fibronectin receptor, the vitronectin receptor proved to be a heterodimer consisting of two glycoproteins of 150 and 115 kDa. This receptor could be incorporated into lipid vesicles, which would then bind to vitronectin but not fibronectin. This binding was prevented by the presence in the binding medium of peptides containing the RGD amino acid sequence. cDNA sequence analysis demonstrated that the 150-kDa subunit of the vitronectin receptor, like the larger subunit of the fibronectin receptor, consists of a heavy and a light chain and that the light chain serves as the membrane anchor (Suzuki et al 1986). The degree of homology between the deduced amino acid sequence of the large subunit from the vitronectin and fibronectin receptors ranges between 34% and 48% depending upon whether or not conservative amino acid replacements are considered (Suzuki et al 1986).

Similar heterodimeric receptors have been implicated in cell-matrix adhesion in other cell types. Using monoclonal antibodies that inhibit attachment of hamster cells to fibronectin but not vitronectin or laminin, Brown & Juliano (1985) have identified a heterodimeric complex that resembles the fibronectin and vinculin receptors just described. Similar bimolecular complexes implicated in fibronectin-mediated cell adhesion have been isolated from mouse fibroblasts (Giancotti et al 1985, 1986a), BHK_{21}/C_{13} cells, and hamster melanoma cells (Knudsen et al 1981, 1982). Finally, heterodimeric receptors with molecular weights of 125–150 kDa have been identified that bind both fibronectin and collagen or collagen alone (Wayne-Carter & Carter 1987). Evidence suggesting that a heteromeric receptor mediates hepatocyte binding in collagen has also been presented (Rubin et al 1986). The picture for collagen receptors is not clear, however. Dedhar et al (1987) have isolated three proteins of 250, 70, and 30 kDa from collagen affinity columns. Lipid vesicles containing these proteins bind only to native collagen I. This binding is inhibited by RGD-containing peptides similar to those found in the triple helical region of collagen. How these peptides interact with one another, if at all, to form a collagen receptor, and how they are related to other proposed collagen receptors has not yet been determined. In general, however, the receptors for extracellular matrix molecules so far identified in mammalian cells appear to be noncovalently associated complexes which, in most cases, interact with a specific ligand, i.e. fibronectin, vitronectin, or collagen. In all cases investigated, these heteromeric receptors adhere to their matrix ligand via an RGD-sensitive mechanism. In most cases, the electrophoretic mobility on SDS-PAGE of one of the members of the complex decreases upon reduction, which indicates that this subunit is rich in intramolecular disulfide bonds.

EXTRACELLULAR MATRIX–LIKE RECEPTORS ON PLATELETS

A glycoprotein complex designated gpIIb-IIIa is found on the surface of platelets and has many properties in common with avian integrin and mammalian extracellular matrix receptors. This complex has been extensively characterized. It binds fibronectin, vitronectin, fibrinogen, and von Willebrand factor in an RGD-sensitive manner (Bennett et al 1983; Ginsberg et al 1983; Gardner & Hynes 1985; Ginsberg et al 1985; Plow et al 1985; Pytela et al 1986). These molecules do not adhere to platelets in the presence of certain monoclonal antibodies specific for gpIIb-IIIa (Bennett et al 1983; Pytela et al 1986) nor to platelets from thrombasthenic patients lacking gpIIb-IIIa. Structurally, gpIIb-IIIa is similar to mammalian extracellular matrix receptors; it consists of a 1:1 heterodimer in which the lower molecular weight subunit (gpIIIa) is highly disulfide cross-linked. The higher molecular weight subunit (gpIIb) consists of two peptides joined by disulfide bonding (Phillips & Agin 1977; McEver et al 1980; Leung et al 1981; Jennings & Phillips 1982; Carrell et al 1985). Interestingly, antibodies specific for platelet gpIIb-IIIa cross-react with glycoproteins on endothelial cell surfaces (Thiagarajan et al 1983; Fitzgerald et al 1985; Leeksma et al 1986; Newman et al 1986; Plow et al 1986). Antibodies against gpIIb-IIIa also react with the β-subunit of the mammalian vitronectin receptor but not with the α-subunit (Ginsberg et al 1987). There is no serological cross-reactivity with other matrix receptors. Receptors related to the platelet gpIIb-IIIa have been designated cytoadhesins (Plow et al 1986), and it has been suggested that these receptors make up one family of a superfamily of glycoproteins (Ginsberg et al 1987).

EXTRACELLULAR MATRIX–LIKE RECEPTORS ON LYMPHOID CELLS

Two groups of receptors found on lymphocytes are similar to extracellular matrix receptors. Each of these receptors is a heterodimer containing a higher molecular weight α-subunit and a lower molecular weight β-subunit.

One group of receptors includes the lymphocyte function–associated antigen LFA-1, which is involved in T-cell helping and killing (Hildreth et al 1982; Sanchez-Madrid et al 1982); Mac-1 (also known as OKM1 or CR3) complement receptors from macrophages (Springer et al 1979; Wright et al 1983); and the p150,95 complex from lymphocytes (Sanchez-Madrid et al 1983). The other group includes the "very late antigens" (VLA-1 through VLA-5), which appear on T cells several days after stimulation by a mitogen or alloantigen (Hemler et al 1983, 1984, 1985a,b, 1987). The LFA-1/Mac-1 and p150,95 heterodimers have identical 95-kDa β-subunits but have unique α-subunits; the same is true of the VLA antigens. Amino acid and cDNA sequence comparisons provide further evidence that these receptors are related to extracellular matrix receptors and cytoadhesins gpIIb-gpIIIa. The first 11 amino acids of the α-chains of LFA-1/Mac-1 and that of the mammalian vitronectin receptor are 63% homologous when conservative amino acid replacements are included (Suzuki et al 1986) and 45% homologous when they are not (Springer et al 1985; Suzuki et al 1986). Comparisons of the N-terminal amino acid sequence of platelet gpIIb with that of the α-subunit of LFA-1 and Mac-1 show that gpIIb is as related to LFA-1 and Mac-1 as they are to one another (Charo et al 1986). Similar conservation of amino-terminal amino acid sequences in the α-subunits of VLA antigen, gpIIb-IIIa, LFA-1/Mac-1, p150,95, and the vitronectin receptor have been noted (Takada et al 1987b).

Comparisons of the nucleotide sequence data for the β-subunits of several receptors have also been made. Remarkable similarities have been noted between the β-subunit of LFA-1/Mac-1 leukocyte adhesion proteins and band 3 glycoprotein of avian integrin (Kishimoto et al 1987). The overall homology is 45%; however, certain sequences of 15–27 amino acids are identical. The cysteine-rich repeats first noted by Tamkun et al (1986) are completely conserved, as are all 56 cysteine residues. There is a 70% conservation of amino acids in the putative transmembrane domains. The nucleotide sequence of platelet gpIIIb shows similar conserved sequences (Fitzgerald et al 1987). Thus proteins expressed independently on human platelets, endothelial cells, and leukocytes, and engaged in entirely distinct adhesive functions, are quite homologous to extracellular matrix receptors expressed on avian cells.

Careful serological comparisons of the VLA antigens and other receptors have revealed further close relationships. Antiserum against the common VLA β-subunit cross-reacts with the β-subunit of the mammalian fibronectin receptor and band 3 of the avian integrin complex (Takada et al 1987a). Anti–human FN receptor antisera immunoprecipitate VLA-5 α-subunits and no others. Anti–avian integrin antisera immunoprecipitate VLA-3 α-subunits and no others (Takada et al 1987a).

SUPERGENE FAMILY OF RECEPTORS

Based upon structural and functional similarities, numerous investigators have suggested that these molecules are all members of a supergene family of cell surface receptors (Charo et al 1986; Cosgrove et al 1986; Leptin 1986; Plow et al 1986; Ruoslahti & Pierschbacher 1986; Ginsberg et al 1987; Kishimoto et al 1987; Takada et al 1987b). Hynes (1987) has proposed a formal nomenclature for this supergene family, which he refers to as "integrins." According to this nomenclature, the integrins can be divided into families each with a common β-subunit and a set of variable α-subunits known to associate with the common β-subunit (Table 1). The different α-chains are denoted by the nature of the ligand (i.e. α_F, fibronectin), the original cell type (α_L, leukocyte), or where no simple designation exists, by the subscript used by the original discoverer (α_1, VLA-1). The nomenclature is tentative and may require revision but should be useful in communicating structural information and demonstrating changes in ligand specificity that accompany the association of different α-subunits with the various β-subunits.

Among the members of the integrin family, avian integrin is somewhat anomolous in that it is comprised of at least three subunits. Avian integrin may be a promiscuous heterotrimer or -dimer capable of binding several ligands, or it may be a mixture of heterodimers each with a common β-subunit with which the CSAT and JG22 monoclonal antibodies react (see above). Hydrodynamic estimations of the molecular weight of the integrin

Table 1   Supergene family of related receptors[a]

β-Subunit (source)	α-Subunit	Receptor (α-β combination)
β_1 (avian integrin band 3, fibronectin receptor, and VLA)	α_0	Avian integrin (band 1)
	α_1	VLA 1
	α_2	VLA 2
	α_3	VLA 3 and avian integrin (band 2)
	α_4	VLA 4
	α_F	VLA 5 and mammalian fibronectin
β_2 (LFA-1/Mac-1 and p150,95)	α_L	LFA^{-1}
	α_M	Mac-1
	α_X	p150,95
β_3 (platelet glycoprotein IIIa and vitronectin receptor)	α_{IIb}	Cytoadhesin glycoprotein IIb-IIIa
	α_V	Vitronectin

[a] The different α-chains are denoted by the nature of the ligand (e.g. α_F, fibronectin), the original cell type (e.g. α_L, leukocyte), or where no simple designation exists, by the subscript used by the original discoverer (e.g. α_1, VLA-1).

complex (Buck et al 1985), plus the fact that all of its ligands (fibronectin, laminin, and vitronectin) compete with one another for binding, suggest that integrin is a promiscuous heterodimer. This model for integrin is illustrated in Figure 1. The subunits have been labeled α and β (*bottom*, Figure 1) to conform to the convention established for other receptors, and as suggested by Hynes (1987). The presence of a second α-like subunit in preparations of integrin remains to be explained. There may be other receptors that share the same β-subunit whose activity has not been detected.

BIOSYNTHESIS AND PROCESSING OF INTEGRINS

Little is known concerning the biosynthesis of integrins. The α-subunit of platelet gpIIb-IIIa is synthesized as one polypeptide chain that is subsequently cleaved into 125-kDa and 25-kDa fragments (Bray et al 1986). There is no evidence that a common message encodes both the α- and β-subunits. Similarly, the subunits of LFA-1/Mac-1 appear to be synthesized

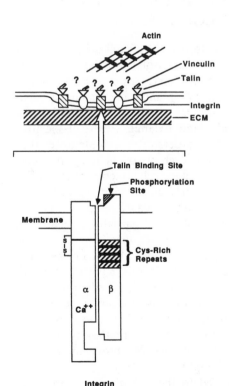

Figure 1 (*Top*) Integrin is shown as a transmembrane molecule acting as a bridge between the extracellular matrix (ECM) and the cytoskeleton-associated molecules talin and vinculin. Question marks designate the undetermined linkages between vinculin and actin. (*Bottom*) Enlarged, schematized view of the structure of integrin.

from mRNAs different from those coding for the β-subunits (Springer et al 1984; Cosgrove et al 1986). Interestingly, however, intracellular processing of the receptors appears to require the association of both α- and β-subunits prior to the appearance of the receptor on the cell surface (Springer et al 1984). For example, in patients with lymphocyte adhesion deficiency disease, in which certain T-cell and macrophage functions are inhibited due to the absence of LFA-1 or Mac-1 from the cell surface, the β-subunit of these receptors is not synthesized and no α-subunits are transported to the cell surface (Springer et al 1984; Springer & Anderson 1986). The α- and β-subunits of the LFA-1/Mac-1 receptors associate before the high-mannose carbohydrate side chains are converted to the complex form in the golgi apparatus (Sanchez-Madrid et al 1983; Springer et al 1984; Sastre et al 1986). Similarly, in the case of mammalian vitronectin receptor the processing of the high-mannose carbohydrate chains occurs at or near the time of association of the α- and β-subunits (Cheresh & Harper 1987).

EXTRACELLULAR MATRIX RECEPTORS NOT BELONGING TO THE INTEGRIN FAMILY

Integrinlike receptors do not appear to control all cell-matrix adhesive interactions. A receptor for laminin has been isolated by laminin affinity chromatography from a rat myoblast cell line (Lesot et al 1983), murine fibrosarcoma cells (Malinoff & Wicha 1983), and human mammary carcinoma cells (Rao et al 1983). This 68-kDa molecule differs from integrin receptors in that it is a monomer, and it binds to laminin with a dissociation constant of 2 nM, a value about three orders of magnitude lower than the measured dissociation constants for receptors of the integrin family. Antibodies against the laminin receptor block cell attachment to laminin (Liotta et al 1985), and preliminary reports suggest that the receptor is able to bind actin (Brown et al 1983). Another collagen receptor, anchorin (Mollenhauer et al 1984), has been isolated and partially sequenced. It contains a hydrophobic domain, which suggests that it is an integral membrane protein (Selmin et al 1986). Cell surface galactosyl transferases have also been implicated in cell-matrix interactions, particularly in the case of neural crest migration (Runyan et al 1986). The substrate with which the surface transferase interacts in this case appears to be laminin. Recent evidence indicates that certain mouse B16 melanoma variants bind to fibronectin in a manner sensitive to the peptide arg-*glu*-asp-val rather than arg-*gly*-asp, which suggests that alternative fibronectin adhesive mechanisms exist on certain cells (Humphries et al 1986).

Recently, attention has been turned to the role of other fibronectin domains in the adhesive process. Mouse fibroblasts will adhere to the cell-binding fragment of fibronectin and form close contact–like structures at the cell-matrix interface, but they will not organize microfilament bundles unless the heparin-binding peptides of fibronectin are also present (Lark et al 1985; Izzard et al 1986; Woods et al 1986). These data suggest that the formation of focal contacts requires the heparin-binding domain of fibronectin. Further, Chinese hamster ovary (CHO) cells unable to produce complex proteoglycans are unable to form focal contacts on intact fibronectin, even though they can form close-contact type adhesions (LeBaron et al 1986). Probably the best candidate for a cell surface protein that could serve as a receptor for the heparin-binding domain of fibronectin is the cell surface proteoglycan from mouse mammary epithelial cells that carries chondroitin sulfate and heparan sulfate glycosaminoglycans. This proteoglycan is thought to span the cytoplasmic membrane and bind to cytoskeletal proteins (Rapraeger & Bernfield 1983; Rapraeger et al 1985, 1986). Another cell surface proteoglycan also has cell adhesion–promoting properties and has been shown to be a receptor involved in the adhesion of neural retina cells via adherons (Schubert & LaCorbiere 1985). Thus the formation of a complex adhesive structure such as a focal contact may begin with the interaction of an integrinlike receptor with the extracellular matrix and the formation of a close contact. These events are probably sufficient for cell motility, but for a more permanent adhesive process to form, additional cell surface receptor interactions likely must occur. In fact, other proteins have been found to be present within the focal contact itself (Oesch & Birchmeier 1982; Rogalski & Singer 1985; Rogalski 1987) and to interact with fibronectin (Aplin et al 1981). These proteins undoubtedly represent still other, as yet poorly characterized, cell-matrix adhesion receptors. Finally, it has been shown that even disialogangliosides can affect the attachment of cells to fibronectin (Kleinman et al 1979; Yamada et al 1981; Perkins et al 1982; Okada et al 1984; Cheresh et al 1986). The fact that glycolipids and antibodies against glycolipids have a limited but reproducible effect on cellular adhesion to fibronectin suggests that these molecules may be involved in secondary adhesion events, such as stabilizing receptors at the adhesive site. Thus a complete adhesive interaction requires that several surface molecules, glycoprotein complexes, proteoglycans, and glycolipids all come together in the proper molecular configuration.

SUMMARY AND CONCLUSIONS

Table 2 lists most of the extracellular matrix and related receptors identified to date. The wide range of binding affinities of these receptors for their

Table 2 Extracellular matrix and related receptors[a]

Receptor designation	Species	Subunit MW (kDa)		Ligand	Source	Distribution	Demonstrated function	Reference
		Nonreduced	Reduced					
Integrin (140-K complex, CSAT ag)	Chick, quail	160/130/110	160–120	FN LM VN Col I, IV(?)	Chick embryos Fibroblasts Myoblasts Dorsal root ganglia	Widely distributed Fibroblasts, Nerve, Muscle, Epithelium, Cytoskeleton-ECM	Cell matrix adhesion Cell migration Neurite outgrowth Cytoskeleton-ECM coupling	Neff et al 1982; Greve & Gottlieb 1982; Knudsen et al 1985; Hasegawa et al 1985
Actosialin	Chick	165/107	170/130	—	Chick fibroblasts	Fibroblasts	Cell-matrix adhesion in focal contacts (?)	Rogalski & Singer 1985
Mammalian fibronectin receptor	Human	160/120	140+25/130	FN	Osteosarcoma, placenta	—	Adhesion to fibronectin	Pytela et al 1985a
	Rat	140/116	—		Rat skeletal muscle (L₆A)			Buck & Horwitz 1987
	Hamster	—	120–160		CHO, C13/B4, BHK21/C13			Brown & Juliano 1985; Knudsen et al 1982
	Mouse	160/135	160/145		SR/Balb3T3 (fibroblasts)	Myeloid cells, erythroid cells, platelets, fibroblasts	Cell adhesion to FN	Giancotti et al 1986a,b
	Mouse	140/115	140		Erythroid precursor cells, pre-B-lymphocytes	Erythroid cells, pre-B-lymphocytes		
Vitronectin receptor	Human	160/100	125/115+25	VN	Osteosarcoma Placenta	—	Adhesion to vitronectin	Pytela et al 1985b
Collagen receptor	Human	147/125	135/130	Col I, VI, FN	Human fibrosarcoma	—	Adhesion to col I and VI and fibronectin	Wayne-Carter & Carter 1987
		145/125	140/135 250/70/30 140/120	Col I, VI Col I Col I	Human fibrosarcoma Osteosarcoma MG-63 Hepatocytes	— — Liver	Adhesion to col I and VI Adhesion to col I Hepatocyte adhesion to col I	Wayne-Carter & Carter 1987 Dedhar et al 1987 Rubin et al 1986
	Rat	—						
Cytoadhesin gpIIB-IIIa	Human and others	140/95	120/110+25	FN, Fgen, vWF	Platelets	Platelets	Platelet aggregation and adhesion, clot retraction	Phillips & Agin 1977; Bennett et al 1983; Ginsberg et al 1983, 1985; Gardner & Hynes 1985
Cytoadhesins	Human	139–145	132–134	—	Human endothelial cells	Endothelium	—	Thiagarajan et al 1983; Fitzgerald et al 1985
	Bovine	93/98	99–107		Bovine endothelial cells			Leeksma et al 1986; Newman et al 1986
LFA-1	Human Mouse	180/95 —	180/95 —	ICAM-1	T-lymphocytes	Leukocytes	T-lymphocyte helping and killing, homeo- and heterotypic interactions of lymphocytes	Hildreth et al 1982; Sanchez-Madrid et al 1982, 1983

Receptor	Species	Subunit mass	Subunit mass	Ligand	Cell source	Tissue distribution	Function	References
Mac-1 (OKM1, CR3)	Human Mouse	— —	170/95 —	Complement C3bi	Monocytes, granulocytes, large granular lymphocytes	Macrophages, tissue macrophages	Complement binding	Springer et al 1979; Wright et al 1983
p150,95	Human	—	150/95	—	Hairy cell leukemia, myeloid cell lines	Granulocytes, monocytes, tissue macrophages	C3bi binding (?)	Sanchez-Madrid et al 1983
VLA-1	Human	200/110	210/130	—	T-lymphocytes	Activated T-cell neuroblastoma, liver sinusoids, kidney mesangium, smooth muscle, skin fibroblasts	—	Hemler et al 1983, 1985a,b
VLA-2	Human	150/110	165/130	—	T-lymphocytes	Platelets, liver sinusoids, kidney mesangium, smooth muscle, skin fibroblasts, neuroblastoma	Collagen binding	Hemler et al 1983, 1985a,b
VLA-3	Human	150/110	135/130	—	T-lymphocytes	Wide distribution	—	Hemler et al 1987
VLA-4	Human	140/110	150/130	—	T-lymphocytes	Most suspension cells	—	Hemler et al 1987
VLA-5	Human	150/110	135/130	FN	T-lymphocytes	Myeloid cell lines, liver sinusoids, kidney mesangium, smooth muscle, skin fibroblasts, neuroblastoma	Adhesion to fibronectin (?)	Hemler et al 1987
Laminin receptor	Human Rat Mouse	—	68	LM	Human mammary tumor (MCF-7) Rat muscle Murine sarcoma	Epithelium Muscle	Adhesion to basement membrane	Rao et al 1983 Lesot et al 1983 Malinoff et al 1983
Membrane-bound heparin sulfate proteoglycan	Mouse Chick	—	—	LM (?), FN (?) LM (?) adherons Col II	Mouse mammary epithelium Epithelium, neural retina	Epithelium Neural retina	Adhesion to interstitial matrix and basement membrane	Raprager et al 1983; Shubert & LaCorbiere 1985
Anchorin I	Chick	31		Col II	Chondrocytes	Chondrocytes	Chondrocyte adhesion to col II	Mollenhauer et al 1984
47K fibronectin receptor	Hamster Mouse	—	47	FN	BHK21/C13 hamster cells 3T3 cells	—	FN binding	Aplin et al 1981

[a]The subunit molecular mass for each receptor is in kilodaltons and was estimated by SDS-polyacrylamide gel electrophoresis under nonreducing and reducing conditions. The precise estimates in different laboratories may vary slightly. Abbreviations: FN, fibronectin; LM, laminin; VN, vitronectin; Col, collagen; TN, talin; Fgen, fibrinogen; vWF, von Willebrand's factor; ECM, extracellular matrix; CSAT ag, cell substratum attachment (integrin); DRG, dorsal root ganglia.

ligands may be important to their function. The affinity of integrins for fibronectin is moderate, with a dissociation constant in the micromolar range. This affinity level leads to relatively rapid dissociation and reformation of receptor-ligand complexes. Thus changes in component concentration can shift binding equilibria within minutes (the time scale of many biologic phenomena) and change the number or organization of adhesive complexes. This type of interaction would be useful in motile cells, in which adhesions must form and dissociate rapidly. In contrast, the affinity of the 68-kDa laminin receptor for its ligand is three orders of magnitude higher. Such levels of affinity would be useful in stabilizing tissue.

Members of the integrin family appear to recognize an RGD sequence on the ligands to which they bind. Since there are many ligands containing the RGD sequence, the question of specificity arises. Avian integrin shows little specificity and appears to behave as a multifunctional, promiscuous receptor for extracellular matrix molecules. Figure 1 summarizes our current view of the structural and functional features of avian integrin. In contrast, the mammalian receptors for vitronectin and fibronectin are specific for their respective ligands. More than one of these receptors may be present simultaneously on a cell surface, e.g. fibroblasts express receptors for fibronectin, laminin, and vitronectin at the same time. This multiplicity of receptors provides potential mechanisms for generating the adhesive differences among cells believed to play a prominent role in morphogenesis. Further adhesive differences may stem from the formation of different combinations of various α- and β-subunits expressed in the cell.

The mechanism of regulation of adhesive interactions with the extracellular matrix is only beginning to be explored. There are several levels at which this regulation might occur. Integrin appears to be more regionalized in more developed cells that are integral parts of tissue structures. Changes in receptor distribution could alter the relative strength of adhesive interactions. In certain instances, avian integrin disappears, or its concentration is reduced, e.g. during the development of embryonic lung (Chen et al 1986) and erythroid cells (Patel & Lodish 1985). Posttranslational modifications provide yet another mechanism for regulating integrin-ligand binding. Avian integrin is phosphorylated on tyrosine upon infection with transforming viruses encoding tyrosine kinases. The phosphorylated receptor shows altered binding to both cytoskeleton-associated and extracellular matrix molecules. These changes could contribute to the altered morphology and adhesive properties seen upon virus-induced transformation.

Probably one of the most unexpected observations concerning these matrix receptors is their resemblance to receptors from lymphoid cells. The remarkably similar structures of these two groups of receptors have been deduced from limited sequence and serological data. On the basis of these similarities, it has been proposed that these receptors all be termed integrins. It has also been suggested that structurally similar antigens, termed position-specific (PS) antigens, expressed in *Drosophila* (Wilcox et al 1984; Wilcox & Leptin 1985), be included in this family of receptors (Leptin 1986) as they are composed of heterodimers and their modulation during embryogenesis suggests they may be involved in morphogenesis. Recent data show that RGD-containing peptides prevent gastrulation in *Drosophila* embryos, which further implicates the integrin family in insect morphogenesis (Naidet et al 1987). Thus, the integrin family is found in several species and cell types and appears to play prominent roles in the development of many organisms.

Diseases in humans involving the integrin family include Glansmann's thrombosthenia in which platelets lack the gpIIb-IIIa receptor (McEver et al 1980; Ginsberg et al 1983) and show abnormal platelet aggregation and clot retraction, resulting in prolonged bleeding times. Leukocyte adhesion deficiency is a disease arising from altered synthesis of the common β-subunit of the LFA-1/Mac-1 receptors. These patients have recurrent infections, retarded wound healing, and other abnormalities arising from altered leukocyte adhesion (Springer & Anderson 1986).

Much of the work presented here is preliminary and incomplete because this is a rapidly expanding field. There are without doubt many more molecules to be identified that are important to the process of cell adhesion. The identification of these additional molecules, characterization of their structures, understanding their regulation, assembly, biosynthesis, their role in developmental biology and various disease processes will be the source of much interesting research in the future.

ACKNOWLEDGMENTS

We would like to thank numerous colleagues for generously sharing their manuscripts and unpublished data. We especially thank our collaborators Keith Burridge, Richard Hynes, and Larry Rohrschneider for their contributions and for allowing us to cite unpublished data. Finally, we are grateful for the tireless assistance of Ms. Marie Lennon in the preparation of the manuscript. This work was supported by National Institutes of Health grants CA-19144, CA-27909, and CA10815 (Dr. Buck) and GM-23244 (Dr. Horwitz) and by the H. M. Watts, Jr. Neuromuscular Disease Research Center (Dr. Horwitz).

200 BUCK & HORWITZ

Literature Cited

Abercrombie, M., Heaysman, E. M., Pegrum, S. M. 1971. The locomotion of fibroblasts in culture. IV. Electron microscopy of the leading lamella. *Exp. Cell Res.* 67: 359–67

Akiyama, S. K., Hasegawa, E., Hasegawa, T., Yamada, K. 1985. The interaction of fibronectin fragments with fibroblastic cells. *J. Biol. Chem.* 260: 13256–60

Akiyama, S. K., Yamada, S. S., Yamada, K. M. 1986. Characterization of a 140 kD avian cell surface antigen as a fibronectin-binding molecule. *J. Cell Biol.* 102: 442–48

Aplin, J. D., Hughes, R. C., Jaffe, C. L., Sharon, N. 1981. Reversible crosslinking of cellular components of adherent fibroblasts to fibronectin and lectin coated substrata. *Exp. Cell Res.* 134: 488–94

Argraves, W. S., Pytela, R., Suzuki, S., Millan, J. L., Pierschbacher, M. D., Ruoslahti, E. 1986. cDNA sequences from the alpha subunit of the fibronectin receptor predict a transmembrane domain and a short cytoplasmic peptide. *J. Biol. Chem.* 261: 12022–24

Bennett, J. S., Hoxie, H., Leitman, S., Valaire, G., Clines, D. 1983. Inhibition of fibrinogen binding to stimulated human platelets by a monoclonal antibody. *Proc. Natl. Acad. Sci. USA* 80: 2417–21

Birchmeier, C., Kreis, T. E., Eppenberger, H. M., Winterhalter, H., Birchmeier, W. 1980. Corrugated attachment membrane on WI-38 fibroblasts: Alternating fibronectin fibers and actin containing focal contacts. *Proc. Natl. Acad. Sci. USA* 77: 4108–12

Boucaut, J. C., Darriebere, T., Poole, T. J., Aoyama, H., Yamada, K. M., Thiery, J. P. 1984. Biological active synthetic peptides as probes of embryonic development: A competitive peptide inhibition of fibronectin function inhibits gastrulation in amphibian embryos and neural crest migration in avian embryos. *J. Cell Biol.* 99: 1822–30

Bozyczko, D., Horwitz, A. 1986. The participation of a putative cell surface receptor for laminin and fibronectin in peripheral neurite extension. *J. Neurosci.* 6: 1241–51

Bray, P. F., Rosa, J.-P., Lingappa, V. R., Kan, Y. W., McEver, R. P., Shuman, M. A. 1986. Biogenesis of platelet receptor for fibrinogen: Evidence for separate precursors for glycoprotein IIb and IIIa. *Proc. Natl. Acad. Sci. USA* 83: 1480–84

Bronner-Fraser, M. 1985. Alterations in neural crest migration by a monoclonal antibody that affects cell adhesion. *J. Cell Biol.* 101: 610–17

Brown, P., Juliano, R. L. 1985. Selective inhibition of fibronectin-mediated cell adhesion by monoclonal antibodies to a cell surface glycoprotein. *Science* 228: 1448–51

Brown, S. S., Malinoff, H. L., Wicha, M. S. 1983. Connectin: Cell surface protein that binds both laminin and actin. *Proc. Natl. Acad. Sci. USA* 80: 5927–30

Buck, C. A., Horwitz, A. F. 1987. Integrin, a transmembrane glycoprotein complex mediating cell-substratum adhesion. 2nd. Abercrombie conf. on cell behaviour. *J. Cell Sci.* In press

Buck, C., Knudsen, K. A., Damsky, C. H., Decker, C., Greggs, R. R., et al. 1985. Integral membrane protein complexes in cell-matrix adhesion. In *The Cell in Contact: Adhesions and Junctions as Morphogenic Determinants*, ed. G. M. Edelman, J. P. Thiery, pp. 345–64. New York: Wiley

Buck, C. A., Shea, E., Duggin, K., Horwitz, A. F. 1986. Integrin the (CSAT antigen): Functionality requires oligomeric integrity. *J. Cell Biol.* 103: 2421–28

Burridge, K., Feramisco, J. 1980. Microinjection and localization of a 130 kD protein in living fibroblasts: A relationship to actin and fibronectin. *Cell* 19: 587–95

Burridge, K., Connell, L. 1983. A new protein of adhesion plaques and ruffling membranes. *J. Cell Biol.* 94: 359–67

Burridge, K. 1987. Substrate adhesions in normal and transformed fibroblasts: Organization and regulation of cytoskeletal, membrane and extracellular matrix components at focal contacts. *Cancer Rev.* 4: 18–78

Carrell, N. A., Fitzgerald, L. A., Steiner, B., Erickson, H. P., Phillips, D. R. 1985. Structure of human platelet membrane glycoproteins IIb and IIIa as determined by electron microscopy. *J. Biol. Chem.* 260: 1743–49

Chapman, A. E. 1984. Characterization of a 140 kD cell surface glycoprotein involved in myoblast adhesion. *J. Cell. Biochem.* 25: 109–21

Charo, I. F., Fitzgerald, L. A., Steiner, B., Rall, S. C. Jr., Bekeart, L. S., Phillips, D. R. 1986. Platelet glycoproteins IIb and IIIa: Evidence for a family of immunologically and structurally related glycoproteins in mammalian cells. *Proc. Natl. Acad. Sci. USA* 83: 8351–55

Chen, W.-T., Greve, J. M., Gottlieb, D. I., Singer, S. J. 1985a. Immunocytochemical localization of 140 kD cell adhesion molecules in cultured chicken fibroblasts and in chicken smooth muscle and intestinal epithelial tissues. *J. Histochem. Cytochem.* 33: 576–86

Chen, W.-T., Hasegawa, E., Hasegawa, T., Weinstock, C., Yamada, K. M. 1985b. Development of cell surface linkage complexes in cultured fibroblasts. *J. Cell Biol.* 100: 1103–14

Chen, W.-T., Chen, J.-M., Mueller, S. C. 1986a. Coupled expression and co-localization of 140 K cell adhesion molecules, fibronectin, and laminin during morphogenesis and cytodifferentiation of chick lung cells. *J. Cell Biol.* 103: 1073–90

Chen, W.-T., Wong, J., Hasegawa, T., Yamada, S., Yamada, K. M. 1986b. Regulation of fibronectin receptor distribution by transformation, exogenous fibronectin, and synthetic peptides. *J. Cell Biol.* 103: 1649–61

Cheresh, D. A., Pierschbacher, M. D., Herzig, M. A., Mujoo, K. 1986. Disialogangliosides GD2 and GD3 are involved in the attachment of human melanoma and neuroblastoma cells to extracellular matrix proteins. *J. Cell Biol.* 102: 688–96

Cheresh, D. A., Harper, J. R. 1987. Arginine-glycine-aspartic acid recognition by a cell adhesion receptor requires its 130 kDa subunit. *J. Biol. Chem.* 262: 1434–37

Cosgrove, L. J., Sandrin, M. S., Rajasekariah, P., McKenzie, I. F. C. 1986. A genomic clone encoding the alpha chain of the OKM-1, LFA-1, and platelet glycoprotein IIb-IIIa molecules. *Proc. Natl. Acad. Sci. USA* 83: 752–56

Damsky, C. H., Knudsen, K. A., Bradley, D., Buck, C. A., Horwitz, A. F. 1985. Distribution of the cell-substratum attachment (CSAT) antigen on myogenic and fibroblastic cells in culture. *J. Cell Biol.* 100: 1528–39

David-Pfeuty, T., Singer, S. J. 1980. Altered distribution of the cytoskeletal proteins vinculin and α-actinin in cultured fibroblasts transformed by Rous sarcoma virus. *Proc. Natl. Acad. Sci. USA* 77: 6687–91

Decker, C., Greggs, R., Duggan, K., Stubbs, J., Horwitz, A. 1984. Adhesive multiplicity in the interaction of embryonic fibroblasts and myoblasts with extracellular matrices. *J. Cell Biol.* 99: 1398–1404

Dedhar, S., Ruoslahti, E., Pierschbacher, M. D. 1987. A cell surface receptor complex for collagen type I recognizes the arg-gly-asp sequence. *J. Cell Biol.* 104: 585–93

Duband, J. L., Rocker, S., Chen, W. T., Yamada, K. M., Thiery, J.-P. 1986. Cell adhesion and migration in the early vertebrate embryo: Location and possible role of the putative fibronectin receptor complex. *J. Cell Biol.* 102: 160–78

Fitzgerald, L. A., Charo, I. F., Phillips, D. R. 1985. Human and bovine endothelial cells synthesize membrane proteins similar to human platelet glycoproteins IIb and IIIa. *J. Biol. Chem.* 260: 10893–96

Fitzgerald, L. A., Steiner, B., Rall, S. C., Lo, S. S., Phillips, D. R. 1987. Protein sequence of endothelial glycoprotein IIIa derived from a cDNA clone; identity with platelet glycoprotein IIIa and similarity to integrin. *J. Biol. Chem.* 262: 3936–39

Gardner, J. M., Hynes, R. O. 1985. Interaction of fibronectin with its receptor on platelets. *Cell* 42: 439–48

Geiger, B., Tokuyasu, K. T., Dutton, A. H., Singer, S. J. 1980. Vinculin, an intracellular protein localized at specialized sites where microfilament bundles terminate at cell membranes. *Proc. Natl. Acad. Sci. USA* 77: 4127–31

Geiger, B., Avnur, Z., Volberg, T., Volk, T. 1985. Molecular domains of adherens junctions. In *The Cell in Contact: Adhesions and Junctions as Morphogenetic Determinants*, ed. G. M. Edelman, J. P. Thiery, pp. 461–89. New York: Wiley

Gerisch, G. 1977. Univalent antibody fragments as tools for the analysis of cell interactions in *Dictyostellium. Curr. Top. Dev. Biol.* 14: 243–70

Giancotti, F. G., Tarone, G., Knudsen, K., Damsky, C., Comoglio, P. M. 1985. Cleavage of a 135 kD cell surface glycoprotein correlated with loss of fibroblast adhesion to fibronectin. *Exp. Cell Res.* 156: 182–90

Giancotti, F. G., Comoglio, P. M., Tarone, G. 1986a. A 135,000 molecular weight plasma membrane glycoprotein involved in fibronectin-mediated cell adhesion. *Exp. Cell Res.* 163: 47–62

Giancotti, F. G., Comoglio, P. M., Tarone, G. 1986b. Fibronectin–plasma membrane interaction in the adhesion of hemopoietic cells. *J. Cell Biol.* 103: 429–37

Ginsberg, M. H., Forsyth, J., Lightsey, A., Chedrak, J., Plow, L. F. 1983. Reduced surface expression and binding of fibronectin by thrombin stimulated thrombasthenic platelets. *J. Clin. Invest.* 71: 619–24

Ginsberg, M. H., Pierschbacher, M. D., Ruoslahti, E., Marguerie, G., Plow, E. 1985. Inhibition of fibronectin binding to platelets by proteolytic fragments and synthetic peptides which support fibroblast adhesion. *J. Biol. Chem.* 260: 3931–36

Ginsberg, M. H., Loftus, J., Ryckwaert, J.-J., Pierschbacher, M., Pytela, R., Ruoslahti, E., Plow, E. F. 1987. Immunochemical and N-terminal sequence comparison of two cytoadhesins indicates they contain similar or identical beta subunits and distinct alpha subunits. *J. Biol. Chem.* 262: 5437–40

Greve, J. M., Gottlieb, D. I. 1982. Monoclonal antibodies which alter the morphology of culture chick myogome cells. *J. Cell Biochem.* 18: 221–30

Hall, D., Neugebauer, K., Reichardt, L. 1987. Embryonic neural retinal cell response to extracellular matrix proteins: Developmental changes and effects of the CSAT antibody. *J. Cell Biol.* 104: 623–34

Hasegawa, T., Hasegawa, E., Chen, W. T., Yamada, K. M. 1985. Characterization of a membrane-associated glycoprotein complex implicated in cell adhesion to fibronectin. *J. Cell Biochem.* 28: 307–18

Heggeness, M. H., Ash, J. F., Singer, S. J. 1978. Transmembrane linkage of fibronectin and intracellular actin containing filaments in cultured human fibroblasts. *Ann. NY Acad. Sci.* 312: 414–17

Hemler, M. E., Ware, C. S., Strominger, J. L. 1983. Characterization of a novel differentiation antigen complex recognized by a monoclonal antibody (A-1A5): Unique activation-specific molecular forms on stimulated T cells. *J. Immunol.* 131: 334–40

Hemler, M. E., Sanchez-Madrid, F., Flotte, T. J., Krensky, A. M., Burakoff, S. J., et al. 1984. Glycoproteins of 210,000 and 130,000 M.W. on activated T cells: Cell distribution and antigenic relation to components on resting cells and T cell lines. *J. Immunol.* 132: 3011–18

Hemler, M. E., Jacobson, J. G., Brenner, M. B., Mann, D., Strominger, J. L. 1985a. VLA-1: A T cell surface antigen which defines a novel late stage of human T cell activation. *Eur. J. Immunol.* 15: 502–8

Hemler, M. E., Jacobson, J. G., Strominger, J. L. 1985b. Biochemical characterization of VLA-1 and VLA-2 cell surface heterodimers on activated T cells. *J. Biol. Chem.* 260: 15246–52

Hemler, M. E., Huang, C., Schwartz, X. 1987. The VLA family. *J. Biol. Chem.* 262: 3300–9

Hildreth, J. E. K., Gotch, F. M., Hildreth, P. D. K., McMichael, A. J. 1982. A human lymphocyte-associated antigen involved in cell-mediated lympholysis. *Eur. J. Immunol.* 13: 202–8

Hirst, R., Horwitz, A., Buck, C., Rohrschneider, L. 1986. Phosphorylation of the fibronectin receptor complex in cells transformed by oncogenes that encode tyrosine kinases. *Proc. Natl. Acad. Sci. USA* 83: 6470–74

Horwitz, A., Neff, N., Sessions, A., Decker, C. 1982. Cellular interactions in myogenesis. In *Muscle Development: Molecular and Cellular Control*, pp. 292–300. Cold Spring Harbor, NY: Cold Spring Harbor Lab.

Horwitz, A., Duggan, K., Greggs, R., Decker, C., Buck, C. 1985. The cell substrate attachment (CSAT) antigen has properties of a receptor for laminin and fibronectin. *J. Cell Biol.* 103: 2134–44

Horwitz, A., Duggan, K., Buck, C., Beckerle, K., Burridge, K. 1986. Interaction of plasma membrane fibronectin with talin—a transmembrane linkage. *Nature* 320: 531–33

Humphries, M. H., Akiyama, S. K., Komoriya, A., Olden, K., Yamada, K. M. 1986. Identification of an alternatively spliced site in human plasma fibronectin that mediates cell type-specific adhesion. *J. Cell Biol.* 103: 2637–48

Hynes, R. O., Destree, A. T. 1978. Relationships between fibronectin (LETS protein) and actin. *Cell* 15: 875–86

Hynes, R. O. 1987. Integrins: A family of cell surface receptors. *Cell* 48: 549–54

Izzard, C. S., Radinsky, R., Culp, L. A. 1986. Substratum contacts and cytoskeletal reorganization of BALB/c 3T3 cells on a cell-binding fragment and heparin-binding fragment of plasma fibronectin. *Exp. Cell Res.* 165: 320–36

Jaffredo, T., Horwitz, A. F., Buck, C. A., Rong, P. M., Dieterlen-Lievre, F. 1986. CSAT antibody interferes with in vivo migration of somitic myoblast precursors into the body wall. In *The Somites in Developing Embryos*, ed. R. Bellaires, J. W. Lash, pp. 225–36. New York: Plenum

Jennings, L. K., Phillips, D. R. 1982. Purification of glycoproteins IIb and IIIa from human platelet plasma membranes and characterization of a calcium-dependent glycoprotein IIb-IIIa complex. *J. Biol. Chem.* 257: 10458–66

Kishimoto, T. K., O'Connor, K., Lee, A., Roberts, T. M., Springer, T. A. 1987. Cloning of the beta subunit of the leukocyte adhesion proteins: Homology to an extracellular matrix receptor defines a novel supergene family. *Cell* 48: 681–90

Kleinman, H. K., Martin, G. R., Fishman, P. H. 1979. Ganglioside inhibition of fibronectin-mediated cell adhesion to collagen. *Proc. Natl. Acad. Sci. USA* 76: 3367–71

Knudsen, K. A., Rao, P., Damsky, C. H., Buck, C. A. 1981. Membrane glycoproteins involved in cell-substratum adhesion. *Proc. Natl. Acad. Sci. USA* 78: 6071–75

Knudsen, K. A., Damsky, C. H., Buck, C. A. 1982. Expression of adhesion-related membrane components in adherent versus nonadherent hamster melanoma cells. *J. Cell. Biochem.* 18: 157–67

Knudsen, K. A., Horwitz, A., Buck, C. A.

1985. A monoclonal antibody identifies a glycoprotein complex involved in cell-substratum adhesion. *Exp. Cell Res.* 157: 218–26

Kohler, G., Milstein, C. 1975. Continuous culture of fused cells secreting antibody of predefined specificity. *Nature* 256: 495–97

Krotoski, D., Domingo, C., Bronner-Fraser, M. 1986. Distribution of a putative cell surface receptor for fibronectin and laminin in the avian embryo. *J. Cell Biol.* 103: 1061–71

Lark, M. W., Laterra, J., Culp, L. A. 1985. Close and focal contact adhesions of fibroblasts to a fibronectin containing matrix. *Fed. Proc.* 44: 394–403

LeBaron, R. G., Esko, J. D., Woods, A., Hook, M. 1986. Chinese hamster cell variants show altered adhesive responses to fibronectin. *J. Cell Biol.* 103: 91a

Leeksma, O. C., Zandberger-Spaargaren, J., Giltay, J. C., van Mourik, J. A. 1986. Cultured human endothelial cells synthesize a plasma membrane protein complex immunologically related to the platelet glycoprotein IIb/IIIa complex. *Blood* 67: 1176–80

Leptin, M., 1986. The fibronectin receptor family. *Nature* 321: 728

Lesot, H., Kihl, U., von der Mark, K. 1983. Isolation of a laminin-binding protein from muscle cell membranes. *EMBO J.* 2: 861–65

Leung, L. L., Kinoshits, T., Nachman, R. L. 1981. Isolation, purification and partial characterization of platelet membrane glycoproteins IIb and IIIa. *J. Biol. Chem.* 256: 1994–97

Liotta, L. A., Hand, P. H., Rao, C. N., Bryant, G., Barksy, S. H., Schlom, J. 1985. Monoclonal antibodies to the human laminin receptor recognize structurally distinct sites. *Exp. Cell Res.* 156: 117–26

Malinoff, H. L., Wicha, M. S. 1983. Isolation of a cell surface receptor protein for laminin from murine fibrosarcoma cells. *J. Cell Biol.* 96: 1475–79

Marchisio, P., Cirillo, D., Teti, A., Zambonin-Zallone, A., Tarone, G. 1987. Rous sarcoma virus transformed fibroblasts and cells of monocytic origin display a peculiar dot-like organization of cytoskeletal proteins involved in microfilament-membrane interactions. *Exp. Cell Res.* 169: 202–14

McEver, R. P., Baenziger, N. L., Majerus, P. W. 1980. Isolation and quantitation of the platelet membrane glycoprotein deficient in the thrombasthenic using a monoclonal hybridoma antibody. *J. Clin. Invest.* 66: 1311–18

Mollenhauer, J., Bee, J. A., Lizarbe, M. A.,

von der Mark, K. 1984. Role of anchorin CII, a 31,000 molecular weight membrane protein, in the interaction of chondrocytes with type II collagen. *J. Cell Biol.* 98: 1572–78

Naidet, C., Semeriva, M., Yamada, K., Thiery, J.-P. 1987. Peptides containing the cell-attachment recognition sequence arg-gly-asp prevent gastrulation in *Drosophila* embryos. *Nature* 325: 248–50

Neff, N. T., Lowrey, C., Decker, C., Tovar, A., Damsky, C., et al. 1982. A monoclonal antibody detaches embryonic skeletal muscle from extracellular matrices. *J. Cell Biol.* 95: 654–66

Newman, P. J., Kawai, Y., Montgomery, R. R., Kunicki, T. J. 1986. Synthesis by cultured human umbilical vein endothelial cells of two proteins structurally and immunologically related to platelet membrane glycoproteins IIb and IIIa. *J. Cell Biol.* 103: 81–85

Oesch, B., Birchmeier, W. 1982. New surface component of fibroblast focal contacts identified by a monoclonal antibody. *Cell* 31: 671–79

Okada, Y., Mugnai, G., Bremer, E. G., Hakomori, S. 1984. Glycosphingolipids in detergent-insoluble substrate attachment matrix (DISAM) prepared from substrate attachment material (SAM): Their possible role in regulating cell adhesion. *Exp. Cell Res.* 155: 448–56

Patel, V., Lodish, H. F. 1985. The fibronectin receptor on mammalian erythroid precursor cells: Characterization and developmental regulation. *J. Cell Biol.* 102: 449–56

Perkins, R. M., Kellie, S., Patel, B., Critchley, D. R. 1982. Gangliosides as receptors for fibronectin. *Exp. Cell Res.* 141: 231–43

Phillips, D. R., Agin, P. 1977. Platelet plasma membrane glycoproteins. *J. Biol. Chem.* 252: 2121–26

Pierschbacher, M. D., Ruoslahti, E. 1984a. Cell attachment activity of fibronectin can be duplicated by small synthetic fragments of the molecule. *Nature* 309: 30–33

Pierschbacher, M. D., Ruoslahti, E. 1984b. Variants of the cell recognition site of fibronectin that retain attachment-promoting activity. *Proc. Natl. Acad. Sci. USA* 81: 5985–88

Plow, E. F., Pierschbacher, M. D., Ruoslahti, E., Marguerie, G. A., Ginsberg, M. H. 1985. The effect of arg-gly-asp-containing peptides on fibrinogen and von Willebrand factor binding to platelets. *Proc. Natl. Acad. Sci. USA* 82: 8057–61

Plow, E. F., Loftus, J. C., Levin, E. G., Fair, D. S., Dixon, D., et al. 1986. Immunologic relationship between platelet membrane glycoprotein gpIIb/IIIa and cell surface molecules expressed by a variety of cells. *Proc. Natl. Acad. Sci. USA* 83: 6002–6

Pytela, R., Pierschbacher, M. D., Ruoslahti, E. 1985a. Identification and isolation of a 140 kd cell surface glycoprotein with properties expected of a fibronectin receptor. *Cell* 40: 191–98

Pytela, R., Pierschbacher, M. D., Ruoslahti, E. 1985b. A 125/115 kDa cell surface receptor specific for vitronectin interacts with the arginine-glycine-aspartic acid adhesion sequence derived from fibronectin. *Proc. Natl. Acad. Sci. USA* 82: 5766–70

Pytela, R., Pierschbacher, M. D., Ginsberg, M. H., Plow, E. F., Ruoslahti, E. 1986. Platelet membrane glycoprotein IIb/IIIa: Member of a family of arg-gly-asp-specific adhesion receptors. *Science* 231: 1559–61

Rao, N. C., Barsky, S. H., Terranova, V. P., Liotta, L. A. 1983. Isolation of a tumor cell laminin receptor. *Biochem. Biophys. Res. Commun.* 111: 804–8

Rapraeger, A. C., Bernfield, M. 1983. Heparan sulfate proteoglycans from mouse mammary epithelial cells, a putative membrane proteoglycan associates quantitatively with lipid vesicles. *J. Biol. Chem.* 258: 3632–36

Rapraeger, A., Jalkanen, M., Endo, E., Koda, J., Bernfield, M. 1985. The cell surface proteoglycan from mouse mammary epithelial cells bears chondroitin sulfate and heparan sulfate glycosaminoglycans. *J. Biol. Chem.* 260: 11046–52

Rapraeger, A., Jalkanen, M., Bernfield, M. 1986. Cell surface proteoglycan associates with the cytoskeleton at the basolateral cell surface of mouse mammary epithelial cells. *J. Cell Biol.* 103: 2683–96

Rogalski, A. A. 1987. Actinsialin (an integral sialoglycoprotein) molecularly distinguished on non-junctional dense plaque sites of microfilament attachment. *J. Cell Biol.* 101: 785–801

Rogalski, A. A., Singer, S. J. 1985. An integral glycoprotein associated with membrane sites of actin with microfilaments. *J. Cell Biol.* 101: 785–801

Rohrschneider, L. 1980. Adhesion plaques of Rous sarcoma virus–transformed cells contain the *src* gene product. *Proc. Natl. Acad. Sci. USA* 77: 3514–18

Rubin, K., Gullberg, D., Borg, T. K., Obrink, B. 1986. Hepatocyte adhesion to collagen isolation of membrane glycoproteins involved in adhesion to collagen. *Exp. Cell Res.* 164: 127–38

Runyan, R., Maxwell, G. D., Shur, B. D. 1986. Evidence for a novel enzymatic mechanism of neural crest cell migration

on extracellular glycoconjugate matrices. *J. Cell Biol.* 102: 432–41

Ruoslahti, E., Pierschbacher, M. D. 1986. Arg-gly-asp: A versatile cell recognition signal. *Cell* 44: 517–18

Sanchez-Madrid, F., Krensky, A. M., Ware, C. F., Robbins, E., Strominger, J. L., et al. 1982. Three distinct antigens associated with human T lymphocyte–mediated cytolysis: LFA-1, LFA-2 and LFA-3. *Proc. Natl. Acad. Sci. USA* 79: 7489–93

Sanchez-Madrid, F., Nagy, J. A., Robbins, E., Simon, P., Springer, T. A. 1983. A human leukocyte differentiation antigen family with distinct alpha subunits and a common beta subunit: The lymphocyte function–associated antigen (LFA-1), the C3bi complement receptor (OKM1/Mac-1), and the p150,95 molecule. *J. Exp. Med.* 158: 1785–1803

Sastre, L., Kishimoto, T. K., Gee, C., Roberts, T., Springer, T. A. 1986. The mouse leukocyte adhesion proteins Mac-1 and LFA-1: Studies on mRNA translation and protein glycosylation with emphasis on Mac-1. *J. Immunol.* 137: 1060–65

Schubert, D., LaCorbiere, M. 1985. Isolation of a cell-surface receptor for chick neural retina adherons. *J. Cell Biol.* 100: 56–63

Selmin, O., Fernandez, M. P., Martin, G. R., Yamada, Y., von der Mark, K. 1986. Isolation and characterization of cDNA clones coding for anchorin, a collagen receptor. *J. Cell Biol.* 103: 259a

Singer, I. 1979. The fibronexus: A transmembrane association of fibronectin-containing fibers and bundles of 5 nm microfilaments in hamster and human fibroblasts. *Cell* 16: 675–85

Singer, I. I., Paradiso, P. R. 1982. A transmembrane relationship between fibronectin and vinculin (130 kD protein) serum modulation in normal and transformed hamster fibroblasts. *Cell* 24: 481–92

Springer, T. A., Galfre, G., Secher, D. S., Milstein, C. 1979. Mac-1: A macrophage differentiation antigen identified by monoclonal antibody. *Eur. J. Immunol.* 9: 301–6

Springer, T. A., Thompson, W. S., Miller, L. J., Schmalstieg, F. C., Anderson, D. C. 1984. Inherited deficiency of the Mac-1, LFA-1, p150,95 glycoprotein family and its molecular basis. *J. Exp. Med.* 160: 1901–18

Springer, T. A., Teplow, D. B., Dreyer, W. J. 1985. Sequence homology of the LFA-1 and Mac-1 leukocyte adhesion glycoproteins and unexpected relation to leukocyte interferon. *Nature* 314: 540–42

Springer, T. A., Anderson, D. C. 1986. The importance of Mac-1, LFA-1 glycoprotein family in monocyte and granulocyte adherence, chemotaxis, and migration into inflammatory sites: insights from an experiment of nature. *Ciba Found. Symp.* 118: 102–26

Suzuki, S., Argraves, W. S., Pytela, R., Arai, H., Krusius, T., et al. 1986. cDNA and amino acid sequences of the cell adhesive protein receptor recognizing vitronectin reveal a transmembrane domain and homologies with other adhesion protein receptors. *Proc. Natl. Acad. Sci. USA* 83: 8614–18

Takada, Y., Huang, C., Hemler, M. E. 1987a. Fibronectin receptor structures within the VLA family of heterodimers. *Nature* 326: 607–9

Takada, Y., Strominger, J. L., Hemler, M. E. 1987b. The VLA family of heterodimers are members of a superfamily of molecules involved in adhesion and embryogenesis. *Proc. Natl. Acad. Sci. USA* 84: 3239–43

Tamkun, J. W., DeSimone, D. W., Fonda, D., Patel, R. S., Buck, C. A., et al. 1986. Structure of integrin, a glycoprotein involved in the transmembrane linkage between fibronectin and actin. *Cell* 46: 271–82

Thiagarajan, P., Sapiro, S. S., Levine, E., DeMarco, L., Yalcin, A. 1983. A monoclonal antibody to human platelet glycoprotein IIIa detects a related protein in cultured human endothelial cells. *J. Clin. Invest.* 75: 896–901

Tomaselli, K., Reichardt, L., Bixby, J. 1986. Distinct molecular interactions mediate neuronal process outgrowth on nonneural cell surface and extracellular matrices. *J. Cell Biol.* 103: 2659–72

Wayne-Carter, E. A., Carter, W. G. 1987. Identification of multiple cell surface receptors for fibronectin and type IV collagen in human fibrosarcoma cells. *J. Cell Biol.* In press

Wehland, J., Osborn, M., Weber, K. 1979. Cell-to-substratum contacts in living cells: A direct correlation between interference

reflection and direct immunofluorescence microscopy using antibodies against α-actin. *J. Cell Sci.* 37: 257–73

Wilcox, M., Brown, N., Piovant, M., Smith, R. J., White, R. A. H. 1984. The *Drosophila* position-specific antigens are a family of cell surface glycoprotein complexes. *EMBO J.* 3: 2307–13

Wilcox, M., Leptin, M. 1985. Tissue-specific modulation of a set of related cell surface antigens in *Drosophila*. *Nature* 316: 351–54

Woods, A., Hook, M., Kjellen, L., Smith, C. G., Rees, D. A. 1984. Relationship of heparin sulfate proteoglycans to the extracellular matrix of cultured fibroblasts. *J. Cell Biol.* 99: 1743–52

Woods, A., Couchman, J. A., Johansson, S., Hook, M. 1986. Adhesion and cytoskeletal organization of fibroblasts in response to fibronectin fragments. *EMBO J.* 5: 665–70

Wright, S. D., Rao, P. E., van Voorhis, W. C., Craigmyle, L. S., Iida, K., et al. 1983. Identification of the C3bi receptor of human monocytes and macrophages by using monoclonal antibodies. *Proc. Natl. Acad. Sci. USA* 80: 5699–5703

Wylie, D. E., Damsky, C. H., Buck, C. A. 1979. Studies on the function of cell surface glycoproteins. *J. Cell Biol.* 80: 385–401

Yamada, K., Kennedy, D. W., Grotendorst, G. R., Momoi, T. 1981. Glycolipids: Receptors for fibronectin? *J. Cell Physiol.* 109: 343–51

Yamada, K., Kennedy, D. W. 1985. Amino acid sequence specificities of an adhesive recognition signal. *J. Cell Biochem.* 281: 99–104

Yamada, K., Akiyama, S., Humphries, M., Yamada, S., Santoro, S. 1986. Platelet and fibroblast interactions with fibronectin and other adhesive proteins. In *Biology and Pathology of the Platelet and Vessel Wall*, ed. G. Jolles, Y. Legrand, A. T. Murden. New York: Academic. In press

Ann. Rev. Cell Biol. 1987. 3 : 207–42

GENERATION OF PROTEIN ISOFORM DIVERSITY BY ALTERNATIVE SPLICING: Mechanistic and Biological Implications

Athena Andreadis, Maria E. Gallego, and Bernardo Nadal-Ginard

Laboratory of Molecular and Cellular Cardiology, Howard Hughes Medical Institute, Department of Cardiology, The Children's Hospital, Departments of Pediatrics, Physiology and Biophysics, Harvard Medical School, Boston, Massachusetts 02115

CONTENTS

PERSPECTIVES AND SUMMARY 208
ALTERNATIVE SPLICING AS A REGULATORY MECHANISM... 208
SELECTION OF SPLICE SITES 209
MODES OF ALTERNATIVE SPLICING 218
   Combinatorial Exons.. 219
   Mutually Exclusive Exons 219
   Internal Donor and Acceptor Sites.. 220
   Retained Introns.. 220
   Multiple 5′ and 3′ Terminal Exons... 220
REGULATORY MECHANISMS OF ALTERNATIVE SPLICING 221
   Possible Categories of Mechanisms 221
   Cis-Regulated Splice-Site Selection 221
   Cis- and Trans-Regulated Splice-Site Selection 226
   Trans-Regulated Splice-Site Selection 230
   Processivity and Directionality of Alternative Splicing.. 232
BIOLOGICAL IMPLICATIONS OF ALTERNATIVE SPLICING 232
   Alternative Splicing Is a Mechanism that Generates Families of Proteins with
      Variable and Constant Domains... 232
   Alternative Splicing Is an Efficient and Reversible Posttranscriptional Mechanism
      for Performing Isoform Switches... 234
   Alternative Splicing May Represent a Primitive Mode of Splicing Important for
      Gene Evolution.. 235

207

0743–4634/87/1115–0207$02.00

PERSPECTIVES AND SUMMARY

RNA splicing is a ubiquitous phenomenon in eukaryotes, and alternative splicing is an important and widespread mechanism for generation of protein diversity made possible by the existence of introns. The differential incorporation of exons into mature mRNAs, often under developmental and/or tissue-specific regulation, allows single genes to produce a variety of related but distinct protein isoforms.

The enzymatic steps involved in constitutive splicing are understood to a first approximation. However, little is known about the mechanisms that distinguish between constitutively versus alternatively spliced exons or about the regulation of alternative splicing where it exists. It has become apparent that the consensus sequence of splice junction sites is not sufficient to specify the nature of an exon. Evidence is accumulating in support of the hypothesis that *cis*-acting sequences, both in introns and in exons, are crucial for correct splicing (constitutive and alternative). Moreover, it seems that the relative affinities of splice junctions can be modified by developmental stage– and tissue–specific *trans*-acting regulatory factors. The existence of these factors has profound significance for the understanding of differentiation, as well as developmental and physiological stage–specific gene expression. Models that use RNA secondary structure to explain alternative splicing have been developed, but their validity has not yet been corroborated by in vivo results.

Alternative splicing increases the coding capacity of the genome; however, in most cases, the selective advantage of encoding different protein isoforms in a single gene is unclear. In terms of evolution, alternative splicing may have been a relatively early event in metazoan genomic development, possibly even predating its constitutive counterpart. Thus alternative RNA splicing, although interesting and important in itself, may also provide a shortcut into clarification of the provenance and role of RNA splicing in general.

ALTERNATIVE SPLICING AS A REGULATORY MECHANISM

The phenotypic expression of structurally and/or functionally distinct, developmentally regulated, and cell type–specific protein isoforms is a fundamental characteristic of metazoan organisms. This isoform diversity has been known for many years, but its widespread occurrence is just beginning to become apparent (Breitbart et al 1987; Leff et al 1986).

The molecular mechanisms responsible for generating this protein diversity belong to two broad categories: those that select a particular gene among the members of a multigene family for expression in a particular cell, developmental, or physiological state (Strehler et al 1986) and those that generate several different proteins from a single gene. The latter mechanism includes DNA rearrangement (Hood et al 1985) and alternative pre-mRNA splicing, the object of this review. These two mechanisms both employ the differential use of DNA sequences that lead to the production, from a single gene, of multiple proteins that have both common and unique domains. DNA rearrangement appears to be restricted to a very limited set of genes coding for components of the immune system (immunoglobulins and T-cell receptors; Hood et al 1985). In contrast, an increasing number of genes in organisms ranging from plants to mammals, including their DNA and RNA viruses, are known to undergo alternative RNA splicing (Table 1). This posttranscriptional mode of switching between isoforms appears to be particularly prevalent in muscle and the nervous system (Hynes 1985; Nadal-Ginard et al 1986). Component cells of both these systems are long-lived but terminally differentiated; they have lost the capacity to undergo DNA replication and cell division. Alternative RNA splicing may be particularly advantageous in cells in which gene reprogramming, which is thought to be most easily accomplished during or immediately following DNA replication, cannot be effected in such a manner (Brown 1984).

SELECTION OF SPLICE SITES

In contrast to their prokaryotic counterparts, most eukaryotic protein-coding genes contain their coding sequences in discontinuous DNA segments (exons) interspersed among sequences (introns) that do not form a part of the mature mRNA. The primary transcripts of these genes also contain the introns, which are precisely excised by a nuclear multistep process known as pre-mRNA splicing. This allows an additional level of posttranscriptional regulation in eukaryotic gene expression (Padgett et al 1986).

The main steps involved in the production of a primary gene transcript are promoter selection, transcription termination, and 3' end formation (Leff et al 1986; Padgett et al 1986; Platt 1986). The primary transcript constitutes the substrate for the splicing apparatus. Each intron forms a complex with specific ribonucleoproteins and small nuclear RNAs (producing a spliceosome; Frendewey & Keller 1985; Padgett et al 1986). Introns are demarcated by invariant consensus sequences at their 5' (donor splice site) and 3' (acceptor splice site) boundaries (Mount 1982). The

Table 1 Cellular genes exhibiting alternative RNA splicing[a]

Gene	Species	Regulation[b]	References
A. Cassette[c]			
Argininosuccinate synthetase	human	—	Freytag et al 1984
c-erb B	chicken	—	Goodwin et al 1986
c-Ki-ras	human, mouse	—	Capon et al 1983; George et al 1985; McCoy et al 1984; McGrath et al 1983; Shimizu et al 1983
αA-Crystallin	mouse, hamster	stochastic	King & Piatigorsky 1983, 1984; van den Heuvel et al 1985
Dopa decarboxylase	Drosophila	tissue	Morgan et al 1986
ERCC-1	human	—	van Duin et al 1986
Fibronectin	human	tissue	Hynes 1985; Kornblihtt et al 1984, 1985; Sekiguchi et al 1986
Gα$_s$ signal transducer	human	—	Bray et al 1986
Histocompatibility antigen H2D[d]	mouse	—	McCluskey et al 1986
Histocompatibility antigen Qa/T1a	mouse	—	Brickell et al 1983
Histocompatibility antigen Q10	mouse	—	Lalanne et al 1985
Ia antigen–associated invariant chain	human	—	Strubin et al 1986
Interleukin-2 receptor	human	—	Leonard et al 1984, 1985
Myelin basic protein	mouse	development	de Ferra et al 1985; Takahashi et al 1985
Myosin heavy chain	Drosophila	development and tissue	Bernstein et al 1986; Rozek & Davidson 1983, 1986
Myosin alkali light chain	Drosophila	development	Falkenthal et al 1985
Nerve growth factor	mouse	tissue	Edwards et al 1986
Polymeric immunoglobulin receptor	rabbit	—	Deitcher & Mostov 1986
Preprotachykinin	bovine	tissue	Nawa et al 1984
Protein kinase C	rat	—	Ono et al 1986
T-cell receptor T3	human	—	Tunnacliffe et al 1986
T-cell receptor chain	mouse	—	Behlke & Loh 1986

α-Tropomyosin	rat	tissue	Ruiz-Opazo et al 1985; Ruiz-Opazo & Nadal-Ginard 1987; Wieczorek & Nadal-Ginard 1987
β-Tropomyosin	rat	tissue	Helfman et al 1986
Tropomyosin I (2)	Drosophila	development and tissue	Basi et al 1984; Basi & Storti 1986; Karlik & Fyrberg 1986
Troponin T (skeletal fast muscle)	rat, quail	development and tissue	Breitbart et al 1985; Breitbart & Nadal-Ginard 1986; Hastings et al 1985; Medford et al 1984
Troponin T (cardiac muscle)	chicken	development and tissue	Cooper & Ordahl 1985
B. Mutually exclusive cassettes[c]			
Myosin light chain 1/3	rat, mouse, chicken	development	Nabeshima et al 1984; Periasamy et al 1984; Robert et al 1984; Strehler et al 1985
α-Tropomyosin	rat	tissue	Ruiz-Opazo et al 1985; Ruiz-Opazo & Nadal-Ginard 1987; Wieczorek & Nadal-Ginard 1987
β-Tropomyosin	rat	tissue	Helfman et al 1986
Tropomyosin II (1)	Drosophila	development and tissue	Karlik & Fyrberg 1986
Troponin T (skeletal fast muscle)	rat, quail	development and tissue	Breitbart et al 1985; Breitbart & Nadal-Ginard 1986; Hastings et al 1985; Medford et al 1984
C. Internal donor sites[c]			
Eip 28/29	Drosophila	—	Schulz et al 1986
Granulocyte colony-stimulating factor	human	—	Nagata et al 1986
HMG-CoA reductase	hamster	—	Reynolds et al 1985
Immunoglobulin heavy chains μ, γ, δ, and ε	mouse	maturation	Alt et al 1980; Early et al 1980; Maki et al 1981; Moore et al 1981; Perlmutter & Gilbert 1984; Rogers et al 1981; Tyler et al 1981; Yaoita et al 1982
Ovomucoid	chicken	stochastic	Stein et al 1980
Plasminogen activator, urokinaselike	porcine	—	Nagamine et al 1985
Prekininogen (bradykinin)	bovine	—	Kitamura et al 1983, 1985

(continued)

Table 1 (*continued*)

Gene	Species	Regulation[b]	References
D. Internal acceptor sites[c]			
Fibronectin	rat	tissue	Hynes 1985; Schwarzbauer et al 1983; Tamkun et al 1984
$G\alpha_s$ signal transducer	human	—	Bray et al 1986
Growth hormone	human	—	DeNoto et al 1981
Histocompatibility antigen *H2K*	mouse	stochastic	Kress et al 1983; Lew et al 1986; Transy et al 1984
Prolactin	rat	—	Maurer et al 1981
Pro-opiomelanocortin	rat	—	Oates & Herbert 1984
Salivary proline-rich protein	human	—	Maeda et al 1985
Serine protease homologue	mouse	—	Cook et al 1985
E. Internal donor and acceptor sites[c]			
Dopa decarboxylase	*Drosophila*	development	Eveleth et al 1986
Fibronectin	human	tissue	Hynes 1985; Kornblihtt et al 1984, 1985; Sekiguchi et al 1986
Gastrin-releasing peptide	human	—	Sausville et al 1986
Histocompatibility antigen *H2K*	mouse	stochastic	Kress et al 1983; Lew et al 1986; Transy et al 1984
Thymidine kinase	chicken	development	Merrill & Tufaro 1986
F. Retained intron[c]			
Dash	*Drosophila*	stochastic	Telford et al 1985
Decay-accelerating factor	human	—	Caras et al 1987
γ-Fibrinogen	rat	tissue (preliminary)	Crabtree & Kant 1982; Homandberg et al 1985
P transposable element	*Drosophila*	tissue	Laski et al 1986
Salivary proline-rich protein	human	—	Maeda et al 1985

G. Multiple 5′ terminal exons[c]

Abl	human	—	Shtivelman et al 1986
Actin 5C	*Drosophila*	stochastic	Bond & Davidson 1986
Alcohol dehydrogenase	*Drosophila*	development	Benyajati et al 1983
Amy 1[a] (α-amylase)	mouse	tissue	Young et al 1981
Antennapedia	*Drosophila*	—	Laughon et al 1986; Schneuwly et al 1986
EGF receptor	*Drosophila*	tissue	Schejter et al 1986; Stroeher et al 1986
Insulinlike growth factor II	rat	development	Soares et al 1986
Myosin light chain 1/3	rat, mouse, chicken	development	Nabeshima et al 1984; Periasamy et al 1984; Robert et al 1984; Strehler et al 1985
Thy-1,2 glycoprotein	mouse	—	Ingraham & Evans 1986

H. Multiple 3′ terminal exons[c]

Antennapedia	*Drosophila*	—	Stroeher et al 1986
Calcitonin	rat, human	tissue	Amara et al 1982; Bovenberg et al 1986; Jonas et al 1985; Rosenfeld et al 1983, 1984
γ-Fibrinogen	human	tissue (preliminary)	Chung & Davie 1984; Fornace et al 1984
Immunoglobulin heavy chains μ, γ, δ, and ε	mouse	maturation	Alt et al 1980; Early et al 1980; Maki et al 1981; Moore et al 1981; Perlmutter & Gilbert 1984; Peterson & Perry 1986; Rogers et al 1981; Tyler et al 1981; Yaoita et al 1982
Insulinlike growth factor I precursor	human	—	Rotwein et al 1986
(2′-5′) Oligo A synthetase	human	tissue	Benech et al 1985; Saunders et al 1985
Prekininogen (bradykinin)	bovine	—	Kitamura et al 1983, 1985
Thymidine kinase	chicken	development	Merrill & Tufaro 1986
α-Tropomyosin	rat	tissue	Ruiz-Opazo et al 1985; Ruiz-Opazo & Nadal-Ginard 1987; Wieczorek & Nadal-Ginard 1987
β-Tropomyosin	rat	tissue	Helfman et al 1986
Tropomyosin II (1)	*Drosophila*	development and tissue	Karlik & Fyrberg 1986

(*continued*)

Table 1 (*continued*)

Gene	Species	Regulation[b]	References
I. Not elucidated			
Brain gene (product unknown)	rat	stochastic	Tsou et al 1986
c-*abl* proto-oncogene	mouse	—	Ben-Neriah et al 1986
Dunce[+]	*Drosophila*	development	Davis & Davidson 1986
EH8	*Drosophila*	development	Vincent et al 1984
FMRFamide neuropeptide	*Aplysia*	—	Schaefer et al 1985
Glucocorticoid receptor	human	—	Hollenberg et al 1985
Neural cell adhesion molecule	chicken	tissue	Hemperly et al 1986; Murray et al 1986
p53 tumor antigen	mouse	—	Arai et al 1986
Protein kinase C	rabbit	—	Ohno et al 1986
T-cell marker *Lyt-2*	mouse	maturation	Zamoyska et al 1985
β-Tropomyosin	human	tissue	MacLeod et al 1985

[a] This list was compiled following an extensive computer-assisted search of the literature and, to the best of the authors' knowledge, was complete at the time of submission. Only those cellular genes in which alternative splice site usage has been clearly documented by cDNA, genomic sequencing, or nuclease protection mapping have been included. Because many genes exhibit more than one pattern of alternative splicing, they appear more than once in the table; within each phenotypic pattern class, genes are listed alphabetically.

[b] Where known, developmental and tissue-specific regulation of splicing is indicated. Documented stochastic (nonregulated) alternative splicing is also specified.

[c] Alternative splicing patterns are as described in the text and shown in Figure 2.

splicing pathway consists of cleavage at the splice donor site, formation of a lariat branch point, and finally cleavage at the acceptor splice site with concomitant ligation of the 5′ and 3′ exons (Figure 1; Padgett et al 1986). Approximately 30 nucleotides upstream from the acceptor splice site is a conserved sequence, where the lariat branch point forms (Keller & Noon 1984).

Because of the size and complexity of mRNA precursors, it has not been possible to investigate rigorously the role of RNA secondary structure in splicing. Tentative results suggesting that hairpins near splice sites can affect splice-site junction choices have emerged from studies with viruses (Munroe 1984; Munroe & Duthie 1986; Nussinov 1980; Seiki et al 1985). Furthermore, recent in vivo and in vitro experiments indicate that sequestering of an exon on a loop of a very stable hairpin leads to its occasional exclusion from the mRNA (Solnick 1985b). It has been postulated that similar stem-and-loop structures may play a role in directing alternative

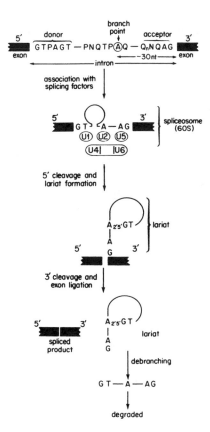

Figure 1 Schematic representation of the steps involved in constitutive splicing. The consensus donor, acceptor, and lariat branchpoint sequences are shown. Long dashes indicate intervening nucleotides; Q = pyrimidine; N = any nucleotide. The A at the branchpoint is circled. The locations of the small nuclear RNAs are extrapolated from primary sequence data, inhibition of splicing subreactions by specific antibodies, and nuclease protection experiments (Padgett et al 1986).

splicing (Breitbart et al 1987; Breitbart & Nadal-Ginard 1986; Ruiz-Opazo & Nadal-Ginard 1987). In general, however, the lack of an effect of large intron deletions (van Santen & Spritz 1985; Wieringa et al 1984) and the wide variation in intron length argue against strict higher order structure requirements. Experiments in which there have been sequence substitutions within introns show that often it is intron length, rather than specific sequence, that affects exon selection (Fu & Manley 1987; Peterson & Perry 1986).

The central problem in pre-mRNA splicing, both constitutive and alternative, is the selection of the correct pair of donor and acceptor sites to be joined. The sequences at the splice sites, though conserved, are multiply repeated elsewhere in the transcript; under the proper conditions, these normally inactive consensus sequences can become functional, as evidenced by the recognition of cryptic equivalents by the splicing complex (Aebi et al 1986; Krainer et al 1984; Metherall et al 1986; Treisman et al 1983). Given the redundancy of splice junction sites, it is imperative that the splicing process be constrained to give rise to only the correct exon combinations. One possible constraint might require that the splicing process be orderly and proceed via a scanning mechanism that starts at one end of the transcript (Sharp 1980). In the β-globin gene, the first intervening sequence is almost always excised before the second, and it appears that intermediates in which the second intron has been removed before the first are not spliced any further (Lang et al 1986). In contrast, the opposite directionality is exhibited in splicing of the E2A pre-mRNA of adenovirus (Gattoni et al 1978). Furthermore, partially spliced molecules with different combinations of persisting introns accumulate in both the nucleus and in vitro systems (Berget & Sharp 1979; Roop et al 1978; Ryffel et al 1980; Tsai et al 1980). From these experiments, one can conclude that there is no strict directional order for the removal of multiple introns from pre-mRNA molecules. However, in the few cases studied carefully in vivo (calcitonin, globin; Bovenberg et al 1986; Zeitlin & Efstratiadis 1984) or in vitro (globin, albumin, interleukin; Lang & Spritz 1986) splicing seems to follow nonobligatory but highly preferred pathways with respect to order of intron excision.

The mechanism by which a selection is made among several putative donor or acceptor splice sites flanking a single intron is not well understood. If a consensus sequence is mutated, the splicing apparatus may select a cryptic consensus within the same intron, thereby creating an exon with different boundaries (Metherall et al 1986; Treisman et al 1983), or it may bypass the exon with the mutated consensus altogether (Mitchell et al 1986). Experiments with chimeric constructs that contain tandem duplications of donors and acceptors have given conflicting results (Kuhne et al 1983; Lang & Spritz 1983) and, together with the results obtained with several exon truncation and substitution experiments, led to the initial conclusion that exon sequences do not play a major role in splicing (Padgett et al 1986). This point of view is contradicted by several recent findings. The pattern of splice-site selection in alternatively spliced viral pre-mRNAs (Somasekhar & Mertz 1985), and possibly in a tropomyosin gene (Karlik & Fyrberg 1985), is altered by mutation within an exon. Moreover, in *cis* competition assays it has been shown that utilization of 5' and 3' splice sites can be significantly affected by the sequences within the flanking exons as well as by the relative proximity of competing splice sites (Eperon et al 1986; Fu & Manley 1987; Reed & Maniatis 1986; van Santen & Spritz 1986). Taken together, these results suggest that intron removal is likely to be kinetically or thermodynamically determined, implying that there is a hierarchy of strengths or affinities among donor and acceptor sites. This phenomenon of intrinsic exon "compatibility" may play a fundamental role in exon selection in differentially spliced transcripts (A. Andreadis, M. Gallego, B. Nadal-Ginard, unpublished observations; Breitbart & Nadal-Ginard 1987). The same results also imply that limited scanning for the appropriate donor and/or acceptor must occur within an intron, once the splicing machinery has attached (Metherall et al 1986).

Despite the lack of experimental proof, the conclusion that splicing operates by some sort of processive mechanism is almost unavoidable. Because donor/acceptor consensus sequences are necessary but not sufficient for correct splicing, the splicing apparatus may recognize a characteristic secondary-structure feature rather than the primary sequence (Gilbert 1985; Solnick 1985b). One of the most compelling arguments for a processive tracking type of mechanism is the rarity of *trans* splicing in vivo. The only exceptions so far described are the spliced common leader of trypanosome RNAs (Milhausen et al 1984; Murphy et al 1986; Sutton & Boothroyd 1986) and *rps*12 of chloroplasts (Koller et al 1986). Within the spliceosome, linearly contiguous donor and acceptor sites may not be closer together than distant sites in the same or other molecules. In this case, the lack of a tracking system would lead to incorrect inter- or intramolecular splicing, which is likely to result in nonfunctional proteins. Such molecules have not been detected in vivo. Although intermolecular splicing can be produced in vitro, it is favored by the formation of hybrid duplexes that link the two reacting substrates, thereby mimicking an intramolecular reaction (Konarska et al 1985; Solnick 1985a).

Whether splicing proceeds via scanning or not, clearly correct exon selection does not require the presence of the entire gene: Chimeric gene constructs are correctly spliced (although this may reflect a bias in selection, since often only correctly spliced species are visible with the probes or functional assays utilized) (A. Andreadis, M. Gallego, B. Nadal-Ginard, unpublished observations; Chu & Sharp 1981; Kaufman & Sharp 1982; Vibe-Pedersen et al 1984; Young et al 1981). Only in the case of a globin intron engineered into the herpes simplex thymidine kinase gene has intron position been correlated with excisability (Greenspan & Weissman 1985). However, the recent experiments suggesting nonequivalence of splice junction sites complicate the interpretation of results from chimeric genes (Breitbart & Nadal-Ginard 1987; Eperon et al 1986; Mitchell et al 1986; Reed & Maniatis 1986).

MODES OF ALTERNATIVE SPLICING

In the majority of instances studied so far, every intron is removed in a way that causes every exon present in a pre-mRNA molecule to be incorporated into a mature mRNA through the invariant ligation of consecutive pairs of donor and acceptor splice sites. This "constitutive splicing" yields a single gene product from each transcription unit. The process appears to be highly precise and efficient but responsive to environmental stimuli: Unspliced RNAs accumulate in cells during heat shock. Interestingly, none of the bona fide heat-shock proteins contain introns in their genes (Yost & Lindquist 1986).

In alternative pre-mRNA splicing, the splicing apparatus can exclude individual exon sequences from the mature mRNA in some transcripts but include them in others, making the distinction between exons and introns ambiguous. Alternative splicing produces mRNAs with different primary structures from a single gene. When the exons involved contain translated sequences, these alternatively spliced mRNAs will encode related but distinct protein variants.

A large, and constantly growing, number of genes are known to generate protein diversity through the use of differential splicing in organisms ranging from *Drosophila* to humans, and including their RNA and DNA viruses. The number of different mRNAs, and hence protein isoforms encoded by a given gene, increases exponentially as a function of the number of exons that participate in alternative splicing (if one assumes that the exons are incorporated independently of each other). This, in turn, augments significantly the phenotypic variability that can be obtained from single genes, such as fibronectin, or from a gene family, such as the contractile proteins, in which the products of alternatively spliced genes assemble to form multimeric complexes (Hynes 1985; Nadal-Ginard et al 1986).

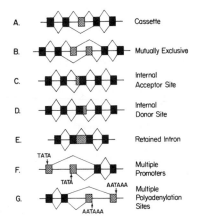

Figure 2   Patterns of alternative splicing. Constitutive exons (*black*), alternative exons (*dotted*), and introns (*heavy solid lines*) are spliced according to different pathways (*fine solid lines*), as described in the text. Alternative promoters (TATA) and polyadenylation signals (AATAAA) are indicated (Breitbart et al 1987).

The proteins produced from alternatively spliced genes fulfill a wide variety of functions, as can be seen by the list in Table 1. Their range of action includes inter- and intracellular communication (neurotransmitters, receptors, transducers, hormones), cell contraction and motility (myofibrillar proteins), cell structural components (cytoskeletal and extracellular matrix proteins), oncogene products and growth factors, enzymes of intermediary metabolism, kinases and proteases that are members of cascade pathways (blood clotting), immune system components (histocompatibility antigens, immunoglobulins), DNA transposition (the P element) and developmental regulators (Antennapedia).

Differentially spliced genes often employ more than one pattern of alternative splice-site usage in combination with conventional constitutive exons. Figure 2 diagrams these patterns, which are explained more extensively below. This classification is descriptive and may not correspond to the underlying molecular mechanisms.

Combinatorial Exons

Many alternatively spliced genes contain "cassette exons," which are defined as exons that can be individually excluded from the mature mRNA. Several genes contain more than one such cassette exon; if n is the number of cassette exons in a gene, the gene can potentially produce 2^n different mRNA species.

Mutually Exclusive Exons

Several contractile protein genes contain pairs of consecutive cassette exons that are included in the mRNA in a mutually exclusive manner. One member of the pair is invariably spliced into a given mRNA, but they are never both included in the same mRNA molecule. This contrasts with the combinatorial pattern described above, in which the splicing of each cassette could theoretically be independent of that of others in the gene. Each mutually exclusive cassette encodes an alternative version of the same protein domain. In troponin T and myosin light chain 1/3, the incompatibility of mutually exclusive exons is enforced by the fact that a frameshift occurs if both exons are simultaneously included or excluded (Medford et al 1984; Periasamy et al 1984). The intron located between two such exons is a pseudointron; although it contains functional 5' and 3' splice sites, these two sites are never joined to each other during splicing.

Internal Donor and Acceptor Sites

Cassette exons may or may not be incorporated into mRNA, but their immediate flanking introns are invariably removed during the splicing process. There exist genes with internal alternative splice sites that lie entirely within potential RNA coding sequence. Splicing at such sites results in the exclusion of a portion of an exon (or in the incorporation of adjacent intron sequences) into the mRNA. As a result, exon/intron boundaries become blurred in this type of splicing. In these exons there is a minimum of two functional donor and/or acceptor splice sites; then the total number of potential mRNA species equals the number of duplicate splice sites.

Retained Introns

Several genes incorporate intron sequences into mRNA by failing to splice a donor-acceptor pair altogether. The retained intron usually maintains an intact translational reading frame and, in effect, creates a longer fusion exon. In this case the intron involved can be viewed as an "optional" intron contained within a larger exon.

Multiple 5' and 3' Terminal Exons

In many systems, alternatively spliced mRNAs are associated with different primary transcripts of the same gene. Heterogeneous sites of transcription initiation and of 3' end formation necessarily result in transcripts with decidedly distinct primary structures. Different promoters and polyadenylation sites may specify alternative 5' and 3' terminal exons, respectively. These exons are not true cassettes, because they contain functional splice sites at their internal boundary only. In some instances, additional alternative splicing occurs in exons internal to these heterogeneous termini.

REGULATORY MECHANISMS OF ALTERNATIVE SPLICING

Possible Categories of Mechanisms

The mechanism of donor/acceptor pair selection is a major problem even in constitutive splicing, since the primary sequence elements necessary for splicing are not sufficiently unique. The logistical complexity increases severely in alternative splicing. It is noteworthy that the donor/acceptor consensus sequences of alternatively spliced exons are identical to those of constitutively spliced ones (62 donors and 62 acceptors involved in alternative splicing have been scored; M. E. Gallego, unpublished observations). Therefore, when one of several alternative splice sites is selected in a spliced RNA, two sites that conform to the consensus are no longer equivalent.

The key to regulated alternative splicing must reside in information encoded in the gene transcript (*cis*) but may require additional control by diffusible factors (*trans*) that may be cell-specific. Certain genes produce heterogeneous primary transcripts (Table 1), with either different promoters or polyadenylation sites. The sequences of such different primary transcripts produced by the same gene may contain enough information in *cis* to dictate a different splicing pathway for each of the transcripts. In this case, specific *trans* factors may not be needed. But other genes produce multiple mRNA variants from a unique primary transcript that is initiated at a single promoter and terminated at a single polyadenylation site (Table 1). The execution of tissue-specific or developmentally regulated splicing in these identical transcripts cannot possibly be encoded in the *cis* sequences exclusively and necessitates additional control by cell-specific *trans* factors.

Before analyzing specific examples, the "null" or "default" splicing mode must be defined. For the alternatively spliced genes studied so far, the mRNAs produced in inappropriate cell types never include all the exons of the pre-mRNA. Instead, the mRNAs are spliced in one or more, but never all, of the possible alternative modes. This splicing pattern defines a ground state that is useful in studies of the regulation of the splicing of that particular gene.

In terms of possible mechanisms, alternative splicing can be divided into three categories that depend on whether the different and specific affinities among splice sites are governed by (*a*) *cis* factors only, (*b*) both *cis* and *trans* factors, or (*c*) *trans* factors only.

Cis-*Regulated Splice-Site Selection*

Cis-regulated splice-site selection involves three overlapping regulatory mechanisms: (*a*) multiple transcriptional promoters and/or terminators, (*b*) pre-mRNA higher-order structure, and (*c*) splice-site hierarchy.

Use of different start- and/or stop-transcription sites results in pre-mRNAs of different length and/or sequence. These pre-mRNAs presumably fold into distinct higher order conformations. If so, RNA conformation could dictate exon selection. For genes in this category, tissue- and development-specific patterns of mRNA production would be only indirectly due to splicing. The true determinants would be the transcriptional factors that bind to, and act upon, the critical nucleic acid sequences, resulting in production of distinct primary transcripts. These factors could influence transcription initiation, promoter activation, 3′ end formation, and polyadenylation. Even in cases of alternatively spliced genes with a unique promoter and terminator, the mRNA might fold into more than one conformation, each making the appropriate splice junctions accessible. Here the relative amounts of the differently spliced mRNA forms would be proportional to the relative free energies of the different pre-mRNA structures. Finally, the multiple splice sites might have different intrinsic affinities for the splicing machinery (for example, splicing factors could bind more or less tightly to specific donor and acceptor consensus sequences), and thus different mRNAs would be produced in proportion to these affinities.

Evidence from a number of systems in vivo and from in vitro experiments gives credence to the existence of all the mechanisms mentioned. For immunoglobulin μ (Figure 3) the relative amounts of secreted versus membrane-bound isoforms are governed by the choice of polyadenylation sites. Immature B cells that have not been challenged with antigen produce the membrane-bound form, whereas mature B cells express the secreted form (Alt et al 1980; Early et al 1980; Rogers et al 1981). Progressive deletions at the (upstream) cleavage/polyadenylation region that specifies the secreted form lead to increasing and finally exclusive production of the membrane-bound product (Danner & Leder 1985). Similar results come from preliminary studies of Ig γ and δ heavy and κ light chains (Flaspohler & Milcarek 1986; Kerr et al 1986; Matis & Milcarek 1986; Rogers et al 1986). The production of the two forms remains regulated only if the two polyadenylation sites are connected in the same molecule (*cis*); the default choice is the secreted form, which utilizes the proximal polyadenylation site. If the two 3′ ends are placed on separate molecules (*trans*), the two IgM mRNAs are coexpressed in approximately equivalent amounts within the same cell. Deletion of various segments of the intron that separates the two polyadenylation sites increases the proportion of the membrane-bound form; the proportion produced correlates inversely with the length of the intron. Regeneration of the intron by insertion of random DNA fragments has the opposite effect on the ratio of membrane-bound to secreted IgM, which indicates that intron length rather than specific

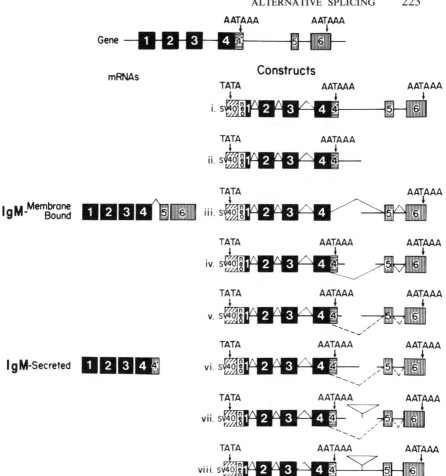

Figure 3 Immunoglobulin μ heavy-chain (mouse) gene organization, mRNAs, and minigene constructs. Exons (*black*, constitutive; *horizontally/vertically striped*, mutually exclusive; *diagonally striped*, heterologous promoters) are diagrammed. The polyadenylation (AATAAA) signals are indicated. In the constructs, exons are spliced according to major (*solid lines*) and minor (*dashed lines*) pathways (Danner & Leder 1985; Peterson & Perry 1986).

sequence determines the processing choice (Peterson & Perry 1986). This finding suggests that the crucial event in these genes is competition between two mutually exclusive events, polyadenylation at the proximal site at the end of exon 4 (which gives rise to the secreted isoform) and splicing of an internal donor of exon 4 to exon 5 and hence to exon 6 (which produces the membrane-bound isoform).

A slightly different picture emerges from studies on the calcitonin gene

(Figure 4). Calcitonin and calcitonin gene–related peptide (CGRP) utilize the same transcription start site but different poly(A) sites, located at the 3′ end of exons 4 and 6, respectively. Thyroid C cells express calcitonin; neuronal cells express CGRP (Amara et al 1982; Jonas et al 1985; Rosenfeld et al 1983, 1984). When the gene is linked to a heterologous promoter that allows gene expression in many different tissues, calcitonin is the default choice in most tissues of a transgenic mouse. But in brain cells, which normally produce CGRP, the heterologous construct gives rise to CGRP only (Rosenfeld et al 1986). However, the calcitonin poly A addition site can be used in the brain if the alternative site is deleted. When the proximal (calcitonin-specific) polyadenylation site is deleted, CGRP is produced in cell lines that normally express CGRP. The same construct, when transfected into cell lines that express calcitonin, remains unspliced, although the distal (CGRP-specific) polyadenylation site is present and functional (M. Rosenfeld, personal communication). It is still unclear whether the tissue-specific regulation occurs at the 3′ cleavage or splicing level (if at the latter, this gene must utilize a neuron-specific splicing factor).

Rat skeletal myosin light chains 1 and 3 are produced by alternative splicing from the same gene. This gene (MLC1/3; Figure 5) seems to utilize all three mechanisms in an integrated fashion: Two widely separated promoters give rise to two transcripts very different in structure and length (Periasamy et al 1984; Strehler et al 1985). Of the two mutually exclusive exons, denoted as 3 and 4 in Figure 5, the one incorporated into MLC1

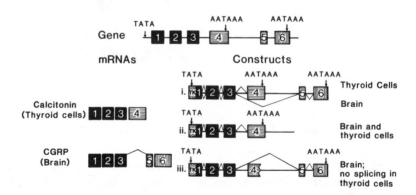

Figure 4   Calcitonin (mouse) gene organization, mRNAs, and minigene constructs. Exons (*black*, constitutive; *horizontally/vertically striped*, mutually exclusive; *diagonally striped*, heterologous promoters) are diagrammed. The promoter (TATA) and polyadenylation (AATAAA) signals are indicated (Rosenfeld et al 1983, 1984; M. Rosenfeld, personal communication).

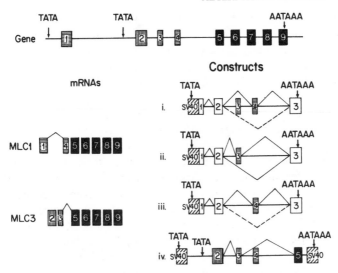

Figure 5   Fast skeletal myosin light-chain 1/3 (rat) gene organization, mRNAs, and minigene constructs. Exons (*black*, constitutive; *horizontally/vertically striped*, mutually exclusive; *diagonally striped*, heterologous promoters; *white*, insulin) are diagrammed. The promoter (TATA) and polyadenylation (AATAAA) signals are indicated. In the constructs, exons are spliced according to major (*solid lines*) and minor (*dashed lines*) pathways.

(4) appears to be the default exon: When either MLC exon 1 or insulin exon 2 is placed upstream from the two exons, exon 4 is used, irrespective of cell type. However, a transcript that starts with the first MLC3-specific exon (2 in Figure 5) uses the MLC3-specific exon (3 in Figure 5) exclusively (M. E. Gallego, unpublished observations). If the 5′ proximal promoter is used, which gives rise to MLC1, exon 1 cannot splice to exon 2 because the latter lacks an acceptor site, and exon 1 is spliced to exon 4, bypassing exon 3. Therefore, the intrinsic affinity of exon 1 for exon 4 must be much higher than its affinity for exon 3. If the 5′ distal promoter is activated, which gives rise to MLC3, exon 2 will select exon 3 over exon 4, and exon 4 will be bypassed. One possibility is that the newly spliced entity consisting of exons 2 and 3 has a donor splice site of low affinity for the acceptor site of exon 4 or folds in such a way that exon 4 becomes inaccessible. Another is that exon 3, whether by itself or spliced to any other upstream exon, has a very low intrinsic affinity for exon 4. For alternative splicing subject only to *cis* control, this would imply that once the appropriate promoter or terminator has been utilized, the subsequent splicing pattern is obligatory.

In vitro studies have demonstrated that sequestering of an exon on the loop of a hairpin stem leads to its frequent exclusion from the final RNA product. However, the same constructs, when spliced in vivo, give only minute proportions of the RNA that lacks the exon; furthermore, the stability of the engineered stems so far tested (Solnick 1985b) is much greater than that of any of the putative stem-and-loop structures detected around naturally occurring exons that are alternatively spliced (Breitbart et al 1987; Breitbart & Nadal-Ginard 1986; Ruiz-Opazo & Nadal-Ginard 1987).

Alternative splicing of duplicated internal donor or acceptor sites and retained introns can be explained satisfactorily by the relative affinities of splice junction sites. A significant body of data on natural and in vitro mutations has demonstrated that cryptic splice sites compete successfully with native sites for splicing factors (Eperon et al 1986; Krainer et al 1984; Treisman et al 1983). Moreover, *cis* competition between splice sites usually results in production of all possible RNAs in ratios that reflect the relative affinities of the splice sites (Eperon et al 1986; Fu & Manley 1987; Reed & Maniatis 1986; van Santen & Spritz 1986). These data suggest that in-frame alternative splice sites could have initially evolved randomly as cryptic sites, and those that coded for selectively advantageous proteins were retained (Craik et al 1983).

Alternative splicing of genes that employ multiple transcriptional initiators/terminators may be determined by secondary structure only, or by a combination of the latter and splice-site hierarchy. Splices executed at one point of the RNA molecule may generate conformational or splice-site affinity changes that affect splicing elsewhere; but this possibility has not yet been carefully examined. If splicing intermediates are differentially processed because of their differing primary sequence, initial splice-site selections could generate a unidirectional cascade of preferential choices and produce complex splicing patterns that result from a series of simple binary decisions.

Cis- *and* Trans-*Regulated Splice-Site Selection*

Genes in this category have *cis* determinants within their primary sequences that specify one of the possible splicing modes, but they also require specific *trans* factors to express their full range of potential mRNAs.

Thus far, the best-studied gene in this category is that encoding the rat skeletal troponin T (TnT; Figure 6), which contains two mutually exclusive exons near its 3′ end. The 5′ proximal exon (α) is adult-specific, whereas the

---

Figure 6 Skeletal troponin T (rat) gene organization, possible mRNAs, and minigene constructs. Exons (*black*, constitutive; *horizontally/vertically striped*, mutually exclusive; *dotted*, cassette; *diagonally striped*, heterologous promoters; *white*, insulin) are diagrammed. The promoter (TATA) and polyadenylation (AATAAA) signals are indicated. In the constructs, exons are spliced according to major (*solid lines*) and minor (*dashed lines*) pathways.

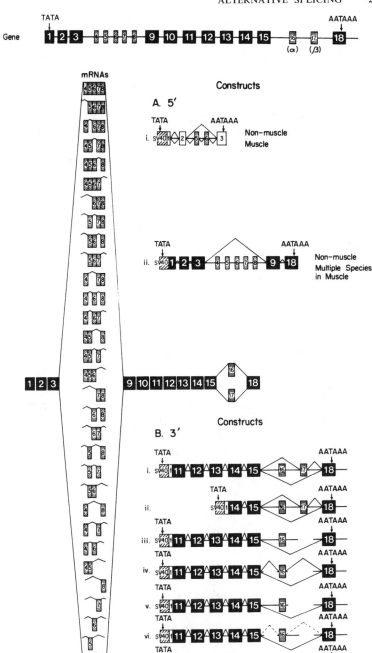

5' distal exon (β) is found in both embryonic and adult tissues and in muscle cell lines (Medford et al 1984). Constructs that contain these two exons, when expressed in muscle or nonmuscle cells, include the β exon, never the α. The α exon is excluded even when β has been deleted, which implies that the exclusion of α is not a simple case of competition between splice sites with different relative affinities. However, when the intron sequences flanking the α exon are almost completely deleted, the exon is present in the mRNA, even in nonmuscle cells. Reconstitution of the intron 5' to the α exon, but not the β exon, leads to exclusion of the α exon from the mRNA (A. Andreadis, unpublished observations). The TnT primary transcript contains a feature that renders the α exon unusable and leads to constitutive retention of the β exon in all tissues except adult muscle. In muscle, a factor must be produced that either alters the conformation of the pre-mRNA or changes the relative kinetics of exon recognition and allows the splicing machinery to recognize the α exon while bypassing the β exon. If this factor was rate-limiting for splicing, its concentration would determine the ratio of α to β mRNA in each tissue.

Similar results have been obtained in preliminary studies with the rat α-tropomyosin gene (Tm; Figure 7), which has alternative polyadenylation sites as well as mutually exclusive exons that encode striated and smooth muscle–specific isoforms (Ruiz-Opazo et al 1985). In the absence of other constraints, the splicing machinery primarily selects the striated muscle–specific exon of the mutually exclusive pair (C. W. J. Smith, unpublished observations). There is some correlation between utilization of a particular polyadenylation site and incorporation of a particular member of the pair. However, polyadenylation cannot be the sole determinant of splicing pattern because Tm produces at least five different mRNAs from a unique primary transcript (Wieczorek & Nadal-Ginard 1987).

Both in TnT and Tm the mutually exclusive exons are flanked by regions of potentially stable secondary structure. Models suggest that each member of the pair could be located in the loop region of a hairpin (Breitbart & Nadal-Ginard 1986; Ruiz-Opazo & Nadal-Ginard 1987). The two hairpins are mutually exclusive because they share one side of their stem. However, there is no experimental evidence yet that these stem-and-loop structures actually form in the cell or that they play a role in splicing.

The cardiac troponin T gene, which has an exon-intron organization similar to its skeletal counterpart (Breitbart & Nadal-Ginard 1986), is also alternatively spliced (Cooper & Ordahl 1985). In contrast to the skeletal gene, the cardiac one has a simpler pattern of splicing, with a single cassette exon that is excluded in approximately half of the mRNA molecules (Cooper & Ordahl 1985). Minigene constructs that contain the cassette exon are alternatively spliced in the same proportions as in the endogenous

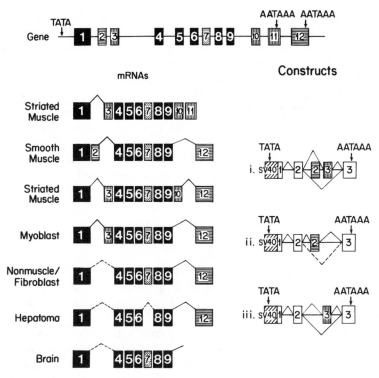

Figure 7 α-Tropomyosin (rat) gene organization, possible mRNAs, and minigene constructs. Exons (*black*, constitutive; *horizontally/vertically striped*, mutually exclusive; *diagonally striped*, heterologous promoters; *white*, insulin) are diagrammed. The promoter (TATA) and polyadenylation (AATAAA) signals are indicated. In the constructs, exons are spliced according to major (*solid lines*) and minor (*dashed lines*) pathways (Ruiz-Opazo & Nadal-Ginard 1987; C. Smith, unpublished observations; Wieczorek & Nadal-Ginard 1987).

gene, even when the construct is expressed in nonmuscle cells (Cooper & Ordahl 1986). Deletion of sequences required to form a hairpin that could potentially sequester this exon does not appear to affect splicing (T. Cooper, personal communication).

When genes employ more than one splicing pattern in separate regions of their pre-mRNA (as do fibronectin, TnT, and Tm; Breitbart et al 1985; Hynes 1985; Ruiz-Opazo & Nadal-Ginard 1987), the splicing in one area of the molecule could in principle influence processing elsewhere. The findings that a mutant *Neurospora* pre-mRNA with an anomalous 3′ extension fails to be spliced (Garriga et al 1984), that mutations within an exon affect splicing patterns both in vivo (Eperon et al 1986; Somasekhar & Mertz 1985) and in vitro (Reed & Maniatis 1986), and that in a flightless

Drosophila mutant with a large insertion in a Tm exon the splicing of other exons is affected (Karlik & Fyrberg 1985) lend validity to the concept that distant intramolecular changes affect local splicing. In addition, the results obtained with the cardiac and skeletal troponin T genes suggest that some of the *cis*-acting elements affecting splicing pattern are located in the introns. In contrast, gene hybrids containing a cassette exon of the fibronectin gene (Vibe-Pedersen et al 1984) or the mutually exclusive exons 3 and 4 of the MLC1/3 gene (M. E. Gallego, unpublished observations) flanked by hybrid introns and heterologous exons, are still alternatively spliced when expressed in HeLa cells. These results, taken together with those from TnT and Tm minigene constructs mentioned above, indicate that alternative splicing that is determined, at least in part, by *cis*-acting elements need not be dependent on the structural integrity of the primary RNA transcript.

Trans-*Regulated Splice-Site Selection*

For many genes that give rise to a unique transcript and exhibit tissue-specific and/or developmentally regulated alternative splicing, the existence of specific *trans*-acting environments must be postulated. Two different types of mechanisms would determine the fate of such gene transcripts: (*a*) varying concentrations of the "constitutive" splicing factors in different cells and (*b*) the presence or absence of cell-specific *trans*-acting factors.

In *cis* competition assays performed in vitro with RNAs containing duplicated donors and/or acceptors of the β-globin gene the resulting exon patterns in the spliced products appear to depend partly on the concentration of the HeLa splicing extract (Reed & Maniatis 1986). However, it is difficult to envision constitutive splicing factors being rate limiting in vivo in a way that determines alternative splicing.

In the skeletal troponin T gene (TnT; Figure 6) exons 4–8 are cassettes that are incorporated into the mRNA in different combinations in vivo (Breitbart et al 1985). As parts of minigene constructs these short exons fail to undergo constitutive splicing, not only in their native muscle context, but also in pre- and nonmyogenic cells. In the absence of the specific *trans*-acting environment provided by the muscle cells, these exons tend to be excluded from the mRNA, although they are flanked by canonical donor and acceptor sequences. During myogenic induction, this pattern changes; there is a progressive increase in the number of TnT species due to differential incorporation of the miniexons (Breitbart & Nadal-Ginard 1987). The most striking finding from experiments with constructs containing the combinatorial TnT exons is that flanking exons 3 and 9, which are constitutively expressed in muscle cells in vivo, are alternatively spliced in nonmuscle cells and myoblasts. All these observations lead to the conclusion that muscle-specific *trans* factors are induced during myogenesis and that these must interact with the primary transcript to modulate the splicing of the various exons (Breitbart & Nadal-Ginard 1987).

Potential candidates for specific *trans* splicing factors are proteins and small nuclear RNAs, alone or in combination (snRNPs), that could act as allosteric modifiers of the splicing machinery or as factors masking specific exons. Small nuclear RNAs are clearly involved in, and crucial to, constitutive splicing (Frendewey & Keller 1985; Kramer et al 1984; Padgett et al 1986; Zhuang & Weiner 1986). SnRNPs would have the potential to produce numerous species by combining relatively few basic components. This would permit an efficient allocation of genetic resources; it could yield final products of high specificity from initial ingredients of low specificity. Small nuclear RNAs alone exhibit enough heterogeneity to make fine tuning possible. In support of this idea, different U1 RNAs (which attach to the donor site of the intron during spliceosome formation) are found to be expressed at different times during *Xenopus* development (Forbes et al 1984).

While much of the RNA splicing machinery is undoubtedly common to all cell types, the muscle cell example shows that certain components are highly specialized. Similar cell-specific factors may well be essential for the appropriate splicing (alternative or constitutive) of many other cell-specific gene transcripts. The processing of these transcripts outside their usual environment, in either heterologous or immature homologous cells, would be expected to produce aberrant splicing. At least in many cases, the onset of transcription of cell-specific genes is not sufficient to produce the differentiated phenotype of a cell; the activation of specific splicing factors is required for correct splice-site selection of alternative, and at least some constitutive, exons in the transcripts expressed in some specialized cells (A. Andreadis, unpublished observations; Breitbart & Nadal-Ginard 1987).

The picture outlined above for splicing of specific gene transcripts contrasts with the behavior of commonly expressed ("housekeeping") genes and a number of cell-specific genes, which are accurately spliced in most cell types. This splicing could be accomplished by a wholly nonspecific splicing apparatus common to all cell types. Alternatively, these genes may have evolved especially strong splice sites that are invariably and efficiently recognized by the various splicing components of different cells. Even these strong splice sites, however, need not all be equivalent: In the dehydrofolate reductase gene, mutation of one splice site disrupts the splice-site hierarchy, and alternative splicing occurs in an otherwise constitutively spliced gene (Mitchell et al 1986).

All alternative splicing might employ specific *trans*-acting factors. In cases where their presence is documented, the mechanism of their regulation is expected to be similar to that of the regulation of transcription factors. Technically, this regulation is difficult to probe due to the current lack of cell type–specific in vitro splicing systems.

Processivity and Directionality of Alternative Splicing

Most molecular mechanisms involved in propagation of genetic information (DNA replication, transcription, translation) are processive and proceed in the 5′ to 3′ direction. However, experiments to determine directionality of splicing have yielded inconsistent results (Kuhne et al 1983; Lang & Spritz 1983). Nevertheless, as discussed above, it is unlikely that splicing proceeds in an entirely random manner.

When mutually exclusive exons were discovered, it was argued that a simple scanning model for splicing would be inadequate, at least for those exons (Medford et al 1984). Yet given the recent results obtained with TnT and MLC1/3 gene constructs (A. Andreadis & M. E. Gallego, unpublished observations), a modified scanning model could still operate: If pre-mRNA high-order structure determines processing, splice junctions destined to be excluded would be either masked by *trans*-acting factors or separated on a hairpin loop; a "generic" splicing system would then indiscriminately process the remaining (default) splice sites. Conversely, if exon selection is dependent on relative splice-site affinities and hence is kinetically determined, the more efficient splicing reaction would completely suppress its competitor. For genes that contain many alternatively spliced exons, ordered events are expected (for example, when MLC3 exons 2 and 3 have been joined, exon 4 must subsequently be bypassed), since causally related splicing decisions would not give equivalent results if reversed in time. This phenomenon could be facilitated if the affinities of donor and acceptor sites change during the splicing process depending on which intermediates are formed. This hypothesis is consistent with experimental data indicating that sequences far from the splice site help determine splicing patterns (Eperon et al 1986; Mitchell et al 1986).

BIOLOGICAL IMPLICATIONS OF ALTERNATIVE SPLICING

Alternative Splicing Is a Mechanism that Generates Families of Proteins with Variable and Constant Domains

Alternative splicing based on the utilization of different promoters or terminators has limited potential to generate protein diversity. In most cases the first or last differentially spliced exon contains the 5′ and 3′ untranslated sequence, respectively. Such alternative splicing may still affect overall protein ratios by altering mRNA stability. Furthermore, different promoters that give rise to untranslated leaders of distinct length and sequence may affect translation. It is a well-known fact that the context in which the initiator AUG is located influences translation efficiency (Kozak 1986a); moreover, in lower eukaryotes mRNA leader length correlates inversely with efficiency of translation (Mueller & Hinnebusch 1986).

In other genes, such as Tm (Helfman et al 1986; Karlik & Fyrberg 1986; Ruiz-Opazo & Nadal-Ginard 1987) and vertebrate MLC1/3 (Nabeshima et al 1984; Periasamy et al 1984), different amino- or carboxy-terminal ends of the protein are encoded by alternative exons. When differentially spliced exons contain translated sequences, their structure needs to be such that, whether they are included or excluded, the resulting mRNAs maintain an intact reading frame that is continuous across exon boundaries (Breitbart et al 1987; Sharp 1980). To date, only two cases of alternative splicing in cellular genes (the *Drosophila* P element and the human decay-accelerating factor) have been described that introduce a frameshift (Caras et al 1987; Laski et al 1986), although this is a common feature of several viral genes (Kozak 1986b; Nevins 1983; Temin 1985; Ziff 1980). The consequence of the constraint imposed by reading frame in most cellular genes is the production of protein isoforms with constant and variable (either additional or interchangeable) domains. From preliminary secondary-structure studies (mainly on the myelin basic protein and the interleukin 2 receptor; de Ferra et al 1985; Leonard et al 1984, 1985), it appears that peptides encoded by cassette exons are often insertions in surface loops. Alternative splicing, therefore, is an efficient way of generating variants with different site affinities and/or functions without disturbing the core structure of the protein.

Mutually exclusive exons (for example, in TnT and Tm) seem to code for protein sequences that interact with other proteins (Pearlstone & Smillie 1982). In the sarcomeric contractile proteins such sequences may be of fundamental importance. Although the mechanisms involved in assembling a sarcomere remain obscure, it is clear that the contractile proteins polymerize in a very precise manner to form a highly ordered paracrystalline structure (Huxley 1953). In this ordered quasi-organelle, different isoforms of the same component are interchangeable in their basic function but interact with a distinct subset of proteins. The existence of constant domains involved in fundamental protein-protein interactions in combination with variable domains that modulate and fine-tune such interactions is likely to have been evolutionarily advantageous in a system such as muscle, where dramatic and sudden changes in performance are common during normal physiological activity. However, the elucidation of the precise functional and/or assembly differences between the numerous contractile protein isoforms remains a challenge hampered by the paucity of sensitive in vitro contractility assays. The recent development of such a functional assay with well-defined components is promising in this respect (Spudich et al 1985).

Beyond the generalizations outlined above, the functional differences between protein isoforms arising from alternative splicing are mostly unknown at present. Notable exceptions are: immunoglobulins, in which alternative heavy-chain 3′ terminal domains are known to target antibodies to a particular subcellular location (Alt et al 1980; Early et al 1980; Rogers et al 1981); fibronectin and neural cell adhesion molecules, in which alternatively spliced exons partly determine cell type–specific adhesion (Humphries et al 1986; Murray et al 1986; Paul et al 1986); the myosin heavy chain in *Drosophila*, in which the inclusion of a cassette exon is imperative for flight capability (Bernstein et al 1986); and the *Drosophila* P element, in which excision of an intron in the germ line, which is otherwise retained in somatic cells, allows production of active transposase (Laski et al 1986).

Alternative Splicing Is an Efficient and Reversible Posttranscriptional Mechanism for Performing Isoform Switches

The existence of multigene families that encode protein isoforms raises the question of the selective evolutionary advantage of these isoforms. One theory is that some gene duplications had the advantage, not of producing two slightly different proteins, but of placing the duplicated genes under different regulatory programs (metabolic, developmental or tissue-specific; McKeown et al 1982). In such cases, differences in the proteins coded by the members of the multigene family would be the consequence of, and not the reason for, the selective advantage.

This scenario is unlikely to apply to most isoforms produced by alternative splicing. With the exception of the genes with multiple promoter and/or terminator sites, in which transcriptional regulation may play a role, alternative isoforms that arise from a unique transcript are likely to have been maintained through evolution because of the selective advantage of the isoform diversity itself. If this is so, the next question is why alternative splicing is so much more widespread than DNA rearrangement (Hood et al 1985), which also yields protein diversity.

The obvious shortcoming of DNA rearrangement is that it irreversibly alters the informational content of the genome. However, it is well suited for the cells of the immune system, which have very specialized functions but also maintain the ability of clonal expansion by extensive cell multiplication. Changes in a single cell can be rapidly magnified by clonal expansion and directly inherited by progeny cells; whenever a different altered protein is required, former clones can be dispensed with, a new DNA rearrangement produced, and the cells stimulated to divide. Clearly, this mechanism is suited only to cell types that can undergo DNA replication and subsequent division.

Gene switches at the transcriptional level among members of a multigene family are an obvious alternative way of generating different protein isoforms (Izumo et al 1986; Strehler et al 1986; Wieczorek et al 1985). This possibility is limited by the number of members in the gene family and by the constraints placed upon transcriptional activation by chromatin conformation (Brown 1984) and methylation (Yisraeli & Szyf 1984). Due to these constraints, changes in the state of a promoter may be particularly difficult to effect in cells that have lost DNA replication capacity, such as nerve and muscle cells. Alternative splicing does not share these shortcomings. It increases the coding capabilities of the genome by expanding the ability of genes to generate protein diversity and places the regulation of this diversity at the posttranscriptional level. Posttranscriptional regulation does not affect genomic information; all the possibilities remain available to the cell and the organism whenever and wherever the gene can be expressed. Alternative splicing may also be able to operate significantly faster than transcriptional switching, particularly in stationary, terminally differentiated cells. In fact, most alternatively spliced exons operate as discrete intragenic duplications that can be recruited at short notice.

In summary, the advantages of alternative splicing are that it is: reversible, in a cellular sense (unlike DNA rearrangement); capable of rapid response to environmental stimuli; a method of maximizing informational output (different proteins) from a minimal input (a single gene); and a system of isoform switching particularly suited to terminally differentiated, long-lived cells that have lost the ability to replicate their DNA.

Alternative Splicing May Represent a Primitive Mode of Splicing Important for Gene Evolution

Splicing is present in archaebacteria (Kaine et al 1983) and T-even bacteriophages (Chu et al 1984, 1986), in addition to eukaryotic cells, and thus appears to be an ancient process. Splicing may have evolved from autocatalyzed RNA ligation, as suggested by the existence of self-splicing introns (Cech 1986; Cech & Bass 1986). However, the role of introns in gene evolution and the selective advantage that has maintained them in eukaryotes remain unexplained. Most hypotheses envision introns as providing opportunities for producing new proteins by exon shuffling. According to these theories, exons are discrete functional units that code for structural or functional domains. In this view the role of introns is to facilitate the creation of complex genes from relatively simple precursors. Metazoan eukaryotic cells, which must adapt as members of multicellular organisms, presumably retained introns to facilitate future specialization (by exon duplication or deletion), whereas prokaryotes have streamlined their genomes by systematic intron deletion (Gilbert 1985; Gilbert et al 1986; Marchionni & Gilbert 1986; Sharp 1980).

Does differential splicing represent a predecessor or a refinement of constitutive splicing? As outlined in previous sections, the distinction between constitutive and alternative exons is a relative one and not exclusively due to intrinsic properties of an exon and its flanking splice sites. The behavior of exons is greatly influenced by the characteristics of the neighboring splice sites and the availability of *trans*-acting factors in the particular cell in which the gene is expressed. Perhaps new exons appearing in a gene through internal duplication or gene shuffling are first alternatively spliced. Such a pattern would have the selective advantage of allowing the generation of a new protein while the old one continued to function, even in an organism homozygous for the mutation. The exons thus generated might remain alternative if they provided a selective advantage. On the other hand, they might evolve into constitutive ones through mutations in the *cis* regulatory elements that produce strong splice sites. According to this hypothesis, alternative splicing would be a very old splicing mechanism that played a crucial role in gene evolution.

From the data presently available, it is clear that most alternative splicing occurs with exons that originated through duplication events. The question is whether these duplications are recent or were present in the ancestral gene, as would be expected if the hypothesis just elaborated is correct. In this connection, the tropomyosin gene (Tm), whose primary structure has been obtained from *Drosophila*, birds, and mammals (Helfman et al 1986; Karlik & Fyrberg 1986; Wieczorek & Nadal-Ginard 1987), provides the best data. Of the twelve exons of this gene, only five are not alternatively spliced in one or another of these organisms (Figure 8). The degree of divergence between pairs of differentially spliced exons within any given Tm gene is greater than that among the same isoform-specific exons in isogenes across the evolutionary ladder (Wieczorek & Nadal-Ginard 1987). This feature suggests that all the alternatively spliced exons so far detected were already present at the time of the *Drosophila* radiation, more than 600 million years ago (Ruiz-Opazo & Nadal-Ginard 1987). If this conclusion is correct, it implies that the primitive Tm gene was significantly more complex than its contemporary counterparts and had the capacity to code for a larger molecule, a larger number of isoforms, or both. The fact that all the Tm genes in existence are differentially spliced

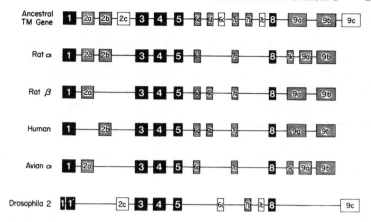

Figure 8 Possible evolutionary path of the tropomyosin gene. Constitutive exons are shown in black; exons within each subgroup that eventually became mutually exclusive are in white or are horizontally/vertically striped (Wieczorek & Nadal-Ginard 1987).

(with the exception of one in *Drosophila* that has lost its introns; Karlik & Fyrberg 1986) suggests that the Tm ancestor was also alternatively spliced and that some of the exons have been selectively lost during evolution.

From these arguments, one can postulate that alternative splicing either preceded or was concomitant with the appearance of constitutive splicing as a means of facilitating rapid gene evolution. This may still be its role today: Genes that remain in flux would retain alternative splicing, whereas those whose product has been optimized for a given cell environment or function would have made all their exons constitutive or eliminated their introns altogether. For those genes that are evolving or require flexibility, alternative splicing would be especially advantageous. Moreover, alternatively spliced exons would necessitate the retention of the splicing apparatus, which would then process the constitutive exons as well. In organisms like trypanosomes, *cis* splicing never arose or was extinguished in favor of *trans* splicing, which better served the particular needs of the organism (Sutton & Boothroyd 1986).

As increasing numbers of genes are found to utilize alternative splicing, the importance of understanding the underpinnings of this phenomenon increases. In vivo and in vitro experiments are currently in progress that attempt to analyze the process of alternative splicing at the molecular level. Detailed knowledge of this mechanism will undoubtedly give invaluable insight into gene regulation and evolution.

ACKNOWLEDGMENTS

We thank Thomas Cooper and Michael Rosenfeld for communicating results prior to publication. The expert assistance of Casserine Toussaint with the figures and Sharon Ward with the table is gratefully acknowledged. This work was supported by grants from the National Institutes of Health, the American Heart Association, and the Muscular Dystrophy Association. AA is a postdoctoral research fellow of the Muscular Dystrophy Association. MEG is a postdoctoral research fellow of the Consejo Superior de Investigaciones Científicas of Spain. BNG is an investigator of the Howard Hughes Medical Institute.

Literature Cited

Aebi, M., Hornig, H., Padgett, R. A., Reiser, J., Weissmann, C. 1986. *Cell* 47: 555–65

Alt, F. W., Bothwell, A. L. M., Knapp, M., Siden, E., Mather, E. 1980. *Cell* 20: 293–301

Amara, S. G., Jonas, V., Evans, R. M. 1982. *Nature* 298: 240–44

Arai, N., Nomura, D., Yokota, K., Wolf, D., Brill, E., et al. 1986. *Mol. Cell. Biol.* 6: 3232–39

Basi, G. S., Boardman, M., Storti, R. V. 1984. *Mol. Cell. Biol.* 12: 2828–36

Basi, G. S., Storti, R. V. 1986. *J. Biol. Chem.* 261: 817–27

Behlke, M. A., Loh, D. Y. 1986. *Nature* 322: 379–82

Benech, P., Mory, Y., Revel, M., Chebath, J. 1985. *EMBO J.* 4: 2249–56

Ben-Neriah, Y., Bernards, A., Paskind, M., Daley, G. Q., Baltimore, D. 1986. *Cell* 44: 577–86

Benyajati, C., Spoerel, N., Haymerle, H., Ashburner, M. 1983. *Cell* 33: 125–33

Berget, S. M., Sharp, P. A. 1979. *J. Mol. Biol.* 129: 547–65

Bernstein, A. I., Hansen, C. J., Becker, K. D., Wassenberg, D. R. II, Roche, E. S., et al. 1986. *Mol. Cell. Biol.* 6: 2511–19

Bond, B. J., Davidson, N. 1986. *Mol. Cell. Biol.* 6: 2080–88

Bovenberg, L. A. R., Van de Meerendonk, M. P. W., Baas, D. R., Steenbergh, M. P., Lips, M. J. C., Jansz, S. H. 1986. *Nucleic Acids Res.* 14: 8785–8803

Bray, P., Carter, A., Simons, C., Guo, V., Puckett, C., et al. 1986. *Proc. Natl. Acad. Sci. USA* 83: 8893–97

Breitbart, R. E., Andreadis, A., Nadal-Ginard, B. 1987. *Ann. Rev. Biochem.* 56: 467–95

Breitbart, R. E., Nadal-Ginard, B. 1986. *J. Mol. Biol.* 188: 313–23

Breitbart, R. E., Nadal-Ginard, B. 1987. *Cell* 49: 793–803

Breitbart, R. E., Nguyen, H. T., Medford, R. M., Destree, A. T., Mahdavi, V., Nadal-Ginard, B. 1985. *Cell* 41: 67–82

Brickell, P. M., Latchman, D. S., Murphy, D., Willison, K., Rigby, W. J. 1983. *Nature* 306: 756–60

Brown, T. D. 1984. *Cell* 37: 359–65

Capon, D. J., Seeburg, P. H., McGrath, J. P., Hayflick, J. S., Edman, U., et al. 1983. *Nature* 304: 507–13

Caras, I. W., Davitz, M. A., Rhee, L., Weddekk, G., Martin, D. W., Nussenzweig, V. 1987. *Nature* 325: 545–49

Cech, T. R. 1986. *Cell* 44: 207–10

Cech, T. R., Bass, B. L. 1986. *Ann. Rev. Biochem.* 55: 599–629

Chu, F. K., Maley, G. F., Maley, F., Belfort, M. 1984. *Proc. Natl. Acad. Sci. USA* 81: 3049–53

Chu, F. K., Maley, G. F., West, D. K., Belfort, M., Maley, F. 1986. *Cell* 45: 157–66

Chu, G., Sharp, P. A. 1981. *Nature* 289: 378–82

Chung, D. W., Davie, E. W. 1984. *Biochemistry* 23: 4232–36

Cook, K. S., Groves, D. L., Min, H. Y., Spiegelman, B. M. 1985. *Proc. Natl. Acad. Sci. USA* 82: 6480–84

Cooper, T. A., Ordahl, C. P. 1985. *J. Biol. Chem.* 260: 11140–48

Cooper, T. A., Ordahl, C. P. 1986. *Cold Spring Harbor Lab. Symp. on RNA Processing*, p. 46. Cold Spring Harbor, NY: Cold Spring Harbor Lab. (Abstr.)

Crabtree, G. R., Kant, J. A. 1982. *Cell* 31: 159–66

Craik, C. S., Rutter, W. J., Fletterick, R. 1983. *Science* 220: 1125–29

Danner, D., Leder, P. 1985. *Proc. Natl. Acad. Sci. USA* 82: 8658–62

Davis, R. L., Davidson, N. 1986. *Mol. Cell. Biol.* 6: 1464–70

de Ferra, F., Engh, H., Hudson, L., Kamholz, J., Puckett, C., et al. 1985. *Cell* 43: 721–27

Deitcher, D. L., Mostov, K. E. 1986. *Mol. Cell. Biol.* 6: 2712–15

DeNoto, F. M., Moore, D. D., Goodman, H. M. 1981. *Nucleic Acids Res.* 9: 3719–30

Early, P., Rogers, J., Davis, M., Calame, K., Bond, M. 1980. *Cell* 20: 313–19

Edwards, R. H., Selby, M. J., Rutter, W. J. 1986. *Nature* 319: 784–87

Eperon, L. P., Estibeiro, J. P., Eperon, I. C. 1986. *Nature* 324: 280–82

Eveleth, D. D., Gietz, D. R., Spencer, H. C., Nargang, E. F., Hodgetts, B. R., Marsh, L. S. 1986. *EMBO J.* 5: 2663–72

Falkenthal, S., Parker, V. P., Davidson, N. 1985. *Proc. Natl. Acad. Sci. USA* 82: 449–53

Flaspohler, J. A., Milcarek, C. 1986. *26th Ann. Meet. Am. Soc. for Cell Biol.*, p. 318a. Washington, DC: Am. Soc. Cell Biol. (Abstr.)

Forbes, D. J., Kirschner, M. W., Caput, D., Dahlberg, J. E., Lund, E. 1984. *Cell* 38: 681–89

Fornace, A. J. Jr., Cummings, D. E., Comeau, C. M., Kant, J. A., Crabtree, G. R. 1984. *J. Biol. Chem.* 259: 12826–30

Frendewey, D., Keller, W. 1985. *Cell* 42: 355–67

Freytag, S. O., Beaudet, A. L., Bock, H. G. O., O'Brien, W. E. 1984. *Mol. Cell. Biol.* 4: 1978–84

Fu, X.-Y., Manley, J. L. 1987. *Mol. Cell. Biol.* 7: 738–48

Garriga, G., Bertrand, H., Lambowitz, A. M. 1984. *Cell* 36: 623–34

Gattoni, R., Stevenin, J., Devilliers, G., Jacob, M. 1978. *FEBS Lett.* 90: 318–23

George, D. L., Scott, A. F., Trusko, S., Glick, B., Ford, E., Dorney, D. J. 1985. *EMBO J.* 4: 1199–1204

Gilbert, W. 1985. *Science* 228: 823–24

Gilbert, W., Marchionni, M., MacKnight, G. 1986. *Cell* 46: 151–54

Goodwin, G. R., Rottman, M. F., Callaghan, T., Kung, H., Maroney, A. P., Nilsen, W. T. 1986. *Mol. Cell. Biol.* 6: 3128–33

Greenspan, D. S., Weissman, S. M. 1985. *Mol. Cell. Biol.* 5: 1894–1900

Hastings, K. E. M., Bucher, E. A., Emerson, C. P. Jr. 1985. *J. Biol. Chem.* 260: 13699–703

Helfman, M. D., Cheley, S., Kuismanen, E., Finn, A. L., Kataoka, Y. Y. 1986. *Mol. Cell. Biol.* 6: 3582–95

Hemperly, J. J., Murray, B. A., Edelman, G. M., Cunningham, B. A. 1986. *Proc. Natl. Acad. Sci. USA* 83: 3037–41

Hollenberg, S. M., Weinberger, C., Ong, E. S., Cerelli, G., Oro, A., et al. 1985. *Nature* 318: 635–41

Homandberg, G. A., Williams, J. E., Evans, D. B., Mosesson, M. W. 1985. *Thromb. Res.* 38: 203–9

Hood, L., Kronenberg, M., Hunkapiller, T. 1985. *Cell* 40: 225–29

Humphries, S. M., Akiyama, K. S., Komoriya, A., Olden, K., Yamada, M. K. 1986. *J. Cell Biol.* 103: 2637–47

Huxley, H. E. 1953. *Proc. R. Soc. London Ser. B* 141: 59–62

Hynes, R. O. 1985. *Ann. Rev. Cell Biol.* 1: 67–90

Ingraham, H. A., Evans, G. A. 1986. *Mol. Cell. Biol.* 6: 2923–31

Izumo, S., Nadal-Ginard, B., Mahdavi, V. 1986. *Science* 231: 597–600

Jonas, V., Lin, C. R., Kawashima, E., Semon, D., Swanson, L. W., et al. 1985. *Proc. Natl. Acad. Sci. USA* 82: 1994–98

Kaine, B. P., Gupta, R., Woese, C. R. 1983. *Proc. Natl. Acad. Sci. USA* 80: 3309–12

Karlik, C. C., Fyrberg, E. A. 1985. *Cell* 41: 57–66

Karlik, C. C., Fyrberg, E. A. 1986. *Mol. Cell. Biol.* 6: 1965–73

Kaufman, R. J., Sharp, P. A. 1982. *Mol. Cell. Biol.* 2: 1304–19

Keller, E. B., Noon, W. A. 1984. *Proc. Natl. Acad. Sci. USA* 81: 7417–20

Kerr, W. G., Burrows, P. D., Hendershot, L. M., Kelley, D., Perry, R. 1986. *26th Ann. Meet. Am. Soc. for Cell Biol.*, p. 38a. Washington, DC: Am. Soc. Cell Biol. (Abstr.)

King, C. R., Piatigorsky, J. 1983. *Cell* 32: 707–12

King, C. R., Piatigorsky, J. 1984. *J. Biol. Chem.* 259: 1822–26

Kitamura, N., Kitagawa, H., Fukushima, B., Takagaki, Y., Miyata, T., Nakanishi, S. 1985. *J. Biol. Chem.* 260: 8610–17

Kitamura, N., Takagaki, Y., Furuto, S., Tanaka, T., Nawa, H., Nakanishi, S. 1983. *Nature* 305: 545–49

Koller, B., Fromm, H., Galun, E., Edelman, M. 1986. *Cell* 48: 111–19

Konarska, M. M., Padgett, R. A., Sharp, P. A. 1985. *Cell* 42: 165–71

Kornblihtt, A. R., Umezawa, K., Vibe-Pedersen, K., Baralle, F. E. 1985. *EMBO J.* 4: 1755–59

Kornblihtt, A. R., Vibe-Pedersen, K., Baralle, F. E. 1984. *Nucleic Acids Res.* 12: 5853–68

Kozak, M. 1986a. *Cell* 44: 283–92

Kozak, M. 1986b. *Cell* 47: 481–83

Krainer, A. R., Maniatis, T., Ruskin, B., Green, M. R. 1984. *Cell* 36: 993–1005

Kramer, A., Keller, W., Appel, B., Luhrmann, R. 1984. *Cell* 38: 299–307

Kress, M., Glaros, D., Khoury, G., Jay, G. 1983. *Nature* 306: 602–4

Kuhne, T., Wieringa, B., Reiser, J., Weissmann, C. 1983. *EMBO J.* 2: 727–33

Lalanne, J.-L., Transy, C., Guerin, S., Darche, S., Meulien, P., Kourilsky, P. 1985. *Cell* 41: 469–78

Lang, K. M., Spritz, R. A. 1983. *Science* 220: 1351–55

Lang, K. M., Spritz, R. A. 1986. *Cold Spring Harbor Lab. Symp. on RNA Processing*, p. 21. Cold Spring Harbor, NY: Cold Spring Harbor Lab. (Abstr.)

Lang, K. M., van Santen, V. L., Spritz, R. A. 1986. *EMBO J.* 4: 1991–96

Laski, F. A., Rio, D. C., Rubin, G. M. 1986. *Cell* 44: 7–19

Laughon, A., Boulet, A. M., Bermingham, J. R., Laymon, R. A., Scott, M. P. 1986. *Mol. Cell. Biol.* 6: 4676–89

Leff, S. E., Rosenfeld, M. G., Evans, R. M. 1986. *Ann. Rev. Biochem.* 55: 1091–1117

Leonard, W. J., Depper, J. M., Crabtree, G. R., Rudikoff, S., Pumphrey, J., et al. 1984. *Nature* 311: 626–31

Leonard, W. J., Depper, J. M., Kanehisa, M., Kronke, M., Peffer, N. J., et al. 1985. *Science* 230: 633–39

Lew, M. A., Margulies, M. D., Maloy, L. W., Lillehoj, P. E., McCluskey, J., Colligan, E. J. 1986. *Proc. Natl. Acad. Sci. USA* 83: 6084–88

MacLeod, A. R., Houlker, C., Reinach, F. C., Smillie, L. B., Talbot, K., et al. 1985. *Proc. Natl. Acad. Sci. USA* 82: 7835–39

Maeda, N., Kim, H. S., Azen, E. A., Smithies, O. 1985. *J. Biol. Chem.* 260: 11123–30

Maki, R., Roeder, W., Traunecker, A., Sidman, C., Wabi, M. 1981. *Cell* 24: 353–65

Marchionni, M., Gilbert, W. 1986. *Cell* 46: 133–41

Matis, S., Milcarek, C. 1986. *26th Ann. Meet. Am. Soc. for Cell Biol.*, p. 185a. Washington, DC: Am. Soc. Cell Biol. (Abstr.)

Maurer, R. A., Erwin, C. R., Donelson, J. E. 1981. *J. Biol. Chem.* 256: 10524–28

McCluskey, J., Boyd, F. L., Maloy, L. W., Colligan, E. J., Margulies, M. D. 1986. *EMBO J.* 5: 2477–83

McCoy, M. S., Bargmann, C. I., Weinberg, R. A. 1984. *Mol. Cell. Biol.* 4: 1577–82

McGrath, J. P., Capon, D. J., Smith, D. H., Chen, E. Y., Seeburg, P. H. 1983. *Nature* 304: 501–6

McKeown, M., MacLeod, C., Firtel, R. 1982. In *Muscle Development: Molecular and Cellular Control*, ed. M. L. Pearson, H. F. Epstein, pp. 61–76. Cold Spring Harbor, NY: Cold Spring Harbor Lab.

Medford, R. M., Nguyen, H. T., Destree, A. T., Summers, E., Nadal-Ginard, B. 1984. *Cell* 38: 409–21

Merrill, F. G., Tufaro, D. F. 1986. *Nucleic Acids Res.* 19: 6281–97

Metherall, J. E., Collins, F. S., Pan, J., Weissmann, S. M., Forget, B. 1986. *EMBO J.* 5: 2551–57

Milhausen, M., Nelson, R. G., Sather, S., Selkirk, M., Agabian, N. 1984. *Cell* 38: 721–29

Mitchell, P. J., Urlaub, G., Chasin, L. 1986. *Mol. Cell. Biol.* 6: 1926–2357

Moore, K. W., Rogers, J., Hunkapiller, T., Early, P., Nottenburg, C., et al. 1981. *Proc. Natl. Acad. Sci. USA* 78: 1800–4

Morgan, A. B., Johnson, A. W., Hirsh, J. 1986. *EMBO J.* 5: 3335–42

Mount, S. M. 1982. *Nucleic Acids Res.* 10: 459–72

Mueller, P. P., Hinnebusch, A. 1986. *Cell* 45: 201–7

Munroe, S. H. 1984. *Nucleic Acids Res.* 12: 8437–56

Munroe, S. H., Duthie, R. S. 1986. *Nucleic Acids Res.* 14: 8447–65

Murphy, W. J., Watkins, K. P., Agabian, N. 1986. *Cell* 47: 517–25

Murray, B. A., Hemperly, J. J., Prediger, E. A., Edelman, G. M., Cunningham, B. A. 1986. *J. Cell Biol.* 102: 189–93

Nabeshima, Y., Fujii-Kuriyama, Y., Muramatsu, M., Ogata, K. 1984. *Nature* 308: 333–38

Nadal-Ginard, B., Breitbart, R. E., Strehler, E. E., Ruiz-Opazo, N., Periasamy, M. 1986. In *Molecular Biology of Muscle Development*, ed. C. Emerson, D. Fischman, B. Nadal-Ginard, M. A. Q. Siddiqui, pp. 387–410. New York: Liss

Nagamine, Y., Pearson, D., Grattan, M. 1985. *Biochem. Biophys. Res. Commun.* 132: 563–69

Nagata, A., Tsuchiya, M., Asano, S., Yamamoto, O., Hirata, Y. 1986. *EMBO J.* 5: 571–81

Nawa, H., Kotani, H., Nakanishi, S. 1984. *Nature* 312: 729–34

Nevins, J. R. 1983. *Ann. Rev. Biochem.* 52: 441–66

Nussinov, R. 1980. *J. Theor. Biol.* 83: 647–62

Oates, E., Herbert, E. 1984. *J. Biol. Chem.* 259: 7421–25

Ohno, S., Kawasaki, H., Imajoh, S., Suzuki, K., Inagaki, M., et al. 1986. *Nature* 325: 161–66

Ono, Y., Kurokawa, T., Fujii, T., Kawahara, K., Igarashi, K., et al. 1986. *FEBS Lett.* 206: 347–52

Padgett, R. A., Grabowski, P. J., Konarska, M. M., Seiler, S., Sharp, P. A. 1986. *Ann. Rev. Biochem.* 55: 1119–50

Paul, I. S., Schwarzbauer, E. J., Tamkun, W. J., Hynes, O. R. 1986. *J. Biol. Chem.* 261: 12258–65

Pearlstone, J. R., Smillie, L. B. 1982. *J. Biol. Chem.* 257: 10587–92

Periasamy, M., Strehler, E. E., Garfinkel, L. I., Gubits, R. M., Ruiz-Opazo, N., Nadal-Ginard, B. 1984. *J. Biol. Chem.* 259: 13595–13604

Perlmutter, A. P., Gilbert, W. 1984. *Proc. Natl. Acad. Sci. USA* 81: 7189–93

Peterson, L. M., Perry, P. R. 1986. *Proc. Natl. Acad. Sci. USA* 83: 8883–87

Platt, T. 1986. *Ann. Rev. Biochem.* 55: 339–72

Reed, R., Maniatis, T. 1986. *Cell* 46: 682–90

Reynolds, G. A., Goldstein, J. L., Brown, M. S. 1985. *J. Biol. Chem.* 260: 10369–77

Robert, B., Daubas, P., Akimenko, M.-A., Cohen, A., Garner, I., et al. 1984. *Cell* 39: 129–40

Rogers, J., Choi, E., Souza, L., Carter, C., Word, C., et al. 1981. *Cell* 26: 19–28

Rogers, J., Fasel, N., Wall, R. 1986. *Mol. Cell. Biol.* 6: 4749–52

Roop, D. R., Nordstrom, J. D., Tsai, S.-Y., Tsai, M. J., O'Malley, B. W. 1978. *Cell* 15: 651–85

Rosenfeld, M. G., Leff, S. E., Russo, A. F., Crenshaw, E. B., Evans, R. M. 1986. *5th Int. Symp. on Calcium Binding Proteins in Health and Disease*, p. 3. Asilomar, Calif: Norman (Abstr.)

Rosenfeld, M. G., Mermod, J.-J., Amara, S. G., Evans, R. M. 1984. *Science* 225: 315–20

Rosenfeld, M. G., Mermod, J.-J., Amara, S. G., Swanson, L. W., Sawchenko, P. E., et al. 1983. *Nature* 304: 129–35

Rotwein, P., Pollock, K. M., Didier, D. K., Krivi, G. G. 1986. *J. Biol. Chem.* 261: 4828–32

Rozek, C. E., Davidson, N. 1983. *Cell* 32: 23–34

Rozek, C. E., Davidson, N. 1986. *Proc. Natl. Acad. Sci. USA* 83: 2128–32

Ruiz-Opazo, N., Nadal-Ginard, B. 1987. *J. Biol. Chem.* 262: 4755–66

Ruiz-Opazo, N., Weinberger, J., Nadal-Ginard, B. 1985. *Nature* 315: 67–70

Ryffel, U., Wyler, T., Muellener, B., Weber, R. 1980. *Cell* 19: 53–61

Saunders, M. E., Gewert, D. R., Tugwell, M. E., McMahon, M., Williams, B. R. G. 1985. *EMBO J.* 4: 1761–68

Sausville, E. A., Lebacq-Verheyden, A. M., Spindel, E. R., Cuttitta, F., Gadzar, A. F., Battey, J. F. 1986. *J. Biol. Chem.* 261: 2451–57

Schaefer, M., Picciotto, M. R., Kreiner, T., Kaldany, R.-R., Taussig, R., Scheller, R. H. 1985. *Cell* 41: 457–67

Schejter, D. E., Segal, D., Glazer, L., Shilo, B. 1986. *Cell* 46: 1091–1101

Schneuwly, S., Kuroiwa, A., Baumgartner, P., Gehring, W. J. 1986. *EMBO J.* 5: 733–39

Schulz, R. A., Cherbas, L., Cherbas, P. 1986. *Proc. Natl. Acad. Sci. USA* 83: 9428–32

Schwarzbauer, J. E., Tamkun, J. W., Lemischka, I. R., Hynes, R. O. 1983. *Cell* 35: 421–31

Seiki, M., Hikikoshi, A., Taniguchi, T., Yoshida, M. 1985. *Science* 228: 1532–34

Sekiguchi, K., Kloss, M. A., Kurachi, K., Yoshitake, S., Hakomori, S. 1986. *Biochemistry* 25: 4936–41

Sharp, P. A. 1980. *Cell* 23: 643–46

Shimizu, K., Birnbaum, D., Ruley, M. A., Fasano, O., Suard, Y., et al. 1983. *Nature* 304: 497–500

Shtivelman, E., Lifshitz, B., Gale, P. R., Roe, A. B., Canaani, E. 1986. *Cell* 47: 277–84

Soares, B. M., Turken, A., Ishii, D., Mills, L., Episkopou, V., et al. 1986. *J. Mol. Biol.* 192: 737–52

Solnick, D. 1985a. *Cell* 42: 157–64

Solnick, D. 1985b. *Cell* 43: 667–76

Somasekhar, M. B., Mertz, J. E. 1985. *Nucleic Acids Res.* 13: 5591–5609

Spudich, J. A., Kron, S. J., Sheetz, M. P. 1985. *Nature* 315: 584–86

Stein, J. P., Catterall, J. F., Kristo, P., Means, A. R., O'Malley, B. W. 1980. *Cell* 21: 681–87

Strehler, E. E., Periasamy, M., Strehler-Page, M.-A., Nadal-Ginard, B. 1985. *Mol. Cell. Biol.* 5: 3168–82

Strehler, E. E., Strehler-Page, M.-A., Perriard, J.-C., Periasamy, M., Nadal-Ginard, B. 1986. *J. Mol. Biol.* 190: 291–317

Stroeher, L. V., Jorgensen, M. E., Garber, L. R. 1986. *Mol. Cell. Biol.* 6: 4667–75

Strubin, M., Berte, C., Mach, B. 1986. *EMBO J.* 5: 3483–88

Sutton, R. E., Boothroyd, J. C. 1986. *Cell* 47: 527–35

Takahashi, N., Roach, A., Teplow, D. B., Prusiner, S. B., Hood, L. 1985. *Cell* 42: 139–48

Tamkun, J. W., Schwarzbauer, J. E., Hynes, R. O. 1984. *Proc. Natl. Acad. Sci. USA* 81: 5140–44

Telford, J., Burckhardt, J., Butler, B., Pirrotta, V. 1985. *EMBO J.* 4: 2609–16

Temin, H. M. 1985. *Mol. Biol. Evol.* 2: 455–68

Transy, C., Lalanne, J.-L., Kourilsky, P. 1984. *EMBO J.* 3: 2383–86

Treisman, R., Orkin, S. H., Maniatis, T. 1983. *Nature* 302: 591–96

Tsai, M.-J., Ting, A. C., Nordstrom, J. L., Zimmer, W., O'Malley, B. W. 1980. *Cell* 22: 219–30

Tsou, A. P., Lai, C., Danielson, P., Noonan, D. J., Sutcliffe, J. G. 1986. *Mol. Cell. Biol.* 6: 768–78

Tunnacliffe, A., Sims, J. E., Rabbitts, T. H. 1986. *EMBO J.* 5: 1245–52

Tyler, B. M., Cowman, A. F., Adams, J. M., Harris, A. W. 1981. *Nature* 293: 406–8

242 ANDREADIS ET AL

van den Heuvel, R., Hendriks, W., Quax, W., Bloemendal, H. 1985. *J. Mol. Biol.* 185: 273–84

van Duin, M., de Wit, J., Odijk, H., Westerveld, A., Yasui, A., et al. 1986. *Cell* 41: 457–67

van Santen, V. L., Spritz, R. A. 1985. *Proc. Natl. Acad. Sci. USA* 82: 2885–89

van Santen, V. L., Spritz, R. A. 1986. *Nucleic Acids Res.* 14: 9911–26

Vibe-Pedersen, K., Kornblihtt, A. R., Baralle, F. E. 1984. *EMBO J.* 3: 2511–16

Vincent, A., O'Connell, P., Gray, M. R., Rosbash, M. 1984. *EMBO J.* 3: 1003–13

Wieczorek, D. F., Nadal-Ginard, B. 1987. *Mol. Cell. Biol.* In press

Wieczorek, D. F., Periasamy, M., Butler-Browne, G. S., Whalen, R. G., Nadal-Ginard, B. 1985. *J. Cell Biol.* 101: 618–29

Wieringa, B., Hofer, E., Weissmann, C. 1984. *Cell* 37: 915–25

Yaoita, Y., Kumagai, Y., Okamura, K., Honjo, T. 1982. *Nature* 297: 697–99

Yisraeli, J., Szyf, M. 1984. In *DNA Methylation*, ed. A. Razin, H. Cedar, A. D. Riggs, pp. 352–78. New York: Springer-Verlag

Yost, H. J., Lindquist, S. 1986. *Cell* 45: 185–93

Young, R. A., Hagenbuchle, O., Schibler, U. 1981. *Cell* 23: 451–58

Zamoyska, R., Vollmer, A. C., Sizer, K. C., Liaw, C. W., Parnes, J. R. 1985. *Cell* 43: 153–63

Zeitlin, S., Efstratiadis, A. 1984. *Cell* 39: 589–602

Zhuang, Y., Weiner, A. M. 1986. *Cell* 46: 827–35

Ziff, E. B. 1980. *Nature* 287: 491–99

Ann. Rev. Cell Biol. 1987. 3:243–93

CONSTITUTIVE AND REGULATED SECRETION OF PROTEINS

Teresa Lynn Burgess and Regis B. Kelly

Department of Biochemistry and Biophysics, University of California, San Francisco, California 94143-0448

CONTENTS

INTRODUCTION.. 243
THE CONSTITUTIVE SECRETORY PATHWAY 245
THE REGULATED SECRETORY PATHWAY... 252
SORTING.. 257
CYTOSKELETON INVOLVEMENT WITH PROTEIN SECRETION 266
SECRETION IN EPITHELIAL CELLS.. 277
RECYCLING OF THE SECRETORY VESICLE MEMBRANE 281

INTRODUCTION

Most cells secrete protein. Proteins destined for secretion are folded, assembled, and glycosylated as they pass through the endoplasmic reticulum (ER) and the cisternae of the Golgi apparatus. In the morphologically complex array of vesicles and tubules that make up the trans Golgi network (TGN), proteins to be secreted are segregated from lysosomal enzymes in animal cells and from vacuolar enzymes in yeast. Both classes of enzyme have sorting signals that determine their intracellular location. When these signals are absent or cannot be read, lysosomal and vacuolar enzymes are secreted rather than sequestered intracellularly. Secretion, in yeast cells at least, may not require sorting information and so can be considered a default pathway in such cells.

Some cells are specialized to store large amounts of protein in cytoplasmic organelles. Because the contents of such organelles usually have

243

0743–4634/87/1115–0243$02.00

high electron density and so appear granular in the electron microscope, storage vesicles are usually called secretory granules. Secretory proteins are at concentrations inside secretory granules ten (exocrine) to two hundred (endocrine) times the concentration in early Golgi cisternae. Secretory granules only release their contents when cells are stimulated by a secretagogue, which in turn changes the level of an intracellular second messenger. Protein secretion that can be regulated by a secretagogue is called regulated secretion.

Regulated secretory cells also have another pathway for protein secretion, the constitutive pathway. Proteins leaving the cell by the constitutive pathway are not concentrated in secretory granules for storage in the cytoplasm. They are constantly, or constitutively, secreted; secretagogue stimulation is not required. The similarity between this pathway and the pathway of secretion in cells lacking secretory granules altogether suggests that the basic, ground-state pathway is the constitutive one. When cells differentiate to acquire the regulated pathway they retain the constitutive one, although it may be vestigial.

Where both regulated and constitutive pathways exist in the same cell, different proteins have a preference for one pathway over another. Whatever the sorting mechanism, it is highly conserved since transfection experiments show that cells correctly sort foreign proteins.

Secretory granule proteins are diverted into the constitutive pathway when sorting is perturbed. By the same argument that was used for vacuolar and lysosomal enzyme sorting, the constitutive pathway in such cells may likewise be considered a default pathway.

Some secretory cells release their secretory granule contents uniformly around their perimeter, while others release them at special domains of the plasma membrane. Association between the Golgi apparatus and the minus ends of microtubules radiating outwards from the microtubule organizing center may explain directional secretion in some cases, but probably not all. Selectivity of secretory granule insertion in the plasma membrane may also be influenced by the subcortical web of actin-associated microfilaments. The subcortical web and microtubule transport of nascent secretory granules may also explain some unusual kinetics of protein secretion.

To describe protein secretion in polarized epithelial cells the simple scheme of one constitutive and one regulated pathway must be modified. Default secretion in this case can be both apical and basolateral. Sorting of secreted proteins is found even in cells that only have constitutive pathways. It remains to be seen if sorting in such cases is facilitated by dedicated cellular machinery or is passive.

Secretion by both constitutive and regulated secretory cells requires massive insertion of new membrane into the plasma membrane. Endocytotic recovery of such membrane is rapid. While some membrane goes to lysosomes, much seems to go to Golgi cisternae. Following the fate of secretory granule–specific membrane proteins after exocytosis suggests the possibility of an endocytotic pathway in regulated secretory cells that is specialized for the reutilization of secretory granule membranes.

THE CONSTITUTIVE SECRETORY PATHWAY

All of the transport steps from cytoplasm to the rough endoplasmic reticulum (RER) and through the Golgi complex appear to be common for secretory proteins, lysosomal enzymes, and plasma membrane proteins. Proteins are segregated away from each other at a late Golgi step that may rely on recognition of specific targeting signals. In this section we summarize the early steps of protein secretion and discuss the segregation of constitutively secreted proteins, membrane proteins, and lysosomal enzymes. Regulated secretion is discussed in detail later.

RER Transport

Proteins destined for export from a cell begin by being translocated into the lumen of the RER. Through the concerted function of the signal peptide, signal-recognition particle (SRP), and the SRP receptor (or docking protein) nascent secretory proteins are selectively targeted to the RER (reviewed by Walter & Lingappa 1986), where protein translocation across the membrane takes place. As the protein gains access to the lumen of the RER, modifications such as signal peptide cleavage, N-linked glycosylation, and disulfide bond formation are carried out by resident RER enzymes. Multimeric membrane proteins and secreted proteins usually complete at least the initial steps of their oligomerization in the RER lumen.

When exit from the RER is examined, major differences are seen in the behavior of different secreted proteins. Both glycosylated and unglycosylated proteins are transported out of the RER and to the Golgi complex at variable rates (e.g. Fitting & Kabat 1982; Lodish et al 1983; Fries et al 1984; Scheele & Tartakoff 1985). One possible explanation for this finding is that transport from the RER to the Golgi is achieved via a carrier-mediated mechanism. An alternate model is that proteins do not bind to specific RER-to-Golgi carrier molecules but are retained in the RER for different periods of time by binding to RER receptors. After release these proteins would be carried by a nonspecific bulk flow mechanism to the

Golgi. Recent data from Wieland et al (1987) provide evidence for bulk flow from RER to Golgi. They show that an acyl-tripeptide, N-acyl-Asn-[^{125}I]Tyr-Thr-NH$_2$, which becomes glycosylated in the RER, is efficiently and rapidly secreted from the cell. It is hard to imagine how such a small peptide could be recognized by an RER-to-Golgi carrier molecule. The presence of bulk flow does not necessarily rule out a carrier-mediated pathway since receptor-mediated endocytosis is also associated with a significant bulk flow uptake of fluid-phase markers (Steinman et al 1983).

Exit from the RER can be inhibited by a variety of treatments. Both inhibitors of ATP production (Jamieson & Palade 1968; Tartakoff et al 1977, 1978) and low temperature (10–15°C) (Saraste & Kuismanen 1984; Saraste et al 1986; Tartakoff 1986) generally block protein exist from the RER. Recent results from Balch et al (1986), based on studies of vesicular stomatitis virus (VSV) G-protein transport, suggest that the ATP requirement and temperature block are two sequential steps in RER-to-Golgi transport. Both of these treatments cause morphological changes in the transitional elements thought to interact with shuttle vesicles transporting proteins between RER and Golgi. In the absence of ATP there are essentially no usual transitional elements visible (Merisko et al 1986a,b), whereas at 10°C there is an increase in the number of transitional elements between the RER and Golgi (Tartakoff 1986). Recently, Lodish et al (1987) isolated cell membranes that have some of the properties expected for such shuttle vesicles.

Protein assembly or conformation may be critical for exit from the RER. A wide variety of different types of protein mutations of both secretory and membrane proteins can block or slow exit from the RER. For example, some deletion-substitution mutations in the cytoplasmic domain of influenza virus hemagglutinin slow or halt RER exit, while other similar mutations do not (Doyle et al 1985). Inhibitors of N-linked glycosylation or glycosylation modification such as glucose trimming can block protein export or have no effect (Lodish & Kong 1984). This result implies that carbohydrate itself is not a required signal for RER export. Inhibition of RER exit by carbohydrate modification or mutation may well be due to the resulting alteration of protein conformation or structure rather than specific recognition of carbohydrate moieties or peptide signals.

Recent data on the apparent requirement for correct membrane protein subunit assembly for RER exit (Copeland et al 1986; Gething et al 1986; Kreis & Lodish 1986) may be a clue in the search for the RER-to-Golgi transport mechanism. The data from these three groups on the transport of the influenza virus hemagglutinin and VSV-G proteins show a strong correlation between correct oligomerization and RER-to-Golgi transport. While the data imply that correct quarternary structure conformation may be necessary for RER exit, it has not yet been shown that correct assembly is sufficient to elicit transport. A further intriguing finding has been the identification of an RER protein, BiP. BiP, originally described as an immunoglobulin heavy-chain binding protein (Bole et al 1986), has also been shown to bind to other proteins in the RER that are aberrant, incorrectly folded or incorrectly oligomerized (Bole et al 1986; Gething et al 1986; Matsuuchi & Kelly 1987). It has not been determined whether BiP participates in normal polymeric protein assembly, recognizes aberrant proteins, or prevents protein exit from the RER. For the moment then, it is not known whether RER exit kinetics are determined by carriers or by dissociation from RER receptors, and the role of protein folding in either is far from clear.

Transport Through the Golgi Apparatus

The Golgi apparatus itself consists of at least three functionally distinct compartments, cis, medial, and trans, through which proteins are transported in an apparently vectorial fashion (reviewed by Pfeffer & Rothman 1987). Evidence for the three compartments comes from subcellular fractionation of enzyme markers, from immunoelectron microscopic localization of specific enzymes and Golgi-specific antigens, and from lectin binding experiments (reviewed by Dunphy & Rothman 1985). As glycoproteins are transported from cis through medial to trans Golgi, sequential carbohydrate modifications are carried out by resident enzymes. A large variety of other modifications, such as acylation and sulfation, also occur during transport through the Golgi. The mechanism by which proteins are transported from one Golgi subcompartment to the next in a vectorial fashion has not been elucidated. The cisternal progression model (see Farquhar 1985 for discussion), in which a cis Golgi stack evolves into first a medial stack then a trans stack, is hard to reconcile with the very clear subcompartment boundaries defined by the marker enzyme localization studies. A more favored model is that each cisterna has resident enzymes and that transient proteins move vectorially through the Golgi stack. Specific budding of shuttle vesicles from one cisterna followed by fusion with the next may occur, but the existence of connecting tubules between cisternae cannot be ruled out. The morphological data demonstrating Golgi-associated vesicles could also be interpreted as reflecting a tubular structure connecting the Golgi subcompartments. However, the data from cell fusion studies (Rothman et al 1984) are more consistent with a vesicular transfer model. They show that transport of VSV G protein from cis to medial Golgi compartments occurs at exactly the same rate within the Golgi of a single cell in culture as it does between the cis Golgi of one cell and the medial Golgi of another cell following their cell-cell fusion. Rothman et al argue that cis to medial Golgi transport must therefore be a dissociative process, presumably involving transport vesicles that dissociate from one cis Golgi cisterna, travel through the cytoplasm, and fuse with a medial Golgi cisterna of either the same or a different Golgi stack.

Trans Golgi Network

After glycosylation in the Golgi cisternae secretory proteins are transported from the trans Golgi to the trans Golgi network (TGN). This compartment (reviewed by Griffiths & Simons 1986) may be the post-Golgi sorting compartment in which protein traffic signals are recognized by specific receptors that route them to their proper destination. Clearly, lysosomal enzymes can be diverted to lysosomes by a mechanism involving a protein sorting signal, which is a modification of an N-linked carbohydrate chain and a membrane-bound receptor (reviewed by Kornfeld 1986) that recognizes specific carbohydrate modification. Lysosomal enzymes bound to the mannose-6-phosphate receptor (MPR) must be segregated from other soluble proteins in the lumen of Golgi elements and from membrane proteins (Figure 1). Clathrin-coated vesicles may be involved in this separation since MPRs and lysosomal enzymes can be detected together in coated buds and vesicles in the TGN (Geuze et al 1985; Griffiths et al 1985).

Clathrin is apparently not involved in constitutive protein release. Morphological evidence shows that the constitutive membrane protein VSV-G is not present in clathrin-coated structures (Griffiths et al 1985). G protein *is* present in "coated" buds in the Golgi, but this coat is not clathrin (Orci et al 1986a). In addition, in yeast it has been shown that the single-copy clathrin heavy-chain gene is not essential for viability (Payne & Schekman 1985); since constitutive release is presumably required for cell growth, the role of clathrin in constitutive secretion must not be essential. There is now direct experimentation addressing the role of clathrin in constitutive secretion. Antibodies known to disrupt clathrin assembly in vitro (Blank & Brodsky 1986) have been delivered into the cytoplasm of living cells, and both endocytosis and constitutive secretion have been assayed (Doxsey et al 1986). While endocytosis can be reproducibly inhibited by 50%, constitutive export is not inhibited at all, which strongly suggests that clathrin is not required for constitutive secretion from mammalian cells.

Although the evidence is still limited, it is likely that membrane proteins

A.

B.

C.

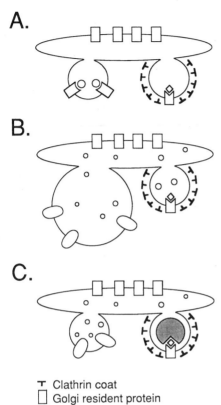

T Clathrin coat
☐ Golgi resident protein
Ⓜ Mannose phosphate receptor

Figure 1   Sorting of lysosomal enzymes from constitutively secreted proteins. Three possible ways of efficient sorting are illustrated, all based on the assumption that lysosomal enzymes (*diamonds*) always bind to mannose-6-phosphate receptors. (A) Constitutively secreted proteins (*circles*) bind to membrane proteins destined for the plasma membrane. (B) Constitutively secreted proteins have no sorting information but are carried by bulk flow. To minimize mis-sorting into lysosomes, the volume of the constitutive transport vesicle has to be larger than that of the coated vesicle, or the number of constitutive vesicles that leave per unit time must be much greater than the number of coated vesicles. (C) If a bulk flow mechanism operates, mis-sorting also can be minimized by excluding soluble proteins from the lumen of the coated vesicle by a space-filling matrix. Note that in all cases, proteins destined for lysosomal membranes and for the plasma membrane (*ovals*) must be segregated from each other and from resident proteins (*rectangles*) of the Golgi apparatus.

and constitutively secreted proteins reach the surface in the same transport vesicle. In addition to the electron microscopic data of Strous et al (1983) from hepatoma cells, data now exist showing that in yeast cells a membrane protein and a secreted protein are in the same secretory vesicle (Holcomb et al 1987). This result is consistent with older observations that mutations blocking yeast secretion also block membrane growth (Tschopp et al 1984).

Figure 1 shows some of the features necessary for segregation of secreted proteins from lysosomal enzymes bound to the MPR. Secreted proteins could also bind to a receptor (Figure 1A). If they do not, a clathrin-coated vesicle carrying bound lysosomal enzymes would also enclose in its contents some soluble constitutive secretory products. To minimize inappropriate bulk flow delivery of constitutive secretory proteins from the TGN to the lysosomes, the vesicles carrying them could be larger than the 50-nm clathrin-coated vesicles containing the bound lysosomal enzymes (Figure 1B). Since the surface-to-volume ratio is higher in the small vesicle, less soluble protein will escape to the lysosome than exists via the large vesicle. Alternatively, the volume of the constitutive route relative to the lysosomal path could be increased by budding many more small constitutive vesicles than clathrin-coated vesicles. A more selective mechanism to thwart inappropriate delivery of soluble proteins to lysosomes (Figure 1C) could involve a gel or matrix material within the forming clathrin-coated vesicle that physically excludes soluble molecules.

Acidification in the Constitutive Pathway

One important feature of the secretory pathway is the role of organelle acidification in transport and/or targeting. Newly synthesized proteins encounter a gradually more acidic environment as they pass through the cell interior (reviewed by Mellman et al 1986). The original evidence for the acidity gradient was pharmacological. Addition of the ionophore monensin, which exchanges a potassium ion for a proton and so neutralizes intravesicular pH, blocks movement of newly synthesized protein through the cell. The weak bases ammonia, chloroquine, and primaquine raise intravesicular pH by accumulating inside acidic compartments in their charged form. Their accumulation neutralizes the pH of the vesicular contents; in addition, it increases the content osmolality, which leads in turn to swelling of the vesicle organelle. In hepatoma cells, inhibition of secretion by weak bases seems to occur in post-Golgi vesicles (Strous et al 1985).

Direct evidence for acidification came from experiments with a weak base, 3-(2,4-dinitroanilino)-3′-amino-N-methyldipropylamine (DAMP), that carries a dinitrophenol (DNP) epitope and so can be recognized by anti-DNP antibodies (Anderson et al 1984). After glutaraldehyde cross-linking of DAMP via its free amino group, the number of DAMP molecules per unit volume could be estimated using colloidal gold coupled to second antibodies or protein A. By this procedure an increasing gradient of DAMP concentration was observed across the Golgi apparatus of constitutively secreting fibroblasts (Anderson & Pathak 1985). The cis Golgi cisternae, the trans Golgi cisternae, and the TGN had 4, 17, and ~ 50 times the background level of gold particles per square micrometer, respectively. If the number of gold particles were proportional to the proton concentration, this would correspond to about one pH unit difference across the Golgi. As expected, the ionophore monensin destroyed the pH gradient, as demonstrated by the lack of a DAMP concentration gradient.

How might low pH compartments facilitate movement of newly synthesized proteins to the surface of the cell? One suggestion (Anderson &

Pathak 1985), based on observations of mannose-6-phosphate receptor recycling (Kornfeld 1986), is that it allows vectorial transport. To give unidirectional movement a pH-sensitive receptor for a specific newly synthesized protein binds a ligand in one cisterna and carries it to another, where the lower pH promotes ligand dissociation. The receptor then recycles back to the donor cisterna; without the difference in pH between compartments vectorial transport would not be possible. A second possibility is that the newly synthesized protein must undergo a structural modification induced by low pH before it can leave the trans Golgi network. Such a requirement would be analogous to the apparent necessity of subunit assembly for exit from the RER. There is evidence that viral membrane proteins undergo some alteration in the Golgi complex. Copeland et al (1986) have shown that the influenza virus hemagglutinin trimer becomes resistant to velocity sedimentation as it is transported through or from the Golgi complex. In addition, several mutant viral glycoproteins have been shown to be slowed or prevented from exiting the Golgi apparatus (see Doyle et al 1985, for example); this could be due to some pH-sensitive structural alteration. A third possibility is that all membrane traffic to the surface is slowed by acidotropic drugs. This could cause major nonspecific internalization of membranes and so could prevent the return of membrane carrying unoccupied receptors to the plasma membrane (Mellman et al 1986). While such a mechanism would account for a general slowing down of all soluble protein secretion, it would not explain why albumin secretion is completely blocked by primaquine, while secretion of orosomucoid and transferrin are only partially blocked (Strous et al 1985).

Is the Constitutive Route from Golgi to Cell Surface a Default Pathway?

Proteins lacking any targeting information probably exit the cell through a bulk flow pathway (Kelly 1985). In mammalian systems, it has been shown that acidotropic drugs such as chloroquine prevent the delivery of nascent lysosomal enzymes to lysosomes. Instead of accumulating in the cells these enzymes are diverted into the constitutive secretion pathway. In addition, two different small molecules, a glycosaminoglycan (Burgess & Kelly 1984) and a glycosylated acyltripeptide (Wieland et al 1987), which probably lack any targeting information, are also apparently secreted via the constitutive route.

Convincing evidence that the constitutive pathway is the default pathway has come from yeast. Overproduction of the yeast vacuolar enzyme procarboxypeptidase Y (proCPY) leads to its secretion (Stevens et al 1986), which is blocked in a yeast secretory mutant in which a late step in constitutive secretion is inhibited. Furthermore, deleting the N-terminus of proCPY (the vacuolar targeting signal) leads to CPY secretion (Valls et al 1987; Johnson et al 1987). Again, inhibitory mutations of the constitutive secretory pathway block this aberrant secretion of CPY. Taken together these data strongly suggest that mislocalized CPY is secreted by default via the constitutive secretory pathway.

In summary, the intracellular transport route from the RER to and through the Golgi apparatus is apparently shared by secretory, lysosomal, and plasma membrane proteins. In addition, resident proteins (e.g. Golgi enzymes) traverse the same pathway until they reach the compartment of their final residence, where they are somehow retained. The specific mechanisms by which vectorial protein transport from one compartment to the next is achieved are unknown. There is some evidence to support a model involving vesicular shuttles, but whether specific carrier molecules or receptors are involved remains to be demonstrated for transport from the RER to and through the Golgi. By the time proteins leave the trans Golgi cisternae or TGN, sorting information for lysosomal, vacuolar, or regulated secretory proteins has been recognized, and these proteins are somehow targeted to the proper organelle. If this information is unreadable or missing, the proteins are instead diverted into a default pathway, probably the constitutive Golgi-to-plasma-membrane route.

THE REGULATED SECRETORY PATHWAY

Hallmarks of Regulated Secretory Cells

The majority of the studies of protein secretion, including the classic work of Palade and coworkers (Palade 1975), were carried out on regulated secretory cells. This type of secretory cell is easily distinguished from constitutive secretory cells by three major characteristics. First, secretion is coupled to an extracellular stimulus. Depending on the specific cell type, exocytosis via vesicle membrane fusion can be triggered by a variety of physiological stimuli, which eventually lead to a transient rise in intracellular Ca^{2+} or another second messenger. The regulation of membrane fusion in secretory exocytosis has been recently reviewed (DeLisle & Williams 1986) and will not be discussed further here. The second characteristic that distinguishes a regulated from a constitutive secretory cell is the concentration and condensation of the secretory products into specialized membrane-bound organelles, the secretory granules. In endocrine cells the concentration factor from the ER to secretory granules may be as high as 200-fold (Salpeter & Farquhar 1981). In contrast, a constitutive secretory cell, such as an immunoglobulin secreting plasma cell shows at most a twofold concentration (Hearn et al 1985) of secretory product between the RER and secretory vesicles. Third, regulated secretory cells store these product-filled secretory granules within their cytoplasm for long periods of time, which means there is a large intracellular pool of mature secretory product. Thus the regulated secretory cell is designed to synthesize and store one or a few secretory products and to discharge rapidly a large fraction thereof in response to physiologically specific stimulation, even in the absence of new protein synthesis.

Because of the characteristic morphology of their secretory granules, regulated secretory cells are easy to identify. These cells include endocrine and exocrine cells, mast cells, platelets, large granular lymphocytes, neutrophils, and neurons. Mammary epithelial cells may fall into this class, but it has not been shown that protein secretion is coupled to an external stimulus (Ortiz & Rocha 1986).

Another characteristic of the regulated pathway is that acidotropic drugs have a dramatic effect on secretion from several cell types. Moore et al (1983a) showed that regulated secretion of newly synthesized adreno-corticotropic hormone (ACTH) was severely inhibited by chloroquine, while exocytosis of pre-existing granules was unaltered. Recently a similar finding was reported by Wagner et al (1986), who showed that regulated secretion of von Willebrand factor (vWf) is also sensitive to acidotropic drugs. These experimental results indicate that there may be a pH-sensitive step in protein targeting to the regulated pathway. Diversion of proteins by acidotropic drugs into the constitutive pathway is found therefore in the regulated secretory pathway as well as in the delivery of lysosomal enzymes.

Morphology of the Regulated Secretory Pathway

The most striking morphological characteristic of regulated secretory cells is the secretory granules themselves. The size of the granules varies, but they always have an electron-opaque content or "dense core." The core may be separated from the membrane by a halo of "space" (as in endocrine cells), or the dense material may fill the entire vesicle (as is common in exocrine granules). The nature of this dense material has been the subject of many investigations, yet very little is known about its structure or assembly. The dense cores of endocrine granules appear to be very stable structures, since even after exocytosis intact cores have been observed (Anderson et al 1973; Tooze & Tooze 1986). In some cases the membrane of isolated endocrine secretory granules can be removed without disturbing the condensed core (Giannattasio et al 1980; Slaby & Farquhar 1980; Zanini et al 1980a). The dense content of exocrine granules appears to be osmotically inert but sensitive to pH > 7.0 (Jamieson & Palade 1971a; Zanini et al 1980b). A few cell types (e.g. polymorphonuclear leukocytes; Bainton & Farquhar 1966, 1968a,b) have more than one type of morphologically and biochemically distinct secretory granules. In poly-morphonuclear leukocytes the two granule types are formed at different developmental stages from different faces of the Golgi apparatus.

Immature secretory granules, or condensing vacuoles, may be connected to the TGN (Tooze & Tooze 1986) and only attain complete separation very late in the maturation pathway. The mechanism by which a condensing vacuole matures into a secretory granule is not known; however, recent immunoelectron microscopic studies have implicated clathrin coats in this process. In two different endocrine cell types, clathrin coats have been found associated with the membranes of these immature granules (Orci et al 1984a, 1985; Tooze & Tooze 1986).

Secretory proteins can be precisely localized along the export pathway using immuno-gold electron microscopy. In insulin-secreting pancreatic cells, antibodies to insulin localize the antigen over the membrane in the Golgi complex. There are also biochemical and cell fractionation data to support the finding that proinsulin is associated with Golgi membranes (Noe & Moran 1984). In contrast, insulin staining is found over the entire dense core of mature and immature secretory granules (Orci et al 1984b, 1985). One explanation for these results is that proinsulin is bound by a membrane receptor in the Golgi and is transferred to the forming secretory granule, where it detaches from the membrane.

Proteolytic Processing

Although many regulated secretory proteins are proteolytically processed prior to secretion, this is not a characteristic unique to regulated secretory proteins. For example, albumin is made in liver cells with an N-terminal "pro" sequence; this peptide is cleaved before albumin is constitutively secreted (Edwards et al 1976; Ikehara et al 1976). The yeast mating pheromone, α-factor, is also processed prior to its constitutive secretion (Julius et al 1984), and many viral membrane glycoproteins that are constitutively exported undergo proteolytic processing (e.g. Wills et al 1984). In addition, there are many regulated secretory proteins that are not processed before secretion, including trypsinogen and growth hormone. (Obviously, we are not considering the removal of a signal peptide, which is an early co- or posttranslational event.)

Proteolytic processing of regulated secretory proteins is a late event. Orci et al (1985) have directly identified the intracellular compartment in which proinsulin is converted to insulin by immunoelectron microscopy using a proinsulin-specific monoclonal antibody. They identified immature secretory granules that are partially clathrin coated as the major, if not the only, site of intracellular prohormone conversion. Somehow proteolysis occurs after the hormone is highly condensed and concentrated. Although this result was implicit in the observations of hormone processing during axonal transport (Gainer et al 1977) and after mature vesicle formation (Gumbiner & Kelly 1981), this is the first direct evidence of proprotein proteolytic processing in immature secretory granules. The proteolytic processing of precursor hormones may not be cell-type or hormone specific. Moore et al (1983b) found that when proinsulin is heterologously expressed in a pituitary cell line, AtT-20, it is proteolytically processed to a form that comigrates with bona fide insulin in SDS-PAGE. Similarly, Comb et al (1985) found that proenkephalin can be correctly processed to Met-enkephalin in AtT-20 cells, and Hellerman et al (1984) showed that preproparathyroid hormone is proteolytically processed in another pituitary cell line. The required protease is likely to be present even in constitutively secreting cells (Hellerman et al 1984; Warren & Shields 1984). The absence of processing in other constitutively secreting cells (Laub & Rutter 1983; Moore et al 1983b) may reflect variation in protease concentration, in substrate preference, or in the kinetics of secretion.

The characteristics of the regulated secretory pathway are summarized in Figure 2.

Nonparallel Secretion and Preferential Release of Newly Synthesized Proteins

In the simplest model of exocytotic secretion, all of the proteins inside secretory vesicles should be released in parallel in response to a stimulation of the secretory cell. There has long been a controversy in the literature concerning the existence of nonparallel secretion, and if it does occur, what mechanism of secretion could explain it. The term nonparallel secretion is confusing because it has been used to describe almost any deviation from the simple model of exocytotic release irrespective of mechanistic implications.

Here we use "nonparallel secretion" to describe cases in which more than one protein is secreted by a tissue. Nonparallel secretion occurs when the ratio of one secreted protein to another is not constant either over time or depends on the secretagogue employed to elicit stimulated release (e.g. Rothman & Wilking 1978; Dagorn et al 1977). In extreme cases (Rothman 1975) data suggesting nonparallel secretion from the exocrine pancreas has been used to argue against the widely held view that secretion occurs exclusively via an exocytotic mechanism (Palade 1975). An enlightening study by Adelson & Miller (1985) seems to reconcile nonparallel secretion and exocytotic release. They compared enzyme secretion from rabbit pancreas under different stimulating conditions both in vivo and in vitro. The ratio of two enzymes, for example, chymotrypsinogen and lipase, does not remain constant under the different stimulation conditions, i.e. secretion

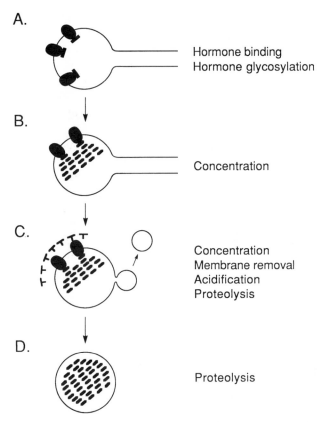

A.

Hormone binding
Hormone glycosylation

B.

Concentration

C.

Concentration
Membrane removal
Acidification
Proteolysis

D.

Proteolysis

Figure 2   Correspondence between steps in the formation of a secretory granule in endocrine cells and molecular events: (A) Hormone binds to membrane in Golgi cisternae; (B) dense core begins to form in rims of Golgi cisternae; (C) immature secretory granule is formed with clathrin coat (further concentration requires membrane removal); (D) the dense core separates from the secretory granule membrane.

is nonparallel. However, enzyme pairs under either stimulating condition are secreted at a constant ratio. There is a strong linear correlation between the secreted amounts of any two enzymes, which demonstrates that the enzymes are not secreted independently of one another. Secretion of enzymes in groups would be expected for enzymes prepackaged into granules and released via exocytosis. The data support the view that nonparallel secretion is due to exocytosis from heterogeneous sources within the pancreas.

The issue of tissue heterogeneity is controversial. Biochemical (Malaisse-

Lagae et al 1975) as well as immunocytochemical (Bendayan & Ito 1979; Posthuma et al 1986) experiments suggest that there are regional differences within the pancreas with regard to the ratios of pancreatic enzymes in peri-insular versus teleinsular acinar cells (but see Bendayan 1985). Others have stressed that the mass of peri-insular cells is small compared to that of the teleinsular cells and may not be sufficient to account for the nonparallelism observed (Posthuma et al 1986).

Convincing evidence for heterogeneity at some level in a glandular system comes from a remarkable analysis of the enzyme content of single, isolated zymogen granules carried out by Mroz & Lechene (1986). They found that the ratio of amylase to chymotrypsinogen within a single granule (taken from the same animal) can vary widely. Quantitative immunoelectron microscopic studies of the digestive enzymes in the exocrine pancreas (e.g. Bendayan et al 1980) show that within an individual cell staining for any one enzyme is similar over all zymogen granules in that cell. Therefore, nonparallel secretion is likely due to heterogeneous cell populations in the exocrine pancreas rather than heterogeneity within individual cells. In neutrophils, however, the evidence suggests that different secretory granules are released at different internal calcium concentrations (Lew et al 1986; see below).

In addition to nonparallel secretion, another observation difficult to reconcile with a simple exocytotic mechanism of release is the preferential secretion of newly synthesized proteins by both endocrine and exocrine cells (e.g. Howell et al 1965; Slaby & Bryan 1976). This could be due simply to biosynthetic heterogeneity within the tissue, for example, it could imply that there is a pool of inert cells. Alternatively, constitutive secretion could account for rapid release of newly synthesized material. Rhodes & Halban (1987) have specifically addressed the question of whether preferential release of proinsulin/insulin from B cells is via a rapid, constitutive secretory pathway. Their results are unambiguous; newly synthesized proinsulin/insulin is preferentially released via a secretagogue (glucose)-regulated pathway. Although their data are consistent with the coexistence of constitutive and regulated secretory pathways in the islet B cell, the constitutive component is quite small and cannot account for the preferential release of newly synthesized proinsulin/insulin. The explanation the authors suggest, which involves cytoskeletal modification, is discussed later.

SORTING

Two Pathways of Secretion in One Cell

We have already mentioned the possibility of regulated secretory cells also secreting proteins via a constitutive pathway. There is now abundant evidence to support this claim obtained in both cell lines and tissues of endocrine and exocrine origin. Gumbiner & Kelly (1982) originally proposed the two-pathway hypothesis to explain their observation that a pituitary tumor cell line (AtT-20) externalizes its endogenous hormone, ACTH, with different kinetics and apparently in a different vesicle than it does an endogenous viral membrane glycoprotein. In another pituitary cell line (GH3) somatostatin inhibits growth hormone secretion but has no effect on the externalization of a membrane protein, VSV G protein (Green & Shields 1984). This result is also consistent with the two-pathway hypothesis, as are the findings on the secretion of von Willebrand factor from cultured human endothelial cells (Sporn et al 1986). As mentioned above, Rhodes & Halban (1987) have provided evidence that isolated islet B cells may export a small fraction of proinsulin and other secretory products via a constitutive pathway. Beaudoin and coworkers (Beaudoin et al 1983, 1986) suggest that some newly synthesized amylase may bypass the zymogen granule compartment and be constitutively secreted from the exocrine pancreas. Arvan & Castle (1987) have provided good evidence for the existence of at least two secretory pathways in the exocrine pancreas. They showed that unstimuilated release of labeled proteins follows a biphasic pattern. About 15% of newly synthesized secretory proteins are released over ~6 hr in an apparently constitutive fashion. A second peak of unstimulated release is observed around 8 hr. On the basis of the effects of secretagogues and some kinetic autoradiographic data this release is thought to be due to basal, zymogen granule–dependent secretion.

Newly synthesized secretory proteins could be segregated into constitutive and regulated pathways or passively distribute between them. Using a quantitative secretion assay, it has been shown that the pituitary cell line AtT-20 can discriminate between different secretory proteins and target them to one or the other pathway. The endogenous hormone pro-opiomelanocortin (POMC)/ACTH is selectively sorted into the regulated secretory pathway at least 30 times more efficiently than either the endogenous extracellular matrix protein, laminin (Burgess et al 1985), or the heterologously expressed secreted viral glycoprotein, truncated G (Moore & Kelly 1985). In fact, both laminin and truncated G appear to be secreted exclusively via the constitutive secretory pathway. These results raise the question of how regulated secretory cells distinguish regulated from constitutive secretory proteins so they can be correctly sorted and targeted to their appropriate intracellular location.

Sorting Signals Are Universal

Intracellular protein targeting may involve the recognition by some cellular sorting machinery of sorting signals common to proteins with the same destination (Blobel 1980). The evidence from signal peptide targeting to the RER suggests that the machinery to read such sorting signals is highly conserved across tissue and species boundaries. Experiments that directly examine this point have been reported. The human endocrine hormones proinsulin and growth hormone are correctly targeted to the regulated secretory pathway in the mouse endocrine cell line AtT-20 (Moore et al 1983b; Moore & Kelly 1985). Other groups have made similar observations for proparathyroid hormone expressed in GH3 cells (Hellerman et al 1984) and for proenkephalin expressed in AtT-20 cells (Comb et al 1985). Growth hormone was also shown to be targeted to the regulated pathway in the neuronal cell line PC-12 (Schweitzer & Kelly 1985). More direct evidence for the existence of a sorting domain for the regulated pathway was recently obtained by Moore & Kelly (1986). Their experiments suggest that human growth hormone contains target information that is sufficient to cause at least partial mislocalization of a constitutive secretory protein into the regulated secretory pathway when the two proteins are fused to give a hybrid molecule. Clearly, this targeting information is not limited to proteins of endocrine or neuroendocrine origin. The rat exocrine protein trypsinogen is recognized and efficiently targeted to the regulated secretory pathway of AtT-20 cells (Burgess et al 1985). Furthermore, growth hormone has been specifically expressed in the exocrine pancreas of transgenic mice using a tissue-specific promoter (Ornitz et al 1985). It has been recently shown that growth hormone can be targeted to zymogen granules (R. J. MacDonald, personal communication). Thus it appears that recognition of the targeting information for regulated secretion is relatively independent of cell type and animal species.

Mechanism of Sorting

In several cases sorting signals have been shown to reside in short, contiguous stretches of the polypeptide chain. Often these are present at the N terminus of the protein; for example, targeting to the RER, mitochondria, and chloroplasts is controlled by such N-terminal signal peptides (Walter & Lingappa 1986; Hurt et al 1984; Horwich et al 1985; Van den Broeck et al 1985). These sequences share little apparent primary sequence homology, but they do have some loosely defined common features (e.g. von Heijne 1986). This type of signal is responsible for targeting proteins from the cytoplasm to a target membrane. It now appears that targeting of proteins from one organelle to another depends on similar mechanisms. In yeast, nascent vacuolar proteins leave the Golgi apparatus (the donor organelle) for the vacuole (the acceptor organelle). A series of experiments from the laboratories of Stevens & Emr have recently shown that such interorganelle targeting information can also be present in a short contiguous amino acid sequence at the N terminus of the targeted protein (Valls et al 1987; Johnson et al 1987). The yeast vacuolar enzyme pro-CPY contains information within the first 50 amino acids of its "pro" sequence that is necessary and sufficient to specify interorganelle targeting to the vacuole.

Sorting information is apparently *not* at the N terminus in the case of regulated secretory proteins. The N-terminal twelve amino acids of rat trypsinogen can be deleted without altering the efficiency of sorting of this mutant trypsinogen into the regulated secretory pathway of AtT-20 cells (Burgess et al 1987). We have also shown that the signal peptide of trypsinogen is not responsible for interorganelle sorting into the regulated secretory pathway. Thus, while it is possible that the sorting domain responsible for targeting to the regulated pathway may be a contiguous stretch of amino acids, such a signal is not at the N terminus of trypsinogen.

A sorting signal or domain could lead to selective protein targeting to the regulated pathway by binding to an appropriate receptor (Figure 3). The receptor could go directly to the mature secretory granule (Figure 3A), or it could be a recycling receptor similar to that proposed for lysosomal enzyme targeting or receptor-mediated endocytosis (see Kornfeld 1986; Mellman et al 1987, for reviews). The features of this model that are relevant to regulated secretory proteins are: (*a*) newly synthesized secretory proteins would be selectively recognized and bound (by their sorting domain) by membrane receptor molecules upon their arrival in the TGN; (*b*) the receptor and ligand would be segregated from constitutive secretory proteins into the condensing vacuole or a region of the TGN (perhaps due to the tendency of occupied receptors to cluster or aggregate); (*c*) the receptor and ligand would dissociate (perhaps due to the low pH of the condensing vacuole or the TGN); (*d*) the receptor would be recycled (perhaps via a clathrin-coated vesicle) back to the TGN, while the ligand would become more concentrated due to membrane removal and/or several rounds of ligand delivery and receptor recycling. In the cases of lysosomal enzyme targeting and receptor-mediated endocytosis the existence of several of these steps is supported by experimental evidence. Receptor molecules have been identified, low pH dependent receptor/ligand dissociation has been demonstrated, and a role for clathrin has been implied by immunoelectron microscopy. Moreover, in the case of receptor-mediated endocytosis anti-clathrin antibodies have been shown to specifically inhibit endocytotic uptake (Doxsey et al 1986). Where data are available for regulated secretory protein targeting, they suggest a homologous mechanism. Weak bases such as chloroquine can disrupt regulated secretory protein targeting (Moore et al 1983a; Wagner et al 1986), which is consistent with a pH-sensitive step; however, there is no direct evidence

A.   B.

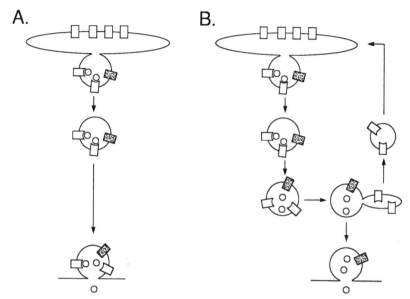

Figure 3   Sorting may or may not involve an intermediate compartment. (A) A receptor may bind a ligand in a donor organelle and transport it directly to an acceptor organelle, which could be the cell surface. (B) A receptor carries the ligand to a low-pH intermediate compartment. Since ligand binding is pH sensitive, the complex dissociates in this compartment. The receptor segregates from other membrane proteins and returns to the Golgi complex. Note that the ligand could be a single molecule or a molecular aggregate in either model. Scheme B is an example of facilitated sorting because it involves a receptor dedicated to recognizing a sorting domain (for discussion, see section on constitutive secretion in epithelia). If the receptor in scheme A is dedicated to sorting, this is also facilitated sorting. If, however, the receptor is a normal surface receptor for the ligand then the sorting is passive.

that pH regulates dissociation from a receptor. There is also good morphological evidence that clathrin is present in the regulated secretory pathway (Orci 1984a, 1985; Tooze & Tooze 1986), although a functional role of clathrin in regulated secretion has not been demonstrated. The question of how sorting into the regulated pathway occurs will probably remain unresolved until an analogue of the MPR is identified. The available evidence is consistent with a recycling receptor mechanism for regulated secretory protein targeting but is in no way definitive. In addition, it is still not known whether secretory granule–specific membrane proteins are somehow cosorted with the soluble content molecules or if their sorting is effected by a completely separate mechanism.

    There are reasons to believe that selective sorting of proteins to the regulated secretory pathway differs dramatically from lysosomal enzyme targeting. The regulated secretory products of a cell may be packaged together into vesicles based on their tendency to form coprecipitates or molecular aggregates with each other. Sorting could be achieved by excluding "nonaggregating" molecules from the forming secretory granules. This condensation-sorting model is particularly attractive because it explains the observed concentration of secretory product into a dense core, as well as de novo secretory granule formation (Kelly 1985).

The electron-dense content or core material is thought to be the concentrated secretory product itself in an osmotically inert state (see Farquhar & Palade 1981 for review). Sulfated proteoglycans, sulfated glycoproteins, the secretogranins, and chromogranins have all been invoked as aggregate-inducing components. Aside from their prevalence in secretory granules, the only direct experiment supporting such a role is that of Reggio & Dagorn (1978), in which they demonstrated that chondroitin sulfate can induce chymotrypsinogen A to form large aggregates. Other factors may play a role in condensation, such as pH (Zanini et al 1980b) or calcium concentration (Herman et al 1973; Clemente & Meldolesi 1975), but there is no consensus on this point.

The morphological and immunocytochemical data are intriguing but provide only circumstantial evidence for the condensation-sorting model. Condensation begins as early as in the Golgi cisternae and proceeds with a concomitant concentration of the secretory product (see Farquhar & Palade 1981). In endocrine cells the concentration factor may be as high as 200-fold over the ER baseline (Salpeter & Farquhar 1981); however, in exocrine cells it may be only 10-fold (Bendayan 1984). It is not known if this difference is significant since two different techniques, autoradiography and immunoelectron microscopy, were used for these experiments.

In most cases, different secretory proteins are evenly intermixed within the secretory granule (e.g. Bendayan et al 1980); however, several reports demonstrate that two secretory products may be segregated into domains within an individual granule (e.g. Ravazzola & Orci 1980; Fumagalli & Zanini 1985). Segregation into domains can easily be interpreted by the condensation-sorting model but not by a simple receptor recycling model since the receptor apparently does not discriminate between different hormones (Moore et al 1983b; Hellerman et al 1984; Burgess et al 1985; Comb et al 1985; Moore & Kelly 1985; Schweitzer & Kelly 1985). Interestingly, the intracellular site of dense core formation (condensation) can be experimentally altered and is likely to be related to the protein concentration within that compartment. Jamieson & Palade (1971b) showed that when exocrine pancreas is hyperstimulated, condensed material begins to form earlier in the secretory pathway. Instead of occurring predominantly in the condensing vacuoles, condensed material is found in abundance throughout the Golgi complex. Furthermore, recent data (S. Hashimoto, G. Fumagalli, A. Zanini, and J. Meldolesi, personal communication) suggest that the amount of prolactin or growth hormone synthesized in an individual cell may determine where the protein begins to aggregate or condense. This apparent concentration dependence of protein condensation is also consistent with a model in which secretory products self-aggregate in Golgi cisternae prior to segregation in the TGN. The apparent lack of an energy requirement (Jamieson & Palade 1971a) for secretory granule maturation and condensation-concentration fits well with this self-assembly model for granule formation.

Both the receptor recycling and condensation sorting models have attractive features, and each model is supported by some experimental evidence. The actual mechanism may be a combination of the two. It is difficult to see how condensation sorting could occur without some association with membrane proteins, since secretory granules have unique membrane proteins.

Efficiency of Sorting to the Regulated Secretory Pathway

It has been noted that sorting into the regulated secretory pathway of the cell line AtT-20 is quite inefficient. Moore & Kelly (1985) estimate that only one-third to one-sixth of newly synthesized POMC/ACTH molecules are actually targeted to secretory granules, the remainder are apparently exported via the constitutive pathway. It has been suggested that this inefficiency could be because AtT-20 cells are rapidly growing (Moore & Kelly 1985) or that these cells maintain only remnants of the regulated secretory pathway. Morphologically they are dedifferentiated relative to a normal pituitary cell. It is hard to imagine why regulated secretion would be inefficient in vivo. If the purpose of this special secretory pathway is to provide a mechanism for releasing a bolus of secretory product over a short time period, it would be very counterproductive to secrete a large fraction of secretory product continuously via the constitutive pathway.

Recent results using either endocrine or exocrine tissues as the experimental system support this view. Arvan & Castle (1987) showed that only 15% of the newly synthesized secretory protein is secreted constitutively. Furthermore, they identified on gels four polypeptides that appear to be bona fide constitutive secretory proteins. Thus the spillover of regulated secretory proteins into the constitutive pathway is likely to be considerably less than 15% of the total. Isolated pancreatic islets exhibit even less spillover of regulated secretory products into the constitutive route. Rhodes & Halban (1987) showed that less than 0.5% of newly synthesized proinsulin/insulin is released from a nonstimulatable (constitutive?)

secretory pool. In addition, the comparison by Gold et al (1984) of insulin secretion from tumor cells to that from normal islet cells suggests that sorting is less efficient in tumor cells. Thus the view is emerging that in vivo, regulated secretion is a highly efficient process. The model systems, such as AtT-20 cells, though useful for sorting studies, do not approach the level of efficiency of regulation observed in vivo.

Are There Multiple Regulated Secretory Pathways in a Single Cell?

Most cells have a constitutive pathway of secretion. A few cell types use the more sophisticated regulated pathway to store secreted protein. Here we examine evidence that cells can have more than one regulated pathway. For example, the studies by Bainton & Farquhar (1966, 1968a,b) on polymorphonuclear leukocytes demonstrated two types of cytoplasmic granules, azurophilic and specific granules. These granules are morphologically and biochemically different and arise at different stages of leukocyte development. The granules appear to be "lysosomelike"; upon stimulation they discharge their content into phagocytic vacuoles (Bainton 1973). In addition, there are morphological and biochemical data demonstrating a third type of regulated secretory granule in neutrophils (e.g. Dewald et al 1982; Gennaro et al 1983). Several studies, including a recent one by Lew et al (1986), show that different granule populations are differentially stimulated to discharge; specific and secretory granules are discharged in parallel under conditions where azurophilic granules are not.

In neurons small, clear synaptic vesicles and dense-core vesicles have been observed together in many different nerve terminals. It is generally believed that the small, clear synaptic vesicles contain classical neurotransmitters, whereas the large dense-core vesicles contain peptide neurotransmitters (Klein et al 1982). Exocytosis of these two types of vesicles can be stimulated in parallel; however, there is also evidence that release can be independently triggered from the two vesicle types (Lundberg & Hokfelt 1983). Low-frequency stimulation of the chorda lingual nerve seems to elicit acetylcholine secretion but little vasoactive intestinal peptide secretion; however, high-frequency stimulation leads to a dramatic increase in vasoactive intestinal peptide output relative to acetylcholine secretion. Recent immunoelectron microscopic evidence suggests that the membrane of secretory granules is qualitatively different from that of synaptic vesicles, even when they are both in the same cell (Navone et al 1986). These data are in favor of a model of two independent regulated pathways in neurons, one generating secretory granules and another, synaptic vesicles.

There is good evidence a single regulated pathway can release several proteins. Colocalization of various neuropeptides and endocrine hormones in dense-core granules is generally observed. For example, glucagon and glicentin colocalize by immunoelectron microscopy to alpha granules of pancreatic A cells (Ravazzola & Orci 1980), and vasopressin and corticotropin-releasing factor colocalize in neurosecretory vesicles of the median eminence (Whitnall et al 1985). It was generally thought that these regulated secretory products were copackaged into a single class of dense-core secretory granules. However, at least in endocrine cells, there may be some mechanism for sorting out different secretory proteins into either topologically distinct regions within a dense vesicle, or even into separate secretory granules. Ravazzola & Orci (1980) have reported that immuno-localization of glucagon and glicentin shows these proteins to be segregated to the core and mantle, respectively, of the same alpha granules. Fumagalli & Zanini (1985) have shown that prolactin and growth hormone can be packaged together in cow somatomammatroph cells; both inter-mixed and topologically segregated granules are observed. Furthermore, within a single cell some granules contain only prolactin or only growth hormone. More recent experimental results suggest that secretogranin II may be packaged into yet a third electron-dense vesicle in these same cells (S. Hashimoto et al, personal communication). Bassetti et al (1986) have localized prolactin and growth hormone in human pituitary adenomas. In most tumor cells the hormones were mixed in the same granules; however, the cells of one tumor had granules that contained only growth hormone or were mixed. In addition, Inoue & Kurosumi (1984) showed that in normal, untreated rat pituitary, FSH and LH are generally distributed into separate granules within each gonadotroph. Whether these heterogeneous granules are the product of possible alternate regulated pathways remains to be demonstrated.

It has been suggested (Fumagalli & Zanini 1985; S. Hashimoto et al, personal communication) that this sorting out may be related to the tendency of various secretory products to form self-aggregates more readily than coaggregates. Since growth hormone and prolactin are intermixed within the Golgi of somatomammotrophs (S. Hashimoto et al, personal communication) selective packaging of condensed hormones into separate granules could provide one way of generating different types of regulated secretory granules within each cell.

Where different secretory granules exist within the same cell, granule size seems to depend on hormone content. In different endocrine cells, the size of endocrine secretory granules is so characteristic that it can be used to distinguish cell types. Granule size does not seem to depend on cell type, but rather on granule content. Inoue & Kurosumi (1984) showed that, within individual cells, granules containing predominantly FSH are about twice the size (~ 500 nm) of those containing predominantly LH

($\sim$250 nm). Hashimoto et al (personal communication) similarly found that prolactin-containing granules in general are larger than growth hormone–containing granules and that the secretogranin II–containing granules are yet smaller. Granules of all three sizes and types are found in single somatomammotroph cells from bovine pituitary. Exactly how condensation of a particular hormone product might determine granule size is a mystery, but perhaps for each of these secretory proteins there are some higher order polymeric structures that are more stable and therefore preferred. A dynamic process by which membrane and secretory protein shuttling achieves uniform granule size was recently discussed by Tooze & Tooze (1986).

A reasonable amount of data suggests that there can be different forms of secretory granules undergoing separately regulated secretion within one cell. However, expression of heterologous proteins in AtT-20 cells by transfection suggests that the mechanisms involved in sorting secretory proteins are conserved across cell types and species boundaries. These two observations can be reconciled by having different organelles appear at different developmental stages as in polymorphonuclear leukocytes (Bainton & Farquhar 1966); or by condensation models of sorting in which self-aggregates form preferentially; or by postulating the existence of a lysosomal enzyme sorting mechanism to fill secretory granules (Taugner et al 1985). Alternatively, more than one type of sorting machinery may exist for targeting regulated secretory proteins.

CYTOSKELETON INVOLVEMENT WITH PROTEIN SECRETION

When a vesicle buds off from the Golgi complex carrying newly synthesized protein, its movement to the cell surface is probably facilitated by microtubular elements, at least some of the time. When the vesicle reaches the cell surface, its ability to fuse with the plasma membrane may be hindered by the actin-associated feltwork that covers the cytoplasmic face of the membrane. Interactions of transport vesicles with microtubules and microfilaments could regulate both the efficiency of secretion and the plasma membrane regions at which secretion occurs.

Association of Secretory Vesicles with Microtubules

Electron micrographs of secretory cells clearly show that secretory vesicles are associated with microtubules (e.g. Orci et al 1973). This by itself offers no great insight since lysosomes (Collot et al 1984), mitochondria (Ball & Singer 1982), the Golgi apparatus (Wehland et al 1983; Rogalski & Singer 1984), the nucleus (Wang et al 1979), and even the endoplasmic reticulum

(Terasaki et al 1986) all show some form of microtubule association. These organelle-microtubule interactions are not all of the same nature. The endoplasmic reticulum (Terasaki et al 1986) and the Golgi apparatus (Wehland et al 1983; Rogalski & Singer 1984; Tassin et al 1985b) associate with microtubules, but the association appears almost static. In contrast, "motile" organelles, e.g. mitochondria and synaptic vesicles, move along microtubules at speeds of up to 5 μm/sec. In living cells it is particularly easy to identify endosomal vesicles labeled with fluorescent ligands and to watch orderly movement to the center of the cell (Pastan & Willingham 1981; Herman & Albertini 1984). To identify secretory vesicles in living cells a marker is needed for the cytoplasmic domain of a specific secretory vesicle membrane protein. Using fluorescently labeled antibodies against the cytoplasmic tails of VSV G protein, Arnheiter et al (1984) demonstrated saltatory movements of constitutive secretory vesicles carrying G proteins. Although direct observation of movement of regulated secretory granules to the surface has not yet been made, the rapid axonal transport of neuropeptides provides indirect evidence for fast transport of secretory granules. Both regulated and constitutive secretory vesicles may be considered "motile" organelles.

The use of video-enhanced contrast techniques to observe small particles and macromolecules has yielded an enormous amount of information about how organelles move along microtubules. We know that small vesicles and mitochondria can move in both directions along a single filament (Allen et al 1985; Vale et al 1985a) that can be shown to consist of a single microtubule (Hayden & Allen 1984; Koonce & Schliwa 1985; Schnapp et al 1985). A microtubule-activated ATPase that has the ability to propagate movement of beads from the minus to the plus end of microtubules has been isolated (Vale et al 1985b; Brady 1985; Scholey et al 1985; Kuznetsov & Gelfand 1986) and called kinesin. Kinesin appears to be widely distributed in cells of organisms as distant as mammals and sea urchins. Evidence exists for a different motor that catalyzes movement in the retrograde direction (plus-to-minus) along microtubules (Vale et al 1985c). Exogenous synaptic vesicles (Schroer et al 1985) and axoplasmic vesicles (Gilbert & Sloboda 1984) show rapid movement when added in vitro to dispersed axoplasmic preparations, unless the vesicles have been treated with proteases beforehand. Although it is an attractive hypothesis that secretory vesicles are moved along microtubules by kinesin, the final proof is not yet in. Most studies of kinesin's action have involved movements of beads along microtubules or of microtubules along glass surfaces. The possibility has been raised that the true motor of membrane-bound organelles may not be kinesin at all (Gilbert & Sloboda 1986).

Motile vesicles can interact with microtubules via ATP-dependent motors. When the viscous drag of the cytoplasm is removed, all organelles move at the same speed (Schnapp et al 1985), which implies there is a single transport mechanism. A single microtubule can support organelle movement in both directions and can carry both mitochondria and smaller organelles. Given the widespread distribution of microtubules throughout the cytoplasm, such a nonselective mechanism would ensure the efficient mixing of motile organelles, whose movement would otherwise be constrained by the significant viscosity of the cytoplasm.

Directed Movement Along Microtubules

Microtubules radiate toward the cell periphery from microtubule-organizing centers centrally located near the nucleus. They all have the same polarity: Their plus ends are oriented away from the nucleus (Bergen et al 1980). The radiating microtubular framework offers an attractive substrate for kinesin-based movement of secretory granules from the center of the cell to its surface. This idea is encouraged by the very tight association between microtubule-organizing centers and the Golgi apparatus. This is perhaps most dramatically shown when myoblasts fuse to become myotubes (Tassin et al 1985a). The microtubule-organizing center can be recognized by antibody to the pericentriolar material of centrosomes and by its ability to nucleate the regrowth of microtubules after their disassembly with the microtubule-destabilizing drug nocodazole. Before myoblast fusion the microtubule-organizing center material appears to be deposited around the centriole, and microtubules spread out asymmetrically from a juxtanuclear site. After myoblast fusion the microtubule-organizing material becomes associated with the external surface of the nuclear envelope, which leads to a perinuclear distribution of microtubule initiation (Tassin et al 1985a). The distribution of the Golgi apparatus during myoblast fusion exactly parallels that of the microtubule-organizing center material (Tassin et al 1985b). When vesicles carrying secretory proteins leave the Golgi apparatus to travel to the cell surface, they do so in the vicinity of a high concentration of microtubules with the correct polarity for kinesin-mediated movement toward the cell periphery.

In many cells, and most cells in culture, the Golgi apparatus is in an asymmetric juxtanuclear position. Long before the function of the Golgi apparatus was known, the polarity of secretion was known to follow the polarity of the Golgi complex. For example, ameloblasts secrete enamel from their basal pole only after the Golgi apparatus has moved from the apical to the basal side of the nucleus (Beams & King 1933). At two different points in the development of the chick corneal epithelium, the Golgi apparatus moves from the apical position it usually assumes in epithelial cells to a basolateral position. Each time there is a concurrent deposition of extracellular matrix on the basolateral side of the epithelium (Trelstad 1970). More recently, the correlation between the polarity of the Golgi apparatus and the polarity of exocytosis has been extended. In migrating cells, for example, the Golgi apparatus orients towards the direction of movement, which is also the major site of membrane insertion (for a review see Singer & Kupfer 1986).

More direct evidence that protein secretion is polarized in the direction of Golgi polarization comes from the study of large granular lymphocytes. The cytoplasm of these lymphocytes is rich in dense secretory granules (Millard et al 1984) that contain a cytolytic pore–forming protein (Henkart et al 1984; Young et al 1986). The apparent function of these natural killer cells, cytotoxic T lymphocytes, and lymphokine-activated killer cells (Henkart et al 1986) is to bind to target cells, secrete the cytolytic factor at the site of contact, and thus lyse the target cell. Cytological studies have demonstrated that when such a granular lymphocyte encounters a target cell there is a rapid reorientation of the Golgi apparatus and microtubule-organizing center toward the target cell. This is perhaps most convincingly shown when the target cell for a cytotoxic T lymphocyte is another cyto-toxic T lymphocyte (Kupfer et al 1986a). Although both cells are T lym-phocytes, the Golgi apparatus in the target cell is randomly oriented, whereas that of the cytotoxic effector cell is appropriately polarized. The observation that the cytoskeletal protein talin is concentrated by the effec-tor cell (but not by the target cell) at the area of the contact provides an intriguing clue as to how such polarity may be established in lymphocytes. Talin is often associated with sites of microfilament attachment to the membrane. Actin is also found to be concentrated at sites of attachment of cytotoxic T cells (Ryser et al 1982) to their targets.

Polarization of the Golgi apparatus is associated with regulated secretion of protein by large granular lymphocytes and with membrane insertion in migrating cells, a process that is associated with constitutive secretion. Direct evidence that helper T cells, which are presumably consti-tutive secretory cells, also demonstrate polarized secretion was obtained by Kupfer et al (1986b). A helper T cell directly interacts with a B cell presenting the appropriate antigen in the context of class II major histo-compatibility antigen and then releases growth factors that stimulate the B cell. Again, the microtubule-organizing center, and presumably the Golgi apparatus, is oriented towards the target B cell, i.e. in the direction of secretion.

The striking polarity of the Golgi apparatus, the origin of secretory vesicles, and the direction of exocytosis, in both regulated and secretory cells, strongly suggest the association of all vesicles carrying newly syn-thesized proteins with the microtubular system radiating from the vicinity of the Golgi apparatus. Although this model is plausible, it is not proven, and even if it is largely correct, it may not be universally applicable. In applying this model to cells in which the distance from the Golgi to the plasma membrane is very small, as it is in lymphocytes, caution seems especially advisable. Polarization of the Golgi itself may give sufficient polarity to exocytosis; diffusion without kinesin-mediated transport may be sufficient to achieve polarized secretion of proteins. Evidence, albeit indirect, that microtubules carry vesicles to the cell surface comes from the use of drugs that perturb the microtubule array.

Microtubule-Disrupting Drugs and Secretion

Given the circumstantial evidence suggesting that transport vesicles associate with microtubules as they travel to the cell surface, one might expect that disruption of microtubules with colchicine, vinblastine, or nocodazole or microtubule stabilization with taxol would interfere with secretion. This is indeed the case for a remarkably wide range of constitutive and regulated secretory cells, including fibroblasts, hepatocytes, and exocrine, endocrine, and mammary gland cells (Busson-Mabillot et al 1982, and references therein). In general, no effect is noticed on the release of secretory granules that are made before drug addition and are presumed to lie at the periphery of the cell. However, secretion of newly synthesized material is inhibited, presumably because microtubules are necessary to transport newly synthesized proteins to the surface. Colchicine inhibits saltatory movements of vesicles in several different systems (Freed & Lebowitz 1970; Wang & Goldman 1978; Bhisey & Freed 1971; Herman & Albertini 1984). Furthermore, in the presence of microtubule-depolymerizing drugs exocytotic events are no longer polarized; for example, VSV G protein is no longer inserted at the growing edge of migrating fibroblasts but instead is randomly inserted into the plasma membrane (Rogalski et al 1984). Whether this change is due to breakdown of a microtubule transport system or fragmentation of the Golgi apparatus is not clear. A controversial result (Rindler et al 1987; but see Salas et al 1986) is that hemagglutinin is delivered to the basolateral rather than the apical membrane of Madin-Darby canine kidney (MDCK) cells after the microtubules have been disrupted. In other cases, autoradiography has been used to show that colchicine blocks apical insertion and causes aberrant localization of basolateral membranes (Blok et al 1981; Bennett et al 1984). Finally, disruption of microtubules allows secretion of albumin from liver cells into the bile (Mullock et al 1980) and casein delivery to basolateral surfaces of the mammary epithelium (Nickerson et al 1980). The pharmacological evidence seems overwhelmingly in favor of the involvement of microtubules in facilitating protein secretion and in directing exocytosis to specific domains of the cell membrane.

It is important, however, to bear in mind some of the limitations of the pharmacological experiments. Not all microtubules transport vesicles; others, perhaps those decorated with 25-nm sidearms or cross-linked to other cytoskeletal elements, serve a structural function (Miller et al 1987). Under the depolymerizing conditions used in these experiments, both structural and transport microtubules are probably destroyed, leading to gross abnormalities in cell organization and perhaps even in the cell-cell interactions required to maintain tight junctions in epithelia. Furthermore, the secretion assay measures many steps between synthesis and exocytosis, any or all of which could be sensitive to disruption of microtubules. When microtubules are disrupted, the ER moves centripetally (Terasaki et al 1986), and the connections between cisternae of adjacent Golgi stacks are broken, although the stacks themselves are left morphologically intact (Wehland et al 1983; Rogalski & Singer 1984; Tassin et al 1986). These disruptions might suffice to inhibit the movement of newly synthesized material to the surface or to change the site of exocytosis on the plasma membrane. Thus, although the pharmacological data suggest that newly formed secretory vesicles move along microtubules to the cell surface, proof of this hypothesis will require studies using inhibitors that are specific for the microtubular transport system and have no effect on structural microtubules.

Selectivity in Microtubule Transport

In the experiments described so far, secretory vesicles seem able to associate with microtubules in both regulated and constitutive secretory cells. In both types of cell the Golgi apparatus is oriented in the direction of selective protein release, and colchicine perturbs secretion. This situation can be described as nonselective cytoskeletal targeting (Figure 4A). Some cells, however, insert different proteins in different regions of the plasma membrane, e.g. MDCK cells, which insert different proteins in their apical and basolateral surfaces (Simon & Fuller 1985; Matlin 1986a). If microtubules direct vesicles containing nascent proteins to the cell surface, then a nonselective vesicle transport system such as that described in Figure 4A could increase the frequency of vesicle collision with a target membrane, but it would not facilitate selective transport to the correct region of plasma membrane. Although the data conflict as to whether microtubules are required for apical insertion of hemagglutinin (Rindler et al 1987; Salas et al 1986), there is agreement that disruption of microtubules has no effect on basolateral insertion of VSV G protein. These data might be explained by selective cytoskeletal targeting (Figure 4B).

According to such a model, one type of vesicle would diffuse randomly from the Golgi complex and collide with the correct plasma membrane by chance, while another type would interact with the cytoskeleton and be delivered to the site of insertion. A third possibility is that there are multiple cytoskeletal targeting mechanisms (Figure 4C). In this case there would

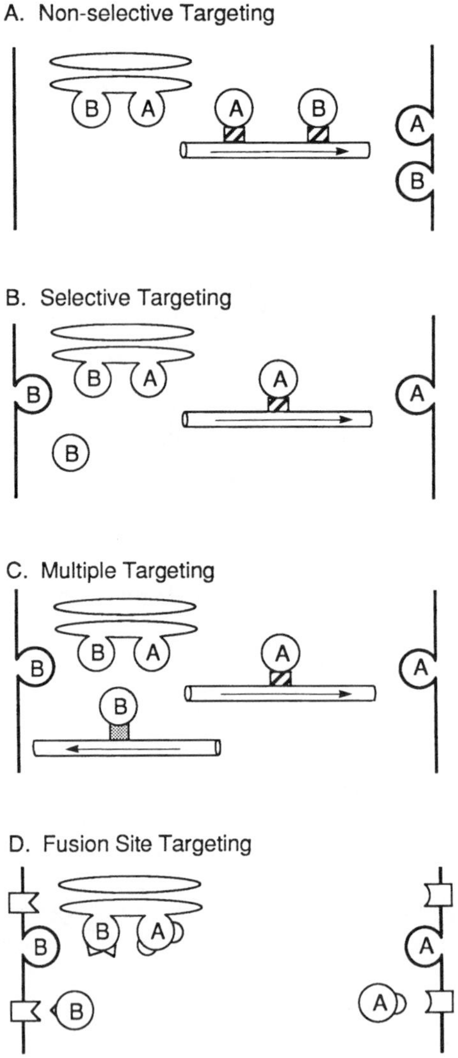

A. Non-selective Targeting

B. Selective Targeting

C. Multiple Targeting

D. Fusion Site Targeting

Figure 4 Exocytosis at restricted domains of the plasma membrane. Mechanisms A–C involve cytoskeletal targeting, while D involves selective membrane fusion sites. (A) Because of the orientation of the microtubules and the Golgi apparatus, vesicles carrying proteins A or B fuse at one end of the cell and release their contents. Because all newly synthesized proteins are secreted together, this is called non-selective cytoplasmic targeting. (B) In selective cytoplasmic targeting only one class of vesicles interacts with the cytoskeleton. A is delivered to a specialized membrane domain, whereas B is secreted from membrane closest to the Golgi complex. (C) In a cell with multiple cytoskeletal pathways, one cytoskeletal structure carries class A vesicles one way and a second structure carries class B vesicles in a different direction. Such multiple pathway targeting may exist in neurons. (D) If there are different docking proteins for different vesicles, and these are segregated from each other, different proteins will be secreted at different domains. Fusion-site targeting mechanisms may exist in epithelial cells and have been well described at the active zones of synapses.

be two different microtubular types, each capable of transporting a different class of membrane vesicle to either the apical or basolateral domain.

The idea that selective or multiple cytoplasmic targeting systems exist has long been popular with neurobiologists. They accept as axiomatic that some transport vesicles must leave the Golgi apparatus and move by fast transport to axonal or nerve-terminal sites. Axonally targeted vesicles must be selectively "shunted away" from membrane vesicles containing proteins destined for neuronal dendrites (e.g. Hammerschlag 1983) and guided to their correct plasma membrane destination by specific cytoskeletal machinery. There is good evidence that the cytoskeletons of the dendrites and axons differ (Peng et al 1986), as predicted by selective-targeting models. There is also good evidence for fast axonal transport of neuropeptides (Lundberg et al 1981; Rasool et al 1981), synaptic vesicle membrane proteins (Lagercrantz 1976; Dahlstrom et al 1985), and presynaptic receptors for neurotransmitters (Young et al 1980; Laduron 1980; Levin 1982), effected perhaps by common vesicles (Laduron 1984). Clearly, vesicles containing components of the nerve-terminal plasma membrane and vesicles containing components of the secretory vesicles interact with the axonal microtubule system and move down it with a velocity consistent with the involvement of a kinesin-type motor. Direct evidence for selectivity is not readily available. On the contrary, mitochondria and lysosomes also move to nerve terminals by what is presumably the same mechanism. To show selectivity, some known neuronal secretory protein or dendritic membrane protein must be shown to be excluded from fast axonal transport. Perhaps because the assumption of selectivity is so widespread, little effort has been made to test it.

Experiments with cells in culture show little evidence for selectivity. When secretory cells in culture extend processes, their secretory vesicles have a tendency to accumulate at the growing tips (Kelly et al 1983), just as they do in growing neurons (Buckley & Kelly 1985). These tips are also enriched in mitochondria and other unidentified membranous organelles, including vesicles containing constitutively secreted proteins (Matsuuchi & Kelly 1987). Clearly, cytoskeletal rearrangement that allows process formation in cultured cells results in nonspecific cytoskeletal targeting of vesicles (Figure 4A). If selective axonal transport exists in neurons, it is a property not shared by the process-extending cells in culture that have been studied to date.

The remarkable structure of neurons offers an insight into how a multiple targeting mechanism (Figure 4C) might use only kinesin (Vale et al 1985b) and a retrograde motor (Vale et al 1985c). In olfactory neurons (Burton 1986) the polarity of dendritic microtubules is opposite that of axonal ones, i.e. the plus ends face the nucleus. If transport vesicles destined for the axon bound a kinesinlike molecule and moved anterogradely, while transport vesicles destined for the dendrites bound the retrograde motor (Vale et al 1985c), multiple cytoskeletal targeting could be achieved without invoking any new cytoskeletal machinery.

Involvement of Actin and Actin-Related Molecules in Secretion

Research on the involvement of microfilaments in protein secretion has as long and as tortuous a history as that on microtubule involvement. A critical recent advance has been the realization that in most animal cells organelles move along microtubules, and actin and myosin are not involved. Now that this point has been clarified, research is focusing on the cortical microfilament web that appears to extend 200–300 nm from the cytoplasmic side of the plasma membrane. It is unlikely that the subplasmalemmal actin and related proteins play a direct role in exocytosis since calcium-dependent exocytosis in an in vitro preparation from sea urchin eggs cannot be inhibited even with huge concentrations of cyto-chalasin B, phalloidin, N-ethylmaleimide-modified myosin subfragment 1, and antibody to actin (Whitaker & Baker 1983). However, calcium entry is likely to change the organization of the subplasmalemmal web. Gelsolin, a Ca^{2+}-activated protein that severs preexisting actin filaments and caps the barbed end, is found in platelets (Nelson & Boyd 1985) and chromaffin cells (Bader et al 1986). Stable complexes of gelsolin and actin form on platelet activation and may disrupt the cortical web (Kurth & Bryan 1984). A calmodulin-regulated actin binding protein, caldesmon, has been shown to bind to secretory granules in a calcium-dependent fashion (Burgoyne et al 1986), which indicates that secretory granules may be sensitive to the state of subplasmalemmal actin. It has been suggested that the subcortical microfilaments serve as a physical barrier excluding secretory granules from the subplasmalemmal region in resting cells (Orci et al 1972; Bur-goyne et al 1982). In the presence of cytochalasin, which binds to free barbed ends of actin and prevents actin polymerization, the web decreased in thickness and the granules were nearer the plasma membrane (Burgoyne et al 1982). Although cytochalasin facilitated secretion in insulin-secreting cells, as predicted by this model, it inhibits secretion in many other cell types (see references in Drenckhahn & Mannherz 1983). Surprisingly, when chromaffin cells are stimulated to exocytose in the presence of tri-fluoperazine, which binds calmodulin and inhibits secretion, secretory vesicles are seen to move into the subplasmalemmal web region from which they had previously been excluded (Burgoyne et al 1982). The suggestion that this is due to a calcium-independent change in the subplasmalemmal web is supported by the finding of a remarkable decrease in filamentous cortical actin in chromaffin cells in response to nicotine (Cheek & Burgoyne 1986). This depolymerization occurred in 15 sec, was reversed by 30 sec, and was insensitive to calcium. Perhaps coincidentally, the appearance of actin-nucleating sites in polymorphonuclear leukocytes in response to chemotactic factors is equally fast and is also calcium-insensitive (Carson et al 1986).

These data indirectly suggest that the actin web provides a physical barrier to exocytosis and that reducing or removing it facilitates exocytosis. The experimental results of Lew et al (1986) on neutrophil secretion support this model. Using Quin 2, they measured the internal calcium concentration required to cause granule exocytosis. They found that under normal conditions a calcium concentration of 6 μM was the EC_{50} for azurophilic granule release, but this figure was reduced to 2.5 μM in the presence of cytochalasin B. They suggested that cytochalasin B may lower the calcium requirement for exocytosis because the cation was no longer needed to activate concomitantly some of the Ca^{2+}-dependent regulatory proteins such as gelsolin.

A second change in the subplasmalemmal cortex has been observed, which involves fodrin. Fodrin is normally found in a continuous $\sim$230-nm-wide band under the plasma membrane of chromaffin cells (Langley et al 1986). When cells are stimulated to secrete in the presence of calcium, the fodrin becomes discontinuous. The time course of this calcium-dependent change is much slower than that described for nicotine in the same cell type; it takes 60 min to return to normal (Perrin & Aunis 1985). Although the authors suggested that fodrin and associated cytoskeletal components move apart to allow granule discharge, the time course is slow compared to the time course of secretion in chromaffin cells (Rojas et al 1986).

The complexity of the rearrangements in the cortical actin web observed in stimulated cells should make us cautious about adopting any one model. Besides interfering with access to the plasma membrane, the web could be involved in endocytotic events (Bendayan 1983), and it could be crucial to tissue organization. For example, actin is associated with the apical domains around the luminal ducts of exocrine cells (Drenckhahn & Mannherz 1983). Disruption of the actin microfilaments could distort the duct itself. A third alternative, suggested by the presence of talin at junction sites between cytotoxic T lymphocytes and their targets (Kupfer et al 1986a) and by the cleavage of talin during platelet activation (O'Halloran et al 1985), is that the microfilament network anchors and stabilizes the plus ends of microtubular arrays at the plasma membrane.

Importance of Cytoskeleton for Secretion

Apparently neither microtubules nor microfilaments are absolutely required for movement of newly synthesized material to the surface (Rogalski et al 1984) or for the exocytotic event itself (Whitaker & Baker 1983). The importance of microtubules in protein secretion stems from the ability of the microtubule arrays to enhance the rate at which newly synthesized material is transported to the surface and to determine to which part of the cell it is delivered. The cortical actin web may act as a physical barrier to secretion, and it may even adsorb and thereby immobilize cytoplasmic organelles. Although the basic elements of protein secretion (Golgi sorting, vesicular transport, and fusion with the plasma membrane) are common to all secretory cells, each class of secretory cell differs from the others as to the site, mode, and regulation of protein discharge. In some cells cytoskeletal regulation may play a major role, whereas in others its role may be insignificant.

Microtubular regulation of secretion may be involved, however, in the preferential release of newly synthesized protein mentioned earlier. Shortly after a pulse-label the specific activity of insulin, for example, is higher in the medium bathing pancreatic B cells than it is in the secretory granules inside the cell. Rhodes & Halban (1987) suggest that newly synthesized vesicles are preferentially released because they preferentially associate with the cytoskeleton. However, more specific mechanisms may operate, such as specific microtubule delivery of newly formed granules to preferred release sites or differential association of old and new vesicles or granules with the cortical web. Preferential release of newly synthesized material is increased if insulin-secreting cells are stimulated while secretory granules are forming (Gold et al 1982). It is even possible that the secretagogue promotes the association of newly formed secretory vesicles with microtubules. The translocation of pigment granules along microtubules is regulated by cAMP via protein kinases (Rozdzial & Haimo 1986; Lynch et al 1986a,b). Since organelle transport can be regulated in melanophores, it would be surprising if similar regulation were not found elsewhere. Such regulation may perhaps be the function of the large amounts of cAMP RII receptor protein and cAMP-dependent protein kinase associated with the trans face of the Golgi stack and centrosomes (DeCamilli et al 1986; Nigg et al 1985).

The Golgi complex and microtubule-organizing center clearly affect the location of exocytosis in migrating cells and T lymphocytes. Microtubules are certainly involved in transporting secretory granules and synaptic vesicle precursors down axonal processes. However, cytoskeletal targeting may not be sufficient to achieve the very high levels of selectivity that are observed in membrane insertion in polarized epithelial cells. Unless there are multiple cytoskeletal pathways (Figure 4C), it is hard to explain why some proteins are inserted apically and others basolaterally. The cortical actin web may also be involved in specifying the site of exocytosis. In exocrine cells, actin and related proteins are concentrated in a terminal web in the apical region (Drenckhahn & Mannherz 1983), while in MDCK cells the actin binding protein fodrin associates with the basolateral membranes after a monolayer is formed (Nelson & Veshnock 1986). Transport vesicles targeted to different regions may have different abilities to penetrate the actin or fodrin web.

Although the cytoskeleton may facilitate delivery of vesicles to the correct plasma membrane domain and the cortical actin web may restrict exocytotic sites, it is unlikely that cytoskeletal targeting is sufficient to explain nonrandom insertion completely. It is reasonable to expect that secretory vesicles recognize different fusion sites on the cytoplasmic face of the plasma membrane (Figure 4D). Direct support for fusion-site targeting, as distinct from cytoskeletal targeting, is not easy to find. Indirect support for this idea comes from the tight association of cortical granules with plasma membrane after disruption of microtubules in sea urchin cells (Whitaker & Baker 1983) and from the striking morphological specializations that hold synaptic vesicles to release sites in the nerve terminal, especially at ribbon synapses. Finding the molecules that control selective fusion between membranes is currently a critical challenge.

SECRETION IN EPITHELIAL CELLS

Constitutive Secretion

In the original model of protein secretion (Gumbiner & Kelly 1982) the constitutive pathway is the route by which membrane proteins are externalized and by which soluble proteins with no other sorting information reach the cell surface. In epithelial cells, the apical and the basolateral domains of the plasma membrane differ in protein composition. After leaving the Golgi complex newly synthesized membrane proteins are segregated from each other and targeted to their appropriate domain. As there are two routes for newly synthesized membrane proteins to reach the surface, soluble proteins lacking any sorting information could take either. However, some constitutively secreting epithelial cells secrete most of their protein in the basolateral direction. If this implies sorting of secreted proteins, then the simple model that the constitutive pathway is the default route and does not require sorting signals on soluble proteins cannot be completely correct, at least in epithelial cells.

In this chapter we review work done on polarized cells in culture that form tight monolayers. They are easier to study than glandular epithelia and have provided a substantial amount of information. Although in such cells protein secretion may be a minor function and may have no obvious physiological significance, much of what we know of protein secretion in polarized epithelia comes from cell lines such as MDCK cells or the human colon carcinoma line, Caco-2 (Pinto et al 1983). Kidney epithelial cells contact a basal lamina on their basal surfaces and secrete basal lamina components selectively on their basolateral surfaces. Laminin and heparan sulfate proteoglycan, two such components, are preferentially secreted from the basolateral surface in MDCK cells (Caplan et al 1986). In support of this result, Rindler et al (1987b) have demonstrated that secretion of apolipoproteins by Caco-2 cells is exclusively basolateral. Thus cell lines resemble hepatocytes in directing the bulk of their secretion to the basolateral surface.

Preferential flow of secreted protein to one surface does not necessarily require sorting. In principle, it could also be explained by appropriate orientation of a nonspecific cytoskeletal targeting system (Figure 4A). There is evidence, however, that not all proteins are secreted from the basolateral surface of MDCK cells. When DNA encoding foreign secretory proteins such as lysozyme (Kondor-Koch et al 1985; Gottlieb et al 1986) is introduced into MDCK cells, the expressed proteins are secreted from both surfaces. When the polyIg receptor is expressed in MDCK cells, it can transport IgA from basolateral to apical membranes (Mostov & Deitcher 1986); when its transmembrane and cytoplasmic domains are deleted, the truncated polyIgA receptor becomes a secreted protein (Mostov & Deitcher 1986), which is preferentially apically discharged. An endogenous MDCK protein, whose function has not yet been identified, is also reported to be secreted apically (Kondor-Koch et al 1985; Gottlieb et al 1986). Although experimental data on directional secretion have come from several different laboratories, it seems likely that nonspecific cytoskeletal targeting (Figure 4A) cannot explain the findings. Protein must be sorted into vesicles with different destinations that are determined by cytoskeletal and/or fusion-site targeting (Figure 4B–D).

How might constitutively secreted proteins be sorted? It is usually helpful first to establish default pathways, if any exist. In MDCK cells proteins apparently lacking specific targeting information, e.g. foreign secretory proteins such as lysozyme, can exit apically and basolaterally. Furthermore, when MDCK cells are treated with ammonium chloride, which inhibits other intracellular sorting mechanisms, laminin is secreted from both surfaces (Caplan et al 1986). Ammonium chloride does not affect the specificity of insertion of membrane proteins, although it reduces the rate (Matlin 1986b). Secretion of unsorted soluble proteins from both surfaces could be due to nonselective bulk flow transport of secretory proteins in vesicles selectively carrying membrane proteins. In contrast, the default pathway in Caco-2 cells appears to be basolateral, since inhibition of sorting of lysosomal enzymes results in their basolateral secretion (M. J. Rindler, personal communication). The difference between these two epithelial cell types could reflect differences in volume flow, i.e. the product of vesicle volume and number of vesicles per second that leave the Golgi complex, for the two surfaces. When the flow to each surface is about the same, as in MDCK cells, default secretion is equally probable in either direction; when a much greater volume is transported to one surface than the other, as may be the case in the apolipoprotein-secreting Caco-2 cells, one surface is the preferred site of secretion.

Since some polypeptides are selectively secreted apically and laminin is secreted basolaterally in MDCK cells, there must be at least two mechanisms to sort soluble proteins out of the common pathway into vesicles exclusively targeted to the apical or basolateral membrane. A domain in laminin, for example, could be recognized by a membrane-bound receptor that is itself targeted to the basolateral membrane. In some way, this sorting mechanism must be sensitive to ammonium ions (Caplan et al 1986). Recent findings on laminin secretion in MDCK cells raise the question of the validity of the default pathway concept. Laminin is secreted constitutively in PC-12 (Schweitzer & Kelly 1985) and AtT-20 cells (Burgess et al 1985) in a way that is consistent with a default pathway. However, the definition of default is essentially negative, that is, no evidence for a sorting domain can be found. The targeting of laminin to the basolateral membrane in MDCK cells suggests that sorting signals present in laminin are utilized by some cells but not others or that the constitutive pathway in AtT-20 cells is not necessarily a default pathway.

In addition to bringing the default pathway concept into question, the data on protein sorting in constitutively secreting epithelial cells force us to define sorting more precisely (Figure 1). In default secretion, the protein has no sorting domain and is carried along by bulk flow in transport vesicles going to the cell surface. When a soluble protein binds to a membrane protein, which carries it to the surface of the cell, it is sorted in the sense that its destination is no longer determined by bulk flow to the cell surface.

It is important to emphasize that a protein sorted by binding to a membrane receptor may or may not itself have a domain specifically and exclusively dedicated to sorting. For example, if a cell makes laminin and the laminin receptor and laminin associates with its receptor inside the cell, then laminin would be transported to whichever cell surface the receptor is delivered to. Laminin would need no sorting domain specifying its eventual destination, and the receptor would have a major cellular function other than sorting. We propose calling such a system passive sorting (Figure 3). Facilitated sorting occurs when a secreted protein has a sorting signal and there is a receptor dedicated to recognizing it. Receptors for facilitated sorting may only go one way, as does the polyIgA receptor involved in transcytosis, or they may be recycled (Figure 3). They can be ligand specific, as the LDL receptor is, or protein-class specific, as are the mannose phosphate receptors that recognize lysosomal enzymes. In the case of basolateral laminin secretion by epithelial cells, it is not clear if sorting is passive or facilitated.

Regulated Secretion

Some epithelial cells are also regulated secretory cells. Indeed, the pancreatic acinar cell is probably the most extensively studied secretory cell. Exocrine cells secrete primarily through their apical domains, whereas sensory epithelial cells and receptor cells in the retina or olfactory neurons (Getchell 1986), for example, secrete basally. Whether endocrine cells exhibit polarized secretion is less clear. Thyroid follicular cells are clearly polarized, but it is not clear whether they should be characterized as endocrine or exocrine. In our hands, antibodies that recognize endocrine but not exocrine cells (Buckley & Kelly 1985) fail to bind to follicular cells, while adjacent parathyroid cells show easily detectable binding. In an attempt to demonstrate membrane polarity in pancreatic endocrine cells Lombardi et al (1985) infected monolayer cultures of pancreatic islet cells with influenza virus and VSV to mark the apical and basolateral surfaces, respectively. By counting the number of virions budding from each face they were able to show that the membrane adjacent to the medium had the characteristics of apical membrane and that adjacent to the dish, basolateral. There is also evidence of polarity in vivo in endocrine cells. The results of physiological experiments suggest that plasma membrane of the islet cell is divided into areas that contain receptors that trigger secretion and areas that are the site of secretion (Kawai et al 1982). Unfortunately, we do not yet know whether these domains correspond to apical and basolateral membranes.

Epithelial cells must have separate pathways for apical and basolateral membrane insertion. In regulated secretory cells, such as the pancreatic acinar cell, the question arises whether the apical membrane proteins reach the cell surface in the zymogen granule membrane. Unfortunately, such data are not available. In the endocrine tumor cell line AtT-20, neither the retroviral basolateral marker gp70 (Gumbiner & Kelly 1982) nor the apical marker hemagglutinin (T. A. Schroer & R. B. Kelly, unpublished) can be found in purified secretory vesicles. This result is not definitive since it is not clear if these cells derived from the pituitary can be considered epithelial. Transformation is known to lead to loss of polarity in endocrine cells (Lombardi et al 1986).

If there are two independent pathways to the apical membrane in exocrine cells, elucidation of their sorting mechanisms will be further complicated. Membrane proteins in the trans Golgi network will need to be segregated from each other and independently targeted to basolateral transport vesicles, apical transport vesicles, and zymogen granules.

RECYCLING OF THE SECRETORY VESICLE MEMBRANE

During exocytotic release of protein, a considerable amount of surface area is added to the plasma membrane of a cell (e.g. Breckenridge & Almers 1987). Since the rate of membrane addition from exocytosis is many times that required for cell growth, membrane must be removed by endocytosis. Although some of the endocytosed membrane is degraded in lysosomes, the cell recycles and reuses considerable amounts of secretory vesicle membrane. Recycling of secretory granule membranes accounts for the observations that the turnover time of granule membrane proteins is much longer than that of the soluble proteins (Meldolesi 1974). Morphological evidence strongly supports the notion that the recycling membrane returns to the Golgi apparatus (Farquhar 1983, 1985), where it is reutilized to enclose newly synthesized secretory proteins. The aim of this section is not to review endocytosis but to ask if any features of endocytosis are specific for cells specialized for protein secretion.

The plasma membrane of many cells is turned over rather rapidly by endocytosis. Cells can internalize about 1–3% of their surface area per minute (Steinman et al 1983). The rate of membrane uptake must be enhanced by exocytosis since the rate of uptake of horseradish peroxidase (HRP) in chromaffin cells increases two- to threefold immediately after stimulation of secretion (von Grafenstein et al 1986). The stimulation of endocytosis begins within 1–2 min of the onset of exocytosis, occurs during exocytosis, and is completed within ~5 min of the end of stimulation (von Grafenstein et al 1986). Temporal overlap of exocytosis and endocytosis in chromaffin cells was observed using capacitance measurements (Neher & Marty 1982). Fusion of a single granule caused about a 0.1% capacitance increase due to the increased membrane area. Interspersed with such increases were drops in capacitance due to endocytosis. Some of these decreases were large, especially after extensive stimulation. They may correspond to the formation of large electron-lucent, endocytic vesicles seen in the electron microscope (Knight & Baker 1983). We can conclude from the capacitance and HRP-uptake measurements that the introduction of new membrane by exocytosis stimulates endocytosis within a short period of time. Minutes after the end of secretion in chromaffin cells, the membrane area has returned to normal. Normal rates of endocytosis are so high there is no need to postulate a specific mechanism (for membrane uptake after exocytosis.

The development of antibodies that bind to the luminal surface of secretory vesicles has allowed exocytosis to be demonstrated directly in chromaffin cells (Patzak & Winkler 1986 and references therein), basophilic leukemia cells (Bonifacino et al 1986), platelets (Stenberg et al 1985) and neurons (von Wedel et al 1981). With this method the fate of individual vesicle components can be followed during recycling. Patzak & Winkler (1986) showed that glycoprotein III (GPIII) of chromaffin granule membranes appears in coated pits and coated vesicles in the first 5 min after exocytosis. After passing through a smooth vesicle compartment, antibody to GPIII reappears in the trans Golgi region within 30–45 min. Some antibody to GPIII was detected in dense-core secretory granules as early as 45 min after exocytosis. This is probably the first convincing evidence that secretory vesicle components return to the Golgi region, where they are used in the production of new secretory granules. These authors stress that label was always over the trans Golgi network and was never associated with the Golgi cisternae. In contrast, when cationized ferritin, dextran, or HRP are used to follow membrane recycling in secretory cells, label is found in the Golgi stack, especially in the medial cisternae (Farquhar 1978; Herzog 1981; Orci et al 1986b). Patzak & Winkler (1986) suggest that there may be two pathways of membrane recycling.

Recycling studies that use antibody to vesicle protein give different results from studies using fluid-phase markers when delivery to mature secretory granules is investigated. The observation that the antibodies eventually reach secretory granules is unambiguous (Patzak & Winkler 1986). However, although HRP taken up by insulin-secreting cells is delivered to organelles that have dense cores, these organelles may not be mature secretory granules (Sawano et al 1986). Core structures are also found in multigranular bodies in which degradation is thought to occur upon crinophagy. During an 85-min chase, most secretory vesicles with HRP label became multigranular bodies. Furthermore, HRP does not associate with newly formed secretory granules. Thus secretory granules destined for degradation probably fuse with an endosomal compartment containing HRP. Since we know that HRP does return to the Golgi complex (Orci et al 1986b), it must be excluded from the regulated pathway. Such a result is predicted for a protein lacking appropriate information for targeting to secretory granules.

Constitutive and regulated secretory cells are alike in some endocytotic properties but perhaps not all. Both constitutively secreting myeloma cells (Ottosen et al 1980) and regulated secreting exocrine and endocrine cells (Farquhar 1983) have a fluid-phase pathway for recycling membrane to the Golgi cisternae. Such recycling seems to be a general feature of cells, not just those whose primary function is protein secretion. The transferrin receptor recycles from the cell surface to a mannosidase I compartment, presumably in the cis Golgi region, in erythroleukemia cells (Snider & Rogers 1986). In regulated cells, evidence suggests that a second recycling pathway exists in addition to the one that returns endocytosed material to the Golgi stacks. In contrast to the rapid uptake of HRP for delivery to the Golgi cisternae (von Grafenstein et al 1986) even at 23°C, the coated vesicle–mediated uptake of secretory granule membrane protein is slow (Patzak & Winkler 1986). As mentioned earlier, the coated vesicle–mediated uptake is to the trans Golgi network, not the Golgi cisternae (Patzak & Winkler 1986). If there are indeed two pathways of internalization, a nonspecific one to the cisternae for bulk membrane flow and a specific one to the trans Golgi network for specific granule membrane proteins, several issues immediately arise: Is either pathway specific to secretory cells? How are different proteins targeted to a particular pathway? These questions are just as challenging as, but perhaps more accessible than, those related to protein sorting in the Golgi apparatus.

ACKNOWLEDGMENTS

The authors thank the members of their laboratory for many useful discussions and helpful ideas. We are grateful to Drs. Trina Schroer (Washington University, School of Medicine) and Ron Vale (University of California, San Francisco) for their assistance on the cytoskeleton section. We are also grateful to our colleagues who sent us their papers and manuscripts in advance of publication. Finally, we acknowledge our indebtedness to Leslie Spector for her patience, accuracy, and speed in preparing the manuscript.

Literature Cited

Adelson, J. W., Miller, P. E. 1985. Pancreatic secretion by nonparallel exocytosis: Potential resolution of a long controversy. *Science* 228: 993–96

Allen, R. D., Weiss, D. G., Hayden, J. H., Brown, D. T., Fujiwake, H., Simpson, M. 1985. Gliding movement of and bidirectional organelle transport along native microtubules from squid axoplasm: Evidence for an active role of microtubules in cytoplasmic transport. *J. Cell Biol.* 100: 1736–52

Anderson, P., Slorach, S. A., Uvnas, B. 1973. Sequential exocytosis of storage granules during antigen-inducing histamine release from sensitized rat mast cells in vitro. An electron microscopic study. *Acta Physiol. Scand.* 88: 359–72

Anderson, R. G. W., Falck, J. R., Goldstein, J. L., Brown, M. S. 1984. Visualization of

284   BURGESS & KELLY

acidic organelles in intact cells by electron microscopy. *Proc. Natl. Acad. Sci. USA* 81: 4838–42

Anderson, R. G. W., Pathak, R. K. 1985. Vesicles and cisternae in the *trans* Golgi apparatus of human fibroblasts are acidic compartments. *Cell* 40: 635–43

Arnheiter, H., Dubois-Dalcq, M., Lazzarini, R. A. 1984. Direct visualization of protein transport and processing in the living cell by microinjection of specific antibodies. *Cell* 39: 99–109

Arvan, P., Castle, J. D. 1987. Phasic release of newly synthesized secretory proteins in the unstimulated rat exocrine pancreas. *J. Cell Biol.* 104: 243–52

Bader, M. F., Trifaro, J. M., Langley, O. K., Thierse, D., Aunis, D. 1986. Secretory cell actin-binding proteins: Identification of a gelsolinlike protein in chromaffin cells. *J. Cell Biol.* 102: 636–46

Bainton, D. F. 1973. Sequential degranulation of the two types of polymorphonuclear leukocyte granules during phagocytosis of microorganisms. *J. Cell Biol.* 58: 249–64

Bainton, D. F., Farquhar, M. G. 1966. Origin of granules in polymorphonuclear leukocytes. *J. Cell Biol.* 28: 277–301

Bainton, D. F., Farquhar, M. G. 1968a. Differences in enzyme content of azurophil and specific granules of PMN leukocytes I. Histochemical staining of bone marrow smears. *J. Cell Biol.* 39: 286–98

Bainton, D. F., Farquhar, M. G. 1968b. Differences in enzyme content of azurophil and specific granules of polymorphonuclear leukocytes II. Cytochemistry and electron microscopy of bone marrow cells. *J. Cell Biol.* 39: 299–317

Balch, W. E., Elliott, M. M., Keller, D. S. 1986. ATP-coupled transport of vesicular stomatitis virus G protein between the endoplasmic reticulum and the Golgi. *J. Biol. Chem.* 261: 14681–89

Ball, E. H., Singer, S. J. 1982. Mitochondria are associated with microtubules and not with intermediate filaments in cultured fibroblasts. *Proc. Natl. Acad. Sci. USA* 79: 123–26

Bassetti, M., Spada, A., Arosio, M., Vallar, L., Brina, M., Giannattasio, G. 1986. Morphological studies on mixed growth hormone (GH)- and prolactin (PRL)-secreting human pituitary adenomas. Coexistence of GH and PRL in the same secretory granule. *J. Clin. Endocrinol. Metab.* 62: 1093–1100

Beams, H. W., King, R. L. 1933. The Golgi apparatus in the developing tooth with special reference to polarity. *Anat. Rec.* 57: 29–39

Beaudoin, A. R., Vachereau, A., St.-Jean, P.

1983. Evidence that amylase is released from two distinct pools of secretory proteins in the pancreas. *Biochim. Biophys. Acta* 757: 302–5

Beaudoin, A. R., St.-Jean, P., Vachereau, A. 1986. Asynchronism between amylase secretion and packaging in the zymogen granules of pig pancreas. *Pancreas* 1: 2–4

Bendayan, M. 1983. Ultrastructural localization of actin in muscle, epithelial and secretory cells by applying the protein A–gold immunocytochemical technique. *Histochem. J.* 15: 39–58

Bendayan, M. 1984. Concentration of amylase along its secretory pathway in the pancreatic acinar cell as revealed by high resolution immunocytochemistry. *Histochem. J.* 16: 85–108

Bendayan, M. 1985. Morphometrical and immunocytochemical characterization of peri-insular and tele-insular acinar cells in the rat pancreas. *Eur. J. Cell Biol.* 36: 263–68

Bendayan, M., Ito, S. 1979. Immunohistochemical localization of exocrine enzymes in normal rat pancreas. *J. Histochem. Cytochem.* 27: 1029–34

Bendayan, M., Roth, J., Perrelet, A., Orci, L. 1980. Quantitative immunocytochemical localization of pancreatic secretory proteins in subcellular compartments of the rat acinar cell. *J. Histochem. Cytochem.* 28: 149–60

Bennett, G., Carlet, E., Wild, G., Parsons, S. 1984. Influence of colchicine and vinblastine on the intracellular migration of secretory and membrane glycoproteins: III. Inhibition of intracellular migration of membrane glycoproteins in rat intestinal columnar cells and hepatocytes as visualized by light and electron microscope radioautography after ^{3}H-fucose injection. *Am. J. Anat.* 170: 545–66

Bergen, L. G., Kuriyama, R., Borisy, G. G. 1980. Polarity of microtubules nucleated by centrosomes and chromosomes of CHO cells in vitro. *J. Cell Biol.* 84: 151–59

Bhisey, A. D., Freed, J. J. 1971. Altered movement of endosomes in colchicine-treated cultured macrophages. *Exp. Cell Res.* 64: 430–38

Blank, G. S., Brodsky, F. M. 1986. Site-specific disruption of clathrin assembly produces novel structures. *EMBO J.* 5: 2087–95

Blobel, G. 1980. Intracellular protein topogenesis. *Proc. Natl. Acad. Sci. USA* 77: 1496–1500

Blok, T., Ginsel, L. A., Mulder-Stapel, A. A., Onderwater, J. J., Daems, W. J. 1981. The effect of colchicine on the intracellular transport of ^{3}H-fucose-labeled glycoproteins in the absorptive cells of cultured human small intestine tissue. *Cell Tissue Res.* 215: 1–12

Bole, D. G., Hendershot, L. M., Kearney, J. F. 1986. Posttranslational association of immunoglobulin heavy chain binding protein with nascent heavy chains in nonsecreting and secreting hybridomas. *J. Cell Biol.* 102: 1558–66

Bonifacino, J. S., Perez, P., Klausner, R. D., Sandoval, I. V. 1986. Study of the transit of an integral membrane protein from secretory granules through the plasma membrane of secreting rat basophilic leukemia cells using a specific monoclonal antibody. *J. Cell Biol.* 102: 516–22

Brady, J. T. 1985. A novel brain ATPase with properties expected for the fast axonal transport motor. *Nature* 317: 73–75

Breckenridge, L. J., Almers, W. 1987. Final steps in exocytosis observed in a cell with giant secretory granules. *Proc. Natl. Acad. Sci. USA* 84: 1945–49

Buckley, K. B., Kelly, R. B. 1985. Identification of a transmembrane glycoprotein specific for secretory vesicles of neural and endocrine cells. *J. Cell Biol.* 100: 1284–94

Burgess, T. L., Kelly, R. B. 1984. Sorting and secretion of adrenocorticotropin in a pituitary tumor cell line after perturbation of the level of a secretory granule-specific proteoglycan. *J. Cell Biol.* 99: 2223–30

Burgess, T. L., Craik, C. S., Kelly, R. B. 1985. The exocrine protein trypsinogen is targeted into the secretory granules of an endocrine cell line: Studies by gene transfer. *J. Cell Biol.* 101: 639–45

Burgess, T. L., Craik, C. S., Matsuuchi, L., Kelly, R. B. 1987. In vitro mutagenesis of pretrypsinogen: The role of the N-terminus in intracellular protein targeting to dense core secretory granules. *J. Cell Biol.* In press

Burgoyne, R. D., Gelsow, M. J., Barron, J. 1982. Dissection of stages in exocytosis in the adrenal chromaffin cell with use of trifluoperazine. *Proc. R. Soc. London Ser. B* 216: 111–15

Burgoyne, R. D., Cheek, T. R., Norman, K. M. 1986. Identification of a secretory granule-binding protein as caldesmon. *Nature* 319: 68–70

Burton, P. R. 1986. Ultrastructure of the olfactory neuron of the bullfrog: The dendrite and its microtubules. *J. Comp. Neurol.* 242: 147–60

Busson-Mabillot, S., Chambaut-Guerin, A.-M., Ovtracht, L., Muller, P., Rossignol, B. 1982. Microtubules and protein secretion in rat lacrimal glands: Localization of short-term effects of colchicine on the secretory process. *J. Cell Biol.* 95: 105–17

Caplan, M. J., Stow, J. L., Newman, A. P., Madri, J. A., Anderson, H. C., et al. 1986. Sorting of newly synthesized secretory and lysosomal proteins in polarized MDCK cells. *J. Cell Biol.* 103: 8a

Carson, M., Weber, A., Zigmond, S. H. 1986. An actin-nucleating activity in polymorphonuclear leukocytes is modulated by chemotactic peptides. *J. Cell Biol.* 103: 2707–14

Cheek, T. R., Burgoyne, R. D. 1986. Nicotine-evoked disassembly of cortical actin filaments in adrenal chromaffin cells. *FEBS Lett.* 207: 110–14

Clemente, F., Meldolesi, J. 1975. Calcium and pancreatic secretion. I. Subcellular distribution of calcium and magnesium in the exocrine pancreas of the guinea pig. *J. Cell Biol.* 65: 88–102

Collot, M., Louvard, D., Singer, S. J. 1984. Lysosomes are associated with microtubules and not with intermediate filaments in cultured fibroblasts. *Proc. Natl. Acad. Sci. USA* 81: 788–92

Comb, M., Liston, D., Martin, M., Rosen, H., Herbert, E. 1985. Expression of the human proenkephalin gene in mouse pituitary cells: Accurate and efficient mRNA production and proteolytic processing. *EMBO J.* 4: 3115–22

Copeland, C. S., Doms, R. W., Bolzau, E. M., Webster, R. G., Helenius, A. 1986. Assembly of influenza hemagglutinin trimers and its role in intracellular transport. *J. Cell Biol.* 103: 1179–91

Dagorn, J. C., Sahel, J., Sarles, H. 1977. Nonparallel secretion of enzymes in human duodenal juice and pure pancreatic juice collected by endoscopic retrograde catheterization of the papilla. *Gastroenterology* 73: 42–45

Dahlstrom, A., Larsson, P. A., Carlson, S. S., Booj, S. 1985. Localization and axonal transport of immunoreactive cholinergic organelles in rat motor neurons—an immunofluorescent study. *Neuroscience* 14: 607–25

DeCamilli, P., Moretti, M., Donini, S. D., Walter, U., Lohmann, S. M. 1986. Heterogeneous distribution of the cAMP receptor protein RII in the nervous system: Evidence for its intracellular accumulation on microtubules, microtubule-organizing centers and in the area of the Golgi complex. *J. Cell Biol.* 103: 189–203

DeLisle, R. C., Williams, J. A. 1986. Regulation of membrane fusion in secretory exocytosis. *Ann. Rev. Physiol.* 48: 225–38

Dewald, B., Bretz, U., Baggiolini, M. 1982. Release of gelatinase from a novel secretory compartment of human neutrophils. *J. Clin. Invest.* 70: 518–25

Doxsey, S., Helenius, A., Blank, G., Brodsky, F. 1986. Inhibition of endocytosis by anti-clathrin antibodies. *J. Cell Biol.* 103: 53a (Abstr. 196)

Doyle, C., Roth, M. G., Sambrook, J., Gething, M.-J. 1985. Mutations in the cytoplasmic domain of the influenza virus hemagglutinin affect different stages of intracellular transport. *J. Cell Biol.* 100: 704–14

Drenckhahn, D., Mannherz, H. G. 1983. Distribution of actin and the actin associated proteins, myosin, tropomyosin, alpha-actinin, vinculin and villin in rat and bovine exocrine glands. *Eur. J. Cell Biol.* 30: 167–76

Dunphy, W. G., Rothman, J. E. 1985. Compartmental organization of the Golgi stack. *Cell* 42: 13–21

Edwards, K., Fleischer, B., Dryburgh, H., Fleischer, S., Schreiber, G. 1976. The distribution of albumin precursor protein and albumin in liver. *Biochem. Biophys. Res. Commun.* 72: 310–18

Farquhar, M. G. 1978. Recovery of surface membrane in anterior pituitary cells. Variations in traffic detected with anionic and cationic ferritin. *J. Cell Biol.* 77: R35–R42

Farquhar, M. G. 1983. Multiple pathways of exocytosis, endocytosis and membrane recycling: Validation of a Golgi route. *Fed. Proc.* 42: 2407–13

Farquhar, M. G. 1985. Progress in unraveling pathways of Golgi traffic. *Ann. Rev. Cell Biol.* 1: 447–88

Farquhar, M. G., Palade, G. E. 1981. The Golgi apparatus (complex)—(1954–1981)—from artifact to center stage. *J. Cell Biol.* 91: 77s–103s

Fitting, T., Kabat, D. 1982. Evidence for a glycoprotein "signal" involved in transport between subcellular organelles. Two membrane glycoproteins encoded by murine leukemia virus reach the cell surface at different rates. *J. Biol. Chem.* 257: 14011–17

Freed, J. J., Lebowitz, M. M. 1970. The association of a class of saltatory movements with microtubules in cultured cells. *J. Cell Biol.* 45: 334–54

Fries, E., Gustafsson, L., Peterson, P. 1984. Four proteins synthesized by hepatocytes are transported from endoplasmic reticulum to Golgi complex at different rates. *EMBO J.* 3: 147–52

Fumagalli, G., Zanini, A. 1985. In cow anterior pituitary, growth hormone and prolactin can be packed in separate granules of the same cell. *J. Cell Biol.* 100:

2019–24

Gainer, H., Sarne, Y., Brownstein, M. J. 1977. Neurophysin biosynthesis: Conversion of a putative precursor during axonal transport. *Science* 195: 1354–56

Gennaro, R., Dewald, B., Horisberger, U., Gubler, H. U., Baggiolini, M. 1983. A novel type of cytoplasmic granule in bovine neutrophils. *J. Cell Biol.* 96: 1651–61

Getchell, T. V. 1986. Functional properties of vertebrate olfactory receptor neurons. *Physiol. Rev.* 66: 772–818

Gething, M.-J., McCammon, K., Sambrook, J. 1986. Expression of wild type and mutant forms of influenza hemagglutinin: The role of folding in intracellular transport. *Cell* 46: 939–50

Geuze, H. J., Slot, J. W., Strous, J. G. A. M., Hasilik, A., Von Figura, K. 1985. Possible pathways for lysosomal enzyme delivery. *J. Cell Biol.* 101: 2253–62

Giannattasio, G., Zanini, A., Rosa, P., Meldolesi, J., Margolis, R. K., Margolis, R. U. 1980. Molecular organization of prolactin granules. III. Intracellular transport of sulfated glycosaminoglycans and glycoproteins of the bovine prolactin granule matrix. *J. Cell Biol.* 86: 273–79

Gilbert, S. P., Sloboda, R. D. 1984. Bidirectional transport of fluorescently labeled vesicles introduced into extruded axoplasm of squid loligo paelei. *J. Cell Biol.* 99: 445–52

Gilbert, S. P., Sloboda, R. D. 1986. Identification of a MAP2-like ATP-binding protein associated with axoplasmic vesicles that translocate on isolated microtubules. *J. Cell Biol.* 103: 947–56

Gold, G., Gishizky, M. L., Grodsky, G. M. 1982. Evidence that glucose "marks" B cells resulting in preferential release of newly synthesized insulin. *Science* 218: 56–58

Gold, G., Gishizky, M. L., Chick, W. L., Grodsky, G. M. 1984. Contrasting patterns of insulin biosynthesis, compartmental storage and secretion. *Diabetes* 33: 556–60

Gottlieb, T. A., Beaudry, G., Rizzolo, L., Colman, A., Rindler, M. J., et al. 1986. Secretion of endogenous and exogenous proteins from polarized MDCK monolayers. *Proc. Natl. Acad. Sci. USA* 83: 2100–4

Green, R., Shields, D. 1984. Somatostatin discriminates between the intracellular pathways of secretory and membrane proteins. *J. Cell Biol.* 99: 97–104

Griffiths, G., Pfeiffer, S., Simons, K., Matlin, K. 1985. Exit of newly synthesized membrane proteins from the *trans* cisterna of the Golgi complex to the plasma membrane. *J. Cell Biol.* 101: 949–64

Griffiths, G., Simons, K. 1986. The *trans* Golgi network: Sorting at the exit site of the Golgi complex. *Science* 234: 438–43

Gumbiner, B., Kelly, R. B. 1981. Secretory granules of an anterior pituitary cell line, AtT-20, contain only mature forms of corticotropin and beta-lipotropin. *Proc. Natl. Acad. Sci. USA* 78: 318–22

Gumbiner, B., Kelly, R. B. 1982. Two distinct intracellular pathways transport secretory and membrane glycoproteins to the surface of pituitary tumor cells. *Cell* 28: 51–59

Hammerschlag, R. 1983. How do neuronal proteins know where they are going? Speculations on the role of molecular address markers. *Dev. Neurosci.* 6: 2–17

Hayden, J., Allen, R. D. 1984. Detection of single microtubules in living cells: Particle transport can occur in both directions along the same microtubule. *J. Cell Biol.* 98: 1785–93

Hearn, S. A., Silver, M. M., Sholdice, J. A. 1985. Immunoelectron microscopic labeling of immunoglobulin in plasma cells after osmium fixation and epoxy embedding. *J. Histochem. Cytochem.* 33: 1212–18

Hellerman, J. G., Cone, R. C., Potts, J. T. Jr., Rich, A., Mulligan, R. C., Kronenberg, H. M. 1984. Secretion of human parathyroid hormone from rat pituitary cells infected with a recombinant retrovirus encoding preproparathyroid hormone. *Proc. Natl. Acad. Sci. USA* 81: 5340–44

Henkart, P. A., Millard, P. J., Reynolds, C. W., Henkart, M. P. 1984. Cytolytic activity of purified cytoplasmic granules from cytotoxic rat large granular lymphocyte tumors. *J. Exp. Med.* 160: 75–93

Henkart, P. A., Yue, C. C., Yang, J., Rosenberg, S. A. 1986. Cytolytic and biochemical properties of cytoplasmic granules of murine lymphokine-activated killer cells. *J. Immunol.* 137: 2611–17

Herman, B., Albertini, D. F. 1984. A time-lapse video image intensification analysis of cytoplasmic organelle movements during endosome translocation. *J. Cell Biol.* 98: 565–76

Herman, L., Sato, T., Hales, N. 1973. The electron microscopic localization of cations to pancreatic islets of Langerhans and their possible role in insulin secretion. *J. Ultrastruct. Res.* 42: 298–311

Herzog, V. 1981. Endocytosis in secretory cells. *Philos. Trans. R. Soc. London Ser. B* 296: 67–72

Holcomb, C. L., Hansen, W. B., Etcheverry, T. E., Schekman, R. 1987. Plasma membrane protein intermediates are present in the secretory vesicles of yeast. *J. Cell. Biochem.* Suppl. 11A, p. 247

Horwich, A. L., Kalousek, F., Melman, I., Rosenberg, L. E. 1985. A leader peptide is sufficient to direct mitochondrial import of a chimeric protein. *EMBO J.* 4: 1129–35

Howell, S. L., Parry, D. G., Taylor, K. W. 1965. Secretion of newly synthesized insulin in vitro. *Nature* 208: 487

Hurt, E. C., Pesold-Hurt, B., Shatz, G. 1984. The amino terminal region of an imported mitochondrial precursor polypeptide can direct cytoplasmic dihydrofolate reductase into the mitochondrial matrix. *EMBO J.* 3: 3149–56

Ikehara, Y., Oda, K., Kato, K. 1976. Conversion of proalbumin into serum albumin in the secretory vesicles of rat liver. *Biochem. Biophys. Res. Commun.* 72: 319–26

Inoue, K., Kurosumi, K. 1984. Ultrastructural immunocytochemical localization of LH and FSH in the pituitary of the untreated male rat. *Cell Tissue Res.* 235: 77–83

Jamieson, J. D., Palade, G. E. 1968. Intracellular transport of secretory proteins in the pancreatic exocrine cell. *J. Cell Biol.* 39: 589–603

Jamieson, J. D., Palade, G. E. 1971a. Condensing vacuole conversion and zymogen granule discharge in pancreatic exocrine cells: Metabolic studies. *J. Cell Biol.* 48: 503–22

Jamieson, J. D., Palade, G. E. 1971b. Synthesis, intracellular transport, and discharge of secretory proteins in stimulated pancreatic exocrine cells. *J. Cell Biol.* 50: 135–58

Johnson, L. M., Bankaitis, V. A., Emr, S. D. 1987. Distinct sequence determinants direct intracellular sorting and modification of a yeast vacuolar protease. *Cell* 48: 875–85

Julius, D., Schekman, R., Thorner, R. 1984. Glycosylation and processing of preproalpha-factor through the yeast secretory pathway. *Cell* 36: 309–18

Kawai, K., Ipp, E., Orci, L., Perrelet, A., Unger, R. H. 1982. Circulating somatostatin acts on islets of Langerhans by way of a somatostatin poor compartment. *Science* 218: 477–78

Kelly, R. B. 1985. Pathways of protein secretion in eukaryotes. *Science* 230: 25–32

Kelly, R. B., Buckley, K. B., Burgess, T. L., Carlson, S. S., Caroni, P., et al. 1983. Membrane traffic in neurons and peptide-secreting cells. *Cold Spring Harbor Symp. Quant. Biol.* 48: 697–705

Klein, R. L., Lagercrantz, H., Zimmerman, H. eds. 1982. *Neurotransmitter Vesicles.* New York: Academic. 384 pp.

288 BURGESS & KELLY

Knight, D. E., Baker, P. F. 1983. Stimulus-secretion coupling in isolated bovine adrenal medullary cells. *Q. J. Exp. Physiol.* 68: 123–42

Kondor-Koch, C., Bravo, R., Fuller, S. D., Cutler, D., Garoff, H. 1985. Protein secretion in the polarized epithelial cell line MDCK. *Cell* 43: 297–306

Koonce, M. P., Schliwa, M. 1985. Bidirectional organelle transport can occur in cell processes that contain single microtubules. *J. Cell Biol.* 100: 322–26

Kornfield, S. 1986. Trafficking of lysosomal enzymes in normal and disease states. *J. Clin. Invest.* 77: 1–6

Kreis, T. E., Lodish, H. F. 1986. Oligomerization is essential for transport of vesicular stomatitis viral glycoprotein to the cell surface. *Cell* 46: 929–37

Kupfer, A., Singer, S. J., Dennert, G. 1986a. On the mechanism of unidirectional killing in mixtures of two cytotoxic T lymphocytes. Unidirectional polarization of cytoplasmic organelles and the membrane-associated cytoskeleton in the effector cell. *J. Exp. Med.* 163: 489–98

Kupfer, A., Swain, S. L., Janeway, C. A. Jr., Singer, S. J. 1986b. The specific direct interaction of helper T cells and antigen-presenting B cells. *Proc. Natl. Acad. Sci. USA* 83: 6080–83

Kurth, M. C., Bryan, J. 1984. Platelet activation induces the formation of a stable gelsolin-actin complex from monomeric gelsolin. *J. Biol. Chem.* 259: 7473–79

Kuznetsov, S. A., Gelfand, V. I. 1986. Bovine brain kinesin is a microtubule activated ATPase. *Proc. Natl. Acad. Sci. USA* 83: 8530–39

Laduron, P. 1980. Axoplasmic transport of muscarinic receptors. *Nature* 286: 287–88

Laduron, P. M. 1984. Axonal transport of muscarinic receptors in vesicles containing noradrenaline and dopamine beta hydroxylase. *FEBS Lett.* 165: 128–32

Lagercrantz, H. 1976. On the composition and function of large dense cored vesicles in sympathetic nerves. *Neuroscience* 1: 81–92

Langley, O. K., Perrin, D., Aunis, D. 1986. Alpha-fodrin in the adrenal gland: Localization by immunoelectron microscopy. *J. Histochem. Cytochem.* 34: 517–25

Laub, O., Rutter, W. J. 1983. Expression of the human insulin gene and cDNA in a heterologous mammalian system. *J. Biol. Chem.* 258: 6043–50

Levin, B. E. 1982. Presynaptic location and axonal transport of β1-adrenoreceptors. *Science* 217: 555–57

Lew, P. D., Monod, A., Waldvogel, F. A., Dewald, B., Baggiolini, M., Pozzan, T. 1986. Quantitative analysis of the cytosolic free calcium dependency of exocytosis from three subcellular compartments in intact human neutrophils. *J. Cell Biol.* 102: 2197–2204

Lodish, H. F., Kong, N. 1984. Glucose removal from N-linked oligosaccharides is required for efficient maturation of certain secretory glycoproteins from the rough endoplasmic reticulum to the Golgi complex. *J. Biol. Chem.* 98: 1720–29

Lodish, H. F., Kong, N., Snider, M., Strous, G. J. A. M. 1983. Hepatoma secretory proteins migrate from the RER to Golgi at characteristic rates. *Nature* 304: 80–83

Lodish, H. F., Kong, N., Hirani, S., Rasmussen, J. 1987. A vesicular intermediate in the transport of hepatoma secretory proteins from the rough endoplasmic reticulum to the Golgi complex. *J. Cell Biol.* 104: 221–30

Lombardi, T., Montesano, R., Wohlwend, A., Amherdt, M., Vassalli, J. D., Orci, L. 1985. Evidence for polarization of plasma membrane domains in pancreatic endocrine cells. *Nature* 313: 694–96

Lombardi, T., Montesano, R., Orci, L. 1986. Loss of polarization of plasma membrane domains in transformed pancreatic endocrine cell lines. *Endocrinology* 119: 502–7

Lundberg, J. M., Hokfelt, T. 1983. Coexistence of peptides and classical neurotransmitters. *Trends Neurosci.* 6: 325–33

Lundberg, J. M., Fahrenkug, J., Brimijoin, S. 1981. Characterization of the axonal transport of vasoactive intestinal polypeptide in nerves of the rat. *Acta Physiol. Scand.* 112: 427–36

Lynch, T. J., Taylor, J. D., Tchen, T. T. 1986a. Regulation of pigment organelle translocation. I. Phosphorylation of the organelle-associated protein p57. *J. Biol. Chem.* 261: 4204–11

Lynch, T. J., Wu, B. Y., Taylor, J. D., Tchen, T. T. 1986b. Regulation of pigment organelle translocation. II. Participation of a cAMP-dependent protein kinase. *J. Biol. Chem.* 261: 4212–16

Malaisse-Lagae, F., Ravazzola, M., Robberecht, P., Vandermeers, A., Malaisse, W. J., Orci, L. 1975. Exocrine pancreas: Evidence for topographic partition of secretory function. *Science* 190: 795–97

Matlin, K. S. 1986a. The sorting of proteins to the plasma membrane in epithelial cells. *J. Cell Biol.* 103: 2565–68

Matlin, K. S. 1986b. Ammonium chloride slows transport of the influenza hemagglutinin but does not lead to missorting in MDCK cells. *J. Biol. Chem.* 261: 15172–78

Matsuuchi, L., Kelly, R. B. 1987. Constitutive secretion of immunoglobulin chains by the endocrine cell line AtT-20: Association of non-secreted chains with a 75 kd protein. In preparation

Meldolesi, J. 1974. Dynamics of cytoplasmic membranes in guinea pig pancreatic acinar cells. Synthesis and turnover of membrane proteins. *J. Cell Biol.* 61: 1–13

Mellman, I., Fuchs, R., Helenius, A. 1986. Acidification of the endocytic and exocytic pathways. *Ann. Rev. Biochem.* 55: 663–700

Mellman, I., Howe, C., Helenius, A. 1987. The control of membrane traffic on the endocytic pathway. *Curr. Top. Membr. Transp.* 29: 255–87

Merisko, E. M., Farquhar, M. G., Palade, G. E. 1986b. Redistribution of clathrin heavy and light chains in anoxic pancreatic acinar cells. *Pancreas* 1: 110–23

Merisko, E. M., Fletcher, M., Palade, G. E. 1986a. The reorganization of the Golgi complex in anoxic pancreatic acinar cells. *Pancreas* 1: 95–109

Millard, P. J., Henkart, M. P., Reynolds, C. W., Henkart, P. A. 1984. Purification and properties of cytoplasmic granules from cytotoxic rat LGL tumors. *J. Immunol.* 132: 3197–3204

Miller, R. H., Lasek, R. J., Katz, M. J. 1987. Preferred microtubules for vesicle transport in lobster axons. *Science* 235: 220–22

Moore, H.-P. H., Kelly, R. B. 1985. Secretory protein targeting in a pituitary cell line: Differential transport of foreign secretory proteins to distinct secretory pathways. *J. Cell Biol.* 101: 1773–81

Moore, H.-P. H., Kelly, R. B. 1986. Rerouting of a secretory protein by fusion with human growth hormone sequences. *Nature* 321: 443–46

Moore, H.-P. H., Gumbiner, B., Kelly, R. B. 1983a. Chloroquine diverts ACTH from a regulated to a constitutive secretory pathway in AtT-20 cells. *Nature* 302: 434–36

Moore, H.-P. H., Walker, M. D., Lee, F., Kelly, R. B. 1983b. Expressing a human proinsulin cDNA in a mouse ACTH-secreting cell. Intracellular storage, proteolytic processing and secretion on stimulation. *Cell* 35: 531–38

Mostov, K. E., Deitcher, D. L. 1986. Polymeric immunoglobulin receptor expressed in MDCK cells transcytoses IgA. *Cell* 46: 613–21

Mroz, E. A., Lechene, C. 1986. Pancreatic zymogen granules differ markedly in protein composition. *Science* 232: 871–73

Mullock, B. M., Jones, R. S., Peppard, J., Hinton, R. H. 1980. Effect of colchicine in

the transfer of IgA across hepatocytes into bile in intact perfused rat livers. *FEBS Lett.* 120: 278–82

Navone, F., Jahn, R., DiGioia, G., Stukenbrok, H., Greengard, P., DeCamilli, P. 1986. Protein p38: An integral membrane protein specific for small vesicles of neurons and neuroendocrine cells. *J. Cell Biol.* 103: 2511–27

Neher, E., Marty, A. 1982. Discrete changes of cell membrane capacitance observed under conditions of enhanced secretion in bovine adrenal chromaffin cells. *Proc. Natl. Acad. Sci. USA* 79: 6712–16

Nelson, T. Y., Boyd, A. E. III 1985. Gelsolin, a Ca^{++}-dependent actin-binding protein in a hamster insulin-secreting cell line. *J. Clin. Invest.* 75: 1015–22

Nelson, W. J., Veshnock, P. J. 1986. Dynamics of membrane-skeleton (fodrin) organization during development of polarity in Madin-Darby canine kidney epithelial cells. *J. Cell Biol.* 103: 1751–65

Nickerson, S. C., Smith, J. J., Keenen, T. W. 1980. Role of microtubules in milk secretion. Action of colchicine on microtubules and exocytosis of secretory vesicles in rat mammary epithelial cells. *Cell Tissue Res.* 207: 361–76

Nigg, E. A., Schafer, G., Hilz, H., Eppenberger, H. M. 1985. cAMP dependent protein kinase type II is associated with the Golgi complex and with centrosomes. *Cell* 45: 1039–51

Noe, B. D., Moran, M. N. 1984. Association of newly synthesized islet prohormones with intracellular membranes. *J. Cell Biol.* 99: 418–24

O'Halloran, T., Beckerle, M. C., Burridge, K. 1985. Identification of talin as a major cytoplasmic protein implicated in platelet activation. *Nature* 317: 449–51

Orci, L., Gabbay, K. H., Malaisse, W. J. 1972. Pancreatic beta-cell web: Its possible role in insulin secretion. *Science* 175: 1128–30

Orci, L., Like, A. A., Amherdt, M., Blondel, B., Kanazawa, Y., et al. 1973. Monolayer cell culture of neonatal rat pancreas: An ultrastructural and biochemical study of functioning endocrine cells. *J. Ultrastruct. Res.* 43: 270–97

Orci, L., Halban, P., Amherdt, M., Ravazzola, M., Vassalli, J.-D., Perrelet, A. 1984a. A clathrin-coated, Golgi-related compartment of the insulin secreting cell accumulates proinsulin in the presence of monensin. *Cell* 39: 39–47

Orci, L., Ravazzola, M., Perrelet, A. 1984b. (Pro)insulin associates with Golgi membranes of pancreatic B cells. *Proc. Natl. Acad. Sci. USA* 81: 6743–46

Orci, L., Ravazzola, M., Amherdt, M.,

Madsen, O., Vassalli, J.-D., Perrelet, A. 1985. Direct identification of prohormone conversion site in insulin-secreting cells. *Cell* 42: 671–81

Orci, L., Glick, B. S., Rothman, J. E. 1986a. A new type of coated vesicular carrier that appears not to contain clathrin: Its possible role in protein transport within the Golgi stack. *Cell* 46: 171–84

Orci, L., Ravazzola, M., Amherdt, M., Brown, D., Perrelet, A. 1986b. Transport of horseradish peroxidase from the cell surface to the Golgi in insulin-secreting cells: Preferential labeling of cisternae located in an intermediate position in the stack. *EMBO J.* 5: 2097–2101

Ornitz, D. M., Palmiter, R. D., Hammer, R. E., Brinster, R. L., Swift, G. H., Mac-Donald, R. J. 1985. Specific expression of an elastase-human growth hormone fusion gene in pancreatic acinar cells of transgenic mice. *Nature* 313: 600–2

Ortiz, C. L., Rocha, V. 1986. Is casein output from mammary epithelia under "standard" stimulation-secretion control? *J. Cell Biol.* 103: 460a

Ottosen, P. D., Courtoy, P. J., Farquhar, M. G. 1980. Pathways followed by membrane recovered from the surface of plasma cells and myeloma cells. *J. Exp. Med.* 152: 1–19

Palade, G. 1975. Intracellular aspects of the process of protein synthesis. *Science* 189: 347–58

Pastan, I. H., Willingham, M. C. 1981. Journey to the center of the cell: Role of the receptosome. *Science* 214: 504–9

Patzak, A., Winkler, H. 1986. Exocytotic exposure and recycling of membrane antigens of chromaffin granules: Ultrastructural evaluation after immuno-labeling. *J. Cell Biol.* 102: 510–15

Payne, G., Schekman, R. 1985. A test of clathrin function in protein secretion and cell growth. *Science* 230: 1009–14

Peng, I., Binder, L. I., Black, M. M. 1986. Biochemical and immunological analyses of cytoskeletal domains of neurons. *J. Cell Biol.* 102: 252–62

Perrin, D., Aunis, D. 1985. Reorganization of alpha-fodrin induced by stimulation in secretory cells. *Nature* 315: 589–92

Pfeffer, S. R., Rothman, J. E. 1987. Biosynthetic protein transport and sorting by the endoplasmic reticulum and Golgi. *Ann. Rev. Biochem.* 56: 829–52

Pinto, M., Robine-Leon, S., Appay, M. D., Kedinger, M., Triadou, N., et al. 1983. Enterocyte-like differentiation and polarization of the human colon carcinoma cell line Caco-2 in culture. *Biol. Cell* 47: 323–30

Posthuma, G., Slot, J. W., Geuze, H. J. 1986.

A quantitative immuno-electron-microscopic study of amylase and chymotrypsin in peri- and tele-insular cells of the rat exocrine pancreas. *J. Histochem. Cytochem.* 34: 203–7

Rasool, C. G., Schwartz, A. L., Bollinger, J. A., Reichlin, S., Bradley, W. G. 1981. Immunoreactive somatostatin distribution and axonal transport in rat peripheral nerve. *Endocrinology* 198: 996–1001

Ravazzola, M., Orci, L. 1980. Glucagon and glicentin immunoreactivity are topologically segregated in the alpha granule of the human pancreatic A cell. *Nature* 284: 66–67

Reggio, H., Dagorn, J. C. 1978. Ionic interactions between bovine chymotrypsinogen A and chondroitin sulfate A.B.C. *J. Cell Biol.* 98: 951–57

Rhodes, C. J., Halban, P. A. 1987. Newly synthesized proinsulin-insulin as well as stored insulin are released from pancreatic B-cells uniquely via a regulated, not constitutive, pathway. *J. Cell Biol.* In press

Rindler, M. J., Ivanov, I. E., Sabatini, D. D. 1987a. Microtubule-acting drugs lead to the non-polarized delivery of the influenza hemagglutinin to the cell surface of polarized Madin-Darby canine kidney cells. *J. Cell Biol.* 104: 231–41

Rindler, M. J., Kayden, H. J., Traber, M. G. 1987b. Secretion of apolipoproteins from cultured epithelial cells. *J. Cell Biochem.* Suppl. 11A, p. 282

Rogalski, A. A., Singer, S. 1984. Association of elements of the Golgi apparatus with microtubules. *J. Cell Biol.* 99: 1092–1100

Rogalski, A. A., Bergman, T. E., Singer, S. 1984. Effect of microtubule assembly status on the intracellular processing and surface expression of an integral protein of the plasma membrane. *J. Cell Biol.* 99: 1101–9

Rojas, E., Pollard, H. B., Heldman, R. 1986. Real-time measurements of acetylcholine-induced release of ATP from bovine medullary chromaffin cells. *FEBS Lett.* 185: 323–27

Rothman, J. E., Miller, R. L., Urbani, L. J. 1984. Intercompartmental transport in the Golgi complex is a dissociative process: Facile transfer of membrane protein between two Golgi populations. *J. Cell Biol.* 99: 260–71

Rothman, S. S. 1975. Protein transport by the pancreas. *Science* 190: 747–53

Rothman, S. S., Wilking, H. 1978. Differential rates of digestive enzyme transport in the presence of cholecystokinin-pancreozymin. *J. Biol. Chem.* 253: 3543–49

Rozdzial, M. M., Haimo, L. T. 1986. Reactivated melanophore motility: Differential regulation and nucleotide requirements of bidirectional pigment granule transport. *J. Cell Biol.* 103: 2755–64

Ryser, J.-E., Rungger-Brandle, E., Chaponnier, C., Gabbiani, G., Vassalli, P. 1982. The area of attachment of cytotoxic T lymphocytes to their target cells shows high mobility and polarization of actin but not myosin. *J. Immunol.* 128: 1159–62

Salas, P. J. I., Misek, D. E., Vega-Salas, D. E., Gundersen, D., Cereijido, M., Rodriguez-Boulan, E. 1986. Microtubules and actin filaments are not critically involved in the biogenesis of epithelial cell surface polarity. *J. Cell Biol.* 102: 1853–67

Salpeter, M. M., Farquhar, M. G. 1981. High resolution analysis of the secretory pathway in mammotrophs of the rat anterior pituitary. *J. Cell Biol.* 91: 240–46

Saraste, J., Kuismanen, E. 1984. Pre- and post-Golgi vacuoles operate in the transport of Semliki Forest virus membrane glycoproteins to the cell surface. *Cell* 38: 535–49

Saraste, J., Palade, G. E., Farquhar, M. G. 1986. Temperature-sensitive steps in transport of secretory proteins through the Golgi complex in exocrine pancreatic cells. *Proc. Natl. Acad. Sci. USA* 83: 6425–29

Sawano, F., Ravazzola, M., Amherdt, M., Perrelet, A., Orci, L. 1986. Horseradish peroxidase uptake and crinophagy in insulin-secreting cells. *Exp. Cell Res.* 164: 174–82

Scheele, G., Tartakoff, A. 1985. Exit of nonglycosylated secretory proteins from the RER is asynchronous in the exocrine pancreas. *J. Biol. Chem.* 260: 926–31

Schnapp, B. J., Vale, R. D., Sheetz, M. P., Reese, T. S. 1985. Single microtubules from squid axoplasm support bidirectional movement of organelles. *Cell* 40: 455–62

Scholey, J. M., Porter, M. E., Grissom, P. M., McIntosh, J. R. 1985. Identification of kinesin in sea urchin eggs and evidence for its location in the mitotic spindle. *Nature* 318: 483–86

Schroer, T. A., Brady, S. T., Kelly, R. B. 1985. Fast axonal transport of foreign synaptic vesicles in squid axoplasm. *J. Cell Biol.* 101: 568–72

Schweitzer, E. S., Kelly, R. B. 1985. Selective packaging of human growth hormone into synaptic vesicles in a rat neuronal (PC12) cell line. *J. Cell Biol.* 101: 667–76

Simons, K., Fuller, S. D. 1985. Cell surface polarity in epithelia. *Ann. Rev. Cell Biol.* 1: 243–88

Singer, S. J., Kupfer, A. 1986. The directed migration of eukaryotic cells. *Ann. Rev. Cell Biol.* 2: 337–65

Slaby, F., Bryan, J. 1976. High uptake of *myo*-inositol by rat pancreatic tissue in vitro stimulates secretion. *J. Biol. Chem.* 251: 5078–86

Slaby, F., Farquhar, M. G. 1980. Characterization of rat somatotrophs and mammotroph secretory granules. Presence of sulfated molecules. *Mol. Cell. Endocrinol.* 18: 33–48

Snider, M. D., Rogers, O. C. 1986. Membrane traffic in animal cells: Cellular glycoproteins return to the site of Golgi mannosidase I. *J. Cell Biol.* 103: 265–75

Sporn, L. A., Marder, V. J., Wagner, D. D. 1986. Inducible secretion of large, biologically potent von Willebrand factor multimers. *Cell* 46: 185–90

Steinberg, R. M., Mellman, I. S., Muller, W. A., Cohn, Z. A. 1983. Endocytosis and the recycling of plasma membrane. *J. Cell Biol.* 96: 1–27

Stenberg, P. E., McEver, R. P., Shuman, M. A., Jacques, Y. V., Bainton, D. F. 1985. A platelet alpha-granule membrane protein (GMP-140) is expressed on the plasma membrane after activation. *J. Cell Biol.* 101: 880–86

Stevens, T. H., Rothman, J. H., Payne, G. S., Schekman, R. 1986. Gene dosage-dependent secretion of yeast vacuolar carboxypeptidase Y. *J. Cell Biol.* 102: 1551–57

Strous, G. J. A. M., Willemsen, R., van Kerkkof, P., Slot, P. W., Geuze, H. J., Lodish, H. F. 1983. Vesicular stomatitis virus glycoprotein, albumin, and transferrin are transported to the cell surface via the same Golgi vesicles. *J. Cell Biol.* 97: 1815–22

Strous, G. J. A. M., Du Maine, A., Zijderhand-Bleekemolen, J. E., Slot, J. W., Schwartz, A. L. 1985. Effect of lysosomotropic amines on the secretory pathway and on the recycling of the asialoglycoprotein receptor in human hepatoma cells. *J. Cell Biol.* 101: 531–39

Tartakoff, A. M. 1986. Temperature and energy dependence of secretory protein transport in the exocrine pancreas. *EMBO J.* 5: 1477–82

Tartakoff, A. M., Vassalli, P., Détraz, M. 1977. Plasma cell immunoglobulin secretion. *J. Exp. Med.* 146: 1332–45

Tartakoff, A. M., Vassalli, P., Détraz, M. 1978. Comparative studies of intracellular transport of secretory proteins. *J. Cell Biol.* 79: 694–707

Tassin, A. M., Maro, B., Bornens, M. 1985a. Fate of microtubule-organizing centers during myogenesis in vitro. *J. Cell Biol.* 100: 35–46

Tassin, A. M., Paintrand, M., Berger, E. G., Bornens, M. 1985b. The Golgi apparatus remains associated with microtubule organizing centers during myogenesis. *J. Cell Biol.* 101: 630–38

Taugner, R., Whalley, A., Angermuller, S., Buhrle, C. P., Hackenthal, E. 1985. Are the renin-containing granules of juxtaglomerular epitheloid cells modified lysosomes? *Cell Tissue Res.* 239: 575–87

Terasaki, M., Chen, L. B., Fujiwara, K. 1986. Microtubules and the endoplasmic reticulum are highly interdependent structures. *J. Cell Biol.* 103: 1557–68

Trelstad, R. L. 1970. The Golgi apparatus in chick corneal epithelium: Changes in intracellular position during development. *J. Cell Biol.* 45: 34–42

Tschopp, J., Esmon, P. C., Schekman, R. 1984. Defective plasma membrane assembly in yeast secretory mutants. *J. Bacteriol.* 160: 966–70

Tooze, J., Tooze, S. A. 1986. Clathrin-coated vesicular transport of secretory proteins during the formation of ACTH-containing secretory granules in AtT-20 cells. *J. Cell Biol.* 103: 839–50

Vale, R. D., Schnapp, B. J., Reese, T. S., Sheetz, M. P. 1985a. Organelle, bead and microtubule translocations promoted by soluble factors from the squid giant axon. *Cell* 40: 559–69

Vale, R. D., Reese, T. S., Sheetz, M. P. 1985b. Identification of a novel force-generating protein, kinesin, involved in microtubule-based motility. *Cell* 42: 39–50

Vale, R. D., Schnapp, B. J., Mitchison, T., Steuer, E., Reese, T. S., Sheetz, M. P. 1985c. Different axoplasmic proteins generate movement in opposite directions along microtubules in vitro. *Cell* 43: 623–32

Valls, L. A., Hunter, C. P., Rothman, J. H., Stevens, T. H. 1987. Protein sorting in yeast: Localization determinant of yeast vacuolar carboxypeptidase Y resides in the propeptide. *Cell* 48: 887–97

Van den Broeck, G., Timko, M. P., Kausch, A. P., Cashmore, A. R., Montagu, M., Herrera-Estrella, L. 1985. Targeting of a foreign protein into chloroplasts by fusion to the transit peptide from the small subunit of ribulose 1,5-bisphosphate carboxylase. *Nature* 313: 358–63

von Grafenstein, H., Roberts, C. S., Baker, P. F. 1986. Kinetic analysis of the triggered exocytosis/endocytosis secretory cycle in cultured bovine adrenal medullary cells. *J. Cell Biol.* 103: 2243–52

von Heijne, G. 1986. A new method for predicting signal sequence cleavage sites. *Nucl. Acids Res.* 14: 4683–90

von Wedel, R. J., Carlson, S. S., Kelly, R. B. 1981. Transfer of synaptic vesicle antigens to the presynaptic plasma membrane during exocytosis. *Proc. Natl. Acad. Sci. USA* 78: 1014–18

Wagner, D. D., Mayadas, T., Marder, V. J. 1986. Initial glycosylation and acidic pH in the Golgi apparatus are required for multimerization of von Willebrand factor. *J. Cell Biol.* 102: 1320–24

Walter, P., Lingappa, V. R. 1986. Mechanisms of protein translocation across the endoplasmic reticulum membrane. *Ann. Rev. Cell Biol.* 2: 499–516

Wang, E., Cross, R. K., Choppin, P. W. 1979. Involvement of microtubules and 10 nm filaments in the movement and positioning of nuclei in syncytia. *J. Cell Biol.* 83: 320–37

Wang, E., Goldman, R. D. 1978. Functions of cytoplasmic fibers in intracellular movements in BHK-21 cells. *J. Cell Biol.* 79: 708–26

Warren, T. G., Shields, D. 1984. Expression of preprosomatostatin in heterologous cells: Biosynthesis, posttranslational processing and secretion of mature somatostatin. *Cell* 39: 547–55

Wehland, J., Henkart, M., Klausner, R., Sandoval, I. V. 1983. Role of microtubules in the distribution of the Golgi apparatus: Effect of taxol and microinjected anti-alpha tubulin antibodies. *Proc. Natl. Acad. Sci. USA* 80: 4286–90

Whitaker, M. J., Baker, P. F. 1983. Calcium-dependent exocytosis and in an in vitro secretory granule plasma membrane preparation from sea urchin eggs and the effects of some inhibitors of cytoskeletal function. *Proc. R. Soc. London Ser. B* 218: 397–413

Whitnall, M. H., Mezey, E., Gainer, H. 1985. Co-localization of corticotropin-releasing factor and vasopressin in median eminence neurosecretory vesicles. *Nature* 317: 248–50

Wieland, F. T., Gleason, M. L., Serafini, T. A., Rothman, J. E. 1987. The rate of bulk flow from the endoplasmic reticulum to the cell surface. *Cell* 50: In press

Wills, J. W., Srinivas, R. V., Hunter, E. 1984. Mutations of the Rous sarcoma virus *env* gene that affect the transport and subcellular location of the glycoprotein products. *J. Cell Biol.* 99: 2011–23

Young, J. D.-E., Hengartner, H., Podack, E. R., Cohn, A. Z. 1986. Purification and characterization of a cytolytic pore-forming protein from granules of cloned lymphocytes with natural killer activity. *Cell* 44: 849–59

Young, W. C. III, Wamsley, J. K., Zarbin, M. A., Kuhar, M. J. 1980. Opioid receptors undergo axonal flow. *Science* 210: 76–77

Zanini, A., Giannattasio, G., Nussdorfer, G., Margolis, R. K., Margolis, R. U., Meldolesi, J. 1980a. Molecular organization of prolactin granules. II. Characterization of glycosaminoglycans and glycoproteins of the bovine prolactin matrix. *J. Cell Biol.* 86: 260–72

Zanini, A., Giannattasio, G., Meldolesi, J. 1980b. Intracellular events in prolactin secretion. In *Synthesis and Release of Adenohypophyseal Hormones*, ed. M. Justisz, K. W. McKerns, pp. 105–23. New York: Plenum

Ann. Rev. Cell Biol. 1987. 3 : 295–317

OLIGOSACCHARIDE SIGNALLING IN PLANTS

Clarence A. Ryan

Institute of Biological Chemistry, Washington State University, Pullman, Washington 99164

CONTENTS

INTRODUCTION.. 295
FUNGAL CELL WALL FRAGMENTS AS DEFENSIVE SIGNALS... 296
   β-Glucans.. 296
   Chitin and Chitosan Derivatives... 302
PLANT CELL WALL FRAGMENTS AS DEFENSIVE SIGNALS... 305
   α-1,4-D-Oligogalacturonides 305
   Unsaturated α-1,4-D-Oligogalacturonides.. 307
   Summary... 309
PLANT CELL WALL FRAGMENTS AS SIGNALS FOR GROWTH AND DEVELOPMENT 310
CONCLUDING REMARKS 311

INTRODUCTION

Within the past decade, research on the biochemistry of plant defense has revealed that fragments from fungal and plant cell walls can act as powerful signalling agents to activate plant defensive genes. Studies of carbohydrates as signalling molecules in plants have expanded from the seminal finding of Albersheim and his associates that β-glucan components of fungal cell walls can elicit synthesis of antibiotic phytoalexins in soybean cells to more recent discoveries in several laboratories that fragments of chitin and chitosan components of fungal cell walls and fragments of pectic components of plant cell walls can also activate genes coding for localized and/or systemic defensive responses. Synthesis of the enzymes chitinase and β-glucanase, which can generate such fragments, is induced in plant tissues during certain pathogenic attacks, and plant and fungal cell walls can act

295

0743–4634/87/1115–0295$02.00

synergistically to elicit defensive responses. These observations support the idea that plants may possess common intracellular mechanisms, responsive to a broad spectrum of oligosaccharide signals, to activate a variety of defensive genes. Recently, fragments of plant cell walls have been found that can profoundly change plant growth and development; such findings greatly expand the significance of oligosaccharide signalling in plants. This article reviews recent research that has established that oligo- and polysaccharides from both pathogens and plants contain structures that are part of signalling systems that can activate plant gene expression and regulate biochemical and physiological processes. The different types of complex carbohydrates involved in signalling responses in plant cells are examined, and the structural features of the simplest active oligosaccharides derived from these polysaccharides are reviewed. Evidence concerning the possible mechanisms involved in oligosaccharide-mediated signal transmission and gene activation is considered together with perspectives for future research.

FUNGAL CELL WALL FRAGMENTS AS DEFENSIVE SIGNALS

β-Glucans

The first observations that cell wall components from fungi can act as signalling agents in plant tissues were made by Albersheim and his associates in the mid 1970s (Anderson-Prouty & Albersheim 1975; Ayres et al 1976a,b; Cline et al 1978) in the course of their study of agents from fungi that induce the accumulation of small antibiotic chemicals in plants. The term phytoalexin (for recent reviews of phytoalexins see Bailey 1982; Darvill & Albersheim 1984; Dixon 1986; Grisebach & Ebel 1978; Ingraham 1982; Kuc 1982; Kuc & Rush 1985; Mansfield 1982; Paxton 1981; Stoessl 1981; West 1981) is used to describe several diverse types of relatively small chemical compounds of plant origin that can inhibit the growth of a broad range of microorganisms. Phytoalexins are not found in healthy plant tissues but are synthesized in diseased tissues in response to chemical signals, called elicitors (Keen et al 1972), that are produced and/or released at the sites of infection. These signals can originate from the attacking pathogens or from the plant itself. An elicitor from the cell walls of the fungus *Phytophthora megasperma* f. sp. soyja was isolated (Ayres et al 1976a,b) using soybean tissues or suspension cultures as assay systems. The phytoalexin glyceollin was induced to accumulate by application of solutions containing the elicitor and was extracted and quantified after ~20 hr. Concentrations of 0.2–1.0 $\mu g/ml$ of crude elicitor were required for maximum accumulation, and as little as 10 ng of the *P. megasperma* β-glucan could induce the biosynthesis of sufficient phytoalexin to inhibit fungal growth (Cline & Albersheim 1981b).

The phytoalexin elicitor was released from *P. megasperma* cell walls by autoclaving at 121°C for 3 hr (Anderson-Prouty & Albersheim 1975; Ayres et al 1976a). The released carbohydrates were heterogenous in size, ranging from an M_r of ~5000 to ~200,000 (the average M_r was ~100,000). The phytoalexin-eliciting polysaccharides were uncharged and were stable even when exposed to extremes of heat (see above) and pH (2 to 10). All of the *P. megasperma*–produced elicitors were shown to be glucans that comprise ~65% of the mycelial walls (Ayres et al 1976a,b; Ebel et al 1976; Albersheim & Valent 1978). The first analyses of the phytoalexin elicitors indicated that they were composed of β-C-3 and β-C-6 linkages (Ebel et al 1976; Albersheim & Valent 1978). Active phytoalexin elicitors could be released by partial acid hydrolysis, by enzymic hydrolysis with an exo-β-1,3-glucanase isolated from *Euglena gracilis* (Ayres et al 1976c), or by heating. Enzymic hydrolysis produced much smaller fragments (with an average M_r of ~10,000) that retained all of the elicitor activity of their larger precursors. When these fragments were completely degraded to glucose with a plant β-1,3-glucanase, the phytoalexin-eliciting activity was lost (Ayres et al 1976c). The terminal glucosyl residues were required for this activity, yet 90% of the glucose could be removed enzymically without any loss of activity. The residues left after enzymic hydrolysis were still primarily C-3 and C-6 linked or C-3/C-6 branched glucose polymers containing about 4% mannosyl residues. The mannose was suspected to have a role in the process since α-D-mannoside inhibited the phytoalexin-eliciting activity; however, to date no small fragments with this activity have been resolved that contain mannose.

A cell wall component of commercial brewers yeast was also shown to possess a phytoalexin elicitor (Hahn & Albersheim 1978). Most of the cell wall polysaccharides of yeast are mannans; however, a glucan fraction was present that was relatively easily separated from the mannans by affinity chromatography using concanavalin A–Sepharose. The elicitor activity was again found to reside in the glucan components (Hahn & Albersheim 1978; Albersheim & Valent 1978) of the cell wall polysaccharide that contained C-3 and C-6 glucosyl linkages.

Elicitor-active cell wall polysaccharides isolated from *P. megasperma*, yeast, and *Colletotrichum lindemuthianum* can induce phytoalexins in red kidney beans and potatoes as well as in soybeans (Cline et al 1978). The partially purified elicitors from *C. lindemuthianum* and *P. megasperma* are now commonly employed to induce the biosynthesis of phytoalexins in many genera from several plant families. The generality of this elicitation of phytoalexin biosynthesis supports the hypothesis that these β-glucan fragments of fungal cell wall polysaccharides function as primary signals to activate genes coding for enzymes in the pathways for synthesis of phytoalexins (Darvill & Albersheim 1984).

The interaction of plants with pathogens is complex and results in the induction of numerous genes in the host that encode not only enzymes required for the synthesis of phytoalexins, but other proteins and enzymes as well, some of which may have roles in plant defense. Many plant genes or gene products have now been shown to be activated by β-glucan elicitors, including genes coding for enzymes of two separate pathways for phytoalexin synthesis, the phenylpropanoid and isoprenoid pathways (Dixon 1986); genes coding for hydroxyproline-rich glycoproteins (HRGPs) (Roby et al 1985; Showalter et al 1985; Showalter & Varner 1987); chitinase and β-glucanase (Boller 1985; Boller et al 1983; Lamb et al 1987; Pegg 1977; Yang & Pratt 1978; Nichols et al 1980; Yoshikawa et al 1981); pathogenesis-related (PR) proteins (Somssich et al 1986; van Loon 1985); protein proteinase inhibitors (Esquerre-Tugaye et al 1985); and ethylene (Pegg 1976; Paradies et al 1980; Toppan & Roby 1982; Roby et al 1985, 1986; Yang & Pratt 1978). As the above systems are further investigated, other genes that are under regulation by carbohydrate fragments will undoubtedly be found. Dixon et al (1983) have summarized data showing that β-glucan elicitors and infection cause marked changes in overall patterns of protein synthesis that are correlated with specific defensive responses and expression of disease resistances. Cramer et al (1985) have shown that levels of numerous mRNA species increase in suspension cultures of bean (*Phaseolus vulgaris* L.) in response to the *C. lindemuthianum* elicitor.

The smallest pure glucan oligosaccharides that still retain elicitor activity have been isolated from polysaccharides of mycelial cell walls of *P. megasperma* var. glycinia and fragmented by mild acid hydrolysis (Sharp et al 1984c). The activity is retained by oligomers with a degree of polymerization (DP) of five or greater. High-resolution chromatography of the most active oligosaccharides on Biogel P2 yielded three glucan fractions that eluted with apparent DPs of 6–8. The terminal glucose of the elicitor-active fractions could be reduced with $NaBH_4$ to a glucitol without affecting elicitor activity; however, periodate oxidation completely destroyed activity. The heptaglucosyl oligosaccharide could be separated into seven components only one of which (fraction AV) was active. Component AV was characterized structurally as a β-glucan with 3-, 6-, and 3-6 linked residues, as shown in Figure 1 (Sharp et al 1984b). This compound is the smallest oligosaccharide elicitor yet reported and is the most potent elicitor known; it has a specific activity of 1800 units/μg glucose. This heptaglucosyl elicitor can elicit phytoalexin biosynthesis in soybean tissues when present at about 1 pg/g tissue (Sharp et al 1984c).

The elicitor activity of the glucoheptaose appears to reside in the unique structure shown in Figure 1 and in no other structural isomers. Calculations indicate that 150–300 structurally distinct β-glucan heptaglucosyl molecules might be present in the acid hydrolysate from mycelial cell wall polysaccharides (Sharp et al 1984c). Of eight structurally different reduced heptaglucosyl molecules analyzed, only the structure shown in Figure 1 was an active elicitor. This glucoheptaose, as well as a derivative with the terminal glucose reduced to glucitol, has been chemically synthesized, as has a glucooctaose (Figure 1) (Ossowski et al 1984). The synthetic heptaglucosyl elicitor possessed the same specific activity in eliciting antimicrobial phytoalexins in soybean as did the mycelial wall–derived heptaose elicitor (Sharp et al 1984a). The synthetic octasaccharide exhibited only about half the potency of the heptaoses (Sharp et al 1984a).

To date, studies with the highly purified heptaglucosyl elicitor have been confined to soybean cells. However, the identification of a specific carbohydrate structure with elicitor activity may provide a conceptual framework that may be applicable to studies of oligo- or polysaccharide elicitors that are present as structural components of cell wall glucans of both pathogenic and nonpathogenic fungi. There may exist within fungal cell walls a variety of polysaccharide structures that can produce different fragments during pathogenic attacks to activate different sets of defensive genes. For example, Keen and coworkers (Keen & Legrand 1980; Keen et al 1983) have shown that mycolaminarin (a β-1,3 storage glucan from *Phytophtera* species) and glucomannan [released from *P. megasperma* cell walls by the action of a β-1,3-endoglucanase from soybean cells (Keen & Yoshikawa, 1983)] can induce the accumulation of the phytoalexin glyceollin in soybeans. The latter elicitor was more active than the cell wall β-glucan fraction. Recently, phytoalexin-eliciting fractions from *C. lindemuthianum* were separated from charged molecules by ion exchange chromatography, further purified by affinity chromatography, and shown

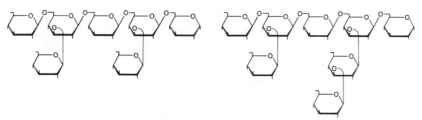

Figure 1   Structures of the glucoheptaose (*left*) and glucooctaose (*right*) elicitors derived from *P. megasperma* cell walls.

to contain a high percentage of galactose and mannose (Dixon 1986). These elicitors induced biosynthesis of early enzymes of the isoflavanoid phytoalexin pathway in beans but did not induce the phytoalexins phaseolin (a pterocarpan) or kievitone (an isoflavone), which are synthesized via these common early steps. Yet, synthesis of phaseolin and kievitone were induced by the cell wall fragments of the fungus (Dixon 1986). Recent research by Lamb et al (1987) has shown that several of the enzymes in beans leading to the biosynthesis of phytoalexins are members of small multigene families. Lamb et al suggest that there is differential activation of different members of plant defensive gene families in response to different elicitors. The β-glucan elicitor from *P. megasperma* has recently been shown to be incapable of inducing synthesis of an isoflavinoid phytoalexin, pisatin, in pea pods (L. Hadwiger, personal communication), although extracts of the mycelial cell walls of the pea pathogen *Fusarium solani* var. pisi are potent elicitors of pisatin in these tissues.

The presence of β-glucanases in plant tissues capable of degrading fungal glucans has generated considerable research regarding the possible roles of β-glucanases in plant defense (Cline & Albersheim 1981a,b; Albersheim & Valent 1978; Nichols et al 1980; Yoshikawa et al 1981; Pegg 1977; Boller 1985). Earlier evidence indicated that the levels of β-glucanases increased in response to pathogen attacks (Ables 1970); thus the activity was thought to aid in lysing cells of the attacking fungi (Pegg & Vessey 1973; Pegg 1977). However, rapid release of β-glucan fragments by these enzymes is now recognized as a possible mechanism for signalling the activation of defensive responses (Nichols et al 1980; Yoshikawa et al 1981). An exo-β-1,3-glucanase was found to be associated with the cell walls of soybean cells (Cline & Albersheim 1981b) that released phytoalexin-eliciting β-glucans from *P. megasperma* cells. The enzyme was purified and found to hydrolyze fungal cell wall β-glucans completely to glucose. Yoshikawa et al (1981) demonstrated that considerable elicitor was released almost immediately from *P. megasperma* cell walls in contact with soybean cotyledons. Keen & Yoshikawa (1983) isolated two endo-β-1,3-glucanases from soybean cells that released phytoalexin elicitors from *P. megasperma* mycelia. Cline & Alberhseim (1981b) have pointed out that the β-glucanases may serve two functions, production of small elicitors that can be internalized as signals and subsequent degradation of these fragments to regulate their effects. However, whether glucan elicitors actually penetrate the target cell has not been determined (Dixon 1986), nor has a requirement for the continued presence of elicitor been firmly established, although some evidence (Dixon et al 1981) suggests that there may be such a requirement.

Signalling by cell wall–derived fragments during interactions between pathogens and host plants may involve synergistic effects of β-glucans and plant cell wall fragments (Darvill & Albersheim 1984), which also have powerful gene-activating properties (see following sections). Fungal or plant cell wall fragments at concentrations too low to elicit phytoalexin synthesis alone, in combination exhibit potent elicitor activity. During pathogen attacks on plants, an array of carbohydrate signals are produced early on. These signals can be amplified by feedback, generated by hydrolase induction and activity, and by synergistic effects of the fragments, thus intensifying the defensive responses.

Complete understanding of these signalling processes will require identification and chemical resolution of the structures within the various fungal cell wall polysaccharides from different pathogens. The determination of the structure of the heptaglucosyl elicitor (Sharp et al 1984b) required a large effort by many scientists and clearly demonstrated a need to develop more sophisticated and efficient, and eventually automated, techniques to successfully dissect and analyze the variety of elicitor polysaccharides available for study. Increased knowledge of the detailed structures of elicitor oligosaccharides will be necessary to elucidate the processes of signal transmission and the biochemical basis of their powerful gene-activating properties.

At present little is known of the signalling process. Lamb et al (1987) provided the first detailed study of the early events of induction of transcriptional events by elicitors, and these data place temporal limits on the biochemical options that might be invoked to explain the signalling mechanism. In cultured bean cells treated with *C. lindemuthianum* elicitor (Anderson-Prouty & Albersheim 1975), the mRNAs coding for three enzymes required for steps early in the phenylpropanoid pathway for phytoalexin synthesis and the mRNA for the enzyme chitinase were synthesized within 10–20 min following addition of elicitor. Transcriptional activation of the mRNAs could be detected within 5 min of elicitor treatment and may occur within 2–3 min (Lamb et al 1987). Transcriptional activation of two of these enzymes by elicitor had been demonstrated previously in parsley cells (Kuhn et al 1984; Chappell et al 1984) but was not investigated at very early times following elicitor addition.

The detection of transcripts in soybean cells 2–5 min after treatment with elicitor (Lamb et al 1987) provides evidence of one of the earliest transcriptional events reported for a plant gene in response to an exogenous signal. Lamb et al speculate that this rapid induction leaves little time for complicated intervening events to occur between elicitor addition and gene activation. This implies that the elicitor activity may involve a direct interaction with *cis*-acting factors or the activation of a short signal transduction cascade modulating *trans*-acting nuclear proteins that would in turn regulate gene expression. These researchers further suggest that the cooperation of rapid but complex signal systems may be necessary to

regulate plant defensive genes under different circumstances and that localized and systemic signal circuits may interact.

The possibility that membrane receptors are involved in elicitor signalling has not been extensively explored, but some recent experiments
suggest that receptors for polysaccharides are present in plant membranes.
Using [^{14}C]mycolaminarin, a β-1,3-glucan storage polysaccharide from
P. megasperma, var. sojae that weakly elicits the phytoalexin glyceollin
accumulation in soybeans, Yoshikawa et al (1983) demonstrated that this
elicitor was specifically bound to membranes from soybean tissues. A
single affinity class of binding sites with a K_d of 11.5 μM was found.
Binding was abolished by treatment with heat and proteolytic enzymes,
which indicates that the interaction was mediated by a protein. The
interpretation that the binding proteins are true receptors was questioned
(Dixon 1986) on grounds that binding was observed throughout membrane fractions and the identity of the sites on the plasma membrane is
therefore still tentative. Also, the possibility that binding of carbohydrates
by these proteins serves a function other than mediating the phytoalexin
response was not ruled out. More experimentation is needed of the type
initiated by Keen and associates, i.e. that utilizing small, well-defined
elicitor molecules such as those isolated and characterized by Sharp et al
(1984a,b). The determination of whether or not a membrane receptor is a
component of carbohydrate signalling is crucial to the understanding of
the phytoalexin response.

Chitin and Chitosan Derivatives

Chitin, an insoluble homopolymer of β-1,4-*N*-acetyl glucosamine (Figure
2a), and chitosan, a polymer of β-1,4-glucosamine (Figure 2b), are components of the cell walls of many fungi (Bartnicki-Garcia 1970), and chitin
is a major component of the cuticle of many members of the Arthropod
family. Neither chitin nor chitosan are normal components of plant tissues.

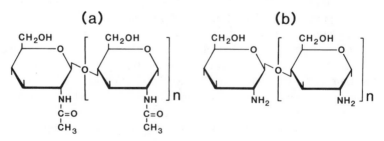

Figure 2 (*a*) Chitin, a β-1,4-D-*N*-acetyl glucosamine polymer, and (*b*) chitosan, a β-1,4-D-
glucosamine polymer.

The enzymes chitinase and chitosanase are endoglycosidases that can hydrolyze chitin and chitosan, respectively, into soluble oligosaccharide fragments (Molano et al 1979; Pegg & Young 1981; Tsukamoto et al 1984; Boller et al 1983) and are present in many species from at least 11 plant families (Prowning & Irzykiewicz 1965). Chitinase activity, along with β-glucanase activity, has been shown to increase in plant tissues in response to pathogen attacks (Ables et al 1970; Pegg & Vessey 1973; Pegg 1977; Toppan & Roby 1982). Because the substrates for chitinase and chitosanase are polysaccharides found in fungal and bacterial cell walls, and the levels of these hydrolases in plants are induced to increase in response to pathogen attacks, these enzymes were thought to be involved in defense by contributing to the lysis of attacking pathogens. However, in 1980 Hadwiger and associates (Nichols et al 1980; Hadwiger & Beckman 1980) reported that extracts of pea endocarp containing chitinase and chitosanase could degrade *F. solani* f. sp. pisi and f. sp. phaseoli cell walls to produce carbohydrate fragments that were powerful elicitors of the isoflavinoid phytoalexin pisatin in the pods. Thus, defensive responses in plants can be activated not only by fragments from the β-glucan components of fungal cell walls but by fragments of the chitin and chitosan components as well. These findings indicate that chitinase may have a role in generating signals as well as in arresting fungal advance.

The activation of plant defense genes by chitin and chitosan fragments has been studied in only a few systems, but the limited data available suggest that these fragments may have broad functions as inducers of defensive responses in plants. The most extensive studies of chitin and chitosan fragments have focused on their possible role(s) in the induction of the phytoalexin pisatin in pea pods in response to attack by the pea pathogen *F. solani* f. sp. phaseoli (Nichols et al 1980; Hadwiger & Beckman 1980; Fritensky et al 1985). Chitosan released during fungal attack was shown to enter the plant cells soon after infection and to induce phytoalexin synthesis and resistance toward the pathogen. Histochemical and immunological evidence indicates that hexosamine-containing fragments accumulate inside plant cells within 15–30 min following fungal inoculation (Hadwiger & Beckman 1980; Hadwiger & Loschke 1981). Purified chitosan, at concentrations of 2 μg/ml extract, induced pisatin accumulation in pea pods and, when applied directly to fungal mycelia at the same levels, chitosan inhibited germination and growth of the fungi.

Chitosan and soluble derivatives of chitin are elicitors or inducers of a number of defensive responses other than those described above, including synthesis of β-glucanase and chitinase in pea pods (Mauch et al 1984), elicitation of the phytoalexin casbene in castor bean (West 1981; Walker-Simmons et al 1984), lignification in wounded wheat leaves (Pearce & Ride

1982), callose synthesis in cultured soybean cells (Kauss 1984), monoterpene accumulation in the phloem of lodgepole pine (Miller et al 1986), proteinase inhibitor I synthesis in detached leaves and suspension cultures of tomato (Walker-Simmons et al 1983, 1984), and synthesis of the Bowman-Birk class of proteinase inhibitors in detached alfalfa leaves (Brown & Ryan 1984).

Several soluble chitin or chitosan derivatives induced synthesis of proteinase inhibitor I in detached young tomato plants (Walker-Simmons & Ryan 1984). Ethylene glycol chitin, nitrous acid–fragmented chitosan, glycol chitosan, and chitosan oligomers all induced synthesis of the small proteinase inhibitor protein. The glycol chitin and nitrous acid–fragmented chitosan were 2–3 times more potent than the other inducers. When supplied to young tomato plants through their cut petioles, nitrous acid–fragmented chitosan was active at concentrations below 100 μg/ml. Since the solutions were supplied for only 30 min, during which time the plants usually imbibed 20–50 μl solution, only about 2–5 μg/plant were required to induce the response (Walker-Simmons & Ryan 1984, and unpublished data).

Soluble fragments of native chitosan were recovered from HCl digests and separated by gel filtration, and the oligosaccharides of DP = 2 through DP = 6 were further purified and assayed for proteinase inhibitor–inducing activity in young tomato plants. Monomeric glucosamine was inactive, and the dimer and trimer were weakly active. Tetrameric glucosamine exhibited moderate activity, whereas the pentamer and hexamer were as active as chitosan.

In a similar study (Kendra & Hadwiger 1984), the smallest glucosamine oligomers shown to inhibit growth of *F. solani* (f. sp. pisi and phaseoli) and elicit pisatin in immature pea pods were the trimer through hexamer; however, they were only moderately active. The monomer and dimer were inactive, but the heptamer was as active as chitosan, both as an antifungal agent and as an elicitor.

The mechanism for signalling gene activation by chitosan or glucosamine oligomers is not understood, but chitosan fragments apparently penetrate the target cells. Hadwiger & Beckman (1980) found that [^{14}C]chitosan entered plant cells and could be detected in the nucleus. Hadwiger et al (1986) suggested that these fragments directly interacted with plant DNA to activate specific defense genes. In pea pods, chitosan enhanced the synthesis of plant mRNA species homologous with several of the 20 or more genes that were activated as part of the resistance response of pea to *F. solani*. The mRNAs appeared to be transcriptionally regulated. These mRNAs included those coding for the enzyme phenylalanine ammonia lyase (PAL), a key enzyme of the biosynthetic pathway for pisatin (Loschke et al 1983). If the chitosan that enters the nucleus regulates pisatin synthesis, then receptors of some type are probably involved to provide specificity in mediating the response. Since a diverse array of defense responses are activated by chitosans, as well as by glucans and plant cell wall fragments (see following section), the genes activated by these diverse oligosaccharides may employ similar signalling mechanisms, either at the receptor level or at the level of transcription.

PLANT CELL WALL FRAGMENTS AS DEFENSIVE SIGNALS

α-1,4-D-Oligogalacturonides

Independent studies of fungal-elicited phytoalexins in beans (a localized reaction) and of wound-inducible proteinase inhibitor proteins in tomato and potato plants (a systemic reaction) indicated that oligosaccharide signals could be generated from plant tissues that were capable of activating the accumulation of these defensive products. In 1978, Hargreaves & Bailey and Hargreaves & Selby reported that autoclaving or freeze-thawing *Phaseolus vulgaris* hypocotyls solubilized dialyzable compounds that induced phytoalexin accumulation in bean cells. They called this putative inducer the "endogenous elicitor." The same year, Stekoll & West (1978) published the first of several reports (Lee & West 1981a,b; Bruce & West 1982; Jin & West 1984; West et al 1985) concerning an endo-polygalacturonase enzyme from the fungal pathogen *Rhizopus stolonifer* that generated pectic fragments from castor bean cell walls. The action of the enzyme induced, in plant tissues, synthesis of the enzyme casbene synthase, which catalyzes the final step in the biosynthesis of the diterpene phytoalexin casbene. In this same year, Ryan (1978) reported that proteinase inhibitor–inducing activity isolated from tomato leaves had the characteristics of a pectic polysaccharide. The inducer of proteinase inhibitors was called the "proteinase inhibitor–inducing factor" or PIIF (Green & Ryan 1972). Subsequent research (Bishop et al 1981, 1984; Ryan et al 1981, 1986) confirmed that the activity of PIIF was associated with pectin and was also present in hydrolysates of tomato leaf cell walls (Bishop et al 1981). PIIF activity was also found to reside (Ryan et al 1981) in a large, well-characterized ($M_r \approx 200,000$) rhamnogalacturonan fragment (McNeil et al 1980; Lau et al 1985) that had previously been isolated from sycamore cell walls (Darvill et al 1978). These experiments conclusively demonstrated that PIIF activity was a component of the plant cell wall.

In 1981, Hahn et al isolated the "endogenous elicitor" activity from soybean cell walls and demonstrated that the elicitor activity resided in the pectic polysaccharide fractions. Similar phytoalexin–eliciting fractions were isolated from cell walls of suspension cultures of tobacco, sycamore, and wheat. The active components of the soybean cell walls were stable to partial acid hydrolysis but were destroyed by treatments with endo-α-1,4-galacturonase. Phytoalexin-eliciting fragments were purified from soybean cell walls and from citrus pectin (Nothnagel et al 1983) that were composed of about 98% α-1,4-linked galacturosyl residues. The most active elicitor fragments from both sources were linear polymers composed of 12 residues of galacturonic acid (Figure 3).

Employing purified oligomers of α-1,4-D-galacturonic acid, from DP = 2 through DP = 15, prepared by partial digestion of poly-galacturonic acid with R. *stolonifer* endopolygalacturonase (Bruce & West 1982), Jin & West (1984) found the casbene synthetase–elicitor activity in linear oligomers of DP = 9 and above, with the highest activity in the DP = 13 oligomer. Little or no activity was found in oligomers with DP < 9. Robertsen (1986) reported that an endopolygalacturonase from *Cladosporium cucumerinum* induced lignification in cucumber (*Cucumus melo*) seedlings and that oligogalacturonides with DPs of 10 and 11, prepared by limited degradation of polygalacturonic acid with the C. *cucumerinum* enzyme, were maximally active in eliciting lignification. Oligomers with DP < 8 or DP > 13 were inactive.

The smallest α-1,4-D-galacturonosyl oligomers that could induce synthesis of proteinase inhibitor proteins in tomato leaves were much smaller than those that elicit phytoalexins and lignification (Bishop et al 1984). Partial acid hydrolysis of a major PIIF-active pectic component of $M_r \approx 5000$, isolated from autoclaved tomato leaves (Bishop et al 1981), produced a series of small PIIF-active α-1,4-D-galacturonic acid oligomers with DPs of 2–6. The dimer exhibited about half the specific activity of PIIF. The specific activities of larger oligomers increased with increasing DP: The activity of DP = 6 was nearly that of the parent PIIF.

Table 1 summarizes data from several recent investigations involving defensive genes or defensive systems activated by plant cell wall fragments or fungal endopolygalacturonase. These reports indicate that a broad spectrum of genera from several plant families respond to cell wall–derived

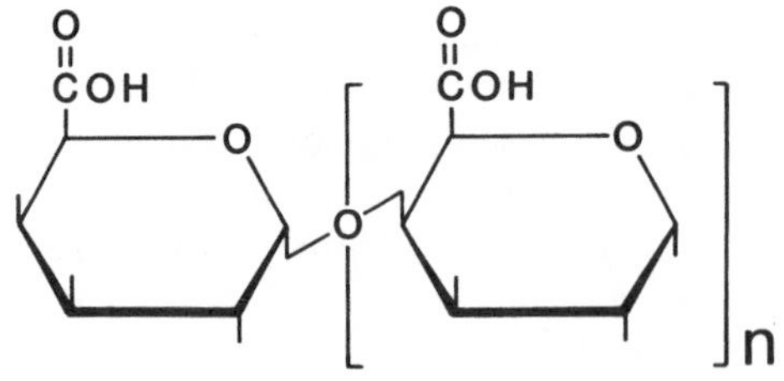

Figure 3 α-1,4-D-Galacturonic acid oligomers.

Table 1   Elicitation or induction of various defensive responses by pectic fragments or by fungal endopolygalacturonases

			Elicitor or Inducer			
Plant	Response[a]	PGase	Pectic fragment	Chitosan	β-Glucan	Ref.[b]
Ricinus communis	P	+	+	+	−	1
Glycine max	P	+	+		+	2
Ipomea batatus	P	+	+			3
Trifolium repens	P		+			4
Phaseolus vulgaris	P	+	+	+	+	5
Pisum sativum	P, E, C, G	+	+	+	+	6
Medicago sativa	PI		+	+		7
Daucus carota	P	+	+			8
Capsicum annum	P	+		+		9
Gossypium herbaceum	N, R	+				10
Cucumis melo	H, E, PI		+		+	11
Pinus contorta	P		+	+		12
Cucumus sativus	L	+	+			13
Nicotiana tabacum	H, P, PI		+	+		14
Lycopersicon esculentum	PI	+	+	+	−	15
Solanum tuberosum	PI	+	+	+		16

[a] Responses: P = phytoalexin; E = ethylene; C = chitinase; G = β-glucanase; N = necrosis; R = resistance; H = hydroxyproline-rich glycoprotein; PI = proteinase inhibitor; L = lignification.
[b] References: 1. Lee & West 1981; Bruce & West 1982; 2. Davis et al 1986; Hahn et al 1981; Sharp et al 1984b,c; 3. Sato et al 1982; 4. West et al 1984; 5. Hargreaves & Selby 1978; Bailey 1986; Wichham, in West et al 1984; C. Lamb, personal communication; 6. Hadwiger & Beckman 1980; Walker-Simmons et al 1984; Walker-Simmons & Ryan 1985; 7. Brown & Ryan 1984; 8. Kurosaki & Nishi 1984; Kurosaki et al 1985; 9. Watson & Brooks 1984; 10. Karban 1985; Venere et al 1984; 11. Roby et al 1985, 1986; Esquerre-Tugaye et al 1985; 12. Miller et al 1986; 13. Robertson 1986; 14. Baker 1985; Modderman et al 1985; M. Walker-Simmons & C. A. Ryan, unpublished data; 15. Walker-Simmons et al 1984; Walker-Simmons & Ryan 1985; Ryan et al 1986; 16. Walker-Simmons & Ryan 1977; R. B. Pearce & C. A. Ryan, unpublished data.

plant and/or fungal oligosaccharides. The increasing number of reports of oligosaccharide signals activating plant defenses emphasizes the need for further examination of the biological roles of oligosaccharides in plants and the mechanism(s) of their signalling activities.

Unsaturated α-1,4-D-Oligogalacturonides

Pectic lyases, enzymes that fragment α-1,4-polygalacturonic acid by way of a trans-eliminase reaction, are secreted by a number of plant pathogens (Bateman & Basham 1976), including the phytopathogenic bacteria *Erwinia carotova* (Moran et al 1968). These bacteria can elicit phytoalexin accumulation in wounded soybean cotyledons (Weinstein et al 1981), apparently by secreting an endopolygalacturonic acid lyase that in turn releases endogenous elicitors from the plant cell wall (Davis et al 1984). The oligosaccharide elicitors thus released have characteristic nonreducing terminal residues of 4,5-unsaturated galacturonic acid that are easily detected by their strong absorption of ultraviolet (UV) light. A lyase activity secreted by *Erwinia* was required for release of the fragments and for elicitation of phytoalexins in soybean cotyledons. Structural analysis of the oligomers released from sodium polypectate by the lyase (Davis et al 1986a,b) indicated that the major phytoalexin-eliciting component was a decasaccharide (DP = 10) composed of α-1,4-D-galacturonic acid residues having a 4,5-unsaturated galacturonosyl nonreducing terminal residue (Figure 4). This oligomer elicited half-maximal phytoalexin accumulation in wounded soybean cotyledons at a concentration of 1 μg/cotyledon (Davis et al 1986a,b). A minor phytoalexin-eliciting oligomer was also isolated and characterized as a linear α-1,4-D-galacturonosyl undecamer (DP = 11) with an unsaturated nonreducing terminus (Davis et al 1986a). This oligomer was about one third as active as the decasaccharide as an elicitor of phytoalexins in soybean tissue.

Two pectin lyases have recently been isolated from culture filtrates of the scab fungus *C. cucumerinum*, a pathogen of cucumber (Robertsen 1986). Both lyases elicited lignification when applied to cucumber hypocotyl segments. A pectic lyase was also purified from *Escherichia coli* strains that contained a cloned lyase gene from *E. crysanthemi*. This enzyme prevented the development of a hypersensitive reaction in tobacco tissues induced by the pathogen *Pseudomonas syringae* pv. pisi. The pure lyase released heat-stable fragments from purified tobacco cell walls that prevented the hypersensitive response and inhibited bacterial multiplication. A pectic lyase purified from the fungus *F. solani* f. sp. pisi (M. Crawford & P. E. Kolattukudy, personal communication), produced fragments that when supplied to leaves of small tomato plants induced accumulation of proteinase inhibitor I (T. Moloshok, G. Pearce, C. A. Ryan, in preparation). Isolation and analysis of oligomeric fragments of a partial digest of polygalacturonic acid revealed that α-1,4-digalacturonic acid with

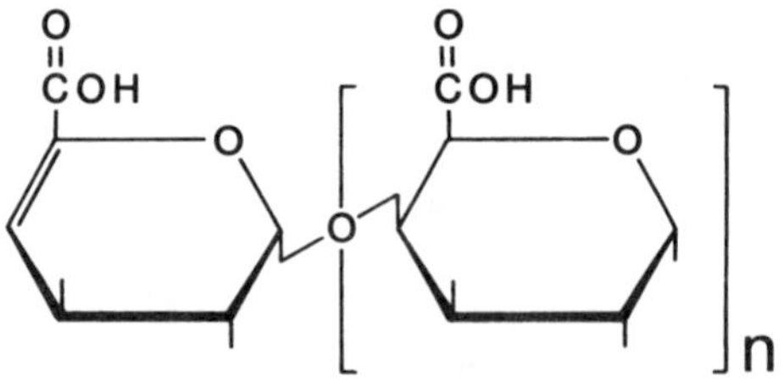

Figure 4   Unsaturated α-1,4-D-oligogalacturonic acid products of pectic lyase on polygalacturonic acid. Oligomers in which n = 1–10 can act as signalling molecules to activate plant defensive genes.

a 4,5-unsaturated nonreducing terminal galacturonosyl residue (Figure 4, $n = 1$) and the unsaturated trimer ($n = 2$) were effective inducers of the synthesis of proteinase inhibitor I when supplied to young tomato plants through their cut petioles.

Summary

The galacturonosyl oligomers, whether saturated or unsaturated, that elicit phytoalexin responses are larger than the oligomers that induce proteinase inhibitors. Accumulation of phytoalexins (Jin & West 1984; Davis et al 1984) and lignin (Robertsen 1986) is elicited by oligomers of DP = 10–13. Proteinase inhibitor synthesis was even induced by dimers (Bishop et al 1984; G. Pearce & C. A. Ryan, in preparation). It is unclear whether this difference indicates a role for oligosaccharides in localized or systemic signalling. No biochemical evidence pertaining to the signalling mechanism exists, nor has it been shown that proteinase inhibitor–inducing oligosaccharides are transported through plants following wounding or pest attacks. Baydoun & Fry (1985) found that radioactively labelled pectic substances added to wounds of tomato leaves were not translocated. These experiments argue against the translocation of oligomers with DP > 6. However, translocation of low levels of smaller, active oligomers was not convincingly ruled out. Campbell (1986) has reported the presence of pectic oligosaccharides in phloem exudates (honeydew) of *Sorghum bicolor* that were produced by the action of endopolygalacturonase activity from aphid saliva on plant cell walls. The way in which pectic oligosaccharides are generated in wounds may be important for their transport as potential signals. The oligosaccharide signals have been collectively termed oligosaccharins (Albersheim et al 1983); this terminology has been accepted by many biochemists and biologists. The name oligosaccharin may prove useful, particularly if the β-glucans, chitosan, or oligogalacturonides are found to employ a common fundamental signalling system. That a number of defensive responses can be activated by some or all of these oligosaccharide signals (see Table 1) suggests that these signals may operate via a common intracellular mechanism throughout the plant kingdom. Such a system would involve receptors that could recognize specific oligosaccharide signals and relay this information to cytoplasmic or nuclear components, which would in turn discriminate among these signals and activate appropriate genes. The recent identification and characterization of several small oligosaccharide molecules with signalling activities, together with current, sophisticated research techniques, should lead to an increased understanding of oligosaccharide signalling and the defensive responses they induce in plants.

PLANT CELL WALL FRAGMENTS AS SIGNALS FOR GROWTH AND DEVELOPMENT

Fragments of plant cell walls have also been identified (and, in at least one instance, purified and structurally characterized) that profoundly affect protein synthesis and viability of plant suspension cultures and growth and development of intact plant tissues. Yamazaki et al (1983) demonstrated that very low concentrations of fragments from partial acid hydrolysates of sycamore cells killed sycamore cells in suspension cultures. The fragments were shown to inhibit incorporation of labelled amino acid into cellular proteins. The active fragments appeared to contain primarily glycosyl residues, but their purification and characterization have not been completed. Fragments released from starch or pectin were inactive. The inhibition of protein synthesis by these fragments may be related to the hypersensitive response that occurs during pathogen attacks and leads to necrosis of nearby cells (Yamazaki et al 1983).

A more clearly defined role for plant cell wall fragments is the inhibition of auxin-stimulated elongation of pea stem segments by a xyloglucan isolated from sycamore cell walls (York et al 1984). A nonasaccharide composed of glucose, xylose, and fucose that possessed potent inhibitory activity was isolated and characterized. Its structure is shown in Figure 5. A corresponding heptasaccharide lacking the galactose and xylose units was inactive. The nonasaccharide was active in the pea system at concentration of 1 to 0.1 μg/ml. This concentration was about 100-fold less than the concentration of auxin required to stimulate growth. The oligomer was hypothesized to act by feedback inhibiting auxin-stimulated growth

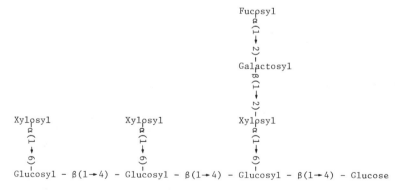

Figure 5 The structure of a xyloglucan oligosaccharide that inhibits auxin-induced stem elongation (York et al 1984).

after its release from the cell wall by auxin-induced glycosidases (York et al 1984). A nonasaccharide with the same structure was isolated from the culture medium of spinach cells (Fry 1986).

Fragments of sycamore and of *Lemna* cell walls generated by hydrolysis with trifluoric acid (TFA) enhanced frond growth of *Lemna gibba* G3 and severely inhibited its flowering (Gollin et al 1984) at concentrations of 50 μg/ml medium. The residue composition of the fragments was not reported, but the DP of the active oligomers was estimated to be greater than four. Although the fragments potently inhibited flowering when added to the medium, there is as yet no evidence that similar fragments have such a function *in situ*.

Oligosaccharide fragments isolated from α-1,4-D-polygalacturonase (PGase) digests of sycamore cell walls have been demonstrated to regulate morphogenesis of thin-layer explants of tobacco (Tran Thanh Van et al 1985). At 1 μg/ml medium at pH 3.8, the fragments inhibited flowering by the explants and induced vegetative bud development. At pH 5.0 the explants normally develop callus; however, at this pH the fragments inhibited callus growth and enhanced vegetative growth. At pH 6.0 the normal explants developed vegetative buds, but the fragments inhibited bud formation and enhanced flowering. The oligosaccharides appear capable of redirecting developmental processes and patterns. Fragments of the cell walls produced by base hydrolysis produce similar results. The concentrations of oligosaccharides that produced these changes in development were estimated to be between 10^{-8} and 10^{-9} M.

The series of experiments described above, almost all of which are from the laboratories of Albersheim and his collaborators, strongly implicates plant cell wall fragments as signal molecules regulating plant growth, development, and morphology.

CONCLUDING REMARKS

The demonstration that oligosaccharides, from dimers to dodecamers, can regulate various cellular and molecular events in plants, including defense against pathogens and plant growth and development, indicates that such molecules may have a central role in plant biology. A systematic study should be initiated to assess the entire spectrum of oligosaccharide messengers present in plants and to determine the structures of these substances. Such knowledge will be essential in developing new probes to study the signalling and recognition systems that regulate gene expression in plants. This type of information could have broad implications, not only for understanding of gene regulation in plants, but for development of new practical applications to improve our natural resources, environment, and eventually the quality of human life, which depends entirely on the health, growth, and productivity of plants for survival.

ACKNOWLEDGMENTS

Research of the author cited in this review was supported by grants from the National Science Foundation; the United States Department of Agriculture, Competitive Grants Program; and the Rockefeller Foundation. I wish to thank Drs. Frank Loewus and Rod Croteau for their critical reviews of, and suggestions for, the manuscript and Drs. Chris Lamb, Lee Hadwiger, P. E. Kolattukudy, and Mark Crawford for the personal communication of results.

Literature Cited

Ables, F. B., Bosshart, R. P., Forrence, L. E., Harbig, W. H. 1970. Preparation and purification of glucanase and chitinase from bean leaves. *Plant Physiol.* 47: 129–34

Albersheim, P., Darvill, A. G., McNeil, M., Valent, B., Sharp, J. K., et al. 1983. Oligosaccharins, naturally occurring carbohydrates with biological regulatory functions. In *Structure and Function of Plant Genomes*, ed. O. Ciferri, L. Dure III, pp. 293–312. New York: Plenum

Albersheim, P., Valent, B. 1978. Host-pathogen interactions in plants: Plants, when exposed to oligosaccharides of fungal origin, defend themselves by accumulating phytoalexins. *J. Cell Biol.* 78: 627–43

Anderson-Prouty, A. J., Albersheim, P. 1975. Host-pathogen interactions: VIII. Isolation of a pathogen-synthesized fraction rich in glucan that elicits a defense response in the pathogenic host. *Plant Physiol.* 56: 286–91

Ayres, A. R., Ebel, J., Finelli, F., Berger, N., Albersheim, P. 1976a. Host-pathogen interactions: IX. Quantitative assays of elicitor activity and characterization of the elicitor present in the extracellular medium of cultures of *Phytophthora megasperma* var. sojae. *Plant Physiol.* 57: 751–59

Ayres, A. R., Ebel, J., Valent, B., Albersheim, P. 1976b. Host-pathogen interactions. X. Fractionation and biological activity of an elicitor isolated from the mycelial walls of *Phytophthora megasperma* var. sojae. *Plant Physiol.* 57: 760–65

Ayres, A., Valent, B., Ebel, J., Albersheim, P. 1976c. Host-pathogen interactions. XI. Composition and structure of wall-released elicitor fractions. *Plant Physiol.* 57: 766–74

Bailey, J. A. 1982. Mechanisms of phytoalexin accumulation. In *Phytoalexins*, ed. J. A. Bailey, J. W. Mansfield, pp. 289–318. Glasgow: Blackie

Baker, G. S., Atkinson, M., Collmer, A. 1985. Effects of cell wall fragments released by pectate lyase on the hypersensitive response in tobacco. *Phytopathology* 75: 1373

Bartnicki-Garcia, S. 1970. Cell wall composition and other biochemical markers in fungal phylogeny. In *Phytochemical Phylogeny*, ed. J. B. Harbourne, pp. 81–101. London: Academic

Bateman, D. F., Basham, H. G. 1976. Degradation of plant cell walls and membranes by microbial enzymes. In *Encyclopedia of Plant Physiology, Physiol. Plant Pathol. New Series*, ed. R. Heitefuss, P. H. Williams, Vol. IV, pp. 316–55. New York: Springer-Verlag

Baydoun, E. A.-H., Fry, S. C. 1985. The immobility of pectic substances in injured tomato leaves and its bearing on the identity of the wound hormone. *Planta* 165: 269–76

Bishop, P. D., Makus, D. J., Pearce, G., Ryan, C. A. 1981. Proteinase inhibitor-inducing activity in tomato leaves resides in oligosaccharides enzymically released from cell walls. *Proc. Natl. Acad. Sci. USA* 78: 3536–40

Bishop, P. D., Pearce, G., Bryant, J. E.,

Ryan, C. A. 1984. Isolation and characterization of the proteinase inhibitor-inducing factor from tomato leaves: Identity and activity of poly- and oligogalacturonide fragments. *J. Biol. Chem.* 259: 13172–77

Boller, T. 1985. Induction of hydrolases as a defense reaction against pathogens. In *Cellular and Molecular Biology of Plant Stress*, ed. J. Key, T. Kosuge, pp. 247–62. New York: Liss

Boller, T., Gehri, A., Mauch, F., Vogeli, U. 1983. Chitinase in bean leaves: Induction by ethylene, purification, properties, and possible function. *Planta* 157: 22–31

Brown, W., Ryan, C. A. 1984. Isolation and characterization of a wound induced trypsin inhibitor from alfalfa leaves. *Biochemistry* 23: 3418–22

Bruce, R. J., West, C. A. 1982. Elicitation of casbene synthetase activity in castor bean: The role of pectic fragments of the plant cell wall in elicitation by a fungal endopolygalacturonase. *Plant Physiol.* 69: 1181–88

Campbell, B. C. 1986. Host-plant oligosaccharins in the honeydew of *Schizaphis graninum* (Rondani) (Insecta, Amphidae). *Experientia* 42: 451–52

Chappell, J., Hahlbrock, K. 1984. Transcription of plant defense genes in response to UV light or fungal elicitor. *Nature* 311: 76–78

Cline, K., Albersheim, P. 1981a. Host-pathogen interactions. XVI. Purification and characterization of a β-glucosyl hydrolase/transferase present in the walls of soybean cells. *Plant Physiol.* 68: 207–20

Cline, K., Albersheim, P. 1981b. Host-pathogen interactions. XVII. Hydrolysis of biologically active fungal glucans from soybean cells. *Plant Physiol.* 68: 221–28

Cline, K., Wade, M., Albersheim, P. 1978. Host-pathogen interactions. XV. Fungal glucans which elicit phytoalexin accumulation in soybean also elicit the accumulation of phytoalexins in other plants. *Plant Physiol.* 62: 918–21

Cramer, C. L., Ryder, T. B., Bell, J. N., Lamb, C. J. 1985. Rapid switching of plant gene expression induced by fungal elicitor. *Science* 227: 1240–43

Darvill, A. G., Albersheim, P. 1984. Phytoalexins and their elicitors—a defense against microbial infection in plants. *Ann. Rev. Plant Physiol.* 35: 234–75

Darvill, A. G., McNeill, M., Albersheim, P. 1978. Structure of plant cell walls. VIII. A new pectic polysaccharide. *Plant Physiol.* 62: 418–22

Davis, K. R., Darvill, A. G., Albersheim, P. 1984. Host-pathogen interactions. XXV.

Endopolygalacturonic acid lyase from *Erwinia carotovora* elicits phytoalexin accumulation by releasing plant cell wall fragments. *Plant Physiol.* 74: 52–60

Davis, K. R., Darvill, A. G., Albersheim, P. 1986a. Host-pathogen interactions: XXX. Characterization of elicitors of phytoalexin accumulation in soybean released from soybean cell walls by endopolygalacturonic acid lyase. *Z. Naturforsch.* 41c: 39–48

Davis, K. R., Darvill, A. G., Albersheim, P., Dell, A. 1986b. Host-pathogen interactions: XXIX. Oligogalacturonides released from sodium polypectate by endopolygalacturonic acid lyase are elicitors of phytoalexins in soybean. *Plant Physiol.* 80: 568–77

Dixon, R. A. 1986. The phytoalexin response: Elicitation, signalling and control of host gene expression. *Biol. Rev.* 61: 239–91

Dixon, R. A., Dey, P. M., Lamb, C. J. 1983. Phytoalexins; enzymology and molecular biology. *Adv. Enzymol. Relat. Areas Mol. Biol.* 53: 1–136

Dixon, R. A., Dey, P. M., Murphy, D. L., Whitehead, I. M. 1981. Dose responses for *Colletotrichem lindemuthianum* elicitor-mediated enzyme induction in French bean cell suspension cultures. *Planta* 151: 272–80

Ebel, J., Ayres, A. R., Albersheim, P. 1976. Host-pathogen interactions. XII. Response of suspension-cultured soybean cells to the elicitor isolated from *Phytophthora megasperma* var. sojae, a fungal pathogen of soybeans. *Plant Physiol.* 57: 775–79

Esquerre-Tugaye, M. T., Mazau, D., Pelissier, B., Roby, D., Rumeau, D., Toppan, A. 1985. In *Cellular and Molecular Biology of Plant Stress*, ed. J. Key, T. Kosuge, pp. 459–73. New York: Liss

Fristensky, B., Riggleman, R. C., Wagoner, W., Hadwiger, L. A. 1985. Gene expression in susceptible and disease resistant interactions of peas induced with *Fusarium solani* pathogens and chitosan. *Physiol. Plant Pathol.* 27: 15–28

Fry, S. C., 1986. In-vivo formation of xyloglucan nonasaccharide: A possible biologically active cell-wall fragment. *Planta* 169: 443–53

Gollin, D. J., Darvill, A. G., Albersheim, P. 1984. Plant cell wall fragments inhibit flowering and promote vegetative growth in *Lemna gibba* G3. *Biol. Cell* 51: 275–80

Green, T., Ryan, C. A. 1972. Wound-induced proteinase inhibitor in plant leaves: A possible defense mechanism against insects. *Science* 175: 776–77

Grisebach, H., Ebel, J. 1978. Phytoalexins,

chemical defense substances of higher plants. *Angew. Chem. Int. Ed. Engl.* 17: 635–47

Hadwiger, L. A., Beckman, J. M. 1980. Chitosan as a component of pea–*Fusarium solani* interactions. *Plant Physiol.* 66: 205–11

Hadwiger, L. A., Daniels, J. W., Fristensky, D. F., Kendra, D. F., Wagoner, W. 1986. Pea genes associated with the non-host resistance to *Fusarium solani* are also induced by chitosan and in race-specific resistance by *Pseudomonas syringae*. In *Biology and Molecular Biology of Plant-Pathogen Interactions*, ed. J. A. Bailey, pp. 263–69. Berlin: Springer Verlag

Hadwiger, L. A., Loschke, D. C. 1981. Molecular communication in host-parasite interactions: Hexosamine polymers (chitosan) as regulator compounds in race-specific and other reactions. *Phytochemistry* 71: 756–62

Hahn, M. G., Albersheim, P. 1978. Host-pathogen interactions. XIV. Isolation and partial characterization of an elicitor from yeast extract. *Plant Physiol.* 62: 107–11

Hahn, M. G., Darvill, A. G., Albersheim, P. 1981. Host-pathogen interactions: XIX. The endogenous elicitor, a fragment of a plant cell wall polysaccharide that elicits phytoalexin accumulation in soybeans. *Plant Physiol.* 68: 1161–69

Hargreaves, J. A., Bailey, J. A. 1978. Phytoalexin production by hypocotyls of *Phaseolus vulgaris* in response to constitutive metabolites released by damaged bean cells. *Physiol. Plant Physiol.* 13: 89–100

Hargreaves, J. A., Selby, C. 1978. Phytoalexin formation in cell suspensions of *Phaseolus vulgaris* in response to an extract of bean hypocotyl. *Phytochemistry* 17: 1099–1102

Ingraham, J. L. 1982. Phytoalexins from the Leguminosae. In *Phytoalexins*, ed. J. A. Bailey, J. W. Mansfield, pp. 21–80. Glasgow: Blackie

Jin, D. J., West, C. A. 1984. Characteristics of galacturonic acid oligomers as elicitors of casbene synthetase activity in castor bean seedlings. *Plant Physiol.* 74: 989–92

Karban, R. 1985. Resistance against spider mites in cotton induced by mechanical abrasion. *Entomol. Exp. Appl.* 37: 137–41

Kauss, H. 1984. Chitosan-elicited synthesis of callose in soybean cells: A rapid and Ca^{++}-dependent process. *6th. John Innes Symp., Cell Surf. in Plant Growth and Dev.* Norwich, p. 53 (Abstr.)

Keen, N. T., Legrand, M. 1980. Surface glycoproteins: Evidence that they may function as the race specific phytoalexin elicitors of *Phytophthora megasperma* f.sp.

glycinea. *Physiol. Plant Pathol.* 17: 175–92

Keen, N. T., Partridge, J. E., Zaki, A. I. 1972. Pathogen-produced elicitor of a chemical defense mechanism in soybeans monogenically resistant to *Phytophthora megasperma* var. sojae. *Phytopathology* 62: 768–72

Keen, N. T., Yoshikawa, M. 1983. β-1,3-Endoglucanase from soybean releases elicitor-active carbohydrates from fungus cell walls. *Plant Physiol.* 71: 460–65

Keen, N. T., Yoshikawa, M., Wang, M. C. 1983. Phytoalexin elicitor activity of carbohydrates from *Phytophthora megasperma* f.sp. glycinea and other sources. *Plant Physiol.* 71: 466–71

Keenan, P., Bryan, I. B., Friend, J. 1985. The elicitation of the hypersensitive response of potato tuber tissue by a component of the culture filtrate of *Phytophthora infestans*. *Physiol. Plant Pathol.* 26: 343–55

Kendra, D. F., Hadwiger, L. A. 1984. Characterization of the smallest chitosan oligomer that is maximally antifungal to *Fusarium solani* and elicits pisatin formation in *Pisum sativum*. *Exp. Mycol.* 8: 276–81

Kuc, J. 1982. Phytoalexins from the Solanaceae. In *Phytoalexins*, ed. J. A. Bailey, J. W. Mansfield, pp. 80–105. Glasgow: Blackie

Kuc, J., Rush, J. S. 1985. Phytoalexins. *Arch. Biochem. Biophys.* 236: 455–72

Kuhn, D. N., Chappell, J., Boudet, A., Hahlbrock, K. 1984. Induction of phenylalanine ammonia-lyase and 4-coumarate: CoA ligase mRNAs in cultured plant cells by UV light of fungal elicitor. *Proc. Natl. Acad. Sci. USA* 81: 1102–6

Kurosaki, F., Futamura, K., Nishi, A. 1985. Factors affecting phytoalexin production in cultured carrot cells. *Plant Cell Physiol.* 26: 693–700

Kurosaki, F., Nishi, A. 1984. Elicitation of phytoalexin production in cultured carrot cells. *Physiol. Plant Pathol.* 24: 169–76

Kurosaki, F., Tsurusawa, Y., Nishi, A. 1985. Partial purification and characterization of elicitors for 6-methoxymellein production in cultured carrot cells. *Physiol. Plant Pathol.* 27: 209–17

Lamb, C. J., Bell, J. N., Corbin, D. R., Lawton, M. A., Mehdy, M. C., et al. 1987. Activation of defense genes in response to elicitor and infection. In *Molecular Strategies for Crop Protection*, ed. C. J. Arntzen, C. A. Ryan, pp. 49–58. New York: Liss

Lau, J. M., McNeil, M., Darvill, A. G., Albersheim, P. 1985. Structure of the backbone of rhamnogalacturonan I, a pectic polysaccharide in the primary cell walls of plants. *Carbohydr. Res.* 137: 111–25

Lee, S.-C., West, C. A. 1981a. Polygalacturonase from *Rhizopus stolonifer* an elicitor of casbene synthetase activity in castor bean (*Ricinus communis* L.) seedlings. *Plant Physiol.* 67: 633–39

Lee, S.-C., West, C. A. 1981b. Properties of *Rhizopus stolonifer* polygalacturonase, an elicitor of casbene synthetase activity in castor bean (*Ricinus communis* L.) seedlings. *Plant Physiol.* 67: 640–45

Loschke, D. C., Hadwiger, L. A., Wagoner, W. 1983. Comparison of mRNA populations coding for phenylalanine ammonia lyase and other peptides from pea tissue treated with biotic and abiotic phytoalexin inducers. *Physiol. Plant Pathol.* 23: 163–73

Mansfield, J. W. 1982. The role of phytoalexins in disease resistance. In *Phytoalexins*, ed. J. A. Bailey, J. W. Mansfield, pp. 253–88. Glasgow: Blackie

Mauch, F., Hadwiger, L. A., Boller, T. 1984. Ethylene: Symptom, not signal for the induction of chitinase and β-1,3-glucanase in pea pods by pathogens and elicitors. *Plant Physiol.* 76: 607–11

McNeil, M., Darwill, A. G., Albersheim, P. 1980. Structure of plant cell walls. X. Rhammogalacturonase I, a structurally complex pectin polysaccharide in the walls of suspension-cultured sycamore cells. *Plant Physiol.* 66: 1128–34

Miller, R. H., Berryman, A. A., Ryan, C. A. 1986. Biotic elicitors of defense reactions in lodgepole pine. *Phytochemistry* 25: 611–12

Modderman, P. W., Schot, C. P., Klis, F. M., Wieringa-Brants, D. H. 1985. Acquired resistance in hypersensitive tobacco against tobacco mosaic virus, induced by plant cell wall components. *Phytopathol. Z.* 113: 165–70

Molano, J., Polacheck, I., Duran, A., Cabib, E. 1979. An endochitinase from wheat germ. *J. Biol. Chem.* 254: 4901–7

Moran, F. S., Nasuno, S., Starr, M. P. 1968. Extracellular and intracellular polygalacturonic acid trans-eliminases of *Erwinia cartovora*. *Arch. Biochem. Biophys.* 123: 298–306

Nichols, E. J., Beckman, J. M., Hadwiger, L. A. 1980. Glycosidic enzyme activity in pea tissue and pea–*Fusarium solani* interactions. *Plant Physiol.* 66: 1068–73

Nothnagel, E. A., McNeil, M., Albersheim, P., Dell, A. 1983. Host-pathogen interactions: XXII. A galacturonic acid oligosaccharide from plant cell walls elicits phytoalexins. *Plant Physiol.* 71: 916–26

Ossowski, P., Pilotte, A., Garegg, P. J., Lindberg, B. 1984. Synthesis of a gluco-

heptaose and a glucooctaose that elicit phytoalexin accumulation in soybean. *J. Biol. Chem.* 259: 11337–40

Paradies, I., Konze, J. R., Elstner, E. I. 1980. Ethylene: Symptom, but not inducer of phytoalexin synthesis in soybean. *Plant Physiol.* 66: 1106–9

Paxton, J. D. 1981. Phytoalexins, a working definition. *Phytopathol. Z.* 101: 106–9

Pearce, R. B., Ride, J. P. 1982. Chitin and related compounds as elicitors of the lignification response in wounded wheat leaves. *Physiol. Plant Pathol.* 20: 119–23

Pegg, G. F. 1976. The involvement of ethylene in plant pathogenesis. In *Encyclopedia of Plant Physiology*, ed. R. Heitefuss, P. H. Williams, Vol. 4, pp. 582–91. Berlin: Springer-Verlag

Pegg, G. F. 1977. Glucanohydrolases of higher plants: A possible defense mechanism against fungi. In *Cell Wall Biochemistry Related to Specificity in Host-Plant Pathogen Interactions*, ed. B. Sherin, J. Raa, pp. 305–45. Tromso: Universitatsforlaget

Pegg, G. F., Vessey, J. C. 1973. Chitinase activity in *Lycopersicon esculentum* and its relationship to the *in vivo* lysis of *Verticillium alboatrum* mycelium. *Physiol. Plant Pathol.* 3: 207–22

Pegg, G. F., Young, D. H. 1981. Changes in glycosidase activity and their relationship to fungal colonisation during infection of tomato by *Verticillium albo-atrum*. *Physiol. Plant Pathol.* 19: 371–82

Prowning, R. F., Irzykiewicz, H. 1965. Studies on the chitinase system in bean and other seeds. *Comp. Biochem. Physiol.* 14: 127–33

Robertson, B. 1986. Do galacturonic acid oligosaccharides have a role in the resistance mechanism of cucumber towards *Cladosporium cucumerinum*? In *Biology and Molecular Biology of Plant-Pathogen Interactions*, ed. J. Bailey, pp. 177–83. Berlin: Springer-Verlag

Roby, D., Toppan, A., Esquerre-Tugaye, M. T. 1985. Cell surfaces in plant-microorganism interactions. V. Elicitors of fungal and of plant origin trigger the synthesis of ethylene and of cell wall hydroxyproline-rich glycoproteins in plants. *Plant Physiol.* 77: 700–4

Roby, D., Toppan, A., Esquerre-Tugaye, M. T. 1986. Cell surfaces in plant-microorganism interactions. VI. Elicitors of ethylene from *Colletotrichum lagenarium* trigger chitinase activity in melon plants. *Plant Physiol.* 81: 228–33

Ryan, C. A. 1978. Proteinase inhibitors in plant leaves: A biochemical model for natural plant protection. *Trends Biochem. Sci.* 5: 148–51

Ryan, C. A., Bishop, P. D., Graham, J. S., Broadway, R. M., Duffey, S. S. 1986. Plant and fungal cell wall fragments activate expression of proteinase inhibitor genes for plant defense. *J. Chem. Ecol.* 12: 1025–36

Ryan, C. A., Bishop, P. D., Pearce, G., Darvill, A. G., McNeil, M., Albersheim, P. 1981. A sycamore cell wall polysaccharide and a chemically related tomato leaf oligosaccharide possess similar proteinase inhibitor-inducing activities. *Plant Physiol.* 68: 616–18

Sato, D., Uritani, I., Saito, T. 1982. Properties of terpene-inducing factor extracted from adults of the sweet potato weevil, *Cylas formicarius* Fabricius (Coleoptera: Brenthidae). *Appl. Entomol. Zool.* 17: 368–74

Sharp, J. K., Albersheim, P., Ossowski, O., Pilotti, A., Garegg, P., Lindberg, B. 1984a. Comparison of the structures and elicitor activities of a synthetic and a mycelial-wall-derived hexa-(β-D-gluopyranosyl)-D-glucitol. *J. Biol. Chem.* 259: 11341–45

Sharp, J. K., McNeil, M., Albersheim, P. 1984b. The primary structures of one elicitor-active and seven elicitor-inactive hexa(β-D-glucopyranosyl)-D-glucitols isolated from the mycelial walls of *Phytophthora megasperma* f.sp. glycinea. *J. Biol. Chem.* 259: 11321–36

Sharp, J. K., Valent, B., Albersheim, P. 1984c. Purification and partial characterization of a β-glucan fragment that elicits phytoalexin accumulation in soybean. *J. Biol. Chem.* 259: 11312–20

Showalter, A. M., Bell, J. N., Cramer, C. L., Bailey, J. A., Varner, J. E., Lamb, C. J. 1985. Accumulation of hydroxyproline-rich glycoprotein mRNAs in response to fungal elicitor and infection. *Proc. Natl. Acad. Sci. USA* 82: 6551–55

Showalter, A. M., Varner, J. E. 1987. Molecular details of plant cell wall hydroxyproline-rich glycoprotein expression during wounding and infection. In *Molecular Stratagies for Crop Protection*, ed. C. J. Arntzen, C. A. Ryan, pp. 375–92. New York: Liss

Somssich, I. E., Schmelzer, E., Bollemann, J., Hahlbrock, K. 1986. Rapid activation by fungal elicitor of genes encoding "pathogenesis-related" proteins in cultured parsley cells. *Proc. Natl. Acad. Sci. USA* 83: 2427–30

Stekoll, M., West, C. A. 1978. Purification and properties of an elicitor of castor bean phytoalexin from culture filtrates of the fungus *Rhizopus stolonifer*. *Plant Physiol.* 61: 38–45

Steossl, A. 1981. Phytoalexins, a biogenetic perspective. *Phytopathol. Z.* 99: 251–72

Toppan, A., Roby, D. 1982. Activité chitinasique de plantes de melon infectées par *Colletotrichum lagenarium* ou traitées par l'ethylene. *Agronomie* 2: 829–34

Tran Thanh Van, K., Touybart, P., Darvill, A. G., Gollin, D. J., Chelf, P., Albersheim, P. 1985. Manipulation of the morphogenic pathways of tobacco explants by oligosaccharins. *Nature* 314: 615–17

Tsukamoto, T., Koga, D., Ide, A., Ishibashi, T., Horino-Matsushige, M., et al. 1984. Purification and some properties of chitinases from yam, *Dioscorea opposita* thumb. *Agric. Biol. Chem.* 48: 931–39

van Loon, L. C. 1985. Pathogenesis-related proteins. *Plant Mol. Biol.* 4: 111–16

Venere, R. J., Brinkerhoff, L. A., Gholson, R. K. 1984. Pectic enzyme: An elicitor of necrosis in cotton inoculated with bacteria. *Proc. Okla. Acad. Sci.* 64: 1–7

Walker-Simmons, M., Hadwiger, L., Ryan, C. A. 1983. Chitosans and pectic polysacharides both induce accumulation of the antifungal phytoalexin pisitin in pea pods and antinutrient proteinase inhibitors in tomato leaves. *Biochem. Biophys. Res. Commun.* 110: 194–99

Walker-Simmons, M., Jin, D., West, C. A., Hadwiger, L., Ryan, C. A. 1984. Comparison of proteinase inhibitor-inducing activities and phytoalexin elicitor activities of a pure fungal endopolygalacturonase, pectic fragments, and chitosans. *Plant Physiol.* 76: 833–36

Walker-Simmons, M., Ryan, C. A. 1977. Wound-induced accumulation of trypsin inhibitor activities in plant leaves: A survey of several plant genera. *Plant Physiol.* 59: 437–39

Walker-Simmons, M., Ryan, C. A. 1984. Proteinase inhibitor synthesis in tomato leaves. Induction by chitosan oligomers and chemically modified chitosan and chitin. *Plant Physiol.* 76: 787–99

Walker-Simmons, M., Ryan, C. A. 1985. Proteinase inhibitor I accumulation in tomato suspension cultures: Induction by plant and fungal cell wall fragments and an extracellular polysaccharide secreted into the medium. *Plant Physiol.* 80: 68–71

Watson, D. G., Brooks, C. J. W. 1984. Formation of caposidol in *Capsicum annum* fruits in response to non-specific elicitors. *Physiol. Plant Pathol.* 24: 331–37

Weinstein, L. I., Hahn, M. G., Albersheim, P. 1981. Host-pathogen interactions: XVIII. Isolation and biological activity of glycinol, a pterocarpan phtyoalexin synthesized by soybean. *Plant Physiol.* 68: 447–57

West, C. A. 1981. Fungal elicitors of the phytoalexin response in higher plants. *Naturwissenschaften* 68: 447–57

West, C. A., Bruce, R. J., Jin, D. F. 1984. Pectic fragments of plant cell walls as mediators of stress responses. In *Structure, Function, and Biosynthesis of Plant Cell Walls*, ed. W. M. Dugger, S. Bartnicki-Garcia, pp. 359–80. Baltimore: Waverly

West, C. A., Moesta, P., Jin, D. F., Lois, A. F., Wickham, K. A. 1985. The role of pectic fragments of the plant cell wall in the response to biological stresses. In *Cellular and Molecular Biology of Plant Stress*, ed. J. L. Key, T. Kosuge, pp. 335–50. New York: Liss

Yamazaki, N., Fry, S. C., Darvill, A. G., Albersheim, P. 1983. Host-pathogen interactions: XXIV. Fragments isolated from suspension-cultured sycamore cell walls inhibit the ability of cells to incorporate [^{14}C]leucine into proteins. *Plant Physiol.* 72: 864–69

Yang, S. F., Pratt, H. K. 1978. The physiology of ethylene in wounded plant tissues. In *Biochemistry of Wounded Plant Tissues*, ed. G. Kahl, pp. 595–622. Berlin: de Gruyter

York, W. S., Darvill, A. G., Albersheim, P. 1984. Inhibition of 2,4-dichlorophenoxyacetic acid–stimulated elongation of pea stem segments by a xyloglucan oligosaccharide. *Plant Physiol.* 72: 295–97

Yoshikawa, M., Keen, N. T., Wang, M. C. 1983. A receptor on soybean membranes for a fungal elicitor of phytoalexin accumulation. *Plant Physiol.* 73: 497–506

Yoshikawa, M., Matama, M., Masago, H. 1981. Release of a soluble phytoalexin elicitor from mycelial walls of *Phytophthora megasperma* var. sojae by soybean tissues. *Plant Physiol.* 67: 1032–35

Ann. Rev. Cell Biol. 1987. 3 : 319–45

CELL ADHESION IN MORPHOGENESIS

David R. McClay and Charles A. Ettensohn

Department of Zoology, Duke University, Durham, North Carolina 27706

CONTENTS

INTRODUCTION 319
MECHANISMS OF CELL ADHESION 320
CELL ADHESION AND MORPHOGENESIS 324
  Aggregation in Dictyostelium 324
  Sea Urchin Gastrulation 327
  Compaction and Early Morphogenesis in the Mouse Embryo 329
  Cell Migration in Vertebrate Embryos 330
  Neurogenesis 333
CONCLUSIONS AND PERSPECTIVES 335

INTRODUCTION

About two billion years ago there occurred a splendid evolutionary event. Cells for the first time began to adhere to one another in clusters. This association of cells provided a basis for the evolution of higher multicellular organisms and created vast opportunities for animal and plant diversification. By the beginning of the Cambrian age some 600 million years ago natural selection had produced many multicellular organisms composed of cells arranged in complex spatial patterns. The evolution of such animals required solutions to (at least) two major problems. First, means of differential gene expression were necessary to provide the organism with specialized cells. Second, developmental mechanisms for cell rearrangements had to arise to organize tissues. The latter mechanisms are known collectively as morphogenesis.

As one component of the morphogenetic process, adhesive interactions play many key roles in the life history of an organism. The initiation of

319

0743–4634/87/1115–0319$02.00

embryogenesis is triggered by a specific adhesive event that brings the sperm and egg into contact at fertilization. As development proceeds, the differentiation of embryonic cells is influenced by their adhesive interactions with other cells and with extracellular substrates (Lash & Vasan 1978; Bissell et al 1982; Ekblom 1984; Hay 1984; Kleinman et al 1984, 1985; Wessel & McClay 1987). In fact, it now appears that many examples of embryonic induction involve this kind of interaction. Cell growth and division are also influenced by cell adhesions (see Glaser 1980; Steinberg 1986). Important cell interactions continue well into adulthood when, for example, cells of the immune system display multiple specific relationships.

This review focuses on a restricted group of cell-cell and cell-substrate interactions: those that participate in morphogenetic cell movements. The discussion excludes many studies on cells in culture, although these have contributed to the detailed understanding of some of the events described below. An ontogenetic approach considers those organisms for which the experimental information concerning adhesion and morphogenesis is most complete. One theme of this review is the importance of adhesive specificity in morphogenetic movements. While the discussion below is distinguished by its emphasis upon events in the embryo, there are many relevant recent reviews of various aspects of cell adhesion and morphogenesis (Edelman 1985, 1986; Ekblom et al 1986; Garrod 1986; Obrink 1986).

MECHANISMS OF CELL ADHESION

The adhesion of a cell to other cells or to an extracellular substratum is a complex phenomenon that may involve a variety of mechanisms, each consisting of several steps. A distinction has been made between "physical" and "chemical" theories of cell adhesion (Bell 1978; Gingell & Vince 1980; Trinkhaus 1984). Although the contribution of nonspecific, physical (e.g. electrostatic) forces to cell adhesion remains unclear, there is now overwhelming evidence that specific molecular (receptor-ligand) interactions are involved in the adherence of cells to one another and to extracellular materials. It is this kind of adhesive interaction that is discussed below.

To analyze a complex process such as the formation of an adhesion by a moving cell, it is necessary to break the process down into its parts. The following model serves this purpose and fits much of the experimental data. At the first encounter between a cell and a substratum (here defined broadly either as another cell or a noncellular substratum, such as an extracellular matrix) binding occurs. This initial binding is brought about by a noncovalent molecular interaction between a component of the substrate and a receptor for that component on the cell surface. The interaction can be homophilic [e.g. NCAM with NCAM (Rutishauser et al 1982), as described below] or heterophilic (e.g. fibronectin with the fibronectin receptor). Once this initial interaction takes place, it can be strengthened by one or more of a series of steps that includes redistribution of adhesive molecules in the plane of the cell membrane (Chow & Poo 1982; Pollerberg et al 1986), coupling of the adhesive components to the cytoskeleton (McClay et al 1981; Horwitz et al 1986), and formation of cell junctions. In most studies of cell adhesion, the initial and strengthening events have not been distinguished, although it is possible to do so experimentally (McClay et al 1981). When antibodies are used to interfere with cell adhesion, it is usually assumed that they block the initial molecular interaction.

A number of molecules associated with adhesion have been identified, but our knowledge of the events underlying molecular interactions remains rudimentary. Many cell surface molecules have been implicated in adhesion (presumably the initial binding). These have been referred to as cell adhesion molecules (CAMs) (Edelman 1985). They include NCAM (reviewed by Edelman 1986), L1 (NgCAM, NILE) (Grumet & Edelman, 1984; Rathjen & Schachner 1984; Bock et al 1985), uvomorulin (also called LCAM, cell-CAM 120/80, and E-cadherin) (Bertolotti et al 1980; Hyafil et al 1980; Yoshida & Takeichi 1982), cadherins (Ogou et al 1983; Yoshida-Noro et al 1984; Nose & Takeichi 1986), R cognin (Trocolli & Hausman 1985; Hausman et al 1987), cell CAM 105 (Ockland et al 1984), and gp80 (contact site A glycoprotein) (Muller & Gerisch 1978; Murray et al 1983), and others (e.g. Hansen et al 1985; Kruse et al 1985). The degree to which these molecules have been characterized varies greatly. In some cases, a molecule is known only by its apparent molecular weight. In contrast, in the cases of NCAM, uvomorulin, and gp80, a great deal is known about the structure of the protein, the regulation of its expression, and the sequence of its gene. (For information on N-CAM see Murray et al 1984a, 1986a, 1986b; Goridis et al 1985; Gennarini et al 1986; Hemperly et al 1986; for uvomorulin/L-CAM see Gallin et al 1985; Schuh et al 1986; for gp80 see Murray et al 1984b; Noegel et al 1985, 1986.)

The original identification of most cell adhesion molecules was based upon the observation that monovalent antibodies (Fabs) directed against them interfere with cell-cell adhesion as measured in in vitro assays (Beug et al 1970, 1973). Such an approach can identify candidates for cell adhesion molecules, but antibody inhibition of adhesion alone is not proof of adhesive function. The binding of antibodies to the cell surface may sterically block access to molecules other than the specific antigen or may affect cell adhesion indirectly by perturbing the function of a regulatory molecule that has global effects on cell adhesion (see Gerisch 1986). In many cases, corroborating (although indirect) evidence for an adhesive function comes from studies demonstrating that the developmental expression of the molecule in question can be correlated with a putative function. In a few cases, additional evidence has come from binding studies using liposome vesicles containing the adhesive component (Hoffman et al 1982; Rutishauser et al 1982) or beads coated with it (Siu et al 1986). In two cases, those of gp80 and discoidin I in *Dictyostelium*, mutants defective in the putative adhesion molecule have been isolated and studied (see below).

Probably only a small proportion of existing CAMs have been described to date; nevertheless, attempts have been made to categorize those already identified. Early classifications subdivided CAMs into calcium-dependent and calcium-independent molecules (calcium-dependent CAMs require calcium for adhesion) (Urushihara et al 1979). More recent classifications are based on the function of the molecules and the extent of their distribution. Some of the molecules function in adhesion in a variety of cell types in many different tissues and appear early in development. These molecules have been termed "primary" by Edelman (1986) to indicate their ubiquitous presence and apparent relative importance. They have been called "general" by Rutishauser (unpublished) to indicate their low specificity. Other molecules have been described as "secondary" (Edelman 1986) or "restricted" to indicate that they are not as widely distributed and/or appear to contribute quantitatively less to adhesive interactions. It should be appreciated that such molecules, despite their restricted distribution or limited contribution to adhesion, may be of great importance in morphogenesis.

In addition to the interest in molecules associated with the cell membrane, considerable attention is being devoted to constituents of the extracellular matrix that function in adhesion. All extracellular matrices contain collagens, other glycoproteins, and proteoglycans. Attachments of cells to these molecules are thought to be important for cell movements and for the maintenance of tissue architecture (see Ekblom et al 1986). Two extracellular matrix glycoproteins that are being studied intensively in this regard are laminin (Kleinman et al 1984; Yamada et al 1985) and fibronectin (Hynes & Yamada 1982; Hynes 1985). These molecules contain short binding domains that interact with receptors in the plasma membrane of cells. For example, the attachment of fibronectin to cells is mediated by the amino acid sequence Arg-Gly-Asp (RGD) located within the cell binding region of the fibronectin molecule (Pierschbacher & Ruoslahti 1984; Ruoslahti & Pierschbacher 1986; see also Yamada & Kennedy 1985). This domain interacts with a complex of integral membrane glycoproteins that collectively constitute the fibronectin receptor also called the CSAT antigen, the 140-kDa complex, or integrin) (Pytela et al 1985; Ruoslahti & Pierschbacher 1986; Tamkun et al 1986; Horwitz & Buck, this volume). The picture is complicated by the fact that other extracellular attachment proteins (such as laminin and vitronectin) also contain the RGD sequence. Several different RGD receptors have been found on different cell types; some exhibit highly specific binding to a single RGD-containing protein, while others are less selective and can bind to several RGD-containing proteins (see Horwitz et al 1985; Ruoslahti & Pierschbacher 1986; Horwitz & Buck, this volume). Also, a cell may have more than one way of binding to a particular extracellular matrix molecule. For example, fibronectin also has a heparan sulfate binding domain, and some cells can interact with this region of the molecule via heparan sulfate–containing proteoglycans located on their surfaces (Kjellan et al 1981; Lattera et al 1983; Rapraeger & Bernfield 1985; Rapraeger et al 1985, 1986; Saunders et al 1986; Woods et al 1985, 1986). Also, a unique peptide on the B1 chain of laminin has cell binding activity (Graf et al 1987).

There is evidence that the initial binding of a cell to a substrate is rapidly stabilized (strengthened). An increase in the local concentration of adhesion molecules (due to insertion of new molecules or translational movement of preexisting molecules in the plane of the membrane) may contribute to adhesive strength; however, the increase in the strength of the coupling between the cell surface and the cytoskeleton probably has a much greater effect. In the presence of cytochalasin B, cells bind with the strength of the initial affinity, but the interaction is not strengthened significantly. In the absence of cytochalasin B (at 37°C and in the presence of ATP) there is a rapid and dramatic strengthening of the adhesion that can make it orders of magnitude stronger in minutes (McClay et al 1981).

Recent studies support these observations at a molecular level. Specific molecules have been shown to link adhesive components on the cell surface to the cytoskeleton (Horwitz et al 1986). Cells in culture adhere to substrates at discrete, specialized regions of their surface known as focal contacts (Chen & Singer 1982). Numerous studies have described colocalization of actin filaments, a number of intracellular proteins, fibonectin and, more recently, the fibronectin receptor at these sites (Chen & Singer 1982; Hynes & Yamada 1982; Chen et al 1986; Grinnell 1986). The 140-kDa receptor complex interacts with talin, a molecule on the cytoplasmic side of the membrane that is known to interact indirectly with actin (Horwitz et al 1986). Thus adhesive interactions with the extracellular matrix appear to involve linking to the cytoskeleton. Much remains to be learned about the dynamics of this linkage during motility.

A final event that can strengthen an adhesion is the formation of specialized cell junctions. This is a slower process than those discussed above, although junctions are dynamic and can begin to form within minutes of an interaction (Overton 1977). Such structures can add mechanical strength to tissues and can provide means of intercellular communication or epithelial impermeability. Our knowledge of the molecular biology of cell junctions is advancing rapidly with descriptions of gap-junction proteins (Unwin & Ennis 1983; Revel 1986), desmosomal proteins (Cowin et al 1985; Yokoo et al 1985; Gorbsky 1986), a tight junction protein known as ZO-1 (Stevenson et al 1986), and proteins associated with the zonula adherens (Boller et al 1985; Cowin et al 1986; Volk & Geiger 1986a,b). Although junction formation during development has been studied extensively (for reviews see Gorbsky 1986; Revel 1986), it is not known to what extent the formation or reorganization of these structures contributes to morphogenetic cell movements.

The various components of cell adhesion are each subject to rapid regulation. The initiation of cell migration in the embryo has been shown to involve several coordinated adhesive changes (McClay & Fink 1982; Fink & McClay 1985). After a cell has begun to move, its translocation involves constant forming and breaking of adhesive contacts with the substratum. In developing epithelia, during cell rearrangement cell-cell junctions are redistributed so that the integrity of the epithelial barrier is constantly maintained (Fristrom 1982; Ettensohn 1985). Thus there is evidence that each of the components of the adhesive process is subject to regulation and that the adhesive process as a whole is highly dynamic.

CELL ADHESION AND MORPHOGENESIS

Ever since the early studies by Wilson (1907), Holtfreter (1943, 1944), Moscona (1952), and others demonstrated tissue- and species-specific sorting out of embryonic cells, it has been believed that specialized adhesive properties of cells play a key role in morphogenesis. This is a reasonable hypothesis in light of what is known about the dramatic cell rearrangements that take place during gastrulation (Keller et al 1985; Gerhart & Keller 1986), neurulation (Jacobson et al 1985), neurogenesis (Le Douarin 1982), and organ formation (Poole & Steinberg 1981, 1982; Zackson & Steinberg 1986). There is little doubt that cell adhesion is important; however, little is known about the extent to which adhesive interactions regulate or direct specific morphogenetic events during development. Below, we examine those experimental systems for which the most detailed information is available.

Aggregation in Dictyostelium

During one phase in the life cycle of slime molds, solitary migrating cells gather to form a multicellular mass by a process known as aggregation.

In response to a depletion of the food supply, individual cells form moving streams that converge on a central point, where they coalesce into a multicellular conglomerate (the pseudoplasmodium or slug).

Early evidence that changes in cell adhesion accompany aggregation came from the observation that the binding of growth-phase (pre-aggregation) cells to one another is abolished by EDTA, while the adhesion of aggregation-stage cells is EDTA resistant (Gerisch 1961). In addition, with the onset of aggregation, slime mold cells acquire the capacity to sort out according to species in mixed aggregates (reviewed in Gerisch 1986). Aggregation is also correlated with a change in the contact behavior of the cells; during this phase they exhibit a preference to adhere to one another in an end-to-end fashion. These observations led to the hypothesis that at least two separate mechanisms of cell-cell adhesion operate in *Dictyostelium*, one that appears at aggregation and is responsible for EDTA-resistant, end-to-end adhesion (via "contact site A") and another that is responsible for EDTA-sensitive adhesion in both growth-phase and aggregation-stage cells (via "contact site B"). The existence of two separate mechanisms was confirmed by immunological experiments showing that Fab fragments of appropriate specificity can disrupt either contact site A or B independently (Beug et al 1970, 1973).

The EDTA-resistant, end-to-end adhesion of aggregating cells is thought to be mediated by an integral membrane glycoprotein of apparent molecular weight 80,000, known as the gp80 or csA glycoprotein. The original identification of this molecule was based on the observation that the ability of Fabs to block adhesion of aggregation-stage cells can be neutralized by absorption with purified gp80 (Muller & Gerisch 1978). Monoclonal antibodies against purified gp80 have been generated, some of which inhibit EDTA-resistant aggregation (Murray et al 1983; Bertholdt et al 1985; Siu et al 1985; Springer & Barondes 1985). Until recently it had not been demonstrated that purified gp80 was active in any cell adhesion assay. Siu et al (1986), however, have reported that polystyrene beads conjugated with gp80 bind specifically to the ends of aggregation-stage cells. These workers also have provided the first evidence that the binding of gp80 may be via a homophilic interaction. The region of the gp80 molecule responsible for binding has not yet been determined.

The timing of the expression of the gp80 glycoprotein is consistent with its proposed role in aggregation-stage adhesion. Neither the gp80 protein nor its mRNA is produced in growth-phase cells, but both accumulate following starvation and then disappear after aggregation is complete (Noegel et al 1986). Levels of both the gp80 mRNA and its cognate protein can be increased experimentally by pulses of exogenous cAMP (Darmon et al 1975; Noegel et al 1985), thus the expression of gp80 may be modulated by the endogenous cAMP pulses normally produced during aggregation. Some steps in the posttranslational processing (glycosylation, phosphorylation, and acylation) of gp80 have been described (reviewed by Gerisch 1986), and the sequence of its gene has been determined (Noegel et al 1986).

The consistent picture that emerges from the studies described above has been complicated by recent analyses of mutations that affect expression of gp80. Mutations in a locus designated mod B have been isolated and shown to prevent the normal glycosylation of gp80 (Murray et al 1984b; Gerisch et al 1985; Loomis et al 1985). These mutants express on the cell surface only very small amounts of an improperly processed form of gp80. In addition, a mutation that prevents the synthesis of the gp80 protein has been identified (Noegel et al 1985). All of these mutants show reduced EDTA-resistant cell adhesion as measured by an agglutination assay. Surprisingly, however, they also show essentially normal phenotypes with regard to streaming, aggregation, and fruiting-body formation. This raises some question about the relevance of EDTA-resistant (end-to-end, contact site A–mediated) adhesion to morphogenesis in *Dictyostelium*. Further work is needed to determine whether streaming is mediated by EDTA-*sensitive* adhesion (independent of gp80), or whether some EDTA-resistant adhesion not detectable by the agglutination assay can be mediated by components other than the gp80 glycoprotein.

Other candidates for cell adhesion molecules have been identified in *Dictyostelium discoideum* based on antibody inhibition and absorption experiments. A 126-kDa glycoprotein has been identified as a candidate for adhesion via contact site B in growth-phase cells (EDTA-sensitive adhesion) (Chadwick & Garrod 1983). Also, Loomis et al (1987) have reported that a 28-kDa protein participates in EDTA-sensitive adhesion. Thus the molecular basis of the adhesion of growth-phase cells remains to be more fully elucidated.

Carbohydrate binding proteins (lectins) have been isolated from slime mold cells and originally were considered as possible mediators of cell-cell adhesion. Despite an intensive search, however, no endogenous ligands for the carbohydrate binding sites of these lectins (known as discoidins) have been found. It now appears that these carbohydrate binding molecules function in the compartmentalization and eventual secretion of undigested bacterial capsular polysaccharides absorbed during feeding (Cooper et al 1986). Recently it has been proposed that the most abundant lectin, discoidin I, may be a fibronectinlike molecule involved in cell-substrate adhesion, not through its lectinlike qualities, but by an RGD-containing cell binding domain similar to that of fibronectin (Springer et al 1984; Gabius et al 1985). RGD-containing peptides, purified discoidin I, and Fabs directed against discoidin I block streaming and aggregation. This is puzzling as mutants with very low levels of discoidin I can form aggregates and fruiting bodies (Springer et al 1984), although they do not show normal streaming movements (Alexander et al 1983; Springer et al 1984).

As noted above, gp80 is not synthesized after aggregation and eventually disappears, so other molecules must mediate cell-cell adhesion in the multicellular slug. One candidate for such a molecule (based on Fab inhibition and absorption experiments) is a 95-kDa membrane glycoprotein, the synthesis of which begins after aggregation and continues throughout development (Steinemann & Parish 1980).

While work with *Dictyostelium* has produced some of the most convincing evidence relating a cell-cell adhesion molecule (gp80) to a specific morphogenetic behavior (end-to-end cell contacts during streaming), these studies illustrate that morphogenesis of even so simple an organism as *Dictyostelium* involves multiple adhesive components. Moreover, the behavior of mutants defective in adhesive proteins indicates that none of the components thus far identified is indispensable. These observations indicate that complex adhesive events underlie even an (apparently) simple morphogenetic process such as aggregation.

Sea Urchin Gastrulation

During cleavage, blastomeres of the sea urchin embryo are in contact with three substrates: an external extracellular matrix called the hyaline layer, a basal lamina, and neighboring cells. During the early stages of development the basal lamina is poorly developed, and the strongest of the three interactions is that with the hyaline layer (Dan 1960). At the mesenchyme blastula stage, primary mesenchyme cells (PMCs) delaminate from the other cells of the blastula and ingress into the blastocoel by passing through the basal lamina. It has been shown experimentally that at the time of ingression PMCs lose their affinity for hyaline and for other cells of the blastula, while at the same time their affinity for the basal lamina increases (McClay & Fink 1982; Fink & McClay 1985). The increased affinity for the basal lamina appears to be specific for fibronectin (Fink & McClay 1985). Presumptive ectoderm and endoderm cells exhibit no changes in their affinities for the three substrates during this period of development.

After ingression, the PMCs leave the vegetal plate and migrate on the basal lamina that lines the wall of the blastocoel. They eventually reach a predictable location near the equator of the embryo, where they reassociate with one another, fuse to form an extensive syncytial network, and begin to synthesize the larval skeleton. Cell transplantation studies have shown that the PMCs are guided to the correct location by environmental cues and that those cues change during development (Ettensohn & McClay 1986). The interaction between the PMCs and the directional cues is a highly specific one, since other cell types placed under the same conditions fail to mimic the migratory behavior of the PMCs (Ettensohn & McClay 1986). Thus the pattern of PMC migration is the result of an interplay between specific properties of the migrating cells and changing signals from their environment. A reasonable working hypothesis is that unique cell surface properties of the PMCs and adhesive cues from the substratum of the cells (the basal lamina and blastocoelic matrix) are involved in this interaction. Both surfaces are known to undergo molecular changes at this time (McClay et al 1983; Solursh & Katow 1982; Wessel & McClay 1984; DeSimone & Spiegel 1986).

During gastrulation the vegetal region of the embryo invaginates, and the archenteron extends across the blastocoel. The invaginating cells undergo rearrangement during this process (Ettensohn 1985; Hardin & Cheng 1986). Monoclonal antibodies have been used to identify several endoderm-specific molecules that appear to participate in the morphogenesis of the archenteron. These antibodies interfere with the initial binding of the cells in vitro, and they recognize antigens that are found only on cells in defined subregions of the endoderm at gastrulation (McClay et al 1987). Expression of these antigens, which are synthesized at the onset of gastrulation, is dependent upon an interaction between the presumptive endoderm and the basal lamina. If collagen is missing from the basal lamina, expression of the endodermal antigens is prevented and invagination does not take place (Wessel & McClay 1987). Renewed collagen deposition is followed by antigen expression and by archenteron formation. Thus interaction between the endoderm cells and the extracellular matrix is necessary both for the differentiation and morphogenesis of the endoderm.

One cell surface antigen has been identified by a monoclonal antibody that inhibits ectodermal adhesion (McClay et al 1987). This antigen is expressed by only two populations of cells, both in the region of the prospective mouth. It is first expressed by ectoderm cells in the presumptive oral field and later by endoderm cells at the tip of the gut rudiment at the time these cells establish continuity with the ectoderm to form the mouth. The relationship between the formation of the mouth and the expression of this antigen is as yet unknown. However, it is clear from these studies and those described above that specific morphogenetic events during sea urchin gastrulation are associated with changes in the spatial and temporal expression of cell surface antigens that appear to serve an adhesive function.

Compaction and Early Morphogenesis in the Mouse Embryo

The first several cleavages in the mouse embryo occur rather slowly, and the blastomeres that result are loosely held together within the vitelline membrane. At the eight-cell stage, the blastomeres become tightly adherent to one another, and their boundaries become less distinct. This process is known as compaction. As development proceeds the blastomeres continue to divide, some more rapidly than others. The more slowly dividing blastomeres give rise to the inner cell mass, from which the embryo proper will arise; the rapidly dividing cells become the trophoblast, which will be involved in implantation and the formation of the placenta.

Antibodies directed against a molecule called uvomorulin have been shown to prevent compaction (Hyafil et al 1980; Vestweber & Kemler 1984; Richa et al 1985). If compaction is inhibited, the blastomeres continue to divide but do not organize themselves into a blastocyst. Other molecules have also been implicated in the compaction process (Bird & Kimber 1984).

It now appears that uvomorulin is homologous to E-cadherin (Yoshida & Takeichi 1982), cell-CAM 120/80 (Damsky et al 1983), and an adhesion molecule originally found in the embryonic chick liver called L-CAM (Bertolotti et al 1980). cDNAs for these molecules have been isolated and sequenced (Gallin et al 1985; Schuh et al 1986), and their distribution and expression has been studied in a variety of tissues (Edelman 1986; Ekblom et al 1986).

As development proceeds, the distribution of uvomorulin becomes progressively more restricted. Initially, uvomorulin is evenly distributed over the surfaces of early blastomeres, but it gradually becomes restricted to the intermediate junctions (zonulae adherentes) of epithelial cells (Boller et al 1985; Damjanov et al 1986). As the distribution of this molecule becomes increasingly restricted, the effects of anti-uvomorulin antibodies become less pronounced (Damsky et al 1985a; Vestweber et al 1985). Thus early in development uvomorulin may be generally distributed on the surfaces of blastomeres and may play a dominant role in cell-cell adhesion. As development proceeds and the distribution of this molecule becomes progressively restricted, its relative contribution may be lessened by the appearance of other adhesive components (Damsky et al 1985a; see also Garrod 1986).

At the time of implantation, trophoblast cells adhere to the uterine wall and invade the epithelial layer of the uterus to initiate the formation of the placenta. The molecular basis of this attachment is not well understood, although it is known that trophoblast cells express a 140-kDa glycoprotein that is recognized by antibodies to integrin (Richa et al 1985). Also, cultured blastocysts containing trophoblasts have been shown to use fibronectin and laminin for attachment and outgrowth (Armant et al 1986a,b). Thus implantation may involve an interaction between the trophoblast cells and fibronectin. Other molecules may be involved as well (e.g. Dutt et al 1987).

As the inner cell mass begins to develop, one of the first morphogenetic events is the delamination of the parietal endoderm (hypoblast). This endoderm eventually lines the inside of the trophoblast layer and forms extraembryonic endoderm. This aspect of morphogenesis has not yet been studied in the embryo; instead, teratocarcinoma stem cell lines have been used as an in vitro model (Adamson et al 1985; Grabel & Casanova 1986). These cells, originally derived from germ cell cancers, can be induced to differentiate into endodermlike cells under appropriate culture conditions. They will adhere to laminin, fibronectin, and other molecules in vitro, and the binding can be disrupted by antibodies of the appropriate specificity (Grabel 1984; Yoshida-Noro et al 1984; Maillet & Buc-Caron 1985). The migration of parietal endoderm cells on substrata coated with laminin or fibronectin can be blocked by an RGD-containing peptide or antibodies that recognize the 140-kDa fibronectin/laminin receptor (Grabel & Watts 1987). Fibronectin and/or laminin appear to serve as preferred substrata for the migration of these cells in vitro and therefore possibly in vivo. Of course, the migration of primitive endoderm cells on an extracellular matrix is just part of the story. It is not known, for example, what triggers the motility of the cells and causes them to leave the inner cell mass.

Gastrulation and organogenesis in the mammal are similar to these processes in the chicken (except for the behavior of the extraembryonic membranes), and many of the fates of the early cells have been described (Gardner 1982; Snow 1985; Lawson et al 1986). The possible adhesive relationships involved in the morphogenesis of the mammalian embryo have been examined by immunolocalization studies using antibodies to known adhesion molecules (Richa et al 1985; Yoshida-Nora et al 1984). In a different approach, a series of mutations in the *t* locus of the mouse have been described that may affect cell adhesion in the embryo (Bennett 1975). Work on these mutants has shown an abnormal galactosyl transferase activity, which may be associated with cell adhesion (Shur 1982, 1983).

Cell Migration in Vertebrate Embryos

NEURAL CREST CELLS   The neural crest cells of the vertebrate embryo originate along the dorsal border of the neural tube. At about the time of neural tube closure they begin to migrate to other sites in the embryo. Fate-map studies using chick-quail chimeras and neural crest cell–specific monoclonal antibodies have described the many pathways and phenotypes of these cells (Le Douarin 1982; Noden 1984; Vincent & Thiery 1984; Bronner-Fraser 1986a). The changes in the neural crest population that initiate migration are not known (Newgreen & Gibbins 1982; Thiery et al 1985b); however, the shift in the levels of NCAM and LCAM that occurs at this time may be involved (Thiery et al 1985b). Once migration is underway, contact inhibition may play a role in directing movement of the cells away from the neural tube (Rovasio et al 1983; Erickson 1985; but see also Davis & Trinkaus 1981). However, contact inhibition alone cannot explain how the neural crest cells are targeted to, and come to rest at, specific locations.

Once migration is initiated, the crest cells follow pathways lined with fibronectin (Newgreen & Thiery 1980; Duband & Thiery 1982). In vitro, fibronectin or fibronectin-containing extracellular matrices provide preferred substrates for the migration of these cells (Greenberg et al 1981; Erickson & Turley 1983). If the interaction between the cells and fibronectin is perturbed with antibodies against fibronectin (Rovasio et al 1983) or the fibronectin receptor (in the cranial region) (Bronner-Fraser 1986b), or with a peptide that contains the cell binding fragment of fibrinonectin (Boucaut et al 1984b), the migration of the neural crest cells is partially blocked. These studies support the idea that fibronectin is used as an adhesive substrate by the migrating cells. However, the migratory environment of the neural crest cells contains matrix components other than fibronectin (Weston et al 1984) that may also play an important role in crest cell migration. For example, laminin can also serve as a substratum for the migration of crest cells in vitro (Goodman & Newgreen 1985). The 140-kDa receptor appears to interact both with fibronectin and laminin, and RGD-containing peptides interfere with the binding of cells to both laminin and fibronectin (Horwitz et al 1985; Horwitz & Buck, this volume). Therefore, some of the perturbation experiments described above may not distinguish between the roles specifically played by fibronectin, laminin, and perhaps other unidentified molecules.

Two lines of evidence suggest that fibronectin (or other extracellular matrix molecules) do not provide all the directional information needed for crest cell migration. When crest cells are implanted into the pathway ahead of the wave of neural crest cell migration, they move radially from the site of implantation (Erickson 1985). Some of the cells move opposite the direction normally followed by the host neural crest cells that later during their migration sweep through the area of the implant. Thus models involving haptotaxis or chemotaxis do not appear to provide a satisfactory explanation for the directionality of crest cell migration. Furthermore, fibronectin is not distributed in a pattern that guides the neural crest cells to specific sites. For example, among their many sites of accumulation in the embryo, crest cells populate specific regions in the wall of the developing gut (Tucker et al 1986). Fibronectin is distributed throughout the wall of the gut, so there is no apparent relationship between the distribution of fibronectin and that of the neural crest cells (Tucker et al 1986). Similarly, it has been reported that the distributions of fibronectin and laminin in the trunk region of the avian embryo are not sufficiently specific to fully account for the pathways of migration followed by the crest cells (Krotoski et al 1986).

It has been suggested that other molecules, including J1 (Kruse et al 1985), cytotactin (which may be equivalent to J1) (Grumet et al 1985), and NCAM (Thiery et al 1985a), participate in the localization of neural crest cells. In addition, monoclonal antibodies have identified specific cell surface determinants on the crest cells (Vincent & Thiery 1984; Ciment & Weston 1985; Ciment et al 1986), but the function of these components is unknown. The migration of neural crest cells to a particular target has not yet been causally linked to a specific mechanism. It appears, however, that once the target is reached the environment provides the neural crest cells with information needed for their differentiation (Howard & Bronner-Fraser 1986; Kalcheim & Le Douarin 1986).

AMPHIBIAN MESODERMAL CELLS   During gastrulation in the amphibian embryo, prospective mesodermal cells that have involuted into the interior of the embryo migrate across the roof of the blastocoel. The filopodia of these cells make attachments to fibrils of the extracellular matrix covering the blastocoel roof. In urodeles, the extracellular matrix fibrils are preferentially aligned along the blastopore–vegetal pole axis, the direction of mesodermal cell migration (Nakatsuji et al 1982). When mesodermal cells are cultured on substrata with this extracellular matrix material, they exhibit oriented movement along the axis of fiber alignment (Nakatsuji & Johnson 1983). If the fiber alignment is experimentally manipulated, the orientation of cell locomotion changes in a corresponding manner (Nakatsuji & Johnson 1984). These observations constitute the best evidence that contact guidance cues (see Trinkaus 1984) operate in vivo in directing the movement of a population of embryonic cells.

The extracellular fibrils on the blastocoel roof contain fibronectin, as shown by immunolocalization studies (Boucaut & Darribere 1983; Darribere et al 1985; Nakatsuji et al 1985). An interaction between the cells and the fibronectin-containing fibrils appears essential for migration, since microinjection of anti-fibronectin Fabs or RGD-containing peptides into the blastocoel cavity blocks the movement of the cells (Boucaut et al

1984a,b). Molecules other than fibronectin may also be involved in the control of migration, since the extracellular fibrils contain components in addition to fibronectin (Nakatsuji et al 1985). Nevertheless, the evidence that these cells use fibronectin for attachment and migration is convincing. Other mesenchymal cell types, such as primordial germ cells (Heasman et al 1981), may also utilize fibronectin as a preferred substratum for migration (see also Lee et al 1984).

Neurogenesis

During morphogenesis, nerve cells extend processes (neurites) to distant target tissues. The pathways followed by these processes often are long and circuitous. Nevertheless, the final anatomy of connections is precise and highly reproducible from individual to individual. Studies of axonal path-finding in fish (Eisen et al 1986), insect (Doe & Goodman 1985a,b), and chick embryos (Lance-Jones & Landmesser 1981; Tosney & Landmesser 1984, 1985, 1986) have shown that the specificity of this path-finding behavior is striking. In several cases it has been shown that individual axons extend along absolutely sterotypical pathways characteristic of that particular neuron.

The mechanisms of neurite outgrowth have been studied intensively in culture. Neurite extension is mediated by the growth cone, the locomotory tip of the extending process (Wessells & Nuttall 1978; Bray 1982). Growth cones migrate preferentially on certain substrates (Letourneau 1975, 1979). For example, laminin (Hammarback et al 1985; Lander et al 1985), fibronectin (Rogers et al 1983), NCAM (Tosney et al 1986), L1 (Fischer et al 1986), and NCalCam (neural calcium-dependent cell adhesion molecule) (Bixby et al 1987) have all been found effective. Neurites interact with laminin and fibronectin by means of the CSAT receptor on the surface of the neuron (Horwitz & Buck, this volume). In the case of laminin, the interaction appears to involve a particular site on one arm of the laminin molecule (Engvall et al 1986). Studies with ciliary ganglion cells have shown that neurite outgrowth can also occur on substrates other than laminin or fibronectin (Sanes 1984; Sanes et al 1986; Kapfhammer et al 1986). Thus a number of substrates can support neurite outgrowth, although some seem to be preferred (e.g. laminin). Similarly, if cells are provided as potential substrates, Schwann cells, astrocytes and muscle cells appear to support outgrowth better than other cell types (Tomaselli et al 1986). Also neurites tend to grow out in bundles or fascicles, an adhesion mediated by a molecule called NILE (or Ng CAM) (Stallcup & Beasley 1985). While these studies have helped to define preferred substrates for neurite outgrowth in vitro, they do not fully address the mechanism of specificity in neuronal pathways in vivo.

Cell ablation studies in insect (Kuwada & Goodman 1985; Bastiani & Goodman 1986; Bastiani et al 1986; du Lac et al 1986), fish (Kuwada 1986), and chick embryos (Lance-Jones & Landmesser 1981; Landmesser & Honig 1986) have shown that there are highly specific recognition cues that direct neurite growth. In the central nervous system of the developing grasshopper, single "guidepost" cells (neurons or glial cells) provide essential information for axonal guidance (Bastiani & Goodman 1986; Bastiani et al 1986; du Lac et al 1986). No information is yet available concerning the molecular basis of these morphogenetic cues, although some monoclonal antibodies have been found that recognize cell surface antigens with highly restricted distributions in the nervous system (McKay et al 1983; Dodd et al 1984; Bastiani et al 1987). Such molecules may be recognized as "adhesive signposts" by extending neurons, although there is no direct evidence that this is so. Other mechanisms could impart directionality to the migration of growth cones. In this context, there is evidence that axonal outgrowth can be directed by chemotactic signals (Lumsden & Davies 1986). In several cases neuronal processes clearly follow pathways established earlier in development by pioneer axons (Bodick & Levinthal 1980; Goodman et al 1984; Landmesser & Honig 1986).

The neural retina has been widely used to study neuronal path-finding. In vertebrate embryos the neural ectoderm of the optic cup separates into two layers that differentiate into the pigmented and neural retinas. On the third day of development in the chick embryo, the neural retina begins a process of differentiation that eventually results in the formation of the sensory layer, several interneuron layers, and a ganglion layer that borders the vitreous humor. Between days three and eleven axons extend from the retina into the optic nerve. These axons ultimately project to the optic tectum, a region of the midbrain. The first axons to project from the retina originate from near the origins of the optic nerve, while later axons project from areas progressively more distant from that region. Following their arrival at the optic tectum, the axons move across the tectal surface to a stereotyped location, where they plunge into the optic tectum and form synapses with other neurons. The final result of this process is that the optic tectum is innervated in a highly ordered pattern that is the reverse of the pattern in the neural retina, with axons from the dorsal neural retina projecting to the ventral tectum, those from the anterior retina projecting to the posterior tectum, etc.

Studies on cell adhesion mechanisms in the neural retina have been of several sorts. Sperry (1963) first showed the precision with which retinal axons project to the optic tectum. He argued that there must be pattern-specifying determinants that somehow guide the retinal axons to the appropriate spots on the optic tectum. Several studies have tested this hypothesis using cultured cells (Marchase 1977; Bonhoeffer & Huf 1985) or regenerating axons (Sperry 1963). These studies consistently show that there are recognition markers on the surface of the tectum that guide the retinal axons to the correct locations. It has also been shown that there is an adhesive relationship between axons in nerve bundles. When certain molecules (NgCAM or NCAM) are blocked with antibodies, the retinal axons are unable to form correct bundles or projections to the tectum (Rutishauser 1984; Silver & Rutishauser 1984; Thanos et al 1984; Stallcup & Beasley 1985). Cell marker studies have established a number of rules for pattern specification on the tectum (Fraser 1980), and monoclonal antibodies have revealed patterns of antigenic determinants across the retinal surface (Trisler et al 1981; Constantine-Paton et al 1986); however, the actual mechanism of cell recognition that establishes the pattern has not yet been established.

Several molecules have been shown to participate in the adhesion of neural retina cells. The best-characterized of these is the neural cell adhesion molecule, NCAM (reviewed by Edelman 1985, 1986). Other molecules also appear to be involved in cell adhesion in the neural retina. These including L1 (Faissner et al 1985); NgCAM or NILE, which is thought to mediate neuron-glial interactions (Stallcup & Beasley 1985; Hoffman et al 1986); R cognin (Ophir et al 1984; Hausman et al 1987); a calcium-dependent cadherin (Hatta et al 1985); and NCalCAM (Bixby et al 1987). There is good evidence from in vitro assays for the participation of these CAMs in cell-to-cell adhesion. How they influence the differentiation, delamination, and integration of the multiple layers of the neural retina is unknown, although antibodies to NCAM and NgCAM disrupt delamination in the brain (Lindner et al 1983; Chuong et al 1987). Clearly, many questions remain unanswered: How do the various layers of the neural retina separate? What directs the ganglionic fibers specifically toward the optic nerve? What directs the axons to resolve their pattern on the surface of the optic tectum? A large number of studies have provided the background needed to answer these questions, and with new technologies molecular explanations of these morphogenetic processes are likely in the near future.

CONCLUSIONS AND PERSPECTIVES

The relationship between the adhesive properties of cells and morphogenetic cell rearrangments remains one of the most challenging areas of research in cell and developmental biology. Modern approaches have established that several molecules provide important substrates for adhesion and motility of a variety of cell types during development. In this as in any expanding field, however, there has been a tendency to accumulate suggestive data at a greater rate than it can be verified and organized into a theoretical framework. Thus, while the number of putative embryonic adhesion molecules is increasing rapidly, the principal challenge that remains is to learn how characteristic patterns of cell movement and tissue organization are specified by these and other molecules.

ACKNOWLEDGMENTS

The authors' research has been funded by NIH Grants HD14483 and EY04480 to DRM, and by NIH Grant HD06644 and a Hargitt post-doctoral fellowship to CAE.

Literature Cited

Adamson, E. D., Strickland, S., Tu, M., Kahan, B. 1985. A teratocarcinoma-derived endoderm cell line (1H5) that can differentiate into extraembryonic endoderm cell types. *Differentiation* 29: 68–76

Alexander, S., Shinnick, T. M., Lerner, R. A. 1983. Mutants of *Dictyostelium discoideum* blocked in expression of all members of the developmentally regulated discoidin multigene family. *Cell* 34: 467–75

Armant, D. R., Kaplan, H. A., Lennarz, W. J. 1986b. Fibronectin and laminin promote in vitro attachment and outgrowth of mouse blastocysts. *Dev. Biol.* 116: 519–23

Armant, D. R., Kaplan, H. A., Mover, H., Lennarz, W. J. 1986a. The effect of a hexapeptide on attachment and outgrowth of mouse blastocysts cultured in vitro. Evidence for the involvement of the cell recognition tripeptide Arg-Gly-Asp. *Proc. Natl. Acad. Sci. USA* 83: 6751–55

Bastiani, M. J., du Lac, S., Goodman, C. S. 1986. Guidance of neuronal growth cones in the grasshopper embryo. I. Recognition of a specific axonal pathway by the pCC neuron. *J. Neurosci.* 6: 3518–31

Bastiani, M. J., Goodman, C. S. 1986. Guidance of neuronal growth cones in the grasshopper embryo. III. Recognition of specific glial pathways. *J. Neurosci.* 6: 3542–51

Bastiani, M. J., Harrelson, A. L., Snow, P. M., Goodman, C. S. 1987. Expression of fasciclin I and II glycoproteins on subsets of axon pathways during neuronal development in the grasshopper. *Cell* 48: 745–55

Bell, G. I. 1978. Models for the specific adhesion of cells to cells. *Science* 200: 618–27

Bennett, D. 1975. The T-locus of the mouse. *Cell* 6: 441–54

Bertholdt, G., Stadler, J., Bozzaro, S., Fichtner, B., Gerisch, G. 1985. Carbohydrate and other epitopes of the contact site A glycoprotein of *Dictyostelium discoideum* as characterized by monoclonal antibodies. *Cell Diff.* 16: 187–202

Bertolotti, R., Rutishauser, U., Edelman, G. 1980. A cell surface molecule involved in aggregation of embryonic liver cells. *Proc. Natl. Acad. Sci. USA* 77: 4831–35

Beug, H., Gerisch, G., Kempff, S., Riedel, V., Cremer, G. 1970. Specific inhibition of cell contact formation in *Dictyostelium* by univalent antibodies. *Exp. Cell Res.* 63: 147–58

Beug, H., Katz, F. E., Gerisch, G. 1973. Dynamics of antigenic membrane sites relating to cell aggregation in *Dictyostelium discoideum*. *J. Cell Biol.* 56: 647–58

Bird, J. M., Kimber, S. J. 1984. Oligosaccharides containing fucose linked a(1–3) and a(1–4) to *N*-acetylglucosamine cause decompaction of mouse morulae. *Dev. Biol.* 104: 449–60

Bissell, M. J., Hall, H. G., Parry, G. 1982. How does the extracellular matrix direct gene expression? *J. Theor. Biol.* 99: 31–68

Bixby, J. L., Pratt, R. S., Lilien, J., Reichardt, L. F. 1987. Neurite outgrowth on muscle cell surfaces involves extracellular matrix receptors as well as Ca^{2+}-dependent and independent cell adhesion molecules. *Proc. Natl. Acad. Sci. USA* 84: 2555–59

Bock, E., Richter-Landsberg, C., Faissner,

A., Schachner, M. 1985. Demonstration of immunochemical identity between the nerve growth factor–inducible large external (NILE) glycoprotein and the cell adhesion molecule-L1. *EMBO J.* 4: 2765–68

Bodick, N., Levinthal, C. 1980. Growing optic nerve fibers follow neighbors during embryogenesis. *Proc. Natl. Acad. Sci. USA* 77: 4374–78

Boller, K., Vestweber, D., Kemler, R. 1985. Cell adhesion molecule uvomorulin is localized in the intermediate junctions of adult intestinal epithelial cells. *J. Cell Biol.* 100: 327–32

Bonhoeffer, F., Huf, J. 1985. Position-dependent properties of retinal axons and their growth cones. *Nature* 315: 409–10

Boucaut, J. C., Darribere, T. 1983. Fibronectin in early amphibian embryos. *Cell Tissue Res.* 234: 135–45

Boucaut, J. C., Darribere, T., Boulekbache, H., Thiery, J. P. 1984a. Antibodies to fibronectin prevent gastrulation but do not perturb neurulation in gastrulated amphibian embryos. *Nature* 307: 364–7

Boucaut, J. C., Darribere, T., Poole, T. J., Aoyama, H., Yamada, K. M., Thiery, J.-P. 1984b. Biologically active synthetic peptides as probes of embryonic development: A competitive peptide inhibitor of fibronectin function inhibits gastrulation in amphibian embryos and neural crest cell migration in avian embryos. *J. Cell Biol.* 99: 1822–30

Bray, D. 1982. Filopodial contraction and growth cone guidance. In *Cell Behavior*, ed. R. Bellairs, A. Curtis, G. Dunn, pp. 299–317. Cambridge, Engl.: Cambridge Univ. Press

Bronner-Fraser, M. 1986a. Analysis of the early stages of trunk neural crest migration in avian embryos using monoclonal antibody HNK-1. *Dev. Biol.* 115: 44–55

Bronner-Fraser, M. 1986b. An antibody to a receptor for fibronectin and laminin perturbs cranial neural crest development in vivo. *Dev. Biol.* 117: 528–36

Buck, C. A., Horwitz, A. F. 1987. Cell surface receptors for extracellular matrix molecules. *Ann. Rev. Cell Biol.* 3: 179–205

Chadwick, C., Garrod, D. R. 1983. Identification of the cohesion molecule, contact sites B, of *Dictyostelium discoideum*. *J. Cell Sci.* 60: 251–66

Chen, W.-T., Singer, S. J. 1982. Immunoelectron microscopic studies of the sites of cell-substratum and cell-cell contacts in cultured fibroblasts. *J. Cell Biol.* 95: 205–22

Chen, W.-T., Wang, J. Hasegawa, T.,

Yamada, S. S., Yamada, K. M. 1986. Regulation of fibronectin receptor distribution by transformation, exogenous fibronectin, and synthetic peptides. *J. Cell Biol.* 103: 1649–61

Chow, I., Poo, M. M. 1982. Redistribution of cell surface receptors induced by cell-cell contact. *J. Cell Biol.* 95: 510–18

Chuong, C.-M., Crossin, K. L., Edelman, G. M. 1987. Sequential expression and differential function of multiple adhesion molecules during the formation of cerebellar cortical layers. *J. Cell Biol.* 104: 331–42

Ciment, G., Glimelius, B., Nelson, D. M., Weston, J. A. 1986. Reversal of a developmental restriction in neural crest–derived cells of avian embryos by a phorbol ester drug. *Dev. Biol.* 118: 392–98

Ciment, G., Weston, J. A. 1985. Segregation of developmental abilities in neural crest–derived cells: Identification of partially restricted intermediate cell types in the branchial arches of avian embryos. *Dev. Biol.* 111: 73–83

Constantine-Paton, M., Blum, A. S., Mendez-Otero, R., Barnstable, C. J. 1986. A cell surface molecule distributed in a dorsoventral gradient in the perinatal rat retina. *Nature* 324: 459–62

Cooper, D. N. W., Haywood-Reid, P. L., Springer, W. R., Barondes, S. H. 1986. Bacterial glycoconjugates are natural ligands for the carbohydrate binding site of discoidin I and influence its cellular compartmentalization. *Dev. Biol.* 114: 416–25

Cowin, P., Kapprell, H.-P., Franke, W. W., Tamkun, J., Hynes, R. O. 1986. Plakoglobin: A protein common to different kinds of intercellular adhering junctions. *Cell* 46: 1063–73

Cowin, P., Mattey, D., Garrod, D. 1985. Identification of desmosomal surface components (desmocollins) and inhibition of desmosome formation by specific Fab's. *J. Cell Sci.* 70: 41–60

Damjanov, I., Damjanov, A., Damsky, C. H. 1986. Developmentally regulated expression of cell-cell adhesion glycoprotein cell-CAM 120–80 in peri-implantation mouse embryos and extraembryonic membranes. *Dev. Biol.* 116: 194–202

Damsky, C. H., Knudson, K. A., Bradley, D., Buck, C. A., Horwitz, A. F. 1985b. Distribution of the cell substratum attachment (CSAT) antigen on myogenic and fibroblastic cells in culture. *J. Cell Biol.* 100: 1528–39

Damsky, C. H., Richa, J., Solter, D., Knudsen, K., Buck, C. A. 1983. Identification and purification of a cell surface glycoprotein mediating intercellular adhesion in embryonic and adult tissue. *Cell* 34: 455–66

Damsky, C. H., Wheelock, M. J., Damjanov, I., Buck, C. 1985a. Two cell adhesion molecules: Characterization and role in early mouse embryo development. In *Molecular Determinants of Animal Form*, ed. G. Edelman, pp. 235–52. New York: Liss

Dan, K. 1960. Cyto-embryology of echinoderms and amphibia. *Int. Rev. Cytol.* 9: 321–67

Darmon, M., Brachet, P., Periera da Silva, L. 1975. Chemotactic signals induce cell differentiation in *Dictyostelium discoideum*. *Proc. Natl. Acad. Sci. USA* 72: 3163–66

Darribere, T., Boulekbache, H., De Li, S., Boucaut, J. C. 1985. Immuno-electron microscopic study of fibronectin in gastrulating amphibian embryos. *Cell Tissue Res.* 239: 75–80

Davis, E. M., Trinkaus, J. P. 1981. Significance of cell-to-cell contacts for the directional movement of neural crest cells within a hydrated collagen lattice. *J. Embryol. Exp. Morphol.* 63: 29–51

DeSimone, D. W., Spiegel, M. 1986. Wheat germ agglutinin binding to the micromeres and primary mesenchyme cells of sea urchin embryos. *Dev. Biol.* 114: 336–46

Dodd, J., Solter, D., Jessell, T. M. 1984. Monoclonal antibodies against carbohydrate differentiation antigens identify subsets of primary sensory neurons. *Nature* 311: 469–72

Doe, C. Q., Goodman, C. S. 1985a. Early events in insect neurogenesis. I. Development and segmental differences in the pattern of neuronal precursor cells. *Dev. Biol.* 111: 193–205

Doe, C. Q., Goodman, C. S. 1985b. Early events in insect neurogenesis. II. The role of cell interactions and cell lineage in the determination of neuronal precursor cells. *Dev. Biol.* 111: 206–19

Duband, J. L., Thiery, J.-P. 1982. distribution of fibronectin in the early phase of avian cephalic neural crest migration. *Dev. Biol.* 93: 308–23

du Lac, S., Bastiani, M. J., Goodman, C. S. 1986. Guidance of neuronal growth cones in the grasshopper embryo. II. Recognition of a specific axonal pathway by the aCC neuron. *J. Neurosci.* 6: 3532–41

Dutt, A., Tang, J.-P., Carson, D. D. 1987. Lactosaminoglycans are involved in uterine epithelial cell adhesion in vitro. *Dev. Biol.* 119: 27–37

Edelman, G. M. 1984. Cell adhesion and morphogenesis: The regulator hypothesis. *Proc. Natl. Acad. Sci. USA* 81: 1460–64

Edelman, G. M. 1985. Cell adhesion and the molecular processes of morphogenesis. *Ann. Rev. Biochem.* 54: 135–69

Edelman, G. M. 1986. Cell adhesion molecules in the regulation of animal form and tissue pattern. *Ann. Rev. Cell Biol.* 2: 81–116

Eisen, J. S., Myers, P. Z., Westerfield, M. 1986. Pathway selection by growth cones of identified motoneurons in live zebra fish embryos. *Nature* 320: 269–71

Ekblom, P. 1984. Basement membrane proteins and growth factors in kidney differentiation. In *42nd Symp. Soc. Dev. Biol.*, ed. R. Trelstad, pp. 173–206. New York: Liss

Ekblom, P., Vestweber, D., Kemler, R. 1986. Cell-matrix interactions and cell adhesion during development. *Ann. Rev. Cell Biol.* 2: 27–47

Engvall, E., Davis, G. E., Dickerson, K., Ruoslahti, E., Varon, S., Manthorpe, M. 1986. Mapping of domains in human laminin using monoclonal antibodies: Localization of the neurite promoting site. *J. Cell Biol.* 103: 2457–66

Erickson, C. A. 1985. Control of neural crest cell dispersion in the trunk of the avian embryo. *Dev. Biol.* 111: 138–57

Erickson, C. A., Turley, E. A. 1983. Substrata formed by combinations of extracellular matrix components alter neural crest cell motility in vitro. *J. Cell Sci.* 61: 299–323

Ettensohn, C. A. 1985. Gastrulation in the sea urchin embryo is accompanied by the rearrangement of invaginating epithelial cells. *Dev. Biol.* 112: 383–90

Ettensohn, C. A., McClay, D. R. 1986. The regulation of primary mesenchyme cell migration in the sea urchin embryo: Transplantations of cells and latex beads. *Dev. Biol.* 117: 380–91

Faissner, A., Teplow, D. B., Kubler, D., Keilhauer, G., Kinzel, V., Schachner, M. 1985. Biosynthesis and membrane topography of the neural cell adhesion molecule L1. *EMBO J.* 4: 3105–13

Fink, R. D., McClay, D. R. 1985. Three cell recognition changes accompany the ingression of sea urchin primary mesenchyme cells. *Dev. Biol.* 107: 66–74

Fischer, G., Kuenemund, V., Schachner, M. 1986. Neurite outgrowth patterns in cerebellar microexplant cultures are affected by antibodies to the cell surface glycoprotein L-1. *J. Neurosci.* 6: 605–12

Fraser, S. E. 1980. A differential adhesion approach to the patterning of nerve connections. *Dev. Biol.* 79: 453–64

Fristrom, D. 1982. Septate junctions in imaginal disks of *Drosophila melanogaster*: A model for the redistribution of septa during cell rearrangement. *J. Cell. Biol.* 94: 77–87

Gabius, H. J., Springer, W. R., Barondes, S. H. 1985. Receptor for the cell binding site of discoidin I. *Cell* 42: 449–56

Gallin, W. J., Prediger, E. A., Edelman, G. M., Cunningham, B. A. 1985. Isolation of a cDNA clone for the liver cell adhesion molecule (L-CAM). *Proc. Natl. Acad. Sci. USA* 82: 2809–13

Gardner, R. L. 1982. Investigation of cell lineage and differentiation in the extra-embryonic endoderm of the mouse embryo. *J. Embryol. Exp. Morphol.* 68: 175–98

Garrod, D. R. 1986. Desmosomes, cell adhesion molecules and the adhesive properties of cells in tissues. *J. Cell Sci. Suppl.* 4: 221–37

Gennarini, G., Hirsch, M. R., He, H. T., Hirn, M., Finne, J., Goridis, C. 1986. Differential expression of mouse neural cell-adhesion molecule mRNA species during brain development and in neural cell lines. *J. Neurosci.* 6: 1983–90

Gerhart, J., Keller, R. 1986. Region-specific cell activities in amphibian gastrulation. *Ann. Rev. Cell Biol.* 2: 201–29

Gerisch, G. 1961. Zellfunktionen und Zellfunktionswechsel in der Entwicklung von *Dictyostelium discoideum*. V. Stadien-spezifische Zellkontaktbildung und ihre quantitative Erfassung. *Exp. Cell Res.* 25: 535–54

Gerisch, G. 1986. Interrelation of cell adhesion and differentiation in *Dictyostelium discoideum*. *J. Cell Sci. Suppl.* 4: 201–9

Gerisch, G., Weinhart, U., Bertholdt, G., Claviez, M., Stadler, J. 1985. Incomplete contact site A glycoprotein in HL220, a modB mutant of *Dictyostelium discoideum*. *J. Cell Sci.* 73: 49–68

Gingell, V., Vince, S. 1980. Long-range forces and adhesion: An analysis of cell-substratum studies. In *Cell Adhesion and Motility*, ed. A. S. G. Curtis, J. D. Pitts, pp. 1–37. Cambridge, Engl.: Cambridge Univ. Press

Glaser, L. 1980. From cell adhesion to growth control. In *38th Symp. Soc. Dev. Biol.*, ed. S. Subtelny, N. K. Wessels, pp. 79–97. New York: Academic

Goodman, C. S., Bastiani, M. J., Doe, C. Q., du Lac, S., Helfand, S. L., et al. 1984. Cell recognition during development. *Science* 225: 1271–79

Goodman, S. L., Newgreen, D. 1985. Do cells show inverse locomotory response to fibronectin and laminin substrates? *EMBO J.* 4: 2769–71

Gorbsky, G. 1986. Desmosomal adhesion in development. *Am. Zool.* 26: 535–40

Goridis, C., Hirn, M., Santoni, M. J., Gennarini, G., Deagostini-Bazin, H., et al. 1985. Isolation of mouse N-CAM-related cDNA—Detection and cloning using monoclonal-antibodies. *EMBO J.* 4: 631–35

Grabel, L. B. 1984. Isolation of a putative cell adhesion mediating lectin from teratocarcinoma stem cells and its possible role in differentiation. *Cell Diff.* 15: 121–24

Grabel, L. B., Casanova, J. E. 1986. The outgrowth of parietal endoderm from mouse teratocarcinoma stem-cell embryoid bodies. *Differentiation* 32: 67–73

Grabel, L. B., Watts, T. D. 1987. The role of extracellular matrix in the migration and differentiation of parietal endoderm from teratocarcinoma embryoid bodies. *J. Cell Biol.* 105: 441–48

Graf, J., Iwamoto, Y., Sasaki, M., Martin, G. R., Kleinman, H. K., et al. 1987. Identification of an amino acid sequence in laminin mediating cell attachment, chemotaxis, and receptor binding. *Cell* 48: 989–96

Greenberg, J. H., Seppa, S., Seppa, H., Hewitt, A. T. 1981. Role of collagen and fibronectin in neural crest cell adhesion and migration. *Dev. Biol.* 87: 259–66

Grinnell, F. 1986. Focal adhesion sites and the removal of substrate-bound fibronectin. *J. Cell Biol.* 103: 2697–2706

Grumet, M., Edelman, G. M. 1984. Heterotypic binding between neuronal membrane vesicles and glial cells is mediated by a specific neuron-glial cell adhesion molecule. *J. Cell Biol.* 98: 1746–56

Grumet, M., Hoffman, S., Crossin, K. L., Edelman, G. M. 1985. Cytotactin, an extracellular matrix protein of neural and non-neural tissues that mediates glia-neuron interaction. *Proc. Natl. Acad. Sci. USA* 82: 8075–79

Hammarback, J. A., Palm, S. L., Furcht, L. T., Letourneau, P. C. 1985. Guidance of neurite outgrowth by pathways of substratum-absorbed laminin. *J. Neurosci. Res.* 13: 213–20

Hansen, O. C., Nybroe, O., Bock, E. 1985. Cell-free synthesis of the D2-cell adhesion molecule—Evidence for three primary translation products. *J. Neurochem.* 44: 712–17

Hardin, J. D., Cheng, L. Y. 1986. The mechanisms and mechanics of archenteron elongation during sea urchin gastrulation. *Dev. Biol.* 115: 490–501

Hatta, K., Okada, T. S., Takeichi, M. 1985. A monoclonal antibody disrupting calcium-dependent cell-cell adhesion of brain tissues: Possible role of its target antigen in animal pattern formation. *Proc. Natl. Acad. Sci. USA* 82: 2789–93

Hausman, R. E., Dobi, E. T., Troccoli, N. M., Christie, T. 1987. Possible roles for cognin in chick embryo neural retina. *Am. Zool.* 27: 171–78

Hay, E. D. 1984. Cell-matrix interactions in the embryo: Cell shape, cell surface, cell skeletons and their role in differentiation. In *42nd Symp. Soc. Dev. Biol.* ed. R. Trelstad, pp. 1–32. New York: Liss

Heasman, J., Hynes, R. O., Swan, A. P., Thomas, J., Wylie, C. C. 1981. Primordial germ cells of *Xenopus* embryos: The role of fibronectin in their adhesion during migration. *Cell* 27: 437–47

Hemperly, J. J., Murray, B. A., Edelman, G. M., Cunningham, B. A. 1986. Sequence of a cDNA clone encoding the polysialic acid–rich and cytoplasmic domains of the neural cell adhesion molecule N-CAM. *Proc. Natl. Acad. Sci. USA* 83: 3037–41

Hoffman, S., Friedlander, D. R., Chuong, C.-M., Grumet, M., Edelman, G. M. 1986. Differential contributions of Ng-CAM and N-CAM to cell adhesion in different neural regions. *J. Cell Biol.* 103: 145–58

Hoffman, S., Sorkin, B. C., White, P. C., Brackenbury, R., Mailhammer, R., et al. 1982. Chemical characterization of a neural cell adhesion molecule purified from embryonic brain membrane. *J. Biol. Chem.* 257: 7720–29

Holtfreter, J. 1943. A study of the mechanics of gastrulation. Part I. *J. Exp. Zool.* 94: 261–318

Holtfreter, J. 1944. A study of the mechanics of gastrulation. Part II. *J. Exp. Zool.* 95: 171–212

Honig, M. G., Lance-Jones, C., Landmesser, L. 1986. The development of sensory projection patterns in embryonic chick hindlimb under experimental conditions. *Dev. Biol.* 118: 532–48

Horwitz, A., Duggan, K., Buck, C., Beckerle, M. C., Burridge, K. 1986. Interaction of plasma membrane fibronectin receptor with talin—A transmembrane linkage. *Nature* 320: 531–33

Horwitz, A., Duggan, K., Greggs, R., Decker, C., Buck, C. 1985. Cell substrate attachment (CSAT) antigen has properties of a receptor for laminin and fibronectin. *J. Cell Biol.* 103: 2134–44

Howard, M. J., Bronner-Fraser, M. 1986. Neural tube–derived factors influence differentiation of neural crest cells in vitro. Effects on activity of neurotransmitter biosynthetic enzymes. *Dev. Biol.* 117: 45–54

Hyafil, F., Morello, D., Babinet, C., Jacob, F. 1980. A cell surface glycoprotein involved in the compaction of embryonal carcinoma cells and cleavage stage embryos. *Cell* 21: 927–34

Hynes, R. O. 1985. Molecular biology of fibronectin. *Ann. Rev. Cell Biol.* 1: 67–90

Hynes, R. O., Yamada, K. M. 1982. Fibronectins: Multifunctional modular glycoproteins. *J. Cell Biol.* 95: 369–77

Jacobson, A. G., Odell, G. M., Oster, G. F. 1985. The cortical tractor model for epithelial folding: Application to the neural plate. In *Molecular Determinants of Animal Form*, ed. G. M. Edelman, pp. 143–66. New York: Liss

Kalcheim, C., Le Douarin, N. M. 1986. Requirement of a neural tube signal for the differentiation of neural crest cells into dorsal root ganglia. *Dev. Biol.* 116: 451–66

Kapfhammer, J. P., Grunbewald, B. E., Raper, J. A. 1986. The selective inhibition of growth cone extension by specific neurites in culture. *J. Neurosci.* 6: 2527–34

Keller, R. E., Danilchik, M., Gimlich, R., Shih, J. 1985. The function and mechanism of convergent extension during gastrulation of *Xenopus laevis*. *J. Embryol. Exp. Morphol.* 89: 185–209 (Suppl.)

Kjellan, L., Pettersson, I., Hook, M. 1981. Cell surface heparan sulfate. Mechanisms of proteoglycan-cell association. *Proc. Natl. Acad. Sci. USA* 78: 5371–75

Kleinman, H. K., Hassell, J. R., Aumailley, M., Terranova, V. P., Martin, G. R., Dubois-Dalq, M. 1985. Biological activities of laminin. *J. Cell Biochem.* 27: 317–25

Kleinman, H. K., McGarvey, M. L., Hassell, J. R., Martin, G. R., van Evercooren, A., Dubois-Dalcq, M. 1984. The role of laminin in basement membranes and in the growth, adhesion and differentiation of cells. In *42nd Symp. Soc. Dev. Biol.*, ed. R. Trelstad, pp. 124–44. New York: Liss

Krotoski, D. M., Domingo, C., Bronner-Fraser, M. 1986. Distribution of a putative cell surface receptor for fibronectin and laminin in the avian embryo. *J. Cell Biol.* 103: 1061–71

Kruse, J., Keilhauer, G., Faissner, A., Timpl, R., Schachner, M. 1985. The J-1 glycoprotein. A novel nervous system cell adhesion molecule of the L-2-HNK-1 family. *Nature* 316: 146–48

Kuwada, J. Y. 1986. Cell recognition by neuronal growth cones in a simple vertebrate embryo. *Science* 233: 740–46

Kuwada, J. Y., Goodman, C. S. 1985. Neuronal determination during embryonic development of the grasshopper nervous system. *Dev. Biol.* 110: 114–26

Lance-Jones, C., Landmesser, L. 1981. Pathway selection by embryonic chick motoneurons in an experimentally altered environment. *Proc. R. Soc. London Ser. B* 214: 19–52

Lander, A. D., Fujii, D. K., Reichardt, L. F. 1985. Laminin is associated with the "neurite outgrowth-promoting factors" found in conditioned media. *Proc. Natl. Acad. Sci. USA* 82: 2183–87

Landmesser, L., Honig, M. G. 1986. Altered sensory projections in the chick hind limb following the early removal of motoneurons. *Dev. Biol.* 118: 511–31

Lash, J. W., Vasan, N. S. 1978. Somite chondrogenesis in vitro. Stimulation by exogenous matrix components. *Dev. Biol.* 66: 151–71

Lattera, J., Silbert, J., Culp, L. 1983. Cell surface heparan sulfate mediates some adhesive responses to glycosaminoglycan-binding matrices, including fibronectin. *J. Cell Biol.* 96: 112–23

Lawson, K. A., Meneses, J. J., Petersen, R. A. 1986. Cell fate and cell lineage in the endoderm of the presomite mouse embryo, studied with an intracellular tracer. *Dev. Biol.* 115: 325–40

Le Douarin, N. M. 1982. *The Neural Crest.* Cambridge, Engl.: Cambridge Univ. Press. 259 pp.

Lee, G., Hynes, R. O., Kirschner, M. 1984. Temporal and spatial regulation of fibronectin in early *Xenopus* development. *Cell* 36: 729–40

Letourneau, P. C. 1975. Possible roles for cell-to-substratum adhesion in neuronal morphogenesis. *Dev. Biol.* 44: 77–91

Letourneau, P. C. 1979. Cell-substratum adhesion of neurite growth cones, and its role in neurite elongation. *Exp. Cell Res.* 124: 127–38

Lindner, J., Rathjen, F. G., Schachner, M. 1983. L1 mono- and polyclonal antibodies modify cell migration in early postnatal mouse cerebellum. *Nature* 305: 427–29

Loomis, W. F., Knecht, D. A., Fuller, D. L. 1987. Adhesion mechanisms and multicellular control of cell-type divergence of *Dictyostelium*. In *Molecular Approaches to Developmental Biology*, ed. E. A. Davidson, R. Firtel, pp. 339–50. New York: Liss

Loomis, W. F., Wheeler, S. A., Springer, W. R., Barondes, S. H. 1985. Adhesion mutants of *Dictyostelium discoideum* lacking the saccharide determinant recognized by two adhesion-blocking monoclonal antibodies. *Dev. Biol.* 109: 111–17

Lumsden, A. G. S., Davies, A. M. 1986. Chemotropic effect of specific target epithelium in the developing mammalian nervous system. *Nature* 323: 538–39

Maillet, L., Buc-Caron, M.-H. 1985. Modulation of specific protein expression in teratocarcinoma cell aggregates by anti-

bodies affecting cell-cell interactions. *Dev. Biol.* 111: 1–7

Marchase, R. B. 1977. Biochemical investigations of retinotectal adhesive specificity. *J. Cell Biol.* 75: 237–57

McClay, D. R., Cannon, G. W., Chambers, S. A., Coffman, J., Burdsal, C., Wessel, G. M. 1987. Cell adhesion determinants in sea urchin morphogenesis. *J. Cell Biol.* In press

McClay, D. R., Cannon, G. W., Wessel, G. M., Fink, R. D., Marchase, R. B. 1983. Patterns of antigenic expression in early sea urchin development. In *Time, Space, and Pattern in Embryonic Development*, ed. W. R. Jeffery, R. A. Raff, pp. 157–69. New York: Liss

McClay, D. R., Fink, R. D. 1982. Sea urchin hyalin: Appearance and function in development. *Dev. Biol.* 92: 285–93

McClay, D. R., Wessel, G. M., Marchase, R. B. 1981. Intercellular recognition: Quantitation of initial binding events. *Proc. Natl. Acad. Sci. USA* 78: 4975–79

McKay, R. D. G., Hockfield, S., Johansen, J., Thompson, I., Frederiksen, K. 1983. Surface molecules identify groups of growing axons. *Science* 222: 788–94

Moscona, A. A. 1952. Cell suspensions from organ rudiments of chick embryos. *Exp. Cell Res.* 3: 536–39

Muller, K., Gerisch, G. 1978. A specific glycoprotein as the target site of adhesion blocking Fab in aggregating *Dictyostelium* cells. *Nature* 274: 445–49

Murray, B. A., Hemperly, J. J., Gallin, W. J., MacGregor, J. S., Edelman, G. M., Cunningham, B. A. 1984a. Isolation of cDNA clones for the chicken neural cell adhesion molecule (N-CAM). *Proc. Natl. Acad. Sci. USA* 81: 5584–88

Murray, B. A., Hemperly, J. J., Prediger, E. A., Edelman, G. M., Cunningham, B. A. 1986a. Alternatively spliced mRNAs code for different polypeptide chains of the chicken neural cell adhesion molecule (N-CAM). *J. Cell Biol.* 102: 189–93

Murray, B. A., Niman, H. L., Loomis, W. F. 1983. Monoclonal antibody recognizing glycoprotein GP-80, a membrane glycoprotein implicated in intercellular adhesion of *Dictyostelium discoideum*. *Mol. Cell Biol.* 3: 863–70

Murray, B. A., Owens, G. C., Prediger, E. A., Crossin, K. L., Cunningham, B. A., Edelman, G. M. 1986b. Cell surface modulation of the neural cell adhesion molecule resulting from alternative mRNA splicing in a tissue-specific developmental sequence. *J. Cell Biol.* 103: 1431–39

Murray, B. A., Wheeler, S., Jongens, T., Loomis, W. F. 1984b. Mutations affecting a surface glycoprotein, gp80 of *Dictyostelium discoideum. Mol. Cell. Biol.* 4: 514–19

Murray, B. A., Yee, L. D., Loomis, W. F. 1981. Immunological analysis of a glycoprotein (contact sites A) involved in intercellular adhesion of *Dictyostelium discoideum. J. Supramol. Struct. Cell. Biochem.* 17: 197–211

Nakatsuji, N., Gould, A. C., Johnson, K. E. 1982. Movement and guidance of migrating mesoderm cells in *Ambystoma maculatum* gastrulae. *J. Cell Sci.* 56: 207–22

Nakatsuji, N., Hashimoto, K., Hayashi, M. 1985. Laminin fibrils in newt gastrulae visualized by immunofluorescent staining. *Dev. Growth Differ.* 27: 639–44

Nakatsuji, N., Johnson, K. E. 1983. Conditioning of a culture substratum by the ectodermal layer promotes attachment and orientated locomotion by amphibian gastrula mesodermal cells. *J. Cell Sci.* 59: 43–60

Nakatsuji, N., Johnson, K. E. 1984. Experimental manipulation of a contact guidance system in amphibian gastrulae by mechanical tension. *Nature* 307: 453–55

Nakatsuji, N., Smolira, M. A., Wylie, C. C. 1985. Fibronectin visualized by scanning electron microscopy immunocytochemistry on the substratum for cell migration in *Xenopus laevis* gastrulae. *Dev. Biol.* 107: 264–68

Newgreen, D. F., Gibbins, I. L. 1982. Factors controlling the time of onset of the migration of neural crest cells in the fowl embryo. *Cell Tissue Res.* 224: 145–60

Newgreen, D. F., Thiery, J.-P. 1980. Fibronectin in early avian embryos: Synthesis and distribution along the migration pathways of neural crest cells. *Cell Tissue Res.* 211: 269–91

Noden, D. 1984. The use of chimeras in analyses of craniofacial development. In *Chimeras in Developmental Biology*, ed. N. M. Le Douarin, A. McLaren, pp. 241–80. London: Academic

Noegel, A., Gerisch, G., Stadler, J., Westphal, M. 1986. Complete sequence and transcript regulation of a cell adhesion protein from aggregating *Dictyostelium* cells. *EMBO J.* 5: 1473–76

Noegel, A., Harloff, C., Hirth, P., Merkl, R., Modersitzki, M., et al. 1985. Probing an adhesion mutant of *Dictyostelium discoideum* with cDNA clones and monoclonal antibodies indicates a specific defect in the contact site A glycoprotein. *EMBO J.* 4: 3805–10

Nose, A., Takeichi, M. 1986. A novel cadherin cell adhesion molecule: Its expression patterns associated with implantation and organogenesis of mouse embryos. *J. Cell Biol.* 103: 2649–58

Obrink, B. 1986. Epithelial cell adhesion molecules. *Exp. Cell Res.* 163: 1–21

Ockland, C., Odin, P., Obrink, B. 1984. Two different cell adhesion molecules, cell CAM 105 and calcium-dependent protein, occur on the surface of rat hepatocytes. *Exp. Cell Res.* 151: 29–45

Ogou, S., Yoshida-Noro, C., Takeichi, M. 1983. Calcium-dependent cell-cell adhesion molecules common to hepatocytes and teratocarcinoma stem cells. *J. Cell Biol.* 97: 944–48

Ophir, J., Moscona, A. A., Ben-Shaud, Y. 1984. Cell disorganization and malformation in neural retina caused by antibodies to R-cognin: Ultrastructural study. *Cell Diff.* 15: 53–60

Overton, J. 1977. Formation of junctions and cell sorting in aggregates of chick and mouse cells. *Dev. Biol.* 55: 103–16

Pierschbacher, M., Ruoslahti, E. 1984. Variants of cell recognition site of fibronectin that retain attachment-promoting activity. *Proc. Natl. Acad. Sci. USA* 81: 5985–88

Pollerberg, G. E., Sadoul, R., Goridis, C., Schachner, M. 1985. Selective expression of the 180-kD component of the neural cell adhesion molecule NCAM during development. *J. Cell Biol.* 101: 1921–29

Pollerberg, G. E., Schachner, M., Davoust, J. 1986. Differentiation state-dependent surface mobilities of two forms of the neural cell adhesion molecule. *Nature* 324: 463–65

Poole, T. J., Steinberg, M. S. 1981. Amphibian pronephric duct morphogenesis: Segregation, cell rearrangement and directed migration of the *Ambystoma* duct rudiment. *J. Embryol. Exp. Morphol.* 63: 1–16

Poole, T. J., Steinberg, M. S. 1982. Evidence for the guidance of pronephric duct migration by a craniocaudally travelling adhesion gradient. *Dev. Biol.* 92: 144–58

Pytela, R., Pierschbacher, M., Ruoslahti, E. 1985. Identification and isolation of a 140 kd cell surface glycoprotein with properties of a fibronectin receptor. *Cell* 40: 191–98

Rapraeger, A., Bernfield, M. 1985. Cell surface proteoglycan of mouse mammary epithelial cells: Protease releases a heparan sulfate–rich ectodomain from a putative membrane-anchored domain. *J. Biol. Chem.* 260: 4103–9

Rapraeger, A., Jalkanen, M., Bernfield, M. 1986. Cell surface proteoglycan associates with the cytoskeleton at the basolateral cell surface of mouse mammary epithelial cells. *J. Cell Biol.* 103: 2683–96

Rapraeger, A., Jalkanen, M., Endo, E.,

Koda, J., Bernfield, M. 1985. The cell surface proteoglycan from mouse mammary epithelial cells bears chondroitin sulfate and heparan sulfate glycosaminoglycans. *J. Biol. Chem.* 260: 11046–52

Rathjen, F. G., Schachner, M. 1984. Immunocytological and biochemical characterization of a new neuronal cell surface component (L1 antigen) which is involved in cell adhesion. *EMBO J.* 3: 1–10

Rathjen, F. G., Wolff, J. M., Frank, R., Bonhoeffer, F., Rutishauser, U. 1987. Membrane glycoproteins involved in neurite fasciculation. *J: Cell Biol.* 104: 343–53

Reh, T. A., Constantine-Paton, M. 1984. Retinal ganglion cell terminals change their projection sites during larval development of *Rana pipiens. J. Neurosci.* 4: 442–57

Revel, J.-P. 1986. Gap junctions in development. In *Developmental Biology, A Comprehensive Synthesis*, ed. L. Browder, Vol. 3, *The Cell Surface in Development and Cancer*, ed. M. Steinberg, pp. 191–204. New York: Plenum

Richa, J., Damsky, C. H., Buck, C. A., Knowles, B. B., Solter, D. 1985. Cell surface glycoproteins mediate compaction, trophoblast attachment, and endoderm formation during early mouse development. *Dev. Biol.* 108: 513–21

Rogers, S. L., Letourneau, P. C., Palm, S. L., McCarthy, J., Furcht, L. T. 1983. Neurite extension by peripheral and central nervous system neurons in response to substratum-bound fibronectin and laminin. *Dev. Biol.* 98: 212–20

Rovasio, R. A., Delouvee, A., Yamada, K. M., Timpl, R., Thiery, J.-P. 1983. Neural crest cell migration: Requirement for exogenous fibronectin and high cell density. *J. Cell Biol.* 96: 462–73

Ruoslahti, E., Pierschbacher, M. D. 1986. Arg-Gly-Asp: A versatile cell recognition signal. *Cell* 44. 517–18

Rutishauser, U. 1984. Developmental biology of a neural cell adhesion molecule. *Nature* 310: 549–54

Rutishauser, U., Hoffman, S., Edelman, G. M. 1982. Binding properties of a cell adhesion molecule from neural tissue. *Proc. Natl. Acad. Sci. USA* 79: 685–89

Sanes, J. 1984. Roles of extracellular matrix in neural development. *Ann. Rev. Physiol.* 45: 581–600

Sanes, J. R., Schachner, M., Covault, J. 1986. Expression of several adhesive macromolecules: Neural cell adhesion molecules L-1, J-1, nerve growth factor–induced large external protein, uvomorulin, laminin, fibronectin, and a heparan sulfate proteoglycan in embryonic adult and denervated adult skeletal muscle. *J. Cell Biol.* 102: 420–31

Saunders, S., Jalkanen, M., Bernfield, M. 1986. The cell surface proteoglycan of mouse mammary epithelial cells is one of multiple receptors for fibronectin. *J. Cell Biol.* 103: 533a

Schuh, R., Vestweber, D., Riede, I., Ringwald, M., Rosenberg, U. B., et al. 1986. Molecular cloning of the mouse cell adhesion molecule uvomorulin: cDNA contains a B1-related sequence. *Proc. Natl. Acad. Sci. USA* 83: 1364–68

Shur, B. D. 1982. Cell surface glycosyltransferase activities during normal and mutant (T/T) mesenchyme migration. *Dev. Biol.* 91: 149–62

Shur, B. D. 1983. Embryonal carcinoma cell adhesion: The role of surface galactosyltransferase and its 90 K lactosaminoglycan substrate. *Dev. Biol.* 99: 360–72

Silver, J., Rutishauser, U. 1984. Guidance of optic axons in vivo by a preformed adhesive pathway on neuroepithelial endfeet. *Dev. Biol.* 106: 485–99

Siu, C. H., Cho, A. S., Choi, A. 1986. Mechanism of action of the contact site A glycoprotein during development of *Dictyostelium discoideum. J. Cell Biol.* 103: 3a

Siu, C. H., Lam, T. Y., Choi, A. H. C. 1985. Inhibition of cell-cell binding at the aggregation stage of *Dictyostelium discoideum* development by monoclonal antibodies directed against an 80,000 dalton surface glycoprotein. *J. Biol. Chem.* 260: 16030–36

Snow, M. H. L. 1985. The embryonic cell lineage of mammals and the emergence of the basic body plan. In *Molecular Determinants of Animal Form*, ed. G. M. Edelman, pp. 73–98. New York: Liss

Solursh, M., Katow, H. 1982. Initial characterization of sulfated macromolecules in the blastocoels of mesenchyme blastulae of *Strongylocentrotus purpuratus* and *Lytechinus pictus. Dev. Biol.* 94: 326–36

Sperry, R. W. 1963. Chemoaffinity in the orderly growth of nerve fiber patterns and connections. *Proc. Natl. Acad. Sci. USA* 50: 703–10

Springer, W. R., Barondes, S. H. 1985. Protein-linked oligosaccharide implicated in cell-cell adhesion in two *Dictyostelium* species. *Dev. Biol.* 109: 102–10

Springer, W. R., Cooper, D. N., Barondes, S. H. 1984. Discoidin I is implicated in cell-substratum attachment and oriented cell migration of *Dictyostelium discoideum* and resembles fibronectin. *Cell* 39: 557–64

Stallcup, W. B., Beasley, L. L. 1985. Involvement of the nerve growth factor–inducible large external glycoprotein (NILE) in neurite fasciculation in primary cultures of rat brain. *Proc. Natl. Acad. Sci. USA* 82: 1276–80

Steinberg, M. S. 1986. Cell surfaces in the control of growth and morphogenesis. In *Developmental Biology, A Comprehensive Synthesis*, ed. L. Browder, Vol. 3, *The Cell Surface in Development and Cancer*, ed. M. Steinberg, pp. 1–14. New York: Plenum

Steinemann, C., Parish, R. W. 1980. Evidence that a developmentally regulated glycoprotein is a target of adhesion-blocking Fab in reaggregating *Dictyostelium*. *Nature* 286: 621–23

Stevenson, B. R., Siliciano, J. D., Mooseker, M. S., Goodenough, D. A. 1986. Identification of ZO-1: A high molecular weight polypeptide associated with the tight junction (zonula occludens) in a variety of epithelia. *J. Cell Biol.* 103: 755–66

Tamkun, J. W., DeSimone, D. W., Fonda, D., Patel, R. S., Buck, C., et al. 1986. Structure of integrin, a glycoprotein involved in the transmembrane linkage between fibronectin and actin. *Cell* 46: 271–82

Terranova, V., Rao, C. N., Kalebic, T. M., Marguiles, T., Liotta, L. A. 1983. Laminin receptor on human breast carcinoma cells. *Proc. Natl. Acad. Sci. USA* 80: 444–48

Thanos, S., Bonhoeffer, F., Rutishauser, U. 1984. Fiber-fiber interaction and tectal cues influence the development of the chicken retinotectal projection. *Proc. Natl. Acad. Sci. USA* 81: 1906–10

Thiery, J.-P., Delouvee, A., Grumet, M., Edelman, G. M. 1985a. Initial appearance and regional distribution of the neuron-glia cell adhesion molecule in the chick embryo. *J. Cell Biol.* 100: 442–56

Thiery, J. P., Duband, J. L., Rutishauser, U., Edelman, G. M. 1982. Cell adhesion molecules in early chicken embryogenesis. *Proc. Natl. Acad. Sci. USA* 79: 6737–41

Thiery, J. P., Duband, J. L., Tucker, G. C. 1985b. Cell migration in the embryo. Role of cell adhesion and tissue environment in pattern formation. *Ann. Rev. Cell Biol.* 1: 91–113

Tomaselli, D. J., Reichardt, L. F., Bixby, J. L. 1986. Distinct molecular interactions mediate neuronal process outgrowth on non-neuronal cell surfaces and extracellular matrices. *J. Cell Biol.* 103: 2659–72

Tosney, K. W., Landmesser, L. T. 1985. Growth cone morphology and trajectory in the lumbosacral region of the chick embryo. *J. Neurosci.* 5: 2345–58

Tosney, K. W., Landmesser, L. T. 1984. Pattern and specificity of axonal outgrowth following varying degrees of chick limb bud ablation. *J. Neurosci.* 4: 2518–27

Tosney, K. W., Landmesser, L. T. 1986. Specificity of motoneuron growth cone outgrowth in the chick hindlimb. *J. Neurosci.* 5: 2336–44

Tosney, K. W., Watanabe, M., Landmesser, L., Rutishauser, U. 1986. The distribution of NCAM in the chick hindlimb during axon outgrowth and synaptogenesis. *Dev. Biol.* 114: 437–52

Trinkaus, J. P. 1984. *Cells into Organs*. New York: Prentice-Hall. 543 pp. 2nd ed.

Trisler, G. D., Schneider, M. D., Nirenberg, M. 1981. A topographic gradient of molecules in retina can be used to identify neuron position. *Proc. Natl. Acad. Sci. USA* 78: 2145–49

Trocolli, N. M., Hausman, R. E. 1985. Vesicle interaction as a model for the retinal cell-cell recognition mediated by R-cognin. *Cell Diff.* 16: 43–49

Tucker, B. C., Ciment, G., Thiery, J. P. 1986. Pathways of avian neural crest cell migration in the developing gut. *Dev. Biol.* 116: 439–50

Unwin, P. N. T., Ennis, P. D. 1983. Two configurations of a channel-forming membrane protein. *Nature* 307: 609–13

Urushihara, H., Ozaki, N. S., Takeichi, M. 1979. Immunological detection of cell surface components related with aggregation of Chinese hamster and chick embryonic cells. *Dev. Biol.* 70: 206–16

Vestweber, D., Kemler, R. 1984. Some structural and functional aspects of the cell adhesion molecule uvomorulin. *Cell Diff.* 15: 269–73

Vestweber, D., Kemler, R., Ekblom, P. 1985. Cell-adhesion molecule uvomorulin during kidney development. *Dev. Biol.* 112: 213–21

Vincent, M., Thiery, J.-P. 1984. A cell surface marker for neural crest and placodal cells. Further evolution in peripheral and central nervous system. *Dev. Biol.* 103: 468–81

Volk, T., Geiger, B. 1986a. A-CAM: A 135-kD receptor of intercellular adherens junctions. I. Immunoelectron microscopic localization and biochemical studies. *J. Cell Biol.* 103: 1441–50

Volk, T., Geiger, B. 1986b. A-CAM: A 135 kD receptor of intercellular adherens junctions. II. Antibody-mediated modulation of junction formation. *J. Cell Biol.* 103: 1451–54

Wessel, G. M., McClay, D. R. 1984. Ontogeny of the basal lamina in the sea urchin embryo. *Dev. Biol.* 103: 235–45

Wessel, G. M., McClay, D. R. 1987. Gastrulation in the sea urchin requires deposition of cross-linked collagen within the extracellular matrix. *Dev. Biol.* 121: 149–65

Wessels, N. K., Nuttall, R. P. 1978. Normal branching, induced branching, and steering of cultured parasympathetic motor neurons in vitro. *Exp. Cell Res.* 115: 111–22

Weston, J. A., Ciment, G., Girdlestone, J. 1984. The role of the extracellular matrix in neural crest development: A reevaluation. In *The Role of Extracellular Matrix in Development*, ed. R. L. Trelstad, pp. 433–60. New York: Liss

Wilson, H. V. 1907. On some phenomena of coalescence and regeneration in sponges. *J. Exp. Zool.* 5: 245–58

Woods, A., Couchman, J. R., Hook, M. 1985. Heparan sulfate proteoglycans of rat embryo fibroblasts. A hydrophobic form may link cytoskeleton and matrix components. *J. Biol. Chem.* 260: 10872–79

Woods, A., Couchman, J. R., Johansson, S., Hook, M. 1986. Adhesion and cytoskeletal organization of fibroblasts in response to fibronectin fragments. *EMBO J.* 5: 665–70

Yamada, K. M., Kennedy, D. W. 1985. Amino acid sequence specificities of an adhesive recognition signal. *J. Cell Biochem.* 28: 99–104

Yamada, Y., Sasaki, M., Kohno, K., Kleinman, H., Kato, S., Martin, G. R. 1985. A novel structure for the protein and gene of the mouse laminin B1 chain. In *Basement Membranes*, ed. S. Shibata, pp. 139–46. Amsterdam: Elsevier

Yokoo, K. M., Jones, J. C. R., Goldman, R. D. 1985. Is the hemidesmosome a half-desmosome? *J. Cell Biol.* 101: 305a

Yoshida, C., Takeichi, M. 1982. Teratocarcinoma cell adhesion: Identification of a cell surface protein involved in calcium-dependent cell aggregation. *Cell* 28: 217–24

Yoshida-Noro, C., Suzuki, N., Takeichi, M. 1984. Molecular nature of the calcium-dependent cell-cell adhesion system in mouse teratocarcinoma and embryonic cells studied with a monoclonal antibody. *Dev. Biol.* 101: 19–27

Zackson, S. L., Steinberg, M. S. 1986. Cranial neural crest cells exhibit directed migration on the pronephric duct pathway: Further evidence for an in vivo adhesion gradient. *Dev. Biol.* 117: 342–53

Ann. Rev. Cell Biol. 1987. 3 : 347–78

INTRACELLULAR TRANSPORT USING MICROTUBULE-BASED MOTORS

Ronald D. Vale

Cell Biology Program, Department of Pharmacology,
University of California, San Francisco, California 94143

CONTENTS

INTRODUCTION.. 347
IN VITRO SYSTEMS FOR STUDYING MICROTUBULE-BASED ORGANELLE TRANSPORT............. 349
    Axonal Transport 349
    Organelle Transport in Nonneuronal Cells 353
CANDIDATES FOR ORGANELLE TRANSPORT MOTORS 355
    Kinesin 355
    Cytoplasmic Dynein and Other Possible Motors 360
BIOCHEMICAL AND STRUCTURAL CHARACTERISTICS OF ORGANELLE TRANSPORT 361
    Association of Motility Proteins with Organelles 361
    Direction of Movement on the Microtubule Lattice..................................... 363
A PERSPECTIVE IN VIVO 364
    Role of Microtubule Polarity and Motility Proteins in Organelle Sorting within the
        Cell 364
    Role of Microtubule-Based Motors in Secretion, Endocytosis, and the Spatial
        Organization of Organelles 368
CONCLUSION 372

INTRODUCTION

Eukaryotic cells contain distinct membrane-bound compartments that serve specialized functions. While such an organization of the intracellular environment has clear advantages, it necessitates a means of transferring material between membrane-bound compartments. This intracellular

347

0743–4634/87/1115–0347$02.00

traffic is mediated by specialized transport organelles that shuttle proteins and lipids between compartments (Farquhar 1985; Kelly 1985).

Both membrane-bound compartments and filamentous polymers that constitute the cytoskeleton are found even in the most primitive eukaryotes. In such simple single-cell organisms, both microtubules and actin filaments are required for cell growth and multiplication (Pringle et al 1986; Novick & Botstein 1985). In addition to their structural functions, actin filaments and microtubules interact with force-generating enzymes such as myosins, dyneins, and kinesins. These motility proteins, together with elements of the cytoskeleton, play an integral role in transport between membrane-bound compartments, in the establishment of cell asymmetry, and in cell division.

A great deal has been learned about the molecular mechanisms of myosin- and dynein-based movement in muscle and ciliated cells, largely because these force-generating proteins are abundant and are organized into ordered arrays in these specialized cells. Much less is known about the mechanism of organelle transport and other ubiquitous forms of microtubule-based motility (Schliwa 1984), in part because the molecules involved in these processes are present in small quantities and because of the difficulties of using intact cells as experimental systems. However, within the last few years, several laboratories have succeeded in reconstituting organelle transport in vitro, thereby providing new opportunities for investigating the mechanism of intracellular microtubule-based transport at a molecular level. Although these studies are quite recent, it is worth examining what such experiments have revealed about how microtubule-based motors function and about their involvement in organelle transport.

This review covers in vitro systems for studying organelle transport and the properties of kinesin, a recently purified microtubule-based force-generating protein that is a good candidate for an organelle transport motor. This review also addresses general questions about cytoplasmic microtubule-based motors: How do they associate with organelles? How do their force-generating mechanisms differ from those of muscle myosin or ciliary dynein? What roles do these proteins play in intracellular sorting of organelles and the spatial organization of organelles and filaments?

This review is not intended as a comprehensive examination of all forms of microtubule-based motility; subjects such as mitosis, for example, are only briefly mentioned. Actin-based systems for organelle transport, which are found in many lower eukaryotes and plant cells, also are not covered in this review. Comprehensive reviews are available on organelle transport (Schliwa 1984; Grafstein & Forman 1980; Rebhun 1972), dynein-powered ciliary movement (Gibbons 1981; Johnson 1985), and microtubules (Kirschner & Mitchison 1986).

IN VITRO SYSTEMS FOR STUDYING MICROTUBULE-BASED ORGANELLE TRANSPORT

Axonal Transport

Although organelle transport occurs in virtually all eukaryotic cells, neurons are particularly well suited to the study of this phenomenon since organelles are transported over great distances and at rapid rates (1–5 μm/ sec) within axons. Historically, components of axonal transport were identified and defined primarily by injecting radiolabeled amino acids into neuronal ganglia and determining the rate at which proteins synthesized in the neuronal cell body are transported down the axon to the nerve terminal (*anterograde* axonal transport). Such experiments showed that a subset of proteins is rapidly transported, at rates of up 400 mm/day (Lasek 1968; Grafstein & Forman 1980). It was later discovered that proteins such as nerve growth factor and toxins are transported at similar rates in the *retrograde* direction (from the nerve terminal toward the cell body) (Bisby 1986). The majority of proteins transported in either direction at such rapid rates are associated with organelles (Tsukita & Ishikawa 1980; Smith 1980). Fast axonal transport of organelles is distinct from transport of cytoskeletal and soluble proteins in the anterograde direction, which occurs at much slower rates (0.2–4 mm/day) (see Lasek 1982 for a discussion of slow transport).

Although microtubules were implicated as essential components for fast axonal transport based upon the inhibitory effects of microtubule-depolymerizing agents (Grafstein & Forman 1980), two fundamental problems hindered further progress toward elucidating the molecular basis of this transport. The first difficulty was that the structures involved in transport (30–200 nm vesicles and 25 nm microtubules) are below the resolution limit of light. These objects can be viewed in the electron microscope (Figure 1), but electron micrographs reveal nothing of the dynamics of movement. A fundamental advance was the development of video-enhanced contrast microscopy (Inoue 1981; Allen et al 1981), which allows 30-nm vesicles and microtubules to be viewed in real time and in their native state (Figure 1). Using the video microscope to study the squid giant axon, Allen et al (1982) observed active transport of numerous small vesicles (30–200 nm diameter), which were not visualized in previous studies of axons by conventional light microscopy. These vesicles moved at velocities of 1–5 μm/sec, which was consistent with previously determined rates of fast transport of labeled proteins.

The complex organization of the intact axonal cytoplasm presented additional difficulties in the study of the mechanism of fast axonal transport. Electron micrographs of axoplasm showed a dense network of cyto-

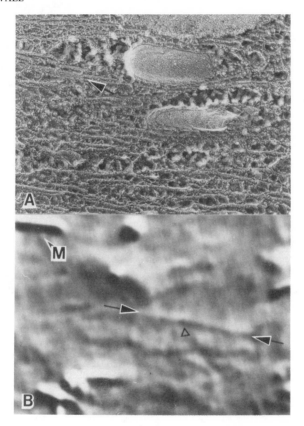

Figure 1 The interior of axons viewed by electron (A) and light (B) microscopy. Panel A shows a turtle optic nerve that was rapid frozen, fractured, shallow-etched, and rotary shadowed, as described by Schnapp & Reese (1982). This electron micrograph (50,000 ×) shows two organelles in a dense network of neurofilaments and cross-bridging elements. The arrow points to a microtubule in the fracture plane. Panel B shows the squid giant axon viewed by video-enhanced differential interference contrast microscopy (10,000 ×), as previously described (Allen et al 1982; Vale et al 1985a). Neurofilaments are not visible by this technique, but individual microtubules can be discerned; one can be seen between the two arrows. Transport of vesicles (*triangle*) along this microtubule was observed in real time. An elongate organelle believed to be a mitochondrion (M) is also indicated. These photographs (courtesy of Dr. B. Schnapp) are examples of morphological information that can be used to approach problems of motility in cells, and they demonstrate the complexity of the axonal cytoskeleton.

skeletal and cross-bridging elements (Schnapp & Reese 1982; Hirokawa 1982; Ellisman & Porter 1980) (Figure 1), but no ordered structural array that might be involved in motility was evident. Such findings emphasized the importance of studying organelle transport outside the confines of the intact cell. The first step in this direction was made by Brady et al (1982), who discovered that fast axoplasmic transport continues in extruded squid axoplasm in much the same fashion as in the intact axon. When attempts were later made to dissociate the highly cross-linked axonal cytoskeleton, it was discovered that organelles continue to move along single filaments separated from the intact axoplasm (Allen et al 1985; Brady et al 1985; Vale et al 1985a). Examination of organelle movement in disrupted axoplasm revealed certain properties of the transport machinery. For example, all organelles, regardless of their size, move continuously and at the same velocity (~ 2 μm/sec) along isolated filaments. In contrast, in intact axoplasm organelles move discontinuously, and large organelles, such as mitochondria, move at a slower rate than smaller vesicular organelles (Vale et al 1985a). These results suggest that differences in transport velocities in the intact axon are most likely due to greater impedance of large organelles by the meshwork of highly interconnected filaments in cytoplasm and are not due to different classes of organelles utilizing different motors.

The finding that organelle transport occurs in dissociated axoplasm indicated that the intact cytoskeleton is not required for this type of movement and that discrete filaments serve as tracks for organelle transport. A combination of video and electron microscopy was used to identify single microtubules as the organelle transport filaments in the squid giant axon (Schnapp et al 1985), in the fresh water amoeba *Reticulomyxa* (Koonce & Schliwa 1985), and in fibroblasts in culture (Hayden & Allen 1984). All of these studies also revealed that single microtubules support bidirectional movement of organelles. This finding was surprising because microtubules have an intrinsic polarity dictated by the repetitive arrangement of asymmetric tubulin dimers in the lattice. (The two ends of the microtubule have been defined as "plus" and "minus" based upon their faster and slower rates of tubulin assembly, respectively.) In contrast to organelle transport, the machineries for muscle contraction, i.e. myosin movement along actin filaments (Adelstein & Eisenberg 1980; Warrick & Spudich, this volume), and for ciliary motility, i.e. dynein movement along microtubule outer doublets (Gibbons 1981), generate unidirectional movement.

Individual organelles, however, only move in one direction on a microtubule and rarely change direction (Allen et al 1985; Schnapp et al 1985). When individual organelles move in opposite directions on a single filament (see Figure 2), they pass one another without colliding. This observation indicates that microtubules have multiple "motility tracks," which perhaps correspond in molecular terms to the protofilaments that comprise each microtubule.

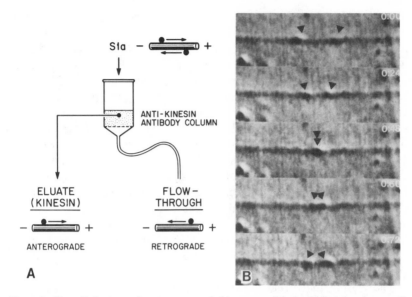

Sla − +

ANTI-KINESIN
ANTIBODY COLUMN

ELUATE
(KINESIN)

FLOW−
THROUGH

− +   − +

ANTEROGRADE   RETROGRADE

A

B

Figure 2   Two distinct translocators are probably responsible for bidirectional organelle movement along microtubules. In dissociated squid axoplasm, single microtubules support organelle movement in both directions, although individual organelles tend to move in only one direction. Panel B shows two organelles (*triangles*) moving in opposite directions along a microtubule and passing one another without colliding (from Schnapp et al 1985). In a crude axoplasmic supernatant (S1a), latex beads also move in both directions along single microtubules, although unlike organelles, single beads move discontinuously and frequently reverse direction. Bead movement toward the plus end of the microtubule (anterograde transport) is induced by kinesin, since this activity is retained by an anti-kinesin monoclonal antibody column (Panel A). In contrast, the translocator that induces retrograde movement does not interact with this antibody, which indicates that it differs from kinesin. This conclusion is also supported by experiments summarized in Table 2.

The identification of molecules that generate microtubule-based movement was aided by the ability to reconstitute organelle movement in vitro. Initial studies showed that organelles isolated from axoplasm (Gilbert & Sloboda 1984) or purified synaptic vesicles (Schroer et al 1985) undergo directed transport when injected into squid giant axons. Two subsequent studies demonstrated that isolated squid axoplasmic organelles, in the presence of ATP, move along salt-extracted axonemes (Gilbert et al 1985) or taxol polymerized squid microtubules (Vale et al 1985b) at velocities similar to those reported in axons. Surprisingly, however, a high-speed axoplasmic supernatant increased the number of organelles moving along purified microtubules and, even in the absence of axoplasmic organelles, caused microtubules to bind to and move along the surface of glass coverslips (Vale et al 1985b). The latter phenomenon was also observed with microtubules in dissociated axoplasm (Allen et al 1985). This unusual ATP-driven movement of microtubules appears to be generated by a soluble protein that binds to glass, since microtubules also moved on coverslips that were exposed to supernatant and subsequently washed (Vale et al 1985b). The soluble microtubule motor also interacts with carboxylated latex beads, causing them to bind to and translocate along microtubules very much as organelles do, although at a threefold slower velocity (Vale et al 1985b). Similar events may also occur in living cells, since carboxylated beads exhibit directed movement when microinjected into axons (Adams & Bray 1983) or cultured fibroblasts (Beckerle 1984). The results from these reconstitution experiments support a model of a soluble enzyme (kinesin) that binds to glass coverslips, charged beads, and organelles and generates a translocating force along microtubules (Figure 3). Low-speed supernatants from axoplasm support bidirectional movement of latex beads along microtubules (Vale et al 1985d), which suggests the presence of a second motor in addition to kinesin, as discussed in a later section.

Organelle Transport in Nonneuronal Cells

Organelle transport along microtubules has also been studied in a variety of nonneuronal cells, and within the last five years, several laboratories

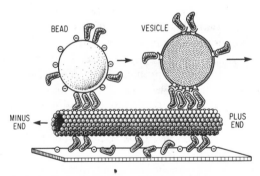

Figure 3 A hypothetical scheme showing how kinesin may induce movement of organelles, carboxylated beads, and microtubules. Kinesin, through a charge interaction, binds to glass coverslips, causing microtubules to translocate along the glass surface, or to carboxylated latex beads, causing these structures to translocate along the surface of the microtubule. Although translocators are presumably oriented randomly on the glass and bead surfaces, only those molecules with the proper orientation relative to the microtubule are capable of generating force. Kinesin interacts with organelles through a membrane-bound protein and causes organelles to translocate along microtubules at three to four times the rate of bead movement. The direction of organelle and bead movement towards the plus end of the microtubule is opposite the direction of microtubule movement on glass, as indicated by the arrows.

have succeeded in reactivating transport in detergent-permeabilized cells (Stearns & Ochs 1982; Clark & Rosenbaum 1982; Forman 1982; Koonce & Schliwa 1986; Rozdzial & Haimo 1986a; Kachar et al 1987; M. A. McNiven & K. R. Porter, personal communication). Chromatophores, which respond to external stimuli by aggregating or dispersing their pigment granules, have been the focus of considerable attention (McNiven & Porter 1984). The centripetal and centrifugal granule movements change the coloration of these neural crest–derived cells. Some unique advantages of studying organelle transport in these cells rather than axons are that the transported pigment granules are large and homogeneous and their movement can be triggered by hormonal or neuronal stimulation, which controls movement using cAMP or Ca^{2+} as second messengers (McNiven & Porter 1984). Because detergent-permeabilized chromatophores require cAMP or Ca^{2+} for activation of granule movement (Clark & Rosenbaum 1982; Rozdzial & Haimo 1986a; Stearns & Ochs 1982; M. A. McNiven & K. R. Porter, personal communication), the activation process can be studied with pharmacological inhibitors, nucleotide analogues, and protein kinases and phosphatases.

The giant freshwater amoeba *Reticulomyxa* is another interesting cell in which to examine reactivated organelle transport (Koonce & Schliwa 1986). After detergent extraction, many organelles in this cell remain attached to a filamentous network comprised of microtubules and actin and undergo rapid (20 μm/sec) linear movements along microtubules after addition of ATP. Organelle movement in this in vitro system does not require soluble components, as movement can be activated even after the network has been washed several times. Bidirectional movement of various types of organelles along microtubules has been observed in crude homogenates of sea urchin eggs (Pryer et al 1986) and *Acanthamoeba* cells (Kacher et al 1987). The latter two studies found that latex beads moved along microtubules and that purified microtubules moved along the surface of glass coverslips in the presence of ATP and crude homogenates, as was observed in squid axon preparations.

Several properties of organelle transport, including nucleotide requirements and transport rates, differ significantly in *Reticulomyxa*, chromatophores, and axons. This variability suggests some divergence and specialization of microtubule-based transport systems and perhaps even reflects different mechanisms. Furthermore, it should be emphasized that organelle transport is not exclusively microtubule based and occurs along actin filaments in some lower eukaryotes and plants (Adams & Pollard 1986; Schliwa 1984). On the other hand, the properties of organelle transport in sea urchin eggs, nonneuronal vertebrate cells, and axons are very similar; a force-generating protein that may be involved in organelle transport has been purified from all of these cell types, as discussed in the next section.

CANDIDATES FOR ORGANELLE TRANSPORT MOTORS

Kinesin

PURIFICATION USING IN VITRO MOTILITY ASSAYS   Microtubule movement on glass and bead movement along microtubules provided powerful assays for purifying the soluble microtubule-based motor. Purification schemes were devised based upon pharmacological experiments that showed that AMP-PNP, a nonhydrolyzable ATP analogue, induces stable associations between organelles and microtubules, a phenomenon that could be explained by a rigorlike attachment of the organelle motor to microtubules (Lasek & Brady 1985). When analogous biochemical experiments were conducted with supernatants that contained the soluble translocator and purified microtubules, a polypeptide of 110–134 kDa from squid, chicken, and bovine neuronal tissues (Vale et al 1985c; Brady 1985) and from sea urchin eggs (Scholey et al 1985) associated with microtubules in the presence, but not the absence, of AMP-PNP. Using microtubule movement as an assay of translocator activity, this protein was demonstrated to be associated with motile activity. It was purified by microtubule affinity followed by gel filtration and hydroxyapatite chromatography and named "kinesin," from the Greek root "kinein," meaning to move (Vale et al 1985c). Squid kinesin has also been purified in an active form by affinity chromatography with a monoclonal antibody that specifically recognizes the 110-kDa squid polypeptide (Vale et al 1985d). More recently, other methods for purifying kinesin have been developed that employ microtubule affinity in the presence of tripolyphosphate (Kuznetsov & Gelfand 1986) or in the absence of nucleotides (R. D. Vale, manuscript in preparation; S. A. Cohn, A. L. Ingold, J. M. Scholey, manuscript in preparation). Although hundreds of micrograms of kinesin can be obtained with these methods, kinesin is much less abundant in neurons than myosin is in muscle or dynein is in ciliated cells, and obtaining milligram quantities of pure protein is currently a difficult task.

BIOCHEMICAL AND MOTILE PROPERTIES   Purified kinesin induces unidirectional movement of microtubules on glass and of beads on microtubules that is indistinguishable from that induced by high-speed axoplasmic supernatants. Thus a completely defined in vitro system exists in which to study the motile properties of this force-generating enzyme. The direction of movement of kinesin-coated beads is always towards the plus end of the microtubules (Vale et al 1985d). Most microtubules in axons are aligned with their plus ends towards the nerve terminal (Burton & Paige 1981; Heidemann et al 1981); hence this direction corresponds to anterograde transport in the axon.

Kinesin isolated from squid neuronal tissue is composed of 110, 65, and 70 kDa polypeptides in a stoichiometric ratio of four 110 kDa polypeptides to one 65 and one 70 kDa polypeptide (Vale et al 1985c). These polypeptides coelute on a variety of columns, including the monoclonal antibody affinity column, and cannot be separated from one another at high salt concentrations, which indicates that they form a tightly associated complex. An 80-kDa polypeptide coelutes with squid kinesin in various types of columns; however, it is found in variable and often substoichiometric amounts and does not appear to be an integral part of the complex. Kinesin from mammalian brain is similar to the squid protein: It consists of 120–130 and 62 kDa polypeptides present in a stoichiometry of two to one (Vale et al 1985c; Kuznetsov & Gelfand 1986; Amos 1987); however, the study by Kuznetsov & Gelfand (1986) has raised the possibility of the presence of additional subunits or associated proteins. The apparent molecular weight of native squid and bovine kinesin, as judged by gel filtration chromatography, is ~600,000; however, an accurate molecular weight determination, which would allow the quaternary structure of kinesin to be ascertained, has not been reported.

Kinesin isolated from sea urchin eggs consists of a prominent 130-kDa polypeptide; the existence of a light chain has not been reported (Scholey et al 1985). Its motile properties in vitro and pharmacological sensitivities to NEM and vanadate are similar to those of squid kinesin (Porter et al 1987), and antibodies prepared against kinesin from either squid or sea urchin will react with kinesins from both species (Scholey et al 1985). A kinesinlike protein has also been partially purified from *Acanthamoeba* that produces movement of latex beads along microtubules at six times the rate seen with squid or bovine kinesin (Kachar et al 1987). Thus there is variability across species in molecular weight and velocity of kinesin movement along microtubules; however, certain domains must be conserved since antibodies against squid kinesin recognize similar polypeptides in many species, including mammals (M. Sheetz, R. D. Vale, unpublished observations). A cDNA clone encoding the 110-kDa polypeptide of *Drosophila* kinesin has been cloned (Yang et al 1987); however, it is not yet clear whether kinesin belongs to a multigene family.

The biochemistry and enzymology of kinesin are in their early stages. Investigations of how this protein generates movement will to some extent draw upon methodologies and ideas that have emerged from over two decades of research on myosin and dynein. However, it is already clear that kinesins are distinctly different from these other force-generating proteins in polypeptide composition and enzymology, as summarized in Table 1. Unlike myosin, kinesin does not generate movement on actin filaments (Pryer et al 1986), and although ciliary dynein and kinesin both utilize microtubules, they generate movement in opposite directions along the polymer (Sale & Satir 1977; Vale et al 1985d). Furthermore, kinesin-induced motility is much less sensitive to the sulfhydryl alkylating reagent N-ethyl-maleimide (NEM) than is myosin- or dynein-induced motility and is less sensitive to vanadate than is dynein-based motility (Vale et al 1985c; Porter et al 1987).

Electron microscopy of kinesin by rotary shadowing or negative staining indicates that it is an elongated molecule ∼ 100 nm long, consisting of a large globular domain connected by a stalk, which may have a flexible joint, to a smaller and possibly forked domain (Amos 1987; Vale et al 1985e). Current evidence suggests that the smaller domain attaches to the micro-tubule in a nucleotide-sensitive manner, and hence the larger domain may be the site of attachment to beads and organelles (Amos 1987). It is unclear, however, how this morphology is related to the polypeptide composition of kinesin.

Table 1 Comparison of motility proteins[a]

	Skeletal muscle myosin	Acanthamoeba myosin I	Ciliary dynein	Cytoplasmic dynein	Kinesin
Substrate	actin	actin	microtubule	microtubule	microtubule
Direction of movement	barbed end	barbed end	minus end	plus end	plus end
Velocity of movement (μm/sec)	2–5	0.03	10	1	0.5
Native MW (kDa)	500	160	1–2,000	—	300–600[b]
Polypeptide composition (kDa)	230; 15–27	135; 17–27	400; 60–130; 14–24	350	110–134; 60–70
NEM sensitive (1 mM)	Yes	Yes	Yes	Yes	No
Vanadate sensitive (25 μM)	No	No	Yes	Yes	No

[a] Information can be found in the following reviews or articles for myosin, Adelstein & Eisenberg 1980, Warrick & Spudich, this volume; *Acanthamoeba* myosin I, Korn 1983, H. Fujisaki, personal communication; dynein, Johnson 1985; a motile cytoplasmic dynein from *C. elegans*, Lye et al 1987; and kinesin, Vale et al 1985c, Scholey et al 1985, Porter et al 1987.

[b] Apparent molecular weight derived from gel filtration chromatography.

Kinesin may have two microtubule binding domains, since kinesin aggregates and moves purified microtubules relative to one another in solution (Vale et al 1985b,c). Recent morphological studies have shown that individual kinesin molecules can cross-bridge two microtubules (Amos 1987; B. J. Schnapp & T. S. Reese, unpublished observations). It is still unclear from these observations whether a single kinesin molecule has two distinct microtubule binding sites (like dynein) or interacts with a second microtubule by charge, which would be analogous to its association with carboxylated beads.

Kinesin requires a nucleotide to produce movement (preferably ATP: K_m for motility ≈ 50 μM ATP; Porter et al 1987; Schnapp et al 1986), and the 110-kDa kinesin polypeptide from squid was demonstrated to bind ATP (Gilbert & Sloboda 1986). Although kinesin has a low ATPase activity in solution (Vale et al 1985c; Scholey et al 1985), the ATPase activities of bovine brain (Kuznetsov & Gelfand 1986) and sea urchin egg kinesins (Cohn et al 1987), as well as that of a kinesinlike protein from *Acanthamoeba* (Kachar et al 1986), are all stimulated by microtubules. Thus the enzymatic activity of kinesin, like those of myosin and dynein, appears to be coupled to motility. Indeed, inhibitors of kinesin-induced movement, such as vanadate, AMP-PNP, and ATP in the absence of magnesium, also decrease the microtubule-activated ATPase activity (Cohn et al 1987). The ATPase activity of kinesin can be activated without microtubules in the presence of millimolar calcium, but it is not activated by high salt and EDTA, conditions that stimulate the myosin ATPase (Kuznetsov & Gelfand 1986).

Microtubule activation of kinesin ATPase activity suggests that hydrolysis is tightly coupled to microtubule binding and force production. However, it is not possible at the present time to put together a complete picture of the steps in kinesin's mechanochemical cycle. Although it has been hypothesized that ATP induces attachment of kinesin to microtubules (Hill 1987; Lasek & Brady 1985), binding experiments show that ATP dissociates kinesin from microtubules and that depletion of ATP (or depletion of magnesium in the presence of ATP) induces a strong binding state of kinesin for microtubules (S. A. Cohn, A. L. Ingold & J. M. Scholey, manuscript submitted; R. D. Vale, manuscript submitted), which could be analogous to the strong binding states of dynein (Johnson 1985) and myosin (see Warrick & Spudich, this volume). Based upon such results, one can imagine a cycle similar to those described for myosin and dynein in which (*a*) ATP dissociates kinesin from microtubules; (*b*) ATP hydrolysis occurs in solution, followed by rebinding of kinesin to a microtubule; (*c*) ADP and P_i are released from kinesin bound to microtubules; and (*d*) ATP binds and dissociates kinesin from microtubules to begin the cycle anew. However, the effects of nonhydrolyzable ATP analogues suggest that perhaps the mechanochemical cycle of kinesin is not as similar to those of myosin and dynein as is suggested by this simple scheme. AMP-PNP weakens the interaction between myosin and actin (Greene & Eisenberg 1980) and between dynein and microtubules (Mitchell & Warner 1981), whereas it induces a stable interaction of kinesin and microtubules. In this respect, kinesin is similar to elongation factor Tu, which forms a stable complex with ribosomes in the presence of GMP-PNP (Kaziro 1978). The AMP-PNP-induced strong binding state may also be distinct from the binding state induced by ATP depletion. One piece of evidence in favor of this idea is that movement of microtubules on glass, which occurs instantaneously when ATP is added to kinesin bound to microtubules by ATP depletion, is observed only after a one minute delay when ATP is introduced to kinesin bound to microtubules with AMP-PNP (Schnapp et al 1986).

BIOLOGICAL ROLES FOR KINESIN   Although a great deal has been learned about kinesin motility in vitro, little is known of how kinesin functions in cells. That kinesin has been detected by immunoblotting and has been purified from nonneuronal and neuronal tissues suggests that it participates in some ubiquitous and fundamental form(s) of intracellular microtubule-based motility. The existence of a soluble pool of kinesin suggests that this protein may be involved in more than one type of motility process; the two most likely are organelle transport and mitosis. Evidence of kinesin's involvement in these processes is discussed below.

Since axons are a rich source of kinesin and kinesin is capable of transporting latex beads along microtubules, one could infer that this protein serves as a motor for axonal transport of organelles. Perhaps the most compelling findings favoring a role for kinesin as an organelle motor are that AMP-PNP induces a rigor attachment of both purified kinesin (Vale et al 1985c; Scholey et al 1985; Brady 1985) and axoplasmic organelles to microtubules (Lasek & Brady 1985). These results implicate kinesin as the molecule responsible for the formation of cross-bridges between organelles and microtubules, although it is possible that other motor proteins also attach to microtubules in the presence of AMP-PNP. Another finding consistent with the hypothesis that kinesin serves as a motor for axonal transport is that vanadate inhibits both kinesin-induced microtubule movement on glass and organelle transport in vitro (Vale et al 1985b,c) and in vivo (Forman et al 1983) at similar concentrations. Since these results suggest, but do not prove, that kinesin is involved in organelle transport, it is still premature to classify kinesin as an organelle motor. A direct demonstration of specific kinesin association with organelles, inhibition of organelle transport using anti-kinesin antibodies, and conclusive reconstitution of organelle motility using purified kinesin must be performed before this conclusion can be reached.

Scholey et al (1985) have shown by light level immunocytochemistry that kinesin is localized in the mitotic spindle of dividing sea urchin eggs, where it may be associated with a spindle matrix (Leslie et al 1987). Currently, there is no other evidence regarding kinesin's function in mitosis, although several possibilities are discussed in recent articles (Mitchison 1986; Vale et al 1986).

Cytoplasmic Dynein and Other Possible Motors

Cytoplasmic proteins with high molecular weight polypeptide chains (~ 400 kDa) have been purified that have dyneinlike characteristics, such as binding to microtubules and high ATPase activity (Pratt 1980; Hisanga & Sakai 1983; Hollenbeck et al 1984; Scholey et al 1984; Asai & Wilson 1985; Vale et al 1985c; Hollenbeck & Chapman 1986). The ATPase activity of these proteins, like that of flagellar dynein, is inhibited by vanadate and can be activated by Triton X-100 (Asai & Wilson 1985). However, these cytoplasmic dyneins are not simply precursors of flagellar dyneins, since proteolytic digest patterns indicate that the two are similar but not identical (Pratt 1986a), and certain antibodies against cytoplasmic dynein do not react with flagellar dynein (Pratt et al 1987). Furthermore, a cytoplasmic dynein has been isolated from Caenorhabditis elegans, an organism that does not produce motile cilia or flagella (Lye et al 1987).

Although it has been widely assumed that cytoplasmic dynein is a mechanochemical enzyme, only the study performed with the dynein from C. elegans has shown that a cytoplasmic dynein produces movement; in this case, microtubule movement along glass coverslips was observed (Lye et al 1987). The rate of microtubule movement on glass induced by this dynein is 2–3 times faster than that induced by kinesin, and this movement exhibits greater sensitivity to NEM and vanadate than does kinesininduced movement (Table 1). Surprisingly, this protein and kinesin generate movement in the same direction on the microtubule lattice whereas ciliary dynein induces movement in the opposite direction (Sale & Satir 1977).

The biological roles of cytoplasmic dyneins, like those of kinesin, are not established. Early studies found cytoplasmic dynein localized in the mitotic apparatus, which suggested that this protein might generate mitotic microtubule-based movements (Piperno 1984). A recent study by Pratt et al (1987) found that sea urchin cytoplasmic dynein copurifies with vesicles in sucrose gradients and appears to be attached to these vesicles, as determined by electron microscope immunocytochemistry. This study raises the possibility that cytoplasmic dynein may be an organelle transport motor, although additional evidence to support this notion has not been obtained.

Another strategy for identifying possible organelle transport motors has been to examine isolated vesicle or microtubule fractions for associated proteins that have characteristics of a force-generating enzyme, such as ATPase activity, and the ability to bind to microtubules. Neither of these characteristics is a sufficient criterion for an organelle transport motor since the major ATPase in cycled microtubule preparations is not a mech-anochemical enzyme but is a vesicle-associated proton pump similar to the F_1 ATPase of mitochondria (Murphy et al 1983). Recently, however, a high–molecular weight protein associated with vesicles from squid axoplasm has been identified that is a promising candidate for a motor. It binds ATP and microtubules, has ATPase activity, and cross-reacts with an antibody to MAP-2 (Gilbert & Sloboda 1986; Pratt 1986b). This protein, however, has not been purified and shown to generate microtubule-based motility.

BIOCHEMICAL AND STRUCTURAL CHARACTERISTICS OF ORGANELLE TRANSPORT

Association of Motility Proteins with Organelles

Several studies have shown that motors for transporting organelles are located on the vesicle surface rather than on the microtubule (Gilbert et al 1985; Vale et al 1985b; Miller & Lasek 1985; Langford et al 1987). Since not all organelles in cells are actively transported, however, there must be mechanisms for regulating the association of motility proteins with organelles and/or their activation once bound to the organelle.

Unlike kinesin's electrostatic interaction with glass coverslips or latex beads, its association with organelles may involve a membrane-associated protein, since pure phospholipid vesicles and trypsinized organelles do not move along microtubules even in the presence of supernatants containing active kinesin (Gilbert et al 1985; Schroer et al 1985; Vale et al 1985b). How could membrane-associated proteins participate in organelle transport? One possibility is that such proteins act as receptors that reversibly bind microtubule motors from a soluble pool. Motility proteins may dissociate from such receptors very slowly, since organelles separated from soluble components in a sucrose gradient exhibit ATP-dependent movement on microtubules (Gilbert et al 1985; Vale et al 1985b). This receptor model predicts saturable and reversible binding of motility proteins to specific receptors present on all organelles transported in the same direction along microtubules. Synapsin I, a protein found in nerve terminals that interacts with synaptic vesicles and cytoskeletal elements (Huttner et al

1983), may bind to an organelle receptor. Alternatively, motility proteins may associate with vesicles by direct insertion into lipid bilayers either by unmasking a hydrophobic domain or through a covalently bound lipid.

Once they are bound to organelles, how many motility proteins are required for movement, and what is their spatial arrangement on the organelle surface? From an energetic standpoint, few motility proteins are required to move organelles. Simple calculations reveal that ATP hydrolysis by a single force-generating enzyme provides sufficient energy to move a 1 μm spherical organelle at a velocity of 2 μm/sec, since such objects experience little viscous drag as they move through an aqueous environment (Sheetz & Spudich 1983). The stepping distance and cycling time of the motor, however, will determine how many motors are actually needed to generate movement at a given velocity. These values are not known for kinesin. Nonetheless, probably less than five motors per organelle are needed to generate organelle movement at 2 μm/sec, since morphological studies reveal less than 5 cross-bridging structures linking squid axoplasmic organelles to microtubules (Gilbert et al 1985; Miller & Lasek 1985; Langford et al 1987). Thus the number of molecules responsible for translocating organelles appears to be much smaller than the large number of myosin heads that interact with actin filaments in muscle or dynein heads that interact with microtubules in ciliary outer doublets.

One can only speculate as to how so few force-generating proteins produce organelle movement. Observations of asymmetric mitochondria translocating with only one end attached to the microtubule suggest that organelles "walk" rather than "roll" on microtubules (Vale et al 1985a). Recent motion analysis of kinesin-induced bead translocation indicates that this movement can follow a single protofilament track (Gelles et al 1987). In order for an organelle or bead to "walk" along a protofilament, either (a) two or more cross-bridges must be involved so that attachment to the microtubule is maintained while a cross-bridge dissociates, or (b) a single cross-bridge maintains attachment during movement. The latter hypothesis is not thought to be true for myosin and dynein, which dissociate from their filaments and reattach in a cyclic fashion.

The characteristics of bidirectional bead movement along microtubules are consistent with the notion that bead movement is generated by a single or a few motors. If individual cross-bridges mediating anterograde or retrograde movement were randomly distributed on the bead, one would predict a "tug of war" resulting in little net displacement of the bead. However, movements of beads for distances of 10 μm in either the anterograde or retrograde directions are often observed. This unidirectional motion must either be generated by several cross-bridging arms of one motor species that are in close proximity and oriented properly with respect to the microtubule lattice or by a single cross-bridge that generates movement until a motor of the opposite polarity is engaged. If multiple cross-bridges are involved, then the force-generating proteins themselves must be composed of more than one cross-bridging arm, like ciliary dynein, or must self-associate to form clusters on the bead surface. It is also possible that these cross-bridges work in some coordinated fashion to maximize contact with the microtubule during movement. The questions raised by these observations point to the need for an examination of the actions of one or a few motor proteins engaged in movement in order to develop an understanding of the molecular mechanism of organelle transport.

Direction of Movement on the Microtubule Lattice

Axoplasmic organelles translocate in both directions along microtubules in vivo, yet purified kinesin generates movement only towards the plus end of the microtubule (anterograde direction). How then are organelles transported along microtubules in the retrograde direction? Is kinesin or a kinesinlike molecule modified to reverse its direction of movement on microtubules, or is another type of microtubule motor protein involved in this process?

Results obtained using the squid giant axon support the latter hypothesis. As discussed previously, bidirectional movement of latex beads along centrosomal microtubules is observed in the presence of a low-speed supernatant from axoplasm to which Triton X-100 has been added to solubilize vesicular components (Vale et al 1985d). The movement of the beads for short distances in alternating directions suggests that two opposing motors are bound simultaneously to the bead surface. These movements are in some respects similar to the oscillatory movements of chromosomes in prometaphase (Tippet et al 1980; Bajer 1982). Bidirectional bead movement has also been described in homogenates of sea urchin eggs (Pryer et al 1986), *Reticulomyxa* (Koonce & Schliwa 1986), and *Acanthamoeba* (Kachar et al 1987). The characteristics of anterograde bead movement in axons suggest that it is powered solely by kinesin (Table 2). In contrast, retrograde movement has distinctly different characteristics, such as a two- to threefold greater velocity and greater sensitivity to NEM and vanadate (Table 2). Furthermore, a monoclonal antibody against kinesin does not recognize the retrograde translocator (Figure 2). Taken together, these data suggest that retrograde movement along microtubules in axons is generated by a protein other than kinesin. Similarly, bidirectional movement of chromatophore pigment granules along unipolar microtubule arrays appears to be driven by different motors that exhibit different nucleotide sensitivities (Rozdzial & Haimo 1986b). The retrograde motor from axons or chromatophores has not been purified; however, inhibition

Table 2   Comparison of anterograde and retrograde bead transport[a]

	Anterograde movement	Retrograde movement
Velocity of beads (μm/sec)	0.5	1.6
Transport of hydroxylated beads	No	Yes
NEM sensitive	No	Yes
Vanadate sensitive (20 μM)	No	Yes
Binding to anti-kinesin monoclonal antibody	Yes	No

[a] Based on retrograde and anterograde bead movement induced by crude axoplasmic supernatants (Vale et al 1985d).

of its motility by NEM or vanadate indicates that it may be similar to dynein.

A PERSPECTIVE IN VIVO

Reconstitution of organelle transport in vitro allows this process to be studied in the presence of a minimal number of components. Ultimately, however, one would like to understand how organelle transport functions in the context of the whole cell. How does the organization of cellular microtubules influence motility? How is the direction of organelle transport in cells determined? What cellular processes require microtubule-based transport? Certain insights into these questions have been gained from a variety of in vitro and in vivo experiments, which are discussed in the following section.

Role of Microtubule Polarity and Motility Proteins in Organelle Sorting within the Cell

CELLULAR MICROTUBULES ARE HETEROGENEOUS, YET UNIFORMLY ORIENTED Microtubules in cells are not randomly oriented. The vast majority of microtubules in most cells are nucleated and assembled from the centrosome, a microtubule-organizing center that consists of a pair of centrioles surrounded by a halo of proteinaceous material (Figure 4). Virtually all microtubules assemble from centrosomes with their minus ends anchored to the centrosome and their plus ends radiating toward the cell periphery (Bergen et al 1980). In axons microtubules are nucleated from unidentified structures that are not centrosomes, yet the polarity of the microtubules is the same: the plus ends point toward the distal nerve terminal (Burton & Paige 1981; Heidemann et al 1981). This uniform polarity of cytoplasmic microtubule networks is found in virtually all eukaryotic cells. An interesting exception, however, are dendrites of olfac-

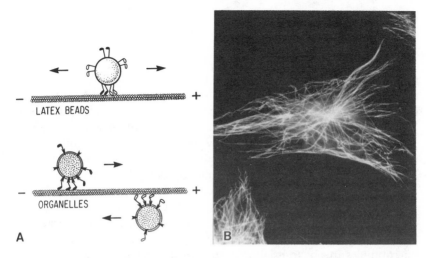

Figure 4 Potential roles of microtubule polarity and motility proteins in cell sorting. Panel A illustrates that while single latex beads move in both directions on microtubules, organelles move in only one direction. This result suggests that latex beads bind both anterograde and retrograde translocators, whereas organelles appear to selectively bind or activate only one polarity-specific motor at a time. Such selectivity could be controlled by receptors on the organelle surface. Panel B shows immunofluorescence staining of a fibroblast cell with an anti-tubulin antibody and a second rhodamine-conjugated antibody. Microtubules originate from the centrosome as a unipolar array with the microtubule plus ends oriented towards the cell periphery. Hence, organelles that bind an anterograde translocator, such as kinesin, will be transported along this unipolar microtubule network to the cell periphery or, in the case of a neuron, down the axon to the nerve terminal. Conversely, organelles that bind or activate a retrograde motor will be transported from the periphery or nerve terminal to the center of the cell. (Photograph courtesy of Dr. M. Kirschner.)

tory neurons, in which microtubules are nucleated by basal bodies in the dendritic bulb and their plus ends extend towards the neuronal cell body (Burton 1986). It is not clear whether dendritic microtubules of other neurons are also organized in this fashion.

The uniform orientation of microtubules indicates that microtubule polarity could serve as a "compass" by which particles could navigate in the cytoplasm (Figure 4). Since the centrosome is generally found in the center of the cell close to the nucleus, microtubules originating from this structure provide tracks from the interior to the periphery of the cell. Thus an object in the cytoplasm, such as an organelle, could be directed toward the interior (minus end) or periphery (plus end) of the cell by the polarity of the microtubule lattice. Actin filaments also have intrinsic polarity, and many are nucleated with the same polarity from sites on the plasma membrane. However, these filaments are not organized in a manner that provides satisfactory cues for directional movement between the center and periphery of the cell (Sanger & Sanger 1980; Heuser & Kirschner 1980).

Although most microtubules in cells have the same polarity, they differ in other ways. Most microtubules are short-lived structures that continuously depolymerize at their plus ends and repolymerize from the centrosome (Schulze & Kirschner 1986; Soltys & Borisy 1985; Saxton et al 1984). Because of this "dynamic instability" of single microtubules (Mitchison & Kirschner 1984), the microtubule network as a whole is constantly changing, which allows cells to alter their morphology rapidly (Kirschner & Mitchison 1986). A subset of cellular microtubules, however, are remarkably stable (half-life of 50 min) and are resistant to depolymerization by nocodazole (Schulze & Kirschner 1987). These stable microtubules are commonly found in cell regions associated with secretion or organelle transport (Schulze & Kirschner 1987; Brady et al 1984). Many stable microtubules are posttranslationally modified along their length by detyrosination (Gunderson et al 1984), acetylation (LeDizet & Piperno 1986), or phosphorylation (Gard & Kirschner 1985). Microtubules can also be comprised of different tubulin isotypes, which are encoded by different tubulin genes (Cleveland & Sullivan 1985), and may interact with a variety of microtubule-associated proteins (Vallee & Bloom 1984). Such posttranslational or genetic variations could give rise to microtubules that behave differently as substrates for organelle transport, thereby providing means of preferentially directing organelle traffic to particular regions of the cell.

POLARITY-SPECIFIC TRANSLOCATORS AS DETERMINANTS OF ORGANELLE TRANS-PORT DIRECTION   The direction of organelle transport is probably specified by molecules on the organelle surface. Evidence for this idea comes from comparison of the movement of latex beads with that of organelles (Figure 4). In the presence of crude axoplasmic supernatant, latex beads move discontinuously and frequently reverse their direction of movement on microtubules. These observations suggest that the beads do not selectively bind anterograde or retrograde translocators and hence have no means of controlling their direction of transport. In contrast, individual organelles in axons tend to move in a single direction (Allen et al 1985; Schnapp et al 1985). Therefore, neuronal vesicles appear to bind and activate only one polarity-specific motor at a time. This specificity may be mediated by receptors on the organelle membrane, as discussed previously. The idea that there is one active motor on a given vesicle may not necessarily hold true in all cells. In neurons, the distances are so great that the direction of organelle transport must be controlled with high fidelity to insure proper delivery to the target. However, some reversal of direction probably does occur, since radiolabelled proteins moving rapidly in the anterograde direction can reverse direction and return to the cell body prior to reaching the terminal (Bisby & Bulgar 1977; Smith 1980). In some nonneuronal cells, however, organelle transport may be used as a means of mixing the contents of the cytoplasm, in which case it would be advantageous to switch the direction of transport of a single vesicle more frequently. Consistent with this idea is the observation that organelles in *Reticulomyxa* reverse their direction of movement along microtubules much more frequently than do those in the squid giant axon (M. Koonce & M. Schliwa, personal communication).

Nonetheless, the presence of (*a*) a unipolar network of microtubules, (*b*) force-generating proteins that recognize microtubule polarity, and (*c*) organelles that can preferentially bind or activate these force-generating proteins provide a potential means of sorting organelles in all types of cells. For example, in a neuron organelles that interact with an anterograde motor such as kinesin will be transported along the unipolar microtubule network to the nerve terminal. Conversely, if a retrograde microtubule motor is bound and active, the organelle will be transported to the cell body. Similar sorting schemes may exist in nonneuronal cells as well. Vesicles derived from the Golgi complex, for example, might bind anterograde motors and be transported to the cell surface, whereas endocytic vesicles might bind retrograde motors and be transported toward the interior of the cell. Specific sorting proteins would then recognize, and mediate fusion with, appropriate membrane targets (Kelly 1985).

Neither binding of motility proteins to neuronal organelles nor their activation has been demonstrated, so this model is speculative at present. In fish chromatophores, however, some progress has been made in elucidating the signals that control pigment granule dispersion and aggregation. In xanthophores and melanophores ACTH and cAMP induce pigment granule dispersion, and removal of these agents results in the reaggregation of granules to the center of the cell (Schliwa 1984). The process of dispersion appears to be mediated by a cAMP-dependent protein kinase that phosphorylates a 57-kDa polypeptide associated with the pigment granules (Lynch et al 196a,b; Rozdzial & Haimo 1986b). It is not clear how phosphorylation of this protein regulates granule motility or whether other cells such as neurons use similar mechanisms for controlling organelle movement. The mechanism of regulation is probably not identical, however, since mitochondria in chromatophore cells do not move in response to the signals that induce granule aggregation and dispersion (Porter & McNiven 1982).

The fate of organelles and their associated microtubule motility proteins upon arrival at their target is unknown. The nerve terminal and cell body are the most probable locations for regulation of axonal transport, i.e. the sites where it is determined which organelles will be transported and in which direction. The organelles moving in the retrograde direction are morphologically distinct from those moving in the anterograde direction (Tsukita & Ishikawa 1980; Smith 1980), which suggests that membranes are processed and sorted in the nerve terminal prior to being transported. Once an anterograde motor on the organelle reaches the terminal, it may be degraded or shuttled back to the cell body in an inactive form on organelles travelling in the retrograde direction (Figure 5).

A more perplexing question is how the retrograde motor arrives at the nerve terminal, since this direction is opposite that of its movement on microtubules. It may reach the terminal simply by diffusion, which is consistent with the seemingly large quantities of soluble motor proteins in the axon. Alternatively, it could be delivered by fast transport as an inactive species on organelles travelling in the anterograde direction (Figure 5). Moving motility proteins as active and inactive species on organelles could allow them to be efficiently shuttled and recycled between the cell body and nerve terminal. Similar recycling pathways have been postulated or described for other proteins involved in shuttling ligands or vesicles between membrane compartments (Steinman et al 1983; Kelly 1985).

Role of Microtubule-Based Motors in Secretion, Endocytosis, and the Spatial Organization of Organelles

GOLGI-DERIVED TRANSPORT AND ENDOCYTOSIS   Microtubules and microtubule-based motility proteins clearly are essential for organelle trans-

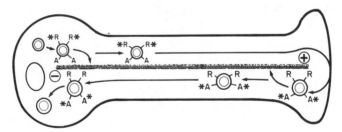

Figure 5 Hypothetical scheme of recycling and sorting of motility proteins. Organelles may bind an anterograde motor (A) such as kinesin in the cell body, which will enable it to be transported to the nerve terminal. Once at the terminal, the motor may be transported by retrograde-moving organelles as an inactive form (A*) back to the cell body. The retrograde transport motor could reach the nerve terminal by slow transport or diffusion (not shown) or by fast axonal transport with the retrograde motor present in an inactive form on organelle surfaces (R*). Activation would occur in the nerve terminal.

port in nerve axons, where the distances between the cell body and nerve terminal can be as large as one meter. Most cells are much smaller, however, and diffusion theoretically could account for movement of small organelles between the cell center and plasma membrane. Nonetheless, there is growing evidence that microtubules and their associated motility proteins may play an important role in organelle transport in nonneuronal cells.

The requirement for microtubules in vesicle-mediated secretion has been studied primarily using microtubule depolymerizing drugs such as colchicine or nocodazole. Numerous studies using various types of cells (for example see Boyd et al 1982; Cho & Garant 1981; Redman et al 1975; Busson-Mabillot et al 1982; Malaisse-Lagae et al 1979; Katz et al 1982; Ehrlich et al 1974) have reported that microtubule depolymerizing agents inhibit Golgi-derived protein secretion. Furthermore, when microtubules are reorganized into the spindle during mitosis, very little transport to the plasma membrane occurs (Warren et al 1983; Hesketh et al 1984), although a great deal of organelle movement is observed on the spindle (Rebhun 1972). Collectively, these results indicate that microtubules may be involved in transport of secretory vesicles from the Golgi complex to the vicinity of the cell surface.

However, there are several reports that microtubule depolymerizing agents do not affect vesicle-mediated protein transport to the cell surface (Salas et al 1986; Carpen et al 1981; Rogalski & Singer 1984). Perhaps not all secretory vesicles require transport on microtubules to reach the cell surface; the smallest ones may diffuse efficiently through the cytoplasm. A requirement for microtubules in directing secretion to a particular region of the cell has also been reported in some studies (Rogalski et al 1984; Rindler et al 1987) but not in others (Salas et al 1986). In fact both Rindler et al (1987) and Salas et al (1986) examined the effects of colchicine on the polarized delivery of hemagglutinin to the apical surface of Madin-Darby canine kidney (MDCK) cells and arrived at different conclusions regarding the role of microtubules in this process. The reason for these contradictory results is unclear; however, these discrepancies draw attention to the pitfalls of using colchicine as the sole means of addressing such questions.

The close proximity of the Golgi apparatus and the centrosome also suggests that microtubules are involved in transport of Golgi-derived vesicles. Golgi elements, which are usually near the center of the cell, become dispersed throughout the cell periphery upon addition of colchicine (Wehland et al 1983; Rogalski & Singer 1984). The Golgi apparatus and centrosome reorient together in migrating fibroblasts and endothelial cells (Kupfer et al 1982; Gottlieb et al 1983), in corneal cells that are depositing an extracellular matrix (Trelstad 1970), and in killer T cells complexed to target cells (Geiger et al 1982; Kupfer et al 1983) such that they are closer to the region of preferential secretion. This reorientation appears to be important for directing membrane insertion at the leading edge of migrating cells, as has been demonstrated using the G protein of vesicular stomatitis virus as a marker (Bergmann et al 1983). Cells that do not migrate have a symmetrical arrangement of their Golgi and microtubule-organizing center (Tassin et al 1985; Spiegelmann et al 1979). Taken together, these findings suggest that microtubules selectively transport secretory vesicles to localized regions of the plasma membrane.

Studies using colchicine indicate that microtubules are important for delivering endocytic vesicles to lysosomes (Wolkoff et al 1984; Kolset et al 1979; Oka & Weigel 1983). Visualization of endosomes with fluorescent ligands and lysosomes by phase microscopy reveals that these organelles first move in a discontinuous, random manner and then migrate centripetally towards the center of the cell along a microtubule network (Pastan & Willingham 1981; Herman & Albertini 1984). The retrograde movement of endosomes and lysosomes towards the centrosome and Golgi complex concentrates these organelles. Such concentration most likely increases the probability of fusion of these two types of organelles (Figure 6), which could explain why delivery of hormone-receptor complexes from endosomes to lysosomes is greatly diminished during mitosis, when few cytoplasmic microtubules are present (Sanger et al 1984).

SPATIAL ARRANGEMENT OF FILAMENTS AND ORGANELLES Microtubule-associated motility proteins may also be involved in positioning of large membrane organelles in the cytoplasm. Membrane tubules of the endoplasmic reticulum (ER), for example, are aligned in a striking manner with microtubules, particularly in the cell periphery (Terasaki et al 1986), and also associate with taxol-polymerized bundles of microtubules (Tokunaka et al 1983). ER tubules gradually retract to the center of the cell after colchicine treatment and migrate out following microtubule tracks when colchicine is washed away (Terasaki et al 1986). On the basis of these results, one might speculate that the ER interacts with an anterograde microtubule motor such as kinesin and is translocated along the microtubule network to the cell periphery. In fact, Franke (1971), has observed cross-bridges between the ER and microtubules that are similar in some respects to the cross-bridges linking vesicles to microtubules in squid axoplasm (Miller & Lasek 1985). Lysosomes, like the ER, can assume a tubular morphology and align themselves with microtubules in a macrophage cell line (Swanson et al 1987). These lysosomes also exhibited saltatory movements characteristic of microtubule-based motility.

The positioning of intermediate filaments may also involve micro-

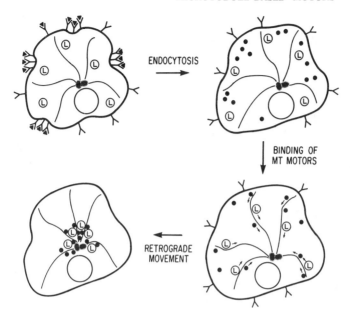

Figure 6   A diagram showing how microtubule-based motility may be involved in delivering endocytic vesicles to lysosomes [based upon studies of Pastan & Willingham (1981) and Herman & Albertini (1984)]. The following sequence of events is shown in this figure; ligand binding to receptors, receptor-mediated endocytosis, binding of a retrograde microtubule (MT) motor to endocytic vesicles and lysosomes (L), and transport of both types of organelles to the center of the cell, where vesicles fuse with lysosomes.

tubule-based motility. A partial correspondence of intermediate filament and microtubule immunofluorescence staining patterns has been observed in some cells (Ball & Singer 1981). The intermediate filament network collapses to a juxtanuclear position in the presence of colchicine (Wang & Goldman 1978; Hynes & Destree 1978; Terasaki et al 1986). When colchicine is removed and microtubules are allowed to repolymerize, intermediate filaments redistribute to the periphery, often aligning themselves with microtubules. The manner in which these filaments redistribute themselves is intriguing in light of the fact that there is no evidence that there is exchange between intermediate filaments and a pool of subunits, as is true of actin filaments and microtubules. One possibility is that intermediate filaments migrate in the anterograde direction along microtubules using motility proteins.

Unlike the ER and intermediate filaments, the membrane stacks of the Golgi are normally found in the center of the cell adjacent to the centrosome and disperse to the periphery when the microtubule network is depolymerized (Wehland et al 1983; Rogalski & Singer 1984). The Golgi membrane probably interacts with microtubules, since Golgi elements in intact cells associate with ends of microtubules formed by taxol polymerization (Rogalski & Singer 1984). Allan & Kreis (1987) have found a 110-kDa protein on the cytoplasmic face of the Golgi membrane that binds to microtubules in vitro and cross-reacts with a monoclonal antibody raised against MAP-2. Whether the Golgi apparatus also moves along microtubules is unknown, but it could acquire its central location by interacting with a retrograde motor and being transported along the unipolar microtubule array toward the centrosome.

Very little is known about how nuclei are positioned in cytoplasm, although it may involve an interaction with, and perhaps movement along, microtubule networks (Aronson 1971). In yeast, for example, nuclear migration during division in *Saccharomyces cerevisiae* (Pringle et al 1986) and localization of nuclei in the center of *Schizosaccharomyces pombe* cells (Hiraoka et al 1984) both require microtubules. Nuclear movements in *Drosophila* embryos may also occur along microtubule tracks (Karr & Alberts 1986).

CONCLUSION

Clearly, a number of important biological processes involve microtubule-based motility; however, deciphering the molecular details of such phenomena is difficult unless these events can be studied outside of the cell. Examination of organelle motility, for example, in the relative simplicity of the test tube has provided important insights into how such movements are generated. This general experimental approach has also led to the identification and purification of two microtubule-based motility proteins, kinesin and cytoplasmic dynein, and additional proteins that generate or regulate motility will likely be purified in the next few years. Nonetheless, it should be apparent from this review that little is known of how these microtubule motors function inside living cells. More concrete information on the biological roles of microtubule-based motility proteins will hopefully be gained through investigations of the protein structure and enzymology of known microtubule translocators such as kinesin and dynein and through the use of reagents such as monoclonal antibodies and cDNA clones for microinjection or genetic manipulation experiments. Future research should combine in vitro and in vivo experimental approaches to elucidate how motility proteins move and position organelles in cytoplasm. It is hoped that this review will stimulate an interest in such subjects among investigators working in a variety of disciplines in cell biology.

ACKNOWLEDGMENTS

I would like to thank Drs. Marc Kirschner, Tim Mitchison, Jim Spudich, and Trina Schroer for their critical evaluation of the manuscript and Drs. Mike Sheetz, Bruce Schnapp, and Tom Reese for helpful discussions.

Literature Cited

Adams, R. J., Bray, D. 1983. Rapid transport of foreign particles microinjected into crab axons. *Nature* 303: 718–20

Adams, R. J., Pollard, T. D. 1986. Propulsion of organelles isolated from *Acanthamoeba* along actin filaments by myosin-I. *Nature* 322: 754–56

Adelstein, R. S., Eisenberg, E. 1980. Regulation and kinetics of actin-myosin-ATP interaction. *Ann. Rev. Biochem.* 49: 921–56

Allan, V. J., Kreis, T. E. 1987. A microtubule-binding protein associated with membranes of the Golgi apparatus. *J. Cell Biol.* 103: 2229–39

Allen, R. D., Allen, N. S., Travis, J. L. 1981. Video-enhanced differential interference contrast (AVEC-DIC) microscopy: A new method capable of analyzing microtubule-related movement in the reticulopodial network of *Allogromia laticollaris*. *Cell Motil.* 1: 291–302

Allen, R. D., Metuzals, J., Tasaki, I., Brady, S. T., Gilbert, S. P. 1982. Fast axonal transport in squid giant axon. *Science* 218: 1127–28

Allen, R. D., Weiss, D. G., Hayden, J. H., Brown, D. T., Fujiwake, H., Simpson, M. 1985. Gliding movement of and bidirectional organelle transport along single native microtubules from squid axoplasm: Evidence for an active role of microtubules in cytoplasmic transport. *J. Cell Biol.* 100: 1736–52

Amos, L. A. 1987. Kinesin from pig brain studied by electron microscopy. *J. Cell Sci.* 87: 105–11

Aronson, J. F. 1971. Demonstration of a colcemid-sensitive attractive force acting between the nucleus and a center. *J. Cell Biol.* 51: 579–83

Asai, D. J., Wilson, L. 1985. A latent activity dynein-like cytoplasmic magnesium adenosine triphosphatase. *J. Biol. Chem.* 260: 699–702

Bajer, A. 1982. Functional anatomy of monopolar spindles and evidence for oscillatory chromosome movements in mitosis. *J. Cell Biol.* 93: 33–48

Ball, R. H., Singer, S. J. 1981. Association

of microtubules and intermediate filaments in normal fibroblasts and its disruption upon transformation by temperature-sensitive mutant of Rous sarcoma virus. *Proc. Natl. Acad. Sci. USA* 78: 6986–90

Beckerle, M. C. 1984. Microinjected fluorescent polystyrene beads exhibit saltatory motion in tissue culture cells. *J. Cell Biol.* 98: 2126–32

Bergen, L. G., Kuriyama, R., Borisy, G. G. 1980. Polarity of microtubules nucleated by centrosomes and chromosomes of CHO cell in vitro. *J. Cell Biol.* 84: 151–59

Bergmann, J. E., Kupfer, A., Singer, S. J. 1983. Membrane insertion at the leading edge of motile fibroblasts. *Proc. Natl. Acad. Sci. USA* 80: 1367–71

Bisby, M. A. 1986. Retrograde axonal transport and nerve regeneration. In *Advances in Neurochemistry*, ed. J. S. Elam, P. Cancelon, Vol. 6, pp. 45–67. New York: Plenum

Bisby, M. A., Bulgar, V. T. 1977. Reversal of axonal transport at a nerve crush. *J. Neurochem.* 29: 313–20

Boyd, A. E. III, Bolton, W. E., Brinkley, B. R. 1982. Microtubules and beta-cell function: Effect of colchicine on microtubules and insulin secretion in vitro by mouse beta cells. *J. Cell Biol.* 92: 425–34

Brady, S. T. 1985. A novel brain ATPase with properties expected for the fast axonal transport motor. *Nature* 317: 73–75

Brady, S. T., Lasek, R. J., Allen, R. D. 1982. Fast axonal transport in extruded axoplasm from squid giant axon. *Science* 218: 1129–31

Brady, S. T., Lasek, R. J., Allen, R. D. 1985. Videomicroscopy of fast axonal transport in extruded axoplasm: A new model for study of molecular mechanisms. *Cell Motil.* 5: 81–101

Brady, S. T., Tytell, M., Lasek, R. J. 1984. Axonal tubulin and axonal microtubules: Biochemical evidence for cold stability. *J. Cell Biol.* 99: 1716–24

Burton, P. R. 1986. Ultrastructure of the olfactory neuron of the bullfrog; the dendrite and its microtubules. *J. Comp. Neurol.* 242: 147–60

Burton, P. R., Paige, J. L. 1981. Polarity of axoplasmic microtubules in the olfactory nerve of the frog. *Proc. Natl. Acad. Sci. USA* 78: 3269–73

Busson-Mabillot, S., Chambaut-Guerin, A.-M., Ovtracht, L., Muller, P., Rossignol, B. 1982. Microtubules and protein secretion in rat lacrimal glands: Localization of short-term effects of colchicine on the secretory process. *J. Cell Biol.* 95: 105–17

Carpen, O., Virtanen, I., Saksela, E. 1981. The cytotoxic activity of human natural killer cells requires an intact secretory apparatus. *Cell. Immunol.* 58: 97–106

Cho, M.-H., Garant, P. R. 1981. Role of microtubules in the organization of the Golgi complex and the secretion of collagen secretory granules by periodontal ligament fibroblast. *Anat. Rec.* 199: 459–71

Clark, T. G., Rosenbaum, J. L. 1982. Pigment particle translocation in detergent-permeabilized melanophores of *Fundulus heteroclitus. Proc. Natl. Acad. Sci. USA* 79: 4655–59

Cleveland, D. W., Sullivan, K. F. 1985. Molecular biology and genetics of tubulin. *Ann. Rev. Biochem.* 54: 331–66

Cohn, S. A., Ingold, A. L., Scholey, J. M. 1987. Correlation between the ATPase and microtubule translocating activities of sea urchin egg kinesin. *Nature* 328: 160–63

Cooke, R. 1986. The mechanism of muscle contraction. *CRC Crit. Rev. Biochem.* 21: 53–118

Ehrlich, H. P., Ross, R., Bornstein, P. 1974. Effects of antimicrotubular agents on the secretion of collagen. *J. Cell Biol.* 62: 390–405

Ellisman, M. H., Porter, K. R. 1980. Microtrabecular structure of the axoplasmic matrix: Visualization of cross-linking structures and their distribution. *J. Cell Biol.* 87: 464–79

Farquhar, M. G. 1985. Progress in unraveling pathways of Golgi traffic. *Ann. Rev. Cell Biol.* 1: 447–88

Forman, D. S. 1982. Vanadate inhibits saltatory organelle movement in a permeabilized cell model. *Exp. Cell Res.* 141: 139–47

Forman, D. S., Brown, K. J., Livengood, D. R. 1983. Fast axonal transport in permeabilized lobster giant axons is inhibited by vanadate. *J. Neurosci.* 3: 1279–88

Franke, W. W. 1971. Cytoplasmic microtubules linked to endoplasmic reticulum with cross-bridges. *Exp. Cell Res.* 66: 486–89

Gard, D. L., Kirschner, M. W. 1985. A

polymer-dependent increase in phosphorylation of B-tubulin accompanies differentiation of a mouse neuroblastoma cell line. *J. Cell Biol.* 100: 764–74

Geiger, B., Rosen, D., Berke, G. 1982. Spatial relationships of microtubule-organizing centers and the contact area of cytotoxic T lymphocytes and target cells. *J. Cell Biol.* 95: 137–43

Gelles, J., Schnapp, B. J., Sheetz, M. P. 1987. Motion analysis of kinesin-induced bead movement. Submitted

Gibbons, I. R. 1981. Cilia and flagella of eukaryotes. *J. Cell Biol.* 91: 107–24

Gilbert, S. P., Sloboda, R. D. 1984. Bidirectional transport of fluorescently labeled vesicles introduced into extruded axoplasm of squid *Loligo pealei. J. Cell Biol.* 99: 445–52

Gilbert, S. P., Sloboda, R. 1986. Identification of a MAP-2-like ATP binding protein associated with axoplasmic vesicles that translocate on isolated microtubule. *J. Cell Biol.* 103: 947–56

Gilbert, S. P., Sloboda, R. D., Allen, R. D. 1985. Translocation of vesicles from squid axoplasm on flagellar microtubules. *Nature* 315: 245–48

Goodenough, U. W., Heuser, J. E. 1984. Structural comparison of purified dynein proteins with in situ dynein arms. *J. Mol. Biol.* 180: 1083–1118

Gottlieb, A., Subrahmanyan, L., Kalnins, V. I. 1983. Microtubule organizing centers and cell migration: Effect of inhibition of migration and microtubule disruption in endothelial cells. *J. Cell Biol.* 96: 1266–72

Grafstein, B., Forman, D. S. 1980. Intracellular transport in neurons. *Physiol. Rev.* 60: 1167–1283

Greene, L. E., Eisenberg, E. 1980. Dissociation of the actin subfragment 1 complex by adenyl-5′-imidodiphosphate, ADP and PPi. *J. Biol. Chem.* 225: 543–48

Gunderson, G. G., Kalnowski, M. H., Bulinsky, J. C. 1984. Distinct populations of microtubules: Tyrosinated and nontyrosinated alpha tubulin are distributed differently in vivo. *Cell* 38: 779–89

Hayden, J., Allen, R. D. 1984. Detection of single microtubules in living cells: Particle transport can occur in both directions along the same microtubule. *J. Cell Biol.* 99: 1785–93

Heidemann, S. R., Landers, J. M., Hamborg, M. A. 1981. Polarity orientation of axonal microtubules. *J. Cell Biol.* 91: 661–65

Herman, B., Albertini, D. F. 1984. A time-lapse video image intensification analysis of cytoplasmic organelle movements during endosome translocation. *J. Cell Biol.* 98: 565–76

Hasketh, T. R., Beaven, M. A., Rogers, J., Burke, B., Warren, G. B. 1984. Stimulated release of histamine by a rat mast cell line is inhibited during mitosis. *J. Cell Biol.* 98: 2250–54

Heuser, J. E., Kirschner, M. W. 1980. Filament organization revealed in platinum replicas of freeze-dried cytoskeletons. *J. Cell Biol.* 86: 212–34

Hill, T. L. 1987. Use of muscle contraction formalism for kinesin in fast axonal transport. *Proc. Natl. Acad. Sci. USA* 84: 474–77

Hiraoka, Y., Toda, T., Yanagida, M. 1984. The NDA3 gene of fission yeast encodes β-tubulin: A cold-sensitive nda3 mutation reversibly blocks spindle formation and chromosome movement in mitosis. *Cell* 39: 349–58

Hirokawa, N. 1982. Cross-linker system between neurofilaments, microtubules and membranous organelles in frog axons by the quick-freeze, deep-etching method. *J. Cell Biol.* 94: 129–42

Hisanga, S., Sakai, H. 1983. Cytoplasmic dynein of the sea urchin. II. Purification, characterization and interactions with microtubules and Ca-calmodulin. *J. Biochem.* 93: 87–98

Hollenbeck, P. J., Chapman, K. 1986. A novel microtubule-associated protein from mammalian nerve shows ATP-sensitive binding to microtubules. *J. Cell Biol.* 103: 1539–45

Hollenbeck, P. J., Suprynowicz, F., Cande, W. Z. 1984. Cytoplasmic dynein-like ATPase cross-links microtubules in an ATP-sensitive manner. *J. Cell Biol.* 99: 1251–58

Huttner, W. B., Schiebler, W., Greengard, P., De Camilli, P. 1983. Synapsin (protein I), a nerve terminal-specific phosphoprotein. III. Its association with synaptic vesicles studied in a highly purified synaptic vesicle preparation. *J. Cell Biol.* 96: 1374–88

Hynes, R. O., Destree, A. T. 1978. 10 nm filaments in normal and transformed cells. *Cell* 13: 151–63

Inoue, S. 1981. Video image processing greatly enhances contrast, quality and speed in polarization-based microscopy. *J. Cell Biol.* 89: 346–56

Johnson, K. A. 1985. Pathway of the microtubule-dynein ATPase and the structure of dynein: A comparison with actomyosin. *Ann. Rev. Biophys. Biophys. Chem.* 14: 161–88

Kachar, B., Fujisake, H., Albanesi, J. P., Korn, E. D. 1987. Soluble microtubule-based translocator protein in *Acanthamoeba castellanii* extracts. *Biophys. J.* 51: 487a

Karr, T. L., Alberts, B. M. 1986. Organization of the cytoskeleton in early *Drosophila* embryos. *J. Cell Biol.* 102: 1494–1509

Katz, P., Zaytoun, A. M., Lee, J. H. 1982. Mechanisms of human cell-mediated cytotoxicity. III. Dependence of natural killing on microtubule and microfilament integrity. *J. Immunol.* 129: 2816–25

Kaziro, Y. 1978. The role of guanosine 5'-triphosphate in polypeptide chain elongation. *Biochim. Biophys. Acta* 505: 95–127

Kelly, R. B. 1985. Pathways of protein secretion in eukaryotes. *Science* 230: 25–32

Kirschner, W. W., Mitchison, T. J. 1986. Beyond self-assembly: From microtubules to morphogenesis. *Cell* 45: 329–42

Kolset, S. O., Tolleshaug, H., Berg, T. 1979. The effects of colchicine and cytochalasin B on uptake and degradation of asialoglycoproteins in isolated rat hepatocytes. *Exp. Cell Res.* 122: 159–67

Koonce, M. P., Schliwa, M. 1985. Bidirectional organelle transport can occur in cell processes that contain single microtubules. *J. Cell Biol.* 100: 322–26

Koonce, M. P., Schliwa, M. 1986. Reactivation of organelle movements along cytoskeletal framework of a giant freshwater amoeba. *J. Cell Biol.* 103: 605–12

Korn, E. D. 1983. *Acanthamoeba castellanii*: Methods and perspectives for the study of cytoskeleton proteins. *Methods Cell Biol.* 25: 313–32

Kupfer, A., Dennert, G., Singer, S. J. 1983. Polarization of the Golgi apparatus and the microtubule-organizing center within cloned natural killer cells bound to their targets. *Proc. Natl. Acad. Sci. USA* 80: 7224–28

Kupfer, A., Louvard, D., Singer, S. J. 1982. Polarization of the Golgi apparatus and the microtubule-organizing center in cultured fibroblasts at the edge of an experimental wound. *Proc. Natl. Acad. Sci. USA* 79: 2603–7

Kuznetsov, S. A., Gelfand, V. I. 1986. Bovine brain kinesin is a microtubule activated ATPase. *Proc. Natl. Acad. Sci. USA* 83: 8350–54

Langford, G., Allen, R. D., Weiss, D. 1987. Substructure of side arms on squid axoplasmic vesicles and microtubules visualized by negative contrast electron microscopy. *Cell Motil. Cytoskeleton* 7: 20–31

Lasek, R. J. 1968. Axoplasmic transport in cat dorsal root ganglion cell as studied with 3H-leucine. *Brain Res.* 7: 360–77

Lasek, R. J. 1982. Translocation of the neuronal cytoskeleton and axonal locomotion. *Philos. Trans. R. Soc. Lond.* 229: 313–27

Lasek, R. J., Brady, S. T. 1985. Attachment of transported vesicles to microtubules in axoplasm is facilitated by AMP-PNP. *Nature* 316: 645–47

LeDizet, M., Piperno, G. 1986. Cytoplasmic microtubules containing acetylated a-tubulin in *Chlamydomonas reinhardti*: Spatial arrangements and properties. *J. Cell Biol.* 103: 13–22

Leslie, R. J., Hird, R. B., Wilson, L., McIntosh, J. R., Scholey, J. M. 1987. Kinesin is associated with a nonmicrotubule component of sea urchin mitotic spindles. *Proc. Natl. Acad. Sci. USA* 84: 2771–75

Lye, R. J., Porter, M. E., Scholey, J. M., McIntosh, J. R. 1987. Identification of a microtubule-based cytoplasmic motor in the nematode *Caenorhabditis elegans*. Submitted

Lynch, T. J., Taylor, J. D., Tchen, T. T. 1986a. Regulation of pigment organelle translocation. I. Phosphorylation of the organelle-associated protein p57. *J. Biol. Chem.* 261: 4204–11

Lynch, T. J., Wu, B. Y., Taylor, J. D., Tchen, T. T. 1986b. Regulation of pigment organelle translocation. II. Participation of a cAMP-dependent protein kinase. *J. Biol. Chem.* 261: 4212–16

Malaisse-Lagae, F., Amherdt, M., Ravazzola, M., Sener, A., Hutton, J. C., et al. 1979. Role of microtubules in the synthesis, conversion, and release of (pro) insulin. *J. Clin. Invest.* 63: 1284–96

McNiven, M. A., Porter, K. R. 1984. Chromatophores—models for studying cytomatrix translocations. *J. Cell Biol.* 99: 152–58

Miller, R. H., Lasek, R. J. 1985. Crossbridges mediate anterograde and retrograde vesicle transport along microtubules in squid axoplasm. *J. Cell Biol.* 101: 2181–93

Mitchell, D. R., Warner, F. D. 1981. Binding of dynein 21S ATPase to microtubules. Effects of ionic conditions and substrate analogs. *J. Biol. Chem.* 256: 12535–44

Mitchison, T. 1986. Role of microtubule polarity in the movement of kinesin and kinetochores. *J. Cell Sci.* 5: 121–28 (Suppl.)

Mitchison, T., Kirschner, M. 1984. Dynamic instability of microtubule growth. *Nature* 312: 237–42

Murphy, D. B., Wallis, K. T., Hiebsch, R. R. 1983. Identity and origin of the ATPase activity associated with neuronal microtubules. II. Identification of a 50,000-dalton polypeptide with ATPase activity similar to F-1 ATPase from mitochondria. *J. Cell Biol.* 96: 1306–15

Novick, D., Botstein, D. 1985. Phenotypic analysis of temperature-sensitive yeast actin mutants. *Cell* 40: 405–16

Oka, J. A., Weigel, P. H. 1983. Microtubule-depolymerizing agents inhibit asialo-orosomucoid delivery to lysosomes but not its endocytosis or degradation in isolated rat hepatocytes. *Biochim. Biophys. Acta* 763: 368–76

Pastan, I. H., Willingham, M. C. 1981. Journey to the center of the cell. Role of the receptosome. *Science* 214: 504–9

Piperno, G. 1984. Monoclonal antibodies to dynein subunits reveal the existence of cytoplasmic antigens in sea urchin egg. *J. Cell. Biol.* 98: 1842–50

Porter, K. R., McNiven, M. A. 1982. The cytoplast: A unit structure in chromatophores. *Cell* 29: 23–32

Porter, M. E., Scholey, J. M., Stemple, D. L., Vigers, G. P. A., Vale, R. D., et al. 1987. Characterization of microtubule movement produced by sea urchin egg kinesin. *J. Biol. Chem.* 262: 2794–2802

Pratt, M. M. 1980. The identification of a dynein ATPase in unfertilized sea urchin eggs. *Dev. Biol.* 74: 364–78

Pratt, M. M. 1986a. Homology of egg and flagellar dynein. *J. Biol. Chem.* 261: 956–64

Pratt, M. M. 1986b. Stable complexes of axoplasmic vesicles and microtubules: Protein composition and ATPase activity. *J. Cell Biol.* 103: 957–68

Pratt, M. M., Barton, N., Betancourt, A., Hammond, C., Schroeder, B. 1987. Cytoplasmic dynein association with vesicular organelles. Submitted

Pringle, J. R., Lillie, A. E. M., Adams, C. W., Jacobs, C. W., Haarer, B. K., et al. 1986. Cellular morphogenesis in the yeast cell cycle. In *Yeast Cell Biology*, ed. J. Hicks, pp. 47–80. New York: Liss

Pryer, N. K., Wadsworth, P., Salmon, E. D. 1986. Polarized microtubule gliding and particle saltations produced by soluble factors from sea urchin eggs and embryos. *Cell Motil.* 6: 537–48

Rebhun, L. I. 1972. Polarized intracellular particle transport: Saltatory movements and cytoplasmic streaming. *Int. Rev. Cytol.* 32: 93–137

Redman, C. M., Banerjee, D., Howell, K., Palade, G. E. 1975. Colchicine inhibition of plasma protein release from rat hepatocytes. *J. Cell Biol.* 66. 42–59

Rindler, M. J., Ivanov, I. E., Sabatini, D. D. 1987. Microtubule-acting drugs lead to the nonpolarized delivery of the influenza hemagglutinin to the cell surface of polarized Madin-Darby canine kidney cells. *J. Cell Biol.* 104: 231–41

Rogalski, A. A., Bergmann, J. E., Singer, S.

J. 1984. Effect of microtubule assembly status of the intracellular processing and surface expression of an integral protein of the plasma membrane. *J. Cell Biol.* 99: 1101–9

Rogalski, A. A., Singer, S. J. 1984. Associations of elements of the Golgi apparatus with microtubules. *J. Cell Biol.* 99: 1092–1100

Rozdzial, M. M., Haimo, L. T. 1986a. Reactivated melanophore motility: Differential regulation and nucleotide requirements of bidirectional pigment granule transport. *J. Cell Biol.* 103: 2755–64

Rozdzial, M. M., Haimo, L. T. 1986b. Bidirectional pigment granule movements of melanophores are regulated by protein phosphorylation and dephosphorylation. *Cell* 47: 1061–70

Salas, A. J. L., Misek, D. E., Vega-Salas, D. E., Gundersen, D., Cereijido, M., Rodriguez-Boulan, E. 1986. Microtubules and actin filaments are not critically involved in the biogenesis of epithelial cell surface polarity. *J. Cell Biol.* 102: 1853–67

Sale, W. S., Satir, P. 1977. The direction of active sliding of microtubules in *Tetrahymena* cilia. *Proc. Natl. Acad. Sci. USA* 74: 2045–49

Sanger, J. M., Sanger, J. W. 1980. Banding and polarity of actin filaments in interphase and cleaving cells. *J. Cell Biol.* 86: 568–75

Sanger, P. R., Brown, P. A., Berlin, R. D. 1984. Analysis of transferrin recycling in mitotic and interphase HeLa cells by quantitative fluorescence microscopy. *Cell* 39: 275–82

Saxton, W. M., Stemple, D. L., Leslie, R. J., Salmon, E. D., Zavortink, M., McIntosh, J. R. 1984. Tubulin dynamics in cultured mammalian cells. *J. Cell Biol.* 99: 2175–86

Schliwa, M. 1984. Mechanism of intracellular organelle transport. In *Cell Muscle Motility*, ed. J. W. Shaw, Vol. 5, pp. 1–81. New York: Plenum

Schnapp, B. J., Kahn, S., Sheetz, M. P., Vale, R. D., Reese, T. S. 1986. Kinesin-driven microtubule sliding involves interacting nucleotide binding sites. *J. Cell Biol.* 103: 551a

Schnapp, B. J., Reese, T. S. 1982. Cytoplasmic structure in rapid-frozen axons. *J. Cell Biol.* 94: 667–79

Schnapp, B. J., Vale, R. D., Sheetz, M. P., Reese, T. S. 1985. Single microtubules from squid axoplasm support bidirectional movement of organelles. *Cell* 40: 455–62

Scholey, J. M., Neighbors, B., McIntosh, J. R., Salmon, E. D. 1984. Isolation of microtubules and a dynein-like MgATPase from unfertilized sea urchin eggs. *J. Biol. Chem.* 259: 6516–25

Scholey, J. M., Porter, M. E., Grissom, P. M., McIntosh, J. R. 1985. Identification of kinesin in sea urchin eggs, and evidence for its localization in the mitotic spindle. *Nature* 318: 483–86

Schroer, T. A., Brady, S. T., Kelly, R. B. 1985. Fast axonal transport of foreign synaptic vesicles in squid axoplasm. *J. Cell Biol.* 101: 568–72

Schulze, E. S., Kirschner, M. W. 1986. Microtubule dynamics in interphase cells. *J. Cell Biol.* 102: 1020–31

Schulze, E. S., Kirschner, M. W. 1987. Dynamic and stable populations of microtubules in cells. *J. Cell Biol.* 104: 277–88

Sheetz, M. P., Spudich, J. A. 1983. Movement of myosin-coated fluorescent beads on actin cables in vitro. *Nature* 303: 31–35

Smith, R. S. 1980. The short term accumulation of axonally transported organelles in the region of localized lesions of single myelinated axons. *J. Neurocytol.* 9: 39–65

Soltys, B. J., Borisy, G. G. 1985. Polymerization of tubulin in vivo: Direct evidence for assembly onto microtubule ends and from centrosomes. *J. Cell Biol.* 100: 1682–89

Spiegelman, B. M., Lopata, M. A., Kirschner, M. W. 1979. Aggregation of microtubule initiation sites preceding neurite outgrowth in mouse neuroblastoma cells. *Cell* 16: 253–63

Stearns, M. E., Ochs, R. L. 1982. A functional in vitro model for studies of intracellular motility in digitonin-permeabilized erythrophores. *J. Cell Biol.* 94: 727–39

Steinman, R. M., Mellman, I. K., Muller, W. A., Cohn, Z. A. 1983. Endocytosis and recycling of plasma membrane. *J. Cell Biol.* 96: 1–27

Swanson, J., Bushnell, A., Silverstein, S. C. 1987. Tubular lysosome morphology and distribution within macrophages depend on the integrity of cytoplasmic microtubules. *Proc. Natl. Acad. Sci. USA* 84: 1921–25

Tassin, A. M., Paintrand, M., Berger, E. G., Bornens, M. 1985. The Golgi apparatus remains associated with microtubule organizing centers during myogenesis. *J. Cell Biol.* 101: 630–38

Terasaki, M., Chen, L. B., Fujiwara, K. 1986. Microtubules and the endoplasmic reticulum are highly interdependent structures. *J. Cell Biol.* 103: 1557–68

Tippet, D. H., Pickett-Heaps, J. D., Leslie, R. 1980. Cell division in two large pennate diatoms *Hantzchia* and *Nitzchia*. III. A

378 VALE

new proposal for kinetochore function during prometaphase. *J. Cell Biol.* 86: 402–16

Tokunaka, S., Friedman, R. M., Tomyama, Y., Pacifici, M., Holtzer, H. 1983. Taxol induces microtubule–rough endoplasmic reticulum complexes and microtubule bundles in cultured chondroblasts. *Differentiation* 24: 39–47

Trelstad, R. L. 1970. The Golgi apparatus in chick corneal epithelium: Changes in intracellular position during development. *J. Cell Biol.* 45: 34–42

Tsukita, S., Ishikawa, H. 1980. The movement of membranous organelles in axons. Electron microscopic identification of anterogradely and retrogradely transported organelles. *J. Cell Biol.* 84: 513–30

Vale, R. D., Reese, T. S., Sheetz, M. P. 1985c. Identification of a novel force-generating protein, kinesin, involved in microtubule-based motility. *Cell* 42: 39–50

Vale, R. D., Schnapp, B. J., Mitchison, T., Steuer, E., Reese, T. S., Sheetz, M. P. 1985d. Different axoplasmic proteins generate movement in opposite directions along microtubules in vitro. *Cell* 43: 623–32

Vale, R. D., Schnapp, B. J., Reese, T. S., Sheetz, M. P. 1985a. Movement of organelles along filaments dissociated from axoplasm of the squid giant axon. *Cell* 40: 449–54

Vale, R. D., Schnapp, B. J., Reese, T. S., Sheetz, M. P. 1985b. Organelle, bead and microtubule translocations promoted by soluble factors from the squid giant axon. *Cell* 40: 559–69

Vale, R. D., Schnapp, B. J., Reese, T. S., Sheetz, M. P. 1985e. Purification and structure of a novel microtubule trans-

locator (kinesin) from squid and bovine neural tissue. *J. Cell Biol.* 101: 37a

Vale, R. D., Scholey, J. M., Sheetz, M. P. 1986. Kinesin: Biological roles for a new microtubule-based motor. *Trends Biochem. Sci.* 11: 464–68

Vallee, R. B., Bloom, G. S. 1984. High molecular weight microtubule-associated proteins. In *Modern Cell Biology*, ed. B. H. Satir, Vol. 3, pp. 21–75. New York: Liss

Wang, E., Goldman, R. D. 1978. Functions of cytoplasmic fibers in intracellular movements in BHK-21 cells. *J. Cell Biol.* 79: 708–26

Warren, G. C., Featherstone, G., Griffiths, G., Burke, B. 1983. Newly synthesized G protein of vesicular stomatitis virus is not transported to the cell surface during mitosis. *J. Cell Biol.* 97: 1623–28

Wehland, J., Henkart, M., Klausner, R., Sandoval, I. V. 1983. Role of microtubules in the distribution of the Golgi apparatus: Effect of taxol and microinjected anti-α-tubulin antibodies. *Proc. Natl. Acad. Sci. USA* 80: 4286–90

Weisenberg, R. C., Allen, R. D., Inoue, S. 1986. ATP-dependent formation and motility of aster-like structures with isolated calf brain microtubule proteins. *Proc. Natl. Acad. Sci. USA* 83: 1728–32

Wolkoff, A. W., Klausner, R. D., Ashwell, G., Harford, J. 1984. Intracellular segregation of asialoglycoproteins and their receptor: A prelysosomal event subsequent to dissociation of the ligand-receptor complex. *J. Cell Biol.* 98: 375–81

Yang, J., Saxton, W., Goldstein, L. S. B. 1987. Isolation and characterization of the gene encoding the heavy chain of *Drosophila* kinesin. *Proc. Natl. Acad. Sci. USA.* In press

Ann. Rev. Cell Biol. 1987. 3 : 379–421

MYOSIN STRUCTURE AND FUNCTION IN CELL MOTILITY

Hans M. Warrick and James A. Spudich

Department of Cell Biology, Stanford University School of Medicine, Stanford, California 94305

CONTENTS

INTRODUCTION 379
MYOSIN: ITS DISCOVERY AND BASIC STRUCTURE 380
CELL BIOLOGY OF MYOSINS 382
MOLECULAR SUBSTRUCTURE OF MYOSIN 388
    The Myosin Head 394
    The Myosin Tail 400
MODELS OF MUSCLE CONTRACTION 402
ASSAYS FOR MYOSIN FUNCTION 403
    In Vitro Movement Assays 403
    Actin-Activated ATPase Assay 406
MOLECULAR GENETIC MANIPULATION OF MYOSIN 408
CONCLUSION AND FUTURE PERSPECTIVES 412

INTRODUCTION

Cell motility refers to three basic forms of cell movement. One is the migration of a cell across a substratum. The second is the variety of changes in shape that a cell undergoes, such as the furrowing that occurs each time a cell divides. The third is movement within a cell. The latter occurs incessantly and can take place in the absence of the first two forms of movement. Inside cells, mitochondria and other organelles constantly undergo directed movement, which sometimes leads to streaming of the entire cytoplasm. Most eukaryotic cells exhibit all three types of cell motility.

The molecular apparatus that drives these movements has been partly characterized. Dynein, first discovered as the motor that drives the beating of eukaryotic cilia and flagella, undoubtedly drives other forms of movement as well. Recently, a new motor, kinesin, has been discovered (Vale et

0743–4634/87/1115–0379$02.00

al 1985a,b,c) that moves along microtubules and may be responsible for vesicle translocations in nerve axons and in other cells. The reader is referred to a recent review by Johnson (1985) for current information on dynein. The biology of kinesin is reviewed by Vale in this volume. This review focuses on the oldest and best-characterized eukaryotic molecular motor, myosin. This protein drives muscle contraction and a variety of other forms of cell motility. There have been numerous recent reviews on myosin that focus primarily on biochemical and biophysical studies of energy transduction (Taylor 1979; Eisenberg & Greene 1980; Adelstein & Eisenberg 1980; Morales et al 1982; Huxley & Faruqi 1983; Goody & Holmes 1983; Harrington & Rodgers 1984; Amos 1985; Eisenberg & Hill 1985; Cooke 1986; Hibberd & Trentham 1986). This review only briefly summarizes major conclusions from those studies and concentrates on recent molecular biological approaches to myosin structure and function that have led to important insights about this molecule.

MYOSIN: ITS DISCOVERY AND BASIC STRUCTURE

In the late nineteenth century, Kühne transformed theories of muscle contraction when he observed a nematode swimming freely within a sample of muscle tissue (Kühne 1863). He concluded that the muscle contents are not solid as had been thought, but instead are a concentrated solution of "albumins." He found that solutions containing high salt concentrations could extract a substance from muscle that he called "myosin" (Kühne 1864). It was well into the twentieth century before Engelhardt & Ljubimowa (1939) made the startling discovery that Kühne's myosin, which had been considered a "structural protein," possessed the ATPase activity believed to be the energy source for force production during muscle contraction. Shortly thereafter, Albert Szent-Gyorgyi and his colleagues (1942) demonstrated that artificial fibrils made from myosin and another protein, which they named "actin" (because it "activated" myosin), contracted when ATP was added. Since that time a great deal of biochemical and biophysical work has contributed to a good understanding of the basic structure of the myosin molecule.

Muscle myosins and the majority of the nonmuscle myosins that have been purified consist of two globular heads joined by an extended α-helical coiled-coil tail (Figure 1). Each head, called subfragment-1 (S1), has an actin binding site and exhibits actin-activated ATPase activity. The tail has the capability of interacting with the tails of other myosin molecules to form a bipolar filament, which has a helical repeat of ~ 140 Å. Myosin

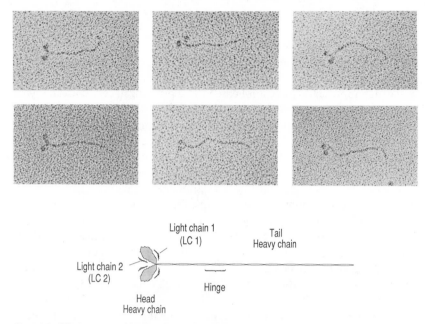

Figure 1 The upper panels show examples of rotary shadowed rabbit myosin viewed with the electron microscope. The lower panel is a representation of a hexameric myosin molecule, showing the two pairs of light chains and the pair of heavy chains. The light chains are known to be associated with the globular heads, but the exact locations have not been established. The bracket denotes the region of the tail that is more flexible and is referred to as the hinge.

consists of three gene products, a heavy chain of about 230 kDa and two light chains of about 20 kDa each. A single myosin molecule is composed of two of each of these products. The carboxy-terminal halves of the two heavy chains assemble into a coiled-coil rod, while their amino-terminal sequences fold to form two globular heads. One of each of the two types of light chain is associated with each S1 domain. Conventional myosin[1] is therefore a hexameric protein (Figure 1).

[1] The broadest enzymatic definition of myosin is a protein that catalyzes translational movement on an actin filament substrate. The term "conventional myosin" refers to a family of myosins that share similarities with those myosins found in muscle. Throughout the text the use of the term myosin refers to this family of conventional myosins. At least one myosin has been cloned and sequenced that is quite different in structure from the conventional myosin in that it contains only one head and lacks the long α-helical tail; it is referred to as myosin I. The conventional myosin in *Acanthamoeba* is referred to as myosin II.

CELL BIOLOGY OF MYOSINS

Given the widespread cellular distribution of actin and myosin, it is generally assumed that these molecules are involved in a variety of cell movements. Immunofluorescence staining, studies of cytochalasin sensitivity, and other approaches have implicated these proteins in cell division (cytokinesis), directed movement of internal membranous organelles and associated cytoplasmic streaming, the poleward movement of chromosomes at anaphase (karyokinesis), cell migration, and other forms of cell motility (Goldman et al 1976). As discussed below, the evidence supporting the role of myosin in some of these motility events is still limited. Upon closer examination, it appears that myosin is not the motor that drives karyokinesis and certain other motility events. Other molecular motors, which include dynein, kinesin, and other forms of myosin, such as the single-headed myosin I (Maruta et al 1979; Korn et al 1987), must drive these forms of movement. For other types of movement, such as cytokinesis and internal organelle movements in certain cells, there is strong evidence that implicates myosin as the motor. These movements are highly regulated, often by phosphorylation of myosin.

The evidence in support of the involvement of myosin in organelle movements is strongest in the case of cytoplasmic streaming in the Characean algae. These simple plants have been studied extensively as the classic example of cytoplasmic streaming. Kühne in the late 1800s pointed out the similarities between cytoplasmic streaming in *Nitella* and muscle contraction (Kühne 1864). Even earlier, Corti's observations (1774) of intracellular movements in *Chara* shocked the scientific community with the discovery of movement in plants. The apparently static behavior of plants had been used as the primary trait to distinguish them from animals. The results from Kamiya and his colleagues laid solid grounds for the idea that cytoplasmic streaming in *Nitella* is driven by an actin and myosin system (for reviews, see Allen & Allen 1978; Kamiya 1981, 1986). More recently, Kersey et al (1976) demonstrated that the actin cables, known to exist on the cytoplasmic surface of the *Nitella* membrane, have the orientation expected for myosin-driven movements, according to the muscle paradigm (Huxley 1963). Indeed, when myosin-coated polystyrene beads are deposited on *Nitella* actin filaments in vitro, they move along the filaments in the same direction as the in vivo streaming occurred before the cell was dissected and its cytoplasm removed (Sheetz & Spudich 1983a,b). The available information suggests that in vivo a myosinlike molecule drives the movement of vesicular elements along the actin filaments attached to the *Nitella* cell membrane. Although myosin has been purified from *Nitella* (Kato & Tonomura 1977), a direct demonstration that a *Nitella* myosin drives cytoplasmic streaming has not been achieved. In the case of cytokinesis, Schroeder (1973) demonstrated the existence of a contractile ring of actin filaments in the furrow of a dividing sea urchin egg, and the division furrow of nonmuscle cells has been shown to contain myosin (Fujiwara & Pollard 1976; for reviews, see Sanger & Sanger 1979; Yumura & Fukui 1985). Myosin antibodies injected into starfish eggs inhibit cytokinesis but not karyokinesis (Mabuchi & Okuno 1977; Kiehart et al 1982). Purified nonmuscle myosin forms bipolar filaments in vitro (for review, see Clarke & Spudich 1977), and myosin filaments have been assumed to be important for the constrictions necessary for cytokinesis. Some models of cytokinesis assume that bipolar myosin filaments pull in both directions on actin filaments that are bound to the cell surface at their ends. These models are analogous to the sarcomere model worked out for striated muscle. It has been difficult, however, to visualize myosin filaments within contractile rings or other motile assemblies of nonmuscle cells. This difficulty probably results from the relative lack of order of these assemblies in nonmuscle cells compared with their nearly crystalline order in skeletal and cardiac muscle. Fukui and coworkers (Yumura & Fukui 1985; Fukui et al 1986) have been able to visualize myosin filaments in *Dictyostelium discoideum* using an improved version of indirect immunofluorescence staining. Their "agar-overlay" technique allows resolution of individual myosin filaments in vitro (Figure 2, *top*) and in situ (Figure 2, *middle*). They have used this technique to observe myosin filaments in a ring that presumably represents the furrowing region of a dividing *Dictyostelium* cell (Figure 2, *bottom*).

The results described above strongly suggest that myosin is the motor that drives cytokinesis. Actin and myosin have also been implicated in karyokinesis by immunofluorescence experiments and electron microscopy observations, and some investigators have suggested that these proteins may drive chromosome movements (for reviews see Inoue 1981; Pickett-Heaps et al 1982). However, anti–myosin antibodies injected into starfish eggs do not inhibit karyokinesis (Mabuchi & Okuno 1977; Kiehart et al 1982), which indicates that myosin may not be involved.

Genetic techniques can provide the strongest evidence for the roles a protein plays in the biology of a cell. The protein of interest must be specifically eliminated from the cell by genetic manipulations, and the resultant phenotype determined. This genetic approach, which provides the ultimate proof of function of a given protein, has been restricted to prokaryotes and a few eukaryotes, such as yeast. Integration of a plasmid by homologous recombination allows one to potentially eliminate the

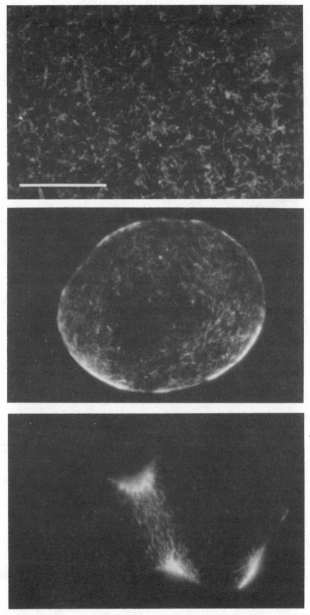

Figure 2 Immunofluorescence micrographs of *Dictyostelium* myosin filaments stained with monoclonal anti-myosin and FITC-labelled goat anti-mouse IgG using the agar overlay method of Fukui et al (1986). The upper panel shows myosin filaments in vitro. The middle panel shows a *Dictyostelium* cell with most of the filamentous myosin near the cell cortex. The lower panel shows a dividing cell with myosin filaments localized in the cleavage furrow. Photographs were kindly provided by Y. Fukui. (Bar = 10 μm)

expression of a particular gene of interest. In yeast, gene targeting has been used to eliminate the single-copy actin gene (Shortle et al 1982). This specific mutation shows that actin is essential for cell growth and survival. The role of myosin in *Dictyostelium* has been examined using gene targeting to interrupt the single-copy *mhc*A myosin gene (De Lozanne 1987; De Lozanne & Spudich 1987). A plasmid was constructed which upon integration creates a cell in which the intact myosin molecule is replaced with an amino-terminal fragment incapable of forming filaments. Cells with this single insertion become very large and multinucleated, and unlike wild-type cells, they cannot divide in suspension culture. When attached to a surface, however, they flatten like wild-type cells but grow extremely large and undergo movements that can result in pieces of the cell being pinched off (Figure 3). Sometimes a nucleus is included in the isolated cell fragment, which can then presumably grow larger. Some fragments lack a nucleus and presumably die. A very similar if not identical phenotype was obtained in experiments by Knecht & Loomis (1987) using antisense *mhc*A RNA to eliminate > 99.6% of the myosin from *Dictyostelium*. These two complementary sets of experiments, the targeting of a plasmid to disrupt the *mhc*A locus and the depletion of myosin by the antisense RNA approach, constitute genetic proof that an intact myosin molecule is required for cytokinesis but not for karyokinesis.

Interestingly, these *Dictyostelium* mutants do not completely lose the ability to move on a substratum or to exhibit chemotaxis toward an external signal. These results suggest that only very low levels of myosin (< 0.4%; Knecht & Loomis 1987) are needed for these movements, or the amino-terminal fragment that is incapable of forming filaments is sufficient for these movements (De Lozanne & Spudich 1987), or other forms of myosin or other molecular motors, such as dynein or kinesin, are involved in these processes.

A second form of myosin has been shown to exist in *Acanthamoeba* (Pollard & Korn 1973; for review, see Korn et al 1987) and in *Dictyostelium* (Cote et al 1985). These unique myosins are called myosin I. They are single-headed species that do not form filaments. A 110-kDa protein from intestinal brush border has many of the characteristics of this myosin (Mooseker 1985). Jung et al (1987) have cloned and sequenced one of the myosin I genes from *Acanthamoeba*. The derived amino acid sequence reveals striking similarities to conventional myosins in the head region. The myosin I protein lacks the extended tail region of conventional myosins (Albanesi et al 1985). Beads coated with myosin I have been shown to move along *Nitella* actin filaments in vitro (Albanesi et al 1985). Evidence has been presented by Adams & Pollard (1986) that this myosin may move vesicles along actin in *Acanthamoeba*. They showed that move-

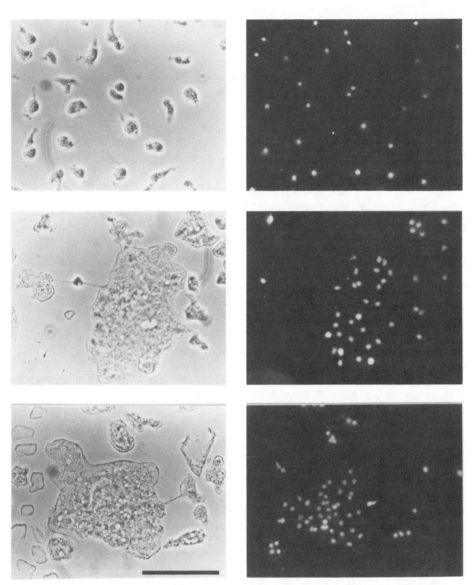

Figure 3 The left-hand panels show phase-contrast micrographs of *Dictyostelium* cells attached to a surface. The right-hand panels show the same fields stained with DAPI to show nuclear morphology. The upper pair of micrographs are of normal *Dictyostelium* cells, and the middle and lower pairs are of *Dictyostelium* transformants in which the expression of myosin heavy chain has been disrupted (De Lozanne & Spudich 1987). Cytokinesis is inhibited in the transformants, resulting in large multinucleated cells. In the middle and lower pairs of micrographs these large cells are in the process of leaving cell fragments behind. If the fragment contains a nucleus it can continue to grow. Photographs were kindly provided by A. De Lozanne. (Bar = 50 μm)

386

ment of *Acanthamoeba* vesicles along *Nitella* actin filaments is inhibited by anti-myosin I antibodies. It is also possible that the *Nitella* cytoplasmic streaming is driven by a myosin I type molecule. The roles played by the various myosins can best be elucidated by the type of gene targeting and antisense RNA experiments described above for the *Dictyostelium* conventional myosin.

Cellular regulation of motility events involving the conventional myosin appears to employ a number of different mechanisms (for reviews, see Kendrick-Jones et al 1976; Adelstein & Eisenberg 1980; Kendrick-Jones & Scholey 1981; Szent-Gyorgyi & Chantler 1986). The most common mechanism involves phosphorylation of the myosin regulatory light chain (for reviews, see Adelstein & Eisenberg 1980; Hartshorne & Siemankowski 1981; Marston 1982; Kamm & Stull 1986). Phosphorylation of a myosin light chain was first discovered in skeletal muscle myosin (Perrie et al 1973). This phosphorylation has a modulatory role in skeletal muscle contraction. Persechini and colleagues (1985) have demonstrated an increase in isometric tension in skinned muscle fibers in which the myosin light chain was phosphorylated by addition of purified light-chain kinase. Phosphorylation of a variety of myosin regulatory light chains on one or more serine residues has been shown to stimulate actin-activated myosin ATPase. This activation was first demonstrated for platelet myosin (Adelstein & Conti 1975). Light-chain phosphorylation also plays a role in the calmodulin-dependent regulation of thymus myosin (Scholey et al 1980, 1982). In smooth muscle the myosin light-chain kinase is also apparently regulated by phosphorylation (Payne et al 1986). The activation of myosin by phosphorylation of its light chain has been documented by myosin-coated bead movement along *Nitella* actin filaments for smooth muscle myosin (Sellers et al 1985) and for *Dictyostelium* myosin (Griffith et al 1987).

Changes in myosin light-chain phosphorylation have been shown to occur in vivo in response to external signals in leukocytes (Fechheimer & Cebra 1982), platelets (Daniel et al 1984), and *Dictyostelium* (Berlot et al 1985). *Dictyostelium* cells migrate across a substratum and show chemotaxis toward cAMP as a normal part of their developmental cycle (for review, see Devreotes et al 1987). A pulse of cAMP induces the cells to put out filopodial and pseudopodial projections, causing the cells to move toward the cAMP source for about 2 min. The cells then stop moving until they receive another pulse of cAMP. There is a fourfold increase in the phosphorylation level of the 18-kDa myosin light chain in vivo within 20 sec of stimulation of *Dictyostelium* cells with cAMP (Berlot et al 1985, 1987). This increase is transient; within 2 min the amount of phosphorylation returns to prestimulus levels. Although the roles played by phosphorylated myosins have not been elucidated, it appears that some aspects of chemotaxis in eukaryotes involve variations in the extent of myosin phosphorylation. The simplest presumption would be that myosin is somehow required for chemotaxis. An alternative hypothesis is that the myosin is not required but, if present, must be phosphorylated to allow chemotaxis to occur. The molecular genetic approach discussed above provides the tools to distinguish between these two hypotheses. In the case of *Dictyostelium*, molecular genetic techniques have shown that an intact conventional myosin, which clearly undergoes light-chain phosphorylation in vivo (Berlot et al 1985, 1987), is not essential for chemotaxis (Knecht & Loomis 1987; De Lozanne & Spudich 1987).

Myosin heavy-chain phosphorylation has been shown to occur in a number of nonmuscle myosins, including those from *Acanthamoeba* (Maruta & Korn 1977; Collins & Korn 1980, 1981; Hammer et al 1983; for review, see Korn et al 1987), *Dictyostelium* (Kuczmarski & Spudich 1980; Kuczmarski 1986), *Physarum* (Ogihara et al 1983), macrophages (Trotter 1982), lymphocytes (Fechheimer & Cebra 1982), and brain cells (Matsamura et al 1982; Murakami et al 1984). For the *Acanthamoeba* myosin II and *Dictyostelium* myosin, this phosphorylation, which occurs on serine and threonine residues, inhibits both filament assembly and actin activation of myosin ATPase (Kuczmarski & Spudich 1980; Korn et al 1987; Peltz et al 1981; Pagh & Gerisch 1986). In the case of *Physarum* myosin (Ogihara et al 1983) and *Acanthamoeba* myosin I (Korn et al 1987), heavy-chain phosphorylation activates the actin-activated myosin ATPase.

Heavy-chain, as well as light-chain, phosphorylation in *Dictyostelium* has been shown to be altered in vivo in response to chemoattractant. The chemotactic response leads to a change in phosphorylation of primarily threonine residues in the light meromyosin (LMM; Figure 4) region of the myosin tail (Berlot et al 1985, 1987). Again, the roles of such phosphorylation in vivo remain obscure. In *Dictyostelium*, this myosin has been shown by genetic manipulation experiments to be nonessential for chemotaxis (De Lozanne & Spudich 1987; Knecht & Loomis 1987).

MOLECULAR SUBSTRUCTURE OF MYOSIN

Knowledge of the detailed molecular substructure of myosin is essential for a complete understanding of its function. To date, studies on myosin structure have been conducted at different levels using a variety of techniques, but a complete three-dimensional characterization of the molecule still proves elusive (Morales et al 1982). The overall shape of the protein has been visualized by electron microscopy (EM) rotary-shadowing techniques, which show myosins, fixed in a number of ways, to be molecules

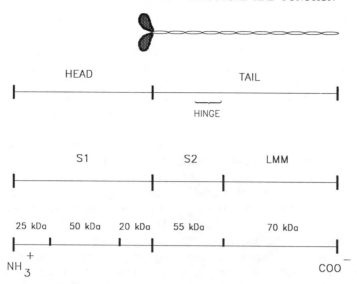

Figure 4 A schematic representation of the organization of a muscle myosin heavy chain. At the top is a representation of the physical appearance of a myosin molecule, showing the globular head region and the straight α-helical coiled-coil tail. The tail region can be further divided by protease digestion into a 55-kDa S2 region and a 70-kDa LMM region. The head or S1 region can be divided by protease digestion into three fragments of 25, 50, and 20 kDa. The hinge region is located in the carboxyl-terminal 20 kDa of the S2 subfragment.

containing globular head domains connected to a rod-shaped tail domain (Lowey & Cohen 1962; Lowey et al 1969; Elliott & Offer 1976, 1978), as discussed above (Figure 1). Low-resolution information has been obtained by electron microscopy of complexes between myosin and other molecules, such as ATP (Sutoh et al 1986). The complex between the myosin head and actin filaments has been productively explored using electron micro-scopy image-processing techniques with negatively stained (Toyoshima & Wakabayashi 1985) and frozen-hydrated unstained samples (Milligan & Flicker 1987). High-resolution structural information from X-ray crystal-lography has been difficult to obtain.

Myosins are relatively easy to isolate from muscle sources in large quantities, which has facilitated the application of biochemical and biophysical techniques to the study of myosin function and structure. As indicated above, all the myosins that have been characterized, except the myosin I types, are hexameric molecules containing a pair of heavy chains and two pairs of light chains. Myosin heavy chains contain regions that are particularly sensitive to protease digestion (Lowey et al 1969; Applegate & Reisler 1983; for review, see Harrington & Rodgers 1984). In rabbit skeletal muscle myosin the globular head (S1) can be separated from the tail and further divided into three subfragments of 20, 25, and 50 kDa. The tail can be cleaved into two fragments of 55 and 70 kDa (Figure 4). Myosins from other species do not show exactly the same digestion pattern; however, among muscle myosins this cleavage pattern is better conserved than among nonmuscle myosins. Areas of protease sensitivity probably correspond to regions of open or flexible structure, which are therefore more accessible to proteases. They should be considered useful substructural mileposts rather than divisions between functional domains.

The primary structure of a number of myosins has recently been determined (Table 1). The protein sequence has been obtained for portions of rabbit and chicken skeletal muscle myosins, which are present in muscle in large amounts (nearly 40% of the total cell protein). Improved molecular biology techniques have enabled the DNA sequences of a number of myosin genes to be determined. DNA sequencing has been the most practical route to deduce the primary structure of nonmuscle myosins, which are present in cells in low amounts (about 1% of the total cell protein). The sequencing of clones obtained from cDNA libraries has, to date, provided information on the carboxyl-terminal region from a number of species. Because full-length cDNAs are very difficult to make, information from genomic DNA is usually required to obtain the complete sequence.

Comparison of the sequences of various myosins from widely divergent organisms shows the evolution of the structure of the myosin molecule. The nucleic acid sequences of myosins show limited conservation, but the predicted amino acids reveal a pattern of homology that is common to all myosins. Considerably more conservation of amino acids appears in the head region than in the tail region of the molecule (Table 2). The pattern of amino acid conservation between, for example, *Dictyostelium* myosin and rat skeletal myosin, visualized using a graphic-matrix comparison technique (Smith & Waterman 1981), shows the two major divisions of the myosin molecule (Figure 5). The head portion (S1) of the molecule shows considerable conservation between these two myosins, and the conserved sequences are at the same relative positions in the head. In contrast, throughout the tail portion (S2 + LMM) of the molecule there is a repeating 28-residue group of charged and hydrophobic amino acids. There is not a single alignment that results in more homology than any other alignment. A very similar graphic-matrix pattern is seen when any two myosin sequences are compared (Saez & Leinwand 1986; Warrick et al 1986; Hammer et al 1987).

Although the primary structure of a number of myosins is now known, there is limited information regarding how the sequence is folded into its

Table 1 Sequenced myosin heavy chains

Organism	Myosin	Molecule sequenced	Amino acid sequence (Length and position)	Reference
Nonmuscle Sources				
Amoeba	type II	DNA	1800; *complete*	Hammer et al 1987
	type I	DNA	1168; *complete*	Jung et al 1987
Slime mold		DNA	2116; *complete*	Warrick et al 1986
Yeast		DNA	783; N terminal	E. Orr, personal communication
Muscle Sources				
Chicken	skeletal	DNA	1940; *complete*	Molina et al 1987
	skeletal	cDNA	438; C terminal	Kavinsky et al 1984
	skeletal	DNA	169; N terminal	Gulick et al 1985
	skeletal	protein	808; N terminal	Maita et al 1987
	smooth	protein	203; internal	Onishi et al 1986
Fruit fly		cDNA	69; C terminal	Rozek & Davidson 1986
		cDNA	95; C terminal	Rozek & Davidson 1986
Hamster	heart	cDNA	310; C terminal	Liew & Jandreski 1986
Human	skeletal	cDNA	876; C terminal	Saez & Leinwand 1986
	skeletal	cDNA	635; C terminal	Saez & Leinwand 1986
Mouse	skeletal	cDNA	165; internal	Weydert et al 1983
	skeletal	cDNA	17; C terminal	Weydert et al 1985
	skeletal	cDNA	114; internal	Weydert et al 1985
	skeletal	cDNA	165; internal	Weydert et al 1985
	skeletal	cDNA	107; internal	Weydert et al 1985
Nematode	body wall	DNA	1966; *complete*	Karn et al 1983
Quail	skeletal	cDNA	25; internal	Hastings & Emerson 1982
Rabbit	skeletal	protein	92; internal	Elzinga & Collins 1977
	skeletal	protein	147; internal	Elzinga & Trus 1980
	skeletal	protein	259; internal	Capony & Elzinga 1981
	skeletal	protein	204; internal	Tong & Elzinga 1983
	skeletal	cDNA	27; internal	Putney et al 1983
	heart	cDNA	832; C terminal	Kavinsky et al 1984
	skeletal	protein	170; internal	Lu & Wong 1985
Rat	skeletal	DNA	1939; *complete*	Strehler et al 1986
	heart	cDNA	428; C terminal	Mahdavi et al 1982

native functional state (Botts et al 1984). Intact myosin molecules will be extremely difficult to crystallize due to their tendency to form semi-ordered macromolecular arrays rather than true crystals. The crystallization of the S1 region of myosin, prepared by proteolytic digestion of myosin from chicken skeletal muscle, has been achieved (Rayment & Winkelmann 1984). EM analysis of these crystals (Winkelmann et al 1985) has revealed

Table 2 Percent amino acid similarity[a] in the S1, S2, and LMM regions of sequenced myosins[b]

	Human S1/S2/LMM	Rabbit S1/S2/LMM	Rat S1/S2/LMM	Chicken S1/S2/LMM	Nematode S1/S2/LMM	Slime mold S1/S2/LMM	Amoeba S1/S2/LMM
Yeast	—/—/—	37†/—/—	38†/—/—	36†/—/—	38†/—/—	42†/—/—	40†/—/—
Amoeba	—/26†/27	40†/21†/28	43/22/33	43/22/19	40/21/28	54/26/29	
Slime mold	—/28†/26	42†/25†/30	42/25/29	45/25/22	44/24/26		
Nematode	—/51†/48	56†/44†/48	54/44/48	54/44/47			
Chicken	—/88†/87	87†/91†/87	85/82/82				
Rat	—/80†/79	80†/82†/82					
Rabbit	—/76†/90						

[a] Amino acid similarities for each region were obtained by dividing the number of identical amino acids in identical positions by the number of amino acids in the region. Gaps were introduced into both sequences to maximize the matches using the "gap" program from the University of Wisconsin Computer Genetics Group (Devereux et al 1984) (gap weight = 5, length weight = 0.3, average match = 1.0, average mismatch = 0). Symbols: — = not available and † = incomplete.
[b] Sequences obtained from Saez & Leinwand 1986, *Homo sapiens (mhc-a)*; E. Orr, personal communication, *Saccharomyces cerevisiae*; Hammer et al 1987, *Acanthamoeba castellanii* (type II); Warrick et al 1986, *Dictyostelium discoideum (mhcA)*; Karn et al 1983, *Caenorhabditis elegans (unc54)*; Molina et al 1987, *Gallus gallus* (embryonic); Strehler et al 1986, *Rat norvegicus* (embryonic); Elzinga & Trus 1980, Tong & Elzinga 1983, Lu & Wong 1985, *Oryctolagus cuniculus*.

RAT

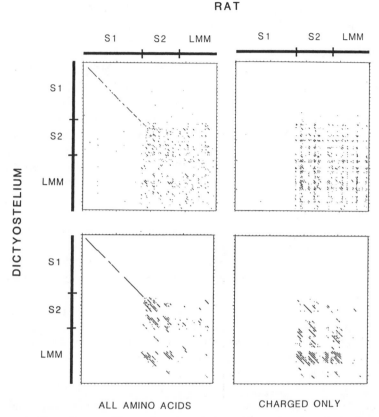

ALL AMINO ACIDS   CHARGED ONLY

Figure 5 These four graphic matrices show amino acid sequence homologies between *Dictyostelium* and rat myosin heavy chains. The diagram was prepared using the "compared" and "dot-plot" programs from the University of Wisconsin Genetics Computer Group (Devereux et al 1984). The upper panels compare the sequences of the two myosins in blocks of 25 amino acids. Each block of 25 amino acids in the *Dictyostelium* sequence was compared with every 25 amino acid block in the rat sequence. When more than 40% of the amino acids matched, a dot was placed at the position in the diagram that corresponds to the location of the amino acid block. The lower panels were constructed using the same technique except a 100–amino acid block was compared each time and more than 30% of the amino acids had to match before a dot was placed. In the left-hand panels all amino acids were considered in the comparisons. In the right-hand panels only the amino acids carrying charged groups (D, E, R, H, K) at pH 7 were considered. In the head (S1) region there is not a sufficient number of conserved charged amino acids to result in a line of homology. The tail (S2+LMM) contains repeating patterns of charged amino acids which result in multiple parallel lines in comparisons between the tail regions. These repeating patterns appear to be important in the assembly of myosin into bipolar thick filaments.

an overall shape of S1 that is consistent with that obtained by three-dimensional reconstructions of S1-decorated actin filaments (Toyoshima & Wakabayashi 1985; Milligan & Flicker 1987).

The Myosin Head

In Figure 6, amino acid sequences from the head region of eight myosin heavy chains have been aligned to illustrate how different regions have been conserved. The four nonmuscle myosins (lines a–d) show more divergence from the muscle myosins (lines e–h) than occurs within the muscle family. With this alignment there are 196 positions ($\sim 22\%$) at which the amino acid is completely conserved in all the sequenced examples except that of *Acanthamoeba* myosin I (line a). At 299 additional positions ($\sim 34\%$) only one myosin has a different amino acid at that particular position. At many other positions the major chemical property of the amino acid is conserved (i.e. charge, hydrophobicity, size). Although the conservation of amino acids in the myosin head region is significantly greater than that in the tail region, it is not as great as the conservation found in other cytoskeletal proteins, particularly actin (Vandekerckhove & Weber 1978). Areas of sequence conservation in myosins can be used as indications that these regions play important roles in myosin functions. In the discussion below we view the amino acid sequence of the S1 domain as a one-dimensional array from its amino to its carboxyl terminus, and correlate the known functions of myosin with its sequence domains. The true functional domains are of course three-dimensional and may involve the contributions of sequence elements that are far apart on the linear sequence. An understanding of myosin function will require the marriage of biochemical information with high-resolution structural information from X-ray diffraction analysis of S1 crystals.

Starting at the amino terminus of the sequence alignment, one can see little sequence conservation in the first 90 amino acids. This area, with the exception of a few aromatic or charged residues, has diverged considerably from the protein sequence of the putative ancestor myosin.

The first area of significant sequence conservation (residues 90–145) contains a hydrophobic stretch followed by an almost completely conserved lysine (residue 136). This lysine is trimethylated in rabbit myosin. The modified lysine creates additional positive charge, which is thought to participate in the ATPase reaction, perhaps by interacting with one of the phosphates of ATP (Tong & Elzinga 1983). All the myosin sequences, except that of yeast, contain lysine at this position. In yeast myosin there is a histidine at this position, which could also be methylated to create additional charge.

Photoactivatable cross-linking reagents that mimic the structure of a nucleotide have been used to examine the region of ATP binding. A tryptophan residue at position 137 in the sequence alignment is frequently cross-linked (Szilagyi et al 1979; Okamoto & Yount 1985), but this residue is not conserved in the nonmuscle family of myosins. A mutation (glycine to arginine) that inserts a charged amino acid into a relatively hydrophobic area immediately preceding the trimethylated lysine (position 126, Figure 6) has been found in the nematode body wall myosin (Dibb et al 1985). Worms with this mutation are unable to move, which indicates that this hydrophobic pocket is important in myosin function. The glycine residue at this position is conserved among all the myosins. At position 90 there is an almost completely conserved lysine that is very susceptible to modification with chemical reagents. The lysine when trinitrated shows absorption spectrum differences in the presence of pyrophosphate (Hozumi & Muhlrad 1981). Taken together, this evidence suggests that this domain of the sequence participates in the ATPase activity of the molecule.

The next region of conservation (residues 170–214) has been identified as part of the binding site for the Mg^{2+}-ATP complex. The sequence in this area has good homology to a region of conservation found among several ATP binding proteins (Walker et al 1982; Higgins et al 1986). This area is only one of three domains, defined by homology with other ATP binding proteins, thought to interact to form a binding site for ATP. The two other domains are not clearly discernible in myosins. Wierenga and associates (1986) have proposed a model for the interaction of nucleotide and proteins suggesting that a β-α-β fold structure is important. In myosins, a related α-β-α structure appears to be used. The structure starts with a basic residue that is replaced in myosins by an absolutely conserved glutamine (residue 192). A short α-helix comes next, followed by a β-sheet containing a G-X-X-G-X-G sequence. All the myosins contain a G-E-S-G-A-G-K-T sequence (residues 198–205). Walker and associates (1982)

Figure 6 (overleaf) The amino acid sequence of the S1 region of the myosin heavy chain from: (a) *Acanthamoeba* type I (Jung et al 1987), (b) yeast (E. Orr, personal communication), (c) *Acanthamoeba* type II (Hammer et al 1987), (d) *Dictyostelium* (Warrick et al 1986), (e) nematode (Karn et al 1983), (f) chicken (Molina et al 1987), (g) rat (Strehler et al 1986), (h) rabbit (Elzinga & Collins 1977; Tong & Elzinga 1983; Karn et al 1985). Letters in bold indicate positions that contain the same amino acid in all or all but one of the sequences. Gaps, symbolized by periods, were inserted into each sequence to obtain the best match. Listed below the eight sequences is the consensus sequence. In this line a dash indicates that no single amino acid is in the majority at that position, a lowercase letter indicates that the amino acid is in the majority at that position, and a capital letter indicates that the amino acid is conserved in all sequences at that position. The *Acanthamoeba* myosin I sequence was not considered in the consensus. The sequence of the rabbit myosin from position 860 to 880 has not been determined.

```
           10         30         50         70         90        110        130        150
            .          .          .          .          .          .          .          .

                    MAYTSKHGVDDMVMLTSISNDAINDNLKKRFAADLIYTYIGHVLISVNPYKQINNLYTERTLKDY
a)
b)  MTGGQSCSSMIVWIPDEKEVFVKGELMSTD..INKNIFTGQEEQIGTVHPLDSTEVSNLSQ.VRISDVFPVNPSTFDKVENMSELIFHENEPSVLYNLEKRYDCDLIYTYSGLFIVAINPYHNLN.LYSEDHINLY
c)  MAAQRRRKGEVESDYIKTLKYKNTGFQVSASDKTLAWPTKDADRAEFCHVEVTKDDGKNFTVRLENGEEKSQPKNE.K..NFLGVNPPKFDGVEDMGELGYLNEPAVLHNLKRYDADLFHTSGLFIVVVNFYKRLP.VYTPEIIDIY
d)  MNPIHDRTSDYHKYLKVKQGDSDLFKLTVSDKRIYWYNPDPKERDSYECGEIVSEDSDFTFKTVGQDRQ.VKKDDANQRNPIKPDGVEDMSELSYLINEPAVFHNLRVRYNQDLIYTYSGLFIVAVNFEKRIP.IYTQEMVDIF
e)  MEHEKDPGWQYLRRTREQVLEDQSKPYDSKKNVWIPDPEEGYLAGEITATKGDQVIVTAREMSVIQVTLKELVQEMNPPKFEKTEDMSNLSFINDASVLHNLRSRYAAMLIYTYSGLFCVVINPYKRLP.IYTDSCARMF
f)  MATDADMAIFGEAAPYLRKSEKERIEAQNKPFDAKSSVFVVHAKESYVKSTIQSKESGKVTVKTEGGETL...TVKEDQIFSMNPPKYDKIEDMAMMTHLHEPAVLYNLKERYAARMIYTYSGLFCVTVNPYKWLP.VYNPEVVLAY
g)  MSSDTEMEVFGIAAPFLRKSEKERIEAQNKPFDAKTYCFVVDSKEEYAKGKIKSSQDGKVTVETEDNRT...LVVKPEDVYAMNPPKFDKIEDMAMLTHLNEPAVLYNLKDRYTSRMIYTYSGLFCVTVNPYKWLP.VYTPEVVDGY
h)  MSSDADMAVFGEAAPYLRKSEKERIEAQNKPFDAKNSVFVVADPKESYVKATVQSREGGKVTVKTEAGAS...VTVKEDQVFPMNPPKYDKIEDMAMMTHLNEPAVLYNLKERYAAWMIYTYSGLFCVTVNPYKWLP.VYNAQVVTAY


                                                                                              Trimethyl lysine
                                                                                                   *
          170        190        210        230        250        270        290
            .          .          .          .          .          .          .

    ------yir-s------e-kp-dak----------d-ke--------t------g-vt--t-----s----vk---v--mNPpkfdk-EdMsnithinepaVLyNLkeRYaa-liyTYSGLFcV-vNPyk-lp-vYt-e-v--y

a)  RGKY.R.........YR.LPPHVYALADDMRTMLSESEDQCVIISGESGAGKTEASKKIMQYIAAVSGATGDVMRVKDV............ILEAFGNAKTIRNNNSSRFGKYMEIQFDLKGDPVGGRISNYLLEKSRVVYQT
b)  HNKHNRLSKSRLDENSHEKLPPHIFAIAEEAYENLLSEGKDQSILVTGESGAGKTENTKKILQYLASITSGSPSNIAPVSG.SSIVESFEKMILQSNPILESFGNACTVRNNNSSRFGKFIKEFNEHGMINGAHIEWYLLEKSRIVHQN
c)  RGRQ.R.........D.KVAPHIFAISDAATRAMLNTRQNQSMLITGESGAGKTENTKKVIQYLTAIA...........GRAEGGL..LEQQLLEFNPILEAFGNAKTIRNNNSSRFGKFIEIQFNAGGQITGANTFYILLEKSRVVFQS
d)  KGRR.R.........NE.VAPHIFAISDVATRSMLDDRQNQSLLITGESGAGKTENTKKVIQYLASVA.........GRNQANGSGVLEQQLQANPILEAFGNAKTVRNNNSSRFGKFEIQFNAGEFISGASIQSTLLEKSRVVFQS
e)  MGKR.K.........TR.MPPHLFAVSDEAYRNMLQDHENQSMLITGESGAGKTENTKKVICYFAAV.GASQQEGGAEVDPNKKKVTLEDQIVTNPVLEAFGNAKTVRNNNSSRFGKFIRIHFNKHGRLASCDIEHYLLEKSRVIRQA
f)  RGKK.R.........QR.APPHIFSISDNAYQFMLTDRENQSILITGESGAGKTVNTKRVIQYFATIAATGDKKKEEQPA.GKMQGTLEDQIISANPLLEAFGNAKTVRNDNSSRFGKFIRIHFGATGKLASADIETYLLEKSRVTFQL
g)  RGKK.R.........QR.APPHIFSISDNAYQFMLTDRENQSILITGESGAGKTVNTKRVIQYFATIAATGDLAKKDSKMK...GTLEDQIISANPLLEAFGNAKTVRNDNSSRFGKFIRIHFGTTGKLASADIETYLLEKSRVTFQL
h)  RGKK.R.........QR.APPHIFSISDNAYQFMLTDRENQSILITGESGAGKTVNTKRVIQYFATIAITGDKKKEEPTPGK.MQGTLEDQIISANPLLEAFGNAKTVRNDNSSRFGKFIRIHFGATGKLASADIETYLLEKSRVTFQL


    rgk--r---------e--pPHiFaisd-AY--mL-drenQSiLiTGESGAGKTeNTKkviqYfa--a-------------gtlEdqi--aNP-LEaFGNAKTvrNnNSSRFGKFirihF---G-lasadie-YLLEKSRvtfQ-

                                *      *      *
                       ATP binding         25-50 kDa junction

          310        330        350        370        390        410        430        450
            .          .          .          .          .          .          .          .

a)  NGERNFHIFYQLLAARARRPEAKFGLQTPDYIFYLNQGKTYIVDGMDDNQB.FQDTNNAMKVIGFTAEEQHEIFRLYTAILLYLGINVQFVDDKGGSIIADSRPVAVETALLYRTIITGEQGRSSVYSCPQDPLGAIYSRDALSKALYS
b)  SKERNYHIFYQLLSGLDDSELKNLRLKSRNVKDYKILSNSNQDIIPGIDVENFKELLSALSIIGFSKDQIRWIFQVVAILLLGNIEFVSDRAEQASFKNDVSAICSNLGVDEKDFQ...TAILRPSRKAGKERVSQSKNSTKLSSLLNA
c)  AGERNFHIFYQILSKAMPELKQKLKLTKPEDYFFLNQNACTVDDMDDAKEFDHMLKAFDILNINEEERLAIFQTISAILHLGNIPFIDVNSETAGLKDEVLNIAEELLGVSAAG.LKAGLLSPRIKAGNEWVTRALNKPKAMASRDA
d)  ETERNYHIFYQLLAGATAEEKK.ALHLAGPESPYLNQSGYYDIKGVSDSEBFKITRQAMDIVGFSQEBQMSIFKIIAGILHLGNIKFEKGACGAVLKDKTALNAASTVFGVNPSV.LEKALMEPRILAGRDLVAQHLNVEKSSSSRDA
e)  PGERCYHIF.QIYSDFRPB.LKKELLLDLPIKDYWFVAQAELIDGIDDVEEFLQTDEAFDILNFSAVEKQDCYRLMSAHMHMGNMKFKQRPREEQAEPDGTVEAEKASNMYGICEEFLKALTKPRVKVGTEWVSKQGNCEQVNWAVGA
f)  KAERSYHIFYQIMSNKKPELIEMLLITTNPYDYQVSQGEITVPSIN.DQBELMATDSAIDILGFTPDEKTAIYKLIGCAVHMYGNMKFKQKQQREQQAEQGGTEVADKTAYLMGLNSADLLKAALCYPRVKVGNEYVTKGQTVQQVYNSVGA
g)  KAERSYHIFYQILSNKKPELIELLLITTNPYDFPYISQGEIVASI.DDRBELLATDSAIDILGFTPEEKSGLYKLTGAVHMYGNMKFKQKQQREQQAEPDGTEVADKTAYLMGLNSSDLLKALCFPRVKVGNEYVTKGQTVDQVYNAVGA
h)  KAERSYHIFYQIMSNKKPELIEMRLITTNPYDYAFVSQGEITVPSI.DDQBELMATDSAVDILGFTSDEKVAIYKILGCAVHMYGNMKFKQKQREQQAEPDGTEVADKAAYLQSLNSADLLKALCYPRVKVGNEYVTKGQTVQQVYNAVGA


    --ER-yHIFyQi-s----pe--k--l--t-p-dy---------q---------dD-eef--tdsA-dilgf----ek--iykl--a-mh-GN-kFkq--rEeqaepdgte-a--ka---l----s---llkal--PRvKvGne-Vtkgqnv-qv---v-A
```

Block 1 (residues ~470–590)

```
        470         490         510         530         550         570         590
a) ......RMFDYII.QRVNDAMYIDDPEALTTGILDIYGFEIFPGKNGFEQLCINFVNEKLQQIFIQLILKABQEEYGAEGIQWENIDYFNNKICC.DLIIEKRPPGLMTILDDVCNFPKGTDKPREKLLGAFPTHAHLAATSQPDEEV.
b) LSRNLYERLFGYIV.DMIRKNLDHGSATLNYIGLLDIAGFEIFPENNSFEQLCINYTNEKLQQFFNHHMFVLEQSEYLKENIQWDYISYGKDLQLITDLIEARGHDRVLPLLVEEAVLPKSLMESFYSKLISFTWDQNSSKFKRSTLKN...
c) LCKALPGRLFLMIV.QKINRILPSHKDKTALWIGVLDIAGFEIFQHNSFEQLINYTNEKLQQFFNHHMFVLEQGEYEREKIDWTFVDYGMDSQDCIDLIEKK.PMGILPLLDEQTVFPDADDTSFTKKLFQTHENHRNFR.PRFDAN...
d) LVKALYGRLFLWIV.KKINNVLCQERK.AYFIGVLDISGFEIFKVNSFEQLCINYTNEKLQQFFNHHMFKLEQEEYLKEKINWTFIDFGLDSQATIDLIDGRQPPGIIALLDEQSVFPNAIDNTLITKLHSHFSKKNAKYEEPRFSK...
e) MAKGLYSRVFNWLVKKCNLTLDOKGIDRDYFIGVLDIAGFEIFDFNSFEQLWINFVNEKLQQFFNHHMFVLEQEEYAREGIQWVFIDFGLDLQACIELIEK..PLGIISMLDECIVPKATDLTLASKLVDQHLGKHPNFEKPKPPKGKQ
f) LAKSVFEKMFLWMVY.VRINQOLDTKQPRQYFIGVLDIAGFEIFDFNSIEQLCINFTNEKLQQFFNHHMFVLEQEEYKKEGIEWEFIDFGMDLAACIELIEK..PMGIFSILEBECMFPKAIDTSFKNKLYDOHLGKSNNFQKPKGKGKL
g) LSKSVTEKLFLWMVY.TRINQOLDTKLPRQHFIGVLDIAGFEIFEYNSIEQLCINFTNEKLQQFFNHHMFVLEQEEYKKEGIEWTFIDFGMDLAACIELIEK..PMGIFSILEBECMFPKAIDTSFKNKLYDOHLGKSNNFQKPKVVKGK.
h) LAKAVTEKMFLWMVY.TRINQOLDTKQPRQYFIGVLDIAGFEIFDFNSIEQLKYNFTNEKLQQFFNHHMFVLEQEEYKKEGIEWCFIDFGMDLAACIELIEK..PMGIFSILEBECMFPKAIDTSFKNKLYEQHLGKHNNFQKPPAGKK.

l-k-ly-r-Fl-V-----iN--ld-k-r-yFIGvLDIAgFEIF--NsfEQlcINftNEKLQQFFNHHMFvLEQeEY-kEgI-W-fIdfG-Dl-acIeLiek--pmgi-s-L-Eec-fPkatdtsf--KL--qhigk-nnf-kpk-kgk-
```

Unidentified function

Block 2 (residues ~610–750)

```
        610         630         650         670         690         710         730         750
a) .....IKHYAGDVVYNVDGFCDKNKD.LLFKDLIGLAECTSSTFFA.GLFPEAKEVAT\s................KKKP\"TAGFKIKESINILVATLSKCTPHYIRCIKPNEKKAANAFNNSLVLHQVKYLGLLENVIRRA
b) ..GFILKDYAGDVNILWKAGYFKT.DPLNDNLLSLLSSQNDIIS.KLFQPEEGGKNLLVCGVEA..........NISNQEVKKSARTSTFKTTSSHREQQI.TLLNQLASTHFHFVRCIIPNNVKVKTFNRSLILDQLRCNGVLEGIRLARE
c) ..NFKIVHYAGEVEYQTSAWLEKRGDEDLSNLCKKSVRFVTG.LFDEDLMPSFKAAFAEEEKAAAGSNRSTGRGKGAQFITVAFQYKEQLAHLMSSTAPHFIRCIIPNLGKKPGVVSDQLVDQLRCNGVLEGIRIARK
d) ..TEPGVTHYAGQVMYEIQDWLEKNKDPLQQDLE.LCFKDSSDNVVTKLFNDP............NIASRAKKGANFITVAAQYKEQLASLMATLETTNPHFVRCIIPNNKQLPAKLEDKVVLDQLRCNGVLEGIRITRK
e) GEAHFAMRHYAGTVRYNCLNWLEKNKDPLNDTVVSAMKQSKGNDLLVEITWQDYTTQEEAAA.........KAKEGGGGKKKKGKSG.SFMTVSMLYRESLNNLMTMLNKTHPHFIRCIIPNEKKQSGMIDAALVLNQLTCNGVLEGIRICRK
f) .EAHFSLVHYAGTVDYNIITGWLEKNKDPLNETVVGL.YQKSSLKTLALLFASVGAGA..........ESGAGGKKGGKKKGSSFQTVSALFRENLNKLMSNLRSTHPHFVRCLLPNETKTPGAMEHELVLHQLRCNGVLEGIRICRK
g) AEAHFSLIHYAGTVDYSVSGWLEKNKDPLNETVVGL.YQKSSNRLLAHLYATFATTDA..........DGGKKKVAKKKGSSFQTVSALFRENLNKLMSNLRSTHPHFVRCLIPNETKTPGAMEHSLVLHQLRCNGVLEGIRICRK
h) AEAHFSLVHYAGTVDYNIITGWLDKNKDPLNETVVGLYQKSMTLLAFLPSGAQAGE..........EGGGGKKGGKKKGSSFQTVSALFRENLNKLMTNLRSTHPHFVRCIIPNETKTPGAMEHELVLHELRCNGVLEGIRICRK

-eaHFsl-hYAGtV-------wleKnkDPLn-tv--l------s-------lf-------------g--k---kkg-sf-Tvsal-re-lnklm--lrsThPhFvRCiIPNe-k-pg-e--ivl-qLrCNGVLEGiRicRk
```

50-20 kDa junction   Actin binding

Block 3 (residues ~770–890)

```
        770         790         810         830         850         870         890
a) GYAYRQSYDKFFYRYRVVCPKTWSGWNGDMVSG.AEAILNHVGMSLGKEYQK.......GKTKIFIRQPESVFSLEELRDRIVFSYANKIQRFLRKTAMRKYYEVKKGGNDALVNKKERRR...........L.SLERPFKTDYINYRQN
b) GYPNRIAFQEFFQRYRILYPRKFNHHDFSSKLK.ASTKQNCRFLLTSL
c) GWPNRLKYDEFLKRYFLLKPGATPTSPSTK...DAVKDLIEHLIAKEPTKVNKDEVRPGVTKIFFRSGQ.LAAIEELREQAISKMVVSIQAGARAFLARRMYDKMRRQTVSAKILQRNIRAWLEKINWANYQLYVKARPLISQRNFQKQI
d) GFPNRIIYADFVKRYYLLAPNV.PRDAED SQK.ATDAVLKHLNID.......PEQYRFGITKIFFRAGQ.LARIEEAREQRISEIIKAIQAATRGWIARKVYKQARERHTVAARIIQKNIRAYIDFKSWFWWKLFSKARPLKKRNFEKEI
e) GFPNRTLHPDFVQRYAILAAKEAKSDDD..KKKCAEAIMSK.LVND..GSLSEEWFRIGLTKVFFKAGV.LAHLEDIRDEKLATILTGFQSQIRWHLGLKDRKRRMEQRAGLLIVQRNVRSWCTLRTWEWFKLYGKVKPWKAGKEAEEL
f) GFPIRIIYADFKQRYKVLNASAIPEGQFIDSKK.ASEKLLG..SID....VDHTQYKFGHTKVFFKAGL.LGLLEEMRDEKLAQLITRTQARCRGFLMRVEFKKMMQRRECIFCIQYNVRAFMNVKHWFWMCLFFKIKPLLKSAESEKEM
g) GFPNRIIYGDFKQRYRVLNASAIPEGQFIDSKK.ACEKLLA..SID....IDHTQYKFGHTKVFFKAGL.LGTLEEMRDERLAKLITRTQAVCRGFLMRVEFQKMMQRRESIFCIQYNIRAFMNVKHWFWWKLFFKIKPLLKSAETEKEM
h) GFPSRILYADFKQRYKVLNASAIPEGQFIDSKK.ASEKLLG..SID....VDHQTYKFGHTKVFFKAGL.LGLLEEMRDDKLAQLITRTQAICRGFLARVEYKKMQERR             MKCLYFKIKPLLKSAETEKEM

GfPnRily-dF-qRY--L-a-a-p-------skk-A--kll----id---------y-fG-TKvfFkaG--L-lEe-Rd---L--lEe-Rd-la---tr-Qa--Rgfl-r---kknmmeri-------q-n-ra----k-w-w-kL--K-kPllk---eke-
```

Amino end of S2   SH2   SH1

3-Methyl histidine

have suggested that in ATP binding proteins the lysine and threonine following the last glycine should be conserved, and they are in the myosins. The lysine is thought to interact with the α-phosphate of ATP (Fry et al 1986). The Wierenga model proposes that another short α-helix follows the cluster of glycines, followed by an acidic residue, which is thought to form a hydrogen bond with the 2'-OH of the ATP ribose. Most of the myosins have acidic residues around the predicted position, but there is not a consensus residue. It is difficult to determine whether the two postulated short α-helices are present in the myosins, since the alternating hydrophobic pattern that is diagnostic for an α-helix is not apparent. Wierenga and associates (1986) have suggested that, as in the primary structures of adenylate kinase and phosphofructokinase, an acidic residue, such as the conserved one at position 179, may interact with the Mg^{2+} cofactor. The ATP binding site has been located on the S1 fragment by electron microscopy using an ATP analog (Sutoh et al 1986).

The region spanning residues 217–240 is striking in its sequence heterogeneity. This region has been associated with the trypsin-sensitive site dividing the 25- and 50-kDa domains in rabbit myosin. Notice the repeated lysine residues in the muscle sequences which may be recognized by trypsin. This relatively polar area could form flexible loops that are near enough to the surface of the myosin head to be accessible to proteases. There are many protease cutting sites predicted by the sequence throughout the entire myosin molecule, but only a few appear to be accessible. The sequences of the nonmuscle myosins, with the exception of that of yeast myosin, appear to be shorter than the muscle sequences in this region. The *Dictyostelium* sequence lacks the loop and does not display a protease sensitivity pattern like that found in rabbit myosin (Warrick et al 1986). Apparently these loops are not required for functional myosin.

The 50-kDa domain contains two highly conserved areas (residues 250–295 and residues 482–537) separated by an extensive stretch of poorly conserved sequence (residues 296–460). Although there is no evidence that assigns a function to the conserved areas, the homologous sequences contain rather unusual concentrations and patterns of amino acids that are generally infrequently used. One must therefore presume that these sequences are important for myosin function. This view is supported by the fact that a mutation in the nematode body wall myosin that converts a conserved arginine residue at position 295 into a cysteine residue results in an inactive myosin (Dibb et al 1985).

Between residues 652 and 687 is another area of extensive heterogeneity in sequence and in length. This region has been correlated with the trypsin-sensitive area separating the 50- and 20-kDa domains in the rabbit myosin. The *Acanthamoeba* myosin II sequence contains more amino acids in its loop than do the muscle myosins. In contrast, this region in *Dictyostelium* myosin is a few amino acids shorter than the loops in the muscle group. Clearly, variation in this area can be tolerated without impairment of myosin function.

A number of studies have used cross-linking reagents to explore the actin-myosin interaction sites (Yamamoto & Sekine 1979; Mornet et al 1981a,b; Sutoh 1982a,b, 1983; Greene 1984; Chen et al 1985a,b, 1987; Katoh et al 1985). One site in S1 that is cross-linked to actin has been mapped to the first 20 amino acids (residues 700–720) of the 20-kDa subfragment (Sutoh 1983). The S1 fragment appears to interact with actin at two sites, a strong binding site in the 20-kDa subfragment and a much weaker one in the 50-kDa subfragment. The 50-20 kDa trypsin site is protected when actin is bound to S1. Labbe et al (1982) demonstrated that with a slightly longer-armed linking reagent, trypsin-nicked S1 can be linked to actin, restoring the ATPase activity, which is normally inactivated by trypsin treatment.

Toward the carboxyl terminus of S1 the muscle myosins have two closely spaced reactive thiol groups called SH1 and SH2. Cross-linking the two thiols inhibits ATPase activity and can trap Mg^{2+}-ATP in the myosin (Wells & Yount 1979, 1980, 1982; Wells et al 1980a,b). In the rabbit sequence the thiols were mapped to cysteine 707 and cysteine 697, respectively, which can be found in positions 749 and 739 in the sequence alignment shown in Figure 6. There is evidence that this region may be rather flexible: The distance between the two thiols appears to decrease when nucleotide binds to the ATP pocket (Dalbey et al 1983). The SH1 group has been localized to a site about two thirds (130 Å) of the distance from the head-tail junction by EM analysis of avidin-biotin-labeled SH1 myosin (Sutoh et al 1984). Conservation of cysteine at these positions does not appear to be essential for myosin function because the nonmuscle myosins contain substitutions at one or both of these locations.

Most myosins contain two light chains, the regulatory light chain and the essential light chain, noncovalently bound with the head region of the heavy chain. Neither light chain is required for actin binding or ATPase activity (Wagner & Giniger 1981; Sivaramakrishnan & Burke 1982). Some light chains seem to be interchangeable between species (Chantler & Szent-Gyorgyi 1980). The regulatory light chain has been localized by electron microscopy on S1 near the head-tail junction (Flicker et al 1983; Yamamoto et al 1985). When a photoreactive analogue of ADP is bound to myosin and then activated, it labels both light chains and the heavy chain in gizzard myosin (Okamoto et al 1986). These results indicate that the regulatory light chain is located in the "neck" region of the myosin but is also close to the ATP binding pocket. A gel overlay technique has been used to identify fragments of S1 that can bind to labeled light chains (Mitchell et al 1986). A 10–12 kDa peptide from the region between SH1 and the carboxyl terminus of S1 appears to bind both types of light chain.

Studies of the events that occur during muscle contraction indicate that there must be some flexibility at the head-tail junction (Huxley & Kress 1985). The α-helical potential of the carboxyl terminus of the S1 fragment is not particularly strong, and there is little sequence conservation in this region. The histidine at position 811 (Figure 6) in rabbit skeletal myosin is modified (3-methyl), but in other rabbit myosin isoforms it is not (Huszar & Elzinga 1972; Huszar 1984). Furthermore, the histidine at this position is not conserved in the sequenced examples of nonmuscle myosins. An additional charge in this area of the S1 fragment is apparently not required for function but may modulate myosin activity. A proline is highly conserved near the S1-S2 junction. This proline is often chosen to mark the end of the head and the beginning of the tail region, but that choice is artificial; it has not been mapped in the structure to show its location. However, the proline is probably near the head-tail junction because measurements of myosin tail length agree with the length predicted from the sequence from the carboxyl terminus to this proline if a constant helical pitch in the tail is assumed. Immediately before the proline there is an unusual concentration of aromatic amino acids.

The analysis of sequence homologies has provided insights into the structure and function of the myosin molecule. However, the mechanisms accounting for the fundamental function of myosin, that of creating motive force, are still obscure. The secrets underlying the ability of myosin to move must lie within S1 since it has been shown to drive the movement of actin filaments in an in vitro assay (Toyoshima et al 1987) (see below). Perhaps the extensive conserved domains contained in the 50-kDa subfragment, which have not been associated with ATP or actin binding, contain the secret to the mechanism underlying the molecular events that result in production of motive force.

The Myosin Tail

The rodlike tail, which allows myosin to self-assemble into filaments, is constructed of two α-helical heavy-chain regions. The structure of the coiled-coil inflexible myosin tail has been extensively reviewed (Parry 1981; McLachlan & Karn 1982, 1983; McLachlan 1983, 1984). In muscle these myosin filaments (thick) are arranged in ordered arrays between actin filaments (thin), which together with accessory proteins form structures known as sarcomeres. A nematode mutant that is not able to form functional myosin filaments (Dibb et al 1985) has two amino acid changes (a glycine to an arginine and a lysine to a methionine), in the sixth and seventh residues from the amino terminus of the tail. The defect found in this mutant indicates that the head-tail junction is involved in some chemical interactions critical for filament assembly. In smooth and nonmuscle cells myosins also form filaments but do not form sarcomeres. As discussed above, the polymerization and depolymerization of myosin appear to play an important role in the regulation of myosin function in nonmuscle cells.

The sequences of a number of myosin tails have been determined (Table 1). Since the tail region is contained in the carboxyl end of the protein, cDNA sequencing has provided examples from several species. As already mentioned, the overall homology between tail regions is less than that of the head region, but even quite distantly related organisms have a >25% homology in the tail (Table 2). The sequences when compared by graphic-matrix analysis reveal homologies that do not fall on a single line as do most of the homologies in the head region. Instead, the homologies apparent in the tail appear in multiple repeating patterns throughout the entire tail (Figure 5).

There are several orders of repeating structure in every filament-forming myosin tail. The smallest repeating structure contains seven amino acids, in which small, generally hydrophobic amino acids are usually found in the first and fourth positions of the repeat. Seven amino acids form two turns of an α-helix and the first and fourth positions fall in the interior coiled-coil locations (Karn et al 1983; McLachlan 1984). This pattern contributes to the formation of the α-helical coiled-coil structure by creating a hydrophobic core.

A group of four seven-residue repeats contains a distinct repeating pattern of alternating charged amino acids. These groups are positioned on the surface of the coiled-coil and create alternating bands of charge, which interact with the tails of adjacent myosin molecules to form the myosin filament. This 28-residue repeating pattern can be seen particularly well in the graphic-matrix analysis in Figure 5 (*bottom right*), where parallel lines spaced 28 residues apart are readily apparent.

With Fourier analysis a 197-residue repeating pattern can be visualized (McLachlan 1984). This unit probably contributes to the determination of the packing of the myosin molecules within the filament. Myosin molecules in muscle filaments are offset by about 140 Å. This distance corresponds to 98 residues, which is about half the length of the 197-residue repeat. From the repeating structures analyzed in the sequence of the tail it is clear that alternating groups of positively and negatively charged amino acids play an important role in forming the myosin filament. Molecular genetic manipulation of the sequence of the tail followed by expression of the modified tail structures will provide insight into the role of each of these patterns in the assembly and stability of the myosin filament.

The α-helix-forming potential appears to be strong and uniform throughout the tail region, as judged by secondary structure predictive methods (Chou & Fasman 1978; Garnier et al 1978). Close examination of the graphic-matrix patterns in Figure 5 shows that the homologies are not uniform throughout the tail region. There are distinct areas where the repeating pattern is less strong. These areas may be the origin of species-specific properties of myosin, such as the bending seen in smooth muscle and nonmuscle myosins (Trybus et al 1982; Craig et al 1983; Umekawa et al 1985). When the tail sequences are aligned into groups of 28 residues it has been observed that additional residues must be occasionally subtracted or added to preserve the overall alignment (McLachlan & Karn 1982; Warrick et al 1986). These irregularities in the tail structure appear to be in different locations in different myosins and may contribute to species-specific properties.

MODELS OF MUSCLE CONTRACTION

In the last 40 years, biochemical and structural studies have led to the development of a number of models for the molecular basis of muscle contraction. According to Huxley's swinging cross-bridge model (Huxley 1969; Huxley & Kress 1985), the myosin head undergoes a change in conformation while bound to the actin filament, resulting in a step (40–100 Å) of relative displacement of the actin and myosin filaments (Figure 7). The S2 region of the tail allows the heads to move away from the thick filament to interact with actin. For each ATP hydrolysis, the myosin binds to actin, undergoes the step in movement, and then releases from the actin filament. Many myosin heads operating asynchronously on one actin filament results in relative displacement of the actin, limited only by the length of the actin filaments.

Other models have been suggested to account for the relative movement of the two filament types. One alternative model has been suggested by Harrington (1971). Harrington and his colleagues (for review, see Harrington & Rodgers 1984; Ueno & Harrington 1986a,b) have demonstrated that a portion of the myosin tail, called the "hinge" (Figures 1 and 4), can undergo a helix-coil transition with very small changes in pH or temperature. Specifically, a 200 Å–long helical fragment can collapse into a random coil spanning about 100 Å. Harrington has suggested that this conformational change occurs in the cell in response to the myosin ATPase cycle and is the shortening step that gives rise to the relative movement of the two filaments. The site where motive force is produced would therefore be in the hinge, and the heads would serve to link the myosin thick filament to the actin filament.

A variety of biophysical and biochemical approaches have been used to obtain evidence for and against various models of motive force production. A large body of important information has been obtained (for reviews, see

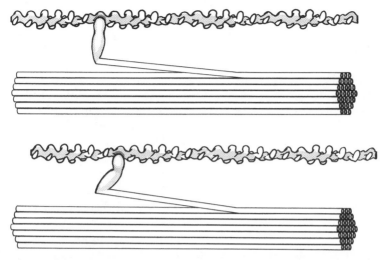

Figure 7 The upper and lower drawings show a schematic representation of a myosin head interacting with an actin filament. The myosin filament contains the LMM region of the heavy chain packed in a cylindrical array. Only one myosin head is shown in each drawing; the head is held away from the myosin filament by the S2 region of the tail, allowing the S1 portion to interact with the actin filament. The upper panel shows the myosin head bound to the actin filament in the putative pre-stroke conformation. The lower panel shows the head after the stroke which resulted in translocation of the actin filament.

Goody & Holmes 1983; Harrington & Rodgers 1984; Cooke 1986) and provides a foundation for future work. Missing from the arsenal of approaches to understanding the molecular basis of this movement has been an in vitro assay for the actual movement event. Such an assay is critical in order to search for the motive force production site(s). Such a search could include a molecular genetic approach to modify and examine areas of interest. Areas of sequence homology in the S1 would be ideal regions to alter. The mutant myosin could then be expressed and its behavior characterized using in vitro biochemical and movement assays. In this way the primary structure could be correlated with important aspects of motive force production.

ASSAYS FOR MYOSIN FUNCTION

In Vitro Movement Assays

In vitro movement of actin and myosin was pioneered by Szent-Gyorgyi and his colleagues in the 1940s. They used muscle actin and myosin to generate artificial threads that contracted when ATP was added (Szent-Gyorgyi 1942, 1947). In the 1950s and 1960s, Kuroda and Kamiya observed movement in cytoplasmic droplets from the alga *Nitella*, which set the stage for the discovery of actin-mediated movements in this organism (for reviews, see Kamiya 1981, 1986). Since those initial seminal discoveries, Kamiya and his colleagues have developed the *Nitella* system extensively. In the last decade, several new methods have been used to observe movements driven by actin and myosin (Yano 1978; Yano et al 1982; Tirosh & Oplatka 1982; Sheetz & Spudich 1983a,b; Yanagida et al 1984; Spudich et al 1985; Higashi-Fujime 1985, 1986; Honda et al 1986; Kron & Spudich 1986). One method that provides quantitative data on the speed of myosin movement on actin filaments (Sheetz & Spudich 1983a,b; Sheetz et al 1984) is currently in use in a number of laboratories. Of primary importance in the development of this assay was the method for laying out a parallel array of oriented actin filaments on which myosin could be observed to move under a light microscope. A *Nitella* cell was cut open longitudinally and the cell wall was pinned flat onto an optically clear silicon substratum, exposing the oriented actin cables (Kersey et al 1976) that are attached to the cytoplasmic face of the cell surface. In order to observe myosin movement, inert beads that are several micrometers in diameter are coated with myosin, and these beads move in an ATP-dependent manner on the *Nitella* actin cables.

Assays have been developed to measure the speed of myosin movement using purified actin and myosin proteins (Higashi–Fujime 1985, 1986; Spudich et al 1985; Honda et al 1986; Kron & Spudich 1986). One, referred to as the myosin-coated surface assay (Kron & Spudich 1986), inverts the arrangement of the *Nitella*-based assay. Myosin is fixed to a glass coverslip, and the ATP-dependent movement of individual fluorescently labeled actin filaments along this two-dimensional array of myosin molecules is observed. This assay was made possible by the finding that individual actin filaments can be observed by fluorescence light microscopy (Yanagida et al 1984).

These assays allow quantitative measurement of the velocity of myosin movement. This velocity has been shown to depend on the myosin type and to be relatively independent of the type of actin used (Sheetz et al 1984; Spudich et al 1985; Kron & Spudich 1986). Skeletal muscle myosin moves at ~ 5 μm/s, smooth muscle myosin moves at 0.4 μm/s, and *Dictyostelium* myosin moves at 1 μm/s. In contrast, skeletal muscle myosin moves at the same velocity on skeletal muscle actin, *Nitella* actin, or *Dictyostelium* actin. The movement is dependent on the concentration of ATP (Sheetz et al 1984; Kron & Spudich 1986) in a manner reminiscent of the ATP concentration dependence of the shortening velocity of skinned muscle fibers (Cooke & Bialek 1979). The ATP concentration needed for half-maximal velocity is nearly an order of magnitude greater than that needed for the half-maximal rate of actin-activated myosin ATPase activity measured in solution (Moos 1973). Half-maximal ATPase activity in solution presumably occurs when half the myosin heads bind ATP. In skinned fibers or in the motility assays, when half the myosin heads have ATP bound, the other half form rigor links with the actin. These rigor links impose a heavy load on those S1s that are actively cycling to pull the actin filament along. This load results in retarded velocities.

The development of these assays has provided a means to localize those regions of the molecule that are critical for movement along actin. Proteolytic fragments of skeletal muscle myosin include short heavy meromyosin (sHMM), which is two-headed and lacks the hinge and LMM regions of the rod, and S1, which lacks the entire rod and is single-headed. The *Nitella*-based assay was used to show that sHMM, attached to formalin-fixed *Staphylococcus aureus* cells by an antibody that recognizes S2 (Figure 8), moves at rates greater than 1 μm/s (Hynes et al 1987). The rates observed in these experiments are similar to that of the maximum velocity of contraction of skinned muscle fibers (Crowder & Cooke 1984). Harada et al (1987) used the myosin-coated surface assay to show that single-headed myosin can promote the movement of actin filaments at velocities very similar to those obtained with double-headed myosin. Their results are in keeping with the results of Cooke & Franks (1978), who used a sensitive tensiometer to demonstrate that single-headed myosin could produce nearly the same force per head as did double-headed myosin in synthetic actin-myosin threads when ATP was added.

The myosin-coated surface assay, modified by using nitrocellulose as

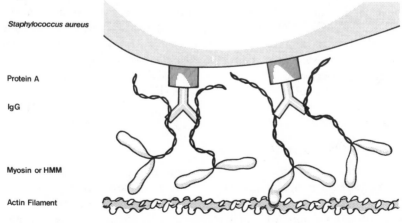

Figure 8 A schematic drawing of the *Nitella*-based motility assay system. A formalin-fixed *Staphylococcus aureus* cell is attached to myosin molecules by a protein A/anti-myosin antibody bridge. Actin arrays of single polarity are obtained from the alga *Nitella*. The rate of myosin movement can be obtained by following the translational progress of the *S. aureus* cell over time.

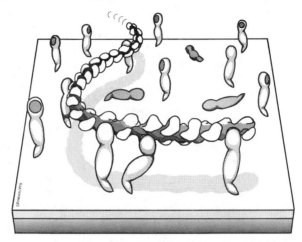

Figure 9 A schematic drawing of the myosin-coated surface assay system. S1 fragments of myosin adhere to a film of nitrocellulose on glass, and the movement of fluorescently-labeled actin filaments can be observed.

the transparent substratum to which myosin fragments were attached, proved particularly useful to demonstrate that S1, prepared either with papain or with chymotrypsin (Figure 9), promotes the movement of actin filaments at rates greater than 1 μm/s (Toyoshima et al 1987). These results show that the movement capability in muscle myosin is localized in the S1 moiety. This conclusion is consistent with the observation of S1-induced movements (Yano et al 1982) of a pinwheel coated with actin filaments and suspended in a beaker. The conclusion is also consistent with the movements of the naturally occurring single-headed myosin I from *Acanthamoeba*, which were observed with the *Nitella*-based assay (Albanesi et al 1985). As described above, myosin I consists of a single head, which resembles the S1 subunit of muscle myosin (Figure 6), and an unusual tail of ~ 35 kDa, which is totally distinct from the tail of muscle myosin (Korn et al 1987).

Actin-Activated ATPase Assay

The velocity of movement of various myosins appears to be closely correlated with their maximum actin-activated ATPase activity. This correlation has been observed both with the in vitro assays described above and with skinned muscle fibers that contain different myosin types (Barany 1967; Stein et al 1982; Sheetz et al 1984). Such a correlation may be expected if we assume that the myosin head moves along actin some unit distance for every ATP hydrolyzed. Extensive study of the ATPase cycle (for reviews, see Adelstein & Eisenberg 1980; Taylor & Amos 1981; Botts et al 1984; Eisenberg & Hill 1985; Brenner & Eisenberg 1986; Cooke 1986) has led to the formulation of diagrams such as that shown in Figure 10. In this scheme, A = actin, M = myosin, T = ATP, D = ADP, and P = inorganic phosphate, and the states indicated are proposed major intermediates. The inner cycle represents states of myosin not associated with actin, whereas the outer cycle represents states of actin-myosin complexes. Several studies (Stein et al 1984; Eisenberg & Hill 1985) have indicated the existence of two distinct states with ADP and P_i bound to myosin (referred to as $X \cdot P_I$ and $X \cdot P_{II}$). In the presence of actin, the conversions shown by dashed lines in Figure 10 probably occur only rarely. Strong binding between actin and myosin occurs in states $A \cdot M \cdot D$ and $A \cdot M$, while in the other states the binding is weak. The $A \cdot M$ state is the so-called rigor complex, which is characterized by its extremely high association constant and presumably represents the end of the work cycle. This rigor complex can be seen in muscles depleted of ATP (Reedy et al 1965, 1983).

The binding of ATP to $A \cdot M$ is extremely rapid and essentially irreversible. According to the kinetic measurements of Eisenberg and associates (Stein et al 1982, 1984; for review, see Eisenberg & Hill 1985), the states $A \cdot M \cdot T$, $A \cdot M \cdot D \cdot P_I$, and $A \cdot M \cdot D \cdot P_{II}$ are in equilibrium with each other and with their actin-free homologues. An important point is that actin accelerates the overall cycle by increasing the rates of the product release steps, which occur only very slowly with myosin alone. Energy considerations suggest that as a result of product release work is performed, presumably in the form of a myosin conformational change (Hibberd et al 1985).

If one could visualize within an actively contracting muscle the structures

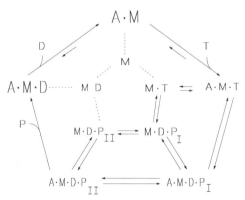

Figure 10   A schematic of a model for the actin-activated ATPase cycle in muscle, in which A = actin, M = myosin, T = ATP, D = ADP, P = phosphate, and P_I and P_{II} represent different putative energy states. The inner five states are the actin-free states and the outer five states are actin bound. $A \cdot M$ and $A \cdot M \cdot D$ are shown in large type because they are strongly bound states.

that the states in the ATPase cycle represent, the myosin heads would be seen to spend most of their time either free of the actin filament or weakly and transiently attached. Almost all of the myosin heads would be in an activated state containing bound ADP and P_i as a result of ATP hydrolysis. When one of these high-energy heads finds the appropriate site on actin and binds strongly, it undergoes its putative conformational change, which results in the movement of the actin filament. The conformational change is associated with product release, followed by the rebinding of ATP, which dissociates the myosin head from the actin filament. The ATP is rapidly hydrolyzed to form the activated state, completing the cycle.

The biochemical and biophysical approaches that have been used to date to study myosin have led to sophisticated analyses of the structure-function relationships in the ATPase cycle. However, there are many questions that remain to be answered, especially those concerning the interrelationships between energy, kinetic, and structural considerations. All of these approaches need to be extended, but new techniques also need to be applied to this fundamental biological phenomenon. Quantitative in vitro assays for myosin movement hold much promise, especially when coupled with molecular genetic techniques, which allow one to express altered forms of the myosin molecule for evaluation.

MOLECULAR GENETIC MANIPULATION OF MYOSIN

A powerful approach that can be used in the analysis of myosin structure-function relationships is to make directed changes in myosin structure and then analyze their effects on myosin activity. The development of molecular genetic techniques provides the tools necessary to create a wide range of changes and to produce the resulting modified protein, the function of which can then be assayed in vitro. By carefully correlating each change to specific aspects of myosin function, it will be possible to begin to understand the important structural features of the molecule.

Reinach and colleagues (1986) used this approach to study the structure and function of chicken skeletal myosin. They cloned the light chain of this myosin, constructed mutants with substitutions within the Ca^{2+}/Mg^{2+} binding site, and expressed them in *Escherichia coli*. These workers showed that an intact Ca^{2+}/Mg^{2+} binding site in the N-terminal region of the light chain is essential for regulation of the interaction of myosin with actin. Using the *Dictyostelium* myosin heavy-chain gene, De Lozanne and co-workers (1987) expressed the LMM region in *E. coli* and showed that the resulting protein is a rod of the expected length. The purified, expressed myosin tail fragment forms filamentous aggregates at low ionic strength and can be phosphorylated in vitro on a threonine residue thought to be important in the *Dictyostelium* chemotactic response. Further constructions will allow the identification of the minimum structure required for each of these functions.

Another important aspect of molecular genetic approaches is that they provide the opportunity to investigate the roles of myosin in vivo. Myosins interact with a number of intracellular components, resulting in particular cellular events. Altered myosins, expressed in vivo, can be tools for the dissection of those interactions. Gene deletions are particularly useful for correlating gene products to cell functions.

E. coli is useful as an expression system to produce altered myosins or myosin fragments for examination in vitro. Myosin is not normally expressed in *E. coli*, therefore this organism is not useful to study in vivo interactions and functions of myosin. Several other organisms have been examined for their potential to provide expression of altered forms of myosin in vitro and in vivo. As discussed below, each has its own particular advantages and limitations.

Some mammalian cells produce large quantities of myosin. Muscle in particular has been used as a source of this protein for biochemical analysis. But myosin gene expression is complex in mammalian cells because they have multiple genes coding for slightly different myosin isoforms, which are expressed in different cell types and at different times in development (Gauthier & Lowey 1979; Shani et al 1981; Nguyen et al 1982; Weydert et al 1983, 1985; Leinwand et al 1983; Winkelmann et al 1983; Shani 1985). To construct a mammalian cell system in which a myosin gene could be modified, expressed, and the resultant protein analyzed, would require techniques to identify and isolate the expressed, altered isoform of myosin from the other forms normally produced by the cell (Liew & Jandreski 1986; Nudel et al 1986; Rovner et al 1986). Since myosins are relatively large proteins, a full-length cDNA clone is very difficult to obtain, and in mammalian cells the coding region is distributed among many exons contained in many tens of kilobases of DNA. Although several mammalian myosin genes have been partially characterized (Umeda et al 1983; Sinha et al 1982; Hastings & Emerson 1982; Friedman et al 1984), only one type (from rat) has been completely sequenced (Strehler et al 1985, 1986). Techniques required for the analysis of a modified mammalian myosin gene product in mammalian cells are at an early stage of development.

The nematode *Caenorhabditis elegans* has provided an opportunity to study muscle myosin structure with classical genetics (for review, see Brenner 1974). These organisms contain two types of muscle, one localized in the body wall and the other involved in feeding. Since each muscle type contains the products of two myosin heavy-chain genes (Schachat et al 1977), it is possible to obtain viable mutants that are defective in motility. By screening for paralysis or impaired movement, a considerble number of mutants have been selected in which one of the two body-wall myosin genes (unc-54) is altered (Epstein et al 1974, 1976; Waterston et al 1980; Zengel & Epstein 1980; MacLeod et al 1981). These mutants can be divided into classes: those that lack myosin, those in which the myosin is of an abnormal size, and those in which the size of the protein appears normal. The first two classes result in nonfunctioning muscles whose structure is grossly disorganized. The latter class contains missense mutations, which result in muscles that have a normal structure but are nonfunctional. The isolation of suppressor mutants is particularly valuable because it gives information on other proteins interacting with myosin or involved in its expression (Wills et al 1983). Other suppressor phenotypes may be the result of intracistronic complementation between different functional domains of the myosin (Moerman et al 1982).

The unc-54 gene was the first myosin gene to be completely sequenced (Karn et al 1983, 1985). Several mutations have now been mapped to the sequence and provide valuable information on how structural changes create functional differences in myosin (Dibb et al 1985). Studies using C. elegans will undoubtedly continue to contribute to our understanding of muscle structure through the analysis of additional mutants as well as through techniques available to express modified myosins in this nematode (Stinchcomb et al 1985; Fire 1986). The complementary biochemistry is underdeveloped due to the difficulty of obtaining large amounts of nematode myosin for analysis.

Physarum presents an opportunity to obtain large amounts of myosin collected from cells in each of several developmental stages. Two of the stages, amoeboid and plasmodial, show very different modes of motility. Stage-specific myosins have been isolated from both amoebae and plasmodia (Kohama & Takano-Ohmuro 1984). A nuclear form of myosin may also exist (Jockusch et al 1973). *Physarum* myosin has been shown to form filaments in vitro (Nachmias & Ingram 1970; Nachmias 1972) and in vivo (Allera et al 1971), and contracting synthetic actomyosin fibers have been formed (Hinssen & D'Haese 1976). *Physarum* myosin is unusually poor in sulfhydryl groups (Nachmias et al 1982), and its ATPase activity has been shown to be regulated by Ca^{2+}, probably through the phosphorylation of one of the light chains (Nachmias 1981; Kohama & Kendrick-Jones 1986). Phosphorylation of the heavy chain (Takahashi et al 1983; Ogihara et al 1983) regulates filament assembly and ATPase activity. The potential for the application of genetic techniques (Cooke & Dee 1975; Haugli & Dove

1972) to the wealth of myosin-related topics in *Physarum* is considerable; however, to date none of the myosin genes have been cloned.

Drosophila has a tremendous potential as an organism in which to manipulate myosin by genetic methods. In *Drosophila* there is only one muscle myosin heavy-chain gene, but it is used to construct up to four different myosin mRNAs (Rozek & Davidson 1986). Two mRNAs are expressed in all stages of development, and two are expressed only in adult and late pupal stages. A series of overlapping clones containing the entire *Drosophila* muscle myosin heavy-chain gene has been described, and portions of the gene have been sequenced (Bernstein et al 1983; Rozek & Davidson 1983, 1986). One of the myosin light chains from *Drosophila* has also been cloned and sequenced (Toffenetti et al 1987). Expression of the myosin heavy-chain gene is required during development since homozygous mutants die in embryonic or larval stages (Mogami et al 1986). A nonmuscle myosin has been isolated from *Drosophila* tissue-culture cells and has been shown to be antigenically distinct from the muscle type (Kiehart & Feghali 1986). The analysis of myosin function in *Drosophila* can take advantage of classical genetic techniques as well as the recently developed P-element transduction, which can be used to modify the adult organism (Rubin & Spradling 1982). Such research has great potential for examination of myosin structure and function. Characterization of *Drosophila* myosin biochemistry has lagged due to the difficulty of isolating sufficient quantities of protein.

In yeast, there is an extensive technology for gene analysis, gene transfer, and gene replacement (Botstein & Davis 1982). Yeast can also be grown in large amounts, so protein can be obtained in biochemical quantities. There are at least two myosinlike genes in *Saccharomyces cerevisiae*, one of which has been partially characterized (Watts et al 1985), and other yeast cytoskeletal protein genes have also been characterized. The single actin gene found in *Saccharomyces cerevisiae* is lethal if present in less than or more than one copy per cell (Shortle et al 1982). Because of the extensive development of yeast genetics, this organism is particularly well suited for molecular genetic manipulations of the myosin molecule. Yeast, however, are not motile, and many aspects of yeast cell biology, such as cell division, are not typical of other eukaryotic cells.

The motile, haploid eukaryote *Dictyostelium discoideum* has a single conventional myosin heavy-chain gene, and this gene, called *mhc*A, has been cloned and sequenced (De Lozanne et al 1987b; Warrick et al 1986). It has no introns, and portions of the gene can be expressed in bacteria as well as reintroduced into *Dictyostelium* and expressed there (De Lozanne & Spudich 1987; De Lozanne et al 1987a). The discovery of homologous recombination in the *Dictyostelium* myosin *mhc*A gene (De Lozanne & Spudich 1987) will allow the construction of a mutant that is devoid of this native myosin, and that mutant should be able to be used as an expression host for modified myosin genes. *Dictyostelium* also contains a small myosin that appears to be related to the single-headed myosin I from *Acanthamoeba* (Cote et al 1985). The ability to obtain homologous integration of a plasmid targeted to a particular gene in *Dictyostelium* provides the methodology to determine the phenotype of a mutant lacking this protein.

CONCLUSION AND FUTURE PERSPECTIVES

Regarding the structure of myosin, many new myosin sequences have appeared in the last year, and comparisons among the various sequences reveal certain regions that are highly conserved. These conserved regions are undoubtedly crucial for myosin function. It is difficult, however, to speculate on the roles of these regions without high-resolution structural information. Such data would show how the protein is folded and which amino acids interact with ATP, which ones interact with actin, and which ones are in a domain that might be expected to undergo the large changes in conformation presumed to be necessary for movement. X-ray crystallographic data have been difficult to obtain for this protein. The crystals obtained by Rayment & Winkelmann (1984) represent an exciting development. Although these crystals may be sufficiently ordered to obtain the necessary information for a high-resolution map, the heterogeneity in S1 preparations made by proteolytic digestion of myosin may present a problem. One way to overcome this heterogeneity would be to express S1 in an appropriate host, using a plasmid containing the gene coding for S1. The expressed S1 would have to be proven to be in its native configuration before it could be used for crystal growth. The recent demonstration that HMM can be expressed in *Dictyostelium* (De Lozanne & Spudich 1987) in a form that moves in the *Nitella*-based motility assay (De Lozanne, unpublished observations) gives hope that homogeneous, functional myosin head fragments may be readily available in the future.

For those interested in the function of myosin in cell motility, the years ahead hold great promise. One can hope to see the application of essentially all of the technologies in biology to the study of myosin function, including high-resolution X-ray crystallography and other biophysical approaches, site-directed mutagenesis and classical physiological genetics, and conventional biochemistry coupled with in vitro quantitative motility assays. The information provided by this multifaceted attack on this fundamental biological problem should take us toward an understanding of the molecular basis of myosin function in myosin-based cell motility.

ACKNOWLEDGMENTS

References to early work on motility and muscle contraction were researched and kindly provided by Harold McGee, with whom we have enjoyed many fruitful discussions. We wish to also thank Roger Cooke, Arturo De Lozanne, and Steve Kron for their constructive comments on early drafts, and Julja Burchard for helping extensively with references.

Literature Cited

Adams, R. J., Pollard, T. D. 1986. Propulsion of organelles isolated from *Acanthamoeba* along actin filaments by myosin I. *Nature* 322: 754–56

Adelstein, R. S., Conti, M. A. 1975. Phosphorylation of platelet myosin increases actin-activated myosin ATPase activity. *Nature* 256: 597–98

Adelstein, R. S., Eisenberg, E. 1980. Regulation and kinetics of the actin-myosin-ATP interaction. *Ann. Rev. Biochem.* 49: 921–56

Albanesi, J. P., Fujisaki, H., Hammer, J. A. III, Korn, E. D., Jones, R., Sheetz, M. P. 1985. Monomeric *Acanthamoeba* myosins I support movement in vitro. *J. Biol. Chem.* 260: 8649–52

Allen, N. S., Allen, R. D. 1978. Cytoplasmic streaming in green plants. *Ann. Rev. Biophys. Bioeng.* 7: 497–526

Allera, A., Beck, R., Wohlfarth-Bottermann, K.-E. 1971. Extensive fibrillar protoplasmic differentiations and their significance for protoplasmic streaming. VII. Identification of the plasma filaments in *Physarum polycephalum* as F-actin by in situ binding of heavy meromyosin. *Cytobiologie* 4: 437–49

Amos, L. A. 1985. Structure of muscle filaments studied by electron microscopy. *Ann. Rev. Biophys. Biophys. Chem.* 14: 291–313

Applegate, D., Reisler, E. 1983. Protease-sensitive regions in myosin subfragment 1. *Proc. Natl. Acad. Sci. USA* 80: 7109–12

Barany, M. 1967. ATPase activity of myosin correlated with the speed of muscle shortening. *J. Gen. Physiol.* 50: 197–218

Berlot, C. H., Devreotes, P. N., Spudich, J. A. 1987. Chemoattractant-elicited increases in *Dictyostelium* myosin phosphorylation are due to changes in myosin localization and increases in kinase activity. *J. Biol. Chem.* 262: 3918–26

Berlot, C. H., Spudich, J. A., Devreotes, P. N. 1985. Chemoattractant-elicited increases in myosin phosphorylation in *Dictyostelium*. *Cell* 43: 307–14

Bernstein, S. I., Mogami, K., Donady, J. J., Emerson, C. P. Jr. 1983. *Drosophila* muscle myosin heavy chain encoded by a single gene in a cluster of muscle mutations. *Nature* 302: 393–98

Botstein, D., Davis, R. W. 1982. In *The Molecular Biology of the Yeast Sac-charomyces*, ed. J. N. Strathern, E. W. Jones, J. R. Broach, Vol. 2, pp. 607–38. Cold Spring Harbor, NY: Cold Spring Harbor Lab.

Botts, J., Takashi, R., Torgerson, P., Hozumi, T., Muhlrad, A., et al. 1984. On the mechanism of energy transduction in myosin subfragment 1. *Proc. Natl. Acad. Sci. USA* 81: 2060–64

Brenner, B., Eisenberg, E. 1986. Rate of force generation in muscle: Correlation with actomyosin ATPase activity in solution. *Proc. Natl. Acad. Sci. USA* 83: 3542–46

Brenner, S. 1974. The genetics of *Caenorhabditis elegans*. *Genetics* 77: 71–94

Capony, J.-P., Elzinga, M. 1981. The amino acid sequence of a 34,000 dalton fragment from S-2 of myosin. *Biophys. J.* 33: 148a

Chantler, P. D., Szent-Gyorgyi, A. G. 1980. Regulatory light-chains and scallop myosin: Full dissociation, reversibility and cooperative effects. *J. Mol. Biol.* 138: 473–92

Chen, T., Applegate, D., Reisler, E. 1985a. Cross-linking of actin to myosin subfragment 1: Course of reaction and stoichiometry of products. *Biochemistry* 24: 137–44

Chen, T., Applegate, D., Reisler, E. 1985b. Cross-linking of actin to myosin subfragment 1 in the presence of nucleotides. *Biochemistry* 24: 5620–25

Chen, T., Liu, J., Reisler, E. 1987. Proteolysis and binding of myosin subfragment 1 to actin. *J. Mol. Biol.* 194: 565–68

Chou, P. Y., Fasman, G. D. 1978. Prediction of the secondary structure of proteins from their amino acid sequence. *Adv. Enzymol.* 47: 45–148

Clarke, M., Spudich, J. A. 1977. Nonmuscle contractile proteins: The role of actin and myosin in cell motility and shape determination. *Ann. Rev. Biochem.* 46: 797–822

Collins, J. H., Korn, E. D. 1980. Actin activation of Ca^{2+}-sensitive Mg^{2+}-ATPase activity of *Acanthamoeba* myosin II is enhanced by dephosphorylation of its heavy chains. *J. Biol. Chem.* 255: 8011–14

Collins, J. H., Korn, E. D. 1981. Purification and characterization of actin-activatable, Ca^{2+}-sensitive myosin II from *Acanthamoeba*. *J. Biol. Chem.* 256: 2586–95

Cooke, D. J., Dee, J. 1975. Methods for the isolation and analysis of plasmodial mutants in *Physarum polycephalum*. *Genet. Res.* 24: 175–87

Cooke, R. 1986. The mechanism of muscle contraction. *CRC Crit. Rev. Biochem.* 21: 53–118

Cooke, R., Bialek, W. 1979. Contraction of glycerinated muscle fibers as a function of the ATP concentration. *Biophys. J.* 28: 241–58

Cooke, R., Franks, K. E. 1978. Generation of force by single-headed myosin. *J. Mol. Biol.* 120: 361–73

Corti, B. 1774. *Osservazione Microscopische sulla Tremella e sulla Circulazione del Fluido in Una Planto Acquaguola.* Lucca, Italy

Cote, G. P., Albanesi, J. P., Ueno, T., Hammer, J. A. III, Korn, E. D. 1985. Purification from *Dictyostelium discoideum* of a low-molecular-weight myosin that resembles myosin I from *Acanthamoeba castellanii*. *J. Biol. Chem.* 260: 4543–46

Craig, R., Smith, R., Kendrick-Jones, J. 1983. Light-chain phosphorylation controls the conformation of vertebrate nonmuscle and smooth muscle myosin molecules. *Nature* 302: 436–39

Crowder, M. S., Cooke, R. 1984. The effect of myosin sulfhydryl modification on the mechanics of fiber contraction. *J. Muscle Res. Cell Motil.* 5: 131–46

Dalbey, R. E., Weiel, J., Yount, R. G. 1983. Forster energy transfer measurements of thiol 1 to thiol 2 distances in myosin subfragment 1. *Biochemistry* 22: 4696–4706

Daniel, J. L., Molish, I. R., Rigmaiden, M., Stewart, G. 1984. Evidence for a role of myosin phosphorylation in the initiation of the platelet shape change response. *J.*

Biol. Chem. 259: 9826–31

De Lozanne, A. 1987. Homologous recombination in *Dictyostelium* as a tool for the study of developmental genes. *Methods Cell Biol.* 28: 489–96

De Lozanne, A., Berlot, C. H., Leinwand, L. A., Spudich, J. A. 1987a. Expression in *E. coli* of a functional *Dictyostelium* myosin heavy chain fragment. *J. Cell Biol.* In press

De Lozanne, A., Spudich, J. A. 1987. Disruption of the *Dictyostelium* myosin heavy chain gene by homologous recombination. *Science* 236: 1086–91

De Lozanne, A., Warrick, H. M., Chasan, R., Leinwand, L. A., Spudich, J. A. 1987b. Molecular genetic approaches to myosin function. In *Proc. UCLA Symp. on Signal Transduction in Cytoplasmic Organization and Cell Motility.* In press

Devereux, J., Haeberli, P., Smithies, O. 1984. A comprehensive set of sequence analysis programs for the VAX. *Nucleic Acids Res.* 12: 387–95

Devreotes, P., Fontana, D., Klein, P., Sherring, J., Theibert, A. 1987. Transmembrane signaling in *Dictyostelium*. *Methods Cell Biol.* 28: 489–96

Dibb, N. J., Brown, D. M., Karn, J., Moerman, D. G., Bolten, S. L., Waterston, R. H. 1985. Sequence analysis of mutations that affect the synthesis, assembly and enzymatic activity of the *unc*-54 myosin heavy chain of *Caenorhabditis elegans*. *J. Mol. Biol.* 183: 543–51

Eisenberg, E., Greene, L. E. 1980. The relation of muscle biochemistry to muscle physiology. *Ann. Rev. Physiol.* 42: 293–309

Eisenberg, E., Hill, T. L. 1985. Muscle contraction and free energy transduction in biological systems. *Science* 227: 999–1006

Elliott, A., Offer, G. 1976. Electron microscopy of myosin molecules from muscle and non-muscle sources. *Proc. R. Soc. Lond. Ser. B* 193: 45–53

Elliott, A., Offer, G. 1978. Shape and flexibility of the myosin molecule. *J. Mol. Biol.* 123: 505–19

Elzinga, M., Collins, J. H. 1977. Amino acid sequence of a myosin fragment that contains SH-1, SH-2, and *N*-methylhistidine. *Proc. Natl. Acad. Sci. USA* 74: 4281–84

Elzinga, M., Trus, B. 1980. Sequence and proposed structure of a 17,000 dalton fragment of myosin. In *Methods in Peptide and Protein Sequence Analysis*, ed. C. Birr, pp. 213-24. Amsterdam: Elsevier/North Holland Biomed.

Engelhardt, W. A., Ljubimowa, M. N. 1939. Myosine and adenosinetriphosphatase. *Nature* 144: 668–69

Epstein, H. F., Harris, H. E., Schachat, F.

H., Suddleson, E. A., Wolff, J. A. 1976. Genetic and molecular studies of nematode myosin. In *Cell Motility—Cold Spring Harbor Conferences on Cell Proliferation*, ed. R. Goldman, T. Pollard, J. Rosenbaum, 3: 203–14. Cold Spring Harbor, NY: Cold Spring Harbor Lab.

Epstein, H. F., Waterston, R. H., Brenner, S. 1974. A mutant affecting the heavy chain of myosin in *Caenorhabditis elegans*. *J. Mol. Biol.* 90: 291–300

Fechheimer, M., Cebra, J. J. 1982. Phosphorylation of lymphocyte myosin catalyzed in vitro and in intact cells. *J. Cell Biol.* 93: 261–68

Fire, A. 1986. Integrative transformation of *Caenorhabditis elegans*. *EMBO J.* 5: 2673–80

Flicker, P. F., Wallimann, T., Vibert, P. 1983. Electron microscopy of scallop myosin: Location of regulatory light chains. *J. Mol. Biol.* 169: 723–41

Friedman, D. J., Umeda, P. K., Sinha, A. M., Hsu, H.-J., Jakovcic, S., Rabinowitz, M. 1984. Characterization of genomic clones specifying rabbit α- and β-ventricular myosin heavy chains. *Proc. Natl. Acad. Sci. USA* 81: 3044–48

Fry, D. C., Kuby, S. A., Mildvan, A. S. 1986. ATP-binding site of adenylate kinase: Mechanistic implications of its homology with *ras*-encoded p21, F1-ATPase, and other nucleotide binding proteins. *Proc. Natl. Acad. Sci. USA* 83: 907–11

Fujiwara, K., Pollard, T. D. 1976. Fluorescent antibody localization of myosin in the cytoplasm, cleavage furrow, and mitotic spindle of human cells. *J. Cell Biol.* 71: 848–75

Fukui, Y., Yumura, S., Yumura, T. K., Mori, H. 1986. Agar overlay method: High-resolution immunofluorescence for the study of the contractile apparatus. *Methods Enzymol.* 134: 573–80

Garnier, J., Osguthorpe, D. J., Robson, B. 1978. Analysis of the accuracy and implications of simple methods for predicting the secondary structure of globular proteins. *J. Mol. Biol.* 120: 97–120

Gauthier, G. F., Lowey, S. 1979. Distribution of myosin isoenzymes among skeletal muscle fiber types. *J. Cell Biol.* 81: 10–25

Goldman, R., Pollard, T., Rosenbaum, J., eds. 1976. *Cell Motility—Cold Spring Harbor Conferences on Cell Proliferation*, Vol. 3. Cold Spring Harbor, NY: Cold Spring Harbor Lab.

Goody, R. S., Holmes, K. C. 1983. Cross-bridges and the mechanism of muscle contraction. *Biochim. Biophys. Acta* 726: 13–39

Greene, L. E. 1984. Stoichiometry of actin·

S-1 cross-linked complex. *J. Biol. Chem.* 259: 7363–66

Griffith, L. M., Downs, S. M., Spudich, J. A. 1987. Myosin light chain kinase and myosin light chain phosphatase from *Dictyostelium*: Effects of reversible phosphorylation on myosin structure and function. *J. Cell Biol.* 104: 1309–23

Gulick, J., Kropp, K., Robbins, J. 1985. The structure of two fast-white myosin heavy chain promoters. *J. Biol. Chem.* 260: 14513–20

Hammer, J. A. III, Albanesi, J. P., Korn, E. D. 1983. Purification and characterization of a myosin I heavy chain kinase from *Acanthamoeba castellanii*. *J. Biol. Chem.* 258: 10168–75

Hammer, J. A. III, Bowers, B., Paterson, B. M., Korn, E. D. 1987. Complete nucleotide sequence and deduced polypeptide sequence of a non-muscle myosin heavy chain gene from *Acanthamoeba*: Evidence of a hinge in the rod-like tail. *J. Cell Biol.* In press

Harada, Y., Noguchi, A., Kishino, A., Yanagida, T. 1987. Sliding movement of single actin filaments on one-headed myosin filaments. *Nature* 326: 805–8

Harrington, W. F. 1971. A mechanochemical mechanism for muscle contraction. *Proc. Natl. Acad. Sci. USA* 68: 685–89

Harrington, W. F., Rodgers, M. E. 1984. Myosin. *Ann. Rev. Biochem.* 53: 35–73

Hartshorne, D. J., Siemankowski, R. F. 1981. Regulation of smooth muscle actomyosin. *Ann. Rev. Physiol.* 43: 519–30

Hastings, K. E. M., Emerson, C. P. Jr. 1982. cDNA clone analysis of six coregulated mRNAs encoding skeletal muscle contractile proteins. *Proc. Natl. Acad. Sci. USA* 79: 1553–57

Haugli, F. B., Dove, W. F. 1972. Mutagenesis and mutant selection in *Physarum polycephalum*. *Mol. Gen. Genet.* 118: 109–24

Hibberd, M. G., Dantzig, J. A., Trentham, D. R., Goldman, Y. E. 1985. Phosphate release and force generation in skeletal muscle fibers. *Science* 228: 1317–19

Hibberd, M. G., Trentham, D. R. 1986. Relationships between chemical and mechanical events during muscular contraction. *Ann. Rev. Biophys. Biophys. Chem.* 15: 119–61

Higashi-Fujime, S. 1985. Unidirectional sliding of myosin filaments along the bundle of F-actin filaments spontaneously formed during superprecipitation. *J. Cell Biol.* 101: 2335–44

Higashi-Fujime, S. 1986. In vitro movements of actin and myosin filaments from muscle. *Cell Motil. Cytoskeleton* 6: 159–62

Higgins, C. F., Hiles, I. D., Salmond, G. P. C., Gill, D. R., Downie, J. A., et al. 1986. A family of related ATP-binding subunits coupled to many distinct biological processes in bacteria. *Nature* 323: 448–50

Hinssen, H., D'Haese, J. 1976. Synthetic fibrils from *Physarum* actomyosin—self assembly, organization and contraction. *Cytobiologie* 13: 132–57

Honda, H., Nagashima, H., Asakura, S. 1986. Directed movement of F-actin in vitro. *J. Mol. Biol.* 191: 131–33

Hozumi, T., Muhlrad, A. 1981. Reactive lysyl of myosin subfragment 1: Location on the 27K fragment and labeling properties. *Biochemistry* 20: 2945–50

Huszar, G. 1984. Methylated lysines and 3-methylhistidine in myosin: Tissue and developmental differences. *Methods Enzymol.* 106: 287–95

Huszar, G., Elzinga, M. 1972. Homologous methylated and nonmethylated histidine peptides in skeletal and cardiac myosins. *J. Biol. Chem.* 247: 745–53

Huxley, H. E. 1963. Electron microscope studies on the structure of natural and synthetic protein filaments from striated muscle. *J. Mol. Biol.* 7: 281–308

Huxley, H. E. 1969. The mechanism of muscular contraction. *Science* 164: 1356–66

Huxley, H. E. 1973. Muscular contraction and cell motility. *Nature* 243: 445–49

Huxley, H. E., Faruqi, A. R. 1983. Time-resolved X-ray diffraction studies on vertebrate striated muscle. *Ann. Rev. Biophys. Bioeng.* 12: 381–417

Huxley, H. E., Kress, M. 1985. Crossbridge behavior during muscle contraction. *J. Muscle Res. Cell Motil.* 6: 153–61

Hynes, T. R., Block, S. M., White, B. T., Spudich, J. A. 1987. Movement of myosin fragments in vitro: Domains involved in force production. *Cell* 48: 953–63

Inoue, S. 1981. Cell division and the mitotic spindle. *J. Cell Biol.* 91: 131s–47s

Jockusch, B. M., Ryser, U., Behnke, O. 1973. Myosin-like protein in *Physarum* nuclei. *Exp. Cell Res.* 76: 464–66

Johnson, K. A. 1985. Pathway of the microtubule-dynein ATPase and the structure of dynein: A comparison with actomyosin. *Ann. Rev. Biophys. Biophys. Chem.* 14: 161–88

Jung, G., Korn, E. D., Hammer, J. A. III. 1987. The heavy chain of *Acanthamoeba* myosin IB is a fusion of myosin-like and nonmyosin-like sequences. *Proc. Natl. Acad. Sci. USA.* In press

Kamiya, N. 1981. Physical and chemical basis of cytoplasmic streaming. *Ann. Rev. Plant Physiol.* 32: 205–36

Kamiya, N. 1986. Cytoplasmic streaming in giant algal cells: A historical survey of experimental approaches. *Bot. Mag. Tokyo* 99: 441–67

Kamm, K. E., Stull, J. T. 1986. Activation of smooth muscle contraction: Relation between myosin phosphorylation and stiffness. *Science* 232: 80–82

Karn, J., Brenner, S., Barnett, L. 1983. Protein structural domains in the *Caenorhabditis elegans* unc-54 myosin heavy chain gene are not separated by introns. *Proc. Natl. Acad. Sci. USA* 80: 4253–57

Karn, J., Dibb, N. J., Miller, D. M. 1985. Cloning nematode myosin genes. In *Cell and Muscle Motility*, ed. J. W. Shay, pp. 185–237. London/New York: Plenum

Kato, T., Tonomura, Y. 1977. Identification of myosin in *Nitella flexilis*. *J. Biochem. Tokyo* 82: 777–82

Katoh, T., Katoh, H., Morita, F. 1985. Actin-binding peptide obtained by the cyanogen bromide cleavage of the 20-kDa fragment of myosin subfragment-1. *J. Biol. Chem.* 260: 6723–27

Kavinsky, C. J., Umeda, P. K., Levin, J. E., Sinha, A. M., Nigro, J. M., et al. 1984. Analysis of cloned mRNA sequences encoding subfragment 2 and part of subfragment 1 of α- and β-myosin heavy chains of rabbit heart. *J. Biol. Chem.* 259: 2775–81

Kendrick-Jones, J., Scholey, J. M. 1981. Myosin-linked regulatory systems. *J. Muscle Res. Cell Motil.* 2: 347–72

Kendrick-Jones, J., Szentkiralyi, E. M., Szent-Gyorgyi, A. G. 1976. Regulatory light chains in myosins. *J. Mol. Biol.* 104: 747–75

Kersey, Y. M., Hepler, P. K., Palevitz, B. A., Wessels, N. K. 1976. Polarity of actin filaments in Characean algae. *Proc. Natl. Acad. Sci. USA* 73: 165–67

Kiehart, D. P., Feghali, R. 1986. Cytoplasmic myosin from *Drosophila melanogaster*. *J. Cell Biol.* 103: 1517–25

Kiehart, D. P., Mabuchi, I., Inoue, S. 1982. Evidence that myosin does not contribute to force production in chromosome movement. *J. Cell Biol.* 94: 165–78

Knecht, D. A., Loomis, W. F. 1987. Antisense RNA inactivation of myosin heavy chain gene expression in *Dictyostelium discoideum*. *Science* 236: 1081–86

Kohama, K., Kendrick-Jones, J. 1986. The inhibitory Ca^{2+}-regulation of the actin-activated Mg-ATPase activity of myosin from *Physarum polycephalum* plasmodia. *J. Biochem.* 99: 1433–46

Kohama, K., Takano-Ohmuro, H. 1984. Stage specific myosins from amoeba and plasmodium of slime mold, *Physarum polycephalum*. *Proc. Jpn. Acad.* 60: 431–34

Korn, E. D., Atkinson, M. A. L., Brzeska,

H., Hammer, J. A. III, Jung, G., Lynch, T. J. 1987. Structure-function studies on *Acanthamoeba* myosins IA, IB, and II. *J. Cell. Biochem.* In press

Kron, S. J., Spudich, J. A. 1986. Fluorescent actin filaments move on myosin fixed to a glass surface. *Proc. Natl. Acad. Sci. USA* 83: 6272–76

Kuczmarski, E. R. 1986. Partial purification of two myosin heavy chain kinases from *Dictyostelium discoideum*. *J. Muscle Res. Cell Motil.* 7: 501–9

Kuczmarski, E. R., Spudich, J. A. 1980. Regulation of myosin self-assembly: Phosphorylation of *Dictyostelium* heavy chain inhibits formation of thick filaments. *Proc. Natl. Acad. Sci. USA* 77: 7292–96

Kühne, W. 1863. Eine lebende Nematode in einer lebenden Muskelfaser beobachtet. *Virchows Archiv.* 26: 222–24

Kühne, W. 1864. *Untersuchungen über das Protoplasma und die Contractilität.* Leipzig: von Wilhelm Engelmann

Labbe, J.-P., Mornet, D. Roseau, G., Kassab, R. 1982. Cross-linking of F-actin to skeletal muscle myosin subfragment 1 with bis(imido esters): Further evidence for the interaction of myosin-head heavy chain with an actin dimer. *Biochemistry* 21: 6897–6902

Leinwand, L. A., Saez, L., McNally, E., Nadal-Ginard, B. 1983. Isolation and characterization of human myosin heavy chain genes. *Proc. Natl. Acad. Sci. USA* 80: 3716–20

Liew, C.-C., Jandreski, M. A. 1986. Construction and characterization of the α form of a cardiac myosin heavy chain cDNA clone and its developmental expression in the Syrian hamster. *Proc. Natl. Acad. Sci. USA* 83: 3175–79

Lowey, S., Cohen, C. 1962. Studies on the structure of myosin. *J. Mol. Biol.* 4: 293–308

Lowey, S., Slayter, H. S., Weeds, A. G., Baker, H. 1969. Substructure of the myosin molecule. I. Subfragments of myosin by enzymic degradation. *J. Mol. Biol.* 42: 1–29

Lu, R. C., Wong, A. 1985. The amino acid sequence and stability predictions of the hinge region in myosin subfragment 2. *J. Biol. Chem.* 260: 3456–61

Mabuchi, I., Okuno, M. 1977. The effect of myosin antibody on the division of starfish blastomeres. *J. Cell Biol.* 74: 251–63

MacLeod, A. R., Karn, J., Brenner, S. 1981. Molecular analysis of the *unc*-54 myosin heavy-chain gene of *Caenorhabditis elegans*. *Nature* 291: 386–90

Mahdavi, V., Periasamy, M., Nadal-Ginard, B. 1982. Molecular characterization of two myosin heavy chain genes expressed in the adult heart. *Nature* 297: 659–64

Maita, T., Hayashida, M., Tanioka, Y., Komine, Y., Matsuda, G. 1987. The primary structure of the myosin head. *Proc. Natl. Acad. Sci. USA* 84: 416–20

Marston, S. B. 1982. The regulation of smooth muscle contractile proteins. *Prog. Biophys. Mol. Biol.* 41: 1–41

Maruta, H., Gadasi, H., Collins, J. H., Korn, E. D. 1979. Multiple forms of *Acanthamobea* myosin I. *J. Biol. Chem.* 254:

Maruta, H., Korn, E. D. 1977. *Acanthamoeba* cofactor protein is a heavy chain kinase required for actin activation of the Mg^{2+}-ATPase activity of *Acanthamoeba* myosin I. *J. Biol. Chem.* 252: 8329–32

Matsumura, S., Muakami, M., Yasuda, S., Kuma, A. 1982. Phosphorylation of the bovine brain myosin. *Biochem. Biophys. Res. Commun.* 108: 1595–1600

McLachlan, A. D. 1983. Analysis of gene duplication repeats in the myosin rod. *J. Mol. Biol.* 169:15–30

McLachlan, A. D. 1984. Structural implications of the myosin amino acid sequence. *Ann. Rev. Biophys. Bioeng.* 13: 167–89

McLachlan, A. D., Karn, J. 1982. Periodic charge distributions in the myosin rod amino acid sequence match cross-bridge spacings in muscle. *Nature* 299: 226–31

McLachlan, A. D., Karn, J. 1983. Periodic features in the amino acid sequence of nematode myosin rod. *J. Mol. Biol.* 164: 605–26

Milligan, R. A., Flicker, P. F. 1987. Structural relationships of actin, myosin, and tropomyosin revealed by cryo-electron microscopy. *J. Cell Biol.* 105: 29–39

Mitchell, E. J., Jakes, R., Kendrick-Jones, J. 1986. Localisation of light chain and actin binding sites on myosin. *Eur. J. Biochem.* 161: 25–35

Moerman, D. G., Plurad, S., Waterston, R. H., Baillie, D. L. 1982. Mutations in the *unc*-54 myosin heavy chain gene of *Caenorhabditis elegans* that alter contractility but not muscle structure. *Cell* 29: 773–81

Mogami, K., O'Donnell, P. T., Bernstein, S. I., Wright, T. R. F., Emerson, C. P. Jr. 1986. Mutations of the *Drosophila* myosin heavy-chain gene: Effects on transcription, myosin accumulation, and muscle function. *Proc. Natl. Acad. Sci. USA* 83: 1393–97

Molina, M. I., Kropp, K. E., Gulick, J., Robbins, J. 1987. The sequence of an embryonic myosin heavy chain gene and isolation of its corresponding cDNA. *J. Biol. Chem.* 262: 6478–88

Moos, C. 1973. Actin activation of heavy meromyosin and subfragment-1 ATPases: Steady state kinetics studies. *Cold Spring Harbor Symp. Quant. Biol.* 37: 137–43

Mooseker, M. S. 1985. Organization, chemistry, and assembly of the cytoskeletal apparatus of the intestinal brush border. *Ann. Rev. Cell Biol.* 1: 209–41

Morales, M. F., Borejdo, J., Botts, J., Cooke, R., Mendelson, R. A., Takashi, R. 1982. Some physical studies of the contractile mechanism in muscle. *Ann. Rev. Phys. Chem.* 33: 319–51

Mornet, D., Bertrand, R., Pantel, P., Audemard, E., Kassab, R. 1981a. Structure of the actin-myosin interface. *Nature* 292: 301–6

Mornet, D., Bertrand, R., Pantel, P., Audemard, E., Kassab, R. 1981b. Proteolytic approach to structure and function of actin recognition site in myosin heads. *Biochemistry* 20: 2110–20

Murakami, N., Matsumura, S., Kumon, A. 1984. Purification and identification of myosin heavy chain kinase from bovine brain. *J. Biochem.* 95: 651–60

Nachmias, V. T. 1972. Filament formation by purified *Physarum* myosin. *Proc. Natl. Acad. Sci. USA* 69: 2011–14

Nachmias, V. T. 1981. *Physarum* myosin light chain one: A potential regulatory factor in cytoplasmic streaming. *Protoplasma* 109: 13–21

Nachmias, V. T., Ingram, W. C. 1970. Actomyosin from *Physarum polycephalum*: Electron microscopy of myosin-enriched preparations. *Science* 170: 743–45

Nachmias, V. T., Rubinstein, N. A., Taylor, T., Cannon, L. E. 1982. Sulfhydryl groups of native myosin and of the myosin heavy chains from *Physarum polycephalum* compared to vertebrate skeletal, smooth, and non-muscle myosins. *Biochim. Biophys. Acta* 700: 198–205

Nguyen, H. T., Gubits, R. M., Wydro, R. M., Nadal-Ginard, B. 1982. Sarcomeric myosin heavy chain is coded by a highly conserved multigene family. *Proc. Natl. Acad. Sci. USA* 79: 5230–34

Nudel, U., Melloul, D., Altoni, B., Greenberg, D., Yaffe, D. 1986. Developmentally regulated expression of chimeric muscle genes transferred into myogenic cells. *Mol. Biol. Dev.* 29: 647–56

Ogihara, S., Ikebe, M., Takahashi, K., Tonomura, Y. 1983. Requirement of phosphorylation of *Physarum* myosin heavy chain for thick filament formation, actin activation of Mg^{2+}-ATPase activity, and Ca^{2+}-inhibitory superprecipitation. *J. Biochem.* 93: 205–23

Okamoto, Y., Sekine, T., Grammer, J., Yount, R. G. 1986. The essential light chains constitute part of the active site of smooth muscle myosin. *Nature* 324: 78–80

Okamoto, Y., Yount, R. G. 1985. Identification of an active site peptide of skeletal myosin after photoaffinity labeling with N-(4-azido-2-nitrophenyl)-2-aminoethyl diphosphate. *Proc. Natl. Acad. Sci. USA* 82: 1575–79

Onishi, H., Maita, T., Miyanishi, T., Watanabe, S., Matsuda, G. 1986. Amino acid sequence of the 203-residue fragment of the heavy chain of chicken gizzard myosin containing the SH1-type cysteine residue. *J. Biochem.* 100: 1433–47

Pagh, K., Gerisch, G. 1986. Monoclonal antibodies binding to the tail of *Dictyostelium discoideum* myosin: Their effects on antiparallel and parallel assembly and actin-activated ATPase activity. *J. Cell. Biol.* 103: 1527–38

Parry, D. A. D. 1981. Structure of rabbit skeletal myosin: Analysis of the amino acid sequences of two fragments from the rod region. *J. Mol. Biol.* 153: 459–64

Payne, M. E., Elzinga, M., Adelstein, R. S. 1986. Smooth muscle myosin light chain kinase: Amino acid sequence at the site phosphorylated by adenosine cyclic 3′,5′-phosphate-dependent protein kinase whether or not calmodulin is bound. *J. Biol. Chem.* 261: 16346–50

Peltz, G., Kuczmarski, E. R., Spudich, J. A. 1981. *Dictyostelium* myosin: Characterization of chymotryptic fragments and localization of the heavy-chain phosphorylation site. *J. Cell Biol.* 89: 104–8

Perrie, W. T., Smillie, L. B., Perry, S. V. 1973. A phosphorylated light-chain component of myosin from skeletal muscle. *Biochem. J.* 135: 151–56

Persechini, A., Stull, J. T., Cooke, R. 1985. The effect of myosin phosphorylation on the contractile properties of skinned rabbit skeletal muscle fibers. *J. Biol. Chem.* 260: 7951–54

Pickett-Heaps, J. D., Tippit, D. H., Porter, K. R. 1982. Rethinking mitosis. *Cell* 29: 729–44

Pollard, T. D., Korn, E. D. 1973. *Acanthamoeba* myosin I: Isolation from *Acanthamoeba castellanii* of an enzyme similar to muscle myosin. *J. Biol. Chem.* 248: 4682–90

Putney, S. D., Herlihy, W. C., Schimmel, P. 1983. A new troponin T and cDNA clones for 13 different muscle proteins, found by shotgun sequencing. *Nature* 302: 718–21

Rayment, I., Winkelmann, D. A. 1984. Crystallization of myosin subfragment 1. *Proc. Natl. Acad. Sci. USA* 81: 4378–80

Reedy, M. K., Goody, R. S., Hofmann, W., Rosenbaum, G. 1983. Co-ordinated electron microscopy and X-ray studies of glycerinated insect flight muscle I. X-ray diffraction monitoring during preparation for electron microscopy of muscle fibers fixed in rigor, in ATP and in AMPPNP. *J. Muscle Res. Cell Motil.* 4: 25–53

Reedy, M. K., Holmes, K. C., Tregear, R. T. 1965. Induced changes in orientation of the crossbridges of glycerinated insect flight muscle. *Nature* 207: 1276–80

Reinach, F. C., Nagai, K., Kendrick-Jones, J. 1986. Site directed mutagenesis of the regulatory light-chain Ca^{2+}/Mg^{2+} binding site and its role in hybrid myosins. *Nature* 322: 80–83

Rovner, A. S., Murphy, R. A., Owens, G. K. 1986. Expression of smooth muscle and nonmuscle myosin heavy chains in cultured vascular smooth muscle cells. *J. Biol. Chem.* 261: 14740–45

Rozek, C. E., Davidson, N. 1983. *Drosophila* has one myosin heavy-chain gene with three developmentally regulated transcripts. *Cell* 32: 23–34

Rozek, C. E., Davidson, N. 1986. Differential processing of RNA transcribed from the single-copy *Drosophila* myosin heavy chain gene produces four mRNAs that encode two polypeptides. *Proc. Natl. Acad. Sci. USA* 83: 2128–32

Rubin, G. M., Spradling, A. C. 1982. Genetic transformation of *Drosophila* with transposable element vectors. *Science* 218: 348–53

Saez, L., Leinwand, L. A. 1986. Characterization of diverse forms of myosin heavy chain expressed in adult human skeletal muscle. *Nucleic Acids Res.* 14: 2951–69

Sanger, J. W., Sanger, J. M. 1979. The cytoskeleton and cell division. *Methods Achievements Exp. Pathol.* 8: 110–42

Schachat, F. H., Harris, H. E., Epstein, H. F. 1977. Two homogeneous myosins in body-wall muscle of *Caenorhabditis elegans*. *Cell* 10: 721–28

Scholey, J. M., Smith, R. E., Drenckhahn, D., Groschel-Stewart, U., Kendrick-Jones, J. 1982. Thymus myosin. Isolation and characterization of myosin from calf thymus and thymic lymphocytes, and studies on the effect of phosphorylation of its $M_r = 20,000$ light chain. *J. Biol. Chem.* 257: 7737–45

Scholey, J. M., Taylor, K. A., Kendrick-Jones, J. 1980. Regulation of non-muscle myosin assembly by calmodulin-dependent light chain kinase. *Nature* 287: 233–35

Schroeder, T. E. 1973. Actin in dividing cells: Contractile ring filaments bind heavy meromyosin. *Proc. Natl. Acad. Sci. USA* 70: 1688–92

Sellers, J. R., Spudich, J. A., Sheetz, M. P. 1985. Light chain phosphorylation regulates the movement of smooth muscle myosin on actin filaments. *J. Cell. Biol.* 101: 1897–1902

Shani, M. 1985. Tissue-specific expression of rat myosin light-chain 2 gene in transgenic mice. *Nature* 314: 283–86

Shani, M., Zevin-Sonkin, D., Saxel, O., Carmon, Y., Katcoff, D., et al. 1981. The correlation between the synthesis of skeletal muscle actin, myosin heavy chain, and myosin light chain and the accumulation of corresponding mRNA sequences during myogenesis. *Dev. Biol.* 86: 483–92

Sheetz, M. P., Chasan, R., Spudich, J. A. 1984. ATP-dependent movement of myosin in vitro: Characterization of a quantitative assay. *J. Cell. Biol.* 99: 1867–71

Sheetz, M. P., Spudich, J. A. 1983a. Movement of myosin-coated fluorescent beads on actin cables in vitro. *Nature* 303: 31–35

Sheetz, M. P., Spudich, J. A. 1983b. Movement of myosin-coated structures on actin cables. *Cell Motil.* 3: 485–89

Shortle, D., Haber, J., Botstein, D. 1982. Lethal disruption of the yeast actin gene by integrative DNA transformation. *Science* 217: 371–73

Sinha, A. M., Umeda, P. K., Kavinsky, C. J., Rajamanickam, C., Hsu, H.-J., et al. 1982. Molecular cloning of mRNA sequences for cardiac α- and β-form myosin heavy chains: Expression in ventricles of normal, hypothyroid, and thyrotoxic rabbits. *Proc. Natl. Acad. Sci. USA* 79: 5847–51

Sivaramakrishnan, M., Burke, B. 1982. The free heavy chain of vertebrate skeletal myosin subfragment 1 shows full enzymatic activity. *J. Biol. Chem.* 257: 1102–5

Smith, T. F., Waterman, M. S. 1981. Identification of common molecular subsequences. *J. Mol. Biol.* 147: 195–97

Spudich, J. A., Kron, S. J., Sheetz, M. P. 1985. Movement of myosin-coated beads on oriented filaments reconstituted from purified actin. *Nature* 315: 584–86

Stein, L. A., Chock, P. B., Eisenberg, E. 1984. The rate-limiting step in the actomyosin adenosinetriphosphatase cycle. *Biochemistry* 23: 1555–63

Stinchcomb, D. T., Shaw, J. E., Carr, S. H., Hirsh, D. 1985. Extrachromosomal DNA transformation of *Caenorhabditis elegans*. *Mol. Cell. Biol.* 5: 3484–96

Strehler, E. E., Mahdavi, V., Periasamy, M., Nadal-Ginard, B. 1985. Intron positions are conserved in the 5′ end region of myosin heavy-chain genes. *J. Biol. Chem.* 260: 468–71

Strehler, E. E., Strehler-Page, M.-A., Parriard, J.-C., Periasamy, M., Nadal-Ginard, B. 1986. Complete nucleotide and encoded amino acid sequence of a mammalian

420   WARRICK & SPUDICH

myosin heavy chain gene: Evidence against intron-dependent evolution of the rod. *J. Mol. Biol.* 190: 291–317

Sutoh, K. 1982a. An actin-binding site on the 20K fragment of myosin subfragment 1. *Biochemistry* 21: 4800–4

Sutoh, K. 1982b. Identification of myosin-binding sites on the actin sequence. *Biochemistry* 21: 3654–61

Sutoh, K. 1983. Mapping of actin-binding sites on the heavy chain of myosin subfragment 1. *Biochemistry* 22: 1579–85

Sutoh, K., Yamamoto, K., Wakabayashi, T. 1984. Electron microscopic visualization of the SH1 thiol of myosin by use of an avidin-biotin system. *J. Mol. Biol.* 178: 323–39

Sutoh, K., Yamamoto, K., Wakabayashi, T. 1986. Electron microscopic visualization of the ATPase site of myosin by photoaffinity labeling with a biotinylated photoreactive ADP analog. *Proc. Natl. Acad. Sci. USA* 83: 212-16

Szent-Gyorgyi, A. 1942. *Studies from the Institute of Medical Chemistry, University Szeged*, Vol. I–III. New York: Karger

Szent-Gyorgyi, A. 1947. *Chemistry of Muscular Contraction*. New York: Academic

Szent-Gyorgyi, A. G., Chantler, P. D. 1986. Control of contraction by myosins. In *Myology*, ed. A. G. Engel, B. Q. Banker, I: 589–612. New York: McGraw-Hill

Szilagyi, L., Balint, M., Sreter, F. A., Gergely, J. 1979. Photoaffinity labelling with an ATP analog of the N-terminal peptide of myosin. *Biochem. Biophys. Res. Commun.* 87: 936–45

Takahashi, K., Ogihara, S., Ikebe, M., Tonomura, Y. 1983. Morphological aspects of thiophosphorylated and dephosphorylated myosin molecules from the plasmodium of *Physarum polycephalum*. *J. Biochem.* 93: 1171–83

Taylor, E. W. 1979. Mechanism of actomyosin ATPase and the problem of muscle contraction. *CRC Crit. Rev. Biochem.* 6: 103–64

Taylor, K. A., Amos, L. A. 1981. A new model for the geometry of the binding of myosin crossbridges to muscle thin filaments. *J. Mol. Biol.* 147: 297–324

Tirosh, R., Oplatka, A. 1982. Active streaming against gravity in glass microcapillaries of solution containing actoheavy meromyosin and native tropomyosin. *J. Biochem. Tokyo* 91: 1435–40

Toffenetti, J., Mischke, D., Pardue, M. L. 1987. Isolation and characterization of the gene for myosin light chain two of *Drosophila melanogaster*. *J. Cell. Biol.* 104: 19–28

Tong, S. W., Elzinga, M. 1983. The sequence of the NH_2-terminal 204-residue fragment of the heavy chain of rabbit skeletal mus-

cle myosin. *J. Biol. Chem.* 258: 13100–10

Toyoshima, C., Wakabayashi, T. 1985. Three-dimensional analysis of the complex of thin filaments and myosin molecules from skeletal muscle. V. Assignment of actin in the actin-tropomyosin-myosin subfragment-1 complex. *J. Biochem.* 97: 245–63

Toyoshima, Y. Y., Kron, S. J., McNally, E. M., Niebling, K. R., Toyoshima, C., Spudich, J. A. 1987. Myosin subfragment-1 is sufficient to move actin filaments in vitro. *Nature* 328: 536–39

Trotter, J. A. 1982. Living macrophages phosphorylate the 20,000 dalton light chains and heavy chains of myosin. *Biochem. Biophys. Res. Commun.* 106: 1071–77

Trybus, K. M., Huaitt, T. D., Lowey, S. 1982. A bent monomeric conformation of myosin from smooth muscle. *Proc. Natl. Acad. Sci. USA* 79: 6151–55

Ueno, H., Harrington, W. F. 1986a. Temperature-dependence of local melting in the myosin subfragment-2 region of the rigor cross-bridge. *J. Mol. Biol.* 190: 59–68

Ueno, H., Harrington, W. F. 1986b. Local melting in the subfragment-2 region of myosin in activated muscle and its correlation with contractile force. *J. Mol. Biol.* 190: 69–82

Umeda, P. K., Kavinsky, C. J., Sinha, A. M., Hsu, H.-J., Jakovcic, S., Rabinowitz, M. 1983. Cloned mRNA sequences for two types of embryonic myosin heavy chains from chick skeletal muscle. *J. Biol. Chem.* 258: 5206–14

Umekawa, H., Naka, M., Inagaki, M., Onishi, H., Wakabayashi, T., Hidaka, H. 1985. Conformational studies of myosin phosphorylated by protein kinase C. *J. Biol. Chem.* 260: 9833–37

Vale, R. D., Reese, T. S., Sheetz, M. P. 1985a. Identification of a novel force generating protein, kinesin, involved in microtubule-based motility. *Cell* 42: 39–50

Vale, R. D., Schnapp, B. J., Reese, T. S., Sheetz, M. P. 1985b. Movement of organelles along filaments dissociated from the axoplasm of the squid giant axon. *Cell* 40: 449–54

Vale, R. D., Schnapp, B. J., Reese, T. S., Sheetz, M. P. 1985c. Organelle, bead, and microtubule translocations promoted by soluble factors from the squid giant axon. *Cell* 40: 559–69

Vandekerckhove, J., Weber, K. 1978. Actin amino acid sequences. Comparison of actins from calf thymus, bovine brain and SV40 transformed mouse 3T3 cells with rabbit skeletal muscle actin. *Eur. J. Biochem.* 90: 451–46

Wagner, P. D., Giniger, E. 1981. Hydrolysis

of ATP and reversible binding to F-actin by myosin heavy chains free of all light chains. *Nature* 292: 560–62

Walker, J. E., Saraste, M., Runswick, M. J., Gay, N. J. 1982. Distantly related sequences in the α- and β-subunits of ATP synthase, myosin, kinases and other ATP-requiring enzymes and a common nucleotide binding fold. *EMBO J.* 1: 945–51

Warrick, H. M., De Lozanne, A., Leinwand, L. A., Spudich, J. A. 1986. Conserved protein domains in a myosin heavy chain from *Dictyostelium discoideum*. *Proc. Natl. Acad. Sci. USA* 83: 9433–37

Waterston, R. H., Thomson, J. N., Brenner, S. 1980. Mutants with altered muscle structure in *Caenorhabditis elegans*. *Dev. Biol.* 77: 271–302

Watts, F. Z., Miller, D. M., Orr, E. 1985. Identification of myosin heavy chain in *Saccharomyces cerevisiae*. *Nature* 316: 83–85

Wells, J. A., Knoeber, C., Sheldon, M. C., Werber, M. M., Yount, R. G. 1980a. Cross-linking of myosin subfragment 1. Nucleotide-enhanced modification by a variety of bifunctional reagents. *J. Biol. Chem.* 255: 11135–40

Wells, J. A., Sheldon, M., Yount, R. G. 1980b. Magnesium nucleotide is stoichiometrically trapped at the active site of myosin and its active proteolytic fragments by thiol cross-linking reagents. *J. Biol. Chem.* 255: 1598–1602

Wells, J. A., Yount, R. G. 1979. Active site trapping of nucleotides by crosslinking two sulfhydryls in myosin subfragment 1. *Proc. Natl. Acad. Sci. USA* 76: 4966–70

Wells, J. A., Yount, R. G. 1980. Reaction of 5,5′-dithiobis(2-nitrobenzoic acid) with myosin subfragment one: Evidence for formation of a single protein disulfide with trapping of metal nucleotide at the active site. *Biochemistry* 19: 1711–17

Wells, J. A., Yount, R. G. 1982. Chemical modification of myosin by active-site trapping of metal-nucleotides with thiol crosslinking reagents. *Methods Enzymol.* 85: 93–115

Weydert, A., Daubas, P., Caravatti, M., Minty, A., Bugaisky, G., et al. 1983. Sequential accumulation of mRNAs encoding different myosin heavy chain isoforms during skeletal muscle development in vivo detected with a recombinant plasmid identified as coding for an adult fast myosin heavy chain from mouse skeletal muscle. *J. Biol. Chem.* 258: 13867–74

Weydert, A., Daubas, P., Lazaridis, I., Barton, P., Garner, I., et al. 1985. Genes for skeletal muscle myosin heavy chains are clustered and are not located on the same mouse chromosome as a cardiac myosin heavy chain gene. *Proc. Natl. Acad. Sci. USA* 82: 7183–87

Wierenga, R. K., Terpstra, P., Hol, W. G. J. 1986. Prediction of the occurrence of the ADP-binding β-α-β-fold in proteins, using an amino acid sequence fingerprint. *J. Mol. Biol.* 187: 101–7

Wills, N., Gesteland, R. F., Karn, J., Barnett, L., Bolten, S., Waterston, R. H. 1983. The genes *sup-7 X* and *sup-5 III* of *C. elegans* suppress amber nonsense mutations via altered transfer RNA. *Cell* 33: 575–83

Winkelmann, D. A., Lowey, S., Press, J. L. 1983. Monoclonal antibodies localize changes on myosin heavy chain isozymes during avian myogenesis. *Cell* 34: 295–306

Winkelmann, D. A., Mekeel, H., Rayment, I. 1985. Packing analysis of crystalline myosin subfragment-1: Implications for the size and shape of the myosin head. *J. Mol. Biol.* 181: 487–501

Yamamoto, K., Sekine, T. 1979. Interaction of myosin subfragment-1 with actin. II. Location of the actin binding site in a fragment of subfragment-1 heavy chain. *J. Biochem.* 86: 1863–68

Yamamoto, K., Tokunaga, M., Sutoh, K., Wakabayashi, T., Sekine, T. 1985. Location of the SH group of the alkali light chain on the myosin head revealed by electron microscopy. *J. Mol. Biol.* 183: 287–90

Yanagida, T., Nakase, M., Nishiyama, K., Oosawa, F. 1984. Direct observation of motion of single F-actin filaments in the presence of myosin. *Nature* 307: 58–60

Yano, M. 1978. Observation of steady streamings in a solution of Mg-ATP and acto-heavy meromyosin from rabbit skeletal muscle. *J. Biochem.* 83: 1203–4

Yano, M., Yamamoto, Y., Shimizu, H. 1982. An actomyosin motor. *Nature* 299: 557–59

Yumura, S., Fukui, Y. 1985. Reversible cyclic AMP-dependent change in distribution of myosin thick filaments in *Dictyostelium*. *Nature* 314: 194–96

Zengel, J. M., Epstein, H. F. 1980. Identification of genetic elements associated with muscle structure of the nematode *Caenorhabditis elegans*. *Cell Motil.* 1: 73–97

Ann. Rev. Cell Biol. 1987. 3: 423–41

GROWTH AND DIFFERENTIATION IN THE HEMOPOIETIC SYSTEM

T. M. Dexter and E. Spooncer

Paterson Institute for Cancer Research, Christie Hospital and Holt Radium Institute, Withington, Manchester M20 9BX, United Kingdom

CONTENTS

INTRODUCTION 423
THE STRUCTURE OF THE HEMOPOIETIC SYSTEM 424
    Hemopoietic Stem Cells 424
    Committed Cells 426
GROWTH FACTORS 427
    Inhibitory Factors 430
    Developmental Factors and the Nature of Commitment 430
    Role of Stromal Cells in Hemopoiesis 432
    Effect of Hemopoietic Growth Factors in Vivo 434
    Role of Growth Factors in Leukemogenesis 435
    Mode of Action of Growth Factors 436
SUMMARY 438

INTRODUCTION

As most other rapidly regenerating systems, such as the intestinal epithelium and the skin epidermis, mature blood cells must be produced continuously throughout life. Once the mature cells are produced, the vast majority are destined to remain functionally active for only a few hours or weeks before being destroyed and broken down (Cronkite & Feinendegen 1976). Indeed, simply to supply the normal steady-state demands, an average human must produce something on the order of 3.7×10^{11} cells a day to replace those cells lost due to natural wastage. This review discusses the various mechanisms controlling growth and development in the hemopoietic system.

423

0743–4634/87/1115–0423$02.00

THE STRUCTURE OF THE HEMOPOIETIC SYSTEM

Functionally mature blood cells are highly specialized: Erythrocytes are specialized for oxygen and carbon dioxide transport; T and B lymphocytes for cell- and antibody-mediated immune responses, respectively; platelets for blood clotting; and the granulocytes and macrophages as general scavengers and accessory cells in the response against invading organisms and their by-products. The granulocytes are further subdivided into various specialized cell types with discrete functions: the neutrophils, eosinophils, basophils, and mast cells (Hardisty & Weatherall 1974) (Figure 1). Clearly, in the hemopoietic system there is an incredible diversity of function. Perhaps more remarkable is that all of these different cell types are continuously being derived from a common set of cells: the pluripotential stem cells. These stem cells arise during early embryogenesis and are first found in the yolk sac. From the yolk sac they migrate to the fetal liver and from there to the bone marrow (Metcalf & Moore 1971). The bone marrow is then the major site of hemopoiesis (or more correctly myelopoiesis, since T-lymphocyte development, for example, occurs in the thymus) in the adult.

Hemopoietic Stem Cells

The stem cells are relatively few in number but can persist throughout life by undergoing proliferation to produce daughter stem cells (Figure 2).

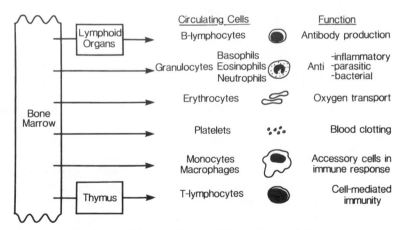

Figure 1   Production of functional mature blood cells from the bone marrow.

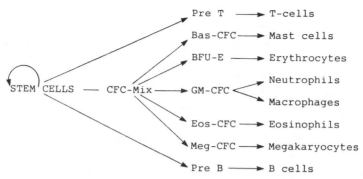

Figure 2   The structure of the hemopoietic system. Proliferation and development require stromal cells or growth factors.

This process of self-renewal is the distinguishing feature of hemopoietic stem cells and is a characteristic shared with stem cells in the other regenerating systems (Potten 1983). Unfortunately, there is no easy assay for stem cells since, by definition, only retrospective analyses can be performed. In other words, the existence of stem cells can only be inferred by the progeny they produce. However, the following lines of evidence indicate that stem cells are discrete biological entities. First, following cytotoxic ablation of an animal's hemopoietic system by radiation or chemicals, normal hemopoiesis can be reestablished if the animal is infused with normal hemopoietic cells from bone marrow, for example (Schofield 1979). This, of course, provides the basis for the clinical use of marrow transplantation in man. Extensive analyses of these transplant chimeras in mice have shown that hemopoiesis is reestablished normally from the donor cells and that relatively few cells are necessary to fully reconstitute the hemopoietic system (Boggs et al 1982). For example, infusion of as few as 10^4–10^5 marrow cells into an irradiated mouse is sufficient to restore hemopoiesis. Since a normal mouse has $\sim 3 \times 10^8$ marrow cells, the infused amount represents only 0.01% of this total. Furthermore, these transplant recipients have a normal life span, and their hemopoietic system shows no evidence of decline with age (Schofield et al 1986). Thus there is a vast proliferative reserve built into the system that would be sufficient to last several times the normal life span.

What evidence is there that the mature cells in these radiation chimeras are derived from stem cells? Abramson et al (1979) approached this problem in transplantation chimeras using unique radiation-induced chromosome markers and showed conclusively that precursor cells from several lineages (myeloid and lymphoid) carried the same karyotype marker. More recently, Lemischka et al (1986) confirmed and extended this finding using retroviruses as insertion markers to follow the progeny of single stem cells. Thus pluripotent stem cells obviously exist: The problem is how to recognize and quantitate them. In this respect, the assay developed by Till & McCulloch (1961) has been of particular significance. These authors showed that when hemopoietic cells are injected into potentially lethally irradiated mice, some of the cells migrate into the spleen, where they lodge and proliferate to form discrete nodules. Subsequent work showed that these nodules were derived from a single cell (i.e. were clonal), contained multiple cell lineages (i.e. were derived from a multipotent cell), and that at least some of the nodules contained primitive cells that could form more spleen colonies when infused into other irradiated mice (i.e. some of the original spleen colony–forming cells could undergo self-renewal). In many respects, therefore, the spleen colony–forming cells (or CFU-S) possess the characteristics of "stem" cells. However, more recent work has shown that the CFU-S are heterogenous; they include not only multipotent stem cells but also more mature cells that have lost their capacity to fully reestablish hemopoiesis, although they often retain a multipotential nature (Magli et al 1982) (Figure 2). This then raises the question of the nature of "commitment": How do stem cells become progressively restricted in their capacity to undergo self-renewal or to produce progeny of several cell lineages?

Committed Cells

Like the stem cells, the committed progenitor cells are distinguished by the progeny they produce. This distinction is usually [but not always, e.g. erythropoietin (Filmanovicz & Gurney 1961)] made on the basis of in vitro analysis in clonogenic soft-gel culture systems (Metcalf 1977). For example, certain cells undergo proliferation and development in vitro to produce colonies containing all the different mature myeloid cells: erythrocytes, neutrophils, megakaryocytes, macrophages, basophils, and erythroid cells (Johnson 1984). Such colonies can be obtained from all mammalian species so far examined (mouse, rat, cat, dog, sheep, cows), including man. Since the colonies are clonal, the mature progeny must be derived from a multipotent cell, called colony–forming cell (mixed) or CFC-Mix. Some of these colonies contain cells that can be replated in soft gels to produce more mixed colonies, which means that some of the CFC-Mix can undergo renewal. In fact, some of the colonies derived from CFC-Mix may contain CFU-S, which would indicate a considerable overlap between these two populations (Metcalf et al 1979). Not all colonies undergo self-renewal; this may simply reflect suboptimal culture conditions but more likely represents true heterogeneity in self-renewal capacity. Thus the multipotential "stem" cell system may best be viewed as one in which the capacity (or probability) of undergoing renewal is progressively lost as the cells become lineage restricted. Indeed, the maintenance or reacquisition of an extended capacity for renewal in the committed cells is representative of a leukemic rather than of a normal cell phenotype.

Lineage-restricted progenitor cells also are classified on the basis of the progeny they produce in clonogenic soft-gel systems. Granulocyte/macrophage colony–forming cells (GM-CFC) undergo proliferation and development to produce neutrophils and/or macrophages, depending upon the growth stimulus used (see below). The most primitive erythroid progenitor cells undergo proliferation, and the cells thus produced migrate through the soft-gel system. These immediate progeny cells then undergo further proliferation and begin to synthesize hemoglobin, thereby generating typical erythroid "bursts". Because of this, the primitive clonogenic cell is often called an erythroid burst–forming unit (BFU-E), and their immediate progeny are called the erythroid colony–forming units (CFU-E). Megakaryocyte, eosinophil, and basophil colony–forming cells (Meg-CFC, Eos-CFC, Bas-CFC, respectively) can also be recognized by their ability to undergo proliferation and development in vitro to produce mature progeny (Metcalf 1977). Of course, all of these lineage-restricted progenitor cells are continuously generated from the multipotential stem cells. In this way, a constant supply of mature cells is ensured, and a balance is maintained throughout the different cell compartments.

GROWTH FACTORS

The development of isolated multipotent stem cells and lineage-restricted progenitor cells in vitro only occurs in the presence of growth factors. In the absence of these growth factors the cells die; in the presence of growth factors they survive, proliferate, differentiate, and mature. Significantly, the growth factors are required throughout development; if they are removed the cells rapidly die, irrespective of their stage of development (Metcalf 1977). Thus the growth factors subserve several functions (survival, proliferation, and development), which are perhaps not mutually exclusive processes.

Many of the growth factors required by hemopoietic cells have now been molecularly cloned and purified to homogeneity. In many cases, both animal and human recombinant material is available for study (Table 1). Interleukin 3 (IL-3) is a multilineage stimulating factor that facilitates the growth and development of multipotential stem cells as well as of progenitor cells restricted to the granulocyte/macrophage, eosinophil, megakaryocyte, erythroid, or mast cell lineages.

Table 1 Growth factors in myelopoiesis

Growth factor[a]	Synonyms	Species	CFU-S	CFC-Mix	GM-CFC	BFU-E	Eos-CFC	Meg-CFC	Mast cells	CFU-E	Mature myeloid cells
IL-3[b]	Multi-CFS, HCGF, BPA	Mouse	+	+	+	+	+	+	+	±	+
		Human, Gibbon	insufficient data								
GM-CSF[c]	MGI-GM, Pluripoietin α	Mouse, Human	-	+	+	+	+	+	+	-	+
G-CSF[d]	MGI-G, Pluripoietin	Mouse	-	-	+(G)	-	-	-	-	-	+
		Human	n/a	(+)	+(G)	(+)	(+)	(+)	-	-	+
M-CSF[e]	MGI-M, CSF-1	Mouse, Human	-	-	+(M)	-	-	-	-	-	+
EOS-DF[f]	IL-4	Mouse	-	-	-	-	+	-	-	-	+
Erythropoietin[g]	—	Human, Monkey, Mouse	n/a	-	-	+	-	-	-	+	-

[a] These growth factors have all been molecularly cloned, from at least one species. Other factors have been described (see text) but have not yet been cloned. [− = no reported effect; + = direct stimulation; (+) = indirect stimulation; G = granulocyte; M = macrophage; n/a = not applicable.] General references: Burgess & Nicola 1983; Metcalf 1985, 1986; Whetton & Dexter 1986.
[b] Hapel et al 1985; Kindler et al 1986; Lord et al 1986; Metcalf et al 1986; Yang et al 1986.
[c] Gabrilove et al 1986; Gough et al 1984; Nicola et al 1983; Sachs 1982.
[d] Nicola et al 1985; Sachs 1982; Souza et al 1986.
[e] Das & Stanley 1982; Kawasaki et al 1985; Sachs 1982; Stanley & Guilbert 1981.
[f] O'Garra et al 1986.
[g] Jacobs et al 1985.

Granulocyte/macrophage colony–stimulating factor (GM-CSF) was initially thought to be restricted in its action to GM-CFC. However, the use of recombinant material has demonstrated that GM-CSF can also influence proliferation and development of eosinophil, megakaryocyte, and (primitive) erythroid progenitor cells. However, high concentrations are required relative to that needed to affect GM-CFC. At least some multipotential stem cells also respond to GM-CSF. Thus IL-3 and GM-CSF have overlapping biological activities, but IL-3 is probably a more powerful stimulator of stem cell growth than is GM-CSF.

Granulocyte colony–stimulating factor (G-CSF) is a somewhat unusual molecule. When whole human bone marrow cells are cultured in G-CSF, a range of colonies typical of those produced by IL-3 or GM-CSF are seen (Souza et al 1986). However, following removal of T cells and accessory cells (T lymphocytes, macrophages, etc) from the bone marrow, only colonies containing mature neutrophils are produced (Metcalf et al 1985). This indicates that G-CSF has a *direct* colony-stimulating effect only on the GM-CFC and that the range of colonies seen upon culture of unfractionated marrow cells results from stimulation of the production by accessory cells of other factors. These factors, and their relationships to established molecules such as GM-CSF, have not yet been characterized.

Macrophage colony–stimulating factor (M-CSF) also appears to be a lineage-restricted growth factor that acts upon GM-CFC to induce proliferation and development into macrophages (Das & Stanley 1982; Stanley & Guilbert 1981). In such colonies, however, neutrophils have also occasionally been observed (Metcalf & Burgess 1982). M-CSF also stimulates early macrophage precursor cells and mature marrow macro-phages to undergo proliferation and enhances the survival of such cells in vitro (Stanley & Guilbert 1981).

A variety of other factors responsible for growth and development of eosinophils and megakaryocytes have also been described. However, with the exception of the eosinophil differentiation factor (or IL-4) (O'Garra et al 1986), these molecules and their target cells are still fairly undefined and are not discussed here. Yet some mention should be made of erythro-poietin, which acts in classical hormone fashion to promote proliferation and hemoglobinization of fairly mature erythroid progenitor cells (CFU-E). Production of this hormone occurs in the kidney and is regulated by the number of mature erythrocytes in the circulating blood and their oxygen carrying capacity (Adamson et al 1978), i.e. via a classical feedback loop control system. This is the *only* hemopoietic cell growth factor for which a system controlling production has been demonstrated.

There is still confusion as to the precise roles of IL-3, M-CSF, GM-CSF, and G-CSF in vivo in marrow myelopoiesis, and there is not enough information about their producer cells, and the systems controlling their production. One reason for this may be that these factors not only stimulate the growth and development of stem cells and their progeny but are also important in maintaining or activating the function of the mature granulocytes and macrophages (Metcalf 1985). They probably play a general systemic role in modulating the overall response of the body to infective organisms (helminths, protozoa, and bacteria) and in stimulating accessory cells in the immune response. Thus high local or circulating levels of these growth factors may represent the outcome of a peripheral immune response rather than having a bearing on marrow myelopoiesis.

Inhibitory Factors

Factors that stimulate growth are more amenable to study than factors that inhibit growth. The latter have proven difficult to characterize biochemically, and their cellular specificity has often been questioned. However, one such factor involved in control of the hemopoietic system has been found in medium conditioned by marrow stromal cells. This factor specifically and reversibly inhibits the entry of multipotential stem cells into S phase (DNA synthesis). Lord and coworkers (1976) first identified this factor on the basis of its ability to decrease the proportion of CFU-S engaged in DNA synthesis. Subsequently, it was shown to be able to prevent the entry into the S phase of CFU-S stimulated by cell-cycle activators (Lord et al 1977). The potential importance of this presently ill-defined factor is that its effects are reversible, dose dependent, and nontoxic. Furthermore, its inhibitory effects are apparently restricted to multipotential stem cells: The factor has no cell-cycle inhibitory effects on the lineage-restricted progenitor cells. The possible relevance of this molecule in the process of leukemogenesis is discussed below. The reader should bear in mind that although the major research emphasis at present is on growth stimulating factors, it is likely that numerous and equally important inhibitory molecules exist.

Developmental Factors and the Nature of Commitment

Clearly hemopoiesis in vitro requires the presence of growth factors that are progressively restricted in their biological activities and target cells. However, one of the fundamental questions that remain to be answered concerns the process by which cells acquire the capacity to respond to growth factors and the nature of the "commitment" implicit in their response. For example, multipotent stem cells can respond to IL-3 but not, as far as is known to M-CSF or to G-CSF. Presumably this means that stem cells have receptors that bind and respond to IL-3 but not to M-CSF. It is possible, then, that "commitment" involves transcriptional activation of the gene(s) coding for the M-CSF receptor, or does control operate at the translational/posttranslational level? An important clue to the answer to this question has recently emerged from a study of a molecule called Hemopoietin 1 or synergistic activity.

Several years ago, Bradley & Hodgson (1979) reported that conditioned media from certain tissues contained a factor that acted synergistically with other known growth factors, i.e. recruited more colony-forming cells. Subsequently, this molecule (Hemopoietin-1 or H-1) was purified from a conditioned medium of a human bladder carcinoma cell line and shown to be distinct from the other growth factors. By itself H-1 has no growth stimulating effect. However, it can enhance the response of macrophage progenitor cells to M-CSF and facilitate the growth and development of primitive marrow cells in the presence of M-CSF or IL-3 (Bartelmez & Stanley 1985; Stanley et al 1986).

In my laboratory we have confirmed and extended these findings. We have used marrow cells that have been highly enriched for CFU-S using fluorescence-activated cell-sorting procedures (Lord & Spooncer 1986). These separated cells contain between 50 and 100% CFU-S and thus represent a fairly homogenous population of multipotent stem cells essentially devoid of other cell types that might influence the assays. When these cells are plated in vitro in IL-3, 4–40% produce multilineage colonies, i.e. they have receptors for, and can respond to, IL-3. When stem cells are cultured in the presence of H-1 they die, which confirms other evidence indicating that H-1 is not a growth factor per se. However, when IL-3 and H-1 are combined, 50–100% more colonies are produced than in the presence of IL-3 alone. In other words, the H-1 recruits more stem cells to an IL-3 responsive state.

The interaction of M-CSF and H-1 is even more interesting. With M-CSF alone, no colonies are produced and the stem cells die; in the presence of M-CSF and H-1 as many colonies develop as in the presence of IL-3 and H-1. Thus H-1 facilitates the development of multipotent stem cells to a stage at which they can respond to M-CSF, i.e. H-1 commits the cells to an M-CSF-responsive state. Moreover, both M-CSF and H-1 must be present *at the same time* for this commitment to occur: The absence of either factor leads to rapid death of the cells. This is puzzling since if H-1 simply induces expression of M-CSF receptors upon the stem cell surface, then preincubation with H-1 should make the cells responsive to M-CSF. The fact that this does not occur indicates that receptor expression per se is not sufficient to maintain a response to M-CSF. Therefore either H-1 acts at the level of gene transcription (i.e. is necessary for the continued transcription of the M-CSF receptor until a certain stage of development is reached) or H-1 plays a posttranslational role in activation of the M-

CSF receptor. Clearly, this point needs to be resolved. Nevertheless, our recent finding (unpublished data) that H-1 also facilitates the ability of multipotent stem cells to respond to GM-CSF indicates an important role for this molecule in the regulation of stem cell differentiation.

Role of Stromal Cells in Hemopoiesis

So far, only growth factors acting on separated hemopoietic cells in vitro have been discussed as putative regulatory elements. But hemopoiesis in vivo is obviously more complex since the cells are in intimate association with a complex stromal cell network. Several lines of evidence suggest that these stromal cells themselves play a major role in hemopoiesis (reviewed in Allen & Dexter 1983; Dexter 1982; Dexter et al 1984) and are not simply acting as mediators for externally generated influences.

The most conclusive data come from work with long-term marrow cultures (Dexter et al 1977, 1984), which provide an in vitro system in which hemopoiesis can be maintained for several months. (In contrast, in the soft-gel system hemopoiesis is sustained for only a few days.) The important characteristic of long-term cultures is the requirement for marrow stromal cells: Stromal cells from spleen, liver, or other tissues do not support hemopoiesis in vitro. A second feature of these cultures is that hemopoiesis is sustained in the absence of added growth factors. Since it is known that isolated hemopoietic cells (grown in soft-gel clonogenic systems) die in the absence of growth factors in vitro, this presumably means that the marrow stromal cells provide the necessary extracellular matrix and growth factors essential for establishment of hemopoiesis. Furthermore, the production of these regulatory molecules is obviously stringently controlled since homeostasis in the long-term cultures is maintained for many months, i.e. a balance is maintained between stem cell self-renewal/differentiation and growth and development of the committed progenitor cells. In most respects, therefore, hemopoiesis in long-term cultures can be compared with hemopoiesis in vivo.

Such long-term cultures have been used to analyze the characteristics of the stromal cells (which comprise a heterogenous collection of various cell types) and the nature of their interactions with hemopoietic cells (Dexter et al 1977, 1984). Several points have emerged. Firstly, the hemopoietic cells must be in direct contact with the stromal cells to facilitate survival, growth, and development: If they are prevented from attaching to the stroma, they die. Secondly, discrete stromal elements show characteristic interactions with the different maturing myeloid cells and with each other. This suggests that different environmental "niches" exist that specifically modulate lineage development. Thirdly, stromal cells from drug-treated animals or congenitally anemic mice may be compromised in their ability

to support hemopoiesis, which indicates that some myeloproliferative disorders may represent a defect in the stromal cell environment rather than
the hemopoietic target cells. The nature of the regulatory interactions
between stroma and hemopoietic cells should be carefully examined with
these points in mind.

GROWTH FACTORS AND OTHER STIMULI PRODUCED FROM MARROW STROMAL
CELLS   The work described above indicates that the marrow stromal cells
produce growth factors. If they do, the factors they produce are not
normally released as diffusible regulatory molecules since (*a*) direct contact
of the hemopoietic cells with the stroma is required to stimulate hemopoiesis and (*b*) media conditioned by the growth of marrow stromal cells
contain little or no (Dexter et al 1977; Quesenberry et al 1984) colony-
stimulating factors, as assessed by the ability of such media to promote
growth of progenitor cells. This further suggests that these growth factors,
presumably produced by stromal cells, represent surface-bound molecules
that exert their effects through direct cell-cell contact. However, the nature
of these growth stimuli remains an enigma. Recent work using molecular
probes has failed to detect mRNA for IL-3, GM-CSF, or G-CSF in
marrow stromal cells in hemopoietically active cultures (T. M. Dexter,
unpublished observations). Thus either the cognate growth factors are not
synthesized by stromal cells (and as a corollary they are not *essential* for
stromal cell–associated hemopoiesis) or they are produced at levels below
the detection limits of our present biological and molecular techniques.
The latter possibility is perhaps likely, since receptors for hemopoietic cell
growth factors on target cells are few in number (100–1000) and relatively
few receptors (5–10%) have to bind a growth factor to elicit a maximal
response (Nicola & Metcalf 1985; Park et al 1986). Obviously, future work
needs to clarify this point. It may be significant that one of the factors
produced and secreted in appreciable amounts by marrow stromal cells is
Hemopoietin-1, which (as described earlier) seems to modulate the ability
of hemopoietic cells to respond to growth factors in situations where the
latter alone initiate no response. However, we should also consider the
possibility that the stromal cells produce (and express on their surface) a
range of as yet uncharacterized growth factors and that IL-3, GM-CSF,
and G-CSF may only be produced by marrow stroma under exceptional
conditions of hemopoietic stress.

The production of cell-cycle modulators by stromal cells has also been
examined; the CFU-S cell-cycle stimulatory and inhibitory molecules
described by Lord and coworkers (Lord & Wright 1980) are included
among such modulators. These factors are produced in long-term cultures,
and their production is directly correlated with CFU-S numbers (Toksoz

et al 1980). For example, when stem cells are depleted from long-term cultures (through mechanical means or drug treatment), relatively greater amounts of stimulatory factors are produced. As CFU-S levels return to normal values, production of stimulatory molecules decreases and that of inhibitory molecules increases. It seems then that stem cell cycling is regulated by the production of these opposing activities and that their production is in turn regulated by the number of stem cells (Lord 1986). Unlike the stromal cell–associated growth factors, the CFU-S cell-cycle modulators are fairly well characterized molecules known to be produced by macrophages.

Effect of Hemopoietic Growth Factors in Vivo

From the previous discussion it is clear that a variety of growth and developmental factors are necessary for stimulating the proliferation and development of hemopoietic stem and progenitor cells removed from their normal stromal cell environment. But in vitro hemopoiesis can also be established in the absence of added growth factors provided that marrow stromal cells are present. As mentioned above, it is difficult to detect either biologically active protein or mRNA for many growth factors in the marrow stromal cells themselves, so it may be assumed that some of the growth factors described (IL-3, GM-CSF, G-CSF) represent in vitro artifacts rather than in vivo biological response inducers or modifiers. However, this does not appear to be the case, since molecularly cloned factors have been shown to be highly active in vivo.

A single injection of a few nanograms of IL-3 into a mouse, for example, can cause rapid recruitment of quiescent stem cells (CFU-S) into the cell cycle (Lord et al 1986). Continued administration of IL-3 leads to a major disturbance in the kinetic behavior of stem cells and progenitor cells in the bone marrow. It also causes the mouse spleen (normally a lymphoid organ) to increase in size and cellularity, become actively myelopoietic, and show a dramatic increase in the numbers of CFU-S and committed progenitor cells (Kindler et al 1986, Metcalf et al 1986). A similar result has been obtained by administering GM-CSF to mice (D. Metcalf, personal communication). The response to G-CSF is even more profound. When administered to mice, this growth factor induces a 10- to 20-fold increase in circulating blood leucocyte levels and can increase the total number of multipotent stem cells (M. A. S. Moore, unpublished observations). Whether or not these effects are mediated directly or indirectly by G-CSF has yet to be determined since (as discussed previously) this growth factor appears to facilitate the production of other growth factors by accessory cells present in the marrow. Nonetheless, G-CSF can obviously play a major role in modulating hemopoiesis.

Studies in primates treated with GM-CSF have reported effects similar to those seen in mice. In normal macaque monkeys, for example, administration of GM-CSF leads to a tenfold rise in circulating leucocyte levels (Donahue et al 1986). Perhaps more interesting, however, were the effects seen on a (single) pancytopenic, immunodeficient monkey that had a simian virus–induced aplasia with no circulating leucocytes detectable. Following administration of GM-CSF, circulating white blood cell levels rose to supranormal values and subsequently fell to undetectable levels upon cessation of treatment (Donahue et al 1986). These data indicate that even the hemopoietic system of animals whose blood cell production has been compromised by viral infection can still mount a response to the growth factor. Such findings have important clinical implications.

Role of Growth Factors in Leukemogenesis

Recent data show that some oncogenes code for parts of growth factors or their receptors; obvious examples are *sis* (B-chain of PDGF), *erb-B* (truncated EGF receptor), and *fms* (M-CSF receptor) (Waterfield et al 1983; Downward et al 1984; Sacca et al 1986). As might be expected, other hemopoietic cell growth factors may also be involved in transformation of hemopoietic cells. When recombinant retroviruses containing the coding sequence for either IL-3 or GM-CSF are used to infect previously immortalized (Lang et al 1985) but growth factor–dependent cells, the infected cells rapidly acquire the ability to grow in the absence of added growth factor. Furthermore, this change is accompanied by the acquisition of the ability to grow in vivo and produce a leukemia. In these cells, therefore, the growth factors are acting as transforming "oncogenes." However, it appears that prior immortalization of the target cells is a prerequisite for malignant transformation. When the same viruses were introduced into normal, freshly isolated, hemopoietic progenitor cells, these cells also acquired the ability to grow in the absence of added growth factor, but in this case the growth was accompanied by terminal maturation, i.e. the cells did not become leukemic (D. Metcalf, unpublished observation). Thus, as in other systems, at least two events are required for acquisition of a malignant phenotype (Land et al 1983; Newbold & Overell 1985; Ruley 1983). These exciting findings deserve further investigation.

Another interesting finding is that some myeloid leukemic cells respond to growth factors by undergoing differentiation and/or clonal extinction. Certain myeloid leukemia cell lines (e.g. murine WEHI-3b cells or M-1 cells, and human HL-60 cells) grow continually in vitro in the absence of added growth factors. When treated with G-CSF, these leukemic cells undergo differentiation into mature progeny (Gabrilove et al 1986; Metcalf 1983; Nicola et al 1985). In some cases, the self-renewal ability of the leukemic stem cell is extinguished, and the cells undergo clonal senescence. Similar effects are seen in many primary leukemias, but since these represent heterogenous cell populations the results are more diverse. The response, for example, may depend upon the presence of appropriate receptors on the cells [M1–M4 myeloid leukemias express the receptor for G-CSF, whereas M5 leukemias do not (M. A. S. Moore, unpublished observations)] and on whether the leukemic cells respond by growing or differentiating or both. In some instances the leukemic cells *require* the growth factor for their survival and proliferation, and treatment with growth factor may lead to enhanced rather than suppressed clonogenic growth (Griffin et al 1986). Therefore, although the growth factors may be useful adjuncts to conventional therapy, their precise therapeutic role still needs to be determined, probably on an individual patient basis.

Another "factor" worthy of comment in this section is the stem cell proliferation inhibitory molecule discussed earlier, which appears to be an important regulatory component of the stem cell system. What is its role in leukemogenesis? The basic mechanism underlying leukemogenesis is perturbation of the growth control system that operates in normal hemopoietic cells. Clearly, loss of growth control may occur through several different mechanisms, one of which may be unresponsiveness to the normal growth-inhibition system. Recent work (Eaves et al 1986) provides some insight into the nature of the defect leading to the hyperproliferative state seen in chronic myeloid leukemia (CML). These workers found that multipotential stem cells from CML patients do not respond to the growth inhibitory molecule produced in vitro in long-term marrow cultures. This finding suggests that CML stem cells have an advantage over normal stem cells and can progressively expand to take over the entire hemopoietic system.

Mode of Action of Growth Factors

A discussion of regulatory networks in hemopoiesis would not be complete without some mention of how growth factors exert their effects upon the target cells to facilitate survival, proliferation, and development. Of particular importance is to understand how the signal generated by the binding of a growth factor to its receptor is transduced, what second messengers this signal elicits, and what biochemical events are initiated in the process.

Progress in this area has been made possible by the development of continuously growing, growth factor–dependent cell lines (Dexter et al 1980), some of which are still capable of multilineage differentiation and of responding to marrow stromal cells (Spooncer et al 1986). The latter ability is of particular significance since the mechanisms underlying survival, growth, and development of multipotent cells can now be followed in vitro using "pure" populations of target cells, in response to "pure" (molecularly cloned) growth factors.

Most work has been done using cell lines dependent on IL-3 for their survival and growth. In the absence of IL-3, the cells die. In investigating this phenomenon, we found that death was associated with a decrease in glucose transport, leading to a critical fall in intracellular ATP levels (Whetton & Dexter 1983, Whetton et al 1984). Thus one of the fundamental processes for which IL-3 is required is the maintenance of the cells' primary metabolism at the level of the glucose transport protein. When IL-3 is removed, the survival of the cells is rapidly compromised. This type of response may well have evolved both in the hemopoietic and in other stem cell systems to protect the organism from metastatic growth of stem cells dislodged from their normal inductive environment.

We then examined the question of how the glucose transport protein is regulated. A clue to the answer came from the finding by Witters et al (1985) that the glucose transporter can be phosphorylated by protein kinase C (PKC). This raised the possibility that glucose transporter activity might be regulated by phosphorylation via activation of PKC. Support for this model is provided by the observation that translocation of PKC from the cytosol to the membrane of cells is one of the earliest events elicited by treatment of cells with IL-3 (Farrar et al 1985). Moreover, this model predicts that tumor-promoting agents (TPA), such as the phorbol diesters, should be able to mimic some of the effects of IL-3 since TPA are potent activators of PKC. This turned out to be the case. If IL-3-dependent cells are treated with TPA, cell viability and clonogenic potential can be maintained for several days in the absence of IL-3. Maintained viability is associated both with translocation (activation) of PKC and with increased glucose transport (Dexter et al 1986; Nishizuka 1984; Whetton et al 1986a). Furthermore, pretreatment of the growth factor–dependent stem cells with TPA down-modulates the number of TPA binding sites and subsequently decreases the ability of the cells to proliferate in response to IL-3 (A. D. Whetton, C. P. Downes, and T. M. Dexter, unpublished observations). Since PKC is the major cellular binding site for TPA (Nishizuka 1984), these data strongly suggest that a decrease in PKC compromises the ability of cells to respond to IL-3. A corollary of these data is that at least some of the metabolic events elicited by IL-3 are mediated via activation of PKC.

These results led us to examine the role of phosphatidylinositol breakdown in the response of cells to IL-3, since the activation of PKC by many other agents is elicited by the receptor-mediated generation of 1,2-diacylglycerol from phosphatidylinositol 4,5-bisphosphate (Berridge &

Irvine 1984). Significantly, we found no hydrolysis of inositol lipid inter-mediates or breakdown into inositol tris-, bis-, or monophosphate (A. D. Whetton, C. P. Downes, and T. M. Dexter, unpublished observations). It appears, therefore, that the activation of PKC occurs via a mechanism not involving inositol lipid breakdown. This effect is also seen with other hemopoietic cell growth factors in primary cell systems. For example, in the response of macrophages to M-CSF (Whetton et al 1986b) PKC is also activated, but there is no evidence that inositol lipid breakdown is involved in this response. Consequently, it is still unclear which second messenger system is involved. However, recent work in other cell systems (Farese et al 1985) suggests that the formation of diacylglycerol in the absence of inositol lipid breakdown is a not uncommon phenomenon.

SUMMARY

Hemopoiesis is regulated by a complex series of interactions, including interactions among hemopoietic cells themselves, hemopoietic cells and the extracellular matrix, hemopoietic cells and marrow stromal cells, and hemopoietic cells and growth factors. In vitro culture systems have allowed a reductionist approach to the solution of these various problems and have facilitated experiments at the mechanistic level. The hemopoietic system is organized hierarchically with multipotential self-renewing stem cells, committed progenitor cells, and mature cells. The various stimuli necessary for growth and development of these cells are rapidly being elucidated. The nature of commitment (or differentiation) remains an enigma, but model systems have been developed in which various aspects of this prob-lem can be investigated. In this respect, growth and differentiation factors obviously have a major role to play. Now that many of these factors have been molecularly cloned (and pure target cell populations are available) their role in vivo and their mode of action can be examined.

ACKNOWLEDGMENTS

T. M. Dexter is a Fellow of the Cancer Research Campaign.

Literature Cited

Abramson, S., Miller, R. G., Phillips, R. A. 1979. The identification in adult bone mar-row of plutipotent and restricted stem cells of the myeloid and lymphoid systems. *J. Exp. Med.* 145: 1567–79

Adamson, J. W., Popovic, W. M., Brown, J. E. 1978. Modulation of in vitro erythro-poiesis: Hormonal interactions and ery-throid colony growth. In *Differentiation*

of Hormonal and Neoplastic Hematopo-ietic Cells, ed. B. Clarkson, P. A. Marks, J. E. Till, pp. 235–48. Cold Spring Harbor, NY: Cold Spring Harbor Lab.

Allen, T. D., Dexter, T. M. 1983. Long-term bone marrow cultures: An ultrastructural review. *Scanning Electron Microsc.* 4: 1851–66

Bartelmez, S. H., Stanley, E. R. 1985. Synergism between hemopoietic growth factors (HGFs) detected by their effect on cells bearing receptors for a lineage specific HCG: Assay of hemopoietin-1. *J. Cell. Physiol.* 122: 370–78

Berridge, M. J., Irvine, R. F. 1984. Inositol triphosphate, a novel second messenger in cellular signal transduction. *Nature* 312: 315–21

Boggs, D. R., Boggs, S. S., Saxe, D. F., Gress, L. A., Carfield, D. R. 1982. Hematopoietic stem cells with high proliferation potential. Assay of their concentration in marrow by the frequency and duration of cure of W/W^v mice. *J. Clin. Invest.* 70: 243–47

Bradley, T. R., Hodgson, G. S. 1979. Detection of primitive macrophage progenitor cells in mouse bone marrow. *Blood* 54: 1446–50

Burgess, A. W., Nicola, N. A. 1983. *Growth Factors and Stem Cells.* London: Academic. 355 pp.

Cronkite, E. P., Feinendegen, L. E. 1976. Notions about human stem cells. *Blood Cells* 2: 263–84

Das, S. K., Stanley, E. R. 1982. Structure-function studies of a colony stimulating factor (CSF-1). *J. Biol. Chem.* 257: 13679–84

Dexter, T. M. 1982. Stromal cell associated haemopoiesis. *J. Cell. Physiol.* Suppl. 1: 87–94

Dexter, T. M., Allen, T. D., Lajtha, L. G. 1977. Conditions controlling the proliferation of haemopoietic stem cells in vitro. *J. Cell. Physiol.* 91: 335–44

Dexter, T. M., Garland, J., Scott, D., Scolnick, E., Metcalf, D. 1980. Growth of factor-dependent hemopoietic precursor cell lines. *J. Exp. Med.* 152: 1036–47

Dexter, T. M., Spooncer, E., Simmons, P., Allen, T. D. 1984. Long-term marrow cultures: An overview of techniques and experience. In *Long-Term Bone Marrow Culture*, ed. D. G. Wright, J. S. Greenberger, pp. 57–96. New York: Liss

Dexter, T. M., Whetton, A. D., Heyworth, C. M. 1986. The relevance of protein kinase C activation, glucose transport and ATP generation in the response of haemopoietic cells to growth factors. In *Oncogenes and Growth Control*, ed. P. Kahn, T. Graf, pp. 163–69. Berlin/Heidelberg: Springer-Verlag

Donahue, R. E., Wang, E. A., Stone, D. K., Kamen, R., Wong, G. G., et al. 1986. Stimulation of haematopoiesis in primates by continuous infusion of recombinant human GM-CSF. *Nature* 321: 872–75

Downward, J., Yarden, Y., Mayes, E., Scrace, G., Totty, N., et al. 1984. Close similarity of epidermal growth factor receptor and v-erb-B oncogene protein sequences. *Nature* 307: 521–27

Eaves, A. C., Cashman, J. D., Gaboury, L. A., Kalousek, D. K., Eaves, C. J. 1986. Unregulated proliferation of primitive chronic myeloid leukemia progenitors in the presence of normal marrow adherent cells. *Proc. Natl. Acad. Sci. USA* 83: 5306–10

Farese, R. V., Davis, J. S., Barnes, D. E., Standaert, M. L., Babischkin, J. S., et al. 1985. The de novo phospholipid effect of insulin is associated with increases in diacylglycerol, but not inositol phosphates or cytosolic Ca^{2+}. *Biochem. J.* 231: 269–78.

Farrar, W. L., Thomas, T. P., Anderson, W. B. 1985. Altered cytosol/membrane enzyme distribution on interleukin-3 activation of protein kinase. *Nature* 315: 235–37

Filmanovicz, E., Gurney, C. W. 1961. Studies on erythropoiesis. XVI. Response to a single dose of erythropoietin in polycythemic mouse. *J. Lab. Clin. Med.* 57: 65–72

Gabrilove, J. L., Welte, K., Harris, P., Platzer, E., Lu, L., et al. 1986. Pluripoietin alpha: A second hematopoietic colony stimulating factor produced by the human bladder carcinoma cell line 5637. *Proc. Natl. Acad. Sci. USA* 83: 2478–82

Gough, N. M., Gough, J., Metcalf, D., Kelso, A., Grail, A., et al 1984. Molecular cloning of cDNA encoding a murine haemopoietic growth regulator granulocyte/macrophage colony stimulating factor. *Nature* 309: 763–67

Griffin, J. D., Young, D., Herrmann, F., Wiper, D., Wagner, K., Sabbath, K. D. 1986. Effects of recombinant human GM-CSF on proliferation of clonogenic cells in acute myeloblastic leukemia. *Blood* 67: 1448–53

Hapel, A. J., Fung, M. C., Johnson, R. M., Young, I. G., Johnson, G., Metcalf, D. 1985. Biologic properties of molecularly cloned and expressed murine interleukin-3. *Blood* 65: 1453–59

Hardisty, R. M., Weatherall, D. J., eds. 1974. *Blood and Its Disorders.* Oxford: Blackwell

Jacobs, K., Shoemaker, C., Rudersdorf, R., Neill, S. D., Kaufman, R. J., et al. 1985. Isolation and characterisation of genomic and cDNA clones of human erythropoietin. *Nature* 313: 806–10

Johnson, G. R. 1984. Haemopoietic multipotential stem cells in culture. *Clin. Haematol.* 13. 309–27

Kawasaki, E. S., Ladner, M. B., Wang, A. M., van Arsdell, J., Waffen, M. K., et al. 1985. Molecular cloning of a complementary DNA encoding human macro-phage-specific colony-stimulating factor (CSF-1). *Science* 230: 291–96

Kindler, V., Thorens, B., de Kossodo, S., Allet, B., Eliason, J. F., 1986. Stimulation of hematopoiesis *in vivo* by recombinant bacterial murine interleukin 3. *Proc. Natl. Acad. Sci. USA* 83: 1001–5

Land, H., Parada, L. F., Weinberg, R. A. 1983. Tumorigenic conversion of primary embryo fibroblasts requires at least two co-operating oncogenes. *Nature* 304: 596–602

Lang, R. A., Metcalf, D., Gough, N. M., Dunn, A. R., Gonda, T. J. 1985. Expression of a hemopoietic growth factor cDNA in a factor dependent cell line results in autonomous growth and tumorigenicity. *Cell* 43: 531–42

Lemischka, I. R., Raulet, D. H., Mulligan, R. C. 1986. Developmental potential and dynamic behaviour of hematopoietic stem cells. *Cell* 45: 917–27

Lord, B. I. 1986. Interactions of regulatory factors in the control of haemopoietic stem cell proliferation. In *Biological Regulation of Cell Proliferation*, ed. R. Baserga, P. Foa, D. Metcalf, E. E. Polli, Sereno Symp. Publ. 34: 167–77. New York: Raven

Lord, B. I., Molineux, G., Testa, N. G., Kelly, M., Spooncer, E., Dexter, T. M. 1986. The kinetic response of haemopoietic precursor cells, in vivo, to highly purified recombinant interleukin-3. *Lymphokine Res.* 5: 97–104

Lord, B. I., Mori, K. J., Wright, E. G. 1977. A stimulator of stem cell proliferation in regenerating bone marrow. *Biomed. Express Paris* 27: 223–26

Lord, B. I., Mori, K. J., Wright, E. G., Lajtha, L. G. 1976. An inhibitor of stem cell proliferation in normal bone marrow. *Br. J. Haematol.* 34: 441–45

Lord, B. I., Spooncer, E. 1986. Isolation of haemopoeitic spleen colony forming cells. *Lymphokine Res.* 5: 59–72

Lord, B. I., Wright, E. G. 1980. Sources of haemopoietic stem cell proliferation: Stimulators and inhibitors. *Blood Cells* 6: 581–93

Magli, M. C., Iscove, N. N., Odartchenko, N. 1982. Transient nature of early haemopoietic spleen colonies. *Nature* 295: 527–29

Metcalf, D. 1977. *Hemopoietic Colonies*. Berlin/Heidelberg: Springer-Verlag. 227 pp.

Metcalf, D. 1983. Clonal analysis of the response of HL60 human myeloid leukaemia cells to biological regulators. *Leuk. Res.* 7. 117–32

Metcalf, D. 1985. The granulocyte-macro-phage colony-stimulating factors. *Science* 229: 16–22

Metcalf, D. 1986. The molecular biology and functions of the granulocyte macrophage colony-stimulating factors. *Blood* 67: 257–67

Metcalf, D., Begley, C. G., Nicola, N. A. 1985. The proliferative effects of human GM-CSF alpha and beta and murine G-CSF in microwell cultures of fractionated human marrow cells. *Leuk. Res.* 9: 521–27

Metcalf, D., Begley, C. G., Nicola, N. A., Lopez, A. F., Williamson, D. J. 1986. Effects of purified bacterially synthesised murine multi-CSF (IL-3) on hematopoiesis in normal adult mice. *Blood* 68: 46–57

Metcalf, D., Burgess, A. W. 1982. Clonal analysis of progenitor cell commitment to granulocyte or macrophage production. *J. Cell. Physiol.* 111: 275–83

Metcalf, D., Johnson, G. R., Mandel, T. E. 1979. Colony formation in agar by multipotential hemopoietic cells. *J. Cell. Physiol.* 98: 401–6

Metcalf, D., Moore, M. A. S. 1971. *Haemopoietic Cells*. Amsterdam: North Holland

Newbold, R. F., Overell, R. W. 1983. Fibroblast immortality is a prerequisite for transformation by EJ c-*Ha-ras* oncogene. *Nature* 304: 648–51

Nicola, N. A., Begley, C. G., Metcalf, D. 1985. Identification of the murine analogue of a regulator that induces differentiation in murine leukaemic cells. *Nature* 314: 625–28

Nicola, N. A., Metcalf, D. 1985. Binding of 125-I-labelled granulocyte colony-stimulating factor to normal murine hemopoietic cells. *J. Cell. Physiol.* 124: 313–21

Nicola, N. A., Metcalf, D., Matsumoto, M., Johnson, G. R. 1983. Purification of a factor inducing differentiation in murine myelomonocytic leukaemia cells. *J. Biol. Chem.* 258: 9017–23

Nishizuka, J. 1984. The role of protein kinase C in cell surface signal transduction and tumour promotion. *Nature* 303: 693–98

O'Garra, A., Warren, D. J., Holman, M., Popham, A. M., Sanderson, C. J., et al. 1986. Interleukin 4 (B-cell growth factor II/eosinophil differentiation factor) is a mitogen and differentiation factor for pre-activated murine B lymphocytes. *Proc. Natl. Acad. Sci. USA* 83: 5228–32

Park, L. S., Friend, D., Gillis, S., Urdal, D. L. 1986. Characterisation of the cell surface receptor for human granulocyte-macrophage colony-stimulating factor. *J. Exp. Med.* 164: 251–62

Potten, C. S., ed. 1983. *Stem Cells*. Edinburgh: Churchill-Livingstone. 304 pp.

Quesenberry, P. J., Song, Z. X., Gualtieri, R. J., Wade, P. M., Alberico, T. A., et al. 1984. Studies on the control of haemopoiesis in Dexter cultures. In *Long-Term Bone Marrow Culture*, ed. D. G. Wright, J. S. Greenberger, pp. 171–93. New York: Liss

Ruley, H. E. 1983. Adenovirus early region 1A enables viral and cellular transforming genes to transform primary cells in culture. *Nature* 304: 602–6

Sacca, R., Stanley, E. R., Sherr, C. J., Rettenmier, C. W. 1986. Specific binding of the mononuclear phagocyte colony-stimulating factor CSF-1 to the product of the v-*fms* oncogene. *Proc. Natl. Acad. Sci. USA* 83: 3331–35

Sachs, L. 1982. Normal development programmes in myeloid leukaemia: Regulatory proteins in the control of growth and differentiation. *Cancer Surv.* 1: 321–42

Schofield, R. 1979. The pluripotent stem cell. *Clin. Haematol.* 8: 221–37

Schofield, R., Dexter, T. M., Lord, B. I., Testa, N. G. 1986. Comparison of haemopoiesis in young and old mice. *Mech. Ageing Dev.* 34: 1–12

Souza, L. M., Boone, T. C., Gabrilove, J., Lai, P. H., Zsebo, K. M., et al 1986. Recombinant human granulocyte colony-stimulating factor: Effects on normal and leukemic myeloid cells. *Science* 232: 61–65

Spooncer, E., Heyworth, C. M., Dunn, A., Dexter, T. M. 1986. Self-renewal and differentiation of interleukin-3-dependent multipotent stem cells are modulated by stromal cells and serum factors. *Differentiation* 31: 111–18

Stanley, E. R., Bartocci, D. P., Rosendaal, M., Bradley, T. R. 1986. Regulation of very primitive multipotent hemopoietic cells by hemopoietin-1. *Cell* 45: 667–74

Stanley, E. R., Guilbert, J. 1981. Methods for the purification, assay, characterisation and target cell binding of a colony stimulating factor (CSF-1). *J. Immunol. Methods* 45: 253–89

Till, J. E., McCulloch, E. A. 1961. A direct measurement of the radiation sensitivity of normal mouse bone marrow cells. *Radiat. Res.* 14: 215–22

Toksoz, D., Dexter, T. M., Lord, B. I., Wright, E. G., Lajtha, L. G. 1980. The regulation of hemopoiesis in long-term bone marrow cultures. II. Stimulation and inhibition of stem cell proliferation. *Blood* 55: 951–36

Waterfield, M. D., Scrace, G. T., Whittle, N., Stroobant, P., Johnsson, A., et al. 1983. Platelet derived growth factor is structurally related to the putative p28 *sis* of simian sarcoma virus. *Nature* 304: 35–39

Whetton, A. D., Bazill, G. W., Dexter, T. M. 1984. Haemopoietic cell growth factor mediates cell survival via its action on glucose transport. *EMBO J.* 3: 409–13

Whetton, A. D., Dexter, T. M. 1983. Effect of haemopoietic cell growth factor on intracellular ATP levels. *Nature* 303: 629–31

Whetton, A. D., Dexter, T. M. 1986. Haemopoietic growth factors. *Trends Biochem. Sci.* 11: 207–11

Whetton, A. D., Heyworth, C. M., Dexter, T. M. 1986a. Phorbol esters activate protein kinase C and glucose transport and can replace the requirement for growth factor in interleukin-3-dependent multipotent stem cells. *J. Cell Sci.* 84: 93–104

Whetton, C. A., Monk, P. N., Consalvey, S. D., Downes, C. P. 1986b. The haemopoietic growth factors interleukin 3 and colony stimulating factor-1 stimulate proliferation but do not induce inositol lipid breakdown in murine bone marrow derived macrophages. *EMBO J.* 5: 3281–86

Witters, L. A., Vater, C. A., Lienhard, G. E. 1985. Phosphorylation of the glucose transporter in vitro and in vivo by protein kinase C. *Nature* 315: 777–79

Yang, Y.-C., Ciarletta, A. B., Temple, P. A., Chung, M. P., Kovacic, S., et al. 1986. Human IL3 (multi-CSF): Identification by expression cloning of a novel hematopoietic growth factor related to murine IL-3. *Cell* 47: 3–10

Ann. Rev. Cell Biol. 1987. 3 : 443–92

POLYPEPTIDE GROWTH FACTORS: Roles in Normal and Abnormal Cell Growth

Thomas F. Deuel

Department of Medicine and Biological Chemistry, Jewish Hospital at Washington University Medical Center, St. Louis, Missouri 63110

CONTENTS

Introduction 444
Platelet-Derived Growth Factor 444
Origins of PDGF 445
Properties of Purified PDGF 445
PDGF Receptor 446
Protein Tyrosine Kinase Activity 447
Activities Induced by PDGF in Cells Bearing the PDGF Receptor 449
PDGF as a Competence Factor 450
PDGF-Inducible Genes 451
Sites of Synthesis and Biological Relevance of PDGF-like Growth Factor Activity 452
Roles of PDGF in Inflammatory and Repair Processes 453
PDGF B-Chain Amino Acid Sequence and Gene Structure 454
PDGF A Chain: Relationship to PDGF B Chain 455
Oncogenes 455
Relationship of PDGF to p28$^{v\text{-sis}}$ 456
Requirement of the PDGF Receptor for Transformation by v-sis 457
Autocrine Growth of SSV-Transformed Cells 457
Internal Autocrine Stimulation of SSV-Transformed Cells 458
The c-sis Gene Product Is Potentially Transforming 459
Origins and Expression of the v-sis Gene 459
Structural Requirements of the v-sis Protein for Transformation 460
Epidermal Growth Factor 461
Structure of EGF and the EGF Precursor 461
EGF-Related Polypeptides 461
EGF Receptor 462
Relationship of the EGF Receptor to the v-erbB Oncogene Product 463
EGF: Potential for Transformation as an Oncogene Product 464
Heparin Binding Growth Factors 464
Transforming Growth Factors, Types α and β 467
Summary and Conclusions 473

443

0743–4634/87/1115–0443$02.00

Introduction

The role of polypeptide growth factors in stimulating the proliferation of cells and in maintaining their viability has been increasingly appreciated. Polypeptide growth factors also play roles in differentiation, development, chemotaxis and activation of inflammatory cells, tissue repair, and disease. The gene structure, cDNA structure, and complete amino acid sequence of several growth factors have been reported. The receptors for several growth factors also have been isolated and characterized. Activation of these receptors by growth factors leads to gene activation, transcription, and ultimately to cell division. Although many polypeptide growth factors have already been identified and characterized, it seems likely that additional growth factors, perhaps including new families of growth factors, will be identified in the next few years.

Polypeptide growth factors differ from hormones in that they usually act through paracrine and autocrine mechanisms. However, polypeptide growth factors have been identified in plasma and may act as hormones as well. Growth factors are synthesized and secreted by both normal and transformed cells. Abnormal secretion of growth factors by normal cells probably results in diseases characterized by a proliferative cellular response or by fibrosis. Cells activated by growth factors share a common phenotypic appearance with cells transformed by retroviruses and appear to stimulate the same metabolic pathways that are stimulated in transformed cells. These observations suggest that similar mechanisms may underlie neoplastic transformation and growth factor-initiated cell proliferation. One growth factor, the platelet-derived growth factor (PDGF), is related to the transforming gene (oncogene) product of the simian sarcoma virus, $p28^{v-sis}$. The v-erbB and v-fms oncogenes were recently established as transposed genes of the receptors for epidermal growth factor (EGF) and colony-stimulating factor 1 (CSF-1), respectively. Four classes of growth factor are reviewed here, each of which has contributed to our understanding of the roles of growth factors in normal and abnormal cellular activities and in neoplastic transformation.

Platelet-Derived Growth Factor

Platelet-derived growth factor (PDGF) was first observed as an activity in media supplemented with serum that supported the growth of chick embryo fibroblasts in culture (Balk 1971; Balk et al 1973). Platelets were identified as the source of the growth-promoting activity (Kohler & Lipton 1974; Ross et al 1974), which later was identified as the principal mitogenic protein in serum for cells of mesenchymal origin (Ross & Vogel 1978). Other activities from platelets observed in serum include an EGF-like molecule (Oka & Orth 1983), and transforming growth factor type β (TGF$_\beta$) (Childs et al 1982). These findings suggest that platelets may be involved in releasing a multitude of cell activators.

Origins of PDGF

PDGF is synthesized in megakaryocytes (Chernoff et al 1980), packaged into platelet α granules, and released from platelets activated by thrombin or at sites of blood vessel injury (Witte et al 1978; D. R. Kaplan et al 1979; K. L. Kaplan et al 1979). Platelet factor 4, another platelet-derived protein localized in α granules, diffuses several cell layers distal to the luminal surface of the vessel wall within 10 min of platelet binding to denuded aortic endothelium (Goldberg et al 1980). PDGF is thought to bind similarly to, and initiate its biological functions on, cellular substrates of the blood vessel wall.

Properties of Purified PDGF

Early efforts at purification (Antoniades et al 1975, 1979; Westermark & Wasteson 1975, 1976; Busch et al 1976; Antoniades & Scher 1977; Heldin et al 1977, 1979; Ross et al 1979) succeeded in defining intrinsic properties of PDGF, which then were used successfully in larger scale purifications (Antoniades 1981; Deuel et al 1981a,b; Heldin et al 1981a; Raines & Ross 1982). PDGF has been purified as two proteins (PDGF I, ~ 31 kDa; PDGF II, ~ 28 kDa) of equal mitogenic activity, the same amino acid composition and immunological reactivity, and with equal affinity for cell surface receptors on Swiss mouse 3T3 cells (Deuel et al 1981a,b; J. S. Huang et al 1982a, 1983). Sedimentation equilibrium analysis defined the molecular mass as 33 kDa (Heldin et al 1981a). PDGF I and PDGF II differ in the extent to which they are glycosylated; PDGF I is $\sim 7\%$ carbohydrate, while PDGF II is $\sim 4\%$ carbohydrate (Deuel et al 1981a). PDGF is very basic (pI ~ 10.2) and has 16 half cystine residues, all of which appear to be linked by disulfide bonds (Deuel et al 1981a). When PDGF is reduced, it is no longer mitogenically active (Antoniades 1981; Deuel et al 1981a,b; Heldin et al 1981a; Raines & Ross 1982; J. S. Huang et al 1982b). Separate chains of M_r 17,000 and 14,000 (PDGF I) and of 15,000 and 14,000 (PDGF II) were found in SDS gels of reduced and carboxymethylated PDGF (Deuel et al 1981a,b) and of M_r 18,000 and 16,000 by high-performance liquid chromatography (HPLC) (Johnsson et al 1982), which suggests that human PDGF is a heterodimer of A and B chains (Johnsson et al 1982). Surprisingly, porcine PDGF is a homodimer of B chains (Stroobant & Waterfield 1984), and it remains to be definitively established that human PDGF is not a mixture of A- and B-chain homodimers. Different researchers have referred to the chains of PDGF as A and B, B and A, or PDGF 1 and 2. In this review, the terms of Johnsson et al (1984) are used: the B chain is defined as the product of the c-*sis* gene.

Antisera that specifically recognize PDGF have been obtained and used to measure PDGF concentrations (J. S. Huang et al 1983). In this way, PDGF in human serum was estimated at between 50 and 60 ng/ml; less than 1 ng/ml was found in serum derived from platelet-poor plasma. PDGF receptor binding was used to measure PDGF concentration, and human clotted blood serum was estimated to contain 15 ng/ml PDGF, or $\sim 30,000$ molecules of PDGF per platelet (Singh et al 1982). Such measurements in plasma and serum are complicated because $[^{125}I]$PDGF binds to a protein that subsequently was identified as α_2-macroglobulin (J. S. Huang et al 1984a). The significance of this binding remains unclear, although some PDGF released locally may escape to circulating plasma and its clearance is then mediated by its interaction with α_2-macroglobulin. The clearance of PDGF from plasma is rapid; PDGF injected intravenously into baboons is cleared from plasma with a $T_{1/2}$ of ~ 2 min (Bowen-Pope et al 1984).

PDGF Receptor

When binding analyses are used to identify PDGF cell surface receptors (Deuel et al 1981b; Heldin et al 1981b, 1982; Huang et al 1982a,b; Bowen-Pope & Ross 1982; Williams et al 1982), a single class of high-affinity binding sites is detected. The K_d ranges from 0.1 to 10×10^{-9} M with 200,000–400,000 binding sites/cell observed on cultured mouse fibroblasts and $\sim 50,000$ sites/cell on cultured vascular smooth muscle cells. The half-maximal concentration of PDGF required for $[^3H]$thymidine uptake (J. S. Huang et al 1982b) and for phosphorylation of ribosomal protein S6 (Nishimura & Deuel 1981) correlates closely with the K_d of PDGF receptors on mouse fibroblasts.

Surface-bound $[^{125}I]$PDGF was internalized and degraded with a $T_{1/2}$ of about 90 min (J. S. Huang et al 1982a). Degradation of $[^{125}I]$PDGF is inhibited by low concentrations of lysosomotrophic agents (Heldin et al 1981b, 1982; J. S. Huang et al 1982a). Incubation of cells with PDGF (50 ng/ml) effectively decreased the number of PDGF receptors, which may reduce the subsequent growth stimulation by PDGF. $[^{125}I]$PDGF bound to human skin fibroblasts also was shown to be internalized and degraded, presumably in lysosomes. The reappearance of the receptor at the cell surface is blocked by cycloheximide, which suggests that new protein synthesis rather than recycling of receptors was required for reexpression of cell surface receptor (Heldin et al 1981b). When cross-linking to [^{125}I]PDGF was used to identify the receptor, a [^{125}I]PDGF binding protein of ~ 164,000 kDa was found in PDGF-responsive cells (Glenn et al 1982). ✳

A tyrosine-specific protein kinase activity stimulated by PDGF was demonstrated in membranes of cells that respond mitogenically to PDGF (Ek et al 1982; Nishimura et al 1982; Ek & Heldin 1984; Frackelton et al 1984). A prominant phosphorylated protein of ~ 180 kDa was tentatively identified as the PDGF receptor. The receptor protein for PDGF was subsequently purified to near homogeneity in small quantities from Swiss 3T3 cell membranes with a PDGF affinity column (S. S. Huang et al 1984) and identified as an autophosphorylating tyrosine kinase of 180 kDa. The kinase activity was stimulated by PDGF, and the protein comigrated with a single band of a 180-kDa ^{35}S-labeled protein, purified from [^{35}S]methionine-labeled Swiss 3T3 cells.

The PDGF receptor was purified to apparent homogeneity from BALB/c 3T3 cells in its activated form (i.e. autophosphorylated) by anti-phosphotyrosine immunoaffinity chromatography (Frackelton et al 1984; Daniel et al 1985). A complete cDNA that encoded the pre-PDGF receptor was cloned and sequenced (Yarden et al 1986). The deduced amino acid sequence revealed a high degree of homology of the PDGF receptor to the v-*kit* oncogene product and to c-*fms*, the putative receptor for CSF-1, the macrophage-specific colony-stimulating factor. These proteins have in common long sequences that interrupt the tyrosine kinase domain. The PDGF and CSF-1 receptors also share a homology in the placement of the extracellular cysteines, a single hydrophobic membrane-spanning domain, and a cytosolic carboxyl terminus. Ubiquitin was found to be covalently bound to the purified PDGF receptor (Yarden et al 1986). The human gene for the PDGF receptor (like that for the CSF-1 receptor) is on chromosome 5 (Yarden et al 1986). This work suggests many new areas of research, which may definitively establish activities mediated by the PDGF receptor, a requirement for this receptor for transformation by the v-*sis* oncogene, and the domains and amino acid residues responsible for its biological activities and their regulation.

Protein Tyrosine Kinase Activity

The tyrosine protein kinase activity of the PDGF receptor is similar to that of other growth factor receptors (EGF, insulin, and IGF-I receptors) (Ushiro & Cohen 1980; Kasuga et al 1982; Jacobs et al 1983; Nishimura & Deuel 1981). These receptors, when activated by ligand, phosphorylate protein substrates, many of which are also phosphorylated in virus-transformed cells (Sefton & Hunter 1984; Hunter & Cooper 1985; Pike & Krebs

1986; Sibley et al 1987). The physiological pathways activated by receptor phosphorylation and other protein kinase activities that induce DNA synthesis and cell division have not been identified. However, many of the same protein substrates are phosphorylated in Rous sarcoma virus (and other) transformed cells and in growth factor–activated cells, which suggests that common pathways are used for growth factor–stimulated and oncogene-mediated cell growth (Radke & Martin 1979; Erikson et al 1981). For example, a 36-kDa protein tentatively identified as a subunit of malic dehydrogenase (Rubsamen et al 1982) may be a substrate both for the viral *src* protein (Radke & Martin 1979) and for the protein kinase associated with the EGF receptor (Erikson et al 1981). Moreover, EGF and PDGF phosphorylate a 39-kDa protein similar to that phosphorylated in virus-transformed cells (Cooper et al 1982), and a 42-kDa protein is phosphorylated both upon transformation by the Rous sarcoma virus or stimulation by PDGF, EGF, or TPA (Nakamura et al 1983; Cooper et al 1984).

Recently, research has focused on two related peripheral membrane proteins, referred to as p35 and p36. p36 is phosphorylated on tyrosine in many transformed cells and in growth factor–stimulated cells and is known as calpactin I. p35 is the major protein phosphorylated in A431 cell membranes stimulated by EGF (Fava & Cohen 1984) and is known as calpactin II. Lipocortins are a class of steroid-inducible proteins that were originally thought to inhibit phospholipase A_2. Lipocortin I appears to be identical to calpactin II (35 kDa) and lipocortin II and p36 are highly homologous. The phosphorylation of these proteins is not known to effect specific cellular activities in transformed or growth factor–stimulated cells (Brugge 1986).

In addition to the activation of tyrosine kinases induced when PDGF binds to its cell surface receptor, serine/threonine kinase activities also are stimulated in cells activated by PDGF (Cooper et al 1982). For example, serum and PDGF stimulate the serine kinase phosphorylation of ribosomal S6 protein (Nishimura & Deuel 1981; Chambard et al 1983). This network of interacting protein kinases among growth factor–stimulated and transformed cells, including protein kinase C, may serve as a common mediator leading to DNA synthesis and cell growth. Activation of protein kinase C by PDGF has been demonstrated (Rozengurt et al 1983b; Rodriguez-Pena & Rozengurt 1985; Blackshear et al 1985) and may serve to generate second messages. Protein kinase C activation appears to mediate many of the changes induced by PDGF, including increased Na^+-H^+ antiport activity, increased intracellular pH, sodium influx, and increased Na^+-K^+ pump activity (Dicker & Rozengurt 1981; Moolenaar et al 1984b;

Rosoff et al 1984; Besterman & Cuatrecasas 1984; Vara & Rozengurt 1985; Vara et al 1985). However, these early events are also elicited by mitogens that do not activate protein kinase C (Vara & Rozengurt 1985; Rozengurt 1986).

Activities Induced by PDGF in Cells Bearing the PDGF Receptor

PDGF initiates highly specific but diverse events in target cells bearing the PDGF receptor. Because DNA synthesis does not begin until 10–15 hr after the initial PDGF signal, attention has been directed to early events to understand how the mitogenic signal is transmitted.

PDGF significantly influences ion fluxes, stimulating entry of Na^+ and efflux of H^+ by an amiloride-sensitive Na^+/H^+ antiport (Rozengurt 1986). Within seconds of PDGF stimulation, there is a transient decrease in intracellular pH, followed by a rise in pH of 0.2–0.4 pH units that is accompanied by an influx of Na^+ (Schuldiner & Rozengurt 1982; Cassel et al 1983). The Na^+/K^+ pump activity also increases, restoring the electrochemical gradient for Na^+ and raising intracellular K^+. Transient alkalinization of cells had previously been shown to induce "competence" in cells (Zetterberg et al 1982), but no evidence is available to establish that the alkalinization induced by PDGF is essential for cell proliferation.

Release of Ca^{2+} from intracellular stores is observed within 15 sec of PDGF stimulation and is associated with increased turnover of phosphatidylinositols (Frantz et al 1980; Shier 1980; Habenicht et al 1981; Frantz 1982; Shier & Durkin 1982; Cassel et al 1983; Rozengurt et al 1983a; Tucker et al 1983; Moolenaar et al 1984a). The time course for release of Ca^{2+} and inositol phosphates strongly suggests that each has a role in the early transduction of the PDGF signal. PDGF also increases levels of arachidonic acid, a precursor of prostaglandins and leukotrienes, and of diacylglycerol, an activator of protein kinase C and a precursor of prostaglandins.

PDGF has also been shown to stimulate amino acid transport (Owen et al 1982), serine phosphorylation of the ribosomal S6 protein (Nishimura & Deuel 1983), the morphological rearrangement of actin filaments (Westermark et al 1983; Bockus & Stiles 1984), and centriole deciliation (Tucker et al 1979).

PDGF stimulates protein synthesis (Canalis 1981; Cochran et al 1981; Owen et al 1982). Of relevance to the putative role of PDGF in wound healing are findings that PDGF enhances synthesis and release of collagen V and regulates the synthesis of collagens III and IV (Narayanan & Page

1983). Collagenase synthesis and secretion also were seen within 12 hr of treatment with PDGF (Bauer et al 1985; Chua et al 1985). PDGF is a potent vasoconstrictor; at low concentrations it induces a dose-dependent contraction of rat aorta strips (Berk et al 1986), an activity that may be important in the early phase of hemostasis.

PDGF also modulates expression of receptors for other ligands. An increase in the expression of interleukin 1 (IL-1) receptors (Clemmons et al 1980) and the secretion of an interleukin 1–like polypeptide have been observed in fibroblasts exposed to PDGF (Clemmons & Shaw 1983). PDGF also induces increased binding, internalization, and degradation of low-density lipoprotein (LDL) (Chait et al 1980; Witte & Cornicelli 1980; Witte et al 1982). The binding of insulin is not influenced by PDGF, nor does EGF modify the binding of PDGF. Preincubation with PDGF, however, decreased the subsequent binding of [^{125}I]EGF to receptors on density-arrested cultures of BALB/c 3T3 cells; this effect was potentiated by cholera toxin (Wrann et al 1980; Heldin et al 1982; Wharton et al 1982).

PDGF reduces binding of EGF (Bowen-Pope et al 1983; Collins et al 1983; Wrann et al 1980) by reducing the affinity of the EGF receptor, an effect similar to that induced by TPA (Magun et al 1980; Wharton et al 1983). These results suggest that protein kinase C may be a common intermediate for the influence of TPA and PDGF on the EGF receptor.

PDGF as a Competence Factor

PDGF alone does not induce cell division in Balb/C 3T3 cells (Pledger et al 1977; Vogel et al 1978; Stiles et al 1979; Clemmons & Van Wyk 1981; Bright & Gaffney 1982), although it can do so at low nanogram concentrations in conjunction with other growth factors or plasma. Cells exposed to PDGF for as little as 2 hr divide if plasma is added subsequently (Pledger et al 1978), but plasma alone has no effect. PDGF thus induces competence for subsequent cell division, and plasma is required for progression through the cell cycle after exposure to PDGF. For cells exposed to PDGF, progression factors are needed throughout the G_1 phase of the cell cycle (Singh et al 1983). Epidermal growth factor (EGF) and IL-1 can be substituted for platelet-poor plasma (Leof et al 1982) and are representative progression factors. Competence may be transferred from one cell to another. When PDGF-treated, competent cells were fused with quiescent Balb/c 3T3 fibroblasts in the presence of platelet-poor plasma, the resultant hybrids were competent to respond to plasma (Smith & Stiles 1981). Competence is blocked by inhibitors of RNA synthesis, which suggests that new protein synthesis is required for the expression of cell competence.

PDGF-Inducible Genes

The pathways by which PDGF induces competence have been studied extensively. Quiescent 3T3 cells stimulated with PDGF preferentially synthesize proteins of 29, 35, 45, 60, and 70 kDa (Pledger et al 1981). Other growth factors induce different patterns of new protein synthesis. Fibroblast growth factor (FGF) induces synthesis of 29- and 35-kDa proteins, whereas insulin and EGF are without significant effect. The 35-kDa protein is similar to, or identical with, a protein secreted by transformed cells (Scher et al 1983). PDGF induces synthesis of high levels of the 35-kDa protein, and spontaneously transformed cells synthesize even greater amounts. The 29-kDa protein induced by PDGF has a long half-life and is localized within the nucleus (Olashaw & Pledger 1983).

Differential colony hybridization was successfully used to identify several of the genes activated shortly after PDGF stimulation of cells (Cochran et al 1984). The *KC* gene seems to code for a 10-kDa polypeptide and the *JE* gene a protein of 19 kDa. A 10–20 fold increase over the steady-state level of *JE* mRNA was observed 1 hr after PDGF stimulation. After stimulation, *JE* mRNA was relatively abundant with ∼3000 copies per cell; *KC* mRNA was somewhat less abundant, with ∼700 copies per cell. Both *JE* and *KC* genes were only transiently activated, and their function is unknown.

A third PDGF-inducible gene was identified as the c-*myc* gene, the normal cellular counterpart (proto-oncogene) of the v-*myc* oncogene (Kelly et al 1983). PDGF stimulation caused a ∼40-fold increase in c-*myc* mRNA levels within 3 hr. Stimulation with FGF or DMA also induced high levels of c-*myc* mRNA. At optimal concentrations of PDGF, c-*myc* was expressed at ∼5–10 copies per cell. The induction of this proto-oncogene by PDGF further establishes the relatedness of growth factor–stimulated pathways with those expressed in cells transformed by retroviruses. Induction of the c-*myc* gene may be a general response to growth stimuli; B lymphocytes stimulated with concanavalin A, T lymphocytes stimulated with lipopolysaccharide, and regenerating liver cells have been shown to express high levels of c-*myc* (Kelly et al 1983; Makino et al 1984). When c-*myc* levels are increased by glucocorticoids after transfection with a c-*myc* gene linked to the inducible mammary tumor virus control region (LTR), the c-*myc* mRNA levels seem to determine the number of cells entering the S phase of the cell cycle (Armelin et al 1984). The activation of oncogenes by growth factors is likely to involve interactions among multiple genes, including proto-oncogenes. Activation of the c-*myc* gene alone probably does not account for all of the stimulatory activity of PDGF. In newly established rat fibroblasts *myc* acts synergistically with the *ras* oncogene (Land et al 1983).

The c-*fos* proto-oncogene also is activated by PDGF (Cochran et al 1984; Greenberg & Ziff 1984; Kruijer et al 1984; Muller et al 1984). The induction of the c-*fos* gene was more rapid than that of the c-*myc* gene; substantial increases of c-*fos* mRNA were seen within 15 min of PDGF addition (Greenberg & Ziff 1984). The increased transcription of both c-*fos* and c-*myc* genes was transient; mRNA levels of c-*fos* had returned to near normal within 3 hr of stimulation by PDGF (Kruijer et al 1984; Muller et al 1984). The mechanisms by which the c-*myc* and c-*fos* gene products induce cell growth are not known, nor is it known what additional pathways, if any, these proto-oncogene products activate. However, research on growth factors and proto-oncogene products is providing clues concerning the central controls of normal and abnormal cell growth.

Sites of Synthesis and Biological Relevance of PDGF-like Growth Factor Activity

Mononuclear phagocytes when activated secrete growth factor activity for fibroblasts, smooth muscle cells, and endothelial cells (Leibovich & Ross 1976; Glenn & Ross 1981; Martin et al 1981), some of which is related to PDGF (Shimokado et al 1985). These cells also secrete a fibroblast growth factor–like activity (Baird et al 1985b) and an IL-1-like activity (Schmidt et al 1982). Cultured endothelial cells also secrete a PDGF-like growth factor activity (Gajdusek et al 1980; Wang et al 1981; Barrett et al 1984; Jaye et al 1985), although these cells lack demonstratable PDGF receptors. PDGF-like molecules also have been identified in cultures of smooth muscle cells; the production of these molecules may be developmentally regulated (Seifert et al 1984). Smooth muscle cells cultured from aortae of two-week-old rats release substantial amounts of PDGF-like growth factor activity, whereas adult aortic smooth muscle cells secrete much less. In contrast, the number of receptors for PDGF is substantially higher in adult rat aortic smooth muscle cells than in those of two-week-old rats (Seifert et al 1984; Bowen-Pope & Seifert 1985). These results suggest autocrine and paracrine roles for PDGF in the development and regulation of the vessel wall and in the migration and mitogenic responses of cellular components involved in wound healing, the proliferative lesions of atherosclerosis in man, and potentially in such diverse lesions as pulmonary fibrosis, myelofibrosis, and rheumatoid arthritis.

Roles of PDGF in Inflammatory and Repair Processes

The migration of neutrophils and mononuclear phagocytes into injured tissues is a hallmark of inflammation. Under appropriate stimulation, these cells destroy foreign debris and microorganisms, remove injured and autolyzed endogenous tissues, and mount an immune response. Both inflammation and tissue repair are complicated responses that involve highly regulated reactions and multiple components. Molecules ordinarily confined within the cell may gain access to the extracellular environment, where they may serve as signals for cell migration (i.e. chemoattractants).

A potential role of platelets or perhaps PDGF in inflammation was suggested when inflammatory cells, such as neutrophils and monocytes, were found in close proximity to platelets in pathological lesions and in experimental models of inflammation, immune complex disease, and atherosclerosis (Angrist & Oka 1963; Henson & Cochrane 1971; Jorgensen et al 1972; Prchal & Blakely 1973; Bachofen et al 1975; Levine et al 1976; Jellinek 1977; Joris & Majno 1979; Bachofen & Weibel 1982). Platelet extracts originally were shown to increase vascular permeability (Packham et al 1968), and cationic proteins isolated from platelet granules were found to increase the permeability of vessels in rabbit skin (Nachman et al 1972). Serum was also shown to have an "intrinsic chemotactic activity" not found in plasma (Hirsch 1960; Weksler & Coupal 1973).

Platelet factor 4 (PF_4) is a well characterized α granule platelet protein. Highly purified PF_4 stimulates the migration of human neutrophils and monocytes at PF_4 concentrations lower than those in human serum (Deuel et al 1981c). PF_4 is also a strong chemoattractant for fibroblasts; maximum fibroblast chemotaxis was observed at PF_4 concentrations of less than one-tenth those eliciting maximal chemotaxis of inflammatory cells (Senior et al 1983). These results suggest that platelet secretory proteins play a role in inflammation apart from the role of PDGF as a mitogenic protein for mesenchymal cells.

PDGF was found to be a strong chemoattractant for human monocytes and neutrophils at concentrations of 20 ng/ml and 1 ng/ml, respectively (Deuel et al 1982); the normal concentration of PDGF in serum is ~50 ng/ml. PDGF is also a powerful chemoattractant for fibroblasts and smooth muscle cells (Grotendorst et al 1981; Seppä et al 1982; Senior et al 1983). Other materials that have been found to elicit fibroblast chemotaxis include β-thromboglobulin, lymphokines, a peptide derived from the fifth component of complement, collagens and collagen-derived peptides, fibronectin, tropoelastin, and elastin-derived peptides (Postlethwaite et al 1976, 1978, 1979; Seppä et al 1981; Senior et al 1982, 1983).

Besides stimulating directed migration, PDGF induces activation of human polymorphonuclear leukocytes (Tzeng et al 1984) and monocytes (Tzeng et al 1985) and causes a dose-dependent, saturatable increase in the release of collagenase by fibroblasts in culture (Bauer et al 1985), an activity essential for tissue repair (Gross 1976; Grillo & Gross 1967; Riley & Peacock 1967; Eisen 1969; Donoff et al 1971). Thus PDGF and perhaps other platelet proteins ordinarily sequestered within the cell are released to the extracellular compartment by platelet activation at sites of injury. They may be recognized by phagocytic cells and cells involved in wound healing, evoking chemotaxis, cell activation, collagenase synthesis and release, and other responses. Recent studies with PDGF in in vivo models support these functional roles for platelet secretory proteins (Grotendorst et al 1985).

Recombinant PDGF (B-chain/B-chain homodimer) has been tested directly in wound healing models. This recombinant PDGF behaves essentially identically to PDGF in chemotactic and mitogenic assays. When it was added directly to linear incisions in rats, a single application enhanced the tensile strength of the healing wounds nearly twofold within 7 days. A marked influx of neutrophils and monocytes and synthesis of new collagen were found in recombinant PDGF–treated wounds (G. F. Pierce, T. A. Mustoe, T. F. Deuel, unpublished data). These results directly establish that PDGF is a potent pharmacological agent capable of significantly accelerating healing of incisional wounds and suggest additional roles of growth factors in chemotaxis and wound healing in general.

PDGF B-Chain Amino Acid Sequence and Gene Structure

The structure of PDGF was initially obtained through partial amino acid sequence analysis (Antoniades & Hunkapiller 1983; Waterfield et al 1983) and the recognition that one chain of PDGF was >90% homologous with the protein product of the v-*sis* oncogene, p28$^{v\text{-}sis}$ (Doolittle et al 1983; Waterfield et al 1983). It was subsequently established that the precursor of the B chain of PDGF is encoded by the c-*sis* gene, the cellular counterpart of the transforming gene v-*sis* of the simian sarcoma virus (SSV) (Johnsson et al 1984; Josephs et al 1984a). The predicted amino acid sequence deduced from DNA sequencing of the human c-*sis* gene corresponded precisely with the amino acid sequence of the B chain of PDGF (Johnsson et al 1984; Josephs et al 1984a). The human c-*sis* gene was identified on the long arm of chromosome 22 (Swan et al 1982). Only a single cellular *sis* gene has been identified with Southern blot analyses of genomic DNA (Wong-Staal et al 1981a; Robbins et al 1982a). The human c-*sis* gene extends over 12 kilobases (kb) of DNA. Regions homologous to the v-*sis* gene of SSV extend over five exons and four introns. However, the first exon with v-*sis* homology in its 5′ flanking region lacks promoter and initiator codons (Josephs et al 1983); these portions of the transcription unit appear upstream of the first exon with homology to the v-*sis* gene (Gazit et al 1984). The sequences of the other v-*sis* exons are highly homologous with those of c-*sis*. RNA transcripts of 4.2 kb have been detected (Eva et al 1982; Westin et al 1982; Graves et al 1984). The 4.2-kb *sis* transcript encodes a 27-kDa polypeptide (Josephs et al 1984b).

PDGF A Chain: Relationship to PDGF B Chain

Biologically active homodimers of the PDGF A chain have been obtained from a human osteosarcoma cell line, U2OS (Betsholtz et al 1983). These homodimers bind to the PDGF receptor, and their chemical fragmentation patterns, chromatographic behavior, and the N-terminal amino acid sequence were used to establish the identity of the homodimers to the A chain of PDGF. The complete amino acid sequence of the precursor of the PDGF A chain was deduced from cDNA clones, and the gene has been localized to chromosome 7. This polypeptide shows extensive homology (~50%) to the precursor of the PDGF B chain. Secretion of a PDGF-like growth factor of ~31 kDa correlated with the expression of A- but not B-chain mRNA (Betsholtz et al 1986).

Rat skeletal myoblasts and arterial smooth muscle cells express the gene for the A chain but not for the B chain. Human melanoma cell lines of primary and metastatic origin also express the A chain, although the primary human melanoma cells also express the B chain (Westermark et al 1986; Sejersen et al 1986).

Oncogenes

A group of important normal cellular genes (proto-oncogenes) were identified when transforming genes (oncogenes) previously identified in acute transforming retroviruses were shown by hybridization and ultimately DNA sequencing to have nearly identical structures. These proto-oncogenes, if inappropriately expressed due to chromosomal translocation, gene amplification, nucleotide substitution, or placement under the control of mobile regulatory elements, may become oncogenes, which can initiate and maintain the transformed phenotype. The proteins encoded by proto-oncogenes were highly conserved in evolution and have been classified by enzymatic activity and by the localization of the protein product. Some oncogene products are localized to the cell nucleus (those of *fos*, *myc*, *myb*), whereas the family of *src* gene products are localized to the plasma membrane and have protein kinase domains (*abl*, *fes/fps*, *fgr*, *src*, *yes*, *erbB*, *fms*, *raf/mil/mht*, *mos*). The family of *ras* oncogenes encode proteins located at the plasma membrane that bind and hydrolyze GTP (*H-ras*, *N-ras*, *R-ras*). The fourth family of oncogenes encode cytosolic proteins for which an identifiable function remains to be established (*erbA*, *ets*, *rel*, *ski*); recent evidence suggests that the *erbA* protein is closely related in sequence to the thyroid hormone binding protein. The fifth family of *sis* oncogenes encode a protein structurally related to a growth factor.

Relationship of PDGF to p28$^{v\text{-}sis}$

A breakthrough in identifying a functional activity with an oncogene product was made when two groups independently demonstrated that PDGF was nearly identical to the protein product of a known oncogene. This observation was followed by the establishment of the near identity of two growth factor receptors (for EGF and CSF-1) with two related onco-gene products. These results reconciled the strikingly similar phenotypic properties of transformed cells and growth factor–stimulated cells at both the biochemical and morphological levels.

A partial amino acid sequence of the B chain of PDGF contained a region of 104 contiguous amino acids with over 90% identity to the amino acid sequence predicted for the product of the v-*sis* gene, p28$^{v\text{-}sis}$ (Waterfield et al 1983). p28$^{v\text{-}sis}$ is the transforming protein of the simian sarcoma virus (SSV), an acute transforming retrovirus, and the v-*sis* oncogene is responsible for initiating and maintaining the transformed state of SSV-transformed cells. Its amino acid sequence had been deduced from se-quencing of the genome of SSV (Devare et al 1983). At the same time, sequence analysis of the N-terminal of PDGF (Antoniades & Hunkapiller 1983) showed it to be nearly identical to part of the predicted amino acid sequence of p28$^{v\text{-}sis}$ (Doolittle et al 1983). These findings led to the identification of the human c-*sis* gene product as the B-chain of PDGF and to the location of c-*sis* on chromosome 22 (Johnsson et al 1984; Josephs et al 1984a). These findings suggested that the c-*sis* gene had been "captured" by, and incorporated into, the genome of the simian sarcoma virus and that the inappropriate expression of the v-*sis* gene mediated cell transformation through its growth factor activity.

The conclusion that the B chain of PDGF had been translocated into the genome of SSV was further established when it was shown that the protein encoded by the v-*sis* gene was nearly identical immunologically and mitogenically to PDGF. Lysates of SSV-transformed NIH-3T3 cells and of control, nontransformed cells were compared (Deuel et al 1983; Robbins et al 1983). SSV-transformed cells contained a potent growth factor activity that was inhibited completely by anti-PDGF antisera and was not found in nontransformed 3T3 cell lysates. This activity in SSV cell lysates had a mitogenic dose-response curve identical to that of purified PDGF. The specific mitogenic activity in SSV-transformed cell lysates was essentially identical with that of purified PDGF. The growth factor activity in SSV-transformed cells was different from PDGF because the processed PDGF-like protein in the SSV-transformed cells after immunoprecipitation with anti-PDGF migrated identically with p28$^{v\text{-}sis}$ in SDS gels (Deuel et al 1983). The product of the v-*sis* oncogene was thus identified as a protein fully capable of activating a tyrosine protein kinase, the PDGF receptor.

Requirement of the PDGF Receptor for Transformation by v-sis

Transformation by the v-*sis* gene appears to require the PDGF cell surface receptor. Only cells of mesenchymal origin have been shown to respond to PDGF, and when transformed cell lines such as glioblastomas, osteosarcomas, and fibrosarcomas were analyzed, only mesenchyme-derived cells expressed *sis* mRNA (Eva et al 1982; Graves et al 1984; Nister et al 1984). Some but not all of these cell lines also secreted a PDGF-like growth factor activity (Heldin et al 1980). PDGF cell surface receptors were down-regulated by PDGF (Heldin et al 1982; J. S. Huang et al 1982a; Williams et al 1982) and by SSV-transformed cell-conditioned medium (Garrett et al 1984). SSV-transformed cells had PDGF cell surface receptors with a dissociation constant identical to that of PDGF receptors on non-transformed cells (J. S. Huang et al 1984b). This finding suggests that the PDGF receptor on transformed cells was down-regulated by the indigenous growth-promoting activity of p28$^{v\text{-}sis}$. Both SSV-transformed cells secreting PDGF-like growth-promoting activity and SSV-transformed cells without detectable secreted activity had markedly reduced numbers of PDGF cell surface receptors (J. S. Huang et al 1984b). Suramin, which blocks binding of PDGF to its receptor when added to SSV-transformed cells, induced the reappearance of new receptors capable of binding PDGF in SSV-transformed cells (Garrett et al 1984). SSV was able to transform only cells that expressed the PDGF receptor (Leal et al 1985). Collectively, these results imply that the PDGF receptor is required for transformation by v-*sis*.

Autocrine Growth of SSV-Transformed Cells

SSV-transformed NIH-3T3 cells and SSV-transformed NRK (normal rat kidney) cells were shown to secrete a protein with PDGF-like growth factor activity. The secreted protein behaved exactly like PDGF in radioimmunoassays and was removed from conditioned media by anti-PDGF antisera. Its behavior was also identical to that of PDGF in dose-response mitogenic assays. The secreted mitogen was competitive with [^{125}I]PDGF for binding. The secreted protein thus has properties of the v-

sis gene product, although this identity has not been confirmed by amino acid sequencing.

Receptors were identified on growth factor–secreting SSV-transformed cells. The number of PDGF receptors on these cells was only 5–10% of the number found on nontransformed cells (J. S. Huang et al 1984b). This is consistent with the secreted growth factor activity acting directly to down-regulate the receptor and thus acting directly as an autocrine stimulator of cell growth in SSV-transformed cells. Also in keeping with the secreted growth factor activating cell growth is the finding that anti-PDGF antisera added to growing SSV-transformed NIH-3T3 cells blocked [^{3}H]thymidine incorporation into DNA. Furthermore, addition of PDGF reversed the inhibition of [^{3}H]thymidine incorporation by antisera. The latter finding confirms that the antisera had reduced the [^{3}H]thymidine incorporation by blocking the secreted growth-promoting activity from interacting with PDGF cell surface receptors.

Internal Autocrine Stimulation of SSV-Transformed Cells

SSV-transformed NPl cells do not secrete detectable growth factor activity, and [^{3}H]thymidine incorporation into the DNA of these cells is not blocked by anti-PDGF antisera (J. S. Huang et al 1984b). Anti-PDGF antisera even at high concentrations only inhibit incorporation of [^{3}H]thymidine into DNA by less than 50% and do not reverse morphological cell transformation. SSV-transformed NPl cells induce tumors in nude mice, and they contain an mRNA that hybridizes with a v-*sis* probe. Transformation, therefore, may be mediated by internally active p28$^{v\text{-}sis}$. An "internal autocrine" hypothesis was suggested to explain these findings (Deuel & Huang 1984a,b; J. S. Huang et al 1984b). p28$^{v\text{-}sis}$ was proposed to activate the PDGF receptor during processing of the receptor in the endoplasmic reticulum and Golgi apparatus. The hypothesis is that as p28$^{v\text{-}sis}$ is transported through the endoplasmic reticulum and the Golgi apparatus, it is free to bind to the luminal domain of the PDGF receptor (its extracellular domain). The hypothesis predicts that the receptor kinase domain in the cytoplasm is activated as the receptor is processed. Thus secretion of p28$^{v\text{-}sis}$ into the surrounding media is not required. This model could also explain the failure of excess anti-PDGF to fully block incorporation of [^{3}H]thymidine into the DNA of those SSV-transformed cells that secrete PDGF-like mitogens and the failure of antisera to cause these transformed cells to revert to a nontransformed phenotype.

Many SSV-transformed cells secrete molecules that are similar to PDGF in their growth stimulatory and receptor binding properties (Deuel et al 1983; Devare et al 1984; Garrett et al 1984; J. S. Huang et al 1984b; Niman 1984; Niman et al 1984; Owen et al 1984; Thiel & Hafenrichter 1984; Wang

& Williams 1984). The original autocrine hypothesis (Sporn & Todaro 1980) suggests that transformed cells secrete a growth factor into the conditioned media and then bind the growth factor through cell surface receptors, thus stimulating their own growth. The experiments outlined above (J. S. Huang et al 1984b) provide convincing evidence that autocrine stimulation of secreting SSV-transformed cells mediates at least some of their stimulated growth and that an internal autocrine mechanism may also operate.

The c-sis *Gene Product Is Potentially Transforming*

The role of PDGF in cell transformation was extended when a cDNA library from the HUT 102 cell line was screened with v-*sis* probes, and a 2.7-kb c-*sis* cDNA that contained the entire v-*sis* homologous region (Clarke et al 1984) was isolated. This clone (pSM-1) was expressed in NIH-3T3 cells, induced colonies in soft agar. These results established that the normal human c-*sis* gene under appropriate regulatory elements was fully transforming in 3T3 cells (Josephs et al 1984b).

Chronic myelogenous leukemia (CML) (Groffen et al 1983; Bartram et al 1984) and Ewing sarcoma (Aurias et al 1983; Turc-Carel et al 1983) cells both have translocations near band 22q11. In CML, the translocation 9 : 22 results in the generation of the characteristic Philadelphia chromosome (Rowley 1973). However, the proximity of the proto-oncogene c-*abl* to the breakpoint compared to the more distant c-*sis* localization (Bartram et al 1984) and the novel c-*abl* mRNAs synthesized (Canaani et al 1984; S. J. Collins et al 1984) suggest that c-*abl* expression but not c-*sis* expression may be important in the inappropriate growth of these tumors. The c-*sis* gene was not expressed in Ewing sarcoma, nor are rearrangements of this gene noted (Bechet et al 1984).

Origins and Expression of the v-sis *Gene*

The simian sarcoma virus was originally isolated from a fibrosarcoma in a woolly monkey (Theilen et al 1971; Wolfe et al 1971). A helper virus, the simian sarcoma–associated virus (SSAV), and a defective transforming component, the simian sarcoma virus (SSV), were identified. The *gag, pol,* and *LTR* regions of SSAV and SSV are homologous, whereas a 1.0-kb sequence is unique to SSV. This gene was termed v-*sis* (Gelmann et al 1981; Robbins et al 1982b) and was found within the envelope region at the 3' end of the virus, where its insertion had resulted in loss of envelope sequences (Gelmann et al 1981, 1982; Robbins et al 1982b). The c-*sis* gene is highly conserved in vertebrates (Robbins et al 1982a; Wong-Staal et al 1981b). The v-*sis* gene is an oncogene, i.e. it mediates cell transformation:

Sequences that include the v-*sis* gene transform NIH-3T3 cells (Robbins et al 1982b). The open reading frame potentially may arise from any of three ATG codons in the envelope sequence of SSV and extend into the v-*sis* region, resulting in polypeptides of 25, 29, or 33 kDa, respectively (Devare et al 1983). The transforming gene product is ∼28 kDa (Robbins et al 1982b; Niman 1984; Thiel & Hafenrichter 1984; Wang & Williams 1984). The normal human *sis* gene contains six regions of homology to the v-*sis* gene over a 12.5-kb region (Chiu et al 1984; Johnsson et al 1984; Josephs et al 1984a). Exon 6 of c-*sis* contains additional sequences not found in v-*sis*.

Structural Requirements of the v-sis Protein for Transformation

The first ATG codon is used in the initial translation product (Hannink et al 1986), which includes a signal peptide sequence between the second and third ATG codons (Hannink & Donoghue 1984). During translation the signal sequence is cleaved and an N-linked oligosaccharide is added at aspargine 93 (Hannink & Donoghue 1984). Approximately 4 kDa is lost from the amino terminus when the signal sequence is removed, resulting in a glycoprotein of ∼28 (Robbins et al 1983) or ∼32 kDa (Hannink et al 1986). This polypeptide then undergoes dimerization after entering the rough endoplasmic reticulum. During transport to the cell surface, the v-*sis* gene product is proteolytically cleaved at a basic dipeptide (Lys-Arg, residues 110 and 111), resulting in the formation of a 42-kDa dimer, and then cleaved at the carboxyl terminus, resulting in a mature 24-kDa dimer (Robbins et al 1983).

Deletion mutagenesis of the v-*sis* gene was used to define the limits at the carboxyl terminus of the v-*sis* gene product required for transformation (Hannink et al 1986). Protein products up to 57 residues shorter than the nondeleted v-*sis* gene product were transforming. The minimal region required for activity defined by these mutagenesis analyses was the 103 residues between residues 112 and 214, a sequence 6 residues shorter than the PDGF-homologous portion of the v-*sis* gene product. Cysteine residues at positions 208 and 210 are necessary for dimerization of the v-*sis* gene product, and products without these residues neither transform nor form dimers under nonreducing conditions (Hannink et al 1986). The amino-terminal signal sequence is required for transformation by v-*sis* as well (Hannink & Donoghue 1984, 1986; Hannink et al 1986). In other studies the minimum transforming region of the v-*sis* gene product was localized between residues 127 and 214, encoding a protein 21 residues smaller than the region homologous to the PDGF B chain gene (residues 112–220) (Sauer et al 1986).

Epidermal Growth Factor

Epidermal growth factor (EGF) was discovered when fractions from sub-maxillary glands not containing nerve growth factor (NGF) were shown to induce premature eyelid opening in newborn mice (Cohen 1962). EGF was soon isolated (Cohen & Elliott 1963) and its amino acid sequence determined (Savage et al 1972). EGF isolated from human urine is effectively identical to β-urogastrone, a polypeptide inhibitor of gastric acid secretion (Carpenter & Cohen 1975; Gregory 1975). EGF is mitogenic for a variety of mesenchymal cells and epithelial cells in culture (Carpenter & Cohen 1975). Its mitogenic activity is stimulated by insulin (Rose et al 1975; Shipley et al 1984).

Structure of EGF and the EGF Precursor

EGF is a 6-kDa polypeptide chain of 53 amino acids with three intrachain disulfide bonds (Taylor et al 1972). The nucleotide sequence of a complete cDNA derived from a male mouse submaxillary gland cDNA library (Gray et al 1983; Scott et al 1983) predicts a large protein precursor of 1217 amino acid residues with mature EGF near the carboxyl terminal at residues 977–1029. The predicted protein precursor contains an internal hydrophobic domain of 20 amino acids between polar regions, which suggests that it has a single membrane-spanning domain. The predicted cytoplasmic domain has 159 amino acids and the extracellular domain has 1010. The precursor contains 8 repeated units of ~40 amino acids with a sequence related to EGF. Each repeat contains 6 cysteine residues and lacks recognizable proteolytic cleavage sites. Residues 565–701 of the EGF precursor are closely related to residues 457–595 of the LDL receptor. The extracellular domain of the LDL receptor also contains 8 cysteine-rich repeats of 40 amino acids each. The EGF precursor may be an integral membrane protein precursor (Gray et al 1983; Scott et al 1983), but how mature EGF is processed from this large membrane-bound protein is not known. In situ hybridization revealed extensive expression of the precursor mRNA in tissues of whole newborn mice. The unprocessed EGF precursor has been found in the region of the distal tubule of mouse kidney (Rall et al 1985).

EGF-Related Polypeptides

The male mouse submaxillary gland has the highest levels of the fully processed EGF polypeptide. In contrast, two related growth factors are expressed in very different tissues. Transforming growth factor type α (TGF_α) has been shown to be expressed in transformed cells, in extracts of human tumors (Roberts et al 1980, 1983a; Marquardt et al 1984; Derynck et al 1984), and in developing fetuses (Proper et al 1982; Matrisian et al 1982; Twardzik et al 1982a; Twardzik 1985). The second EGF-like polypeptide was identified in cells infected with vaccinia virus (Blomquist et al 1984; Brown et al 1985; Reisner 1985). The vaccinia virus EGF-like polypeptide is 77 amino acids long (Stroobant et al 1985) and like EGF is derived from a large precursor protein, which after processing has 18 additional residues at the amino terminus and 9 added residues at the carboxyl terminus (Twardzik et al 1985a).

EGF Receptor

$[^{125}\text{I}]$EGF binds to a specific cell surface receptor with an apparent K_d of 10^{-9}–10^{-10} M (Hollenberg & Cuatrecasas 1973). It was later shown that TGF_α competes with EGF for this binding site. The human epidermoid carcinoma cell line (A-431), which has $> 10^6$ EGF receptor sites per cell (Fabricant et al 1977; Haigler et al 1978; Stoscheck & Carpenter 1984a), was used to identify phosphoproteins of 150 and 170 kDa. EGF stimulates tyrosine phosphorylation of these proteins (Carpenter et al 1979; King et al 1980). The EGF receptor was purified from A431 cells as the most abundant source of the receptor and shown to be a tyrosine-specific protein kinase with capacity for autophosphorylation and activity for exogenous substrates (Cohen et al 1980, 1982; Ushiro & Cohen 1980; Ehrhart et al 1981; Hunter & Cooper 1981). It is an integral membrane protein of 170 kDa (Cohen et al 1982); the nonglycosylated precursor polypeptide is ~ 130 kDa. In addition to glycosylation, posttranslational modifications include serine, threonine, and tyrosine phosphorylations (Hunter & Cooper 1981, 1985; Sefton & Hunter 1984; Pike & Krebs 1986). The endogenous kinase mediates tyrosine phosphorylation, whereas other kinases are responsible for phosphorylation of serine and threonine residues, including protein kinase C (Cochet et al 1984), which preferentially phosphorylates threonine 654 (Hunter et al 1984; Davis & Czech 1985). Phosphorylation of the EGF receptor by protein kinase C inhibits the intrinsic tyrosine kinase activity, reduces the apparent affinity of the receptor for EGF, and stimulates internalization of the receptor (Lee & Weinstein 1978; Shoyab et al 1979; Downward et al 1984a; Hunter et al 1984; Iwashita & Fox 1984; Bequinot et al 1985; Davis & Czech 1985). Thus protein kinase C has a negative influence on the EGF receptor. The EGF receptor is internalized when activated by EGF and is subsequently degraded (Carpenter & Cohen 1976). The EGF receptor has a half-life ($T_{1/2}$) of ~ 10 hr in human fibroblasts, but in the presence of EGF this value is reduced to ~ 1 hr (Stoscheck & Carpenter 1984b). Recycling of the

EGF receptor is insignificant (Stoscheck & Carpenter 1984b); thus the interaction of EGF with its receptor provides a negative feedback loop that reduces ("down-regulates") the potential for additional signaling by EGF.

Relationship of the EGF Receptor to the v-erbB Oncogene Product

Amino acid sequence analyses of tryptic digests of the EGF receptor demonstrated a striking homology between the EGF receptor and the deduced amino acid sequence of the v-*erbB* oncogene product (Downward et al 1984b), the product of the *erbB* gene of avian erythroblastosis virus (Yamamoto et al 1983). This remarkable finding established that the receptor protein itself has the potential for transformation. The complete cDNA of the EGF receptor was then obtained and shown to have close identity to known tyrosine kinases, including those of the oncogene products of *src*, *abl*, and *fms*, and a typical transmembrane hydrophobic sequence (Ullrich et al 1984). The external domain has 622 amino acids, 12 N-linked glycosylation sites, and clusters of cysteine-rich sequences. Tyrosines 1173, 1148, and 1058 were identified as potential autophosphorylation sites (Downward et al 1984a). The genes for c-*erbB* (Merlino et al 1984; Ullrich et al 1984) and the EGF receptor were located on human chromosome 7 (Kondo & Shimizu 1983), further confirming the identity of the c-*erbB* and EGF receptor genes.

A tyrosine kinase activity was subsequently demonstrated for the v-*erbB* gene product (Decker 1985; Gilmore et al 1985; Kris et al 1985). The v-*erbB* protein is truncated for the extracellular ligand binding domain of the EGF receptor and lacks either 32 or 71 amino acids in the carboxyl terminus (Ullrich et al 1984), but the shortened cytoplasmic domain retains the protein tyrosine kinase site (Yamamoto et al 1983; Privalsky et al 1984). Only a small fraction of the v-*erbB* product reaches the plasma membrane (Hayman & Beug 1984; Privalsky & Bishop 1984; Schmidt et al 1985); much of the protein remains within the Golgi complex. Similar results have been observed with the v-*fms* oncogene protein product (Anderson et al 1984; Manger et al 1984; Rettenmier et al 1985a,b). When activated by ligand the EGF receptor is internalized, whereas the v-*erbB* product, in the absence of a ligand binding site, remains at the cell surface. The kinase activity associated with the v-*erbB* protein is constitutively expressed. The v-*erbB* protein must reach the plasma membrane to be transforming (Beug & Hayman 1984; Roussel et al 1984; Schmidt et al 1985).

EGF: Potential for Transformation as an Oncogene Product

EGF has not been identified by homology to a specific oncogene product as might be expected based on the relationship of PDGF and the v-*sis* oncogene product. It is not clear whether the constitutive production of any of the polypeptide growth factors by cells that have the capacity to respond to the mitogen is sufficient to induce transformation, although this hypothesis has a strong logical base. In one test of this hypothesis, an expression vector was used to induce the constitutive production and secretion of human EGF. Upon transfection into FR 3T3 fibroblasts or Rat-1 fibroblasts, the EGF expression vector induced focus formation with the same efficiency as a plasmid carrying a *Ha-ras* oncogene. Clonal cell lines derived expressed high levels of EGF mRNA associated with secretion of an EGF-like activity. These cells induced tumors in nude mice (Stern et al 1987). The results demonstrate that EGF, like PDGF, when inappropriately expressed in cells bearing the EGF receptor, leads to transformation, and they support the model that the constitutive elaboration of normal growth factors in cells with appropriate receptors may be oncogenic.

Heparin Binding Growth Factors

Endothelial cells line the luminal surface of the vascular system and are believed to be essential for the integrity of blood vessels and to serve other functions as well, such as transport of substances to and from the circulation. Abnormalities in endothelial cell function are associated with atherosclerosis, thrombosis, and tumor metastases (Denekemp & Hobson 1984; Folkman & Cotran 1976). Angiogenesis, i.e. new blood vessel formation, is required for wound healing and for the growth of solid tumors (Folkman et al 1971; Folkman 1972). In some instances tumor cells appear to induce their own capillary formation, and thus tumor angiogenesis factors have been sought which mediate endothelial cell proliferation and which perhaps are needed for support of tumor growth and development (Ziche & Gullino 1982; Maciag 1984; Thomas 1985; Folkman 1985, 1986).

Based upon the use of heparin for purification, two classes of factors supporting endothelial cell growth have been identified (Lobb et al 1986a; Gospodarowicz 1987) and designated as heparin binding growth factors. These factors, e.g. endothelial cell growth factor (ECGF) and fibroblast growth factor (FGF), however, were originally identified on the basis of the range of activities they induced (Maciag et al 1984; Baird et al 1985a; Lobb et al 1986a; Gospodarowicz 1987). Each of these mitogens induces angiogenesis in vivo (Lobb & Fett 1984; Lobb et al 1985; Folkman 1985, 1986) or supports endothelial cell growth in culture. Class 1 heparin binding growth factors (HBGFs) have an acidic pI (4.8–6.0); the following growth factors are either identical or very closely related: acidic fibroblast growth factor (aFGF) (Lobb & Fett 1984; Thomas et al 1984; Lobb et al 1986a,b; Strydom et al 1986), endothelial cell growth factor (ECGF) (Maciag et al 1984), eye-derived growth factor II (EDGF-II) (Courty et al 1985), α-retina-derived growth factor (α-RDGF) (D'Amore & Klagsbrun 1984), anionic hypothalamus-derived growth factor (aHDGF) (Klagsbrun & Shing 1985), brain-derived growth factor (BDGF) (J. S. Huang et al 1986), and the astroglial growth factor I (AGF-I) (Pettmann et al 1985).

Class 2 HBGFs (Lobb & Fett 1984; Lobb et al 1986a) have a basic pI (8–10), and this group of identical or very closely related factors includes basic FGF (Gospodarowicz 1975; Gospodarowicz et al 1984), tumor angiogenesis factor (TAF) (Folkman 1985), EDGF-I (Courty et al 1985), β-RDGF (D'Amore & Klagsbrun 1984), cationic HDGF (Klagsbrun & Shing 1985), cartilage-derived growth factor (CDGF) (Sullivan & Klagsbrun 1985), astroglial growth factor (AGF) (Pettmann et al 1985), hepatoma-derived growth factor (HDGF) (Klagsbrun et al 1986), a portion of the macrophage-derived growth factor (MDGF) (Baird et al 1985b; Shimokado et al 1985), and perhaps others.

Amino acid sequence analyses of class 1 and class 2 HBGFs identified an approximately 50% homology between these mitogens from bovine neural tissues (Gimenez-Gallego et al 1985; Esch et al 1985a,b; Strydom et al 1986). Both the class 1 and class 2 HBGFs appeared to bind to the same cell surface receptors (Neufeld & Gospodarowicz 1986). A specific high-affinity receptor(s) for the HBGF has been identified on endothelial cell surfaces (Schreiber et al 1985; Friesel et al 1986; S. S. Huang & J. S. Huang 1986). The receptor(s) appears to have intrinsic, autophosphorylating tyrosine kinase activity when activated (S. S. Huang & J. S. Huang 1986).

Both class 1 and 2 HBGFs are found in neural tissues. Class 2 HBGFs have a broader distribution, however, and may be found in adrenal gland, corpus luteum, and kidney (Gospodarowicz 1987). The class 1 HBGFs have enhanced activity with heparin (Thornton et al 1983; Schreiber et al 1985; Orlidge & D'Amore 1986). This activation is unique among growth factors and may serve not only to activate class 1 HBGFs in tissues but perhaps also to localize class 1 HBGFs to heparinlike molecules on cell surfaces. The class 1 HBGFs were shown to elicit chemotaxis of endothelial cells (Terranova et al 1985), a characteristic that could be of major importance for repair of vascular endothelium. Heparin also elicits endothelial cell chemotaxis.

A human cDNA clone encoding the precursor of class 1 HBGFs (ECGF) was isolated from human brain cDNA (Jaye et al 1986). A bovine cDNA clone encoding a class 2 HBGF (basic FGF) was isolated from a pituitary cDNA library (Abraham et al 1986). A human class 1 HBGF mRNA transcript was 4.8 kb, whereas two transcripts of class 2 HBGF were 5 and 2.2 kb. The predicted amino acid sequences in both cases were 155 amino acids and agreed well with previously determined sequences (Burgess et al 1986; Ueno et al 1986). Lower molecular weight forms of both class 1 and class 2 HBGFs were identified as truncated forms resulting from proteolysis (Burgess et al 1985, 1986; Ueno et al 1986). These forms include the 140–amino acid polypeptide class 1 HBGF (Thomas et al 1985; Gimenez-Gallego et al 1985; Esch et al 1985a) and the 146–amino acid polypeptide of class 2 HBGF (basic FGF) (Gospodarowicz et al 1984; Esch et al 1985b). Yet to be resolved but of potential importance is whether different tissues process HBGFs differently or whether the different species obtained result from proteolysis during purification (Gospodarowicz et al 1986; Baird et al 1985c; Burgess et al 1985, 1986; Ueno et al 1986). A second issue of importance is the apparent lack of the signal peptide required for cotranslational transport through the membrane of the rough endoplasmic reticulum; thus it is not known how HBGF is secreted. Such a signal peptide also appears to be absent in the sequence of the IL-1 cDNA (Auron et al 1984). IL-1 is active in stimulating macrophages (Dexter 1986) and shares significant homology with the HBGFs (Gimenez-Gallego et al 1985; Thomas et al 1985).

As noted, the complete amino acid sequences of bovine class 2 HBGF (Esch et al 1985a,b) and class 1 HBGF (Gimenez-Gallego et al 1985) were 50% homologous, despite the striking difference in their isoelectric points. An antibody generated from a synthetic peptide from a conserved region of class 2 HBGFs recognized class 1 HBGF (Esch et al 1985a,b). Clusters of basic amino acids are highly conserved between the two proteins and likely represent heparin binding sites (Schwartzbauer et al 1983). Nine amino acids at the carboxyl terminus are conserved in both polypeptides. The structural gene for the class 1 HBGF is localized to human chromosome 5 (Zabel et al 1985) in a region (q31.3–33.2) deleted from the chromosomes of some patients with acute nonlymphocytic leukemia.

HBGFs are potent mitogens; they stimulate proliferation of endothelial cells at concentrations of 1–10 ng/ml. Both classes of HBGF stimulate the proliferation of the same mesoderm- or neuroectoderm-derived cells, which is consistent with their observed binding to the same cell surface receptor (Esch et al 1985a; Gimenez-Gallego et al 1985; Lobb et al 1986a; Gospodarowicz et al 1986). These factors also induce angiogenesis in vivo in chicken embryo allantoric membranes (Esch et al 1985b; Lobb et al

1985; Shing et al 1985; Thomas et al 1985) and in cornea (Lobb et al 1985; Shing et al 1985).

They also induced vascular granulation tissue when implanted subcutaneously (Gospodarowicz et al 1982; Davidson et al 1985). Type 2 HBGF is required for proliferation of cultured myoblasts (Gospodarowicz et al 1976) but not of embryonal rhabdomyosarcoma cells. Embryonal rhabdomyosarcomas are richly vascularized tumors (Hajdu 1979). Cultured human embryonal rhabdomyosarcoma cells were shown to express the class 2 (basic FGF) gene and two microheterogeneous forms of class 2 HBGFs, which suggests there is autocrine regulation of proliferation of these cells by class 2 HBGF (Schweigerer et al 1987). If class 2 HBGFs are secreted by human embryonal rhabdomyosarcoma cells, this appears to be an exception; little or no secretion of class 2 HBGFs (basic FGF) by cultured cells has been otherwise observed (Klagsbrun et al 1986; Vlodavsky et al 1986). This latter result is consistent with the absence of a signal peptide and suggests that the HBGFs may be released only under very special circumstances or perhaps in association with cell damage.

However, secretion of class 2 HBGF was demonstrated in extracts of stimulated peritoneal macrophages (Baird et al 1985b); the mechanism of release was not established. Macrophages secrete additional growth factors, including a PDGF-like mitogen and other as yet uncharacterized growth factors (Shimokado et al 1985). These activities may be of major importance since circulating monocytes can bind to endothelial cells, and the activated monocyte may release mitogenic activities involved in the repair of damaged vessels. Such a process could be important in fibrotic diseases and in the genesis of atherosclerosis (Faggiotto et al 1984; Pawlowski et al 1985). Other roles for HBGFs are likely; the class 2 HBGFs have been shown to support the survival of cerebral cortical neurons in primary culture (Morrison et al 1986), a finding consistent with a special role for HBGFs in tissues with high concentrations of these mitogens.

Transforming Growth Factors, Types α and β

Transforming growth factors (TGFs) were identified when transformed cell lines were shown to have reduced numbers of binding sites for EGF (Todaro et al 1976). An acid- and heat-stable activity that was competitive with EGF for binding was secreted by the transformed cells and was termed the sarcoma growth factor (De Larco & Todaro 1978) based upon its secretion from murine sarcoma virus–transformed cells. Ultimately this activity was isolated and is presently known as TGF_α. The sarcoma growth factor also was shown to induce the reversible transformed phenotype in normal rat kidney (NRK) cells grown in soft agar. Careful quantitation of colony growth was ultimately used to purify this second activity from nonneoplastic cells (Roberts et al 1981). The polypeptide responsible for the reversible transforming activity was designated TGF_β. It did not compete with EGF but required EGF or TGF_α for colony formation of NRK cells in soft agar (Roberts et al 1983a). A second polypeptide called TGF_α was independently isolated from chemically transformed cells; it also reversibly transformed AKR-2B mouse embryo fibroblasts in culture and was called TGF_γ (Moses et al 1981). It was later shown that TGF_γ and TGF_β are the same factor (Moses et al 1984).

TRANSFORMING GROWTH FACTOR α   Purification of the sarcoma growth factor(s) resulted in the separation of two very different polypeptides (Anzano et al 1983). The EGF-like activity of the sarcoma growth factor(s) was designated TGF_α; together with TGF_β it induces colony formation of NRK cells in soft agar. TGF_α was purified to homogeneity from conditioned media from human melanoma cells (Marquardt & Todaro 1982) and from fibroblasts transformed with retroviruses (Twardzik et al 1982b; Massagué 1983a). Amino acid sequencing of TGF_α (Marquardt et al 1983, 1984) showed that rodent and human TGF_α (Derynck et al 1984) are single-chain polypeptides of 50 amino acids with three disulfide bonds in homologous positions. Human (Derynck et al 1984) and rat TGF_α cDNAs (Gray et al 1983; Scott et al 1983; Lee et al 1985) are encoded by mRNAs of ~4.5–4.8 kb. The predicted precursor contains 160 residues with extensions at the amino- and carboxyl-terminal regions; the mature processed product is a 50–amino acid polypeptide. A hydrophobic, cysteine-rich extension of TGF_α was found in the nucleotide sequence; the sequence predicts that a transmembrane protein will be produced after the cleavage that releases mature TGF_α. The function of this transmembrane protein is not known (Derynck et al 1984; Todaro et al 1985), but it resembles the transmembrane extension of the EGF precursor and may be a feature of the growth factor receptors.

A plasmid directing the constitutive synthesis of the human TGF_α precursor was used to study the processing and posttranslational modification of TGF_α. TGF_α was synthesized as part of a glycosylated transmembrane precursor, which was cleaved to produce the 50–amino acid TGF_α and two larger glycosylated peptides. The cytoplasmic domain (carboxyl terminus) of the precursor remained attached to the membrane and was modified by the addition of palmitate. The nucleotide sequences encoding the transmembrane extension are very highly conserved between human and rat TGF_α; only a single conservative amino acid change was found. This conservation suggests that this portion of the molecule may also serve an essential biological function. The recombinant TGF_α polypeptide retains full biological activity (Derynck et al 1984). The fibroblasts that expressed the cDNA clone lost their capacity for anchorage-dependent growth and induced tumors in nude mice. Anti–human TGF_α monoclonal antibodies prevented TGF_α-expressing cells from forming colonies in soft agar. The TGF_α gene thus is similar to the human c-*sis* and EGF genes in that it can cause cell transformation if inappropriately expressed. It is likely that any growth factor will be transforming if it is inappropriately expressed in cells that also express its receptor. This hypothesis can be tested experimentally.

Evidence to support a role of TGFs in cell transformation was obtained by showing that the extent of growth of human tumor cell lines in soft agar increases as more EGF-like polypeptides are secreted (Todaro et al 1980). Temperature-sensitive mutants of the Kirsten sarcoma virus secreted EGF-like (i.e. TGF_α) polypeptides only at permissive temperatures (Ozanne et al 1980; De Larco et al 1981). The loss of transforming potential correlated with failure to secrete TGF_α. Anti–EGF receptor antisera blocked DNA synthesis and TGF_α-dependent, anchorage-independent growth of NRK cells (Carpenter et al 1983).

TGF_α species of different molecular weights have been identified (De Larco & Todaro 1978; Todaro et al 1980; Marquardt & Todaro 1982). A 10-kDa polypeptide from conditioned media in SDS gels may represent the 50–amino acid, mature TGF_α (De Larco & Todaro 1978; Marquardt & Todaro 1982; Linsley et al 1985; Marquardt et al 1984), whereas a second polypeptide of ~ 20 kDa may represent the 160–amino acid precursor predicted from the cDNA structure (Derynck et al 1984; Lee et al 1985). This 20-kDa species reacted with antibodies developed from the carboxyl-terminal 17 amino acids of TGF_α, although at a greatly reduced level (Linsley et al 1985).

EGF and TGF_α have been shown to directly compete for binding to the EGF receptor (Massagué 1983b; Derynck et al 1984; Tam et al 1984), despite the limited ($\sim 50\%$) homology between EGF and TGF_α. The three disulfide bonds of each, however, are in highly homologous positions and the conformation in this region of each protein may be recognized by the EGF receptor (Marquardt et al 1983, 1984; Derynck et al 1984). Four of the eight amino acids in the third disulfide loop are identical in all EGF-like structures. TGF_α produces all the physiological responses expected of EGF, including precocious eyelid opening in newborn mice (Smith et al 1985).

TGF_α and vaccinia growth factor (VGF) have extensive sequence homology with EGF and bind to the EGF receptor. Topical administration of TGF_α or VGF in antibiotic cream to second-degree burns accelerates epidermal regeneration in comparison with untreated or control treated burns (Schultz et al 1987). Regenerated epithelium from burns treated with TGF$_\alpha$ or VGF appeared normal histologically. Thus the topical application of selective growth factors may be useful in accelerating healing of second-degree burns. EGF present in saliva and applied to wounds by licking is thought to accelerate healing of cutaneous injuries in mice (Niall et al 1982). Topical application of EGF accelerated epidermal regeneration of middermal skin injuries (Brown et al 1986) and corneal abrasions (Brightwell et al 1985) and increased the tensile strength of healing corneal incisions (Brightwell et al 1985; Woost et al 1985).

TRANSFORMING GROWTH FACTOR β TGF$_\beta$ influences the rate of proliferation of many cell types, acting as a growth-inhibiting substance in most cell types (Holly et al 1980; Massagué 1984; Tucker et al 1984b; Shipley et al 1985; Roberts et al 1985b; Leof et al 1986). It has inhibitory activity in the adipogenic differentiation of 3T3 fibroblasts and influences myogenesis, chondrogenesis, osteogenesis, epithelial cell differentiation, and immune cell function (Ignotz & Massagué 1985; Massagué et al 1986; Seyedin et al 1985; Centrella et al 1986; Masui et al 1986; Rook et al 1986). TGF$_\beta$ thus may be the prototype of a family of homologous polypeptides that regulate control and development of tissues and organisms. This family includes various inhibins and activins, which, for example, stimulate the ability of cultured pituitary cells to release follicle-stimulating hormone (Mason et al 1985; Vale et al 1986; Ling et al 1986); the Müllerian inhibitory substance, which inhibits development of the Müllerian duct in male mammalian embryos (Cate et al 1986); and the products of the decapentaplegic gene complex, which support the development of *Drosophila* (Padgett et al 1987) and inhibit IL-2 dependent T-cell proliferation (Kehrl et al 1986).

The Müllerian inhibitory substance is a glycoprotein of ~ 140 kDa expressed early in gonadal differentiation (Cate et al 1986). The highly conserved C-terminal domain of this protein shows extensive homology with TGF$_\beta$ and is able to inhibit the growth of Müllerian derived tumors (Fuller et al 1982, 1984). The inhibins, protein inhibitors of the secretion of follicle stimulating hormones, are also structurally related to TGF$_\beta$. These proteins show significant homology to TGF$_\beta$, including the placement of cysteine residues, which suggests structural similarity (Mason et al 1985), although their biological activities appear to be unrelated (Ying et al 1986). Furthermore, cartilage-inducing factors A and B (CIF-A, CIF-B) are factors isolated from bovine demineralized bone that induce the synthesis of cartilage. CIF-A is the major species isolated from bone. Both molecules induce fetal cells to undergo differentiation and to synthesize cartilage-specific macromolecules in vitro (Seyedin et al 1986). CIF-B also induces the anchorage-independent proliferation of NRK-49F cells when these cells are simultaneously treated with EGF. CIF-B competes with

CIF-A for the same membrane receptors on NRK-49F cells, but partial amino acid sequencing reveals that CIF-B is a distinct molecule. Comparison of the sequence of the N-terminal 30 amino acids of CIF-A from bovine demineralized bone revealed 100% identity with the corresponding sequence of human TGF_β. CIF-A stimulates normal rat kidney fibroblasts to become anchorage independent and to form colonies in soft agar in the presence of EGF, as does TGF_β. In addition, TGF_β from human platelets induces rat muscle mesenchymal cells to differentiate and to synthesize cartilage-specific macromolecules in a manner equivalent to CIF-A.

TGF_β is ubiquitously expressed in embryonic and adult tissues (Anzano et al 1985a; Moses et al 1981; Tucker et al 1983, 1984a). Its secretion increases in transformed cells and in cells activated by mitogens (Anzano et al 1985b; Derynck et al 1985). Platelets have been used as a source for its purification (Childs et al 1982; Assoian et al 1983). The fact that TGF_β is released from platelets at sites of blood vessel injury suggests that it plays a role in wound healing (Assoian et al 1983; Frolik et al 1983; Roberts et al 1983b; Massagué 1984). TGF_β is a homodimeric polypeptide of ~ 25 kDa with nine half-cystine residues per 12.5-kDa monomer. A precursor polypeptide of 391 amino acids was predicted by cDNA cloning of TGF_β encoded by a 2.4-kb mRNA (Derynck et al 1985).

The amino acid sequence of TGF_β has been highly conserved during evolution (Derynck et al 1985; Roberts et al 1983b), and it is active at picomolar concentrations (Frolik et al 1984; Tucker et al 1984b; Massagué & Like 1985). TGF_β receptors also appear to be ubiquitously expressed, are not subject to down-regulation by high levels of TGF_β (Roberts et al 1984, 1985a), and appear to lack tyrosine kinase activity (Fanger et al 1985). The TGF_β receptor is a disulfide-linked homodimer of 280–300 kDa (Massagué 1985) and is found on cells from normal tissues and tumors (Tucker et al 1984b).

A separate homodimeric form of TGF_β, $TGF_{\beta2}$, has been recently identified in porcine blood platelets. $TGF_{\beta2}$ is homologous to TGF_β, which now is called $TGF_{\beta1}$. $TGF_{\beta1-2}$ dimers have also been isolated; these are heterodimers containing one $TGF_{\beta1}$ chain and one $TGF_{\beta2}$ chain. $TGF_{\beta1}$ and $TGF_{\beta2}$ were shown to interact differently with a family of receptors in target cells. A 280-kDa receptor displayed high affinity for both $TGF_{\beta1}$ and $TGF_{\beta2}$, and its occupancy by $TGF_{\beta1}$ or $TGF_{\beta2}$ correlated directly with the inhibition of cell proliferation. Receptors of 65 and 85 kDa have also been found with high affinity for $TGF_{\beta1}$ but lower affinity for $TGF_{\beta2}$. These distinct forms of TGF_β receptors are postulated to provide flexibility for the regulation of growth and differentiation by the TGF_βs (Cheifetz et al 1987).

A serious problem in elucidation of the function of TGF_β has been the apparently disparate activities it mediates. Together with EGF or TGF_α, it supports colony formation of NRK cells in soft agar and the anchorage-independent growth of ARK-2B fibroblasts and of rat embryo fibroblasts bearing the *myc* oncogene (Moses et al 1984; Roberts et al 1985a,b). However, TGF_β inhibits growth of BSC-1 monkey kidney cells in conditioned media (Holly et al 1980; Tucker et al 1984a), of human foreskin keratinocytes (Moses et al 1985), of human bronchial epithelial cells (Masui et al 1985), and of lines derived from human tumors (Roberts et al 1985b; Moses et al 1985). Although TGF_β and EGF together stimulate anchorage-independent growth of NRK fibroblasts, the two activities are antagonistic in anchorage-dependent growth of NRK cells (Anzano et al 1982; Roberts et al 1985b).

PDGF with TGF_β induces reversible transformation of NRK cells (Assoian et al 1984b). TGF_α and TGF_β were found secreted together in conditioned media of malignant cells. Neither TGF_α nor TGF_β is specific for the transformed cell but the secretory activity of TGF_α in human tumor cell lines (Todaro et al 1980) may serve as a marker for neoplastic activity. High–molecular weight TGF_α has been identified in the urine of cancer patients and may be useful in assessing growth of tumors in humans (Twardzik et al 1982c, 1985b; Sherwin et al 1983; Kimball et al 1984).

The mechanisms by which TGF_β influences cells remain to be elucidated. TGF_β has been shown to increase the following: the number of EGF binding sites on NRK fibroblasts (Assoian et al 1984a), glucose uptake (Inman & Colowick 1985), amino acid transport (Boerner et al 1985; Racker et al 1985), the secretion of collagenase (Chua et al 1985), and the release of calcium from mouse calvaria cells (Tashjian et al 1985). The total protein, collagen, and DNA content increases when TGF_β is placed within chambers implanted subcutaneously in experimental animals (Sporn et al 1983).

TGF_β is a potent desmoplastic agent. When injected subcutaneously into newborn mice, it causes a rapid increase in connective tissue formation (Roberts et al 1986). TGF_β also is active in stimulating fibroblast chemotaxis and production of collagen and fibronectin (Roberts et al 1986; Varga & Jimenez 1986). TGF_β preferentially stimulates the synthesis of fibronectin and procollagen I chains 3–5 fold, as shown by polypeptide analysis. Exposure to TGF_β caused elevation of the steady-state levels of mRNAs coding for type 1 procollagen and fibronectin without increasing the rate of transcription of either of these genes. These findings suggest that TGF_β selectively stabilizes the mRNAs for procollagen I and fibronectin.

TGF_β has a biphasic influence on bone cell replication that depends both on the TGF_β concentration and on the cell density in monolayer culture. After 24 hr of treatment with TGF$_\beta$, DNA synthesis in confluent cells is progressively enhanced, but to achieve this effect in subconfluent cells, higher concentrations of TGF$_\beta$ are needed. Moreover, in sparse cell cultures, TGF$_\beta$ inhibits the growth of osteoblast cell lines. At all cell densities, however, TGF$_\beta$ stimulates collagen synthesis. Since TGF$_\beta$ is found in medium conditioned by bone explants and in bone tissue extracts, these results support the idea that TGF$_\beta$ is an important and multifunctional autocrine regulator of bone formation (Centrella et al 1987).

Summary and Conclusion

An increasing number of polypeptide growth factors have been identified that regulate not only cell proliferation but an extraordinary range of cell activities, including matrix protein deposition and resolution, the maintenance of cell viability, cell differentiation, inflammation, and tissue repair. Normal cells appear to require growth factors for proliferation and for maintenance of viability. Cells that secrete a polypeptide growth factor have an advantage in growth. These factors can act either externally through cell surface receptors or perhaps internally during the transport of receptors and growth factors through the ER and Golgi, causing autocrine stimulation of cell growth. Depending on the cell type, growth factors can also be potent inhibitors of cell growth rather than stimulating growth, and the effects can depend on the presence or absence of other growth factors. Platelet-derived growth factor has been shown to be nearly identical to the product of the v-*sis* gene of the simian sarcoma virus, which appears to cause cell transformation through its interactions with the PDGF receptor activating the tyrosine kinase activity of the PDGF receptor. Similarly, two proto-oncogenes, c-*erbB* and c-*fms*, encode growth factor receptors. The EGF receptor activity of the v-*erb* oncogene product appears to be constitutively activated without the need for growth factor, perhaps because of the truncation at the amino terminus deleting the EGF binding domain. The induction of the *myc* and the *fos* proteins by growth factor stimulation of quiescent cells, as well as the potential for the p21 product of the *ras* oncogene to act as an intermediate in transducing adrenergic signals, provide direct evidence that these pathways are important for stimulation of cell growth. Cells transformed by the v-*sis* oncogene always appear to bear PDGF cell surface receptors, which suggests that this oncogene has a specific requirement of the PDGF receptor for transformation. In contrast, cells transformed by the v-*erbB* and v-*fms* oncogenes are not stimulated by EGF or by CSF-1. Thus it seems likely that the tyrosine kinase activity of the corresponding receptor is ubiquitously expressed in these cases.

Major questions remain unanswered. In particular, what are the mechanisms by which growth factors initiate pathways leading to DNA synthesis? What are the physiological substrates of the growth factor receptor tyrosine kinase? Considerable effort also is needed to further define the cellular specificity of the different growth factors, particularly within intact tissues, and to determine how the various growth factors interact. Four growth factors have been discussed in this review, each of which has contributed to our understanding of neoplasia through different mechanisms. Similarly, these growth factors have led to important conclusions concerning pathways not only of mitogenesis but of differentiation, development, inflammation, and tissue repair. Further research should provide important new insights concerning the relationship between the overexpression of growth factors and a variety of pathological conditions, including atherosclerosis, pulmonary fibrosis, bone marrow fibrosis, and neoplasia.

ACKNOWLEDGMENTS

I thank Drs. Philip Majerus, Laura Shawver, Glen Pierce, and Rodney Kawahara for helpful discussions and review in the preparation of this manuscript. TFD is supported by grants awarded by the National Institutes of Health (HL31102 and HL14147) and by a grant from the Monsanto Corporation.

Literature Cited

Abraham, J. A., Mergia, A., Whang, J. L., Tumolo, A., Friedman, J., et al. 1986. Nucleotide sequence of a bovine clone encoding the angiogenic protein, basic fibroblast growth factor. *Science* 233: 545–48

Anderson, S. J., Gonda, M. A., Rettenmier, C. W., Sherr, C. J. 1984. Subcellular localization of glycoproteins encoded by the viral oncogene *v-fms. J. Virol.* 51: 730–41

Angrist, A., Oka, M. 1963. Pathogenesis of bacterial endocarditis. *J. Am. Med. Assoc.* 183: 249–52

Antoniades, H. N. 1981. Human platelet-derived growth factor (PDGF): Purification of PDGF-I and PDGF-II and separation of their reduced subunits. *Proc. Natl. Acad. Sci. USA* 78: 7314–17

Antoniades, H. N., Hunkapiller, M. W. 1983. Human platelet-derived growth factor (PDGF) amino-terminal amino acid sequence. *Science* 220: 963–65

Antoniades, H. N., Scher, C. D., Stiles, C. D. 1979. Purification of human platelet-derived growth factor. *Proc. Natl. Acad. Sci. USA* 76: 1809–13

Antoniades, H. N., Stathakos, D., Scher, C. D. 1975. Isolation of a cationic polypeptide from human serum that stimulates proliferation of 3T3 cells. *Proc. Natl. Acad. Sci. USA* 72: 2635–39

Anzano, M. A., Roberts, A. B., DeLarco, J. E., Wakefield, L. M., Assoian, R. K., et al. 1985b. Increased secretion of type beta transforming growth factor accompanies viral transformation of cells. *Mol. Cell. Biol.* 5: 242–47

Anzano, M. A., Roberts, A. B., Meyers, C. A., Komoriya, A., Lamb, L. C., et al. 1982. Synergistic interaction of two classes of transforming growth factors from murine sarcoma cells. *Cancer Res.* 42: 4776–78

Anzano, M. A., Roberts, A. B., Smith, J. M., Sporn, M. B., DeLarco, J. E. 1983. Sarcoma growth factor from conditioned medium is composed of both type alpha and type beta transforming growth factors. *Proc. Natl. Acad. Sci. USA* 80: 6264–68

Anzano, M. A., Roberts, A. B., Sporn, M. B. 1985a. Anchorage-independent growth of primary rat embryo cells is induced by platelet-derived growth factor and inhibited by type beta transforming growth factor. *Fed. Proc.* 44: 694

Armelin, H. A., Armelin, M. C. S., Kelly, K., Stewart, T., Leder, P., et al. 1984. Functional role for c-*myc* in mitogenic response to platelet-derived growth factor. *Nature* 310: 655–60

Assoian, R. K., Frolik, C. A., Roberts, A. B., Miller, D. M., Sporn, M. B. 1984a. Transforming growth factor-beta controls receptor levels for epidermal growth factor in NRK fibroblasts. *Cell* 36: 35–41

Assoian, R. K., Grotendorst, G. R., Miller, D. M., Sporn, M. B. 1984b. Three growth factors from human platelets coordinating phenotype transformation. *Nature* 309: 804–6

Assoian, R. K., Komoriya, A., Myers, C. A., Miller, D. M., Sporn, M. B. 1983. Transforming growth factor-beta in human platelets. *J. Biol. Chem.* 258: 7155–60

Aurias, A., Rimbaut, C., Buffe, D., Dubousset, J., Mazabraud, A. 1983. Chromosomal translocations in Ewing's sarcoma. *N. Engl. J. Med.* 309: 496–97

Auron, P. E., Webb, A. C., Rosenwasser, L. J., Mucci, S. F., Rich, A., et al. 1984. Nucleotide sequence of human monocyte interleukin 1 precursor cDNA. *Proc. Natl. Acad. Sci. USA* 81: 7907–11

Bachofen, M., Weibel, E. R. 1982. Structural alterations of lung parenchyma in the adult respiratory distress syndrome. *Clin. Chest Med.* 3: 35–56

Bachofen, M., Weibel, E. R., Roos, B. 1975. Postmortem fixation of human lungs for electron microscopy. *Am. Rev. Respir. Dis.* 111: 247–56

Baird, A., Esch, F., Böhlen, P., Ling, N., Gospodarowicz, D. 1985c. Isolation and partial characterization of an endothelial cell growth factor from the bovine kidney: Homology with basic fibroblast growth factor. *Regulatory Peptides* 12: 201–13

Baird, A., Mormède, P., Böhlen, P. 1985b. Immunoreactive fibroblast growth factor in cells of peritoneal exudate suggests its identity with macrophage derived growth factor. *Biochem. Biophys. Res. Commun.* 126: 358–64

Baird, A., Mormède, P., Ying, S.-Y., Wehrenberg, W. B., Ueno, N., et al. 1985a. A nonmitogenic pituitary function of fibroblast growth factor: Regulation of thyrotropin and prolactin secretion. *Proc. Natl. Acad. Sci. USA* 82: 5545–49

Balk, S. D. 1971. Calcium as a regulator of the proliferation of normal, but not of transformed, chicken fibroblasts in a plasma-containing medium. *Proc. Natl. Acad. Sci. USA* 68: 271–75

Balk, S. D., Whitefield, J. F., Youdale, T.,

Braun, A. C. 1973. Roles of calcium, serum, plasma, and folic acid in the control of proliferation of normal and Rous sarcoma virus–infected chicken fibroblasts. *Proc. Natl. Acad. Sci. USA* 70: 675–79

Ballester, R., Rosen, O. M. 1985. Fate of immunoprecipitable protein kinase C in GH_3 cells treated with phorbol 12-myristate 13-acetate. *J. Biol. Chem.* 260: 15194–99

Barrett, T. B., Gajdusek, J. C. M., Schwartz, S. M., McDougall, J. K., Benditt, E. P. 1984. Expression of the *sis* gene by endothelial cells in culture and in vivo. *Proc. Natl. Acad. Sci. USA* 81: 6772–74

Bartram, C. R., de Klein, A., Hagemeijer, A., Grosveld, G., Heisterkamp, N., et al. 1984. Localization of the human c-*sis* oncogene in Ph[1]-positive and Ph[1]-negative chronic myelocytic leukemia by in situ hybridization. *Blood* 63: 223–25

Bauer, E. A., Cooper, T. W., Huang, J. S., Altman, J., Deuel, T. F. 1985. Stimulation of in vitro human skin collagenase expression by platelet-derived growth factor. *Proc. Natl. Acad. Sci. USA* 82: 4132–36

Bechet, J.-M., Bornkamm, G., Freese, U.-K., Lenoir, G. M. 1984. The c-*sis* oncogene is not activated in Ewing's sarcoma. *N. Engl. J. Med.* 310: 393

Bequinot, L., Hanover, J. A., Ito, S., Richert, N. D., Willingham, M. C. 1985. Phorbol esters induce transient internalization without degradation of unoccupied epidermal growth factor receptors. *Proc. Natl. Acad. Sci. USA* 82: 2774–78

Berk, B. C., Alexander, R. W., Brock, T. A., Gimbrone, M. A., Webb, C. R. 1986. Vasoconstriction: A new activity for platelet-derived growth factor. *Science* 232: 87–90

Besterman, J. M., Cuatrecasas, P. 1984. Phorbol esters rapidly stimulate amiloride-sensitive Na^+/H^+ exchange in a human leukemic cell line. *J. Cell Biol.* 99: 340–43

Betsholtz, C., Heldin, C.-H., Nister, M., Ek, B., Wasteson, A., et al. 1983. Synthesis of a PDGF-like growth factor in human glioma and sarcoma cells suggests the expression of the cellular homologue to the transforming protein of simian sarcoma virus. *Biochem. Biophys. Res. Commun.* 117: 176–82

Betsholtz, C., Johnsson, A., Heldin, C.-H., Westermark, B., Lind, P., et al. 1986. cDNA sequence and chromosomal localization of human platelet-derived growth factor A-chain and its expression in tumor cell lines. *Nature* 320: 695–99

Beug, H., Hayman, M. J. 1984. Temperature-sensitive mutants of avian erythroblastosis virus: Surface expression of the *erbB* product correlates with transformation. *Cell* 36: 963–72

Blackshear, P. J., Wen, L., Glynn, B. P., Witters, L. A. 1986. Protein kinase C–stimulated phosphorylation in vitro of a M_r 80,000 protein phosphorylated in response to phorbol esters and growth factors in intact fibroblasts: Distinction from protein kinase C and prominence in brain. *J. Biol. Chem.* 261: 1459–69

Blackshear, P. J., Witters, L. A., Girard, P. R., Kuo, J. F., Quamo, S. N. 1985. Growth factor-stimulated protein phosphorylation in 3T3-L1 cells: Evidence for protein kinase C-dependent and -independent pathways. *J. Biol. Chem.* 260: 13304–15

Blomquist, M. C., Hunt, L. T., Barker, W. C. 1984. Vaccinia virus 19–kilodalton protein: Relationship to several mammalian proteins including two growth factors. *Proc. Natl. Acad. Sci. USA* 81: 7363–67

Bockus, B. J., Stiles, C. D. 1984. Regulation of cytoskeletal architecture by platelet-derived growth factor, insulin, and epidermal growth factor. *Exp. Cell Res.* 153: 186–97

Boerner, P., Resnick, R. J., Racker, E. 1985. Stimulation of glycolysis and amino acid uptake in NRK-49F cells by transforming growth factor beta and epidermal growth factor. *Proc. Natl. Acad. Sci. USA* 82: 1350–53

Bowen-Pope, D. F., Dicorleto, P. E., Ross, R. 1983. Interactions between the receptors for platelet-derived growth factor and epidermal growth factor. *J. Cell Biol.* 96: 679–83

Bowen-Pope, D. F., Malpass, T. W., Foster, D. M., Ross, R. 1984. Platelet-derived growth factor in vivo: Levels, activity, and rate of clearance. *Blood* 64: 458–69

Bowen-Pope, D. F., Ross, R. 1982. Platelet-derived growth factor. II. Specific binding to cultured cells. *J. Biol. Chem.* 257: 5161–71

Bowen-Pope, D. F., Seifert, R. A. 1985. Exogenous and endogenous sources of PDGF-like molecules and their potential roles in vascular biology. In *Cancer Cells*, ed. J. Ferramisco, B. Ozanne, C. Stiles, pp. 183–88

Bright, M. D., Gaffney, E. V. 1982. Demonstration of competence and progression activities for human fibroblasts. *Exp. Cell Res.* 137: 309–16

Brightwell, J. R., Riddle, S. L., Eiferman, R. A., Valenzuela, P., Barr, P. J., et al. 1985. Biosynthetic human EGF accelerates healing of neodecadron-treated primate corneas. *Invest. Ophthalmol. Vis. Sci.* 26: 105–6

Brown, G. L., Curtsinger, L. III, Brightwell, J. R., Ackerman, D. M., Tobin, G. R., et al. 1986. Enhancement of epidermal regeneration by biosynthetic epidermal growth factor. *J. Exp. Med.* 163: 1319–24

Brown, J. P., Twardzik, D. R., Marquardt, H., Todaro, G. J. 1985. Vaccinia virus encodes a polypeptide homologous to epidermal growth factor and transforming growth factor. *Nature* 313: 491–92

Brugge, J. S. 1986. The p35/p36 substrates of protein-tyrosine kinases as inhibitors of phospholipase A_2. *Cell* 46: 149–50

Burgess, W. H., Mehlman, T., Friesel, R., Johnson, W. V., Maciag, T. 1985. Multiple forms of endothelial cell growth factor: Rapid isolation and biological and chemical characterization. *J. Biol. Chem.* 260: 11389–92

Burgess, W. H., Mehlman, T., Marshak, D. R., Fraser, B. A., Maciag, T. 1986. Structural evidence that endothelial cell growth factor B is the precursor of both endothelial cell growth factor a and acidic fibroblast growth factor. *Proc. Natl. Acad. Sci. USA* 83: 7216–20

Busch, C., Wasteson, A., Westermark, B. 1976. Release of a cell growth promoting factor from human platelets. *Thromb. Res.* 8: 493–500

Canaani, E., Steiner-Saltz, D., Aghai, E., Gale, R. P., Berrebi, A., et al. 1984. Altered transcription of an oncogene in chronic myeloid leukaemia. *Lancet* I: 593–95

Canalis, E. 1981. Effect of platelet-derived growth factor on DNA and protein synthesis in cultured rat calvaria. *Metabolism* 30: 970–75

Carpenter, G., Cohen, S. 1975. Human epidermal growth factor and the proliferation of human fibroblasts. *J. Cell. Physiol.* 88: 227–38

Carpenter, G., Cohen, S. 1976. ^{125}I-labeled human epidermal growth factor: Binding, internalization, and degradation in human fibroblasts. *J. Cell Biol.* 71: 159–71

Carpenter, G., King, L. Jr., Cohen, S. 1979. Rapid enhancement of protein phosphorylation in A-431 cell membrane preparations by epidermal growth factor. *J. Biol. Chem.* 254: 4884–91

Carpenter, G., Stoscheck, C. M., Preston, Y. A., DeLarco, J. E. 1983. Antibodies to the epidermal growth factor receptor block the biological activities of sarcoma growth factor. *Proc. Natl. Acad. Sci. USA* 80: 5627–30

Cassel, D., Rothenberg, P., Zhuang, Y.-X., Deuel, T. F., Glaser, L. 1983. Platelet-derived growth factor stimulates Na^+/H^+

and induces cytoplasmic alkalinization in NR6 cells. *Proc. Natl. Acad. Sci. USA* 80: 6224–28

Cate, R. L., Mattaliano, R. J., Hession, C., Tizard, R., Farber, N. M., et al. 1986. Isolation of the bovine and human genes for Müllerian inhibiting substance and expression of the human gene in animal cells. *Cell* 45: 685–98

Centrella, M., Massagué, J., Canalis, E. 1986. Human platelet-derived transforming growth factor-beta stimulates parameters of bone growth in fetal rat calvariae. *Endocrinology* 119: 2306–12

Centrella, M., McCarthy, T. L., Canalis, E. 1987. Transforming growth factor *β* is a bifunctional regulator of replication and collagen synthesis in osteoblast-enriched cell cultures from fetal rat bone. *J. Biol. Chem.* 262: 2869–74

Chait, A., Ross, R., Albers, J. J., Bierman, E. L. 1980. Platelet-derived growth factor stimulates activity of low density lipoprotein receptors. *Proc. Natl. Acad. Sci. USA* 77: 4084–88

Chambard, J. C., Franchi, A., LeCam, A., Pouyssegur, J. 1983. Growth factor-stimulated protein phosphorylation in B_0/G_1-arrested fibroblasts. Two distinct classes of growth factors with potentiating effects. *J. Biol. Chem.* 258: 1706–13

Cheifetz, S., Weatherbee, J. A., Tsang, M. L.-S., Anderson, J. K., Mole, J. E., et al. 1987. The transforming growth factor-*β* system, a complex pattern of cross-reactive ligands and receptors. *Cell* 48: 409–15

Chernoff, A., Levine, R. F., Goodman, D. S. 1980. Origin of platelet-derived growth factor in megakaryocytes in guinea pigs. *J. Clin. Invest.* 65: 926–30

Childs, C. B., Proper, J. A., Tucker, R. F., Moses, H. L. 1982. Serum contains a platelet-derived transforming growth factor. *Proc. Natl. Acad. Sci. USA* 79: 5312–16

Chiu, I.-M., Reddy, E. P., Givol, D., Robbins, K. C., Tronick, S., et al. 1984. Nucleotide sequence analysis identifies the human c-*sis* proto-oncogene as a structural gene for platelet-derived growth factor. *Cell* 37: 123–29

Chua, C. C., Gieman, D. E., Keller, G. H., Ladda, R. L. 1985. Induction of collagenase secretion in human fibroblast cultures by growth promoting factors. *J. Biol. Chem.* 260: 5213–16

Clarke, M. F., Westin, E., Schmidt, D., Josephs, S. F., Ratner, L., et al. 1984. Transformation of NIH/3T3 cells by a human c-*sis* cDNA clone. *Nature* 308: 464–67

Clemmons, D. R., Shaw, D. S. 1983. Variables controlling somatomedin production by cultured human fibroblasts. *J.*

Cell. Physiol. 115: 137–42

Clemmons, D. R., Van Wyk, J. J. 1981. Somatomedin-C and platelet derived growth factor stimulate human fibroblast replication. *J. Cell. Physiol.* 106: 361–67

Clemmons, D. R., Van Wyk, J. J., Pledger, W. J. 1980. Sequential addition of platelet factor and plasma to BALB/c 3T3 fibroblast cultures stimulates somatomedin-C binding early in cell cycle. *Proc. Natl. Acad. Sci. USA* 77: 6644–48

Cochet, C., Gill, G. N., Meisenhelder, J., Cooper, J. A., Hunter, T. 1984. C-kinase phosphorylates the epidermal growth factor receptor and reduces its epidermal growth factor-stimulated tyrosine protein kinase activity. *J. Biol. Chem.* 259: 2553–58

Cochran, B. H., Lillquist, J. S., Stiles, C. D. 1981. Post-transcriptional control of protein synthesis in BALB/c 3T3 cells by platelet-derived growth factor and platelet-poor plasma. *J. Cell. Physiol.* 109: 429–38

Cochran, B. H., Zullo, J., Verma, I. M., Stiles, C. D. 1984. Expression of the c-*fos* gene and the *fos*-related gene as stimulated by platelet-derived growth factor. *Science* 226: 1080–82

Cohen, S. 1962. Isolation of a mouse submaxillary gland protein accelerating incisor eruption and eyelid opening in the new-born animal. *J. Biol. Chem.* 237: 1555–62

Cohen, S., Carpenter, G., King, L. Jr. 1980. Epidermal growth factor-receptor–protein kinase interactions. Co-purification of receptor and epidermal growth factor-enhanced phosphorylation activity. *J. Biol. Chem.* 255: 4834–42

Cohen, S., Elliott, G. A. 1963. The stimulation of epidermal keratinization by a protein isolated from the submaxillary gland of a mouse. *J. Invest. Dermatol.* 40: 1–5

Cohen, S., Ushiro, H., Stoscheck, C., Chinkers, M. 1982. A native 170,000 epidermal growth factor receptor–kinase complex from shed plasma membrane vesicles. *J. Biol. Chem.* 257: 1523–31

Collins, M. K. L., Rozengurt, E. 1982a. Binding of phorbol esters to high-affinity sites on murine fibroblastic cells elicits a mitogenic response. *J. Cell. Physiol.* 112: 42–50

Collins, M. K. L., Rozengurt, E. 1982b. Stimulation of DNA synthesis in murine fibroblasts by the tumour promoter teleocidin: Relationship to phorbol esters and vasopressin. *Biochem. Biophys. Res. Commun.* 104: 1159–66

Collins, M. K. L., Rozengurt, E. 1984. Homologous and heterologous mitogenic desensitization of Swiss 3T3 cells to phorbol esters and vasopressin: Role of receptor and postreceptor steps. *J. Cell. Physiol.* 118: 133–42

Collins, M. K. L., Sinnett-Smith, J. W., Rozengurt, E. 1983. Platelet-derived growth factor treatment decreases the affinity of the epidermal growth factor receptors of Swiss 3T3 cells. *J. Biol. Chem.* 258: 11689–93

Collins, S. J., Kubonishi, I., Miyoshi, I., Groudine, M. T. 1984. Altered transcription of the c-*abl* oncogene in K-562 and other chronic myelogenous leukemia cells. *Science* 225: 72–74

Cooper, J. A., Bowen-Pope, D. F., Raines, E., Ross, R., Hunter, T. 1982. Similar effects of platelet-derived growth factor and epidermal growth factor on the phosphorylation of tyrosine in cellular proteins. *Cell* 31: 263–73

Cooper, J. A., Sefton, B. M., Hunter, T. 1984. Diverse mitogenic agents induce the phosphorylation of two related 42,000-dalton proteins on tyrosine in quiescent chick cells. *Mol. Cell. Biol.* 4: 30–37

Courty, J., Loret, C., Moenner, M., Chevallier, B., Lagente, O., et al. 1985. Bovine retina contains three growth factor activities with different affinity to heparin: Eye derived growth factor I, II, III. *Biochimie* 67: 265–69

Dalla-Favera, R., Gelmann, E. P., Gallo, R. C., Wong-Staal, F. 1981. A human onc gene homologous to the transforming gene (v-*sis*) of simian sarcoma virus. *Nature* 292: 31–35

D'Amore, P. A., Klagsbrun, M. 1984. Endothelial cell mitogens derived from retina and hypothalamus: Biochemical and biological similarities. *J. Cell Biol.* 99: 1545–49

Daniel, T. O., Tremble, P. M., Frackelton, A. R. Jr., Williams, L. T. 1985. Purification of the platelet-derived growth factor receptor by using an anti-phosphotyrosine antibody. *Proc. Natl. Acad. Sci. USA* 82: 2684–87

Davidson, J. M., Klagsbrun, M., Hill, K. E., Buckley, A., Sullivan, R., et al. 1985. Accelerated wound repair, cell proliferation, and collagen accumulation are produced by a cartilage-derived growth factor. *J. Cell Biol.* 100: 1219–27

Davis, R. J., Czech, M. P. 1985. Tumor-promoting phorbol diesters cause the phosphorylation of epidermal growth factor receptors in normal fibroblasts at threonine-654. *Proc. Natl. Acad. Sci. USA* 82: 1974–78

Decker, S. J. 1985. Phosphorylation of the *erbB* gene product from an avian erythroblastosis virus-transformed chick fibroblast cell line. *J. Biol. Chem.* 260: 2003–7

De Larco, J. E., Preston, Y. A., Todaro, G. J. 1981. Properties of a sarcoma-growth-factor-like peptide from cells transformed by a temperature-sensitive sarcoma virus. *J. Cell. Physiol.* 109: 143–52

De Larco, J. E., Todaro, G. J. 1978. Growth factors from murine sarcoma virus–transformed cells. *Proc. Natl. Acad. Sci. USA* 75: 4001–5

Denekemp, J., Hobson, B. 1984. Endothelial proliferation in normal and tumour blood vessels. *Microvasc. Res.* 27: 388 (Abstr.)

Derynck, R., Jarrett, J. A., Chen, E. Y., Eaton, D. H., Bell, J. R., et al. 1985. Human transforming growth factor-beta cDNA sequence and expression in tumour cell lines. *Nature* 316: 701–5

Derynck, R., Roberts, A. B., Winkler, M. E., Chen, E. Y., Goeddel, D. V. 1984. Human transforming growth factor-α: Precursor structure and expression in *E. coli. Cell* 38: 287–97

Deuel, T. F., Huang, J. S. 1984a. Platelet derived growth factor: Structure, function and roles in normal and transformed cells. *J. Clin. Invest.* 74: 669–76

Deuel, T. F., Huang, J. S. 1984b. Roles of growth factor activities in oncogenesis. *Blood* 64: 951–58

Deuel, T. F., Huang, J. S., Huang, S. S., Stroobant, P., Waterfield, M. D. 1983. Expression of a platelet-derived growth factor-like protein in simian sarcoma virus transformed cells. *Science* 221: 1348–50

Deuel, T. F., Huang, J. S., Proffitt, R. T., Baenziger, J. U., Chang, D., et al. 1981a. Human platelet-derived growth factor: Purification and resolution into two active protein fractions. *J. Biol. Chem.* 256: 8896–99

XDeuel, T. F., Huang, J. S., Proffitt, R. T., Chang, D., Kennedy, B. B. 1981b. Platelet derived growth factor: Preliminary characterization. *J. Supramol. Struct. Cell Biochem.* Suppl. 5, p. 128

Deuel, T. F., Senior, R. M., Chang, D., Griffin, G. L., Heinrikson, R. L., et al. 1981c. Platelet factor 4 is chemotactic for neutrophils and monocytes. *Proc. Natl. Acad. Sci. USA* 78: 4584–87

Deuel, T. F., Senior, R. M., Huang, J. S., Griffin, G. L. 1982. Chemotaxis of monocytes and neutrophils to platelet-derived growth factor. *J. Clin. Invest.* 69: 1046–49

Devare, S. G., Reddy, E. P., Law, J. D., Robbins, K. C., Aaronson, S. A. 1983. Nucleotide sequence of the simian sarcoma virus genome: Demonstration that its acquired cellular sequences encode the transforming gene product p28^{sis}. *Proc. Natl. Acad. Sci. USA* 80: 731–35

Devare, S. G., Shatzman, A., Robbins, K. C., Rosenberg, M., Aaronson, S. A. 1984.

Expression of the PDGF-related transforming protein of simian sarcoma virus in *E. coli. Cell* 36: 43–49

Dexter, M. 1986. Growth factors. From the laboratory to the clinic. *Nature* 321: 198

Dicker, P., Rozengurt, E. 1978. Stimulation of DNA synthesis by tumour promoter and pure mitogenic factors. *Nature* 276: 723–26

Dicker, P., Rozengurt, E. 1981. Phorbol ester stimulation of Na influx and Na-K pump activity in Swiss 3T3 cells. *Biochem. Biophys. Res. Commun.* 100: 433–41

Donoff, R. B., McLennan, J. E., Grillo, H. ✴ C. 1971. Preparation and properties of collagenases from epithelium and mesenchyme of healing mammalian wounds. *Biochim. Biophys. Acta* 227: 639–53

Doolittle, R. F., Hunkapiller, M. W., Hood, L. E., Devare, S. G., Robbins, K. C., et al. 1983. Simian sarcoma virus onc gene, v-*sis*, is derived from the gene (or genes) encoded in platelet-derived growth factor. *Science* 221: 275–76

Downward, J., Parker, P., Waterfield, M. D. 1984a. Autophosphorylation sites on the epidermal growth factor receptor. *Nature* 311: 483–85

Downward, J., Yarden, Y., Mayes, E., Scrace, G., Totty, N., et al. 1984b. Close similarity of epidermal growth factor receptor and v-*erb* B oncogene protein sequences. *Nature* 307: 521–27

Ehrhart, J.-C., Creuzet, C., Rollet, E., Loeb, J. 1981. Epidermal growth factor–stimulated phosphorylation of tyrosine residues on a 120,000 dalton protein in mouse liver plasma membrane subfractions. *Biochem. Biophys. Res. Commun.* 102: 602–9

Eisen, A. Z. 1969. Skin collagenase: Localization and distribution in normal human skin. *J. Invest. Dermatol.* 52: 442–48

Ek, B., Heldin, C.-H. 1984. Use of an antiserum against phosphotyrosine for the identification of phosphorylated components in human fibroblasts stimulated by platelet-derived growth factor. *J. Biol. Chem.* 259: 11145–52

Ek, B., Westermark, B., Wasteson, A., Heldin, C.-H. 1982. Stimulation of tyrosine-specific phosphorylation by platelet-derived growth factor. *Nature* 295: 419–20

Erikson, E., Shealy, D. J., Erikson, R. L. 1981. Evidence that viral transforming gene products and epidermal growth factor stimulate phosphorylation of the same cellular protein with similar specificity. *J. Biol. Chem.* 256: 11381–84

Esch, F., Baird, A., Ling, N., Ueno, N., Hill, F., et al. 1985b. Primary structure of bovine pituitary basic fibroblast growth factor (FGF) and comparison with the amino-terminal sequence of bovine brain acidic FGF. *Proc. Natl. Acad. Sci. USA* 82: 6507–11

Esch, F., Ueno, N., Baird, A., Hill, F., Denoroy, L., et al. 1985a. Primary structure of bovine brain acidic fibroblast growth factor (FGF). *Biochem. Biophys. Res. Commun.* 133: 554–62

Eva, A., Robbins, K. C., Andersen, P. R., Srinivasan, A., Tronick, S. R., et al. 1982. Cellular genes analogous to retroviral onc genes are transcribed in human tumour cells. *Nature* 295: 116–19

✴ Fabricant, R. N., De Larco, J. E., Todaro, G. J. 1977. Nerve growth factor receptors on human melanoma cells in culture. *Proc. Natl. Acad. Sci. USA* 74: 565–69

Faggiotto, A., Ross, R., Harker, L. 1984. Studies of hypercholesterolemia in the nonhuman primate. I. Changes that lead to fatty streak formation. *Arteriosclerosis* 4: 323–40

Fanger, B. O., Wakefield, L. M., Sporn, M. B. 1985. Properties and actions of transforming growth factor-beta (TGF-beta) receptors. *Fed. Proc.* 44: 1241

Fava, R. A., Cohen, S. 1984. Isolation of a calcium-dependent 35-kilodalton substrate for the epidermal growth factor receptor/kinase from A-431 cells. *J. Biol. Chem.* 259: 2636–45

Folkman, J. 1972. Anti-angiogenesis: New concept for therapy of solid tumors. *Ann. Surg.* 175: 409–16

Folkman, J. 1985. Tumor angiogenesis. *Adv. Cancer Res.* 43: 175–203

Folkman, J. 1986. How is blood vessel growth regulated in normal and neoplastic tissue?—G. H. A. Clowes memorial award lecture. *Cancer Res.* 46: 467–73

Folkman, J., Cotran, R. 1976. Relation of vascular proliferation to tumor growth. *Int. Rev. Exp. Pathol.* 16: 207–48

Folkman, J., Merler, E., Abernathy, C., Williams, G. 1971. Isolation of a tumor factor responsible for angiogenesis. *J. Exp. Med.* 133: 275–88

Frackelton, A. R. Jr., Tremble, P. M., Williams, L. T. 1984. Evidence for the platelet-derived growth factor–stimulated tyrosine phosphorylation of the platelet-derived growth factor receptors in vivo: Immunopurification using a monoclonal antibody to phosphotyrosine. *J. Biol. Chem.* 259: 7909–15

Frantz, C. N. 1982. Univalent cation concentration and regulation of the BALB/c-3T3 growth cycle. In *Genetic Expression in the Cell Cycle*, ed. G. M. Padilla, K. S. McCarty, Sr., pp. 411–46. New York: Academic

Frantz, C. N., Stiles, C. D., Pledger, W. J., Scher, C. D. 1980. Effect of ouabain on growth regulation by serum components in BALB/c-3T3 cells: Inhibition of entry into S phase by decreased protein synthesis. *J. Cell. Physiol.* 105: 439–49

Friesel, R., Burgess, W. H., Mehlman, T., Maciag, T. 1986. The characterization of the receptor for endothelial cell growth factor by covalent ligand attachment. *J. Biol. Chem.* 261: 7581–84

Frolik, C. A., Dart, L. L., Meyers, C. A., Smith, D. M., Sporn, M. B. 1983. Purification and initial characterization of a type beta transforming growth factor from human placenta. *Proc. Natl. Acad. Sci. USA* 80: 3676–80

Frolik, C. A., Wakefield, L. A., Smith, D. M., Sporn, M. B. 1984. Characterization of a membrane receptor for transforming growth factor-beta in normal rat kidney fibroblasts. *J. Biol. Chem.* 259: 10995–11000

Fuller, A. F. Jr., Budzik, G. P., Krane, I. M., Donahoe, P. K. 1984. Müllerian inhibiting substance inhibition of a human endometrial carcinoma cell line xenographed in nude mice. *Gynecol. Oncol.* 17: 124–32

Fuller, A. F. Jr., Guy, S., Budzik, G. P., Donahoe, P. K. 1982. Müllerian inhibiting substance inhibits colony growth of a human ovarian carcinoma cell line. *J. Clin. Endocrinol. Metab.* 54: 1051–55

Gajdusek, C., DiCorleto, P. Ross, R., Schwartz, S. M. 1980. An endothelial cell-derived growth factor. *J. Cell Biol.* 85: 467–72

Garrett, J. S., Coughlin, S. R., Niman, H. L., Tremble, P. M., Giels, G. M., et al. 1984. Blockade of autocrine stimulation in simian sarcoma virus–transformed cells reverses down-regulation of platelet-derived growth factor receptors. *Proc. Natl. Acad. Sci. USA* 81: 7466–70

Gazit, A., Igarashi, H., Chiu, I., Srinivasan, A., Yaniv, A., et al. 1984. Expression of the normal human *sis*/PDGF-2 coding sequences includes cellular transformation. *Cell* 39: 89–97

Gelmann, E. P., Petri, E., Cetta, A., Wong-Staal, F. 1982. Deletions of specific regions of the simian sarcoma-associated virus genome are found in defective viruses and in the simian sarcoma virus. *J. Virol.* 41: 593–604

Gelmann, E. P., Wong-Staal, F., Kramer, R. A., Gallo, R. C. 1981. Molecular cloning and comparative analyses of the genomes of simian sarcoma virus and its associated helper virus. *Proc. Natl. Acad. Sci. USA* 78: 3373–77

Gilmore, T., DeClue, J. E., Martin, G. S. 1985. Protein phosphorylation at tyrosine is induced by the v-*erbB* gene product in vivo and in vitro. *Cell* 40: 609–18

Gimenez-Gallego, G., Rodkey, J., Bennett,

C., Rios-Candelore, M., DiSalvo, J., et al. 1985. Brain-derived acidic fibroblast growth factor: Complete amino acid sequence and homologies. *Science* 230: 1385–88

Glenn, K., Bowen-Pope, D. F., Ross, R. 1982. Platelet-derived growth factor. III. Identification of a platelet-derived growth factor receptor by affinity labeling. *J. Biol. Chem.* 257: 5172–76

Glenn, K., Ross, R. 1981. Human monocyte-derived growth factor(s) for mesenchymal cells: Activation of secretion by endotoxin and concanavalin A. *Cell* 25: 603–15

Goldberg, I. D., Stemerman, M. B., Handin, R. I. 1980. Vascular permeation of platelet factor 4 after endothelial injury. *Science* 209: 611–12

Gospodarowicz, D. 1975. Purification of a fibroblast growth factor from bovine pituitary. *J. Biol. Chem.* 250: 2515–20

Gospodarowicz, D. 1987. Isolation and characterization of acidic and basic fibroblast growth factor. *Methods Enzymol.* In press

Gospodarowicz, D., Baird, A., Cheng, J., Lui, G. M., Esch, F., et al. 1986. Isolation of fibroblast growth factor from bovine adrenal gland: Physicochemical and biological characterization. *Endocrinology* 118: 82–90

Gospodarowicz, D., Cheng, J., Lui, G.-M., Baird, A., Böhlen, P. 1984. Isolation of brain fibroblast growth factor by heparin-Sepharose affinity chromatography: Identity with pituitary fibroblast growth factor. *Proc. Natl. Acad. Sci. USA* 81: 6963–67

Gospodarowicz, D., Lui, G.-M., Gonzalez, R. 1982. High-density lipoproteins and the proliferation of human tumor cells maintained on extracellular matrix–coated dishes and exposed to defined medium. *Cancer Res.* 42: 3704–13

Gospodarowicz, D., Weseman, J., Moran, J. S., Lindstrom, J. 1976. Effect of fibroblast growth factor on the division and fusion of bovine myoblasts. *J. Cell Biol.* 70: 395–405

Graves, D. T., Owen, A. J., Barth, R. K., Tempst, P., Winoto, A., et al. 1984. Detection of c-*sis* transcripts and synthesis of PDGF-like proteins by human osteosarcoma cells. *Science* 226: 972–74

Gray, A., Dull, T. J., Ullrich, A. 1983. Nucleotide sequence of epidermal growth factor cDNA predicts a 128,000-molecular weight protein precursor. *Nature* 303: 722–25

Greenberg, M. E., Ziff, E. B. 1984. Stimulation of 3T3 cells induces transcription of the c-*fos* proto-oncogene. *Nature* 311: 433–38

Gregory, H. 1975. Isolation and structure of urogastrone and its relationship to epidermal growth factor. *Nature* 257: 325–27

Grillo, H. C., Gross, J. 1967. Collagenolytic activity during mammalian wound repair. *Dev. Biol.* 15: 300–17

Groffen, J., Heisterkamp, N., Stephenson, J. R., Geurts van Kessel, A., de Klein, A., et al. 1983. c-*sis* is translocated from chromosome 22 to chromosome 9 in chronic myelocytic leukemia. *J. Exp. Med.* 158: 9–15

Gross, J. 1976. Aspects of the animal collagenases. In *Biochemistry of Collagen*, ed. G. N. Ramachandran, A. H. Reddi, pp. 275–317. New York: Plenum

Grotendorst, G. R., Martin, G. R., Pencev, D., Sodek, J., Harvey, A. K. 1985. Stimulation of granulation tissue formation by platelet-derived growth factor in normal and diabetic rats. *J. Clin. Invest.* 76: 2323–29

Grotendorst, G. R., Seppä, H. E. J., Kleinman, H. K., Martin, G. R. 1981. Attachment of smooth muscle cells to collagen and their migration toward platelet-derived growth factor. *Proc. Natl. Acad. Sci. USA* 78: 3669–72

Habenicht, A. J., Glomset, J. A., King, W. C., Nist, C., Mitchell, C. D., et al. 1981. Early changes in phosphatidylinositol and arachidonic acid metabolism in quiescent Swiss 3T3 cells stimulated to divide by platelet-derived growth factor. *J. Biol. Chem.* 256: 12329–35

Haigler, H. T., Ash, J. F., Singer, S. J., Cohen, S. 1978. Visualization by fluorescence of the binding and internalization of growth factor in human carcinoma cells A-431. *Proc. Natl. Acad. Sci. USA* 75: 3317–21

Hajdu, S. I. 1979. *Pathology of Soft Tissue Tumors.* Philadelphia: Lea & Febiger. 599 pp.

Hannink, M., Donoghue, D. J. 1984. Requirement for a signal sequence in the biological expression of the v-*sis* oncogene. *Science* 226: 1197–99

Hannink, M., Donoghue, D. J. 1986. Biosynthesis of the v-*sis* gene product: Signal sequence cleavage, glycosylation and proteolytic processing. *Mol. Cell. Biol.* 6: 1343–48

Hannink, M., Sauer, M. K., Donoghue, D. J. 1986. Deletions in the C-terminal coding region of the v-*sis* gene: Dimerization is required for transformation. *Mol. Cell. Biol.* 6: 1304–14

Hayman, M. J., Beug, H. 1984. Identification of a form of the avian erythroblastosis virus *erb-B* gene product at the cell surface. *Nature* 309: 460–62

Heldin, C.-H., Johnsson, A., Wennergren, S., Wernstedt, C., Betsholtz, C., et al. 1986. A human osteosarcoma cell line secretes a growth factor structurally related to a homodimer of PDGF A-chains. *Nature* 319: 511–14

Heldin, C.-H., Wasteson, A., Westermark, B. 1977. Partial purification and characterization of platelet factors stimulating the multiplication of normal human glial cells. *Exp. Cell Res.* 109: 429–37

Heldin, C.-H., Wasteson, A., Westermark, B. 1982. Interaction of platelet-derived growth factor with its fibroblast receptor: Demonstration of ligand degradation and receptor modulation. *J. Biol. Chem.* 257: 4216–21

Heldin, C.-H., Westermark, B., Wasteson, A. 1980. Chemical and biological properties of a growth factor from human-cultured osteosarcoma cells: Resemblance with platelet-derived growth factor. *J. Cell. Phys.* 105: 235–46

Heldin, C.-H., Westermark, B., Wasteson, A. 1981a. Platelet-derived growth factor. Isolation by a large scale procedure and analysis of subunit composition. *Biochem. J.* 193: 907–13

Heldin, C.-H., Westermark, B., Wasteson, A. 1981b. Specific receptors for platelet-derived growth factor on cells derived from connective tissue and glia. *Proc. Natl. Acad. Sci. USA* 78: 3664–68

Henson, P. M., Cochrane, C. G. 1971. Acute immune complex disease in rabbits. The role of complement and of a leukocyte-dependent release of vasoactive amines from platelets. *J. Exp. Med.* 133: 554–71

Hirsch, J. G. 1960. Comparative bactericidal activities of blood serum and plasma serum. *J. Exp. Med.* 112: 15–22

Hollenberg, M. D., Cuatrecasas, P. 1973. Epidermal growth factor receptors in human fibroblasts and modulation of action by cholera toxin. *Proc. Natl. Acad. Sci. USA* 70: 2964–68

Holly, R. W., Bohlen, P., Fava, R., Baldwin, J. H., Kleeman, G., et al. 1980. Purification of kidney epithelial cell growth inhibitors. *Proc. Natl. Acad. Sci. USA* 77: 5989–92

Huang, J. S., Huang, S. S., Deuel, T. F. 1983. Human platelet derived growth factor: Radioimmunoassay and discovery of a specific plasma binding protein. *J. Cell Biol.* 97: 383–88

Huang, J. S., Huang, S. S., Deuel, T. F. 1984a. Specific covalent binding of platelet derived growth factor to human plasma α_2-macroglobulin. *Proc. Natl. Acad. Sci. USA* 81: 342–46

Huang, J. S., Huang, S. S., Deuel, T. F. 1984b. Transforming protein of simian sarcoma virus stimulates autocrine cell growth of SSV-transformed cells through platelet-derived growth factor cell surface receptors. *Cell* 39: 79–87

Huang, J. S., Huang, S. S., Kennedy, B. B., Deuel, T. F. 1982a. Platelet derived growth factor: Specific binding to target cells. *J. Biol. Chem.* 257: 8130–36

Huang, J. S., Huang, S. S., Kuo, M.-D. 1986. Bovine brain-derived growth factor: Purification and characterization of its interaction with responsive cells. *J. Biol. Chem.* 261: 11600–7

Huang, J. S., Proffitt, R. T., Baenziger, J. U., Chang, D., Kennedy, B. B., et al. 1982b. Human platelet-derived growth factor: Purification and initial characterization. In *Differentiation and Function of Hematopoietic Cell Surfaces*, ed. V. Marchesi, R. Gallo, P. Majerus, pp. 225–30. New York: Liss

Huang, S. S., Huang, J. S., 1986. Association of bovine brain-derived growth factor receptor with protein tyrosine kinase activity. *J. Biol. Chem.* 261: 9568–71

Huang, S. S., Huang, J. S., Deuel, T. F. 1984. The platelet derived growth factor receptor protein is a tyrosine specific protein kinase. In *Cold Spring Harbor Conferences on Cell Proliferation and Cancer: The Cancer Cell*, 1: 43–49. Cold Spring Harbor, NY: Cold Spring Harbor Lab.

Hunter, T., Cooper, J. A. 1981. Epidermal growth factor induces rapid tyrosine phosphorylation of proteins in A431 human tumor cells. *Cell* 24: 741–52

Hunter, T., Cooper, J. A. 1985. Protein-tyrosine kinases. *Ann. Rev. Biochem.* 54: 897–930

Hunter, T., Ling, N., Cooper, J. A. 1984. Protein kinase C phosphorylation of the EGF receptor at a threonine residue close to the cytoplasmic face of the plasma membrane. *Nature* 311: 480–83

Ignotz, R. A., Massagué, J. 1985. Type β transforming growth factor controls the adipogenic differentiation of 3T3 fibroblasts. *Proc. Natl. Acad. Sci. USA* 82: 8530–34

Inman, W. H., Colowick, S. P. 1985. Stimulation of glucose uptake by transforming growth factor-beta: Evidence for the requirement of epidermal growth factor-receptor activation. *Proc. Natl. Acad. Sci. USA* 82: 1346–49

Iwashita, S., Fox, C. F. 1984. Epidermal growth factor and potent phorbol tumor promoters induce epidermal growth factor receptor phosphorylation in a similar but distinctly different manner in human epidermoid carcinoma A431 cells. *J. Biol. Chem.* 259: 2559–67

Jacobs, S., Kull, F. C. Jr., Earp, H. S., Svobada, M. E., Van Wyk, J. J., et al. 1983. Somatomedin-C stimulates the phosphorylation of the β-subunit of its own receptor. *J. Biol. Chem.* 258: 9581–84

Jaye, M., Howk, R., Burgess, W., Ricca, G. A., Chiu, I.-M., et al. 1986. Human endo-

thelial cell growth factor: Cloning, nucleotide sequence, and chromosome localization. *Science* 233: 541–45

Jaye, M., McConathy, E., Drohan, W., Tong, B., Deuel, T., et al. 1985. Modulation of the *sis* gene transcript during endothelial cell differentiation in vitro. *Science* 228: 882–85

Jellinek, H. 1977. Arterial surface changes examined by scanning (SEM) and transmission (TEM) electron microscope. *Adv. Exp. Biol.* 82: 324–27

Johnsson, A., Heldin, C.-H., Wasteson, A., Westermark, B., Deuel, T. F., et al. 1984. The c-*sis* gene encodes a precursor of the B chain of platelet-derived growth factor. *EMBO J.* 3: 921–28

Johnsson, A., Heldin, C.-H., Westermark, B., Wasteson, A. 1982. Platelet-derived growth factor: Identification of constituent polypeptide chains. *Biochem. Biophys. Res. Commun.* 104: 66–74

Jorgensen, L., Packham, M. A., Rowsell, H. C., Mustard, J. F. 1972. Deposition of formed elements of blood on the intima and signs of intimal injury in the aorta of rabbit, pig and man. *Lab. Invest.* 27: 341–50

Joris, I., Majno, G. 1979. Inflammatory components of atherosclerosis. In *Advances in Inflammation Research*, ed. G. Weissman, B. Samuelsson, R. Paoletti, 1: 71–85. New York: Raven

Josephs, S. F., Dalla-Favera, R. D., Gelmann, E. P., Gallo, R. C., Wong-Staal, F. 1983. 5' viral and human cellular sequences corresponding to the transforming gene in simian sarcoma virus. *Science* 219: 503–5

Josephs, S. F., Guo, C., Ratner, L., Wong-Staal, F. 1984a. Human-proto-oncogene nucleotide sequences corresponding to the transforming region of simian sarcoma virus. *Science* 223: 487–91

Josephs, S. F., Ratner, L., Clarke, M. F., Westin, E. H., Reitz, M. S., et al. 1984b. Transforming potential of human c-*sis* nucleotide sequences encoding platelet-derived growth factor. *Science* 225: 636–39

Kaplan, D. R., Chao, F. C., Stiles, C. D., Antoniades, H. N., Scher, C. D. 1979. Platelet α granules contain a growth factor for fibroblasts. *Blood* 53: 1043–52

Kaplan, D. R., Broekman, M. J., Chernoff, A., Lesznik, G. R., Drillings, M. 1979. Platelet α-granule proteins: Studies on release and subcellular localization. *Blood* 53: 604–18

Kasuga, M., Zick, Y., Blithe, D. L., Crettaz, M., Kahn, C. R. 1982. Insulin stimulates tyrosine phosphorylation of the insulin receptor in a cell-free system. *Nature* 298: 667–69

Kehrl, J. H., Wakefield, L. M., Roberts, A. B., Jakowlew, S., Alvarez-Mon, M., et al. 1986. Production of transforming growth factor β by human T lymphocytes and its potential role in the regulation of T cell growth. *J. Exp. Med.* 163: 1037–50

Kelly, K., Cochran, B. H., Stiles, C. D., Leder, P. 1983. Cell-specific regulation of the c-*myc* gene by lymphocyte mitogens and platelet-derived growth factor. *Cell* 35: 603–10

Kimball, E. S., Bohn, W. H., Cockley, K. D., Warren, T. C., Sherwin, S. A. 1984. Distinct high performance liquid chromatography pattern of transforming growth factor activity in urine of cancer patients as compared with normal individuals. *Cancer Res.* 44: 3616–19

King, L. E. Jr., Carpenter, G., Cohen, S. 1980. Characterization by electrophoresis of epidermal growth factor stimulated phosphorylation using A-431 membranes. *Biochemistry* 19: 1524–28

Klagsbrun, M., Sasse, J., Sullivan, R., Smith, J. A. 1986. Human tumor cells synthesize an endothelial cell growth factor that is structurally related to basic fibroblast growth factor. *Proc. Natl. Acad. Sci. USA* 83: 2448–52

Klagsbrun, M., Shing, Y. 1985. Heparin affinity of anionic and cationic capillary endothelial cell growth factors: Analysis of hypothalamus-derived growth factors and fibroblast growth factors. *Proc. Natl. Acad. Sci. USA* 82: 805–9

Kohler, N., Lipton, A. 1974. Platelets as a source of fibroblast growth-promoting activity. *Exp. Cell Res.* 87: 297–301

Kondo, I., Shimizu, N. 1983. Mapping of the human gene for epidermal growth factor receptor (EGFR) on the p13 → q22 region of chromosome 7. *Cytogenet. Cell Genet.* 35: 9–14

Kris, R. M., Lax, I., Gullick, W., Waterfield, M. D., Ullrich, A., et al. 1985. Antibodies against a synthetic peptide as a probe for the kinase activity of the avian EGF receptor and v-*erbB* protein. *Cell* 40: 619–25

Kruijer, W., Cooper, J. A., Hunter, T., Verma, I. M. 1984. Platelet-derived growth factor induces rapid but transient expression of the c-*fos* gene and protein. *Nature* 312: 711–16

Land, H., Parada, L. F., Weinberg, R. A. 1983. Tumorigenic conversion of primary embryo fibroblasts requires at least two cooperating oncogenes. *Nature* 304: 596–602

Leal, F., Williams, L. T., Robbins, K. C., Aaronson, S. A. 1985. Evidence that the v-*sis* gene product transforms by interaction with the receptor for platelet-derived growth factor. *Science* 230: 327–30

Lee, D. C., Rose, T. M., Webb, N. R., Todaro, G. J. 1985. Cloning and sequence analysis of a cDNA for rat transforming growth factor-alpha. *Nature* 313: 489–91

Lee, L. S., Weinstein, I. B. 1978. Tumor-promoting phorbol esters inhibit binding of epidermal growth factor to cellular receptors. *Science* 202: 313–15

Leibovich, S. J., Ross, R. 1976. A macrophage-dependent factor that stimulates the proliferation of fibroblasts in vitro. *Am. J. Pathol.* 84: 501–13

Leof, E. B., Proper, J. A., Goustin, A. S., Shipley, G. D., DiCorleto, P. E., et al. 1986. Induction of c-*sis* mRNA and activity similar to platelet-derived growth factor by transforming growth factor β: A proposed model for indirect mitogenesis involving autocrine activity. *Proc. Natl. Acad. Sci. USA* 83: 2453–57

Leof, E. B., Wharton, W., Van Wyk, J. J., Pledger, W. J. 1982. Epidermal growth factor (EGF) and somatomedin C regulate G_1 progression in competent BALB/c-3T3 cells. *Exp. Cell Res.* 141: 107–15

Levine, P. H., Weinger, R. S., Simon, J., Scoon, K. L., Krinsky, N. I. 1976. Leukocyte-platelet interaction release of hydrogen peroxide by granulocytes as a modulator of platelet reactions. *J. Clin. Invest.* 57: 955–63

Ling, N., Ying, S.-Y., Ueno, N., Shimasaki, S., Esch, F., et al. 1986. Pituitary FSH is released by a heterodimer of the β-subunits from the two forms of inhibin. *Nature* 321: 779–82

Linsley, P. S., Hargreaves, W. R., Twardzik, D. R., Todaro, G. J. 1985. Detection of larger polypeptides structurally and functionally related to type I transforming growth factor. *Proc. Natl. Acad. Sci. USA* 82: 356–60

Lobb, R. R., Alderman, E. M., Fett, J. W. 1985. Induction of angiogenesis by bovine brain derived class 1 heparin-binding growth factor. *Biochemistry* 24: 4969–73

Lobb, R. R., Fett, J. W. 1984. Purification of two distinct growth factors from bovine neural tissue by heparin affinity chromatography. *Biochemistry* 23: 6295–99

Lobb, R. R., Harper, J. W., Fett, J. W. 1986a. Review. Purification of heparin-binding growth factors. *Anal. Biochem.* 154: 1–14

Lobb, R. R., Sasse, J., Sullivan, R., Shing, Y., D'Amore, P., et al. 1986b. Purification and characterization of heparin-binding endothelial cell growth factors. *J. Biol. Chem.* 261: 1924–28

Maciag, T. 1984. Angiogenesis. *Prog. Hemostasis Thromb.* 7: 167–82

Maciag, T., Mehlman, T., Friesel, R., Schreiber, A. B. 1984. Heparin binds endothelial cell growth factor, the principal endothelial cell mitogen in bovine brain. *Science* 225: 932–35

Magun, B. E., Matrisian, L. M., Bowden, G. T. 1980. Epidermal growth factor. *J. Biol. Chem.* 255: 6373–81

Makino, R., Hayashi, K., Sigamura, T. 1984. c-*myc* transcript is induced in rat liver at a very early stage of regeneration or by cycloheximide treatment. *Nature* 310: 697–700

Manger, R., Najita, L., Nichols, E. J., Hakomori, S., Rohrschneider, L. 1984. Cell surface expression of the McDonough strain of feline sarcoma virus *fms* gene product (gp140^{fms}). *Cell* 39: 327–37

Marquardt, H., Hunkapiller, M. W., Hood, L. E., Todaro, G. J. 1984. Rat transforming growth factor type 1: Structure and relation to epidermal growth factor. *Science* 223: 1079–81

Marquardt, H., Hunkapiller, M. W., Hood, L. E., Twardzik, D. R., De Larco, J. E., et al. 1983. Transforming growth factors produced by retrovirus-transformed rodent fibroblasts and human melanoma cells: Amino acid sequence homology with epidermal growth factor. *Proc. Natl. Acad. Sci. USA* 80: 4684–88

Marquardt, H., Todaro, G. J. 1982. Human transforming growth factor. Production by a melanoma cell line, purification and initial characterization. *J. Biol. Chem.* 257: 5220–25

Martin, B. M., Gimbrone, M. A., Unanue, E. R., Cotran, R. S. 1981. Stimulation of nonlymphoid mesenchymal cell proliferation by a macrophage-derived growth factor. *J. Immunol.* 126: 1510–15

Mason, A. J., Hayflick, J. S., Ling, N., Esch, F., Ueno, N., et al. 1985. Complementary DNA sequences of ovarian follicular fluid inhibin show precursor structure and homology with transforming growth factor-β. *Nature* 318: 659–63

Massagué, J. 1983a. Epidermal growth factor-like transforming growth factor. I. Isolation, chemical characterization, and potentiation by other transforming factors from feline sarcoma virus-transformed rat cells. *J. Biol. Chem.* 258: 13606–13

Massagué, J. 1983b. Epidermal growth factor-like transforming growth factor. II. Interaction with epidermal growth factor receptors in human placenta membranes and A-431 cells. *J. Biol. Chem.* 258: 13614–20

Massagué, J. 1984. Type beta transforming growth factor from feline sarcoma virus-transformed rat cells. *J. Biol. Chem.* 259: 9756–61

Massagué, J. 1985. Subunit structure of a high-affinity receptor for type β trans-

forming growth factor: Evidence for a disulfide-linked glycosylated receptor complex. *J. Biol. Chem.* 260: 7059–66

Massagué, J., Cheifetz, S., Endo, T., Nadal-Ginard, B. 1986. Type β transforming growth factor is an inhibitor of myogenic differentiation. *Proc. Natl. Acad. Sci. USA* 83: 8206–10

Massagué, J., Like, B. 1985. Cellular receptors for type beta transforming growth factor. *J. Biol. Chem.* 260: 2636–45

Masui, T., Wakefield, L. M., Lechner, J. F., LaVeck, M. A., Sporn, M. B., et al. 1985. Type beta transforming growth factor is the primary differentiation-inducing serum factor for normal human bronchial epithelial cells. *Fed. Proc.* 44: 4019 (Abstr.)

Masui, T., Wakefield, L. M., Lechner, J. F., LaVeck, M. A., Sporn, M. B., et al. 1986. Type beta transforming growth factor is the primary differentiation-inducing serum factor for normal human bronchial epithelial cells. *Proc. Natl. Acad. Sci. USA* 83: 2438–42

Matrisian, L. M., Pathak, M., Magun, B. E. 1982. Identification of an epidermal growth factor-related transforming growth factor from rat fetuses. *Biochem. Biophys. Res. Commun.* 107: 761–69

Merlino, G. T., Xu, Y.-H., Ishii, S., Clark, A. J. L., Semba, K., et al. 1984. Amplification and enhanced expression of the epidermal growth factor receptor gene in A431 human carcinoma cells. *Science* 224: 417–20

Moonlenaar, W. H., Tertoolen, L. G. J., de Laat, S. W. 1984a. Growth factors immediately raise cytoplasmic free Ca^{2+} in human fibroblasts. *J. Biol. Chem.* 259: 8066–69

Moonlenaar, W. H., Tertoolen, L. G. J., de Laat, S. W. 1984b. Phorbol ester and diacylglycerol mimic growth factors in raising cytoplasmic pH. *Nature* 312: 371–74

Morrison, R. S., Sharma, A., de Vellis, J., Bradshaw, R. A. 1986. Basic fibroblast growth factor supports the survival of cerebral cortical neurons in primary culture. *Proc. Natl. Acad. Sci. USA* 83: 7537–41

Moses, H. L., Branum, E. L., Proper, J. A., Robinson, R. A. 1981. Transforming growth factor production by chemically transformed cells. *Cancer Res.* 41: 2842–48

Moses, H. L., Childs, C. B., Halper, J., Shipley, G. D., Tucker, R. F. 1984. Role of transforming growth factors in neoplastic transformation. In *Control of Cell Growth and Proliferation*, ed. C. M. Veneziale, pp. 147–67. New York: Van Nostrand Reinhold

Moses, H. L., Tucker, R. F., Leof, E. B.,

Coffey, R. J., Halper, J., et al. 1985. Type-beta transforming growth factor is a growth stimulator and a growth inhibitor. In *Cancer Cells*, ed. J. Feramisco, B. Ozanne, C. Stiles, 3: 65–71. Cold Spring Harbor, NY: Cold Spring Harbor Lab.

Muller, R., Bravo, R., Burchkardt, J., Lurran, T. 1984. Induction of a c-*fos* gene and protein by growth factors precedes activation of c-*myc*. *Nature* 312: 716–20

Nachman, R. L., Weksler, B., Ferris, B. 1972. Characterization of human platelet vascular permeability-enhancing activity. *J. Clin. Invest.* 51: 549–56

Nakamura, K. D., Martinez, R., Weber, M. J. 1983. Tyrosine phosphorylation of specific proteins after mitogen stimulation of chicken embryo fibroblasts. *Mol. Cell. Biol.* 3: 380–90

Narayanan, A. S., Page, R. C. 1983. Biosynthesis and regulation of type V collagen in diploid human fibroblasts. *J. Biol. Chem.* 258: 11694–99

Neufeld, G., Gospodarowicz, D. 1986. Basic and acidic fibroblast growth factors interact with the same cell surface receptors. *J. Biol. Chem.* 261: 5631–37

Niall, M., Ryan, G. B., O'Brien, B. J. 1982. The effect of epidermal growth factor on wound healing in mice. *J. Surg. Res.* 33: 164–69

Niman, H. L. 1984. Antisera to a synthetic peptide of the *sis* viral oncogene product recognize human platelet-derived growth factor. *Nature* 307: 180–83

Niman, H. L., Houghten, R. A., Bowen-Pope, D. F. 1984. Detection of high molecular weight forms of platelet-derived growth factor by sequence-specific antisera. *Science* 226: 701–3

Nishimura, J., Deuel, T. F. 1981. Stimulation of protein phosphorylation in Swiss mouse 3T3 cells by human platelet derived growth factor. *Biochem. Biophys. Res. Commun.* 103: 355–61

Nishimura, J., Deuel, T. F. 1983. Platelet derived growth factor stimulates the phosphorylation of ribosomal protein S6. *FEBS Lett.* 156: 130–34

Nishimura, J., Huang, J. S., Deuel, T. F. 1982. Platelet-derived growth factor stimulates tyrosine-specific protein kinase activity in Swiss mouse 3T3 cell membranes. *Proc. Natl. Acad. Sci. USA* 79: 4303–7

Nishizuka, Y. 1984. The role of protein kinase C in cell surface signal transduction and tumour promotion. *Nature* 308: 693–98

Nister, M., Heldin, C.-H., Wasteson, A., Westermark, B. 1984. A glioma-derived analog to platelet-derived growth factor: Demonstration of receptor competing

activity and immunological crossreactivity. *Proc. Natl. Acad. Sci. USA* 81: 926–30

Oka, Y., Orth, D. N. 1983. Human plasma epidermal growth factor/β-urogastrone is associated with blood platelets. *J. Clin. Invest.* 72: 249–59

Olashaw, N. E., Pledger, W. J. 1983. Association of platelet-derived growth-factor-induced protein with nuclear material. *Nature* 306: 272–74

Orlidge, A., D'Amore, P. A. 1986. Cell specific effects of glycosaminoglycans on the attachment and proliferation of vascular wall components. *Microvasc. Res.* 31: 41–53

Owen, A. J., Geyer, R. P., Antoniades, H. N. 1982. Human platelet-derived growth factor stimulates amino acid transport and protein synthesis by human diploid fibroblasts in plasma-free media. *Proc. Natl. Acad. Sci. USA* 79: 3203–7

Owen, A. J., Pantazis, P., Antoniades, H. N. 1984. Simian sarcoma virus–transformed cells secrete a mitogen identical to platelet-derived growth factor. *Science* 225: 54–56

Ozanne, B., Fulton, R. J., Kaplan, P. I. 1980. Kirsten murine sarcoma virus transformed cell lines and a spontaneously transformed rat cell line produce transforming growth factors. *J. Cell. Physiol.* 105: 163–80

Packham, M. A., Nishizawa, E. E., Mustard, J. F. 1968. Response of platelets to tissue injury. *Biochem. Pharmacol.* 17: 171–84 (Suppl.)

Padgett, R. W., St. Johnston, R. D., Gelbart, W. M. 1987. A transcript from a *Drosophila* pattern gene predicts a protein homologous to the transforming growth factor β family. *Nature* 325: 81–84

Pawlowski, N. A., Abraham, E. L., Pontier, S., Scott, W. A., Cohn, Z. A. 1985. Human monocyte-endothelial cell interaction in vitro. *Proc. Natl. Acad. Sci. USA* 82: 8208–12

Pettmann, B., Weibel, M., Sensenbrenner, M., Labourdette, G. 1985. Purification of two astroglial growth factors from bovine brain. *FEBS Lett.* 189: 102–8

Pike, L. J., Krebs, E. G. 1986. Protein tyrosine kinase activity of hormone and growth factor receptors. In *Receptors*, ed. P. M. Conn, Vol. 3, pp. 93–134. New York: Academic

Pledger, W. J., Hart, C. A., Locatell, K. L., Scher, C. D. 1981. Platelet-derived growth factor modulated proteins: Constitutive synthesis by a transformed-cell line. *Proc. Natl. Acad. Sci. USA* 78: 4358–62

Pledger, W. J., Stiles, C. D., Antoniades, H. N., Scher, C. D. 1977. Induction of DNA synthesis in BALB/c 3T3 cells by serum complements: Reevaluation of the commitment process. *Proc. Natl. Acad. Sci. USA* 74: 4481–85

Pledger, W. J., Stiles, C. D., Antoniades, H. N., Scher, C. D. 1978. An ordered sequence of events is required before BALB/c 3T3 cells become committed to DNA synthesis. *Proc. Natl. Acad. Sci. USA* 75: 2839–43

Postlethwaite, A. E., Seyer, J. M., Kang, A. H. 1978. Chemotactic attraction of human fibroblasts to type I, II, and III collagens and collagen-derived peptides. *Proc. Natl. Acad. Sci. USA* 75: 871–75

Postlethwaite, A. E., Snyderman, R., Kang, A. H. 1976. The chemotactic attraction of human fibroblasts to a lymphocyte-derived factor. *J. Exp. Med.* 144: 1188–1203

Postlethwaite, A. E., Snyderman, R., Kang, A. H. 1979. Generation of a fibroblast chemotactic factor in serum by activation of complement. *J. Clin. Invest.* 64: 1379–85

Prchal, J. T., Blakely, J. 1973. Granulocyte platelet rosettes. *N. Engl. J. Med.* 289: 1146

Privalsky, M. L., Bishop, J. M. 1984. Subcellular localization of the v-erb-B protein, the product of a transforming gene of avian erythroblastosis virus. *Virology* 135: 356–69

Privalsky, M. L., Ralston, R., Bishop, J. M. 1984. The glycoprotein encoded by the retroviral oncogene v-erb-B is structurally related to tyrosine-specific protein kinases. *Proc. Natl. Acad. Sci. USA* 81: 704–7

Proper, J. A., Bjornson, C. L., Moses, H. L. 1982. Mouse embryos contain polypeptide growth factor(s) capable of inducing a reversible neoplastic phenotype in nontransformed cells in culture. *J. Cell. Physiol.* 110: 169–74

Racker, E., Resnick, R. J., Feldman, R. 1985. Glycolysis and methylaminoisobutyrate uptake in rat-1 cells transfected with ras or myc oncogenes. *Proc. Natl. Acad. Sci. USA* 82: 3535–38

Radke, K., Martin, G. S. 1979. Transformation by Rous sarcoma virus: Effects of src gene expression on the synthesis and phosphorylation of cellular polypeptides. *Proc. Natl. Acad. Sci. USA* 76: 5212–16

Raines, E. W., Ross, R. 1982. Platelet-derived growth factor. I. High yield purification and evidence for multiple forms. *J. Biol. Chem.* 257: 5154–60

Rall, L. B., Scott, J., Bell, G. I., Crawford, R. J., Penschow, J. D., et al. 1985. Mouse prepro–epidermal growth factor synthesis by the kidney and other tissues. *Nature* 313: 228–31

Reisner, A. H. 1985. Similarity between the vaccinia virus 19 K early protein and epidermal growth factor. *Nature* 313: 801–3

Rettenmier, C. W., Chen, J. H., Roussel, M. F., Sherr, C. J. 1985a. The product of the c-fms proto-oncogene: A glycoprotein with associated tyrosine kinase activity. *Science* 228: 320–22

Rettenmier, C. W., Roussel, M. F., Quinn, C. O., Kitchingman, G. R., Look, A. T., et al. 1985b. Transmembrane orientation of glycoproteins encoded by the v-fms oncogene. *Cell* 40: 971–81

Riley, W. B., Peacock, E. E. 1967. Identification, distribution, and significance of collagenolytic enzyme in human tissues. *Proc. Exp. Biol. Med.* 124: 207–10

Robbins, K. C., Antoniades, H. N., Devare, S. G., Hunkapiller, M. W., Aaronson, S. A. 1983. Structural and immunological similarities between sarcoma virus gene product(s) and human platelet-derived growth factor. *Nature* 305: 605–8

Robbins, K. C., Devare, S. G., Reddy, E. P., Aaronson, S. A. 1982b. In vivo identification of the transforming gene product of simian sarcoma virus. *Science* 218: 1131–33

Robbins, K. C., Hill, R. L., Aaronson, S. A. 1982a. Primate origin of the cell-derived sequences of simian sarcoma virus. *J. Virol.* 41: 721–25

Roberts, A. B., Anzano, M. A., Lamb, L. C., Smith, J. M., Sporn, M. B. 1981. New class of transforming growth factors potentiated by epidermal growth factor: Isolation from non-neoplastic tissues. *Proc. Natl. Acad. Sci. USA* 78: 5339–43

Roberts, A. B., Anzano, M. A., Meyers, C. A., Wideman, J., Blacher, R., et al. 1983b. Purification and properties of a type beta transforming growth factor from bovine kidney. *Biochemistry* 22: 5692–98

Roberts, A. B., Anzano, M. A., Wakefield, L. M., Roche, N. S., Stern, D. F., et al. 1985b. Type beta transforming growth factor: A bifunctional regulator of cellular growth. *Proc. Natl. Acad. Sci. USA* 82: 119–23

Roberts, A. B., Frolik, C. A., Anzano, M. A., Assoian, R. K., Sporn, M. B. 1984. Purification of type beta transforming growth factors from nonneoplastic tissues. In *Methods for Preparation of Media, Supplements, and Substrate for Serum-free Animal Cell Culture*, ed. D. Barnes, G. Sato, D. Sirbasku, pp. 181–94. New York: Liss

Roberts, A. B., Frolik, C. A., Anzano, M. A., Sporn, M. B. 1983a. Transforming growth factors from neoplastic and nonneoplastic tissues. *Fed. Proc.* 42: 2621–26

Roberts, A. B., Lamb, L. C., Newton, D. L.,

Sporn, M. B., De Larco, J. E., et al. 1980. Transforming growth factors: Isolation of polypeptides from virally and chemically transformed cells by acids/ethanol extraction. *Proc. Natl. Acad. Sci. USA* 77: 3494–98

Roberts, A. B., Roche, N. S., Sporn, M. B. 1985a. Selective inhibition of the anchorage-independent growth of *myc*-transfected fibroblasts by retinoic acid. *Nature* 315: 237–39

Roberts, A. B., Sporn, M. B., Assoian, R. K., Smith, J. M., Roche, N. S., et al. 1986. Transforming growth factor type-β: Rapid induction of fibrosis and angiogenesis in vivo and stimulation of collagen formation in vitro. *Proc. Natl. Acad. Sci. USA* 83: 4167–71

Rodriguez-Pena, A., Rozengurt, E. 1984. Disappearance of Ca²⁺-sensitive, phospholipid-dependent protein kinase activity in phorbol ester-treated 3T3 cells. *Biochem. Biophys. Res. Commun.* 120: 1053–59

Rodriguez-Pena, A., Rozengurt, E. 1985. Serum, like phorbol esters, rapidly activates protein kinase C in intact quiescent fibroblasts. *EMBO J.* 4: 71–76

Rodriguez-Pena, A., Rozengurt, E. 1986. Phosphorylation of an acidic mol. wt. 80,000 cellular protein in a cell-free system and intact Swiss 3T3 cells: A specific marker of protein kinase activity. *EMBO J.* 5: 77–83

Rook, A. M., Kehrl, J. H., Wakefield, L. M., Roberts, A. B., Sporn, M. B., et al. 1986. Effects of transforming growth factor-β on the functions of natural killer cells: Depressed cytolytic activity and blunting of interferon responsiveness. *J. Immunol.* 136: 3916–20

Rose, S. P., Pruss, R. M., Herschman, H. R. 1975. Initiation of 3T3 fibroblast cell division by epidermal growth factor. *J. Cell. Physiol.* 86: 593–98

Rosoff, P. M., Stein, L. F., Cantley, L. C. 1984. Phorbol esters induce differentiation in a pre-B-lymphocyte cell line by enhancing Na⁺/H⁺ exchange. *J. Biol. Chem.* 259: 7056–60

Ross, R., Glomset, J., Kariya, B., Harker, L. 1974. A platelet-dependent serum factor that stimulates the proliferation of arterial smooth muscle cells in vitro. *Proc. Natl. Acad. Sci. USA* 71: 1207–10

Ross, R., Vogel, A. 1978. The platelet-derived growth factor. Review. *Cell* 14: 203–10

Ross, R., Vogel, A., Davies, P., Raines, E., Kariya, B., et al. 1979. The platelet-derived growth factor and plasma control cell proliferation. In *Hormones and Cell Culture*, ed. G. H. Sato, R. Ross, Book A,

pp. 3–16. Cold Spring Harbor, NY: Cold Spring Harbor Lab.

Roussel, M. F., Rettenmier, C. W., Look, A. T., Sherr, C. J. 1984. Cell surface expression of v-*fms*-coded glycoproteins is required for transformation. *Mol. Cell. Biol.* 4: 1999–2009

Rowley, J. D. 1973. A new consistent chromosomal abnormality in chronic myelogenous leukaemia identified by quinacrine fluorescence and Giemsa staining. *Nature* 243: 290–93

Rozengurt, E. 1986. Early signals in the mitogenic response. *Science* 234: 161–66

Rozengurt, E., Rodriguez-Pena, A., Coombs, M., Sinnett-Smith, J. 1984. Diacylglycerol stimulates DNA synthesis and cell division in mouse 3T3 cells: Role of Ca²⁺-sensitive phospholipid-dependent protein kinase. *Proc. Natl. Acad. Sci. USA* 81: 5748–52

Rozengurt, E., Rodriguez-Pena, M., Smith, K. A. 1983b. Phorbol esters, phospholipase C, and growth factors rapidly stimulate the phosphorylation of a M_r 80,000 protein in intact quiescent 3T3 cells. *Proc. Natl. Acad. Sci. USA* 80: 7244–48

Rozengurt, E., Stroobant, P., Waterfield, M. D., Deuel, T. F., Keehan, M. 1983a. Platelet-derived growth factor elicits cyclic AMP accumulation in Swiss 3T3 cells: Role of prostaglandin production. *Cell* 34: 265–72

Rubsamen, H., Saltenberger, K., Friis, R. R., Eigenbrodt, E. 1982. Cytosolic malic dehydrogenase activity is associated with a putative substrate for the transforming gene product of Rous sarcoma virus. *Proc. Natl. Acad. Sci. USA* 79: 228–32

Sauer, M. K., Hannink, M., Donoghue, D. J. 1986. Deletions in the N-terminal coding region of the v-*sis* gene: Determination of the minimal transforming region. *J. Virol.* 59: 292–300

Savage, C. R., Inagami, T., Cohen, S. 1972. The primary structure of epidermal growth factor. *J. Biol. Chem.* 247: 7612–21

Scher, C. D., Dick, R. L., Whipple, A. P., Locatell, K. L. 1983. Identification of a BALB/c-3T3 cell protein modulated by platelet-derived growth factor. *Mol. Cell. Biol.* 3: 70–81

Schmidt, J. A., Beug, H., Hayman, M. J. 1985. Effects of inhibitors of glycoprotein processing on the synthesis and biological activity of erb-B oncogene. *EMBO J.* 4: 105–12

Schmidt, J. A., Mizel, S. B., Cohen, D., Green, I. 1982. Interleukin-1, a potential regulator of fibroblast proliferation. *J. Immunol.* 128: 2177–82

Schreiber, A. B., Kenney, J., Kowalski, W. J., Friesel, R., Mehlman, T., et al. 1985. Interaction of endothelial cell growth factor with heparin: Characterization by receptor and antibody recognition. *Proc. Natl. Acad. Sci. USA* 82: 6138–42

Schuldiner, S., Rozengurt, E. 1982. Na^+/H^+ antiport in Swiss 3T3 cells: Mitogenic stimulation leads to cytoplasmic alkalinization. *Proc. Natl. Acad. Sci. USA* 79: 7778–82

Schultz, G. S., White, M., Mitchell, R., Brown, G., Lynch, J., et al. 1987. Epithelial wound healing enhanced by transforming growth factor-α and vaccinia growth factor. *Science* 235: 350–52

Schwarzbauer, J. E., Tamkun, J. W., Lemischka, I. R., Hynes, R. O. 1983. Three different fibronectin mRNAs arise by alternative splicing within the coding region. *Cell* 35: 421–31

Schweigerer, L., Neufeld, G., Mergia, A., Abraham, J. A., Fiddes, J. C., et al. 1987. Basic fibroblast growth factor in human rhabdomyosarcoma cells: Implications for the proliferation and neovascularization of myoblast-derived tumors. *Proc. Natl. Acad. Sci. USA* 84: 842–46

Scott, J., Urdea, M., Quiroga, M., Sanchez-Pescador, R., Fong, N., et al. 1983. Structure of a mouse submaxillary messenger RNA encoding epidermal growth factor and seven related proteins. *Science* 221: 236–40

Sefton, B. M., Hunter, T. 1984. Tyrosine protein kinases. *Adv. Cyclic Nucleotides Protein Phosphorylation Res.* 18: 195–226

Seifert, R. A., Schwartz, S. M., Bowen-Pope, D. F. 1984. Developmentally regulated production of platelet-derived growth factor-like molecules. *Nature* 311: 669–71

Sejersen, T., Betsholtz, C., Sjölund, M., Heldin, C.-H., Westermark, B., et al. 1986. Rat skeletal myoblasts and arterial smooth muscle cells express the gene for the A-chain but not the gene for the B-chain (c-sis) of platelet derived growth factor (PDGF) and produce a PDGF-like protein. *Proc. Natl. Acad. Sci. USA* 83: 6844–48

Senior, R. M., Griffin, G. L., Huang, J. S., Walz, D. A., Deuel, T. F. 1983. Chemotactic activity of platelet alpha granule proteins for fibroblasts. *J. Cell Biol.* 96: 382–85

Senior, R. M., Griffin, G. L., Mecham, R. P. 1982. Chemotactic response of fibroblasts to tropoelastin and elastin-derived peptides. *J. Clin. Invest.* 70: 614–18

Seppä, H., Grotendorst, G., Seppä, S., Schiffmann, E., Martin, G. R. 1982. Platelet-derived growth factor is chemotactic

for fibroblasts. *J. Cell Biol.* 92: 584–88

Seppä, H. E. J., Yamada, K. M., Seppä, S. T., Silver, M. H., Kleinman, H. K., et al. 1981. The cell binding fragment of fibronectin is chemotactic for fibroblasts. *Cell Biol. Int. Rep.* 5: 813–19

Seyedin, S. M., Thomas, T. C., Thompson, A. Y., Rosen, D. M., Piez, K. A. 1985. Purification and characterization of two cartilage-inducing factors from bovine demineralized bone. *Proc. Natl. Acad. Sci. USA* 82: 2267–71

Seyedin, S. M., Thompson, A. Y., Bentz, H., Rosen, D. M., McPherson, J. M., et al. 1986. Cartilage-inducing factor-A: Apparent identity to transforming growth factor-β. *J. Biol. Chem.* 261: 5693–95

Sherwin, S. A., Twardzik, D. R., Bohn, W. H., Cockley, K. D., Todaro, G. J. 1983. High–molecular weight transforming growth factor activity in the urine of patients with disseminated cancer. *Cancer Res.* 43: 403–7

Shier, W. T. 1980. Serum stimulation of phospholipase A_2 and prostaglandin release in 3T3 cells is associated with platelet-derived growth-promoting activity. *Proc. Natl. Acad. Sci. USA* 77: 137–41

Shier, W. T., Durkin, J. P. 1982. Role of stimulation of arachidonic acid release in the proliferative response of 3T3 mouse. *J. Cell. Physiol.* 112: 171–81

Shimokado, K., Raines, E. W., Madtes, D. K., Barrett, T. B., Benditt, E. P., et al. 1985. A significant part of macrophage-derived growth factor consists of at least two forms of PDGF. *Cell* 43: 277–86

Shing, Y., Folkman, J., Haudenschild, C., Lund, D., Crum, R., et al. 1985. Angiogenesis is stimulated by a tumor-derived endothelial cell growth factor. *J. Cell. Biochem.* 29: 275–87

Shipley, G. D., Childs, C. B., Volkenant, M. E., Moses, H. L. 1984. Differential effects of epidermal growth factor, transforming growth factor, and insulin on DNA and protein synthesis and morphology in serum-free cultures of AKR-2B cells. *Cancer Res.* 44: 710–16

Shipley, G. D., Tucker, R. F., Moses, H. L. 1985. Type β transforming growth factor/growth inhibitor stimulates entry of monolayer cultures of AKR-2B cells into S phase after a prolonged prereplicative interval. *Proc. Natl. Acad. Sci. USA* 82: 4147–51

Shoyab, M., De Larco, J. E., Todaro, G. J. 1979. Biologically active phorbol esters specifically alter affinity of epidermal growth factor membrane receptors. *Nature* 279: 387–92

Sibley, D. R., Benovic, J. L., Caron, M. C., Lefkowitz, R. J. 1987. Regulation of transmembrane signaling by receptor phosphorylation. *Cell* 48: 913–22

Singh, J. P., Chaikin, M. A., Pledger, W. J., Scher, C. D., Stiles, C. D. 1983. Persistence of the mitogenic response to platelet-derived growth factor (competence) does not reflect a long-term interaction between growth-factor and the target cell. *J. Cell Biol.* 96: 1497–1502

Singh, J. P., Chaikin, M. A., Stiles, C. D. 1982. Phylogenetic analysis of platelet-derived growth factor by radioreceptor assay. *J. Cell Biol.* 95: 667–71

Smith, J. C., Stiles, C. D. 1981. Cytoplasmic transfer of mitogenic response to platelet-derived growth factor. *Proc. Natl. Acad. Sci. USA* 78: 4363–67

Smith, J. M., Sporn, M. B., Roberts, A. B., Derynck, R., Winkler, M. E. 1985. Human transforming growth factor-alpha causes precocious eyelid opening in newborn mice. *Nature* 315: 515–16

Sporn, M. B., Roberts, A. B., Shull, J. H., Smith, J. M., Ward, J. M., et al. 1983. Polypeptide transforming growth factors isolated from bovine sources and used for wound healing in vivo. *Science* 219: 1329–31

Sporn, M. B., Todaro, G. J. 1980. Autocrine secretion and malignant transformation of cells. *N. Engl. J. Med.* 303: 878–80

Stern, D. F., Hare, D. L., Cecchini, M. A., Weinberg, R. A. 1987. Construction of a novel oncogene based on synthetic sequences encoding epidermal growth factor. *Science* 235: 321–24

Stiles, C. D., Pledger, W. J., Van Wyk, J. J., Antoniades, H. N., Scher, C. D. 1979. Hormonal control of early events in the BALB/c-3T3 cell cycle: Commitment to DNA synthesis. In *Hormones and Cell Culture*, ed. G. H. Sato, R. Ross, Book A, pp. 425–29. Cold Spring Harbor, NY: Cold Spring Harbor Lab.

Stoscheck, C. M., Carpenter, G. 1984a. Characterization of the metabolic turnover of epidermal growth factor receptor protein in A-431 cells. *J. Cell. Physiol.* 120: 296–302

Stoscheck, C. M., Carpenter, G. 1984b. Down regulation of epidermal growth factor receptors: Direct demonstration of receptor degradation in human fibroblasts. *J. Cell Biol.* 98: 1048–53

Stroobant, P., Rice, A. P., Gullick, W. J., Cheng, D. J., Kerr, I. M., et al. 1985. Purification and characterization of vaccinia virus growth factor. *Cell* 42: 383–93

Stroobant, P., Waterfield, M. D. 1984. Purification and properties of porcine platelet-derived growth factor. *EMBO J.* 3: 2963–67

Strydom, D. J., Harper, J. W., Lobb, R. R. 1986. Amino acid sequence of bovine brain derived class 1 heparin-binding growth factor. *Biochemistry* 25: 945–51

Sullivan, R., Klagsbrun, M. 1985. Purification of cartilage-derived growth factor by heparin affinity chromatography. *J. Biol. Chem.* 260: 2399–2403

Swan, D. C., McBride, O. W., Robbins, K. C., Keithley, D. A., Reddy, E. P., et al. 1982. Chromosomal mapping of the simian sarcoma virus onc gene analogue in human cells. *Proc. Natl. Acad. Sci. USA* 79: 4691–95

Tam, J. P., Marquardt, H., Rosberger, D. F., Wong, T. W., Todaro, G. J. 1984. Synthesis of biologically active rat transforming growth factor I. *Nature* 309: 376–78

Tashjian, A. H., Voelkel, E. F., Lazzaro, M., Singer, F. R., Roberts, A. B., et al. 1985. Human transforming growth factors alpha and beta stimulate prostaglandin production and bone resorption in cultured mouse calvaria. *Proc. Natl. Acad. Sci. USA* 82: 4535–38

Taylor, J. M., Mitchell, W. M., Cohen, S. 1972. Epidermal growth factor: Physical and chemical properties. *J. Biol. Chem.* 247: 5928–34

Terranova, V. P., DiFlorio, R., Lyall, R. M., Hic, S., Friesel, R., et al. 1985. Human endothelial cells are chemotactic to endothelial cell growth factor and heparin. *J. Cell Biol.* 101: 2330–34

Theilen, G. H., Gould, D., Fowler, M., Dungworth, D. L. 1971. C-type virus in tumor tissue of a woolly monkey (Lagothrix spp.) with fibrosarcoma. *J. Natl. Cancer Inst.* 47: 881–89

Thiel, H.-J., Hafenrichter, R. 1984. Simian sarcoma virus transformation-specific glycopeptide: Immunological relationship to human platelet-derived growth factor. *Virology* 136: 414–24

Thomas, K. A. 1985. Mechanisms of action of mitogenic growth factors. *Comments Mol. Cell. Biophys.* 3: 1–13

Thomas, K. A., Rios-Candelore, M., Fitzpatrick, S. 1984. Purification and characterization of acidic fibroblast growth factor from bovine brain. *Proc. Natl. Acad. Sci. USA* 81: 357–61

Thomas, K. A., Rios-Candelore, M., Gimenez-Gallego, G., DiSalvo, J., Bennett, C., et al. 1985. Pure brain-derived acidic fibroblast growth factor is a potent angiogenic vascular endothelial cell mitogen with sequence homology to interleukin 1. *Proc. Natl. Acad. Sci. USA* 82: 6409–13

Thornton, S. C., Mueller, S. N., Levine, E. M. 1983. Human endothelial cells: Use of heparin in cloning and long-term serial cultivation. *Science* 222: 623–25

Todaro, G. J., De Larco, J. E., Cohen, S. 1976. Transformation by murine and feline sarcoma viruses specifically blocks binding of epidermal growth factor to cells. *Nature* 264: 26–31

Todaro, G. J., Fryling, C., De Larco, J. E. 1980. Transforming growth factors produced by certain human tumor cells: Polypeptides that interact with epidermal growth factor receptors. *Proc. Natl. Acad. Sci. USA* 77: 5258–62

Todaro, G. J., Lee, D. C., Webb, N. R., Rose, T. M., Brown, J. P. 1985. Rat type-α transforming growth factor: Structure and possible function as a membrane receptor. In *Cancer Cells*, ed. J. Feramisco, B. Ozanne, C. Stiles, 3: 51–58. Cold Spring Harbor, NY: Cold Spring Harbor Lab.

Tucker, R. F., Branum, E. L., Shipley, G. D., Ryan, R. J., Moses, H. L. 1984b. Specific binding to cultured cells of ^{125}I-labeled type beta transforming growth factor from human platelets. *Proc. Natl. Acad. Sci. USA* 81: 6757–61

Tucker, R. F., Shipley, G. D., Moses, H. L., Holly, R. W. 1984a. Growth inhibitor from BSC-1 cells closely related to platelet type beta transforming growth factor. *Science* 226: 705–7

Tucker, R. F., Scher, C. D., Stiles, C. D. 1979. Centriole deciliation associated with the early response of 3T3 cells to growth factors but not to SV40. *Cell* 18: 1065–72

Tucker, R. F., Snowdowne, K. W., Borle, A. B. 1983. Platelet derived growth factor produces transient increases in the intracellular concentration of free calcium in Balb/c 3T3 cells. *J. Cell Biol.* 97: 343a (Abstr.)

Turc-Carel, C., Philip, I., Berger, M.-P., Philip, T., Lenoir, G. M. 1983. Chromosomal translocations in Ewing's sarcoma. *N. Engl. J. Med.* 309: 497–98

Twardzik, D. R. 1985. Differential expression of transforming growth factor-α during prenatal development of the mouse. *Cancer Res.* 45: 5413–16

Twardzik, D. R., Brown, J. P., Ranchalis, J. E., Todaro, G. J., Moss, B. 1985a. Vaccinia virus-infected cells release a novel polypeptide functionally related to transforming and epidermal growth factors. *Proc. Natl. Acad. Sci. USA* 82: 5300–4

Twardzik, D. R., Kimball, E. S., Sherwin, S. A., Ranchalis, J. E., Todaro, G. J. 1985b. Comparison of growth factors functionally related to epidermal growth factor in the urine of normal and human tumor-bearing athymic mice. *Cancer Res.* 45: 1934–39

Twardzik, D. R., Ranchalis, J. E., Todaro, G. J. 1982a. Mouse embryonic transforming growth factors related to those isolated from tumor cells. *Cancer Res.* 42: 590–93

Twardzik, D. R., Sherwin, S. A., Ranchalis, J. E., Todaro, G. J. 1982c. Transforming growth factors in the urine of normal, pregnant, and tumor-bearing humans. *J. Natl. Cancer Inst.* 69: 793–98

Twardzik, D. R., Todaro, G. J., Marquardt, H., Reynolds, F. H., Stephenson, J. R. 1982b. Transformation induced by Abelson murine leukemia virus involves production of a polypeptide growth factor. *Science* 216: 894–97

Tzeng, D. Y., Deuel, T. F., Huang, J. S., Baehner, R. L. 1985. Platelet-derived growth factor promotes human peripheral monocyte activation. *Blood* 66: 179–83

Tzeng, D. Y., Deuel, T. F., Huang, J. S., Senior, R. M., Boxer, L. A., et al. 1984. Platelet-derived growth factor promotes polymorphonuclear leukocyte activation. *Blood* 64: 1123–28

Ueno, N., Baird, A., Esch, F., Ling, N., Guillemin, R. 1986. Isolation of an amino terminal extended form of basic fibroblast growth factor. *Biochem. Biophys. Res. Commun.* 138: 580–88

Ullrich, A., Coussens, L., Hayflick, J. S., Dull, T. J., Gray, A., et al. 1984. Human epidermal growth factor receptor cDNA sequence and aberrant expression of the amplified gene in A431 epidermoid carcinoma cells. *Nature* 309: 418–25

Ushiro, H., Cohen, S. 1980. Identification of phosphotyrosine as a product of epidermal growth factor-activated protein kinase in A-431 cell membranes. *J. Biol. Chem.* 255: 8363–65

Vale, W., Rivier, J., Vaughan, J., McClintock, R., Corrigan, A., et al. 1986. Purification and characterization of an FSH releasing protein from porcine ovarian follicular fluid. *Nature* 321: 776–79

Vara, F., Rozengurt, E. 1985. Stimulation of Na^+/H^+ antiport activity by epidermal growth factor and insulin occurs without activation of protein kinase C. *Biochem. Biophys. Res. Commun.* 130: 646–53

Vara, F., Schneider, J. A., Rozengurt, E. 1985. Ionic responses rapidly elicited by activation of protein kinase C in quiescent Swiss 3T3 cells. *Proc. Natl. Acad. Sci. USA* 82: 2384–88

Varga, J., Jimenez, S. A. 1986. Stimulation of normal human fibroblast collagen production and processing by transforming growth factor-β. *Biochem. Biophys. Res. Commun.* 138: 974–80

Vlodavsky, I., Sullivan, R., Fridman, R., Sasse, J., Folkman, J., et al. 1986. Heparin binding endothelial cell growth factor produced by endothelial cells and sequestered by the subendothelial extracellular matrix. *J. Cell Biol.* 103: 98a (Abstr.)

Vogel, A., Raines, E., Kariya, B., Rivest, M. J., Ross, R. 1978. Coordinate control of 3T3 cell proliferation by platelet-derived growth factor and plasma components. *Proc. Natl. Acad. Sci. USA* 75: 2810–14

Wang, C.-H., Largis, E. E., Schaffer, S. A. 1981. The effects of endothelial cell-conditioned media on the proliferation of aortic smooth muscle cells and 3T3 cells in culture. *Artery* 9: 358–71

Wang, J. Y. J., Williams, L. T. 1984. A v-*sis* oncogene protein produced in bacteria competes for platelet-derived growth factor binding to its receptor. *J. Biol. Chem.* 259: 10645–48

Waterfield, M. D., Scrace, G. T., Whittle, N., Stroobant, P., Johnsson, A., et al. 1983. Platelet-derived growth factor is structurally related to the putative transforming protein p28^{sis} of simian sarcoma virus. *Nature* 304: 35–39

Weksler, B. B., Coupal, C. E. 1973. Platelet-dependent generation of chemotactic activity in serum. *J. Exp. Med.* 137: 1419–30

Westermark, B., Heldin, C.-H., Ek, B., Johnsson, A., Mellstrom, K., et al. 1983. Biochemistry and biology of platelet-derived growth factor. In *Growth and Maturation Factors*, ed. G. Guroff, pp. 75–114. New York: Wiley

Westermark, B., Johnsson, A., Paulsson, Y., Betsholtz, C., Heldin, C.-H., et al. 1986. Human melanoma cell lines of primary and metastatic origin express the genes encoding the chains of platelet-derived growth factor (PDGF) and produce a PDGF-like growth factor. *Proc. Natl. Acad. Sci. USA* 83: 7197–7200

Westermark, B., Wasteson, A. 1975. The response of cultured human normal glial cells to growth factor. In *Advances in Metabolic Disorders*, ed. R. Luft, K. Hall, 8: 85–100. New York: Academic

Westermark, B., Wasteson, A. 1976. A platelet factor stimulating human normal glial cells. *Exp. Cell Res.* 98: 170–74

Westin, E. H., Wong-Staal, F., Gelmann, E. P., Dalla-Favera, R. D., Papas, T. S., et al. 1982. Expression of cellular homologues of retroviral onc genes in human hematopoietic cells. *Proc. Natl. Acad. Sci. USA* 79: 2490–94

Wharton, W., Leof, E., Olashaw, N., O'Keefe, E. J., Pledger, W. J. 1983. Mitogenic response to epidermal growth factor (EGF) modulated by platelet-derived growth factor in cultured fibroblasts. *Exp. Cell Res.* 147: 443–48

Wharton, W., Leof, E., Pledger, W. J., O'Keefe, E. J. 1982. Modulation of the epidermal growth factor receptor by platelet-derived growth factor and choleragen: Effects on mitogenesis. *Proc. Natl. Acad. Sci. USA* 79: 5567–71

Williams, L. T., Tremble, P., Antoniades, H. N. 1982. Platelet-derived growth factor binds specifically to receptors on vascular smooth muscle cells and the binding becomes nondissociable. *Proc. Natl. Acad. Sci. USA* 79: 5867–70

Witte, L. D., Cornicelli, J. A. 1980. Platelet-derived growth factor stimulates low density lipoprotein receptor activity in cultured human fibroblasts. *Proc. Natl. Acad. Sci. USA* 77: 5962–66

Witte, L. D., Cornicelli, J. A., Miller, R. W., Goodman, D. S. 1982. Effects of platelet-derived and endothelial cell–derived growth factors on the low density lipoprotein receptor pathway in cultured human fibroblasts. *J. Biol. Chem.* 257: 5392–5401

Witte, L. D., Kaplan, K. L., Nossel, H. L., Lages, B. A., Weiss, H. J., et al. 1978. Studies of the release from human platelets of the growth factor for cultured human arterial smooth muscle cells. *Circ. Res.* 42: 402–9

Wolfe, L. G., Deinhardt, F., Theilen, G. H., Rabin, H., Kawakami, T., et al. 1971. Induction of tumors in marmoset monkeys by simian sarcoma virus, type 1 (Lagothrix): A preliminary report. *J. Natl. Cancer Inst.* 47: 1115–20

Wong-Staal, F., Dalla-Favera, R., Franchini, G., Gelmann, E. P., Gallo, R. C. 1981b. Three distinct genes in human DNA related to the transforming genes of mammalian sarcoma retroviruses. *Science* 213: 226–28

Wong-Staal, F., Dalla-Favera, R., Gelmann, E. P., Manzari, V., Szala, S., et al. 1981a. The v-*sis* transforming gene of simian sarcoma virus is a new onc gene of primate origin. *Nature* 294: 273–75

Woost, P. G., Brightwell, J., Eiferman, R. A., Schultz, G. S. 1985. Effect of growth factors with dexamethasone on healing of rabbit corneal stromal incisions. *Exp. Eye Res.* 40: 47–60

Wrann, M., Fox, C. F., Ross, R. 1980. Modulation of epidermal growth factor receptors on 3T3 cells by platelet-derived growth factor. *Science* 210: 1363–65

Yamamoto, T., Nishida, T., Miyajima, N., Kawai, S., Ooi, T., et al. 1983. The *erb* B gene of avian erythroblastosis virus is a member of the *src* gene family. *Cell* 35: 71–78

Yarden, Y., Escobedo, J. A., Kuang, W.-J., Yang-Feng, T. L., Daniel, T. O., et al. 1986. Structure of the receptor for platelet-derived growth factor helps define a family

492   DEUEL

of closely related growth factor receptors. *Nature* 323: 226–32

Ying, S.-Y., Becker, A., Baird, A., Ling, N., Ueno, N., et al. 1986. Type beta transforming growth factor (TGF-β) is a potent stimulator of the basal secretions of follicle stimulating hormone (FSH) in a pituitary monolayer system. *Biochem. Biophys. Res. Commun.* 135: 950–56

Zabel, B. U., Kronenberg, H. M., Bell, G. I., Shows, T. B. 1985. Chromosome mapping of genes on the short arm of human chromosome 11: Parathyroid hormone gene is at 11p15 together with the genes for insulin, c-Harvey-ras 1, and β-hemoglobin. *Cytogenet. Cell Genet.* 39: 200–5

Zetterberg, A., Engstrom, W., Larsson, O. 1982. Growth activation of resting cells: Induction of balanced and unbalanced growth. *Ann. NY Acad. Sci.* 397: 130–47

Ziche, M., Gullino, P. M. 1982. Angiogenesis and neoplastic progression in vitro. *J. Natl. Cancer Inst.* 69: 483–87

SUBJECT INDEX

A

Acanthamoeba
  myosin in, 385-88
Acquired immunodeficiency syndrome
  helper T cells and, 161
Acrosin
  sperm-egg interactions and, 133
Acrosome reaction, 111-13
  site of, 113-18
Actin
  laminin receptor and, 72
  membrane immunoglobulin and, 154
  protein secretion and, 274-75
Actin filaments
  platelet-derived growth factor and, 449
  polarity of, 365-66
α-Actinin
  membrane immunoglobulin and, 154
Activins
  follicle-stimulating hormone and, 470
Adrenocorticotropic hormone
  secretion of
    inhibition of, 253
Albumin
  acrosome reaction and, 112
  secretion of
    primaquine and, 251
Amino acid transport
  platelet-derived growth factor and, 449
Ammonia
  lysosomal proteolysis and, 5
Amphibian mesodermal cells
  migration of, 332-33
Anionic hypothalamus-derived growth factor, 465
Arachidonic acid
  platelet-derived growth factor and, 449
Archaebacteria
  RNA splicing in, 235
Astroglial growth factor, 465
Astroglial growth factor I, 465
Atherosclerosis
  platelet-derived growth factor and, 452
ATP
  intracellular proteolysis and, 3, 12

Autophagy
  proteolysis and, 3
Avian fibronectin receptor, 180-88
Avian leukosis virus, 33
Avian sarcoma viruses
  src proteins and, 33-35, 42
Axonal transport
  microtubules and, 349-53

B

Bacterial antigens
  antibody production and, 161-68
Bacteriophages
  T-even
    RNA splicing in, 235
Basement membrane proteins
  laminin and, 65-69
Basement membranes
  laminin and, 57-58
  metastasis and, 75-76
  structure of
    molecular models of, 76-78
B lymphocyte activation, 143-70
  antigen receptors and, 145-56
  bacterial antigens and, 161-68
  c-myc proto-oncogene and, 168-69
  helper T cells and, 156-61
B lymphocyte activators
  polyclonal, 162-64
B lymphocyte antigen receptors
  B lymphocyte activation and, 145-56
  Fc receptors and, 149-51
  transmembrane signaling by, 146-49
B lymphocyte growth factor
  lipopolysaccharide and, 162
B lymphocytes
  Epstein-Barr virus and, 90
  intracellular calcium and, 147
  phosphoinositide breakdown and, 147-48
  ubiquitin and, 9
Bone marrow
  hemopoiesis and, 424
Bovine papilloma virus type 1
  replication of, 87-105
    cellular factors and, 103-4
    copy number control and, 97-98
    plasmid segregation and, 96-97

sites of initiation and, 99-102
  trans-acting factors and, 93-95, 102
Brain-derived growth factor, 465
Burkitt's lymphoma
  Epstein-Barr virus and, 90, 164

C

Cadherins
  cell adhesion and, 321
Caenorhabditis elegans
  myosin in, 409-10
Calcitonin
  thyroid C cells and, 224
Calcitonin gene-related peptide
  neuronal cells and, 224
Calcium
  platelet-derived growth factor and, 449
Caldesmon
  secretory granule binding and, 274
Calmodulin
  membrane immunoglobulin and, 154-55
  ubiquitination of, 12-13
Cancer
  Epstein-Barr virus and, 90
Carbohydrate
  laminin and, 61
Cartilage-derived growth factor, 465
α-Casein
  ubiquitin conjugation and, 19
Catecholamines
  acrosome reaction and, 112
Cell adhesion, 319-36
  mechanisms of, 320-24
  morphogenesis and, 324-35
Cell adhesion molecules, 321-22, 335
Cell membranes
  src protein and, 39-40
Cell migration
  laminin and, 73-74
  vertebrate embryos and, 330-33
Cell motility
  eukaryotic cells and, 379-80
Cell transformation
  avian sarcoma viruses and, 33-35

platelet-derived growth factor and, 457-60
Rous sarcoma virus and, 31-33
transforming growth factor α and, 469
Centriole deciliation
platelet-derived growth factor and, 449
Chagas disease
laminin antibodies and, 64
Chitin
plant defense and, 302-5
Chitosan
plant defense and, 302-5
Chloroquine
adrenocorticotropic hormone secretion and, 253
lysosomal enzymes and, 251
lysosomal proteolysis and, 4-5
Chondrogenesis
transforming growth factor β and, 470
Chondroitin sulfate
chymotrypsinogen A and, 262
Chondroitin sulfate proteoglycan
basement membranes and, 68
Chromaffin cells
secretory vesicles and, 274-75
Chromatin
interphase
ubiquitinated histones and, 21
Chromogranins
secretory granules and, 262
Chymotrypsinogen A
chondroitin sulfate and, 262
Clathrin
regulated protein secretory pathway and, 261
c-myc protó-oncogene
B lymphocyte activation and, 168-69
platelet-derived growth factor and, 451-52
Colchicine
microtubules and, 270
Collagen
cell transformation and, 32-33
extracellular matrix and, 322
fibroblast chemotaxis and, 453
neural cells and, 74
platelet-derived growth factor and, 449
Collagen I
chemotaxis and, 73
Collagen IV
laminin binding to, 58, 68
Complement
fibroblast chemotaxis and, 453
laminin and, 69

Congenital thymic aplasia
helper T cells and, 161
Corticotropin-releasing factor
localization in neurosecretory cells, 265
Cyclic nucleotides
acrosome reaction and, 112
Cytochalasin B
calcium-dependent exocytosis and, 274
cell adhesion and, 323
Cytoskeleton
protein secretion and, 266-77
Cytotactin
neural crest cell migration and, 332

D

Diacylglycerol
platelet-derived growth factor and, 449
Dictyostelium discoideum
aggregation in
cell adhesion and, 324-27
myosin of, 383-88
myosin heavy-chain gene of, 411
Di George's syndrome
helper T cells and, 161
Discoidin I
cell adhesion and, 322, 326-27
DNA sequencing
platelet-derived growth factor and, 454-55
Drosophila
homeotic proteins of
PEST sequences in, 9
myosin of, 411
Dynein
cell motility and, 379-80
organelle transport and, 360-61

E

Egg binding protein, 132-34
Elastin-derived peptides
fibroblast chemotaxis and, 453
Endocrine hormones
localization in dense-core granules, 264-65
Endocytosis
Golgi-derived transport and, 368-70
Endoplasmic reticulum
microtubules and, 266-67
protein secretion and, 243
ubiquitin activation and, 22
Endothelial cell growth factor, 464-65

Entactin
laminin binding to, 58
Enzymes
hormone processing
intracellular proteolysis and, 3
ubiquitin metabolism and, 16-20
Epidermal growth factor, 461-64
platelet-derived growth factor and, 450
structure of, 461
Epidermal growth factor receptor, 462-63
transforming growth factor α and, 469
v-erb oncogene product and, 463
Epithelial cells
Epstein-Barr virus and, 90
laminin and, 74
protein secretion and, 277-81
transforming growth factor β and, 470
Epstein-Barr virus
B lymphocyte CR2 and, 164-65
replication of, 87-105
cellular factors and, 103-4
copy number control and, 97-98
plasmid segregation and, 96-97
sites of initiation and, 99-102
trans-acting factors and, 93-95, 102
Erythrocyte agglutination
laminin and, 73
Escherichia coli
myosin expression in, 408-9
N-Ethylmaleimide-modified
myosin subfragment 1
calcium-dependent exocytosis and, 274
Eukaryotic cells
motility of, 379-80
RNA splicing in, 208, 235-36
Extracellular fluid
laminin and, 69
Extracellular matrix receptors, 179-99
lymphoid cells and, 190-91
mammalian cells and, 188-90
nonintegrin, 194-95
platelets and, 190
supergene family of, 192-93
Eye-derived growth factor II, 465

F

Fibroblast growth factor, 464-65
Fibroblasts

platelet-derived growth factor and, 454
Fibronectin
amphibian mesodermal cells and, 332-33
attachment to cells
RGD sequence and, 322
cell transformation and, 32-33
chemotaxis and, 73
fibroblast chemotaxis and, 453
integrin binding and, 184
laminin binding and, 72
morphogenesis in mouse embryo and, 330
neural cells and, 74
neural crest cell migration and, 331-32
neurite outgrowth and, 333
Fibronectin receptor
avian, 180-88
mammalian, 188-89
Fodrin
chromaffin cells and, 275
Follicle-stimulating hormone
transforming growth factor β and, 470
Fungal cell wall fragments
plant defense and, 296-305

G

Galactosyltransferase
sperm-egg interactions and, 132-33
Gangliosides
laminin binding and, 72-73
Gastric acid secretion
β-urogastrone and, 461
Glicentin
localization in pancreatic A cells, 265
Glucagon
localization in pancreatic A cells, 265
β-Glucanases
plant defense and, 300
β-Glucans
plant defense and, 296-302
Glucose transport protein
protein kinase C and, 437-38
Glyceollin
plant defense and, 296-302
Glycolipid
cell transformation and, 32-33
Glycoproteins
extracellular matrix and, 322
sulfated
secretory granules and, 262
Glycosaminoglycans
acrosome reaction and, 112

Glycosyltransferases
cell-cell interactions and, 132-33
Golgi apparatus
microtubules and, 266-67
polarity of, 268-70
protein secretion and, 243, 247-50
ubiquitin activation and, 22
gp80
cell adhesion and, 321-22
Granulocyte colony-stimulating factor
myelopoiesis and, 429-30
Growth factors
hemopoietic system and, 427-38
leukemogenesis and, 435-36
mode of action of, 436-38
See also specific type
Growth hormone
immunolocalization of, 265
secretion of
somatostatin and, 258

H

Heat-shock proteins
ubiquitin and, 23-24
Heat-shock response
ubiquitin and, 2
Helper T cells
B lymphocyte activation and, 156-61
B lymphocyte antigens and, 151-52
Hemopoiesis
stromal cells and, 432-34
Hemopoietic stem cells, 424-26
Hemopoietic system, 423-38
growth factors and, 427-38
structure of, 424-27
Hemostasis
platelet-derived growth factor and, 449
Heparan sulfate proteoglycan
laminin binding to, 58, 68-69
Heparin
endothelial cell chemotaxis and, 465
Heparin binding growth factors, 464-67
Hepatocytes
adhesion of
laminin and, 72
Hepatoma-derived growth factor, 465
Histone H2A
ubiquitination of, 18
Histones
ubiquitin and, 2
ubiquitinated

interphase chromatin and, 21
Hormone processing enzymes
intracellular proteolysis and, 3
Hyaluronic acid
cell transformation and, 32-33

I

Immune cells
transforming growth factor β and, 470
Immune complex disease
platelet-derived growth factor and, 452
Immunoglobulin
membrane
antigen uptake via, 153-55
transmembrane signaling by, 148-49
Infectious mononucleosis
Epstein-Barr virus and, 90
Inflammation
platelet-derived growth factor and, 452-53
Influenza virus hemagglutinin
transport of, 247
Inhibins
follicle-stimulating hormone and, 470
Integrin, 180-88
biosynthesis and processing of, 193-94
Interferon
B lymphocytes and, 157
Interleukin-1
lipopolysaccharide and, 162
Interleukin-1 receptor
platelet-derived growth factor and, 450
Interleukin-2
B lymphocytes and, 157
Interleukin-3
cell survival and, 437
multipotential stem cells and, 427
myelopoiesis and, 429-30
Intracellular proteolysis, 1-25
properties of, 2-9
Introns
gene evolution and, 235
Ionophores
acrosome reaction and, 112

K

Kinesin
cell motility and, 379-80
organelle transport and, 355-60

L

β-Lactalbumin
  ubiquitin conjugation and, 19
β-Lactoglobulin
  acrosome reaction and, 112
Laminin, 57-78
  B1 chain of, 73
  basement membrane proteins and, 65-69
  basement membranes and, 57-58
  biosynthesis and regulation of, 69-70
  cell binding and, 71-76
  cell migration and, 73-74
  cell surface ligands and, 72-73
  integrin binding and, 184
  metastasis and, 75-76
  model for, 63-64
  morphogenesis in mouse embryo and, 330
  neural crest cell migration and, 332
  neurite outgrowth and, 333
  neurite process formation and, 74-75
  RGD sequence in, 323
  self-assembly of, 65-67
  structure of, 58-65
Laminin receptor, 71-72
Leader peptides
  intracellular proteolysis and, 3-4, 8
Leishmaniasis
  laminin antibodies and, 64
Leukemogenesis
  growth factors and, 435-36
Leukocytes
  laminin and, 74
Leukotrienes
  platelet-derived growth factor and, 449
Lipopolysaccharide
  macrophage activation and, 162-63
Low-density lipoprotein
  platelet-derived growth factor and, 450
Lymphocyte homing receptor
  ubiquitin and, 2, 21-22
Lymphoid cells
  extracellular matrix-like receptors on, 190-91
Lymphokines
  fibroblast chemotaxis and, 453
Lymphoma
  Burkitt's
    Epstein-Barr virus and, 90, 164

Lysolecithin
  acrosome reaction and, 112
Lysosomal enzymes
  chloroquine and, 251
Lysosomes
  intracellular proteolysis and, 4-5
  microtubules and, 266
Lysozyme
  ubiquitin conjugation and, 19

M

Macrophage activation
  lipopolysaccharide and, 162-63
Macrophage colony-stimulating factor
  myelopoiesis and, 429-30
Macrophage-derived growth factor, 465
Mammalian cells
  extracellular matrix receptors of, 188-90
Mammalian fertilization, 109-35
  acrosome reaction and, 111-18
  egg binding protein and, 132-34
  sperm binding to zona pellucida and, 118-24
  sperm capacitation and, 111
  sperm receptor and, 124-32
Megakaryocytes
  platelet-derived growth factor and, 445
Metastasis
  laminin and, 75-76
4-Methylumbelliferyl-*p*-guanidinobenzoate
  sperm-egg interactions and, 133
Microtubule protein transport
  drugs disrupting, 270-71
  selectivity in, 271-74
Microtubules
  axonal transport and, 349-53
  directed movement along, 268-70, 363-64
  organelle transport and, 349-55
  secretory vesicles and, 266-68
Mitochondria
  microtubules and, 266
Mitochondrial DNA
  replication of, 98-99
Monoclonal antibodies
  laminin receptor and, 71
Mononuclear phagocytes
  platelet-derived growth factor and, 452-53
Morphogenesis
  cell adhesion and, 324-35

Motility proteins
  organelles and, 361-63
  organelle transport and, 364-68
Mouse embryo
  morphogenesis in
    compaction and, 329-30
Muscle contraction
  models of, 402-3
Myogenesis
  transforming growth factor β and, 470
Myosin
  cell biology of, 382-88
  function of
    assays for, 403-8
  head, 394-400
  membrane immunoglobulin and, 154
  molecular-genetic manipulation of, 408-12
  molecular substructure of, 388-402
  structure of, 380-81
  tail, 400-2

N

Nasopharyngeal carcinoma
  Epstein-Barr virus and, 90
Nerve growth factor, 461
Neural cells
  laminin and, 74
Neural crest cells
  migration of, 330-32
Neurite processes
  laminin and, 74-75, 333
Neurogenesis
  cell adhesion and, 333-35
Neuronal cells
  calcitonin gene-related peptide and, 224
Neuropeptides
  localization in dense-core granules, 264-65
Neutrophils
  platelet-derived growth factor and, 452-53
Nicotine
  chromaffin cells and, 275
Nidogen
  laminin binding to, 58, 67
Nitella
  myosin in, 382-88
Nocadozole
  microtubules and, 270

O

α-1,4-D-Oligogalacturonides
  plant defense and, 305-9
Oligosaccharide signaling
  plants and, 295-312

Oncogenes
  platelet-derived growth factor
    and, 455-57
  viral *src*, 31-49
Organelle acidification
  protein transport and, 250-51
Organelles
  spatial organization of
    microtubule-based motors
    and, 368-72
Organelle transport
  characteristics of, 361-64
  microtubule-based, 349-55
  microtubule polarity and, 364-
    68
  motility proteins and, 364-68
  nonneuronal cells and, 353-55
Organelle transport motors, 355-
  61
Orosomucoid
  secretion of
    blocking of, 251
Osteogenesis
  transforming growth factor β
    and, 470

P

Phalloidin
  calcium-dependent exocytosis
    and, 274
Phorbol esters
  lipopolysaccharide and, 167-
    68
Phosphoinositide
  B lymphocytes and, 147-48
Physarum
  myosin in, 410-11
Phytoalexins
  plant defense and, 296-302
Plant defense, 295-312
  fungal cell wall fragments
    and, 296-305
  plant cell wall fragments and,
    305-11
Plasmids
  replication of, 87-105
Plasminogen activator
  cell transformation and, 32-
    33
Platelet-derived growth factor,
    444-46
  activities induced by, 449-50
  cell transformation and, 457-
    60
  competence and, 450-52
  origins of, 445
  properties of, 445-46
Platelet-derived growth factor
  receptor, 446-47
  tyrosine protein kinase activ-
    ity of, 447-48

Platelets
  extracellular matrix-like recep-
    tors on, 190
Polyclonal antibodies
  laminin and, 64
Polylysine
  neural cells and, 74
Polymorphonuclear leukocytes
  platelet-derived growth factor
    and, 454
Polyornithine
  neural cells and, 74
Polypeptide growth factors, 443-
    74
Preproparathyroid hormone
  proteolytic processing of, 255
Primaquine
  albumin secretion and, 251
Proacrosin
  sperm-egg interactions and,
    133
Procarboxypeptidase Y
  secretion of, 251-52
Proenkephalin
  proteolytic processing of, 255
Proinsulin
  proteolytic processing of, 255
Prolactin
  immunolocalization of, 265
Pro-opiomelanocortin
  sorting of, 258
Prostaglandins
  lipopolysaccharide and, 162
  platelet-derived growth factor
    and, 449
Proteases
  laminin and, 60-61
Proteinases
  sperm-egg interactions and,
    133
Protein kinase C
  glucose transport protein and,
    437-38
  lipopolysaccharide and, 167-
    68
  platelet-derived growth factor
    and, 449-50
Protein secretion, 243-83
  constitutive pathway for, 245-
    52
  cytoskeleton and, 266-77
  epithelial cells and, 277-81
  regulated pathway for, 252-57
  sorting and, 257-66
Protein synthesis
  platelet-derived growth factor
    and, 449
Proteoglycans
  extracellular matrix and, 322
  sulfated
    secretory granules and, 262
Proteolysis
  basal rates of, 3

intracellular, 1-25
  properties of, 2-9
  ubiquitin and, 9-24
  secretory proteins and, 254-55

R

R cognin
  cell adhesion and, 321
α-Retina-derived growth factor,
  465
Retroviruses
  src-containing, 33
Ribosomal S6 protein
  platelet-derived growth factor
    and, 449
RNase
  ubiquitin conjugation and, 19
RNA splicing, 207-37
  alternative, 208-9
    biological implications of,
    232-37
  modes of, 218-20
  regulatory mechanisms of,
    221-32
  site selection for, 209-18
Rough endoplasmic reticulum
  protein secretion and, 245-47
Rous sarcoma virus
  cell transformation and, 31-33

S

Saccharomyces cerevisiae
  myosinlike genes of, 411
Sarcoma viruses
  See Avian sarcoma viruses;
    Rous sarcoma virus
Schwann cells
  laminin and, 75
Sea urchin gastrulation
  cell adhesion and, 327-28
Secretogranins
  secretory granules and, 262
Secretory vesicles
  membrane recycling of, 281-
    83
  microtubules and, 266-68
Somatostatin
  growth hormone secretion
    and, 258
Sperm
  capacitation of, 111
Sperm receptor, 124-32
src proteins
  activation of, 46-49
  avian sarcoma viruses and,
    33-35, 42
  cell membranes and, 39-40
  cellular substrates of, 45-46
  membrane-binding domains
    of, 40-41
  modulation of, 42-44

phosphorylation of, 44-45
Staphylococcus aureus
    B lymphocytes and, 165
Steroids
    acrosome reaction and, 112
Stromal cells
    hemopoiesis and, 432-34
Sulfatides
    laminin binding and, 72-73
SV40 DNA
    replication of, 98-99

T

Taxol
    microtubules and, 270
β-Thromboglobulin
    fibroblast chemotaxis and,
        453
Thyroid C cells
    calcitonin and, 224
Transferrin
    secretion of
        blocking of, 251
Transforming growth factor α,
    461-62, 468-70
Transforming growth factor β,
    470-73
Transit peptides
    intracellular proteolysis and,
        3-4, 8
Trifluoperazine
    chromaffin cells and, 274
Tropoelastin
    fibroblast chemotaxis and,
        453

α-Tropomyosin
    gene encoding, 228
Troponin T
    gene encoding, 226-30
Tumor angiogenesis factor, 465
Tumor cells
    laminin and, 73
Tumor necrosis factor
    lipopolysaccharide and, 162
Tyrosine protein kinase
    platelet-derived growth factor
        receptor and, 447-48

U

Ubiquitin
    conjugates of, 21-22
    covalent bonding of, 2
    intracellular proteolysis and,
        9-24
    metabolism of
        enzymes of, 16-20
    properties of, 9-12
β-Urogastrone
    gastric acid secretion and,
        461
Uvomorulin
    cell adhesion and, 321
    morphogenesis in mouse
        embryo and, 329

V

Vaccinia growth factor
    transforming growth factor α
        and, 469-70

Vaccinia virus EGF-like
    polypeptide, 462
Vasopressin
    localization in neurosecretory
        cells, 265
Vertebrate embryos
    cell migration in, 330-33
Vesicular stomatitis virus G-
    protein
        secretory vesicles and, 267
        transport of, 246-47
Vinblastine
    microtubules and, 270
Viral *src* oncogene, 31-49
    See also *src* proteins
Vitronectin
    integrin binding and, 184
    RGD sequence in, 323
Vitronectin receptor
    mammalian, 189

W

Wound healing
    platelet-derived growth factor
        and, 449

Y

Yeast cells
    protein secretion in, 243-44

Z

Zona pellucida
    sperm binding to, 114-24

CUMULATIVE INDEXES

CONTRIBUTING AUTHORS, VOLUMES 1–3

A

Al-Awqati, Q., 2:179–99
Anderson, R. G. W., 1:1–39
Andreadis, A., 3:207–42

B

Beckwith, J., 2:315–36
Bourne, H. R., 2:391–419
Brinkley, B. R., 1:145–72
Brown, M. S., 1:1–39
Buck, C. A., 3:179–205
Burgess, T. L., 3:243–93

C

Chaponnier, C., 1:353–402

D

DeFranco, A. L., 3:143–78
Deuel, T. F., 3:443–92
Dexter, T. M., 3:423–41
Dingwall, C., 2:367–90
Dreyfuss, G., 2:459–98
Duband, J. L., 1:91–113

E

Edelman, G. M., 2:81–116
Ekblom, P., 2:27–47
Ettensohn, C. A., 3:319–45
Ezzell, R., 1:353–402

F

Farquhar, M. G., 1:447–88
Finer-Moore, J., 1:317–51
Fujiki, Y., 1:489–530
Fuller, S., 1:243–88

G

Gallego, M. E., 3:207–42
Garoff, H., 1:403–45
Gerhart, J., 2:201–29
Goldstein, J. L., 1:1–39

H

Hanafusa, H., 3:31–56
Hartwig, J. H., 1:353–402

Horwitz, A. F., 3:179–205
Hynes, R. O., 1:67–90

J

Janmey, P., 1:353–402
Jove, R., 3:31–56

K

Keller, R., 2:201–29
Kelly, R. B., 3:243–93
Kemler, R., 2:27–47
Kikkawa, U., 2:149–78
Kupfer, A., 2:337–65
Kwiatkowski, D., 1:353–402

L

Laskey, R. A., 2:367–90
Lazarow, P. B., 1:489–530
Lee, C., 2:315–36
Lefebvre, P. A., 2:517–46
Lind, S., 1:353–402
Lingappa, V. R., 2:499–516

M

MacDonald, H. R., 2:231–53
Marchesi, V. T., 1:531–61
Martin, G. R., 3:57–85
McClay, D. R., 3:319–45
Mecsas, J., 3:87–108
Mooseker, M. S., 1:209–41
Murray, A., 1:289–315

N

Nabholz, M., 2:231–53
Nadal-Ginard, B., 3:207–42
Nishizuka, Y., 2:149–78

O

O'Farrell, P. H., 2:49–80
Olmsted, J. B., 2:421–57

P

Parry, D. A. D., 1:41–65
Pederson, D. S., 2:117–47

R

Rechsteiner, M., 3:1–30
Rosenbaum, J., 2:517–46
Russell, D. W., 1:1–39
Ryan, C. A., 3:295–317

S

Schekman, R., 1:115–43
Schneider, W. J., 1:1–39
Scott, M. P., 2:49–80
Semenza, G., 2:255–313
Shapiro, L., 1:173–207
Simons, K., 1:243–88
Simpson, R. T., 2:117–47
Singer, S. J., 2:337–65
Smith, D., 1:353–402
Southwick, F. S., 1:353–402
Spooncer, E., 3:423–41
Spudich, J. A., 3:379–421
Steinert, P. M., 1:41–65
Stossel, T. P., 1:353–402
Stroud, R. M., 1:317–51
Stryer, L., 2:391–419
Sugden, B., 3:87–108
Szostak, J. W., 1:289–315

T

Thiery, J., 1:91–113
Thoma, F., 2:117–47
Timpl, R., 3:57–85
Trimmer, J. S., 2:1–26
Tucker, G. C., 1:91–113

V

Vacquier, V. D., 2:1–26
Vale, R. D., 3:347–78
Vestweber, D., 2:27–47

W

Walter, P., 2:499–516
Warrick, H. M., 3:379–421
Wassarman, P. M., 3:109–42

Y

Yin, H. L., 1:353–402

Z

Zaner, K. S., 1:353–402

CHAPTER TITLES, VOLUMES 1–3

CELL-EXTRACELLULAR MATRIX INTERACTIONS
Cell-Matrix Interactions and Cell Adhesion
During Development — Peter Ekblom, Dietmar Vestweber, and Rolf Kemler — 2:27–47
Cell Surface Receptors for Extracellular
Matrix Molecules — Clayton A. Buck and Alan F. Horwitz — 3:179–205

CELL GROWTH AND DIFFERENTIATION
Growth and Differentiation in the
Hemopoietic System — T. M. Dexter and E. Spooncer — 3:423–41
Polypeptide Growth Factors: Roles in Normal
and Abnormal Cell Growth — Thomas F. Deuel — 3:443–92

CELL TRANSFORMATION
Cell Transformation by the Viral *src*
Oncogene — Richard Jove and Hidesaburo Hanafusa — 3:31–56
Replication of Plasmids Derived from Bovine
Papilloma Virus Type 1 and Epstein-Barr
Virus in Cells in Culture — Joan Mecsas and Bill Sugden — 3:87–108
Polypeptide Growth Factors: Roles in Normal
and Abnormal Cell Growth — Thomas F. Deuel — 3:443–92

CELLULAR IMMUNOLOGY
T-Cell Activation — H. Robson MacDonald and Markus Nabholz — 2:231–53
Molecular Aspects of B-Lymphocyte
Activation — Anthony L. DeFranco — 3:143–78

CENTRIOLES
Microtubule Organizing Centers — B. R. Brinkley — 1:145–72

CHROMATIN
Core Particle, Fiber, and Transcriptionally
Active Chromatin Structure — D. S. Pederson, F. Thoma, and R. T. Simpson — 2:117–47

CHROMOSOMES
Chromosome Segregation in Mitosis and
Meiosis — Andrew W. Murray and Jack W. Szostak — 1:289–315

CILIA AND FLAGELLA
Regulation of the Synthesis and Assembly of
Ciliary and Flagellar Proteins During
Regeneration — Paul A. Lefebvre and Joel L. Rosenbaum — 2:517–46

CONTRACTILE PROTEINS AND ASSEMBLIES
Organization, Chemistry, and Assembly of
the Cytoskeletal Apparatus of the Intestinal
Brush Border — Mark S. Mooseker — 1:209–41
Nonmuscle Actin-Binding Proteins — T. P. Stossel, C. Chaponnier, R. M. Ezzell, J. H. Hartwig, P. A. Janmey, D. J. Kwiatkowski, S. E. Lind, D. B. Smith, F. S. Southwick, H. L. Yin, and K. S. Zaner — 1:353–402

500

The Directed Migration of Eukaryotic Cells   S. J. Singer and Abraham Kupfer   2:337–65
Intracellular Transport Using
   Microtubule-Based Motors   Ronald D. Vale   3:347–78
Myosin Structure and Function in Cell
   Motility   Hans M. Warrick and James A.
     Spudich   3:379–421

CYTOSKELETON
   Intermediate Filaments   Peter M. Steinert and David A. D.
     Parry   1:41–65
   Microtubule-Associated Proteins   J. B. Olmsted   2:421–57
   Intracellular Transport Using
     Microtubule-Based Motors   Ronald D. Vale   3:347–78

DEVELOPMENTAL BIOLOGY
   Cell Migration in the Vertebrate Embryo   Jean Paul Thiery, Jean Loup
     Duband, and Gordon C. Tucker   1:91–113
   Activation of Sea Urchin Gametes   James S. Trimmer and Victor D.
     Vacquier   2:1–26
   Cell-Matrix Interactions and Cell Adhesion
     During Development   Peter Ekblom, Dietmar Vestweber,
     and Rolf Kemler   2:27–47
   Spatial Programming of Gene Expression in
     Early *Drosophila* Embryogenesis   Matthew P. Scott and Patrick H.
     O'Farrell   2:49–80
   Cell Adhesion Molecules in the Regulation of
     Animal Form and Tissue Pattern   Gerald M. Edelman   2:81–116
   Region-Specific Cell Activities in Amphibian
     Gastrulation   John Gerhart and Ray Keller   2:201–29
   Early Events in Mammalian Fertilization   Paul M. Wassarman   3:109–42
   Cell Adhesion in Morphogenesis   David R. McClay and Charles A.
     Ettensohn   3:319–45

ENDOCYTOSIS
   Receptor-Mediated Endocytosis   Joseph L. Goldstein, Michael S.
     Brown, Richard G. W. Anderson,
     David W. Russell, and Wolfgang
     J. Schneider   1:1–39

EXOCYTOSIS
   Constitutive and Regulated Secretion of
     Proteins   Teresa Lynn Burgess and Regis B.
     Kelly   3:243–93

EXTRACELLULAR MATRIX
   Molecular Biology of Fibronectin   Richard Hynes   1:67–90
   Laminin and Other Basement Membrane
     Components   George R. Martin and Rupert Timpl   3:57–85

GENES
   Structure and Function of Nuclear and
     Cytoplasmic Ribonucleoprotein Particles   Gideon Dreyfuss   2:459–98

INTERCELLULAR COMMUNICATION
   Oligosaccharide Signalling in Plants   Clarence A. Ryan   3:295–317

INTRACELLULAR MEMBRANE SYSTEMS
   Progress in Unraveling Pathways of Golgi
     Traffic   Marilyn Gist Farquhar   1:447–88
   Constitutive and Regulated Secretion of
     Proteins   Teresa Lynn Burgess and Regis B.
     Kelly   3:243–93

INTRACELLULAR PROTEOLYSIS
Ubiquitin-Mediated Pathways for Intracellular
Proteolysis   Martin Rechsteiner   3:1–30

mRNA
Generation of Protein Isoform Diversity by
Alternative Splicing: Mechanistic and
Biological Implications   Athena Andreadis, Maria E.
                                                Gallego, and Bernardo
                                                Nadal-Ginard   3:207–42

PEROXISOMES
Biogenesis of Peroxisomes   P. B. Lazarow and Y. Fujiki   1:489–530

PLASMALEMMA
Receptor-Mediated Endocytosis   Joseph L. Goldstein, Michael S.
                                                Brown, Richard G. W. Anderson,
                                                David W. Russell, and Wolfgang
                                                J. Schneider   1:1–39

Generation of Polarity During Caulobacter
Cell Differentiation   Lucille Shapiro   1:173–207
Cell Surface Polarity in Epithelia   Kai Simons and Stephen D. Fuller   1:243–88
Acetylcholine Receptor Structure, Function,
and Evolution   Robert M. Stroud and Janet
                                                Finer-Moore   1:317–51
Stabilizing Infrastructure of Cell Membranes   V. T. Marchesi   1:531–61
The Role of Protein Kinase C in
Transmembrane Signalling   Ushio Kikkawa and Yasutomi
                                                Nishizuka   2:149–78
Proton-Translocating ATPases   Qais Al-Awqati   2:179–99
Anchoring and Biosynthesis of Stalked Brush
Border Membrane Proteins: Glycosidases
and Peptidases of Enterocytes and Renal
Tubuli   Giorgio Semenza   2:255–313
G Proteins: A Family of Signal Transducers   Lubert Stryer and Henry R. Bourne   2:391–419

PROTEIN TRAFFIC CONTROL
Protein Localization and Membrane Traffic   Randy Schekman   1:115–43
Cell Surface Polarity in Epithelia   Kai Simons and Stephen D. Fuller   1:243–88
Using Recombinant DNA Techniques to
Study Protein Targeting in the Eucaryotic
Cell   Henrik Garoff   1:403–45
Biogenesis of Peroxisomes   P. B. Lazarow and Y. Fujiki   1:489–530
Cotranslational and Posttranslational Protein
Translocation in Prokaryotic Systems   Catherine Lee and Jon Beckwith   2:315–36
Protein Import into the Cell Nucleus   Colin Dingwall and Ronald A.
                                                Laskey   2:367–90
Mechanism of Protein Translocation Across
the Endoplasmic Reticulum   Peter Walter and Vishwanath R.
                                                Lingappa   2:499–516